Progress in Optical Science

Volume 4

Series editor

Javid Atai

The purpose of the series Progress in Optical Science and Photonics is to provide a forum to disseminate the latest research findings in various areas of Optics and its applications. The intended audience is physicists, electrical and electronic engineers, applied mathematicians, and advanced graduate students.

More information about this series at http://www.springer.com/series/10091

Psang Dain Lin

Advanced Geometrical Optics

Psang Dain Lin
Department of Mechanical Engineering
National Cheng Kung University
Tainan
Taiwan

ISSN 2363-5096 ISSN 2363-510X (electronic)
Progress in Optical Science and Photonics
ISBN 978-981-10-9586-3 ISBN 978-981-10-2299-9 (eBook)
DOI 10.1007/978-981-10-2299-9

Printed on acid-free paper

This Springer imprint is published by Springer Nature
The registered company is Springer Science+Business Media Singapore Pte Ltd.

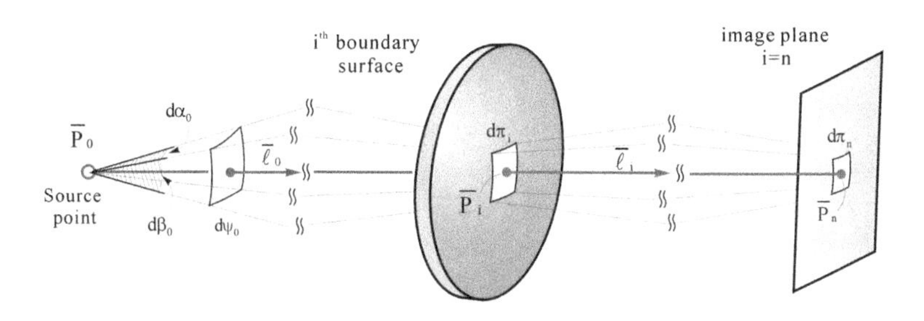

i^th boundary surface
image plane
i=n
$\overline{P}_0$
Source point
$d\alpha_0$
$d\beta_0$
$d\psi_0$
$\overline{\ell}_0$
$d\pi_i$
$\overline{P}_i$
$\overline{\ell}_i$
$d\pi_n$
$\overline{P}_n$

Preface

The study of geometrical optics dates back to ancient Greek and Egyptian times. However, geometrical optics remains firmly rooted in the use of paraxial optics and skew-ray tracing equations. For many years, it has been known that the first- and second-order derivative matrices (i.e., Jacobian and Hessian matrices) of merit functions provide highly effective tools for the analysis and design of optical systems. However, computing these derivative matrices analytically is extremely challenging since ray-tracing equations are inherently recursive functions. To overcome this limitation, this book proposes a straightforward computational scheme for deriving the Jacobian and Hessian matrices of a ray and its optical path length using homogeneous coordinate notation.

The book represents both a modernization and an extension of my last book, *New Computation Methods for Geometrical Optics*, published in 2013 by Springer Singapore. Part I of the book reviews the basic principles and theories of skew-ray tracing, paraxial optics and primary aberrations. Much of the material is likely to be known to the readers. However, it serves as essential reading in laying down a solid foundation for the modeling work presented in Parts II and III of the book. Part II derives the Jacobian matrices of a ray and its optical path length. Although this issue is also addressed in other books and publications, the authors generally fail to consider all of the variables of a non-axially symmetrical system. The modeling work presented in Part II thus provides a more robust framework for the analysis and design of non-axially symmetrical systems such as prisms and head-up display. Importantly, Part II also presents a new method for determining the point spread function and modulation transfer function of an optical system such that the image quality can be evaluated accurately. Part III of the book proposes a computational scheme for deriving the Hessian matrices of a ray and its optical path length. The validity of the proposed method is demonstrated using various optical systems for illustration purposes. It is shown that the Hessian matrix approach overcomes the limitations of traditional finite difference methods and provides an effective means of determining an appropriate search direction when tuning the system variables in the system design process.

This book is intended to be used as a reference book for introductory graduate and senior undergraduate geometrical optics courses. With this in mind, the text contains numerous illustrative examples aimed at helping the reader understand the underlying theories and concepts of the related modeling work and proposed methods. It is noted that while aspherical lenses are common in the geometrical optics field nowadays, the book focuses deliberately on the simpler case of flat or spherical boundary surfaces in order to more clearly convey the main concepts and ideas. However, once students and self-taught practitioners have mastered the fundamentals described in this edition, they will find no problem in applying the related equations to the more complex case of aspherical lenses.

This book is dedicated to all the faculty and staff at the Department of Mechanical Engineering, National Cheng Kung University, Taiwan. Without their support and encouragement, this book would never have been possible. Special thanks are also extended to the Ministry of Science and Technology of Taiwan for the generous financial support provided every year to the author in developing the methodologies and underlying concepts presented in this book.

I am indebted to Dr. Chung Yu Tsai of Formosa University in Taiwan for his many stimulating discussions on the subject of prism analysis and design. My thanks also go to Dr. Chien-Sheng Liu, Che-Wei Chang, Ying-Yan Hsu, and Chia-Kuei Hsu for their help in verifying the equations, figures, and notations used throughout the text. Any shortcomings or errors in the book are my responsibility, and mine alone. Finally, I would like to thank Patrick Wyton for his efforts in proofreading the text.

Tainan, Taiwan Psang Dain Lin

Acknowledgements

To my former advisor

Kornel F. Ehmann

To my wife

Chiung-Jung Huang

and

In memory of my past wife

Su-Chin Wang

Contents

Part I A New Light on Old Geometrical Optics (Raytracing Equations of Geometrical Optics)

1 Mathematical Background 3
1.1 Foundational Mathematical Tools and Units 3
1.2 Vector Notation 5
1.3 Coordinate Transformation Matrix 7
1.4 Basic Translation and Rotation Matrices 9
1.5 Specification of a Pose Matrix by Using Translation and Rotation Matrices 15
1.6 Inverse Matrix of a Transformation Matrix 16
1.7 Flat Boundary Surface 17
1.8 RPY Transformation Solutions 19
1.9 Equivalent Angle and Axis of Rotation 20
1.10 The First- and Second-Order Partial Derivatives of a Vector 22
1.11 Introduction to Optimization Methods 26
References 28

2 Skew-Ray Tracing of Geometrical Optics 29
2.1 Source Ray 29
2.2 Spherical Boundary Surfaces 32
2.2.1 Spherical Boundary Surface and Associated Unit Normal Vector 32
2.2.2 Incidence Point 34
2.2.3 Unit Directional Vectors of Reflected and Refracted Rays 37

2.3 Flat Boundary Surfaces. 44
2.3.1 Flat Boundary Surface and Associated Unit Normal Vector 44
2.3.2 Incidence Point 46
2.3.3 Unit Directional Vectors of Reflected and Refracted Rays 47
2.4 General Aspherical Boundary Surfaces. 55
2.4.1 Aspherical Boundary Surface and Associated Unit Normal Vector. 55
2.4.2 Incidence Point 57
2.5 The Unit Normal Vector of a Boundary Surface for Given Incoming and Outgoing Rays 64
2.5.1 Unit Normal Vector of Refractive Boundary Surface 65
2.5.2 Unit Normal Vector of Reflective Boundary Surface 67
References. 68

3 Geometrical Optical Model. 71
3.1 Axis-Symmetrical Systems 71
3.1.1 Elements with Spherical Boundary Surfaces 76
3.1.2 Elements with Spherical and Flat Boundary Surfaces. 77
3.1.3 Elements with Flat and Spherical Boundary Surfaces. 78
3.1.4 Elements with Flat Boundary Surfaces 79
3.2 Non-axially Symmetrical Systems. 87
3.3 Spot Diagram of Monochromatic Light 97
3.4 Point Spread Function 99
3.5 Modulation Transfer Function. 104
3.6 Motion Measurement Systems 109
References. 113

4 Raytracing Equations for Paraxial Optics. 115
4.1 Raytracing Equations of Paraxial Optics for 3-D Optical Systems 115
4.1.1 Transfer Matrix 117
4.1.2 Reflection and Refraction Matrices for Flat Boundary Surface 118
4.1.3 Reflection and Refraction Matrices for Spherical Boundary Surface 119
4.2 Conventional 2 × 2 Raytracing Matrices for Paraxial Optics . . . 123
4.2.1 Refracting Boundary Surfaces 124
4.2.2 Reflecting Boundary Surfaces 125

4.3 Conventional Raytracing Matrices for Paraxial Optics Derived from Geometry Relations 128
4.3.1 Transfer Matrix for Ray Propagating Along Straight-Line Path 129
4.3.2 Refraction Matrix at Refractive Flat Boundary Surface 131
4.3.3 Reflection Matrix at Flat Mirror 133
4.3.4 Refraction Matrix at Refractive Spherical Boundary Surface 135
4.3.5 Reflection Matrix at Spherical Mirror 138
References. 142

5 Cardinal Points and Image Equations 143
5.1 Paraxial Optics 143
5.2 Cardinal Planes and Cardinal Points 145
5.2.1 Location of Focal Points 146
5.2.2 Location of Nodal Points. 148
5.3 Thick and Thin Lenses 149
5.4 Curved Mirrors 151
5.5 Determination of Image Position Using Cardinal Points 153
5.6 Equation of Lateral Magnification. 154
5.7 Equation of Longitudinal Magnification 155
5.8 Two-Element Systems 156
5.9 Optical Invariant 159
5.9.1 Optical Invariant and Lateral Magnification. 160
5.9.2 Image Height for Object at Infinity 161
5.9.3 Data of Third Ray 162
5.9.4 Focal Length Determination 164
References. 165

6 Ray Aberrations 167
6.1 Stops and Aperture 167
6.2 Ray Aberration Polynomial and Primary Aberrations 169
6.3 Spherical Aberration 171
6.4 Coma 173
6.5 Astigmatism 177
6.6 Field Curvature 179
6.7 Distortion 180
6.8 Chromatic Aberration 181
References. 183

Part II New Tools for Optical Analysis and Design (First-Order Derivative Matrices of a Ray and its OPL)

7 Jacobian Matrices of Ray $\bar{R}_i$ with Respect to Incoming Ray $\bar{R}_{i-1}$ and Boundary Variable Vector $\bar{X}_i$ . . . 187

7.1 Jacobian Matrix of Ray . . . 188
7.2 Jacobian Matrix $\partial\bar{R}_i/\partial\bar{R}_{i-1}$ for Flat Boundary Surface . . . 189
7.2.1 Jacobian Matrix of Incidence Point . . . 190
7.2.2 Jacobian Matrix of Unit Directional Vector of Reflected Ray . . . 191
7.2.3 Jacobian Matrix of Unit Directional Vector of Refracted Ray . . . 191
7.2.4 Jacobian Matrix of $\bar{R}_i$ with Respect to $\bar{R}_{i-1}$ for Flat Boundary Surface . . . 192
7.3 Jacobian Matrix $\partial\bar{R}_i/\partial\bar{R}_{i-1}$ for Spherical Boundary Surface . . . 195
7.3.1 Jacobian Matrix of Incidence Point . . . 196
7.3.2 Jacobian Matrix of Unit Directional Vector of Reflected Ray . . . 197
7.3.3 Jacobian Matrix of Unit Directional Vector of Refracted Ray . . . 198
7.3.4 Jacobian Matrix of $\bar{R}_i$ with Respect to $\bar{R}_{i-1}$ for Spherical Boundary Surface . . . 198
7.4 Jacobian Matrix $\partial\bar{R}_i/\partial\bar{X}_i$ for Flat Boundary Surface . . . 201
7.4.1 Jacobian Matrix of Incidence Point . . . 202
7.4.2 Jacobian Matrix of Unit Directional Vector of Reflected Ray . . . 203
7.4.3 Jacobian Matrix of Unit Directional Vector of Refracted Ray . . . 203
7.4.4 Jacobian Matrix of $\bar{R}_i$ with Respect to $\bar{X}_i$ . . . 204
7.5 Jacobian Matrix $\partial\bar{R}_i/\partial\bar{X}_i$ for Spherical Boundary Surface . . . 206
7.5.1 Jacobian Matrix of Incidence Point . . . 207
7.5.2 Jacobian Matrix of Unit Directional Vector of Reflected Ray . . . 208
7.5.3 Jacobian Matrix of Unit Directional Vector of Refracted Ray . . . 208
7.5.4 Jacobian Matrix of $\bar{R}_i$ with Respect to $\bar{X}_i$ . . . 209
7.6 Jacobian Matrix of an Arbitrary Ray with Respect to System Variable Vector . . . 210
Appendix 1 . . . 213
Appendix 2 . . . 215
References . . . 218

8 Jacobian Matrix of Boundary Variable Vector $\bar{X}_i$ with Respect to System Variable Vector $\bar{X}_{sys}$ 219
8.1 System Variable Vector 219
8.2 Jacobian Matrix $d\bar{X}_0/d\bar{X}_{sys}$ of Source Ray 220
8.3 Jacobian Matrix $d\bar{X}_i/d\bar{X}_{sys}$ of Flat Boundary Surface 221
8.4 Jacobian Matrix $d\bar{X}_i/d\bar{X}_{sys}$ of Spherical Boundary Surface 226
Appendix 1 233
Appendix 2 236
Appendix 3 238
Appendix 4 241
References 243

9 Prism Analysis 245
9.1 Retro-reflectors 245
9.1.1 Corner-Cube Mirror 245
9.1.2 Solid Glass Corner-Cube 247
9.2 Dispersing Prisms 248
9.2.1 Triangular Prism 249
9.2.2 Pellin-Broca Prism and Dispersive Abbe Prism 250
9.2.3 Achromatic Prism and Direct Vision Prism 251
9.3 Right-Angle Prisms 253
9.4 Porro Prism 254
9.5 Dove Prism 255
9.6 Roofed Amici Prism 256
9.7 Erecting Prisms 257
9.7.1 Double Porro Prism 257
9.7.2 Porro-Abbe Prism 259
9.7.3 Abbe-Koenig Prism 260
9.7.4 Roofed Pechan Prism 261
9.8 Penta Prism 262
Appendix 1 263
References 264

10 Prism Design Based on Image Orientation 267
10.1 Reflector Matrix and Image Orientation Function 267
10.2 Minimum Number of Reflectors 274
10.2.1 Right-Handed Image Orientation Function 275
10.2.2 Left-Handed Image Orientation Function 277
10.3 Prism Design Based on Unit Vectors of Reflectors 277
10.4 Exact Analytical Solutions for Single Prism with Minimum Number of Reflectors 282
10.4.1 Right-Handed Image Orientation Function 284
10.4.2 Left-Handed Image Orientation Function 284

10.4.3 Solution for Right-Handed Image Orientation Function . . . 285
10.4.4 Solution for Left-Handed Image Orientation Function . . . 288
10.5 Prism Design for Given Image Orientation Using Screw Triangle Method . . . 291
References . . . 294

11 Determination of Prism Reflectors to Produce Required Image Orientation . . . 295
11.1 Determination of Reflector Equations . . . 295
11.2 Determination of Prism with n = 4 Boundary Surfaces to Produce Specified Right-Handed Image Orientation . . . 298
11.3 Determination of Prism with n = 5 Boundary Surfaces to Produce Specified Left-Handed Image Orientation . . . 302
Reference . . . 307

12 Optically Stable Systems . . . 309
12.1 Image Orientation Function of Optically Stable Systems . . . 309
12.2 Design of Optically Stable Reflector Systems . . . 312
12.2.1 Stable Systems Comprising Two Reflectors . . . 312
12.2.2 Stable Systems Comprising Three Reflectors . . . 313
12.2.3 Stable Systems Comprising More Than Three Reflectors . . . 314
12.3 Design of Optically Stable Prism . . . 316
Reference . . . 318

13 Point Spread Function, Caustic Surfaces and Modulation Transfer Function . . . 319
13.1 Infinitesimal Area on Image Plane . . . 320
13.2 Derivation of Point Spread Function Using Irradiance Method . . . 322
13.3 Derivation of Spot Diagram Using Irradiance Method . . . 326
13.4 Caustic Surfaces . . . 327
13.4.1 Caustic Surfaces Formed by Point Source . . . 328
13.4.2 Caustic Surfaces Formed by Collimated Rays . . . 330
13.5 MTF Theory for Any Arbitrary Direction of OBDF . . . 333
13.6 Determination of MTF for Any Arbitrary Direction of OBDF Using Ray-Counting and Irradiance Methods . . . 336
13.6.1 Ray-Counting Method . . . 336
13.6.2 Irradiance Method . . . 337
Appendix 1 . . . 344
Appendix 2 . . . 345
Appendix 3 . . . 346
Appendix 4 . . . 346
References . . . 349

14 Optical Path Length and Its Jacobian Matrix ... 353
14.1 Jacobian Matrix of OPL_i Between (i−1)th and ith Boundary Surfaces ... 353
14.1.1 Jacobian Matrix of OPL_i with Respect to Incoming Ray $\bar{R}_{i-1}$ ... 354
14.1.2 Jacobian Matrix of OPL_i with Respect to Boundary Variable Vector $\bar{X}_i$ ... 355
14.2 Jacobian Matrix of OPL Between Two Incidence Points ... 357
14.3 Computation of Wavefront Aberrations ... 362
14.4 Merit Function Based on Wavefront Aberration ... 368
References ... 369

Part III A Bright Light for Geometrical Optics (Second-Order Derivative Matrices of a Ray and its OPL)

15 Wavefront Aberration and Wavefront Shape ... 373
15.1 Hessian Matrix $\partial^2\bar{R}_i/\partial\bar{R}_{i-1}^2$ for Flat Boundary Surface ... 374
15.1.1 Hessian Matrix of Incidence Point $\bar{P}_i$ ... 375
15.1.2 Hessian Matrix of Unit Directional Vector $\bar{\ell}_i$ of Reflected Ray ... 375
15.1.3 Hessian Matrix of Unit Directional Vector $\bar{\ell}_i$ of Refracted Ray ... 375
15.2 Hessian Matrix $\partial^2\bar{R}_i/\partial\bar{R}_{i-1}^2$ for Spherical Boundary Surface ... 376
15.2.1 Hessian Matrix of Incidence Point $\bar{P}_i$ ... 376
15.2.2 Hessian Matrix of Unit Directional Vector $\bar{\ell}_i$ of Reflected Ray ... 377
15.2.3 Hessian Matrix of Unit Directional Vector $\bar{\ell}_i$ of Refracted Ray ... 377
15.3 Hessian Matrix of $\bar{R}_i$ with Respect to Variable Vector $\bar{X}_0$ of Source Ray ... 378
15.4 Hessian Matrix of OPL_i with Respect to Variable Vector $\bar{X}_0$ of Source Ray ... 380
15.5 Change of Wavefront Aberration Due to Translation of Point Source $\bar{P}_0$ ... 382
15.6 Wavefront Shape Along Ray Path ... 387
15.6.1 Tangent and Unit Normal Vectors of Wavefront Surface ... 389
15.6.2 First and Second Fundamental Forms of Wavefront Surface ... 390
15.6.3 Principal Curvatures of Wavefront ... 392
Appendix 1 ... 399
Appendix 2 ... 400
References ... 403

16 Hessian Matrices of Ray $\bar{R}_i$ with Respect to Incoming Ray $\bar{R}_{i-1}$ and Boundary Variable Vector $\bar{X}_i$ 405
16.1 Hessian Matrix of a Ray with Respect to System Variable Vector 405
16.2 Hessian Matrix $\partial^2\bar{R}_i/\partial\bar{X}_i^2$ for Flat Boundary Surface 407
16.2.1 Hessian Matrix of Incidence Point $\bar{P}_i$ 407
16.2.2 Hessian Matrix of Unit Directional Vector $\bar{\ell}_i$ of Reflected Ray 408
16.2.3 Hessian Matrix of Unit Directional Vector $\bar{\ell}_i$ of Refracted Ray 408
16.3 Hessian Matrix $\partial^2\bar{R}_i/\partial\bar{X}_i\partial\bar{R}_{i-1}$ for Flat Boundary Surface 409
16.3.1 Hessian Matrix of Incidence Point $\bar{P}_i$ 410
16.3.2 Hessian Matrix of Unit Directional Vector $\bar{\ell}_i$ of Reflected Ray 411
16.3.3 Hessian Matrix of Unit Directional Vector $\bar{\ell}_i$ of Refracted Ray 411
16.4 Hessian Matrix $\partial^2\bar{R}_i/\partial\bar{X}_i^2$ for Spherical Boundary Surface 412
16.4.1 Hessian Matrix of Incidence Point $\bar{P}_i$ 412
16.4.2 Hessian Matrix of Unit Directional Vector $\bar{\ell}_i$ of Reflected Ray 413
16.4.3 Hessian Matrix of Unit Directional Vector $\bar{\ell}_i$ of Refracted Ray 413
16.5 Hessian Matrix $\partial^2\bar{R}_i/\partial\bar{X}_i\partial\bar{R}_{i-1}$ for Spherical Boundary Surface 414
16.5.1 Hessian Matrix of Incidence Point $\bar{P}_i$ 415
16.5.2 Hessian Matrix of Unit Directional Vector $\bar{\ell}_i$ of Reflected Ray 415
16.5.3 Hessian Matrix of Unit Directional Vector $\bar{\ell}_i$ of Refracted Ray 416
Appendix 1 417
Appendix 2 420
Reference 423

17 Hessian Matrix of Boundary Variable Vector $\bar{X}_i$ with Respect to System Variable Vector $\bar{X}_{sys}$ 425
17.1 Hessian Matrix $\partial^2\bar{X}_0/\partial\bar{X}_{sys}^2$ of Source Ray 425
17.2 Hessian Matrix $\partial^2\bar{X}_i/\partial\bar{X}_{sys}^2$ for Flat Boundary Surface 426
17.3 Design of Optical Systems Possessing Only Flat Boundary Surfaces 430
17.4 Hessian Matrix $\partial^2\bar{X}_i/\partial\bar{X}_{sys}^2$ for Spherical Boundary Surface 433
17.5 Design of Retro-reflectors 437

Appendix 1 441
Appendix 2 443
Appendix 3 445
Appendix 4 446
References 449

18 Hessian Matrix of Optical Path Length 451
18.1 Determination of Hessian Matrix of OPL 451
18.1.1 Hessian Matrix of OPL_i with Respect to Incoming Ray $\bar{R}_{i-1}$ 453
18.1.2 Hessian Matrix of OPL_i with Respect to $\bar{X}_i$ and $\bar{R}_{i-1}$ 453
18.1.3 Hessian Matrix of OPL_i with Respect to Boundary Variable Vector $\bar{X}_i$ 453
18.2 System Analysis Based on Jacobian and Hessian Matrices of Wavefront Aberrations 454
18.3 System Design Based on Jacobian and Hessian Matrices of Wavefront Aberrations 456
Reference 457

VITA 459

Symbols and Notation

$\Vert\ \Vert$	Magnitude of a vector
$\vert\ \vert$	Absolute value of a scalar quantity
$\det(\bar{A})$	Determinant of a square matrix $\bar{A}$
(1, 2, 3)	Sequence has components 1, 2, and 3
$G = \{1, 2, 3\}$	G is a set with components 1, 2, and 3
$3 \in G$	3 is a component of set G
$4 \notin G$	4 is not a component of set G
$\bar{I}_{3\times 3}$	3×3 identity matrix
$\bar{0}_{2\times 3}$	2×3 zero matrix
$C\theta$	$C\theta = \cos\theta$
$S\theta$	$S\theta = \sin\theta$
$\bar{i}, \bar{j}, \bar{k}$	Unit directional vectors of x, y, and z axes of a coordinate frame
$(xyz)_0$	World coordinate frame
${}^{g}\bar{r}_i$	The ith boundary surface expressed w.r.t. $(xyz)_g$
$\bar{r}_i$	The ith boundary surface expressed w.r.t. $(xyz)_0$
$(xyz)_i$	Boundary coordinate frame imbedded in $\bar{r}_i$
$(xyz)_{ej}$	Element coordinate frame imbedded in jth element
${}^{h}\bar{A}_g$	Pose matrix of $(xyz)_g$ w.r.t. $(xyz)_h$
${}^{0}\bar{A}_i$	Pose matrix of $(xyz)_i$ w.r.t. $(xyz)_0$
$\mathrm{tran}(t_x, t_y, t_z)$	Pose matrix corresponding to the translation along $t_x\bar{i} + t_y\bar{j} + t_z\bar{k}$
$\mathrm{rot}(\bar{x}, \theta)$	Rotation matrix about the unit direction vector of x axis
$\mathrm{rot}(\bar{y}, \theta)$	Rotation matrix about the unit direction vector of y axis
$\mathrm{rot}(\bar{z}, \theta)$	Rotation matrix about the unit direction vector of z axis
$\mathrm{rot}(\bar{\kappa}, \theta)$	Rotation matrix around a unit vector $\bar{\kappa} = \kappa_x\bar{i} + \kappa_y\bar{j} + \kappa_z\bar{k}$

$RPY(\omega_z, \omega_y, \omega_x)$	Pose matrix by roll, pitch, and yaw motions, $RPY(\omega_z, \omega_y, \omega_x) = rot(\bar{z}, \omega_z) rot(\bar{y}, \omega_y) rot(\bar{x}, \omega_x)$
n	Total number of boundary surfaces of an optical system
k	Total number of elements of an optical system
i	Boundary surface index, i = 1,2,…,n
j	Element index, j = 1,2,…,k
${}^g\bar{n}_i$	Unit normal vector (expressed w.r.t. $(xyz)_g$) of ${}^g\bar{r}_i$, ${}^g\bar{n}_i = [{}^gn_{ix} \quad {}^gn_{iy} \quad {}^gn_{iz} \quad 0]^T$
$\bar{n}_i$	Unit normal vector (expressed w.r.t. $(xyz)_0$) of $\bar{r}_i$, $\bar{n}_i = [n_{ix} \quad n_{iy} \quad n_{iz} \quad 0]^T$
$\overline{X}_0$	Variable vector of source ray, $\overline{X}_0 = [P_{0x} \quad P_{0y} \quad P_{0z} \quad \alpha_0 \quad \beta_0]^T$
$\overline{X}_i$	Boundary variable vector of $\bar{r}_i$ $\overline{X}_i = [J_{ix} \quad J_{iy} \quad J_{iz} \quad e_i \quad \xi_{i-1} \quad \xi_i]^T$ for a flat boundary surface $\overline{X}_i = [t_{ix} \quad t_{iy} \quad t_{iz} \quad \omega_{ix} \quad \omega_{iy} \quad \omega_{iz} \quad \xi_{i-1} \quad \xi_i \quad R_i]^T$ for a spherical boundary surface
$\overline{X}_{ej}$	Element variable vector of the jth element
$\overline{X}_{sys}$	System variable vector of an optical system
${}^g\bar{P}_i$	Incidence point (expressed w.r.t. $(xyz)_g$) at ${}^g\bar{r}_i$
${}^g\bar{\ell}_i$	Unit directional vector (expressed w.r.t. $(xyz)_g$) of reflected/or refracted ray at ${}^g\bar{r}_i$
${}^g\bar{R}_i$	Reflected or refracted ray (expressed w.r.t. $(xyz)_g$) at ${}^g\bar{r}_i$, ${}^g\bar{R}_i = [{}^g\bar{P}_i \quad {}^g\bar{\ell}_i]^T$
$\bar{P}_i$	Incidence point (expressed w.r.t. $(xyz)_0$) at $\bar{r}_i$
$\bar{\ell}_i$	Unit directional vector (expressed w.r.t. $(xyz)_0$) of reflected/or refracted ray at $\bar{r}_i$
$\bar{R}_0$	Source ray (expressed w.r.t. $(xyz)_0$) originating from source point $\overline{P}_0$, $\bar{R}_0 = [\bar{P}_0 \quad \bar{\ell}_0]^T$
$\bar{\ell}_0$	Unit directional vector (expressed w.r.t. $(xyz)_0$) of a source ray, $\bar{\ell}_0 = [C\beta_0 C(90° + \alpha_0) \quad C\beta_0 S(90° + \alpha_0) \quad S\beta_0 \quad 0]^T$
(α_0, β_0)	Spherical coordinates of $\bar{\ell}_0$
$\bar{R}_i$	Reflected or refracted ray (expressed w.r.t. $(xyz)_0$) at $\bar{r}_i$, $\bar{R}_i = [\bar{P}_i \quad \bar{\ell}_i]^T$
${}^i\bar{q}_i$	Generating curve of the ith boundary surface
${}^i\bar{\eta}_i$	Unit normal vectors of ${}^i\bar{q}_i$
β_i	Parameter of ${}^i\bar{q}_i$
α_i	Parameters of ${}^i\bar{r}_i$ with $0 \leq \alpha_i < 2\pi$
s_i	Sign parameter, $s_i = +1$ or $s_i = -1$
λ_i	Geometrical length from point $\overline{P}_{i-1}$ to $\overline{P}_i$

θ_i	Incidence angle (=reflection angle) at $\bar{r}_i$
$\underline{\theta}_i$	Refraction angle at $\bar{r}_i$
ξ_i	Refractive index of medium i
ξ_{air}	Refractive index of air
$N_i = \xi_{i-1}/\xi_i$	Refractive index of medium i – 1 relative to that of medium i
$\bar{m}_i$	Common unit normal vector of active unit normal vector $\bar{n}_i$ and $\bar{\ell}_{i-1}$ at $\bar{r}_i$
q_{ej}	Thickness of element j
ξ_{ej}	Refractive index of element j
$\bar{\ell}_{0/chief}$	Unit directional vector of the chief ray of a source point. $\bar{\ell}_{0/chief} = [C\beta_{0/chief}C(90^\circ + \alpha_{0/chief}) \quad C\beta_{0/chief}S(90^\circ + \alpha_{0/chief}) \quad S\beta_{0/chief} \quad 0]^T$
$\alpha_0(\beta_{0/chief})$	Sagittal cone generated by sweeping $\bar{\ell}_0$ with $\beta_0 = \beta_{0/chief}$
${}^{ej}\bar{A}_i$	Pose matrix of $(xyz)_i$ w.r.t. element coordinate frame $(xyz)_{ej}$
R_i	Radius of ith spherical boundary surface
κ_i	Curvature of $\bar{r}_i$, $R_i = 1/\kappa_i$
L_{ej}	Total number of boundary surfaces of element j
m_{ej}	$i = m_{ej}$ is the last boundary surface when a ray passed through the jth element
q_{sys}	Dimension of $\bar{X}_{sys} = [\bar{X}_0^T \quad \bar{X}_\xi^T \quad \bar{X}_\kappa^T \quad \bar{X}_{rest}^T]^T$
q_0	Dimension of $\bar{X}_0$, $q_0 = 5$
$\bar{X}_\xi$	Variable vector constituted by all refractive indices of a system
q_ξ	Dimension of $\bar{X}_\xi$
$\bar{X}_R$	Variable vector constituted by all radii of spherical boundary surfaces in a system
q_R	Dimension of $\bar{X}_R$
$\bar{X}_{rest}$	$\bar{X}_{rest} = \bar{X}_{sys} - \bar{X}_0 - \bar{X}_\xi - \bar{X}_R$
q_{rest}	Dimension of $\bar{X}_{rest}$
rms	Root-mean-square value of the radius of a spot
(x_n, z_n)	In-plane coordinates of an image plane
$B(x_n, z_n)$	Point spread function (PSF) at (x_n, z_n)
ϖ	Phase shift
OBDF	Object brightness distribution function
FD	Finite difference
PSF	Point spread function
MTF	Modulation transfer function
PSD	Position sensitive detector
LSF	Line spread function
I	Luminous intensity

$I(x_0)$	OBDF in the sagittal direction, $I(x_0) = b_0 + b_1 C(2\pi \nu x_0)$
$I(x_n, z_n)$	Luminous intensity at (x_n, z_n) of an image plane
ν	Frequency of OBDF
M_0	Modulation of OBDF at $\bar{P}_0$
M_n	Modulation of $I(x_n, z_n)$ on an image plane
$\bar{\Phi}$	Merit function
$J_i(u, v)$	Jacobian matrix between $\bar{X}_i$ and $\bar{X}_{sys}$, $\partial\bar{X}_i/\partial\bar{X}_{sys} = [J_i(u, v)]$
$d\pi_n$	Infinitesimal area in an image plane
ψ_0	Solid angle subtended by a ray cone with its apex at $\bar{P}_0$
$\bar{S}_i = \partial\bar{R}_i/\partial\bar{X}_i$	Jacobian matrix of ray $\bar{R}_i$ w.r.t. $\overline{X}_i$
$\bar{M}_i = \partial\bar{R}_i/\partial\bar{R}_{i-1}$	Jacobian matrix of ray $\bar{R}_i$ w.r.t. incoming ray $\bar{R}_{i-1}$
OPL	Optical path length
OPL_i	The OPL between points $\bar{P}_{i-1}$ and $\bar{P}_i$,$OPL_i = \xi_{i-1}\lambda_i$
$OPL(\bar{P}_g, \bar{P}_h)$	The OPL between two incidence points $\bar{P}_g$ and $\bar{P}_h$
$\bar{\Omega}$	Wavefront
$\bar{r}_{ref}$	Reference sphere for computation of wavefront aberration
$\bar{P}_\Omega$	Incidence point of a ray at wavefront $\bar{\Omega}$
$W(\bar{X}_0)$	Wavefront aberration
P-V	Peak-to-valley wavefront aberration
$\bar{\Phi} = \begin{bmatrix} \bar{a} & \bar{b} & \bar{c} \end{bmatrix}$	Image orientation function, which is a merit function
n_{mini}	Minimum number of reflectors a prism needs to produce a required image orientation
$\bar{m}$	Auxiliary unit vector in discussing image orientation, $\bar{m} = \begin{bmatrix} m_x & m_y & m_z & 0 \end{bmatrix}^T$
μ	Auxiliary angle in discussing image orientation
$\bar{t}_\Omega$	A tangent vector of wavefront $\bar{\Omega}$
$\bar{n}_\Omega$	The unit normal vector of wavefront $\bar{\Omega}$
I_Ω	First fundamental form of the wavefront $\bar{\Omega}$
II_Ω	Second fundamental form of the wavefront $\bar{\Omega}$
κ_Ω	Principal curvatures at wavefront $\bar{\Omega}$
$\bar{H}_\Omega = \begin{bmatrix} h_{\Omega 1} & h_{\Omega 2} \end{bmatrix}^T$	Principal direction at wavefront $\bar{\Omega}$

Part I
A New Light on Old Geometrical Optics (Raytracing Equations of Geometrical Optics)

Although the basics of geometrical optics have changed very little since their original inception, their applications have undergone rapid, extensive, dynamic and fascinating changes. Thus, while Snell's law continues to hold, anyone wishing to "practice optics" is well advised to acquire a solid grounding in ray tracing. Accordingly, Chap. 1 of this book provides a comprehensive overview of the notations and modeling conventions required to analyze optical systems using a ray-tracing approach based on homogeneous coordinate notation. The principles of skew-ray tracing at flat and spherical boundary surfaces are then addressed in Chap. 2. Chapter 3 introduces the modeling techniques required to analyze the behavior and performance of axial-symmetrical and non-axially symmetrical optical systems. Chapter 4 first describes a paraxial optics using 6×6 matrix raytracing equations. These equations are then simplified to 2×2 ray-tracing matrices of conventional paraxial optics. Finally, Chap. 5 presents an approach for determining the cardinal points of axis symmetrical optical systems. The contents of Chaps. 1–5 are likely to be known to the readers. However, they serve as essential reading in providing a solid foundation for the modeling work presented later in Parts II and III of this book.

"*Give me a place to stand and I will move the Earth.*" —Archimedes, Greek mathematician, physicist, engineer, inventor, and astronomer

Chapter 1
Mathematical Background

The homogeneous coordinate notation is a powerful mathematical tool used in a wide range of fields, including the motion of rigid bodies [1, 2], robotics [3], gearing theory [4], and computer graphics [5]. Previous publications of geometrical optics used vector notation, which is comparatively awkward for computations for non-axially symmetrical systems. In order to circumvent its limitations, this book employs homogeneous coordinate notation. Accordingly, this chapter briefly reviews the basic principles of the homogeneous coordinate notation in order to set the mathematical modeling presented in the rest of the book in proper context.

1.1 Foundational Mathematical Tools and Units

Sets, and their associated theories, are important foundational tools in such diverse fields as mathematics and philosophy. In general, any collection of objects that can be either listed or described by some predicate can be regarded as constituting a set. Table 1.1 lists some of the most commonly used symbols in set theory. As shown, the components of a set are enclosed within two curly brackets. Furthermore, the symbol $\in$ indicates the membership of an object x to a particular set. By contrast, the negation of membership is indicated by "$x \notin G$" (i.e., object x is not a member of set G). Given two sets G and F, $G = F$ if and only if the two sets contain exactly the same components. Furthermore, the difference between the two sets, denoted as "G minus F" (or $G - F$) indicates the result obtained when 'subtracting' all of the components of F from those of G. Note that in this book, the notation $x \in G$ indicates that x is a single particular component of set G. By contrast, the notation "$x = 1$ to $x = 50$" indicates that x runs continuously from 1 to 50.

P.D. Lin, *Advanced Geometrical Optics*, Progress in Optical Science
and Photonics 4, DOI 10.1007/978-981-10-2299-9_1

Table 1.1 Commonly used symbols in set theory

Symbol	Symbol name	Meaning/definiton	Example
{ }	Set	A collection of components	G = {3, 7, 9, 14}, F = {9, 14, 28}
$a \in G$	Component of	Set membership	G = {3, 9, 14}, $3 \in G$
$X \notin G$	Not component of	No set membership	G = {3, 9, 14}, $1 \notin G$
G = F	Equality	Both sets have the same members	G = {3, 9, 14}, F = {3, 9, 14}, G = F
G − F	Relative complement	Objects that belong to G and not to F	G = {3, 9, 14}, F = {1, 2, 3}, G − F = {9, 14}

In mathematics, a sequence is an ordered list of objects or events (known as components), and is indicated by the placement of rounded brackets around the constituent objects (or events). The number of ordered components within the sequence (possibly infinite) is referred to as the sequence dimension. Unlike a set, the same components can appear multiple times in a sequence. Therefore, (1, 2, 3), (3, 2, 1) and (1, 1, 2, 3) are three different sequences, while {1, 2, 3} and {3, 2, 1} are the same set.

In mathematics, a 2-D matrix is a rectangular array of numbers, symbols or expressions, arranged in rows and columns and enclosed within square brackets. The individual items in a matrix are known as components. Moreover, the component located at the intersection of the wth row and vth column of the matrix is referred to as the (w, v)th component of the matrix. A matrix is sometimes referred to using a formula expressed in terms of its (w, v)th component. For example, if the (w, v)th component of matrix $\bar{A}$ is denoted as a_{wv}, $\bar{A}$ may be referred to as $\bar{A} = [a_{wv}]$. Furthermore, a column matrix $\bar{X}_{sys}$ is usually expressed in a horizontal form as $\bar{X}_{sys} = \begin{bmatrix} \bar{X}_0^T & \bar{X}_\xi^T & \bar{X}_\kappa^T & \bar{X}_{rest}^T \end{bmatrix}^T$ in order to save space. By convention, a matrix is denoted by the use of an overbar over its symbol. Thus, $\bar{R}$ represents a matrix, whereas R denotes a scalar. A scalar can also be represented by an underbar under its symbol (e.g., $\underline{\theta}$) or a caret over it (e.g., $\hat{m}$). Consequently, R, R_i, $\hat{R}$ and $\underline{R}$ all represent scalar properties. Finally, a vector can be expressed using either a sequence notation, e.g., (1, 2, 3), or a column matrix notation, e.g., $\begin{bmatrix} 1 & 2 & 3 & 1 \end{bmatrix}^T$).

The notations described above are widely used in the engineering field. However, there are insufficient symbols available to uniquely describe all of the optical properties and parameters of interest. As a result, some symbols are reused in different chapters of this book to represent different physical quantities. Typical examples of such "temporary" symbols include D, E, F, G, H, L, M, P, Q, T, U, f, g, h, q, η, and ψ. Note, however, the symbols in the Symbol and Notation list are always used consistently (i.e., without change in meaning) throughout the entire book.

In this book, the notation $\| \ \|$ (i.e., two bars on either side of a vector) is used to denote the magnitude of the vector, i.e., the magnitude of vector $\bar{P} = P_x\bar{i} + P_y\bar{j} + P_z\bar{k}$ is denoted as

$$\|\bar{P}\| = \sqrt{P_x^2 + P_y^2 + P_z^2}. \tag{1.1}$$

By contrast, the notation $| \ |$ (i.e., one bar on either side of a symbol) is used to denote the absolute value of a scalar quantity. Finally, the determinant of a matrix, $\bar{A}$, is denoted as $\det(\bar{A})$.

Unless stated otherwise, the units of length and time used in this book are millimeters and seconds, respectively. However, in measuring angles, both degrees (e.g., $\theta_i = 90°$) and radians (e.g., $\theta_i = \pi/2$) are used. Within each chapter (and in the appendices), the equations are placed within rounded brackets and are numbered sequentially starting from the chapter. Moreover, the equations presented in the appendices are referred to in the main body of the text using both the equation number and the corresponding chapter number (e.g., "Eq. (5.3) of Appendix 2 in Chap. 5" or "Eq. (5.3) of Appendix 2 in this chapter"). Finally, each reference is cited using a unique reference number enclosed by square brackets (e.g., [3]).

1.2 Vector Notation

In homogeneous coordinate notation of 3-Dimensional (3-D) space, the ith position vector of coordinate frame $(xyz)_g$, i.e., ${}^g\bar{P}_i = P_{ix}\bar{i} + P_{iy}\bar{j} + P_{iz}\bar{k}$, in which $\bar{i}$, $\bar{j}$ and $\bar{k}$ are unit vectors along the x_g, y_g and z_g axes, respectively, is written in the following column matrix form:

$${}^g\bar{P}_i = \begin{bmatrix} P_{ix} \\ P_{iy} \\ P_{iz} \\ 1 \end{bmatrix}, \tag{1.2}$$

where the post-script "i" of symbol ${}^g\bar{P}_i$ indicates that the position vector refers to the ith boundary surface of the optical system, while the pre-superscript "g" indicates that the vector is defined with respect to coordinate frame $(xyz)_g$. For simplicity, the pre-superscript "g" is omitted from the individual components within the ith position vector since the frame to which the vector (and its components) refers is readily induced from the leading symbol of the vector, i.e., ${}^g\bar{P}_i$. Thus, the position vector is represented simply by the row matrix, ${}^g\bar{P}_i = \begin{bmatrix} P_{ix} & P_{iy} & P_{iz} & 1 \end{bmatrix}^T$, where the post-superscript "T" indicates the transpose of the row vector as a column vector. The same notation rules are also applicable to the ith unit directional vector, i.e.,

$$^{g}\bar{\ell}_{i} = \begin{bmatrix} \ell_{ix} \\ \ell_{iy} \\ \ell_{iz} \\ 0 \end{bmatrix}, \tag{1.3}$$

where $\ell_{ix}^2 + \ell_{iy}^2 + \ell_{iz}^2 = 1$.

The vector $[0 \quad 0 \quad 0 \quad 0]^T$ is undefined. Moreover, the dot product of two vectors, e.g., ${}^{g}\bar{P}_{i} = [P_{ix} \quad P_{iy} \quad P_{iz} \quad 1]^T$ and ${}^{g}\bar{P}_{j} = [P_{jx} \quad P_{jy} \quad P_{jz} \quad 1]^T$, is a scalar with the form

$${}^{g}\bar{P}_{i} \cdot {}^{g}\bar{P}_{j} = P_{ix}P_{jx} + P_{iy}P_{jy} + P_{iz}P_{jz}. \tag{1.4}$$

Note that ${}^{g}\bar{P}_{i}$ and ${}^{g}\bar{P}_{j}$ are both expressed with respect to the same coordinate frame, i.e., $(xyz)_g$. The cross product of ${}^{g}\bar{P}_{i}$ and ${}^{g}\bar{P}_{j}$, denoted as ${}^{g}\bar{P}_{i} \times {}^{g}\bar{P}_{j}$, is another vector perpendicular to the plane formed by ${}^{g}\bar{P}_{i}$ and ${}^{g}\bar{P}_{j}$, and is computed as

$${}^{g}\bar{P}_{i} \times {}^{g}\bar{P}_{j} = \begin{bmatrix} P_{iy}P_{jz} - P_{iz}P_{jy} \\ P_{iz}P_{jx} - P_{ix}P_{jz} \\ P_{ix}P_{jy} - P_{iy}P_{jx} \\ 1 \end{bmatrix}. \tag{1.5}$$

Note that ${}^{g}\bar{P}_{i}$ and ${}^{g}\bar{P}_{j}$ are both expressed with respect to coordinate frame $(xyz)_g$.

Example 1.1 The position vector of point $\bar{P}_1$ in Fig. 1.1 can be represented as ${}^{4}\bar{P}_{1} = [3 \quad 4 \quad 2 \quad 1]^T$ or ${}^{5}\bar{P}_{1} = [-4 \quad 3 \quad 2 \quad 1]^T$ with respect to coordinate frame $(xyz)_4$ or $(xyz)_5$, respectively.

Example 1.2 The unit directional vector $\bar{\ell}_4$ shown in Fig. 1.2 can be represented as ${}^{4}\bar{\ell}_{4} = [1 \quad 0 \quad 0 \quad 0]^T$ or ${}^{5}\bar{\ell}_{4} = [1/\sqrt{2} \quad 1/\sqrt{2} \quad 0 \quad 0]^T$ with respect to coordinate frame $(xyz)_4$ or $(xyz)_5$, respectively.

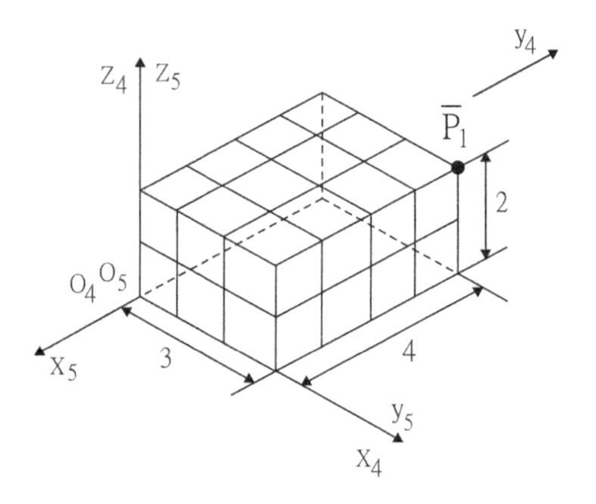

Fig. 1.1 Position vector of point $\bar{P}_1$

1.3 Coordinate Transformation Matrix

In this book, the orientations of the three axes of a coördinate frame relative to one another are assumed to comply with the right-hand rule, unless stated otherwise. Moreover, the homogeneous coordinate transformation matrix ${}^{h}\bar{A}_{g}$ (referred to as the pose matrix hereafter) is used to describe the relative position and orientation of frame $(xyz)_g$ with respect to another coordinate frame $(xyz)_h$ in 3-D space, and has the form

$$ {}^{h}\bar{A}_{g} = \begin{bmatrix} {}^{h}\bar{I} & {}^{h}\bar{J} & {}^{h}\bar{K} & {}^{h}\bar{t} \end{bmatrix} = \begin{bmatrix} I_x & J_x & K_x & t_x \\ I_y & J_y & K_y & t_y \\ I_z & J_z & K_z & t_z \\ 0 & 0 & 0 & 1 \end{bmatrix} = \begin{bmatrix} \bar{R}_{3\times3} & \bar{t}_{3\times1} \\ \bar{0}_{1\times3} & 1 \end{bmatrix}. \tag{1.6} $$

Equation (1.6) is a 4 × 4 matrix in which each column corresponds to a vector. Moreover, ${}^{h}\bar{t}$ describes the position of the origin o_g of coordinate frame $(xyz)_g$ with respect to coordinate frame $(xyz)_h$ (see Fig. 1.3). Note that no restriction is imposed on the value of ${}^{h}\bar{t}$, provided that the vector can reach the desired position. Vectors ${}^{h}\bar{I}$, ${}^{h}\bar{J}$ and ${}^{h}\bar{K}$ in Eq. (1.6) describe the orientations of the three unit vectors of coordinate frame $(xyz)_g$ with respect to coordinate frame $(xyz)_h$. It is noted that ${}^{h}\bar{I}$, ${}^{h}\bar{J}$ and ${}^{h}\bar{K}$ are not independent of one another. More specifically, their nine components are related via the following equations:

(a) ${}^{h}\bar{I}$ is the vector cross product of ${}^{h}\bar{J}$ and ${}^{h}\bar{K}$, i.e.,

$$ {}^{h}\bar{I} = {}^{h}\bar{J} \times {}^{h}\bar{K} = \begin{bmatrix} J_yK_z - J_zK_y \\ J_zK_x - J_xK_z \\ J_xK_y - J_yK_x \\ 0 \end{bmatrix}. \tag{1.7} $$

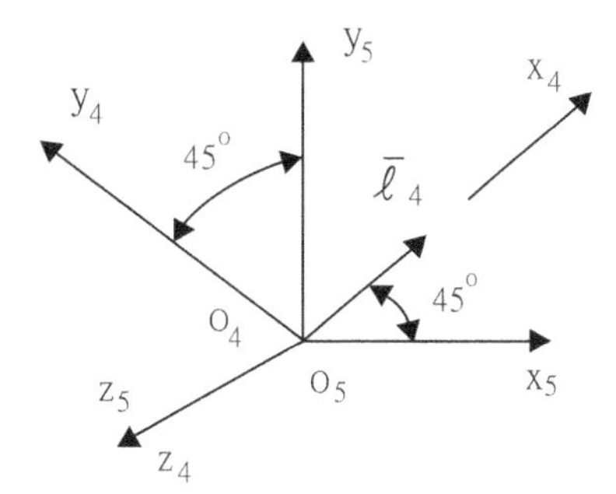

Fig. 1.2 Unit directional vector $\bar{\ell}_4$

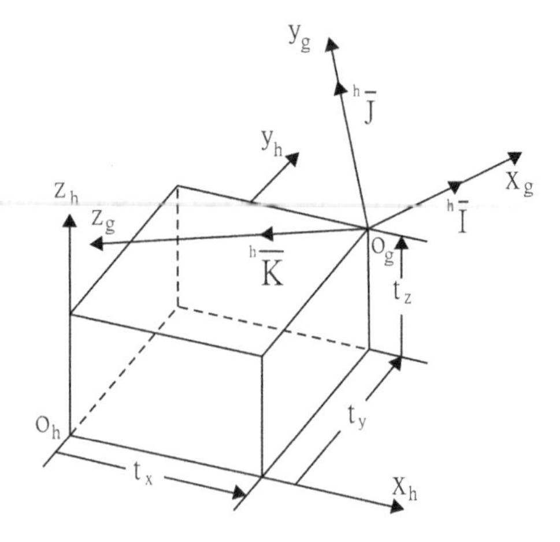

Fig. 1.3 Schematic interpretation of homogeneous coordinate transformation matrix $^h\bar{A}_g$

(b) The vectors $^h\bar{J}$ and $^h\bar{K}$ are of unit magnitude and perpendicular to each other, i.e.,

$$J_X^2 + J_Y^2 + J_Z^2 = 1, \tag{1.8}$$

$$K_X^2 + K_Y^2 + K_Z^2 = 1, \tag{1.9}$$

$$J_X K_X + J_Y K_Y + J_Z K_Z = 0. \tag{1.10}$$

Given a vector $^g\bar{P}_i$, its transformation $^h\bar{P}_i$ can be obtained by the matrix product $^h\bar{P}_i = {}^h\bar{A}_g {}^g\bar{P}_i$.

Furthermore, successive coordinate transformation can be performed by applying the concatenation rule, i.e.,

$$^h\bar{A}_g = {}^h\bar{A}_a {}^a\bar{A}_b \ldots {}^e\bar{A}_f {}^f\bar{A}_g, \tag{1.11}$$

where Eq. (1.11) shows successive coordinate transformation from $(xyz)_h$ to $(xyz)_a$, $(xyz)_a$ to $(xyz)_b$,..., and $(xyz)_f$ to $(xyz)_g$.

Example 1.3 The pose matrix $^4\bar{A}_5$ of the two coordinate frames shown in Fig. 1.4 can be expressed as

$$^4\bar{A}_5 = \begin{bmatrix} -1 & 0 & 0 & 3 \\ 0 & 0 & -1 & 2 \\ 0 & -1 & 0 & 1 \\ 0 & 0 & 0 & 1 \end{bmatrix}.$$

Example 1.4 Given a vector $^5\bar{P}_1 = [-4 \quad 3 \quad 2 \quad 1]^T$ and pose matrix $^4\bar{A}_5$ in Example 1.3, the transformation $^4\bar{P}_1$ is obtained as $^4\bar{P}_1 = {}^4\bar{A}_5 {}^5\bar{P}_1 = [7 \quad 0 \; -2 \quad 1]^T$.

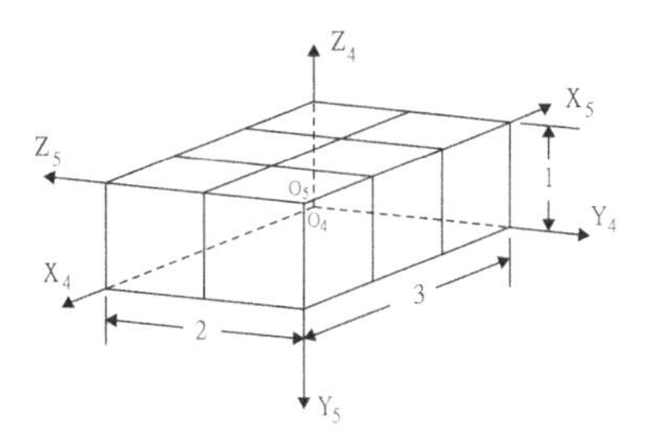

Fig. 1.4 Schematic illustration of pose matrix ${}^{4}\bar{A}_{5}$

1.4 Basic Translation and Rotation Matrices

The transformation matrices corresponding to translations along vectors $t_x\bar{i}$, $t_y\bar{j}$ and $t_z\bar{k}$ with respect to coordinate frame $(xyz)_h$ are given respectively by (see Figs. 1.5, 1.6 and 1.7).

$$ {}^{h}\bar{A}_{g} = \text{tran}(t_x, 0, 0) = \begin{bmatrix} 1 & 0 & 0 & t_x \\ 0 & 1 & 0 & 0 \\ 0 & 0 & 1 & 0 \\ 0 & 0 & 0 & 1 \end{bmatrix}, \tag{1.12} $$

$$ {}^{h}\bar{A}_{g} = \text{tran}(0, t_y, 0) = \begin{bmatrix} 1 & 0 & 0 & 0 \\ 0 & 1 & 0 & t_y \\ 0 & 0 & 1 & 0 \\ 0 & 0 & 0 & 1 \end{bmatrix}, \tag{1.13} $$

$$ {}^{h}\bar{A}_{g} = \text{tran}(0, 0, t_z) = \begin{bmatrix} 1 & 0 & 0 & 0 \\ 0 & 1 & 0 & 0 \\ 0 & 0 & 1 & t_z \\ 0 & 0 & 0 & 1 \end{bmatrix}. \tag{1.14} $$

Note that $\text{tran}(t_x, 0, 0)$, $\text{tran}(0, t_y, 0)$ and $\text{tran}(0, 0, t_z)$ possess a commutative property, i.e., the order of multiplication does not change the result.

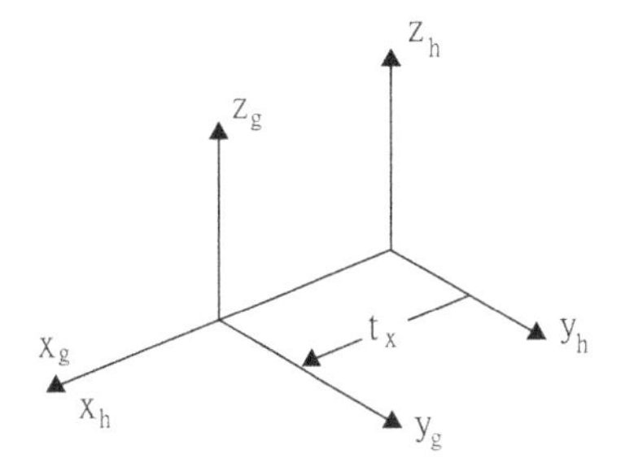

Fig. 1.5 Translation along x_h axis by distance t_x

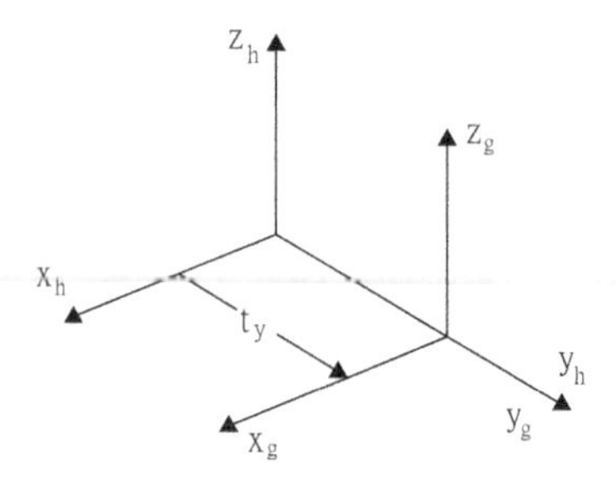

Fig. 1.6 Translation along y_h axis by distance t_y

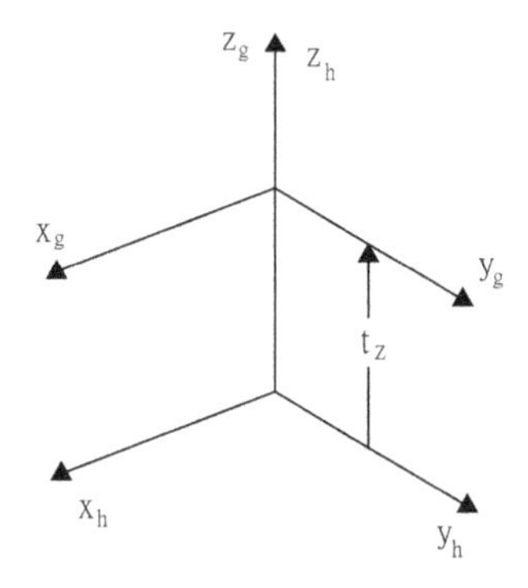

Fig. 1.7 Translation along z_h axis by distance t_z

The transformation matrices corresponding to rotations about the x_h, y_h and z_h axes of coordinate frame $(xyz)_h$ through an angle θ each time are given respectively by (see Figs. 1.8, 1.9 and 1.10)

$$ {}^{h}\bar{A}_g = \text{rot}(\bar{x}, \theta) = \begin{bmatrix} 1 & 0 & 0 & 0 \\ 0 & \cos\theta & -\sin\theta & 0 \\ 0 & \sin\theta & \cos\theta & 0 \\ 0 & 0 & 0 & 1 \end{bmatrix} = \begin{bmatrix} 1 & 0 & 0 & 0 \\ 0 & C\theta & -S\theta & 0 \\ 0 & S\theta & C\theta & 0 \\ 0 & 0 & 0 & 1 \end{bmatrix}, \quad (1.15) $$

$$ {}^{h}\bar{A}_g = \text{rot}(\bar{y}, \theta) = \begin{bmatrix} C\theta & 0 & S\theta & 0 \\ 0 & 1 & 0 & 0 \\ -S\theta & 0 & C\theta & 0 \\ 0 & 0 & 0 & 1 \end{bmatrix}, \quad (1.16) $$

$$ {}^{h}\bar{A}_g = \text{rot}(\bar{z}, \theta) = \begin{bmatrix} C\theta & -S\theta & 0 & 0 \\ S\theta & C\theta & 0 & 0 \\ 0 & 0 & 1 & 0 \\ 0 & 0 & 0 & 1 \end{bmatrix}, \quad (1.17) $$

where S and C denote sine and cosine, respectively. It should be noted that $\bar{x}$, $\bar{y}$ and $\bar{z}$ of Eqs. (1.15), (1.16) and (1.17) are unit directional vectors of the x_h, y_h and z_h

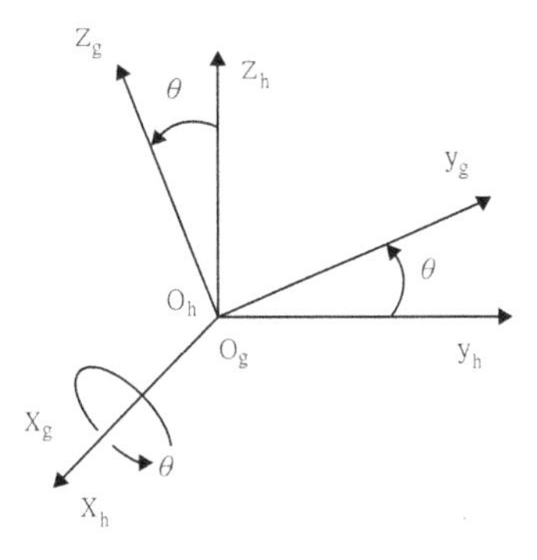

Fig. 1.8 Schematic illustration of $\mathrm{rot}(\bar{x}, \theta)$

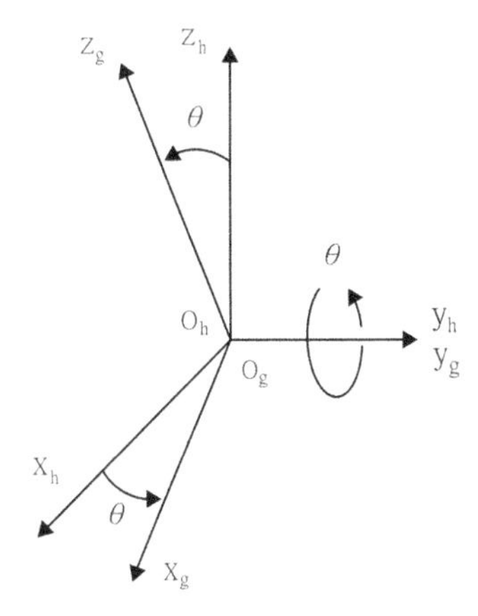

Fig. 1.9 Schematic illustration of $\mathrm{rot}(\bar{y}, \theta)$

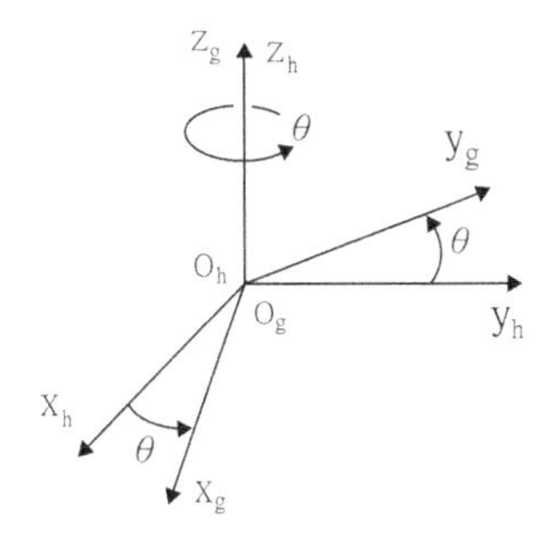

Fig. 1.10 Schematic illustration of $\mathrm{rot}(\bar{z}, \theta)$

axes, not of the x_g, y_g and z_g axes. Furthermore, the transformation matrices have a simple geometric interpretation. For example, in the transformation matrix ${}^h\bar{A}_g = \mathrm{rot}(\bar{z}, \theta)$, the third column represents the z axis and remains constant during rotation, while the first and second columns represent the x and y axes, and vary as shown in Fig. 1.10.

When performing matrix manipulations, the following properties are of great practical use:

$$\mathrm{tran}(t_x, t_y, t_z)\mathrm{tran}(p_x, p_y, p_z) = \mathrm{tran}(t_x + p_x, t_y + p_y, t_z + p_z), \tag{1.18}$$

$$\mathrm{rot}(\bar{x}, \theta)\mathrm{rot}(\bar{x}, \Phi) = \mathrm{rot}(\bar{x}, \theta + \Phi), \tag{1.19}$$

$$\mathrm{rot}(\bar{y}, \theta)\mathrm{rot}(\bar{y}, \Phi) = \mathrm{rot}(\bar{y}, \theta + \Phi), \tag{1.20}$$

$$\mathrm{rot}(\bar{z}, \theta)\mathrm{rot}(\bar{z}, \Phi) = \mathrm{rot}(\bar{z}, \theta + \Phi) \tag{1.21}$$

$$\mathrm{rot}(\bar{x}, \theta)\mathrm{tran}(t_x, 0, 0) = \mathrm{tran}(t_x, 0, 0)\mathrm{rot}(\bar{x}, \theta), \tag{1.22}$$

$$\mathrm{rot}(\bar{y}, \theta)\mathrm{tran}(0, t_y, 0) = \mathrm{tran}(0, t_y, 0)\mathrm{rot}(\bar{y}, \theta), \tag{1.23}$$

$$\mathrm{rot}(\bar{z}, \theta)\mathrm{tran}(0, 0, t_z) = \mathrm{tran}(0, 0, t_z)\mathrm{rot}(\bar{z}, \theta). \tag{1.24}$$

The transformation matrix representing rotation around an arbitrary unit vector ${}^h\bar{\kappa} = \begin{bmatrix} \kappa_X & \kappa_Y & \kappa_Z & 0 \end{bmatrix}^T$, where $\kappa_X^2 + \kappa_Y^2 + \kappa_Z^2 = 1$, located at the origin of coordinate frame $(xyz)_h$ (see Fig. 1.11) has the form

$${}^h\bar{A}_g = \mathrm{rot}(\bar{\kappa}, \theta) = \begin{bmatrix} \kappa_x^2(1 - C\theta) + C\theta & \kappa_x\kappa_y(1 - C\theta) - \kappa_z S\theta & \kappa_x\kappa_z(1 - C\theta) + \kappa_y S\theta & 0 \\ \kappa_x\kappa_y(1 - C\theta) + \kappa_z S\theta & \kappa_y^2(1 - C\theta) + C\theta & \kappa_y\kappa_z(1 - C\theta) - \kappa_x S\theta & 0 \\ \kappa_x\kappa_z(1 - C\theta) - \kappa_y S\theta & \kappa_y\kappa_z(1 - C\theta) + \kappa_x S\theta & \kappa_z^2(1 - C\theta) + C\theta & 0 \\ 0 & 0 & 0 & 1 \end{bmatrix}. \tag{1.25}$$

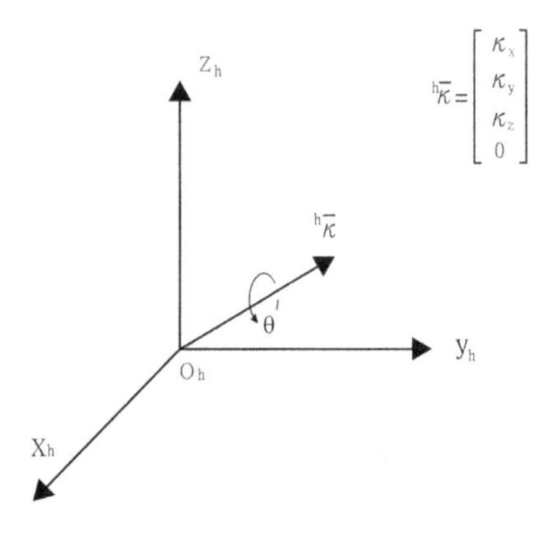

Fig. 1.11 $\mathrm{rot}(\bar{\kappa}, \theta)$ is a rotation around an arbitrary unit vector ${}^h\bar{\kappa}$

Note that the unit vector of Eq. (1.25) is ${}^{h}\bar{\kappa}$, not ${}^{g}\bar{\kappa}$, and $\mathrm{rot}(\bar{\kappa}, -\theta) = \mathrm{rot}(\bar{\kappa}, \theta)^{T}$. Notably, this general rotation matrix enables each of the elementary rotation matrices to be individually obtained. For example, $\mathrm{rot}(\bar{x}, \theta)$ is $\mathrm{rot}(\bar{\kappa}, \theta)$ with ${}^{h}\bar{\kappa} = [1 \quad 0 \quad 0 \quad 0]^{T}$. Substituting these values of ${}^{h}\bar{\kappa}$ into Eq. (1.25) yields the results shown in Eqs. (1.15) to (1.17). Equation (1.25) provides an alternative method for specifying the orientation of a pose matrix. Unfortunately, however, the axis of rotation ${}^{h}\bar{\kappa}$ is not intuitively obvious when searching for a particular orientation.

The pose matrix is also commonly specified using the notation $\mathrm{RPY}(\omega_z, \omega_y, \omega_x)$, in which the orientation is described in terms of three angles (roll, pitch and yaw), i.e.,

$$
\begin{aligned}
{}^{h}\bar{A}_{g} &= \mathrm{RPY}(\omega_z, \omega_y, \omega_x) = \mathrm{rot}(\bar{z}, \omega_z)\mathrm{rot}(\bar{y}, \omega_y)\mathrm{rot}(\bar{x}, \omega_x) \\
&= \begin{bmatrix} C\omega_y C\omega_z & S\omega_x S\omega_y C\omega_z - C\omega_x S\omega_z & C\omega_x S\omega_y C\omega_z + S\omega_x S\omega_z & 0 \\ C\omega_y S\omega_z & C\omega_x C\omega_z + S\omega_x S\omega_y S\omega_z & -S\omega_x C\omega_z + C\omega_x S\omega_y S\omega_z & 0 \\ -S\omega_y & S\omega_x C\omega_y & C\omega_x C\omega_y & 0 \\ 0 & 0 & 0 & 1 \end{bmatrix}.
\end{aligned}
\tag{1.26}
$$

As shown more clearly in Fig. 1.12, $\mathrm{RPY}(\omega_z, \omega_y, \omega_x)$ describes the pose matrix ${}^{h}\bar{A}_{g}$ by a rotation of angle ω_z about z_h to reach an intermediate coordinate frame $(xyz)_1$, followed by a rotation of angle ω_y about y_1 (not y_h) to reach another intermediate coordinate frame $(xyz)_2$, and finally, a rotation ω_x about x_2 (neither x_h nor x_1) to reach the target coordinate frame $(xyz)_g$. It is noted that the two intermediate coordinate frames, $(xyz)_1$ and $(xyz)_2$, are important only in the interpretation (i.e., not the calculation) of the three angles.

The relative orientation of two coordinate frames can also be described by the following Euler transformation (Fig. 1.13):

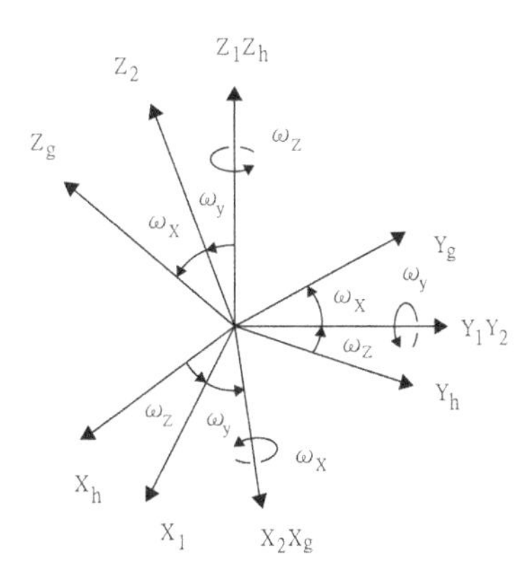

Fig. 1.12 Roll, pitch and yaw angles

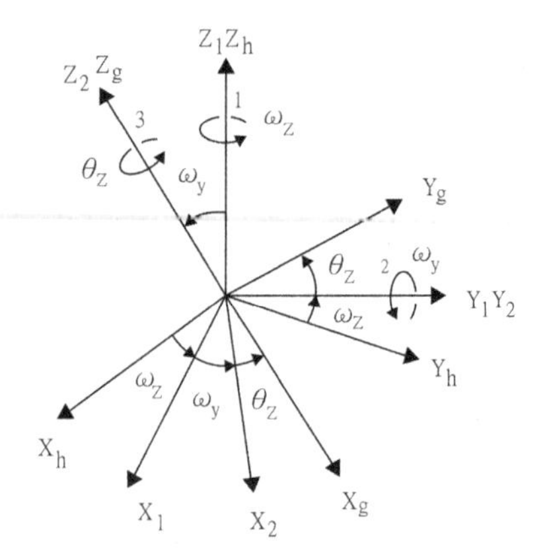

Fig. 1.13 Euler angles

$$
\begin{aligned}
{}^{h}\bar{A}_g &= \mathrm{Euler}(\omega_Z, \omega_y, \theta_Z) = \mathrm{rot}(\bar{z}, \omega_Z)\mathrm{rot}(\bar{y}, \omega_y)\mathrm{rot}(\bar{z}, \theta_Z) \\
&= \begin{bmatrix} C\omega_Z C\omega_y C\theta_Z - S\omega_Z S\theta_Z & -C\omega_Z C\omega_y S\theta_Z - S\omega_Z C\theta_Z & C\omega_Z S\omega_y & 0 \\ S\omega_Z C\omega_y C\theta_Z + C\omega_Z S\theta_Z & -S\omega_Z C\omega_y S\theta_Z + C\omega_Z C\theta_Z & S\omega_Z S\omega_y & 0 \\ -S\omega_y C\theta_Z & S\omega_y S\theta_Z & C\omega_y & 0 \\ 0 & 0 & 0 & 1 \end{bmatrix}.
\end{aligned}
\tag{1.27}
$$

However, it is noted from Eq. (1.27) that the Euler transformation involves two rotations about two different z axes. Consequently, it is not used in this book in order to avoid confusion.

Example 1.5 For the two coordinate frames shown in Fig. 1.14, the pose matrix ${}^4\bar{A}_5$ is given by ${}^4\bar{A}_5 = \mathrm{tran}(1, 0, 0)\mathrm{tran}(0, 2, 0)\mathrm{tran}(0, 0, 3) = \mathrm{tran}(1, 2, 3)$.

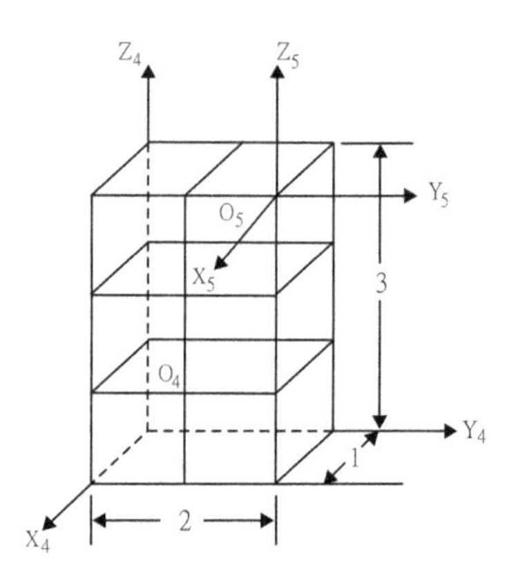

Fig. 1.14 Coordinate frames $(xyz)_4$ and $(xyz)_5$

1.5 Specification of a Pose Matrix by Using Translation and Rotation Matrices

As discussed in the previous section, the pose matrix, ${}^{h}\bar{A}_{g}$, can be specified in various ways using rotation and translation matrices. To describe the procedure for obtaining ${}^{h}\bar{A}_{g}$, several intermediate coordinate frames are required, say $(xyz)_1$, $(xyz)_2$, $(xyz)_3$ … (see Fig. 1.15). As shown in Fig. 1.15, ${}^{h}\bar{A}_{1} = \mathrm{tran}(t_x, 0, 0)$, ${}^{1}\bar{A}_{2} = \mathrm{rot}(\bar{z}, \Phi_z)$, ${}^{2}\bar{A}_{3} = \mathrm{tran}(0, t_y, 0)$, ${}^{3}\bar{A}_{4} = \mathrm{rot}(\bar{x}, \Phi_x)$, ${}^{4}\bar{A}_{5} = \mathrm{tran}(p_x, 0, 0)$, ${}^{5}\bar{A}_{6} = \mathrm{rot}(\bar{z}, \omega_z)$, ${}^{6}\bar{A}_{7} = \mathrm{tran}(0, 0, t_z)$, and ${}^{7}\bar{A}_{g} = \mathrm{rot}(\bar{x}, \omega_x)$. Therefore, ${}^{h}\bar{A}_{g}$ can be obtained as

$$ {}^{h}\bar{A}_{g} = {}^{h}\bar{A}_{1}\,{}^{1}\bar{A}_{2}\,{}^{2}\bar{A}_{3}\,{}^{3}\bar{A}_{4}\,{}^{4}\bar{A}_{5}\,{}^{5}\bar{A}_{6}\,{}^{6}\bar{A}_{7}\,{}^{7}\bar{A}_{g}, $$

or simply

$$ {}^{h}\bar{A}_{g} = \mathrm{tran}(t_x, 0, 0)\mathrm{rot}(\bar{z}, \Phi_z)\mathrm{tran}(0, t_y, 0)\mathrm{rot}(\bar{x}, \Phi_x)\mathrm{tran}(p_x, 0, 0)\mathrm{rot}(\bar{z}, \omega_z)\mathrm{tran}(0, 0, t_z)\mathrm{rot}(\bar{x}, \omega_x). \tag{1.28} $$

One can simulate the processes of Eq. (1.28) by using a coordinate frame starting from the pose of $(xyz)_h$ (i.e., the coordinate frame defined by the leading super-script of ${}^{h}\bar{A}_{g}$) and then followed by the translation and rotation operators to reach the final pose of $(xyz)_g$ (i.e., the coordinate frame defined by the post sub-script of ${}^{h}\bar{A}_{g}$). It is noted in Eq. (1.28) that, as in the case of all matrix multiplication operations, the order in which the rotation and translation operations are performed is important since different combinations of translation and rotation

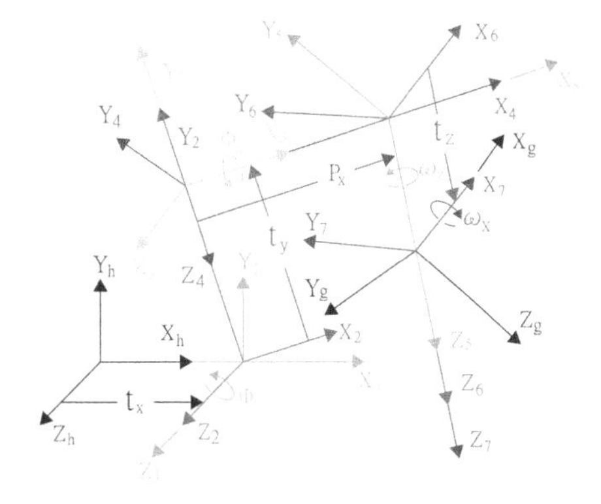

Fig. 1.15 Pose matrix of $(xyz)_g$ with respect to $(xyz)_h$ can be specified by a sequence of rotations and translations

operations may lead to different pose matrices. Alternatively, the same pose matrix, ${}^{h}\bar{A}_{g}$, can be obtained via different combinations of translation and rotation motions.

Example 1.6 Referring to Fig. 1.4 again, the pose matrix ${}^{4}\bar{A}_{5}$ can be obtained by all of the following procedures:

$$\begin{aligned}
{}^{4}\bar{A}_{5} &= \text{tran}(3,0,0)\text{tran}(0,2,0)\text{tran}(0,0,1)\text{rot}(\bar{y},180^{\circ})\text{rot}(\bar{x},90^{\circ})\\
&= \text{tran}(3,0,0)\text{tran}(0,0,1)\text{tran}(0,2,0)\text{rot}(\bar{x},90^{\circ})\text{rot}(\bar{z},180^{\circ})\\
&= \text{tran}(3,0,0)\text{tran}(0,2,0)\text{tran}(0,0,1)\text{rot}(\bar{z},180^{\circ})\text{rot}(\bar{x},-90^{\circ})\\
&= \text{tran}(0,2,0)\text{tran}(0,0,1)\text{tran}(3,0,0)\text{rot}(\bar{z},180^{\circ})\text{rot}(\bar{x},-90^{\circ})\\
&= \text{rot}(\bar{y},180^{\circ})\text{rot}(\bar{x},90^{\circ})\text{tran}(-3,0,0)\text{tran}(0,-1,0)\text{tran}(0,0,-2)\\
&= \text{rot}(\bar{x},90^{\circ})\text{rot}(\bar{z},180^{\circ})\text{tran}(-3,0,0)\text{tran}(0,0,-2)\text{tran}(0,-1,0)\\
&= \text{rot}(\bar{z},180^{\circ})\text{rot}(\bar{x},-90^{\circ})\text{tran}(0,0,-2)\text{tran}(-3,0,0)\text{tran}(0,-1,0)\\
&= \text{rot}(\bar{x},90^{\circ})\text{tran}(3,0,0)\text{rot}(\bar{z},180^{\circ})\text{tran}(0,-1,0)\text{tran}(0,0,-2).
\end{aligned}$$

1.6 Inverse Matrix of a Transformation Matrix

The inverse matrix of ${}^{h}\bar{A}_{g}$, denoted as ${}^{g}\bar{A}_{h}$, carries the transformed coordinate frame $(xyz)_{g}$ back to the original frame $(xyz)_{h}$. Given the pose matrix described in Eq. (1.6), the inverse matrix has the form

$$\begin{aligned}
\left({}^{h}\bar{A}_{g}\right)^{-1} &= {}^{g}\bar{A}_{h}\\
&= \begin{bmatrix} I_x & I_y & I_z & -(I_x t_x + I_y t_y + I_z t_z)\\ J_x & J_y & J_z & -(J_x t_x + J_y t_y + J_z t_z)\\ K_x & K_y & K_z & -(K_x t_x + K_y t_y + K_z t_z)\\ 0 & 0 & 0 & 1 \end{bmatrix} = \begin{bmatrix} I_x & I_y & I_z & f_x\\ J_x & J_y & J_z & f_y\\ K_x & K_y & K_z & f_z\\ 0 & 0 & 0 & 1 \end{bmatrix},
\end{aligned} \tag{1.29}$$

where

$$f_x = -(I_x t_x + I_y t_y + I_z t_z) \tag{1.30}$$

$$f_y = -(J_x t_x + J_y t_y + J_z t_z), \tag{1.31}$$

$$f_z = -(K_x t_x + K_y t_y + K_z t_z) \tag{1.32}$$

For the case where the pose matrix ${}^{h}\bar{A}_{g}$ is given by a sequence of rotation and translation motions (e.g.,

$$ {}^{h}\bar{A}_{g} = \mathrm{tran}(t_x, 0, 0)\mathrm{tran}(0, t_y, 0)\mathrm{tran}(0, 0, t_z)\mathrm{rot}(\bar{z}, \omega_z)\mathrm{rot}(\bar{y}, \omega_y)\mathrm{rot}(\bar{x}, \omega_x)), $$

then, from basic matrix theory, the inverse matrix is given as

$$ \left({}^{h}\bar{A}_{g}\right)^{-1} = {}^{g}\bar{A}_{h} = \mathrm{rot}(\bar{x}, -\omega_x)\mathrm{rot}(\bar{y}, -\omega_y)\mathrm{rot}(\bar{z}, -\omega_z)\mathrm{tran}(-t_x, -t_y, -t_z). $$

Note that this result can be easily verified by post-multiplying ${}^{h}\bar{A}_{g}$ by ${}^{g}\bar{A}_{h}$.

Example 1.7 Consider the pose matrix ${}^{4}\bar{A}_{5} = \mathrm{tran}(0, 0, 1)\mathrm{tran}(0, 2, 0)\mathrm{tran}(3, 0, 0)$ $\mathrm{rot}(\bar{y}, 180^\circ)\mathrm{rot}(\bar{x}, 90^\circ)$ given in Example 1.3. Its inverse matrix is given by

$$ \begin{aligned} \left({}^{4}\bar{A}_{5}\right)^{-1} &= {}^{5}\bar{A}_{4} = \mathrm{rot}(\bar{x}, -90^\circ)\mathrm{rot}(\bar{y}, -180^\circ)\mathrm{tran}(-3, 0, 0)\mathrm{tran}(0, -2, 0)\mathrm{tran}(0, 0, -1) \\ &= \mathrm{rot}(\bar{x}, -90^\circ)\mathrm{rot}(\bar{y}, -180^\circ)\mathrm{tran}(-3, -2, -1) = \begin{bmatrix} -1 & 0 & 0 & 3 \\ 0 & 0 & -1 & 1 \\ 0 & -1 & 0 & 2 \\ 0 & 0 & 0 & 1 \end{bmatrix}. \end{aligned} $$

1.7 Flat Boundary Surface

In general, any flat boundary surface can be specified uniquely in 3-D space in a coordinate frame $(xyz)_g$ using its unit normal vector ${}^{g}\bar{n} = \begin{bmatrix} {}^{g}n_x & {}^{g}n_y & {}^{g}n_z & 0 \end{bmatrix}^T$ and a point by $({}^{g}n_x)x + ({}^{g}n_y)y + ({}^{g}n_z)z + {}^{g}e = 0$. Consequently, in the homogeneous coordinate notation, a flat boundary surface can be described in terms of its unit normal vector ${}^{g}\bar{n}$ and the parameter ${}^{g}e$ by means of the following column matrix:

$$ {}^{g}\bar{r} = \begin{bmatrix} {}^{g}n_x \\ {}^{g}n_y \\ {}^{g}n_z \\ {}^{g}e \end{bmatrix}. \tag{1.33} $$

Equation (1.33) indicates that the flat boundary surface is located at a distance $-({}^{g}e)$ from the origin of coordinate frame $(xyz)_g$ in the unit normal direction ${}^{g}\bar{n}$ (see Fig. 1.16). Again, the pre-superscript "g" of the leading symbol ${}^{g}\bar{r}$ in Eq. (1.33) indicates that the flat boundary surface is defined with respect to coordinate frame $(xyz)_g$. For any point ${}^{g}\bar{P}$ lying on flat boundary surface ${}^{g}\bar{r}$, the matrix product ${}^{g}\bar{r} \cdot {}^{g}\bar{P} = 0$ is fulfilled.

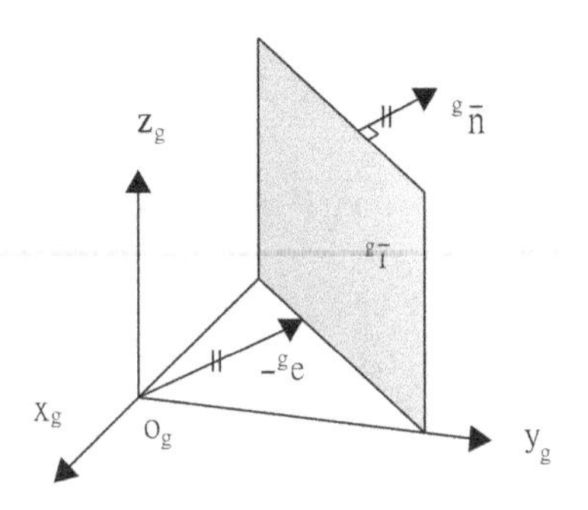

Fig. 1.16 Definition of a flat boundary surface in 3-D space by means of four parameters

Given a flat boundary surface ${}^g\bar{r}$ and two coordinate frames $(xyz)_g$ and $(xyz)_h$, the transformation of flat boundary surface ${}^h\bar{r}$ can be computed via the following matrix product:

$$ {}^h\bar{r} = \begin{bmatrix} {}^h n_x \\ {}^h n_y \\ {}^h n_z \\ {}^h e \end{bmatrix} = \left(\left({}^h\bar{A}_g\right)^{-1}\right)^T {}^g\bar{r} = {}^h\bar{A}'_g \begin{bmatrix} {}^g n_x \\ {}^g n_y \\ {}^g n_z \\ {}^g e \end{bmatrix}, \tag{1.34} $$

where ${}^h\bar{A}_g$ describes the pose of coordinate frame $(xyz)_g$ with respect to coordinate frame $(xyz)_h$, and is given by Eq. (1.6). It is noted from Eq. (1.34) that ${}^h\bar{A}'_g$ is the transpose of the inverse matrix of ${}^h\bar{A}_g$, and has the form

$$ {}^h\bar{A}'_g = \begin{bmatrix} I_x & J_x & K_x & 0 \\ I_y & J_y & K_y & 0 \\ I_z & J_z & K_z & 0 \\ f_x & f_y & f_z & 1 \end{bmatrix}, \tag{1.35} $$

where f_x, f_y and f_z are given in Eqs. (1.30), (1.31) and (1.32), respectively.

Example 1.8 As will be described later in Chap. 2, the ith flat boundary surface ${}^i\bar{r}_i$ can be written as ${}^i\bar{r}_i = [0 \quad 1 \quad 0 \quad 0]^T$ in the boundary coordinate frame $(xyz)_i$. If the pose matrix of $(xyz)_i$ with respect to the world coordinate frame $(xyz)_0$ is given by

$$ {}^0\bar{A}_i = \begin{bmatrix} I_{ix} & J_{ix} & K_{ix} & t_{ix} \\ I_{iy} & J_{iy} & K_{iy} & t_{iy} \\ I_{iz} & J_{iz} & K_{iz} & t_{iz} \\ 0 & 0 & 0 & 1 \end{bmatrix}, $$

then ${}^{i}\bar{r}_i$ can be transferred to $(xyz)_0$ using Eq. (1.34) to give

$$
{}^{0}\bar{r}_i = \begin{bmatrix} J_{ix} \\ J_{iy} \\ J_{iz} \\ -(J_{ix}t_{ix} + J_{iy}t_{iy} + J_{iz}t_{iz}) \end{bmatrix}.
$$

1.8 RPY Transformation Solutions

As described in Sect. 1.5, the pose matrix, ${}^{h}\bar{A}_g$, can be specified in various ways, including $RPY(\omega_z, \omega_y, \omega_x)$, $Euler(\omega_z, \omega_y, \theta_z)$, $rot(\bar{\kappa}, \theta)$, and so on. In this section, an additional post-subscript "i" is added to each of the variables in these formulations in order to indicate that they refer to the ith boundary surface of an optical system. The solutions of t_{ix}, t_{iy}, t_{iz}, ω_{ix}, ω_{iy} and ω_{iz} for a given numeric value of ${}^{h}\bar{A}_g$ are obtained for the case where ${}^{h}\bar{A}_g$ is specified as

$$
{}^{h}\bar{A}_g = tran(t_{ix}, 0, 0)tran(0, t_{iy}, 0)tran(0, 0, t_{iz})rot(\bar{z}, \omega_{iz})rot(\bar{y}, \omega_{iy})rot(\bar{x}, \omega_{ix}), \tag{1.36}
$$

since this is the transformation most commonly used in this book. (Note that for the case where the orientation part of ${}^{h}\bar{A}_g$ is specified as $Euler(\omega_{iz}, \omega_{iy}, \theta_{iz})$, refer to pp. 65–72 of [3]).

If the numeric value of ${}^{h}\bar{A}_g$ is known and

$$
\begin{aligned}
{}^{h}\bar{A}_g &= tran(t_{ix}, 0, 0)tran(0, t_{iy}, 0)tran(0, 0, t_{iz})rot(\bar{z}, \omega_{iz})rot(\bar{y}, \omega_{iy})rot(\bar{x}, \omega_{ix}) \\
&= \begin{bmatrix} I_{ix} & J_{ix} & K_{ix} & t_{ix} \\ I_{iy} & J_{iy} & K_{iy} & t_{iy} \\ I_{iz} & J_{iz} & K_{iz} & t_{iz} \\ 0 & 0 & 0 & 1 \end{bmatrix} \\
&= \begin{bmatrix} C\omega_{iy}C\omega_{iz} & S\omega_{ix}S\omega_{iy}C\omega_{iz} - C\omega_{ix}S\omega_{iz} & C\omega_{ix}S\omega_{iy}C\omega_{iz} + S\omega_{ix}S\omega_{iz} & t_{ix} \\ C\omega_{iy}S\omega_{iz} & C\omega_{ix}C\omega_{iz} + S\omega_{ix}S\omega_{iy}S\omega_{iz} & -S\omega_{ix}C\omega_{iz} + C\omega_{ix}S\omega_{iy}S\omega_{iz} & t_{iy} \\ -S\omega_{iy} & S\omega_{ix}C\omega_{iy} & C\omega_{ix}C\omega_{iy} & t_{iz} \\ 0 & 0 & 0 & 1 \end{bmatrix},
\end{aligned} \tag{1.37}
$$

then the three angles, ω_{iz}, ω_{iy} and ω_{ix}, and three position pose variables, t_{ix}, t_{iy} and t_{iz}, can be obtained sequentially as

$$\begin{cases} \omega_{iz} = \text{atan2}(I_{iy}, I_{ix}) \text{ or } \omega_{iz} = \text{atan2}(-I_{iy}, -I_{ix}) & \text{, if } I_{ix}^2 + I_{iy}^2 \neq 0 \\ \omega_{iz} = 0 & \text{, if } I_{ix}^2 + I_{iy}^2 = 0, \end{cases} \tag{1.38}$$

$$\omega_{iy} = \text{atan2}(-I_{iz}, I_{ix}C\omega_{iz} + I_{iy}S\omega_{iz}), \tag{1.39}$$

$$\omega_{ix} = \text{atan2}(K_{ix}S\omega_{iz} - K_{iy}C\omega_{iz}, -J_{ix}S\omega_{iz} + J_{iy}C\omega_{iz}), \tag{1.40}$$

$$t_{ix} = t_{ix}, \tag{1.41}$$

$$t_{iy} = t_{iy}, \tag{1.42}$$

$$t_{iz} = t_{iz}. \tag{1.43}$$

Equation (1.38) shows that there are two solutions, separated by 180°, for ω_{iz} in the case of $I_{ix}^2 + I_{iy}^2 \neq 0$. The function atan2 returns the arctangent in the range of $-\pi$ to π. However, it is usually necessary to add 2π to the solution in order to solve a variable in its required domain. Equations (1.41), (1.42) and (1.43) show that the three position pose variables t_{ix}, t_{iy} and t_{iz} can be extracted directly from the (1, 4) th, (2, 4)th and (3, 4)th elements of ${}^h\bar{A}_g$.

Example 1.9 $\text{atan2}(1,1) = \text{atan2}(3,3) = 45°$, $\text{atan2}(-1,-1) = \text{atan2}(-3,-3) = -135°$. $\text{atan2}(-1,1) = \text{atan2}(-3,3) = -45°$, $\text{atan2}(1,-1) = \text{atan2}(3,-3) = 135°$. It is noted from this example that when $\eta = \text{atan2}(y, x)$, it does not indicate $C\eta = x$ and $S\eta = y$.

Example 1.10 Again, consider the pose matrix ${}^4\bar{A}_5$ of Example 1.3. The values of the six pose variables are $\omega_Z = \text{atan2}(0,-1) = 180°$, $\omega_Y = \text{atan2}(0,1) = 0°$, $\omega_x = \text{atan2}(-1,0) = -90°$, $t_x = 3$, $t_y = 2$, and $t_z = 1$. Therefore, ${}^4\bar{A}_5$ can be obtained by ${}^4\bar{A}_5 = \text{tran}(3,0,0)\text{tran}(0,2,0)\text{tran}(0,0,1)\text{rot}(\bar{z}, 180°)\text{rot}(\bar{x}, -90°)$, which is shown in the third row of Example 1.6.

1.9 Equivalent Angle and Axis of Rotation

When discussing the orientation of an image, the equivalent angle and axis of rotation are both required to define the pose matrix ${}^h\bar{A}_g = \text{tran}(t_{ix}, t_{iy}, t_{iz})\text{rot}(\bar{\kappa}_i, \theta_i)$, with $0° \leq \theta_i \leq 180°$. In this case, θ_i, ${}^h\bar{\kappa}_i = \begin{bmatrix} \kappa_{ix} & \kappa_{iy} & \kappa_{iz} & 0 \end{bmatrix}^T$ and (t_{ix}, t_{iy}, t_{iz}) can be obtained respectively as follows (p. 29 of [3]):

$$\theta_i = \operatorname{atan2}\left(\sqrt{(J_{iz} - K_{iy})^2 + (K_{ix} - I_{iz})^2 + (I_{iy} - J_{ix})^2}, I_{ix} + J_{iy} + K_{iz} - 1\right), \tag{1.44}$$

$$\kappa_{ix} = \frac{0.5(J_{iz} - K_{iy})}{S\theta_i}, \tag{1.45}$$

$$\kappa_{iy} = \frac{0.5(K_{ix} - I_{iz})}{S\theta_i}, \tag{1.46}$$

$$\kappa_{iz} = \frac{0.5(I_{iy} - J_{ix})}{S\theta_i}, \tag{1.47}$$

$$t_{ix} = t_{ix}, \tag{1.48}$$

$$t_{iy} = t_{iy}, \tag{1.49}$$

$$t_{iz} = t_{iz}. \tag{1.50}$$

When θ_i is very small, ${}^h\bar{\kappa}_i$ is physically not well defined due to the small magnitude of both the numerator and the denominator in Eqs. (1.45), (1.46) and (1.47). Thus, when θ_i is small, the unit vector should be normalized to ensure that ${}^h\bar{\kappa}_i$ is a unit vector (i.e., $\left|{}^h\bar{\kappa}_i\right| = 1$).

Furthermore, Paul [3] suggests that if $\theta_i \geq 90°$, then a different approach must be followed for determining ${}^h\bar{\kappa}_i$ by first computing the following three terms:

$$\kappa_{ix} = \operatorname{sign}(J_{iz} - K_{iy})\sqrt{\frac{I_{ix} - C\theta_i}{1 - C\theta_i}}, \tag{1.51}$$

$$\kappa_{iy} = \operatorname{sign}(K_{ix} - I_{iz})\sqrt{\frac{J_{iy} - C\theta_i}{1 - C\theta_i}}, \tag{1.52}$$

$$\kappa_{iz} = \operatorname{sign}(I_{iy} - J_{ix})\sqrt{\frac{K_{iz} - C\theta_i}{1 - C\theta_i}}, \tag{1.53}$$

where $\operatorname{sign}(*) = +1$ if $* \geq 0$ and $\operatorname{sign}(*) = -1$ if $* < 0$. However, only the largest component of ${}^h\bar{\kappa}_i$ can be determined from Eqs. (1.51), (1.52) and (1.53); corresponding to the most positive components of I_{ix}, J_{iy} and K_{iz}, respectively. The remaining components are more accurately determined via the following equations:

If κ_{ix} is the largest, then

$$\kappa_{iy} = \frac{0.5(I_{iy} + J_{ix})}{\kappa_{ix}(1 - C\theta_i)}, \tag{1.54}$$

$$\kappa_{iz} = \frac{0.5(K_{ix} + I_{iz})}{\kappa_{ix}(1 - C\theta_i)}. \tag{1.55}$$

If κ_{iy} is the largest, then

$$\kappa_{ix} = \frac{0.5(I_{iy} + J_{ix})}{\kappa_{iy}(1 - C\theta_i)}, \tag{1.56}$$

$$\kappa_{iz} = \frac{0.5(J_{iz} + K_{iy})}{\kappa_{iy}(1 - C\theta_i)}. \tag{1.57}$$

If κ_{iz} is the largest, then

$$\kappa_{ix} = \frac{0.5(K_{ix} + I_{iz})}{\kappa_{iz}(1 - C\theta_i)}, \tag{1.58}$$

$$\kappa_{iy} = \frac{0.5(J_{iz} + K_{iy})}{\kappa_{iz}(1 - C\theta_i)}. \tag{1.59}$$

1.10 The First- and Second-Order Partial Derivatives of a Vector

In vector calculus, the Jacobian matrix is a matrix containing all of the first-order partial derivatives of a vector-valued function with respect to another vector. If vector $\bar{F} = \bar{F}(\bar{X}_i, \bar{Y}_i) = \begin{bmatrix} f_1(\bar{X}_i, \bar{Y}_i) & f_2(\bar{X}_i, \bar{Y}_i) \end{bmatrix}^T$ is a function of $\bar{X}_i = \begin{bmatrix} x_{i1} & x_{i2} & x_{i3} \end{bmatrix}^T$ and $\bar{Y}_i = \begin{bmatrix} y_{i1} & y_{i2} \end{bmatrix}^T$, the Jacobian matrix of component f_p, $p \in \{1, 2\}$, with respect to $\bar{X}_i$ is a row matrix with the form

$$\frac{\partial f_p}{\partial \bar{X}_i} = \frac{\partial f_p}{\partial(x_{i1}, x_{i2}, x_{i3})} = \begin{bmatrix} \frac{\partial f_p}{\partial x_{i1}} & \frac{\partial f_p}{\partial x_{i2}} & \frac{\partial f_p}{\partial x_{i3}} \end{bmatrix}, \tag{1.60}$$

where the symbol ∂x_{iv} indicates partial differentiation with respect to the vth variable of $\bar{X}_i$. The Jacobian matrix of a vector with respect to a scalar is a column matrix. For example, $\partial \bar{F}/\partial x_{iv}$, $v \in \{1, 2, 3\}$, is written as

$$\left[\frac{\partial \bar{F}}{\partial x_{iv}}\right] = \begin{bmatrix} \partial f_1/\partial x_{iv} \\ \partial f_2/\partial x_{iv} \end{bmatrix}. \tag{1.61}$$

It can be deduced from Eqs. (1.60) and (1.61) that $\partial \bar{F}/\partial \bar{X}_i$ is a 2×3 matrix with the form

$$\frac{\partial \bar{F}}{\partial \bar{X}_i} = \frac{\partial(f_1, f_2)}{\partial(x_{i1}, x_{i2}, x_{i3})} = \begin{bmatrix} \partial f_1/\partial x_{i1} & \partial f_1/\partial x_{i2} & \partial f_1/\partial x_{i3} \\ \partial f_2/\partial x_{i1} & \partial f_2/\partial x_{i2} & \partial f_2/\partial x_{i3} \end{bmatrix}. \tag{1.62}$$

The Hessian matrix of a scalar function f_p is a matrix containing all of the second-order partial derivatives of f_p with respect to its variable vectors, i.e., $\partial^2 f_p/\partial \bar{Y}_i \partial \bar{X}_i$. The Hessian matrix can be determined directly by differentiating Eq. (1.60) to give

$$\begin{aligned} \frac{\partial^2 f_p}{\partial \bar{Y}_i \partial \bar{X}_i} &= \left[\frac{\partial^2 f_p}{\partial y_{iw} \partial x_{iv}}\right] = \frac{\partial^2 f_p}{\partial(y_{i1}, y_{i2})\partial(x_{i1}, x_{i2}, x_{i3})} \\ &= \begin{bmatrix} \partial^2 f_p/\partial y_{i1}\partial x_{i1} & \partial^2 f_p/\partial y_{i1}\partial x_{i2} & \partial^2 f_p/\partial y_{i1}\partial x_{i3} \\ \partial^2 f_p/\partial y_{i2}\partial x_{i1} & \partial^2 f_p/\partial y_{i2}\partial x_{i2} & \partial^2 f_p/\partial y_{i2}\partial x_{i3} \end{bmatrix}. \end{aligned} \tag{1.63}$$

Note that $\partial^2 f_p/\partial \bar{Y}_i \partial \bar{X}_i$ is different from $\partial^2 f_p/\partial \bar{X}_i \partial \bar{Y}_i$. The Hessian matrix $\partial^2 \bar{F}/\partial \bar{Y}_i \partial \bar{X}_i$ is a 3-D matrix. Hessian matrices are used in large-scale optimization problems within Newton-type methods since they contain all of the coefficients of the quadratic term in the local Taylor expansion of the function of interest, $\bar{F}$. That is,

$$\begin{aligned} \bar{F}(\bar{X}_i + \Delta\bar{X}_i, \bar{Y}_i + \Delta\bar{Y}_i) &\approx \bar{F}(\bar{X}_i, \bar{Y}_i) + \frac{\partial \bar{F}}{\partial \bar{X}_i}\Delta\bar{X}_i + \frac{\partial \bar{F}}{\partial \bar{Y}_i}\Delta\bar{Y}_i \\ &+ \frac{1}{2}\left(\Delta\bar{X}_i^T \frac{\partial^2 \bar{F}}{\partial \bar{X}_i^2}\Delta\bar{X}_i + \Delta\bar{Y}_i^T \frac{\partial^2 \bar{F}}{\partial \bar{Y}_i \partial \bar{X}_i}\Delta\bar{X}_i + \Delta\bar{X}_i^T \frac{\partial^2 \bar{F}}{\partial \bar{X}_i \partial \bar{Y}_i}\Delta\bar{Y}_i + \Delta\bar{Y}_i^T \frac{\partial^2 \bar{F}}{\partial \bar{Y}_i^2}\Delta\bar{Y}_i\right). \end{aligned} \tag{1.64}$$

The following discussions consider the Jacobian and Hessian matrices of the matrix $\bar{F} = \bar{G}\bar{H}$, where $\bar{G} = \left[g_{pm}\right] = \bar{G}(\bar{X}_i, \bar{Y}_i)$ and $\bar{H} = \left[h_{mq}\right] = \bar{H}(\bar{X}_i, \bar{Y}_i)$ are two matrices with dimensions of $P \times M$ and $M \times Q$, respectively. The production differentiation rule for $\partial \bar{F}/\partial \bar{X}_i$ is commonly given as

$$\frac{\partial \bar{F}}{\partial \bar{X}_i} = \frac{\partial \bar{G}}{\partial \bar{X}_i}\bar{H} + \bar{G}\frac{\partial \bar{H}}{\partial \bar{X}_i}. \tag{1.65}$$

Proof By definition, the (p, q)th component of matrix $\bar{F}$ is described by

$$f_{pq} = \sum_{m=1}^{M} g_{pm} h_{mq}. \tag{1.66}$$

The production differentiation rule then yields the vth component of $\partial\bar{F}/\partial\bar{X}_i$ as

$$\frac{\partial f_{pq}}{\partial x_{iv}} = \sum_{m=1}^{M} \left(\frac{\partial g_{pm}}{\partial x_{iv}} h_{mq} + g_{pm} \frac{\partial h_{mq}}{\partial x_{iv}} \right). \tag{1.67}$$

Hence Eq. (1.65) follows; thereby completing the proof.

Mathematically, Eq. (1.65) violates the matrix multiplication rule since the number "3" in the dimensions of $\partial\bar{G}/\partial\bar{X}_i$ (i.e., $P \times M \times 3$) may not be equal to the number "M" in the dimensions of $\bar{H}$ (i.e., $M \times Q$). However, Eq. (1.65) is still widely accepted as the chain rule for matrix multiplication when its interpretation is based on Eq. (1.67). Moreover, Eq. (1.65) possesses an important advantage in that it intuitively represents the Jacobian matrix of $\bar{F}$ with respect to $\bar{X}_i$. Therefore, in this book, the Jacobian matrix is assigned the form shown in Eq. (1.65). However, in writing computer code, it is important to recognize that Eq. (1.65) is based on an interpretation of Eq. (1.67), which is obtained by replacing $\partial\bar{X}_i$ in Eq. (1.65) with ∂x_{iv}.

The Hessian matrix $\partial^2\bar{F}/\partial\bar{Y}_i\partial\bar{X}_i$ is obtained simply by differentiating Eq. (1.65) with respect to $\bar{Y}_i$ to give

$$\frac{\partial^2\bar{F}}{\partial\bar{Y}_i\partial\bar{X}_i} = \frac{\partial^2\bar{G}}{\partial\bar{Y}_i\partial\bar{X}_i}\bar{H} + \frac{\partial\bar{G}}{\partial\bar{X}_i}\frac{\partial\bar{H}}{\partial\bar{Y}_i} + \frac{\partial\bar{G}}{\partial\bar{Y}_i}\frac{\partial\bar{H}}{\partial\bar{X}_i} + \bar{G}\frac{\partial^2\bar{H}}{\partial\bar{Y}_i\partial\bar{X}_i}. \tag{1.68}$$

Again, it is important to note that Eq. (1.68) is valid only based on the interpretation of the following equation:

$$\frac{\partial^2 f_{pq}}{\partial y_{iw}\partial x_{iv}} = \sum_{m=1}^{M} \left(\frac{\partial^2 g_{pm}}{\partial y_{iw}\partial x_{iv}} h_{mq} + \frac{\partial g_{pm}}{\partial x_{iv}}\frac{\partial h_{mq}}{\partial y_{iw}} + \frac{\partial g_{pm}}{\partial y_{iw}}\frac{\partial h_{mq}}{\partial x_{iv}} + g_{pm}\frac{\partial^2 h_{mq}}{\partial y_{iw}\partial x_{iv}} \right). \tag{1.69}$$

Note that Eq. (1.69) is obtained by replacing $\partial\bar{Y}_i$ and $\partial\bar{X}_i$ in Eq. (1.68) with ∂y_{iw} and ∂x_{iv}, respectively. In this book, $\bar{F} = \bar{G}\bar{H}$ will be given before Eq. (1.68) to serve as a reminder that Eq. (1.68) is based on Eq. (1.69).

However, confusion may still arise in presenting the Hessian matrix $\partial^2\bar{F}/\partial\bar{X}_i^2$ of a matrix function $\bar{F} = \bar{G}\bar{H}$, where $\bar{G} = \left[g_{pm}\right] = \bar{G}(\bar{X}_i)$ and $\bar{H} = \left[h_{mq}\right] = \bar{H}(\bar{X}_i)$ are $H \times M$ and $M \times Q$ matrices, respectively. In this case, the Jacobian matrix $\partial\bar{F}/\partial\bar{X}_i$ of $\bar{F}$ is given in Eq. (1.65) with $\bar{G}$ and $\bar{H}$ being functions only of $\bar{X}_i$.

Intuitively, Eq. (1.65) can be differentiated directly to give the Hessian matrix $\partial^2\bar{F}/\partial\bar{X}_i^2$ of $\bar{F}$, yielding

$$\frac{\partial^2\bar{F}}{\partial\bar{X}_i^2}=\frac{\partial^2\bar{G}}{\partial\bar{X}_i^2}\bar{H}+\frac{\partial\bar{G}}{\partial\bar{X}_i}\frac{\partial\bar{H}}{\partial\bar{X}_i}+\bar{G}\frac{\partial^2\bar{H}}{\partial\bar{X}_i^2}+\frac{\partial\bar{G}}{\partial\bar{X}_i}\frac{\partial\bar{H}}{\partial\bar{X}_i}. \tag{1.70}$$

However, Eq. (1.70) can be mistakenly rewritten as

$$\frac{\partial^2\bar{F}}{\partial\bar{X}_i^2}=\frac{\partial^2\bar{G}}{\partial\bar{X}_i^2}\bar{H}+\bar{G}\frac{\partial^2\bar{H}}{\partial\bar{X}_i^2}+2\frac{\partial\bar{G}}{\partial\bar{X}_i}\frac{\partial\bar{H}}{\partial\bar{X}_i}. \tag{1.71}$$

This mathematical error can be observed from the alternative form of $\partial^2\bar{F}/\partial\bar{X}_i^2$ obtained by directly differentiating Eq. (1.67) to give

$$\frac{\partial^2 f_{pq}}{\partial x_{iw}\partial x_{iv}}=\sum_{m=1}^{M}\left(\frac{\partial^2 g_{pm}}{\partial x_{iw}\partial x_{iv}}h_{mq}+\frac{\partial g_{pm}}{\partial x_{iv}}\frac{\partial h_{mq}}{\partial x_{iw}}+g_{pm}\frac{\partial^2 h_{mq}}{\partial x_{iw}\partial x_{iv}}+\frac{\partial g_{pm}}{\partial x_{iw}}\frac{\partial h_{mq}}{\partial x_{iv}}\right). \tag{1.72}$$

In Eq. (1.72), the second and fourth terms are generally not equal if $p \neq q$ or $w \neq v$ (i.e., Equation (1.71) is valid only if $p = q$ and $w = v$). In order to prevent confusion, the following equation is thus used instead of Eq. (1.70) to describe the Hessian matrix $\partial^2\bar{F}/\partial\bar{X}_i^2$:

$$\frac{\partial^2\bar{F}}{\partial\bar{X}_i^2}=\frac{\partial^2\bar{G}}{\partial\bar{X}_i^2}\bar{H}+\frac{\partial\bar{G}}{\partial\bar{X}_i}\frac{\partial\bar{H}}{\partial\bar{X}_{\underline{i}}}+\bar{G}\frac{\partial^2\bar{H}}{\partial\bar{X}_i^2}+\frac{\partial\bar{G}}{\partial\bar{X}_{\underline{i}}}\frac{\partial\bar{H}}{\partial\bar{X}_i}. \tag{1.73}$$

It is noted in Eq. (1.73) that term $\partial\bar{X}_{\underline{i}}$ (in which the post-subscript "i" of $\bar{X}_i$ is underlined) indicates partial differentiation with respect to the wth variable of $\bar{X}_i$. Consequently, $\partial\bar{X}_{\underline{i}}$ and $\partial\bar{X}_i$ in Eq. (1.73) can be replaced with ∂x_{iw} and ∂x_{iv}, respectively, in order to obtain the (w, v)th component of $\partial^2\bar{F}/\partial\bar{X}_i^2$ (i.e., $\partial^2 f_{pq}/\partial x_{iw}\partial x_{iv}$). It is noted that this notation is of particular importance in Part 3 of this book, which addresses the problem of developing computer code for numerical computations.

Example 1.11 If $\bar{F} = [f_1 \quad f_2]^T$ with

$$f_1 = 3x_1 + 4x_2^3 + 5x_2 + 6x_3 + 4x_3^2 + 5,$$
$$f_2 = 6x_1 + 4x_2^2 + 3x_2 + x_3^3 + 7x_3 + 5,$$

and $\bar{X} = [x_1 \quad x_2 \quad x_3]^T$, determine (1) $\partial f_1/\partial x_1$, (2) $\partial f_1/\partial\bar{X}$, (3) $\partial\bar{F}/\partial x_1$, (4) $\partial\bar{F}/\partial\bar{X}$, (5) $\partial^2 f_1/\partial\bar{X}^2$, and (6) $\partial^2\bar{F}/\partial\bar{X}^2$.

Answer:

(1) $\partial f_1/\partial x_1 = 3$, i.e., a scalar.

(2) $\partial f_1/\partial \bar{X} = \begin{bmatrix} 3 & (12x_2^2+5) & (8x_3+6) \end{bmatrix}$, i.e., a row matrix.

(3) $\partial \bar{F}/\partial x_1 = \begin{bmatrix} 3 \\ 6 \end{bmatrix}$, i.e., a column matrix.

(4) $\frac{\partial \bar{F}}{\partial \bar{X}} = \begin{bmatrix} 3 & 12x_2^2+5 & 8x_3+6 \\ 6 & 8x_2+3 & 3x_3^2+7 \end{bmatrix}$, i.e., a 2×3 matrix.

(5) $\frac{\partial^2 f_1}{\partial \bar{X}^2} = \begin{bmatrix} 0 & 0 & 0 \\ 0 & 24x_2 & 0 \\ 0 & 0 & 8 \end{bmatrix}$, i.e., a 3×3 matrix.

(6) $\partial^2 \bar{F}/\partial \bar{X}^2 = \begin{bmatrix} \partial^2 f_1/\partial \bar{X}^2 \\ \partial^2 f_2/\partial \bar{X}^2 \end{bmatrix}_{2\times 3\times 3}$, i.e., a multi-dimensional matrix with $\partial^2 f_1/\partial \bar{X}^2$ given in (5) and $\frac{\partial^2 f_2}{\partial \bar{X}^2} = \begin{bmatrix} 0 & 0 & 0 \\ 0 & 8 & 0 \\ 0 & 0 & 6x_3 \end{bmatrix}$.

1.11 Introduction to Optimization Methods

In optical systems design, optimization methods play a key role in tuning the system variables so as to maximize the system performance. Numerous optimization methods have been developed in recent decades. Generally speaking, these methods can be classified as either direct search methods or derivatives-based methods. The term "direct search methods" refers to a class of methods which do not calculate, use or approximate the derivatives of a merit function. In other words, only the function values of a merit function $\bar{\Phi}$ are used in the search process. By contrast "derivatives-based methods", as the name implies, use derivatives of the merit function $\bar{\Phi}$ to perform the search for the optimum solution. In other words, the merit function $\bar{\Phi}$ is assumed to be smooth and at least twice continuously differentiable everywhere in the feasible design space.

In practice, the performance of optimization methods depends on the search direction $\Delta\bar{X}_{sys}$ and step size f, where $\bar{X}_{sys}$ is the system variable vector. During the optimization process, the search direction, $\Delta\bar{X}_{sys}$, is used to update the estimated system variable vector $\bar{X}_{sys}$ in accordance with (Fig. 1.17)

$$\bar{X}_{sys/next} = \bar{X}_{sys/current} + f\Delta\bar{X}_{sys}, \tag{1.74}$$

where $\bar{X}_{sys/next}$ and $\bar{X}_{sys/current}$ are the estimated values of $\bar{X}_{sys}$ in the next and current iterations, respectively. Furthermore, a convergence parameter g is defined to check the magnitude of $\Delta\bar{X}_{sys}$, i.e., $\|\Delta\bar{X}_{sys}\|$. If $\|\Delta\bar{X}_{sys}\| \leq g$, the iterative optimization process is terminated. Otherwise, the iteration procedure continues, and new values of the merit function $\bar{\Phi}$, search direction $\Delta\bar{X}_{sys}$, step size f, and system vector $\bar{X}_{sys/next}$ are determined.

In this book, the importance of the Jacobian and Hessian matrices of a merit function is demonstrated using four common optimization methods, namely the steepest-descent method (p. 431 of [6]), the classical Newton method (p. 460 of [6]), the modified Newton's method (p. 461 of [6]), and the quasi-Newton method (p. 466 of [6]). The steepest-descent method is one of the simplest and most commonly used optimization methods since the search direction $\Delta\bar{X}_{sys}$ is simply evaluated by taking the Jacobian value of the merit function $\bar{\Phi}$, i.e.,

$$\Delta\bar{X}_{sys} = -\frac{\partial\bar{\Phi}}{\partial\bar{X}_{sys}}. \tag{1.75}$$

However, the steepest-descent method has a poor rate of convergence since it uses only the Jacobian matrix. In the classical Newton method, the search direction $\Delta\bar{X}_{sys}$ is thus evaluated in terms of both the Jacobian matrix and the Hessian matrix, i.e.,

$$\Delta\bar{X}_{sys} = -\left(\frac{\partial^2\bar{\Phi}}{\partial\bar{X}_{sys}^2}\right)^{-1}\frac{\partial\bar{\Phi}}{\partial\bar{X}_{sys}}. \tag{1.76}$$

However, in the classical Newton method, the step size is taken as one. Therefore, there is no way to ensure that the merit function will be reduced in each iteration, and hence the method is not guaranteed to converge to a local minimum point even with the use of the classical Hessian matrix. This problem can be corrected if a step size is incorporated in the calculation of the search direction $\Delta\bar{X}_{sys}$ by using any one-dimensional search methods to calculate the step size f in the search direction. This is called the modified Newton's method.

For some optical design problems (e.g., the distortion of image), calculating the Hessian matrix may be tedious or even impossible. Moreover, problems can arise in the classical and modified Newton's methods if the Hessian matrix is singular in any iteration. Quasi-Newton methods overcome these drawbacks by generating an approximation of the Hessian matrix or its inverse in each iteration. Importantly, these approximations are generated using only the Jacobian matrix of a merit function. Therefore, such methods retain the desirable features of both the steepest-descent method and the two Newton methods.

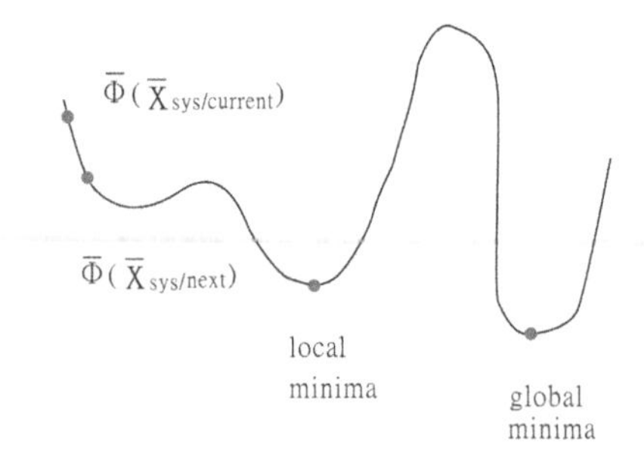

Fig. 1.17 Representation of local and global optimum points

References

1. Uicker JJ (1965) On the dynamic analysis of spatial linkages using 4 × 4 matrices. PhD dissertation, Northwestern University, Evanston, ILL, USA
2. Denavit J, Hartenberg RS (1955) A kinematic notation for lower pair mechanisms based on matrices, Trans. ASME. J Appl Mech 77:215–221
3. Paul RP (1982) Robot manipulators-mathematics, programming and control. MIT press, Cambridge, Mass
4. Litvin FL (1989) Theory of gearing. NASA Reference Publication
5. Foley JD, Dam AV, Feiner SK, Hughes JF (1981), Computer graphics, principles and practices, 2nd edn. Addision-Wesley Publishing Company
6. Arora JS (2012) Introduction to optimum design, 3rd edn. Elservier Inc

Chapter 2
Skew-Ray Tracing of Geometrical Optics

In geometrical optics (or ray optics), light propagation is described in terms of "rays", where each ray is regarded as an idealized narrow bundle of light with zero width [1]. This is different from beam, which is a concept used in almost all fields of physics. Geometrical optics provides equations for predicting the paths followed by the rays through an optical system. These equations are somewhat simplistic, and cannot therefore accurately describe such effects as diffraction and polarization. However, they nevertheless provide a powerful tool for investigating the performance of optical systems during the initial design and analysis stage. Geometrical optics in this book are applied to perform sequential raytracing based on Snell's law and a homogeneous coordinate notation, and then compute the first- and second-order derivative matrices of various optical quantities. Notably, the equations of geometrical optics expressed in terms of homogeneous coordinate notation are often far simpler than their vector counterparts; particularly for more complex optical elements such as prisms.

2.1 Source Ray

In geometrical optics, rays are usually, but not necessarily, assumed to move from left to right. In accordance with the homogeneous coordinate notation used in this book, the incidence point ${}^{g}\bar{P}_{i}$ and unit directional vector ${}^{g}\bar{\ell}_{i}$ at the ith boundary surface are designated as ${}^{g}\bar{P}_{i} = [\,P_{ix} \quad P_{iy} \quad P_{iz} \quad 1\,]^{T}$ and ${}^{g}\bar{\ell}_{i} = [\,\ell_{ix} \quad \ell_{iy} \quad \ell_{iz} \quad 0\,]^{T}$, respectively, where the pre-superscript "g" of the leading symbols ${}^{g}\bar{P}_{i}$ and ${}^{g}\bar{\ell}_{i}$

P.D. Lin, *Advanced Geometrical Optics*, Progress in Optical Science and Photonics 4, DOI 10.1007/978-981-10-2299-9_2

indicates that the components of the respective vectors are referred with respect to coordinate frame $(xyz)_g$. The ray at the ith boundary surface can thus be denoted as

$$ {}^{g}\bar{R}_i = \left[{}^{g}\bar{P}_i \quad {}^{g}\bar{\ell}_i \right]^T = \left[P_{ix} \quad P_{iy} \quad P_{iz} \quad \ell_{ix} \quad \ell_{iy} \quad \ell_{iz} \right]^T. \tag{2.1} $$

Note that when ${}^{0}\bar{R}_i$ (or ${}^{0}\bar{P}_i$ or ${}^{0}\bar{\ell}_i$) is referred to the world coordinate frame $(xyz)_0$, the pre-superscript "0" is omitted for reasons of simplicity.

An optical lens or prism (referred to simply as an optical element hereafter) is a block of optical material possessing multiple boundary surfaces and having a constant refractive index. To trace the path of a ray through an optical system with k elements and n boundary surfaces, it is first necessary to label the individual elements within the system from $j = 0$ to $j = k$ and the boundary surfaces from $i = 0$ to $i = n$. By convention, labels $j = 0$ and $i = 0$ are assigned to the source ray $\bar{R}_0$, which originates at point source

$$ \bar{P}_0 = \begin{bmatrix} P_{0x} \\ P_{0y} \\ P_{0z} \\ 1 \end{bmatrix} \tag{2.2} $$

and travels along the unit directional vector

$$ \bar{\ell}_0 = \begin{bmatrix} \ell_{0x} \\ \ell_{0y} \\ \ell_{0z} \\ 0 \end{bmatrix} = \begin{bmatrix} C\beta_0 C(90^\circ + \alpha_0) \\ C\beta_0 S(90^\circ + \alpha_0) \\ S\beta_0 \\ 0 \end{bmatrix}, \tag{2.3} $$

where β_0 is the angle between $\bar{\ell}_0$ and the projection of $\bar{\ell}_0$ on the horizontal plane passing through the origin of a notional unit sphere centered at $\bar{P}_0$ (see Fig. 2.1). Furthermore, α_0 is the angle between the meridional plane and $\bar{\ell}_0$, and is measured along the zenith direction (i.e., the z_0 axis direction) of the unit sphere. For convenience, the cone shown in Fig. 2.1 is referred to hereafter as the $\alpha_0(\beta_0)$ cone, and is generated by sweeping $\bar{\ell}_0$ with a constant value of β_0 around the zenith direction of the unit sphere. It is noted that $\bar{\ell}_0$ is parallel with the y_0 axis when $\alpha_0 = 0^\circ$ and $\beta_0 = 0^\circ$. From Eqs. (2.2) and (2.3), the variable vector $\bar{X}_0$ of the source ray $\bar{R}_0$ is obtained as

$$ \bar{X}_0 = \left[P_{0x} \quad P_{0y} \quad P_{0z} \quad \alpha_0 \quad \beta_0 \right]^T. \tag{2.4} $$

When a point source $\bar{P}_0$ of an axis-symmetrical system is confined to the y_0z_0 plane (where y_0 points along the optical axis of the system), then a meridional ray (or tangential ray) is the ray lying on that plane. Furthermore, the y_0z_0 plane is referred to as the meridional plane. It is noted that in an axis-symmetrical system,

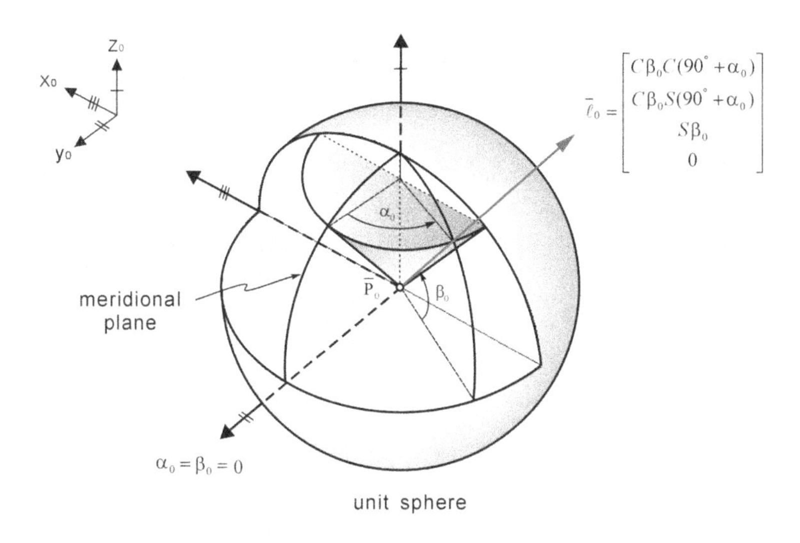

Fig. 2.1 Schematic representation of unit directional vector $\bar{\ell}_0$ originating from point source $\bar{P}_0$

any source ray originating from point source $\bar{P}_0 = [0 \quad P_{0y} \quad P_{0z} \quad 1]^T$ with $\alpha_0 = 0$ always travels on the meridional plane.

The discussions above consider a single ray originating at a single point located at an arbitrary location. Such a point source provides a good approximation for finite but small sources. However, in many practical cases, an expression for the path of perfectly collimated light within the optical system is required. This can most conveniently be achieved by placing a number of source points, each emitting a single ray with a fixed unit directional vector $\bar{\ell}_0$, at several discrete locations in the optical system (Fig. 2.2).

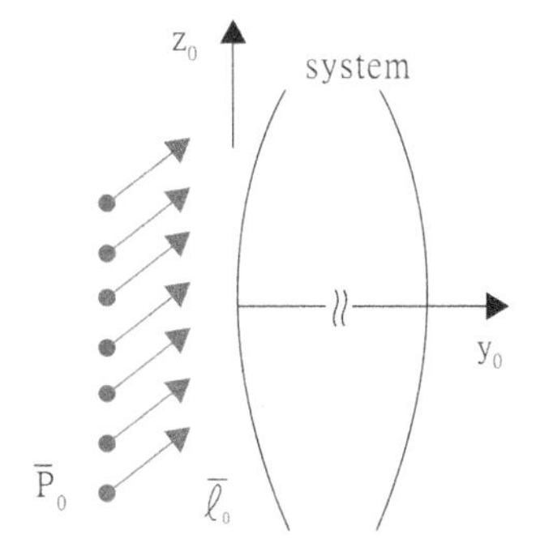

Fig. 2.2 Collimated light comprising multiple parallel light rays

2.2 Spherical Boundary Surfaces

Raytracing is a commonly used technique in geometrical optics for the design, analysis and synthesis of optical systems. The basic foundations of raytracing were presented originally by Hamilton [2] in 1830 and were then further developed by Silverstein [3] in 1918. Raytracing forms the basis of many optical models and techniques [2–33]; with outstanding basic documents written by Spencer et al. [4] and Stavroudis [5] with vector analysis, but most of them use paraxial, meridional or skew ray-tracing with approximations. Accordingly, this chapter presents an analytical skew-ray tracing methodology which is more easily understandable and applied.

2.2.1 Spherical Boundary Surface and Associated Unit Normal Vector

Spherical lenses are one of the most common elements in optical systems. In such lenses, the two boundary surfaces are formed by the partial surfaces of two notional spheres, while the lens axis is ideally perpendicular to both boundary surfaces. The boundary surfaces may be convex (i.e., the sphere center and incoming ray are located on opposite sides of the vertex), concave (i.e., the sphere center and incoming ray are located on the same side of the vertex), or planar (i.e., flat). The line joining the centers of the two spheres making up the lens surfaces represents the axis of the lens. As shown in Fig. 2.3, the generating curve ${}^{i}\bar{q}_i$ of a spherical boundary surface, denoted as the ith boundary surface in an optical system, is a curve in the x_iz_i plane with the form

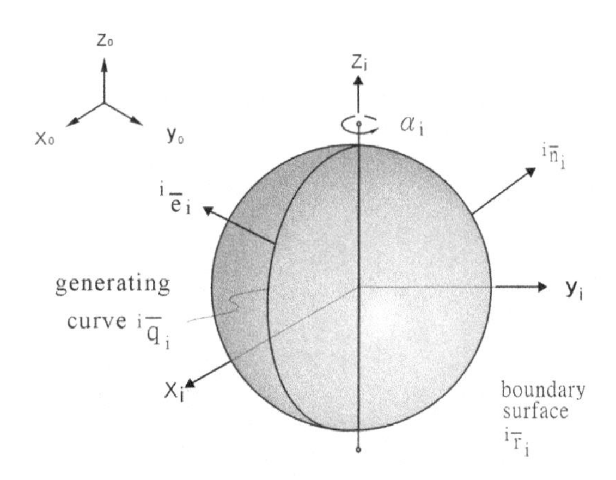

Fig. 2.3 Generating curve and associated unit normal vector of spherical boundary surface

$$ {}^{i}\bar{q}_i = \begin{bmatrix} x_i \\ 0 \\ z_i \\ 1 \end{bmatrix} = \begin{bmatrix} |R_i|C\beta_i \\ 0 \\ |R_i|S\beta_i \\ 1 \end{bmatrix}, \frac{-\pi}{2} \le \beta_i \le \frac{\pi}{2}, \tag{2.5} $$

where $|R_i|$ denotes the absolute value of the radius R_i, and R_i can be either positive or negative, depending on whether the surface is convex or concave, respectively. For any generating curve, there exist two possible unit normal vectors, which point in opposite directions from one another. These unit normal vectors can be formulated as

$$ {}^{i}\bar{\eta}_i = \begin{bmatrix} \eta_{ix} \\ \eta_{iy} \\ \eta_{iz} \\ 0 \end{bmatrix} = \frac{s_i}{\sqrt{(dx_i/d\beta_i)^2 + (dz_i/d\beta_i)^2}} \begin{bmatrix} dz_i/d\beta_i \\ 0 \\ -dx_i/d\beta_i \\ 0 \end{bmatrix} = s_i \begin{bmatrix} C\beta_i \\ 0 \\ S\beta_i \\ 0 \end{bmatrix}, \tag{2.6} $$

where s_i is set to either +1 or −1 to indicate the two possible directions of the vector. Having formulated the two unit normal vectors of the generating curve, the spherical boundary surface ${}^{i}\bar{r}_i$ and associated unit normal vectors ${}^{i}\bar{n}_i$ can be obtained by rotating ${}^{i}\bar{q}_i$ and ${}^{i}\bar{\eta}_i$, respectively, about the z_i axis through an angle α_i $(0 \le \alpha_i < 2\pi)$, i.e.,

$$ {}^{i}\bar{r}_i = \begin{bmatrix} x_i \\ y_i \\ z_i \\ 1 \end{bmatrix} = \mathrm{rot}(\bar{z}, \alpha_i)\,{}^{i}\bar{q}_i = \begin{bmatrix} C\alpha_i & -S\alpha_i & 0 & 0 \\ S\alpha_i & C\alpha_i & 0 & 0 \\ 0 & 0 & 1 & 0 \\ 0 & 0 & 0 & 1 \end{bmatrix} \begin{bmatrix} |R_i|\,C\beta_i \\ 0 \\ |R_i|\,S\beta_i \\ 1 \end{bmatrix} = \begin{bmatrix} |R_i|\,C\beta_i C\alpha_i \\ |R_i|\,C\beta_i S\alpha_i \\ |R_i|\,S\beta_i \\ 1 \end{bmatrix}, \tag{2.7} $$

$$ {}^{i}\bar{n}_i = \mathrm{rot}(\bar{z}, \alpha_i)\,{}^{i}\bar{\eta}_i = s_i \begin{bmatrix} C\alpha_i & -S\alpha_i & 0 & 0 \\ S\alpha_i & C\alpha_i & 0 & 0 \\ 0 & 0 & 1 & 0 \\ 0 & 0 & 0 & 1 \end{bmatrix} \begin{bmatrix} C\beta_i \\ 0 \\ S\beta_i \\ 0 \end{bmatrix} = s_i \begin{bmatrix} C\beta_i C\alpha_i \\ C\beta_i S\alpha_i \\ S\beta_i \\ 0 \end{bmatrix}. \tag{2.8} $$

Equations (2.7) and (2.8) describe a spherical boundary surface ${}^{i}\bar{r}_i$ and its unit normal vector ${}^{i}\bar{n}_i$ in terms of two parameters, α_i and β_i. Both equations are expressed with respect to an arbitrary coordinate frame $(xyz)_i$. However, many derivations in this book are built relative to the world coordinate frame $(xyz)_i$. The following pose matrix of $(xyz)_i$ with respect to $(xyz)_0$ is thus required:

$$
\begin{aligned}
{}^{0}\bar{A}_i &= \mathrm{tran}(t_{ix},0,0)\mathrm{tran}(0,t_{iy},0)\mathrm{tran}(0,0,t_{iz})\mathrm{rot}(\bar{z},\omega_{iz})\mathrm{rot}(\bar{y},\omega_{iy})\mathrm{rot}(\bar{x},\omega_{ix}) \\
&= \begin{bmatrix} C\omega_{iz}C\omega_{iy} & C\omega_{iz}S\omega_{iy}S\omega_{ix} - S\omega_{iz}C\omega_{ix} & C\omega_{iz}S\omega_{iy}C\omega_{ix} + S\omega_{iz}S\,\omega_{ix} & t_{ix} \\ S\omega_{iz}C\omega_{iy} & S\omega_{iz}S\omega_{iy}S\omega_{ix} + C\omega_{iz}C\omega_{ix} & S\omega_{iz}S\omega_{iy}C\omega_{ix} - C\omega_{iz}S\omega_{ix} & t_{iy} \\ -S\omega_{iy} & C\omega_{iy}S\omega_{ix} & C\omega_{iy}C\omega_{ix} & t_{iz} \\ 0 & 0 & 0 & 1 \end{bmatrix} \\
&= \begin{bmatrix} I_{ix} & J_{ix} & K_{ix} & t_{ix} \\ I_{iy} & J_{iy} & K_{iy} & t_{iy} \\ I_{iz} & J_{iz} & K_{iz} & t_{iz} \\ 0 & 0 & 0 & 1 \end{bmatrix},
\end{aligned}
\tag{2.9}
$$

where $t_{ix}, t_{iy}, t_{iz}, \omega_{ix}, \omega_{iy}$ and ω_{iz} are the pose variables of the spherical boundary surface. The unit normal vectors $\bar{n}_i$ of the boundary surface with respect to frame $(xyz)_o$ can then be obtained via the following transformation:

$$
\bar{n}_i = \begin{bmatrix} n_{ix} \\ n_{iy} \\ n_{iz} \\ 0 \end{bmatrix} = {}^{0}\bar{A}_i\,{}^{i}\bar{n}_i = s_i \begin{bmatrix} I_{ix}C\beta_iC\alpha_i + J_{ix}C\beta_iS\alpha_i + K_{ix}S\beta_i \\ I_{iy}C\beta_iC\alpha_i + J_{iy}C\beta_iS\alpha_i + K_{iy}S\beta_i \\ I_{iz}C\beta_iC\alpha_i + J_{iz}C\beta_iS\alpha_i + K_{iz}S\beta_i \\ 0 \end{bmatrix}. \tag{2.10}
$$

2.2.2 *Incidence Point*

In Fig. 2.4, a ray originating from incidence point $\bar{P}_{i-1}$ located on the previous boundary surface $\bar{r}_{i-1}$ is directed along the unit directional vector $\bar{\ell}_{i-1}$ and is reflected or refracted at boundary surface $\bar{r}_i$. Any intermediate point $\bar{P}'_{i-1}$ along this ray is given by $\bar{P}'_{i-1} = \bar{P}_{i-1} + \lambda\bar{\ell}_{i-1}$. The parameter $\lambda = \lambda_i$, for which the incoming ray $\bar{R}_{i-1}$ hits the current boundary surface $\bar{r}_i$ at incidence point

$$
\bar{P}_i = \begin{bmatrix} P_{ix} \\ P_{iy} \\ P_{iz} \\ 1 \end{bmatrix} = \begin{bmatrix} P_{i-1x} + \lambda_i\,\ell_{i-1x} \\ P_{i-1y} + \lambda_i\,\ell_{i-1y} \\ P_{i-1z} + \lambda_i\,\ell_{i-1z} \\ 1 \end{bmatrix} = \bar{P}_{i-1} + \lambda_i\bar{\ell}_{i-1}, \tag{2.11}
$$

can be obtained by equating Eq. (2.7) to ${}^{i}\bar{P}_i = {}^{i}\bar{A}_0\,\bar{P}_i = ({}^{0}\bar{A}_i)^{-1}\bar{P}_i$. That is,

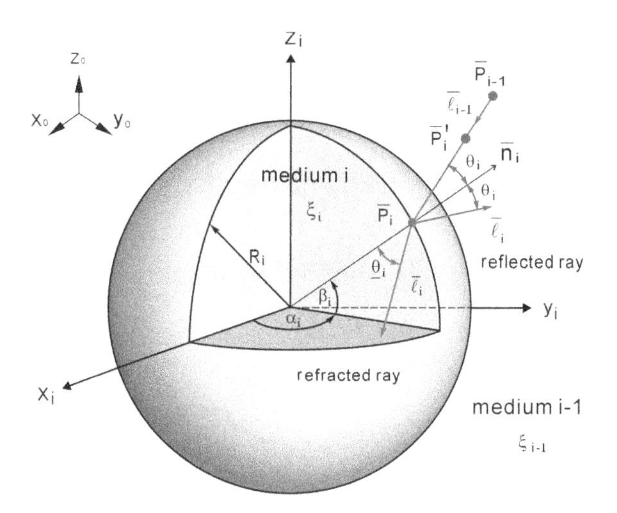

Fig. 2.4 Raytracing at spherical boundary surface $\bar{r}_i$

$$
{}^{i}\bar{P}_i = \begin{bmatrix} I_{ix} & I_{iy} & I_{iz} & -(I_{ix}t_{ix} + I_{iy}t_{iy} + I_{iz}t_{iz}) \\ J_{ix} & J_{iy} & J_{iz} & -(J_{ix}t_{ix} + J_{iy}t_{iy} + J_{iz}t_{iz}) \\ K_{ix} & K_{iy} & K_{iz} & -(K_{ix}t_{ix} + K_{iy}t_{iy} + K_{iz}t_{iz}) \\ 0 & 0 & 0 & 1 \end{bmatrix} \begin{bmatrix} P_{i-1x} + \lambda_i \, \ell_{i-1x} \\ P_{i-1y} + \lambda_i \, \ell_{i-1y} \\ P_{i-1z} + \lambda_i \, \ell_{i-1z} \\ 1 \end{bmatrix}
$$
$$
= \begin{bmatrix} \sigma_i \\ \rho_i \\ \tau_i \\ 1 \end{bmatrix} = {}^{i}\bar{r}_i = \begin{bmatrix} |R_i| \; C\beta_i C\alpha_i \\ |R_i| \; C\beta_i S\alpha_i \\ |R_i| \; S\beta_i \\ 1 \end{bmatrix}, \tag{2.12}
$$

where σ_i, ρ_i, and τ_i are coordinates of incidence point expressed in boundary coordinate frame $(xyz)_i$, given by

$$
\begin{aligned} \sigma_i &= I_{ix}(P_{i-1x} + \ell_{i-1x}\lambda_i) + I_{iy}\left(P_{i-1y} + \ell_{i-1y}\lambda_i\right) \\ &\quad + I_{iz}(P_{i-1z} + \ell_{i-1z}\lambda_i) - (I_{ix}t_{ix} + I_{iy}t_{iy} + I_{iz}t_{iz}) \\ &= I_{ix}P_{ix} + I_{iy}P_{iy} + I_{iz}P_{iz} - (I_{ix}t_{ix} + I_{iy}t_{iy} + I_{iz}t_{iz}), \end{aligned} \tag{2.13}
$$

$$
\begin{aligned} \rho_i &= J_{ix}(P_{i-1x} + \ell_{i-1x}\lambda_i) + J_{iy}\left(P_{i-1y} + \ell_{i-1y}\lambda_i\right) \\ &\quad + J_{iz}(P_{i-1z} + \ell_{i-1z}\lambda_i) - (J_{ix}t_{ix} + J_{iy}t_{iy} + J_{iz}t_{iz}) \\ &= J_{ix}P_{ix} + J_{iy}P_{iy} + J_{iz}P_{iz} - (J_{ix}t_{ix} + J_{iy}t_{iy} + J_{iz}t_{iz}), \end{aligned} \tag{2.14}
$$

$$\begin{aligned}\tau_i &= K_{ix}(P_{i-1x} + \ell_{i-1x}\lambda_i) + K_{iy}(P_{i-1y} + \ell_{i-1y}\lambda_i) \\ &\quad + K_{iz}(P_{i-1z} + \ell_{i-1z}\lambda_i) - (K_{ix}t_{ix} + K_{iy}t_{iy} + K_{iz}t_{iz}) \\ &= K_{ix}P_{ix} + K_{iy}P_{iy} + K_{iz}P_{iz} - (K_{ix}t_{ix} + K_{iy}t_{iy} + K_{iz}t_{iz}).\end{aligned} \tag{2.15}$$

From the sum of σ_i^2, ρ_i^2 and τ_i^2, λ_i is obtained as

$$\lambda_i = -D_i \pm \sqrt{D_i^2 - E_i}, \tag{2.16}$$

where

$$D_i = \ell_{i-1x}(P_{i-1x} - t_{ix}) + \ell_{i-1y}(P_{i-1y} - t_{iy}) + \ell_{i-1z}(P_{i-1z} - t_{iz}), \tag{2.17}$$

$$E_i = P_{i-1x}^2 + P_{i-1y}^2 + P_{i-1z}^2 - R_i^2 + t_{ix}^2 + t_{iy}^2 + t_{iz}^2 - 2(t_{ix}P_{i-1x} + t_{iy}P_{i-1y} + t_{iz}P_{i-1z}). \tag{2.18}$$

The non-negative parameter λ_i represents the geometrical path length from point $\bar{P}_{i-1}$ to point $\bar{P}_i$. Also note that the $\pm$ sign in Eq. (2.16) indicates the two possible intersection points of the ray with a complete sphere. More specifically, $\lambda_i = -D_i - \sqrt{D_i^2 - E_i}$ refers to the nearer intersection point while $\lambda_i = -D_i + \sqrt{D_i^2 - E_i}$ refers to the further point. Clearly, only one of these points is useful in practical systems, and thus the appropriate sign must be chosen. When $\rho_i^2 + \sigma_i^2 \neq 0$, α_i $(0 \le \alpha_i < 2\pi)$ and β_i $(-\pi/2 \le \beta_i \le \pi/2)$ at the incidence point $\bar{P}_i$ can be solved from the following equations:

$$\alpha_i = \text{atan2}(\rho_i, \sigma_i), \tag{2.19}$$

$$\beta_i = \text{atan2}\left(\tau_i, \sqrt{\sigma_i^2 + \rho_i^2}\right). \tag{2.20}$$

It is noted from the third component of Eq. (2.12) that β_i can also be determined as $\beta_i = \arcsin(\tau_i/|R_i|)$. However, β_i cannot be obtained using this equation when the incoming ray $\bar{R}_{i-1}$ does not intersect with the complete sphere of $\bar{r}_i$ when $|R_i| < |\tau_i|$. Notably, Eq. (2.20) avoids this problem since it always yields a solution provided that $\sigma_i^2 + \rho_i^2 + \tau_i^2 \neq 0$.

The points located at $\beta_i = \pm\pi/2$ on a spherical boundary surface are pseudo-singular points (or irregular points), at which $\partial^i\bar{r}_i/\partial\beta_i$ and $\partial^i\bar{r}_i/\partial\alpha_i$ are not linearly independent. As a result, the cross product $\partial^i\bar{r}_i/\partial\beta_i \times \partial^i\bar{r}_i/\partial\alpha_i$ cannot be performed. To avoid this problem, an assumption is made throughout this book that the y_i axis of the spherical boundary surface always coincides with the optical axis of the system.

2.2.3 *Unit Directional Vectors of Reflected and Refracted Rays*

To trace a reflected or refracted ray at the current boundary surface $\bar{r}_i$, the incidence angle θ_i (defined in geometrical optics as a non-obtuse angle, i.e., $0° \le \theta_i \le 90°$) must first be known. In practice, $C\theta_i$ is determined via the dot product of $\bar{\ell}_{i-1}$ and $\bar{n}_i$, i.e., $C\theta_i = \left|\bar{\ell}_{i-1} \cdot \bar{n}_i\right|$ (Figs. 2.5 and 2.6). For every incidence point, there exist two possible unit normal vectors, and thus for each particular application, it is necessary to choose the correct one (referred to hereafter as the active unit normal vector, $\bar{n}_i$). By default, this book always chooses the active unit normal vector which forms an obtuse angle $90° < \eta < 180°$ with $\bar{\ell}_{i-1}$. Consequently,$C\theta_i$ can be computed without the absolute symbol as

$$
\begin{aligned}
C\theta_i &= -\bar{\ell}_{i-1} \cdot \bar{n}_i = -\left(\ell_{i-1x}n_{ix} + \ell_{i-1y}n_{iy} + \ell_{i-1z}n_{iz}\right) \\
&= -s_i\left[\ell_{i-1x}(I_{ix}C\beta_iC\alpha_i + J_{ix}C\beta_iS\alpha_i + K_{ix}S\beta_i) + \ell_{i-1y}\left(I_{iy}C\beta_iC\alpha_i + J_{iy}C\beta_iS\alpha_i + K_{iy}S\beta_i\right)\right. \\
&\quad \left. + \ell_{i-1z}(I_{iz}C\beta_iC\alpha_i + J_{iz}C\beta_iS\alpha_i + K_{iz}S\beta_i)\right].
\end{aligned}
\tag{2.21}
$$

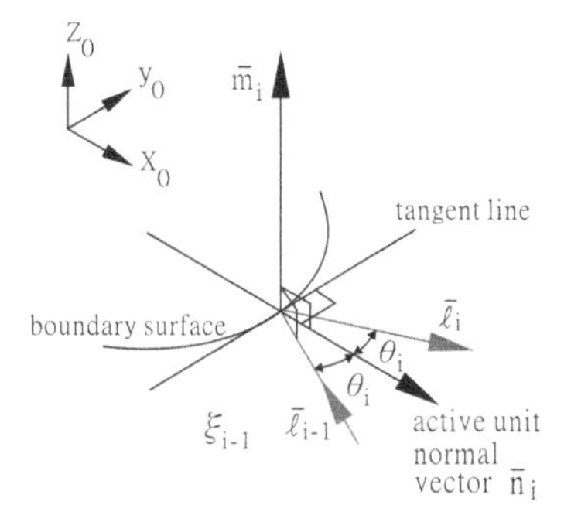

Fig. 2.5 Reflected unit directional vector obtained by rotating active unit normal vector $\bar{n}_i$ about $\bar{m}_i$ by angle θ_i

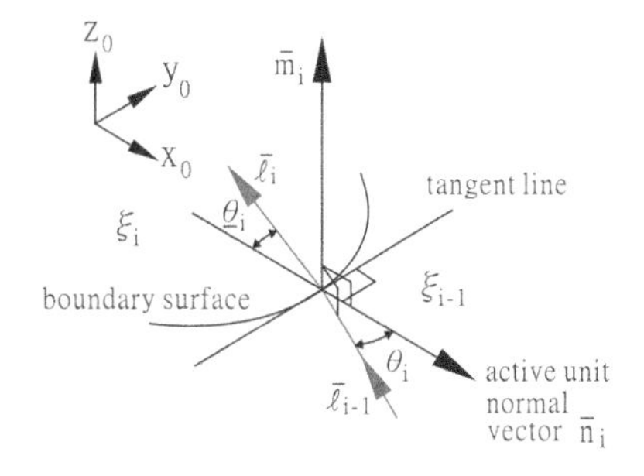

Fig. 2.6 Refracted unit directional vector obtained by rotating active unit normal vector $\bar{n}_i$ about $\bar{m}_i$ by angle $\pi - \underline{\theta}_i$

The refraction angle $\underline{\theta}_i$ between two optical media satisfies Snell's law, that is

$$S\underline{\theta}_i = \frac{\xi_{i-1}}{\xi_i} S\theta_i = N_i S\theta_i, \tag{2.22}$$

where ξ_i is the refractive index of medium i and $N_i = \xi_{i-1}/\xi_i$ is the refractive index of medium i − 1 relative to that of medium i. Total internal reflection is a phenomenon that the ray will not cross the boundary and instead be totally reflected back internally. It happens when a ray strikes the boundary surface at an incident angle θ_i larger than the critical angle $\theta_{i/critical}$ (where $S\theta_{i/critical} = N_i$) if $\xi_i < \xi_{i-1}$.

To trace the reflected or refracted ray at the boundary surface, the common unit normal vector $\bar{m}_i$ of the active unit normal vector $\bar{n}_i$ and $\bar{\ell}_{i-1}$ (see Fig. 2.5) is required, i.e.,

$$\bar{m}_i = \begin{bmatrix} m_{ix} & m_{iy} & m_{iz} & 0 \end{bmatrix}^T = \frac{\bar{n}_i \times \bar{\ell}_{i-1}}{S\theta_i}. \tag{2.23}$$

It is useful to have the following equation, which is derived from Eq. (2.23), when we determine the unit directional vectors $\bar{\ell}_i$ of the reflected and refracted rays:

$$S\theta_i(\bar{m}_i \times \bar{n}_i) = (\bar{n}_i \times \bar{\ell}_{i-1}) \times \bar{n}_i = \bar{\ell}_{i-1} - (\bar{n}_i \cdot \bar{\ell}_{i-1})\bar{n}_i = \bar{\ell}_{i-1} + \bar{n}_i\, C\theta_i. \tag{2.24}$$

According to the reflection law of optics, the reflected unit directional vector $\bar{\ell}_i$ can be obtained by rotating the active unit normal vector $\bar{n}_i$ about $\bar{m}_i$ through an angle θ_i (Fig. 2.5 and Eq. (1.25)). In other words, $\bar{\ell}_i$ is obtained as

$$\bar{\ell}_i = \begin{bmatrix} \ell_{ix} & \ell_{iy} & \ell_{iz} & 0 \end{bmatrix}^T = \mathrm{rot}(\bar{m}_i, \theta_i)\bar{n}_i$$
$$= \begin{bmatrix} m_{ix}^2(1-C\theta_i)+C\theta_i & m_{iy}m_{ix}(1-C\theta_i)-m_{iz}S\theta_i & m_{iz}m_{ix}(1-C\theta_i)+m_{iy}S\theta_i & 0 \\ m_{ix}m_{iy}(1-C\theta_i)+m_{iz}S\theta_i & m_{iy}^2(1-C\theta_i)+C\theta_i & m_{iz}m_{iy}(1-C\theta_i)-m_{ix}S\theta_i & 0 \\ m_{ix}m_{iz}(1-C\theta_i)-m_{iy}S\theta_i & m_{iy}m_{iz}(1-C\theta_i)+m_{ix}S\theta_i & m_{iz}^2(1-C\theta_i)+C\theta_i & 0 \\ 0 & 0 & 0 & 1 \end{bmatrix}$$
$$\begin{bmatrix} n_{ix} \\ n_{iy} \\ n_{iz} \\ 0 \end{bmatrix}. \tag{2.25}$$

Further simplification of Eq. (2.25) is possible by utilizing Eq. (2.24), resulting in

$$\bar{\ell}_i = \begin{bmatrix} \ell_{ix} \\ \ell_{iy} \\ \ell_{iz} \\ 0 \end{bmatrix} = \begin{bmatrix} \ell_{i-1x} + 2C\theta_i\, n_{ix} \\ \ell_{i-1y} + 2C\theta_i\, n_{iy} \\ \ell_{i-1z} + 2C\theta_i\, n_{iz} \\ 0 \end{bmatrix} = \bar{\ell}_{i-1} + 2C\theta_i\, \bar{n}_i. \tag{2.26}$$

Notably, $\bar{\ell}_i$ can also be obtained by rotating $\bar{\ell}_{i-1}$ about $\bar{m}_i$ through an angle $\pi + 2\theta_i$, i.e.,

$$\bar{\ell}_i = \text{rot}(\bar{m}_i, \pi + 2\theta_i)\bar{\ell}_{i-1}. \tag{2.27}$$

According to the refraction law of optics, the refracted unit directional vector $\bar{\ell}_i$ can be obtained by rotating the active unit normal vector $\bar{n}_i$ about $\bar{m}_i$ through an angle $\pi - \underline{\theta}_i$ (Fig. 2.6), i.e.,

$$\bar{\ell}_i = \begin{bmatrix} \ell_{ix} & \ell_{iy} & \ell_{iz} & 0 \end{bmatrix}^T = \text{rot}(\bar{m}_i, \pi - \underline{\theta}_i)\bar{n}_i$$
$$= \begin{bmatrix} m_{ix}^2(1 + C\underline{\theta}_i) - C\underline{\theta}_i & m_{iy}m_{ix}(1 + C\underline{\theta}_i) - m_{iz}S\underline{\theta}_i & m_{iz}m_{ix}(1 + C\underline{\theta}_i) + m_{iy}S\underline{\theta}_i & 0 \\ m_{ix}m_{iy}(1 + C\underline{\theta}_i) + m_{iz}S\underline{\theta}_i & m_{iy}^2(1 + C\underline{\theta}_i) - C\underline{\theta}_i & m_{iz}m_{iy}(1 + C\underline{\theta}_i) - m_{ix}S\underline{\theta}_i & 0 \\ m_{ix}m_{iz}(1 + C\underline{\theta}_i) - m_{iy}S\underline{\theta}_i & m_{iy}m_{iz}(1 + C\underline{\theta}_i) + m_{ix}S\underline{\theta}_i & m_{iz}^2(1 + C\underline{\theta}_i) - C\underline{\theta}_i & 0 \\ 0 & 0 & 0 & 1 \end{bmatrix}$$
$$\begin{bmatrix} n_{ix} \\ n_{iy} \\ n_{iz} \\ 0 \end{bmatrix}. \tag{2.28}$$

Again, further simplification of Eq. (2.28) is possible by utilizing Eq. (2.24) and Snell's law, $S\underline{\theta}_i = N_i S\theta_i$. The following refracted unit directional vector $\bar{\ell}_i$ is thus obtained:

$$\bar{\ell}_i = \begin{bmatrix} \ell_{ix} \\ \ell_{iy} \\ \ell_{iz} \\ 0 \end{bmatrix} = \begin{bmatrix} \left(-\sqrt{1 - N_i^2 + (N_i C\theta_i)^2}\right) n_{ix} + N_i(\ell_{i-1x} + C\theta_i\ n_{ix}) \\ \left(-\sqrt{1 - N_i^2 + (N_i C\theta_i)^2}\right) n_{iy} + N_i(\ell_{i-1y} + C\theta_i\ n_{iy}) \\ \left(-\sqrt{1 - N_i^2 + (N_i C\theta_i)^2}\right) n_{iz} + N_i(\ell_{i-1z} + C\theta_i\ n_{iz}) \\ 0 \end{bmatrix}$$
$$= \left(-\sqrt{1 - N_i^2 + (N_i C\theta_i)^2}\right)\bar{n}_i + N_i\left(\bar{\ell}_{i-1} + C\theta_i\bar{n}_i\right), \tag{2.29}$$

where $\bar{n}_i = \begin{bmatrix} n_{ix} & n_{iy} & n_{iz} & 0 \end{bmatrix}^T$ and $C\theta_i$ are given by Eqs. (2.10) and (2.21), respectively. Of course, the refracted unit directional vector $\bar{\ell}_i$ can also be determined by rotating $\bar{\ell}_{i-1}$ about $\bar{m}_i$ through an angle $\pi - \underline{\theta}_i + \theta_i$, i.e., $\bar{\ell}_i = \text{rot}(\bar{m}_i, \pi - \underline{\theta}_i + \theta_i)\bar{\ell}_{i-1}$.

It is important to point out that total internal reflection occurs when $1 - N_i^2 + (N_i C\theta_i)^2 < 0$.

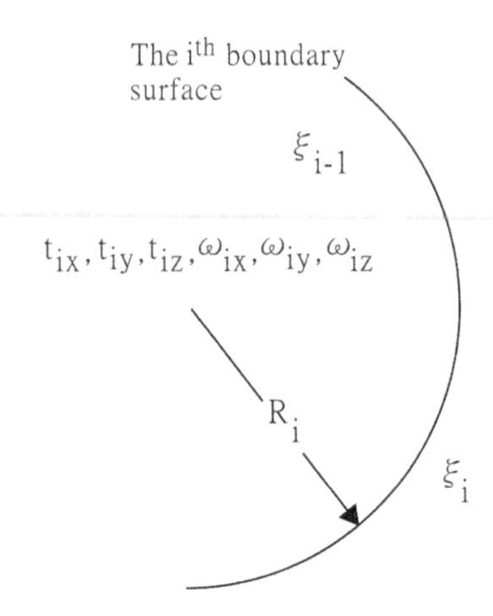

Fig. 2.7 Boundary variables of spherical boundary surface

Referring to Fig. 2.7, designate

$$\bar{X}_i = [\, t_{ix} \quad t_{iy} \quad t_{iz} \quad \omega_{ix} \quad \omega_{iy} \quad \omega_{iz} \quad \xi_{i-1} \quad \xi_i \quad R_i \,]^T \tag{2.30}$$

as the boundary variable vector of a spherical boundary surface, where this vector comprises the six pose variables of Eq. (2.9), the refractive indices ξ_{i-1} and ξ_i, and the radius R_i of the boundary surface.

It is noted from Eqs. (2.11), (2.26) and (2.29) that the ray $\bar{R}_i$ is a function of the incoming ray $\bar{R}_{i-1}$ with the given source ray $\bar{R}_0$ (Fig. 2.8). In other words, $\bar{R}_i$ is a recursive function, i.e., a function which operates in turn on another function (or functions). It is like a Russian nesting doll. Each doll has a smaller and smaller doll inside it. To evaluate a recursive function, it is first necessary to evaluate the internal functions, and to then determine the outer function based on the results of these internal functions.

Example 2.1 The raytracing equations for a meridional ray traveling through an axis-symmetrical system can be obtained by setting the x_0 components of Eqs. (2.11), (2.26) and (2.29) to zero. However, in the following, these equations are derived independently for the system shown in Fig. 2.9, in which a spherical

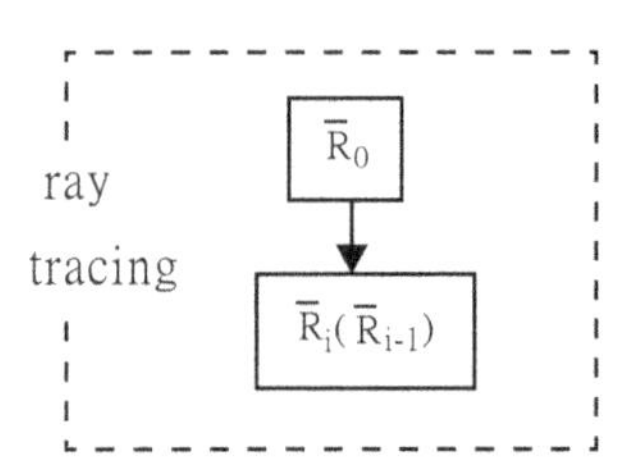

Fig. 2.8 Ray $\bar{R}_i$ is function of incoming ray $\bar{R}_{i-1}$ by given source ray $\bar{R}_0$

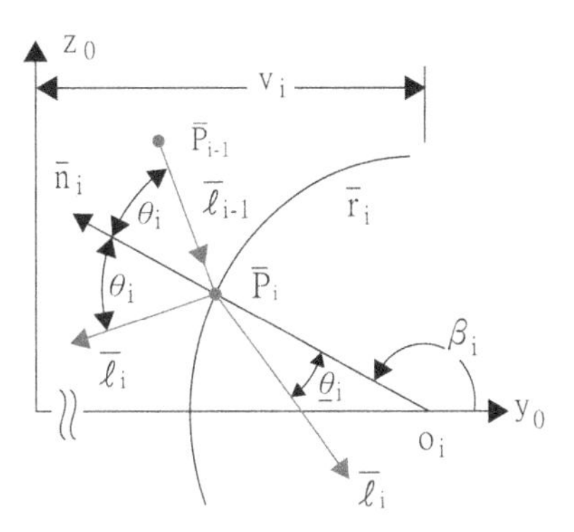

Fig. 2.9 Tracing meridional ray at spherical boundary surface in axis-symmetrical system

boundary surface $\bar{r}_i = [0 \quad |R_i|C\beta_i + v_i \quad |R_i|S\beta_i \quad 1]^T$ with center o_i is located at point $y_0 = v_i$ along the optical axis of an axis-symmetrical system. Assume that a meridional ray $\bar{R}_{i-1}$ originating at point $\bar{P}_{i-1} = [0 \quad P_{i-1y} \quad P_{i-1z} \quad 1]^T$ and directed along $\bar{\ell}_{i-1} = [0 \quad \ell_{i-1y} \quad \ell_{i-1z} \quad 0]^T$ ($\ell_{i-1y}^2 + \ell_{i-1z}^2 = 1$ with $0 < \ell_{i-1y}$ and $\ell_{i-1z} < 0$) is reflected or refracted at boundary surface $\bar{r}_i$. The incidence point $\bar{P}_i$ at which the ray strikes $\bar{r}_i$ is given by

$$\bar{P}_i = \begin{bmatrix} 0 \\ P_{iy} \\ P_{iz} \\ 1 \end{bmatrix} = \bar{P}_{i-1} + \lambda_i \bar{\ell}_{i-1} = \begin{bmatrix} 0 \\ P_{i-1y} + \lambda_i \ell_{i-1y} \\ P_{i-1z} + \lambda_i \ell_{i-1z} \\ 1 \end{bmatrix},$$

where the parameter λ_i is obtained by setting $\bar{P}_i = \bar{r}_i$, yielding $\lambda_i = -D_i \pm \sqrt{D_i^2 - E_i}$, with $D_i = \ell_{i-1y}(P_{i-1y} - v_i) + \ell_{i-1z}P_{i-1z}$ and $E_i = (P_{i-1y} - v_i)^2 + P_{i-1z}^2 - R_i^2$.

Since the active unit normal vector is $\bar{n}_i = [0 \quad C\beta_i \quad S\beta_i \quad 0]^T$, the incidence angle θ_i can be computed as

$$C\theta_i = |\bar{\ell}_{i-1} \cdot \bar{n}_i| = |\bar{\ell}_{i-1}^T \bar{n}_i| = (-\bar{\ell}_{i-1}) \cdot \bar{n}_i = -(\ell_{i-1y}C\beta_i + \ell_{i-1z}S\beta_i).$$

It is noted from Fig. 2.9 that the reflected unit directional vector $\bar{\ell}_i$ can be obtained by rotating the active unit normal vector $\bar{n}_i$ about the x_0 axis through an angle θ_i. Thus, the reflected unit directional vector $\bar{\ell}_i$ is obtained as

$$\bar{\ell}_i = \begin{bmatrix} 0 \\ \ell_{iy} \\ \ell_{iz} \\ 0 \end{bmatrix} = \text{rot}(\bar{x}, \theta_i)\bar{n}_i = \begin{bmatrix} 0 \\ C(\theta_i + \beta_i) \\ S(\theta_i + \beta_i) \\ 0 \end{bmatrix}.$$

According to the refraction law of optics, the refracted unit directional vector $\bar{\ell}_i$ can be obtained by rotating the active unit normal vector $\bar{n}_i$ about the x_0 axis through an angle $\pi - \underline{\theta}_i$. In other words,

$$\bar{\ell}_i = \begin{bmatrix} 0 \\ \ell_{iy} \\ \ell_{iz} \\ 0 \end{bmatrix} = \mathrm{rot}(\bar{x}, \pi - \underline{\theta}_i)\bar{n}_i = \begin{bmatrix} 0 \\ -C(\underline{\theta}_i - \beta_i) \\ S(\underline{\theta}_i - \beta_i) \\ 0 \end{bmatrix},$$

where the refraction angle $\underline{\theta}_i$ satisfies Snell's law, i.e., $S\underline{\theta}_i = (\xi_{i-1}/\xi_i)S\theta_i = N_i S\theta_i$.

It is noted that for the case where the source ray $\bar{R}_{i-1}$ travels in the upward direction (i.e., $0 < \ell_{i-1z}$), the unit directional vectors of the reflected and refracted rays are determined as $\bar{\ell}_i = \mathrm{rot}(-\bar{x}, \theta_i)\bar{n}_i$ and $\bar{\ell}_i = \mathrm{rot}(-\bar{x}, \pi - \underline{\theta}_i)\bar{n}_i$, respectively. The example above shows that the raytracing equations for a meridional ray traveling through an axis-symmetrical system can be easily obtained. However, caution should be exercised when applying the rotation matrix since the rotation axis depends on both $\bar{\ell}_{i-1}$ and the active unit normal vector $\bar{n}_i$ (see Eq. (2.23)).

Example 2.2 Skew-ray tracing in any element containing spherical boundary surfaces is comparatively difficult. As a result, only numerical examples are provided here for illustration purposes. Consider the bi-convex element shown in Fig. 2.10, with $v_1 = 5$, thickness $q_{e1} = 10$, refractive index $\xi_{e1} = 1.5$, and surface radii $R_1 = 50$ and $R_2 = -100$. (1) Assign boundary coordinate frames $(xyz)_1$ and $(xyz)_2$ to the two spherical boundary surfaces, $\bar{r}_1$ and $\bar{r}_2$. (2) Find the unit normal vectors ${}^1\bar{n}_1$ and ${}^2\bar{n}_2$. (3) Determine the pose matrix ${}^0\bar{A}_1$ and boundary variable vector $\bar{X}_1$. (4) Assign the pose matrix ${}^0\bar{A}_2$ and determine boundary variable vector $\bar{X}_2$. (5) Find the unit normal vectors $\bar{n}_1$ and $\bar{n}_2$. (6) Write out the source ray $\bar{R}_0$ when $\bar{P}_0 = [0 \quad -5 \quad 5 \quad 1]^T$ and $\alpha_0 = 0°$, $\beta_0 = 5°$. (7) Find λ_1, α_1, β_1, unit normal vectors $\bar{n}_1$, the value of s_1 for the active unit normal vector, the incidence angle θ_1, the refractive index N_1, and the refracted ray $\bar{R}_1$ when ray $\bar{R}_0$ is refracted at the first boundary surface, $\bar{r}_1$. (8) Find λ_2, α_2, β_2, unit normal vectors $\bar{n}_2$, the value of s_2 for the active unit normal vector, the incidence angle θ_1, the refractive index N_2, and the refracted ray $\bar{R}_2$ when ray $\bar{R}_1$ is refracted at the second boundary surface, $\bar{r}_2$.

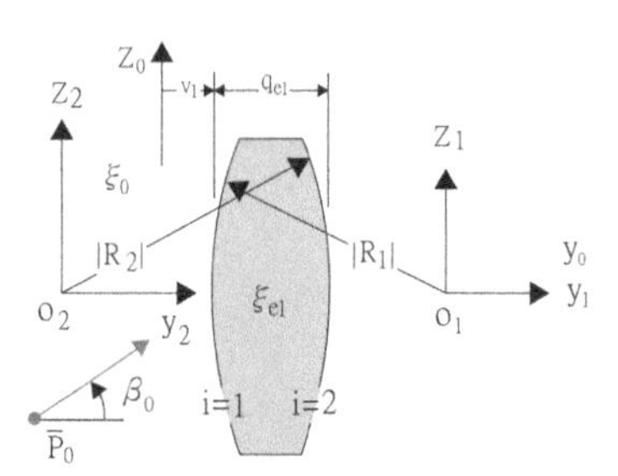

Fig. 2.10 Assigned coordinate frames $(xyz)_1$ and $(xyz)_2$ for bi-convex lens

Solution

(1) The assigned coordinate frames $(xyz)_1$ and $(xyz)_2$ are shown in Fig. 2.10. Note that their origins, o_1 and o_2, are located at the centers of $\bar{r}_1$ and $\bar{r}_2$, respectively, while the y_1 and y_2 axes coincide with the y_0 axis.

(2) ${}^1\bar{n}_1 = s_1[C\beta_1 C\alpha_1 \quad C\beta_1 S\alpha_1 \quad S\beta_1 \quad 0]^T$,
${}^2\bar{n}_2 = s_2[C\beta_2 C\alpha_2 \quad C\beta_2 S\alpha_2 \quad S\beta_2 \quad 0]^T$.

(3) $${}^0\bar{A}_1 = \text{tran}(0, v_1 + R_1, 0) = \text{tran}(0, 5 + 50, 0) = \begin{bmatrix} 1 & 0 & 0 & 0 \\ 0 & 1 & 0 & 55 \\ 0 & 0 & 1 & 0 \\ 0 & 0 & 0 & 1 \end{bmatrix},$$

$$\begin{aligned}\bar{X}_1 &= [t_{1x} \quad v_1 + R_1 \quad t_{1z} \quad \omega_{1x} \quad \omega_{1y} \quad \omega_{1z} \quad \xi_0 \quad \xi_1 \quad R_1]^T \\ &= [0 \quad 55 \quad 0 \quad 0 \quad 0 \quad 0 \quad 1 \quad 1.5 \quad 50]^T.\end{aligned}$$

(4) $$\begin{aligned}{}^0\bar{A}_2 &= \text{tran}(0, v_1 + q_{e1} + R_2, 0) = \text{tran}(0, 5 + 10 - 100, 0) \\ &= \begin{bmatrix} 1 & 0 & 0 & 0 \\ 0 & 1 & 0 & -85 \\ 0 & 0 & 1 & 0 \\ 0 & 0 & 0 & 1 \end{bmatrix},\end{aligned}$$

$$\begin{aligned}\bar{X}_2 &= [t_{2x} \quad v_1 + q_{e1} + R_2 \quad t_{2z} \quad \omega_{2x} \quad \omega_{2y} \quad \omega_{2z} \quad \xi_1 \quad \xi_2 \quad R_2]^T \\ &= [0 \quad -85 \quad 0 \quad 0 \quad 0 \quad 0 \quad 1.5 \quad 1 \quad -100]^T.\end{aligned}$$

(5) The two unit normal vectors of $\bar{r}_1$ are given as $\bar{n}_1 = {}^0\bar{A}_1\,{}^1\bar{n}_1 = s_1[C\beta_1 C\alpha_1 \quad C\beta_1 S\alpha_1 \quad S\beta_1 \quad 0]^T$. Similarly, the unit normal vectors of $\bar{r}_2$ are given as $\bar{n}_2 = {}^0\bar{A}_2\,{}^2\bar{n}_2 = s_2[C\beta_2 C\alpha_2 \quad C\beta_2 S\alpha_2 \quad S\beta_2 \quad 0]^T$.

(6) $\bar{R}_0 = [0 \quad -5 \quad 5 \quad C5°C90° \quad C5°S90° \quad S5°]^T$.

(7) $\lambda_1 = 10.38951$,
$\alpha_1 = -90°$,
$\beta_1 = 6.78304°$.
The active unit normal vector of $\bar{r}_1$ is $\bar{n}_1 = [C\beta_1 C\alpha_1 \quad C\beta_1 S\alpha_1 \quad S\beta_1 \quad 0]^T$, leading to $s_1 = 1$,
$\theta_1 = 11.78304°$,
$N_1 = \xi_0/\xi_1 = 1/1.5$,
$\bar{R}_1 = [0 \quad 5.34997 \quad 5.90551 \quad 0 \quad 0.99983 \quad 0.01817]^T$.

(8) $\lambda_2 = 9.46673$,
$\alpha_2 = 90°$,
$\beta_2 = 3.48433°$.
The active unit normal vector of $\bar{r}_2$ is $\bar{n}_2 = [-C\beta_2 C\alpha_2 \quad -C\beta_2 S\alpha_2 - S\beta_2 \quad 0]^T$, leading to $s_2 = -1$,
$\theta_2 = 2.44297°$,
$N_2 = \xi_1/\xi_2 = 1.5$,
$\bar{R}_2 = [0 \quad 14.81515 \quad 6.07756 \quad 0 \quad 0.99999 \quad -0.00317]^T$.

2.3 Flat Boundary Surfaces

Many optical systems contain elements with flat boundary surfaces. Typical examples include plano-convex lenses, plano-concave lenses, optical flats, beam-splitters, and flat first-surface mirrors. Raytracing at a flat boundary surface is thus of great practical interest. Most studies treat flat boundary surfaces as a spherical surface with zero curvature (e.g., p. 312 of [34]). However, certain ray-tracing equations for prisms cannot be obtained using such an approach. Accordingly, a more different methodology for dealing with flat boundary surfaces is required.

2.3.1 *Flat Boundary Surface and Associated Unit Normal Vector*

As stated in Sect. 2.2, the y_i axis of boundary coordinate frame $(xyz)_i$ in an axis-symmetrical optical system is assumed to coincide with the optical axis of the system. If β_i is the length parameter along y_i axis measured from the origin of $(xyz)_i$, a flat boundary surface can be obtained simply by rotating its generating line (Fig. 2.11)

$$ {}^{i}\bar{q}_i = \begin{bmatrix} 0 \\ 0 \\ \beta_i \\ 1 \end{bmatrix} (0 \le \beta_i) \tag{2.31} $$

about the y_i axis through an angle α_i $(0 \le \alpha_i < 2\pi)$, i.e.,

$$ {}^{i}\bar{r}_i = \begin{bmatrix} x_i \\ y_i \\ z_i \\ 1 \end{bmatrix} = \mathrm{rot}(\bar{y}, \alpha_i)\,{}^{i}\bar{q}_i = \begin{bmatrix} C\alpha_i & 0 & S\alpha_i & 0 \\ 0 & 1 & 0 & 0 \\ -S\alpha_i & 0 & C\alpha_i & 0 \\ 0 & 0 & 0 & 1 \end{bmatrix} \begin{bmatrix} 0 \\ 0 \\ \beta_i \\ 1 \end{bmatrix} = \begin{bmatrix} \beta_i S\alpha_i \\ 0 \\ \beta_i C\alpha_i \\ 1 \end{bmatrix}. \tag{2.32} $$

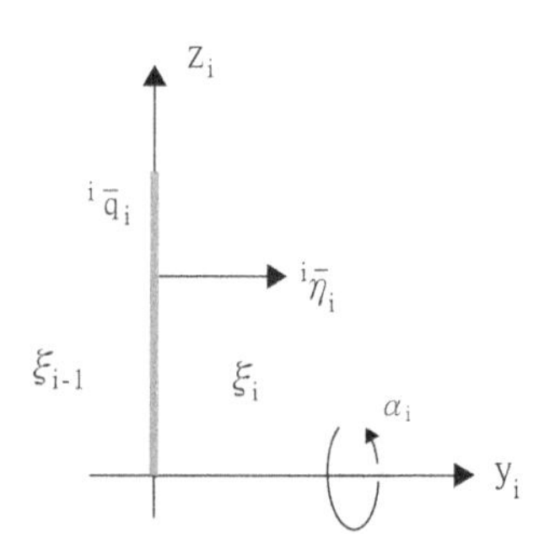

Fig. 2.11 Generating line and associated unit normal vector of flat boundary surface

The two possible unit normal vectors ${}^{i}\bar{\eta}_i$ of the generating curve ${}^{i}\bar{q}_i$ are given by

$$ {}^{i}\bar{\eta}_i = \begin{bmatrix} \eta_{ix} \\ \eta_{iy} \\ \eta_{iz} \\ 0 \end{bmatrix} = s_i \begin{bmatrix} 0 \\ -1 \\ 0 \\ 0 \end{bmatrix}, \tag{2.33} $$

where, again, s_i is set to +1 or −1 to show that there exist two possible unit normal vectors pointing in opposite directions from one another. The unit normal vectors ${}^{i}\bar{n}_i$ of boundary surface ${}^{i}\bar{r}_i$ are then determined by rotating ${}^{i}\bar{\eta}_i$ about the y_i axis through an angle α_i, i.e.,

$$ {}^{i}\bar{n}_i = \mathrm{rot}(\bar{y}, \alpha_i)\, {}^{i}\bar{\eta}_i = s_i \begin{bmatrix} C\alpha_i & 0 & s\alpha_i & 0 \\ 0 & 1 & 0 & 0 \\ -S\alpha_i & 0 & C\alpha_i & 0 \\ 0 & 0 & 0 & 1 \end{bmatrix} \begin{bmatrix} 0 \\ -1 \\ 0 \\ 0 \end{bmatrix} = s_i \begin{bmatrix} 0 \\ -1 \\ 0 \\ 0 \end{bmatrix}. \tag{2.34} $$

Equations (2.32) and (2.34) describe the flat boundary surface ${}^{i}\bar{r}_i$ and its unit normal vector ${}^{i}\bar{n}_i$ with respect to the boundary coordinate frame $(xyz)_i$. However, as discussed earlier, many derivations in this book are built relative to the world coordinate frame $(xyz)_o$. The following pose matrix of frame $(xyz)_i$ with respect to $(xyz)_0$ is thus required:

$$ \begin{aligned} {}^{0}\bar{A}_i &= \mathrm{tran}(t_{ix}, 0, 0)\mathrm{tran}(0, t_{iy}, 0)\mathrm{tran}(0, 0, t_{iz})\mathrm{rot}(\bar{z}, \omega_{iz})\mathrm{rot}(\bar{y}, \omega_{iy})\mathrm{rot}(\bar{x}, \omega_{ix}) \\ &= \begin{bmatrix} C\omega_{iz}C\omega_{iy} & C\omega_{iz}S\omega_{iy}S\omega_{ix} - S\omega_{iz}C\omega_{ix} & C\omega_{iz}S\omega_{iy}C\omega_{ix} + S\omega_{iz}S\omega_{ix} & t_{ix} \\ S\omega_{iz}C\omega_{iy} & S\omega_{iz}S\omega_{iy}S\omega_{ix} + C\omega_{iz}C\omega_{ix} & S\omega_{iz}S\omega_{iy}C\omega_{ix} - C\omega_{iz}S\omega_{ix} & t_{iy} \\ -S\omega_{iy} & C\omega_{iy}S\omega_{ix} & C\omega_{iy}C\omega_{ix} & t_{iz} \\ 0 & 0 & 0 & 1 \end{bmatrix} \\ &= \begin{bmatrix} I_{ix} & J_{ix} & K_{ix} & t_{ix} \\ I_{iy} & J_{iy} & K_{iy} & t_{iy} \\ I_{iz} & J_{iz} & K_{iz} & t_{iz} \\ 0 & 0 & 0 & 1 \end{bmatrix}, \end{aligned} \tag{2.35} $$

where $t_{ix}, t_{iy}, t_{iz}, \omega_{ix}, \omega_{iy}$ and ω_{iz} are the pose variables of the flat boundary surface. The unit normal vector $\bar{n}_i$ can then be obtained with respect to $(xyz)_0$ as

$$ \bar{n}_i = \begin{bmatrix} n_{ix} \\ n_{iy} \\ n_{iz} \\ 0 \end{bmatrix} = {}^{0}\bar{A}_i {}^{i}\bar{n}_i = s_i \begin{bmatrix} I_{ix} & J_{ix} & K_{ix} & t_{ix} \\ I_{iy} & J_{iy} & K_{iy} & t_{iy} \\ I_{iz} & J_{iz} & K_{iz} & t_{iz} \\ 0 & 0 & 0 & 1 \end{bmatrix} \begin{bmatrix} 0 \\ -1 \\ 0 \\ 0 \end{bmatrix} = -s_i \begin{bmatrix} J_{ix} \\ J_{iy} \\ J_{iz} \\ 0 \end{bmatrix}. \tag{2.36} $$

2.3.2 Incidence Point

Assume in Fig. 2.12 that a ray originating at point $\bar{P}_{i-1}$ on the previous boundary surface $\bar{r}_{i-1}$ is directed along the unit directional vector $\bar{\ell}_{i-1}$ and is reflected or refracted at current flat boundary surface $\bar{r}_i$. Any intermediate point $\bar{P}'_{i-1}$ lying along this ray as it travels from $\bar{P}_{i-1}$ is given by $\bar{P}'_{i-1} = \bar{P}_{i-1} + \lambda\bar{\ell}_{i-1}$. Moreover, the incidence point $\bar{P}_i$ at which the ray hits the flat boundary surface is given by

$$\bar{P}_i = \begin{bmatrix} P_{ix} \\ P_{iy} \\ P_{iz} \\ 1 \end{bmatrix} = \begin{bmatrix} P_{i-1x} + \lambda_i \ell_{i-1x} \\ P_{i-1y} + \lambda_i \ell_{i-1y} \\ P_{i-1z} + \lambda_i \ell_{i-1z} \\ 1 \end{bmatrix} = \bar{P}_{i-1} + \lambda_i \bar{\ell}_{i-1}, \tag{2.37}$$

where the parameter λ_i can be obtained by equating Eq. (2.32) to ${}^i\bar{P}_i = {}^i\bar{A}_0\bar{P}_i = ({}^0\bar{A}_i)^{-1}\bar{P}_i$, i.e.,

$$\begin{aligned} {}^i\bar{P}_i &= \begin{bmatrix} I_{ix} & I_{iy} & I_{iz} & -(I_{ix}t_{ix} + I_{iy}t_{iy} + I_{iz}t_{iz}) \\ J_{ix} & J_{iy} & J_{iz} & -(J_{ix}t_{ix} + J_{iy}t_{iy} + J_{iz}t_{iz}) \\ K_{ix} & K_{iy} & K_{iz} & -(K_{ix}t_{ix} + K_{iy}t_{iy} + K_{iz}t_{iz}) \\ 0 & 0 & 0 & 1 \end{bmatrix} \begin{bmatrix} P_{i-1x} + \lambda_i \ell_{i-1x} \\ P_{i-1y} + \lambda_i \ell_{i-1y} \\ P_{i-1z} + \lambda_i \ell_{i-1z} \\ 1 \end{bmatrix} \\ &= \begin{bmatrix} \sigma_i \\ \rho_i \\ \tau_i \\ 1 \end{bmatrix} = {}^i\bar{r}_i = \begin{bmatrix} \beta_i S\alpha_i \\ 0 \\ \beta_i C\alpha_i \\ 1 \end{bmatrix}, \end{aligned} \tag{2.38}$$

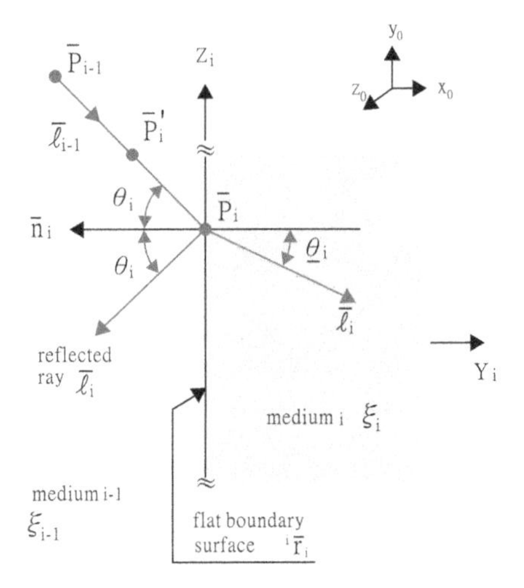

Fig. 2.12 Raytracing at flat boundary surface

in which σ_i, ρ_i and τ_i are given in Eqs. (2.13), (2.14) and (2.15), respectively. λ_i can be solved from the second component of Eq. (2.38) (i.e., $\rho_i = 0$) as

$$\lambda_i = \frac{-\left[J_{ix}P_{i-1x} + J_{iy}P_{i-1y} + J_{iz}P_{i-1z} - (J_{ix}t_{ix} + J_{iy}t_{iy} + J_{iz}t_{iz})\right]}{J_{ix}\ell_{i-1x} + J_{iy}\ell_{i-1y} + J_{iz}\ell_{i-1z}} = \frac{-D_i}{E_i}, \quad (2.39)$$

where

$$D_i = J_{ix}P_{i-1x} + J_{iy}P_{i-1y} + J_{iz}P_{i-1z} + e_i, \quad (2.40)$$

$$E_i = J_{ix}\ell_{i-1x} + J_{iy}\ell_{i-1y} + J_{iz}\ell_{i-1z}. \quad (2.41)$$

Note that parameter e_i in Eq. (2.40) is defined as $e_i = -(J_{ix}t_{ix} + J_{iy}t_{iy} + J_{iz}t_{iz})$ and is adopted in this book as a means of lumping six parameters together, thereby reducing the total number of variables to be considered. As discussed in Sect. (1.7, $-e_i$ represents the distance from the origin of coordinate frame $(xyz)_o$ to the origin of coordinate frame $(xyz)_i$ along the direction of the unit normal $\bar{n}_i$ (Fig. 1.16 with $g = 0$).

The parameters α_i $(0 \le \alpha_i < 2\pi)$ and β_i $(0 \le \beta_i)$, where incidence point $\bar{P}_i$ hits the $x_i z_i$ plane, are determined respectively by

$$\beta_i = \sqrt{\sigma_i^2 + \tau_i^2}, \quad (2.42)$$

$$\alpha_i \begin{cases} = \text{atan2}(\sigma_i, \tau_i) & \text{when } \beta_i \ne 0 \\ = \text{any value} & \text{when } \beta_i = 0 \end{cases}. \quad (2.43)$$

Notably, even though the point at $\beta_i = 0$ of a flat boundary surface is a pseudo-singular point, it poses no computational problem.

2.3.3 Unit Directional Vectors of Reflected and Refracted Rays

The incidence angle θ_i, whose domain is $0° \le \theta_i \le 90°$, can be computed as $C\theta_i = \left|\bar{\ell}_{i-1} \cdot \bar{n}_i\right|$. Again, at every incidence point, there exist two possible unit normal vectors aligned in opposite directions to one another. To choose the correct unit normal vector, let the active unit normal vector $\bar{n}_i$ be once again defined as the vector possessing an obtuse angle $90° < \eta < 180°$ with $\bar{\ell}_{i-1}$. Having chosen the active unit normal vector $\bar{n}_i$, $C\theta_i$ can be computed without the absolute symbol as

$$C\theta_i = \left|\bar{\ell}_{i-1} \cdot \bar{n}_i\right| = -\bar{\ell}_{i-1} \cdot \bar{n}_i = s_i\left(J_{ix}\ell_{i-1x} + J_{iy}\ell_{i-1y} + J_{iz}\ell_{i-1z}\right) = s_i E_i. \quad (2.44)$$

The refraction angle $\underline{\theta}_i$ between two optical media must satisfy Snell's law, i.e., $S\underline{\theta}_i = (\xi_{i-1}/\xi_i)S\theta_i = N_i S\theta_i$. As for the case of a spherical boundary surface, the reflected unit directional vector $\bar{\ell}_i$ at a flat boundary surface can be obtained by rotating the active unit normal vector $\bar{n}_i$ about the unit common normal vector $\bar{m}_i$ (given in Eq. (2.23)) through an angle θ_i, yielding

$$\bar{\ell}_i = \begin{bmatrix} \ell_{ix} \\ \ell_{iy} \\ \ell_{iz} \\ 0 \end{bmatrix} = \begin{bmatrix} \ell_{i-1x} + 2C\theta_i\, n_{ix} \\ \ell_{i-1y} + 2C\theta_i\, n_{iy} \\ \ell_{i-1z} + 2C\theta_i\, n_{iz} \\ 0 \end{bmatrix} = \bar{\ell}_{i-1} + 2C\theta_i\, \bar{n}_i. \quad (2.45)$$

Similarly, the refracted unit directional vector $\bar{\ell}_i$ can be obtained by rotating the active unit normal vector $\bar{n}_i$ about the unit common vector $\bar{m}_i$ through an angle $\pi - \underline{\theta}_i$ (its detailed derivations were presented in Eq. (2.28)), giving

$$\bar{\ell}_i = \begin{bmatrix} \ell_{ix} \\ \ell_{iy} \\ \ell_{iz} \\ 0 \end{bmatrix} = \begin{bmatrix} -n_{ix}\sqrt{1 - N_i^2 + (N_i C\theta_i)^2} + N_i(\ell_{i-1x} + n_{ix}C\theta_i) \\ -n_{iy}\sqrt{1 - N_i^2 + (N_i C\theta_i)^2} + N_i(\ell_{i-1y} + n_{iy}C\theta_i) \\ -n_{iz}\sqrt{1 - N_i^2 + (N_i C\theta_i)^2} + N_i(\ell_{i-1z} + n_{iz}C\theta_i) \\ 0 \end{bmatrix} \quad (2.46)$$
$$= \left(N_i C\theta_i - \sqrt{1 - N_i^2 + (N_i C\theta_i)^2}\right)\bar{n}_i + N_i\bar{\ell}_{i-1}.$$

Note that total internal reflection occurs at this boundary surface when $1 - N_i^2 + (N_i C\,\theta_i)^2 < 0$.

It is noted from Eqs. (2.37), (2.45) and (2.46) that only six variables (i.e., $J_{ix}, J_{iy}, J_{iz}, e_i, \xi_{i-1}, \xi_i$) are needed to completely describe the effects of a flat boundary surface. The boundary variable vector of a flat boundary surface can thus be defined as

$$\bar{X}_i = [J_{ix} \quad J_{iy} \quad J_{iz} \quad e_i \quad \xi_{i-1} \quad \xi_i]^T. \quad (2.47)$$

It is noted that not all of the variables in $\bar{X}_i$ are independent since $J_{ix}^2 + J_{iy}^2 + J_{iz}^2 = 1$. Another example is $\xi_{i-1} = \xi_i$ for a reflective boundary surface. However, the inter-dependence of some of the components in $\bar{X}_i$ presents no computational difficulties if a system variable vector $\bar{X}_{sys}$ containing all of the independent variables of interest of the optical system is defined.

The discussions above consider the raytracing problem for a reflected or refracted ray $\bar{R}_i$ at a single boundary surface. However, the same approach can be applied successively to trace rays in an optical system containing n boundary

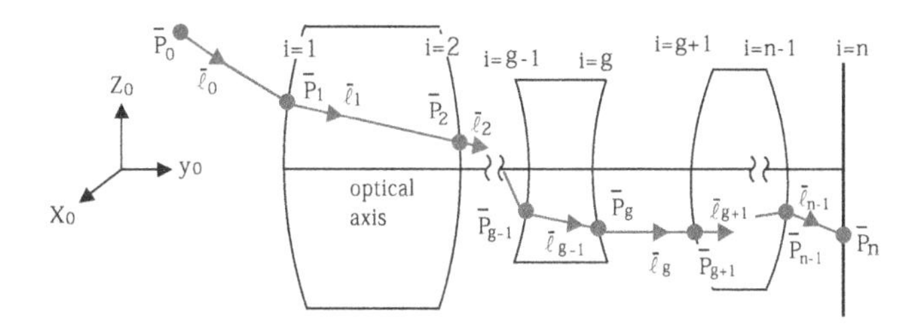

Fig. 2.13 Sequential tracing of ray $\bar{R}_g$ through optical system using proposed approach

surfaces. To trace a ray reflected or refracted at the gth boundary surface, it is first necessary to label the boundary surfaces of the system sequentially from 1 to n. The preceding raytracing methodology can then be applied sequentially with i = 1, i = 2, …, until i = g to obtain the ray $\bar{R}_g$ refracted or reflected at the gth boundary surface (see Fig. 2.13), where

$$\bar{P}_g = \begin{bmatrix} P_{gx} \\ P_{gy} \\ P_{gz} \\ 1 \end{bmatrix} = \begin{bmatrix} P_{g-1x} + \lambda_g \ell_{g-1x} \\ P_{g-1y} + \lambda_g \ell_{g-1y} \\ P_{g-1z} + \lambda_g \ell_{g-1z} \\ 1 \end{bmatrix} = \begin{bmatrix} P_{g-2x} + \lambda_{g-1} \ell_{g-2x} + \lambda_g \ell_{g-1x} \\ P_{g-2y} + \lambda_{g-1} \ell_{g-2y} + \lambda_g \ell_{g-1y} \\ P_{g-2z} + \lambda_{g-1} \ell_{g-2z} + \lambda_g \ell_{g-1z} \\ 1 \end{bmatrix} = \ldots$$
$$= \begin{bmatrix} P_{0x} + \lambda_1 \bar{\ell}_{0x} + \ldots + \lambda_g \bar{\ell}_{g-1x} \\ P_{0y} + \lambda_1 \bar{\ell}_{0y} + \ldots + \lambda_g \bar{\ell}_{g-1y} \\ P_{0z} + \lambda_1 \bar{\ell}_{0z} + \ldots + \lambda_g \bar{\ell}_{g-1z} \\ 1 \end{bmatrix} = \bar{P}_0 + \lambda_1 \bar{\ell}_0 + \ldots + \lambda_g \bar{\ell}_{g-1}. \tag{2.48}$$

Example 2.3 Skew-ray tracing in an optical element possessing only flat boundary surfaces is much simpler than in the case of an optical element containing spherical boundary surfaces. Consider the rectangular optical flat shown in Fig. 2.14 with separation v_1, thickness q_{e1}, and refractive index ξ_{e1}. (1) Assign boundary coordinate frames $(xyz)_1$ and $(xyz)_2$ to the two flat boundary surfaces, $\bar{r}_1$ and $\bar{r}_2$. (2) Find their unit normal vectors ${}^1\bar{n}_1$ and ${}^2\bar{n}_2$. (3) Determine the pose matrix ${}^0\bar{A}_1$ and boundary variable vector $\bar{X}_1$. (4) Assign the pose matrix ${}^0\bar{A}_2$ and determine boundary variable vector $\bar{X}_2$. (5) Find unit normal vectors $\bar{n}_1$, D_1, E_1, λ_1, β_1, α_1, $C\theta_1$, N_1, the value of s_1 for the active unit normal vector, and $\bar{\ell}_1$ when source ray $\bar{R}_0 = [0 \quad 0 \quad 0 \quad 0 \quad C\beta_0 \quad S\beta_0]^T, 0 < \beta_0$, is refracted at the first boundary surface, $\bar{r}_1$. (6) Find unit normal vectors $\bar{n}_2$, D_2, E_2, λ_2, $C\theta_2$, N_2, the value of s_2 for the active unit normal vector, and $\bar{\ell}_2$ when ray $\bar{R}_1$ is refracted at the second boundary surface, $\bar{r}_2$. (7) Determine the ray displacement D.

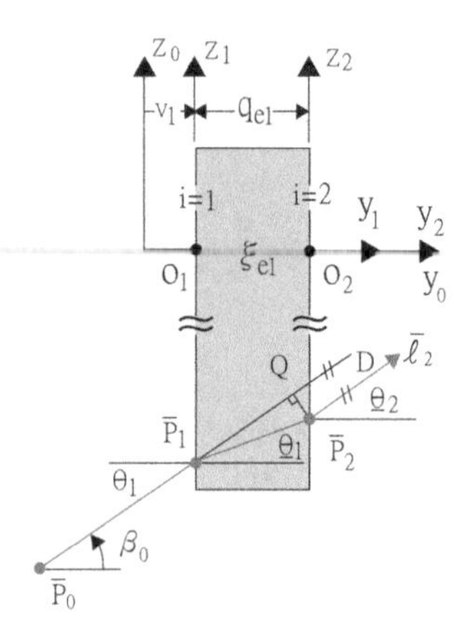

Fig. 2.14 Assigned coordinate frames $(xyz)_1$ and $(xyz)_2$ for rectangular optical flat

Solution

(1) The assigned coordinate frames $(xyz)_1$ and $(xyz)_2$ are shown in Fig. 2.14 with their origins, o_1 and o_2, located at any convenient in-plane points on boundary surfaces $\bar{r}_1$ and $\bar{r}_2$, respectively. Note that the y_1 and y_2 axes are both aligned with the y_0 axis of the world coordinate frame $(xyz)_0$.

(2) ${}^1\bar{n}_1 = s_1[0 \quad -1 \quad 0 \quad 0]^T$,
${}^2\bar{n}_2 = s_2[0 \quad -1 \quad 0 \quad 0]^T$.

(3) If o_1 lies on the y_0 axis, then

$$ {}^0\bar{A}_1 = \text{tran}(0, v_1, 0) = \begin{bmatrix} 1 & 0 & 0 & 0 \\ 0 & 1 & 0 & v_1 \\ 0 & 0 & 1 & 0 \\ 0 & 0 & 0 & 1 \end{bmatrix}, $$

$$ \bar{X}_1 = [J_{1x} \quad J_{1y} \quad J_{1z} \quad e_1 \quad \xi_0 \quad \xi_1]^T = [0 \quad 1 \quad 0 \quad -v_1 \quad 1 \quad \xi_{e1}]^T. $$

(4) If o_2 lies on the y_0 axis, then

$$ {}^0\bar{A}_2 = \text{tran}(0, v_1 + q_{e1}, 0) = \begin{bmatrix} 1 & 0 & 0 & 0 \\ 0 & 1 & 0 & v_1 + q_{e1} \\ 0 & 0 & 1 & 0 \\ 0 & 0 & 0 & 1 \end{bmatrix}, $$

$$ \bar{X}_2 = [J_{2x} \quad J_{2y} \quad J_{2z} \quad e_2 \quad \xi_1 \quad \xi_2]^T = [0 \quad 1 \quad 0 \quad -(v_1 + q_{e1}) \quad 1 \quad 1]^T. $$

(5) $\bar{n}_1 = {}^0\bar{A}_1{}^1\bar{n}_1 = s_1[0 \quad -1 \quad 0 \quad 0]^T$,
$D_1 = -v_1$,
$E_1 = C\beta_0$,
$\lambda_1 = v_1/C\beta_0$,
$\bar{P}_1 = [0 \quad v_1 \quad v_1S\beta_0/C\beta_0 \quad 1]^T$,
$\beta_1 = v_1S\beta_0/C\beta_0$,
$\alpha_1 = \text{atan2}(0, v_1S\beta_0/C\beta_0) = 0°$,
$C\theta_1 = C\beta_0$.
$N_1 = \xi_0/\xi_1 = 1/\xi_{e1}$,
The active unit normal vector of $\bar{r}_1$ is $\bar{n}_1 = [0 \quad -1 \quad 0 \quad 0]^T$, yielding $s_1 = 1$,
$\bar{\ell}_1 = \left[0 \quad \sqrt{1 - N_1^2 + (N_1C\beta_0)^2} \quad N_1S\beta_0 \quad 0\right]^T$.

(6) $\bar{n}_2 = {}^0\bar{A}_2{}^2\bar{n}_2 = s_2[0 \quad -1 \quad 0 \quad 0]^T$,
$D_2 = -q_{e1}$,
$E_2 = \sqrt{1 - N_1^2 + (N_1C\beta_0)^2}$,
$\lambda_2 = q_{e1}/\sqrt{1 - N_1^2 + (N_1C\beta_0)^2}$,
$C\theta_2 = \sqrt{1 - N_1^2 + (N_1C\beta_0)^2}$.
$N_2 = \xi_1/\xi_2 = \xi_{e1} = 1/N_1$,
The active unit normal vector of $\bar{r}_2$ is $\bar{n}_2 = [0 \quad -1 \quad 0 \quad 0]^T$, yielding $s_2 = 1$,
$\bar{\ell}_2 = [0 \quad C\beta_0 \quad S\beta_0 \quad 0]^T$.
It is thus proven that the exit ray $\bar{R}_2$ is parallel to $\bar{R}_0$.

(7) From Fig. 2.14, it follows that
$D = [q_{e1} \tan \beta_0 - (P_{2z} - P_{1z})]C\beta_0 = q_{e1}S\beta_0 - \frac{q_{e1}S\beta_0C\beta_0}{\sqrt{1-N_1^2+(N_1C\beta_0)^2}}$.
After expansion of $S(\theta_1 - \underline{\theta}_1)$ and substitution of $\underline{\theta}_1$ from $S\theta_1 = N_1S\underline{\theta}_1$ and $\theta_1 = \beta_0$, D can be reformulated as

$$D = q_{e1}S\beta_0\left(1 - \frac{C\beta_0}{\sqrt{N_1^2 - (S\beta_0)^2}}\right) = q_{e1}S\beta_0 - \frac{q_{e1}S\beta_0C\beta_0}{\sqrt{1 - N_1^2 + (N_1C\beta_0)^2}}.$$

Example 2.4 Consider the triangular prism shown in Fig. 2.15 with vertex angle η_{e1} and refractive index $\xi_{e1} = 1.5$. (1) Assign boundary coordinate frames $(xyz)_1$ and $(xyz)_2$ to the two flat boundary surfaces, $\bar{r}_1$ and $\bar{r}_2$. (2) Determine the pose matrices ${}^0\bar{A}_1$ and ${}^0\bar{A}_2$. (3) Find the unit normal vectors $\bar{n}_1$, D_1, E_1, λ_1, β_1, α_1, $C\theta_1$, N_1, the value of s_1 for the active unit normal vector, and $\bar{\ell}_1$ when ray $\bar{R}_0 =$

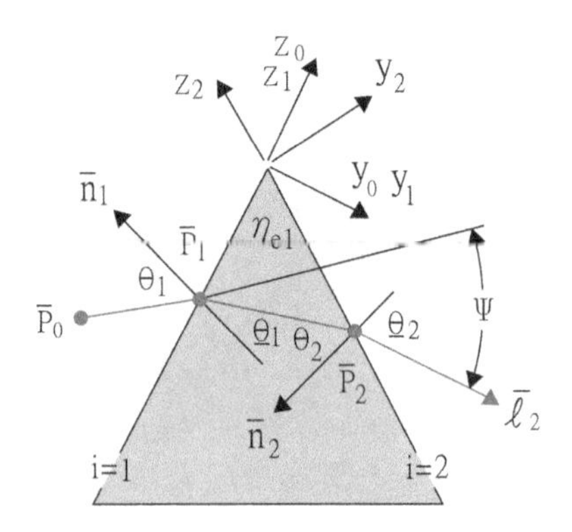

Fig. 2.15 Assigned coordinate frames $(xyz)_1$ and $(xyz)_2$ for triangular prism

$[0 \quad P_{0y} \quad P_{0z} \quad 0 \quad C\beta_0 \quad S\beta_0]^T$ ($P_{0y} < 0$, $P_{0z} < 0, P_{0z} - P_{0y}S\beta_0/C\beta_0 < 0$, and $0 < \beta_0$) is refracted at the first boundary surface, $\bar{r}_1$. (4) Find the unit normal vectors $\bar{n}_2$, D_2, E_2, λ_2, β_2, α_2, $C\theta_2$, the value of s_2 for the active unit normal vector when ray $\bar{R}_1$ is refracted at the second boundary surface, $\bar{r}_2$. (5) Determine the ray deviation angle ψ $(0 \le \psi \le 180°)$ in terms of θ_1, η_{e1} and N_1.

Solution

(1) The assigned coordinate frames $(xyz)_1$ and $(xyz)_2$ are shown in Fig. 2.15 with their origins both coinciding with the origin of $(xyz)_0$.

(2) $${}^0\bar{A}_1 = \bar{I}_{4\times4} = \begin{bmatrix} 1 & 0 & 0 & 0 \\ 0 & 1 & 0 & 0 \\ 0 & 0 & 1 & 0 \\ 0 & 0 & 0 & 1 \end{bmatrix},$$

$${}^0\bar{A}_2 = \text{rot}(\bar{x}, \eta_{e1}) = \begin{bmatrix} 1 & 0 & 0 & 0 \\ 0 & C\eta_{e1} & -S\eta_{e1} & 0 \\ 0 & S\eta_{e1} & C\eta_{e1} & 0 \\ 0 & 0 & 0 & 1 \end{bmatrix}.$$

(3) $\bar{n}_1 = s_1[0 \quad -1 \quad 0 \quad 0]^T$ with $s_1 = \pm 1$

$D_1 = P_{0y}$,

$E_1 = C\beta_0$,

$\lambda_1 = -P_{0y}/C\beta_0$,

$\bar{P}_1 = [0 \quad 0 \quad P_{0z} - P_{0y}S\beta_0/C\beta_0 \quad 1]^T$,

$\beta_1 = -(P_{0z} - P_{0y}S\beta_0/C\beta_0)$,

$\alpha_1 = \text{atan2}(0,\, P_{0z} - P_{0y}S\beta_0/C\beta_0) = 180°$,

$C\theta_1 = C\beta_0$.

$N_1 = \xi_0/\xi_1 = 1/\xi_{e1}$,

The active unit normal vector of $\bar{r}_1$ is $\bar{n}_1 = [0 \quad -1 \quad 0 \quad 0]^T$, leading to $s_1 = 1$,

$$\bar{\ell}_1 = \left[0 \quad \sqrt{1 - N_1^2 + (N_1 C\beta_0)^2} \quad N_1 S\beta_0 \quad 0\right]^T.$$

(4) $\bar{n}_2 = s_2[0 \quad -C\eta_{e1} \quad -S\eta_{e1} \quad 0]^T$ with $s_2 = \pm 1$,

$$D_2 = S\eta_{e1}(P_{0z} - P_{0y} S\beta_0 / C\beta_0),$$

$$E_2 = C\eta_{e1}\sqrt{1 - N_1^2 + (N_1 C\beta_0)^2} + N_1 S\eta_{e1} S\beta_0,$$

$$\lambda_2 = \frac{-S\eta_{e1}(P_{0z} - P_{0y} S\beta_0 / C\beta_0)}{C\eta_{e1}\sqrt{1 - N_1^2 + (N_1 C\beta_0)^2} + N_1 S\eta_{e1} S\beta_0},$$

$$C\theta_2 = C\eta_{e1}\sqrt{1 - N_1^2 + (N_1 C\beta_0)^2} + N_1 S\beta_0 S\eta_{e1}.$$

The active unit normal vector of $\bar{r}_2$ is $\bar{n}_2 = [0 \quad -C\eta_{e1} \quad -S\eta_{e1} \quad 0]^T$, leading to $s_2 = 1$.

(5) It is possible to obtain $\bar{\ell}_2$ from Eq. (2.46) so as to determine ψ as $C\psi = \bar{\ell}_0 \cdot \bar{\ell}_2$. Alternatively, Snell's law can be applied successively at the two boundary surfaces to give $S\theta_1 = N_1 S\underline{\theta}_1$ and $S\theta_2 = N_2 S\underline{\theta}_2 = S\underline{\theta}_2 / N_1$, respectively. Applying $\theta_2 = \eta_{e1} - \underline{\theta}_1$, it follows that

$$\begin{aligned}\psi &= \theta_1 - \underline{\theta}_1 + \underline{\theta}_2 - \theta_2 = \theta_1 + \underline{\theta}_2 - \eta_{e1} \\ &= \theta_1 - \eta_{e1} + \sin^{-1}\left(S\eta_{e1}\sqrt{\xi_{e1}^2 - (S\theta_1)^2} - C\eta_{e1} S\theta_1\right).\end{aligned}$$

In general, the refraction index is higher for short wavelengths (blue light) than for long wavelengths (red light). Therefore, the deviation angle ψ will be greater for blue light than for red.

The minimum deviation angle $\psi_{\min i}$ is found to occur when $\theta_1 = \underline{\theta}_2$ by setting $\partial\psi / \partial\theta_1 = 0$. Under this condition, the refractive index of the prism is given as

$$\xi_{e1} = \frac{S((\psi_{\min i} + \eta_{e1})/2)}{S(\eta_{e1}/2)}.$$

It is noted that this formulation provides a precise and convenient approach for measuring the refractive index ξ_{e1} since the minimum deviation angle $\psi_{\min i}$ can be readily determined using a spectrometer.

Example 2.5 The following example derives the raytracing equations for a meridional ray $\bar{R}_{i-1}$ incident on a flat boundary surface $\bar{r}_i$ in an axis-symmetrical system. As shown in Fig. 2.16, the flat boundary surface $\bar{r}_i$ is located at $y_0 = v_i$ and is oriented perpendicularly to the optical axis. The meridional ray $\bar{R}_{i-1}$ originates at point $\bar{P}_{i-1} = [0 \quad P_{i-1y} \quad P_{i-1z} \quad 1]^T$, travels along $\bar{\ell}_{i-1} = [0 \quad \ell_{i-1y} \quad \ell_{i-1z} \quad 0]^T$ ($\ell_{i-1y}^2 + \ell_{i-1z}^2 = 1$ with $0 < \ell_{i-1y}$ and $\ell_{i-1z} < 0$), and is then reflected or refracted at $\bar{r}_i$ with a unit normal vector $\bar{n}_i = s_i[0 \quad -1 \quad 0 \quad 0]^T$ ($s_i = \pm 1$). The incidence point $\bar{P}_i$ at which the ray strikes $\bar{r}_i$ is given by

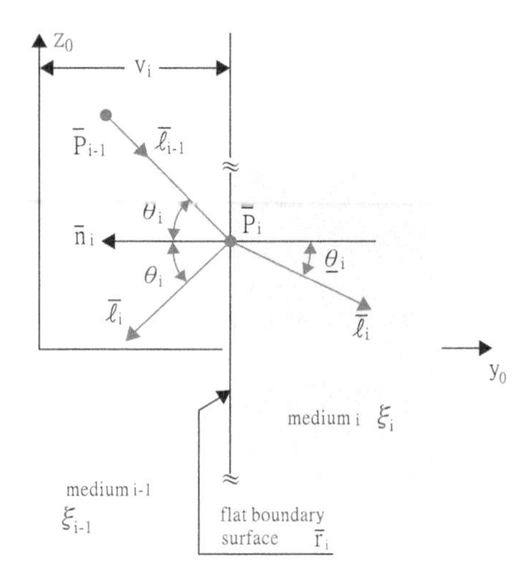

Fig. 2.16 Tracing meridional ray at flat boundary surface of axis-symmetrical system

$$\bar{P}_i = \begin{bmatrix} 0 \\ P_{iy} \\ P_{iz} \\ 1 \end{bmatrix} = \bar{P}_{i-1} + \lambda_i \bar{\ell}_{i-1} = \begin{bmatrix} 0 \\ P_{i-1y} + \lambda_i \ell_{i-1y} \\ P_{i-1z} + \lambda_i \ell_{i-1z} \\ 1 \end{bmatrix},$$

where the parameter λ_i is obtained by setting $P_{i-1y} + \lambda_i \ell_{i-1y} = v_i$, yielding

$$\lambda_i = \frac{v_i - P_{i-1y}}{\ell_{i-1y}}.$$

The incidence angle θ_i is given by $C\theta_i = \left|\bar{\ell}_{i-1} \cdot \bar{n}_i\right| = (-\bar{\ell}_{i-1}) \cdot \bar{n}_i = \ell_{i-1y}$, where the active unit normal vector has the form $\bar{n}_i = \begin{bmatrix} 0 & -1 & 0 & 0 \end{bmatrix}^T$. The reflected unit directional vector $\bar{\ell}_i$ is then obtained by rotating the active unit normal vector $\bar{n}_i$ about the x_0 axis through an angle θ_i. Imposing $\ell_{i-1y}^2 + \ell_{i-1z}^2 = 1$, the reflected unit directional vector $\bar{\ell}_i$ is thus obtained as

$$\bar{\ell}_i = \begin{bmatrix} 0 \\ \ell_{iy} \\ \ell_{iz} \\ 0 \end{bmatrix} = \text{rot}(\bar{x}, \theta_i)\, \bar{n}_i = \begin{bmatrix} 0 \\ -C\theta_i \\ -S\theta_i \\ 0 \end{bmatrix} = \begin{bmatrix} 0 \\ -\ell_{i-1y} \\ -\sqrt{1 - \ell_{i-1y}^2} \\ 0 \end{bmatrix}.$$

The refracted unit directional vector $\bar{\ell}_i$ can be obtained by rotating the active unit normal vector $\bar{n}_i$ about the x_0 axis through an angle $\pi - \underline{\theta}_i$, i.e.,

$$\bar{\ell}_i = \begin{bmatrix} 0 \\ \ell_{iy} \\ \ell_{iz} \\ 0 \end{bmatrix} = \text{rot}(\bar{x}, \pi - \underline{\theta}_i)\bar{n}_i = \begin{bmatrix} 0 \\ C\underline{\theta}_i \\ -S\underline{\theta}_i \\ 0 \end{bmatrix}.$$

The refraction angle $\underline{\theta}_i$ must satisfy Snell's law, i.e., $S\underline{\theta}_i = (\xi_{i-1}/\xi_i)S\theta_i = N_i S\theta_i$. Consequently, the refracted unit directional vector $\bar{\ell}_i$ can be formulated as

$$\bar{\ell}_i = \begin{bmatrix} 0 \\ \ell_{iy} \\ \ell_{iz} \\ 0 \end{bmatrix} = \begin{bmatrix} 0 \\ \sqrt{1 - N_i^2 + (N_i \ell_{i-1y})^2} \\ -N_i\sqrt{1 - \ell_{i-1y}^2} \\ 0 \end{bmatrix}.$$

It is noted that when the source ray $\bar{R}_{i-1}$ travels in the upward direction (i.e., $0 < \ell_{i-1z}$), the unit directional vectors of the reflected and refracted rays are computed as $\bar{\ell}_i = \text{rot}(-\bar{x}, \theta_i)\bar{n}_i$ and $\bar{\ell}_i = \text{rot}(-\bar{x}, \pi - \underline{\theta}_i)\bar{n}_i$, respectively, where $\bar{n}_i = [0 \quad -1 \quad 0 \quad 0]^T$ is the active unit normal vector.

2.4 General Aspherical Boundary Surfaces

Most optical systems comprise flat or spherical boundary surfaces since such surfaces are easily manufactured with low cost. However, certain systems also contain aspherical boundary surfaces. Accordingly, this section extends the raytracing methodology described above to the case of aspherical boundary surfaces ([10, 35–37], p. 312 of [34]). In practice, the ability to trace rays at aspherical boundary surfaces is highly important since even though such surfaces are difficult and expensive to manufacture, there are cases where elements with aspherical boundary surfaces have significant advantages over those with spherical boundary surfaces.

2.4.1 Aspherical Boundary Surface and Associated Unit Normal Vector

While in principle an aspherical surface can take a wide variety of forms, any aspherical boundary surface can be designed with a generating curve ${}^i\bar{q}_i$ of the form

$${}^i\bar{q}_i = \begin{bmatrix} 0 \\ y_i(\beta_i) \\ z_i(\beta_i) \\ 1 \end{bmatrix} (0 \le z_i(\beta_i)), \tag{2.49}$$

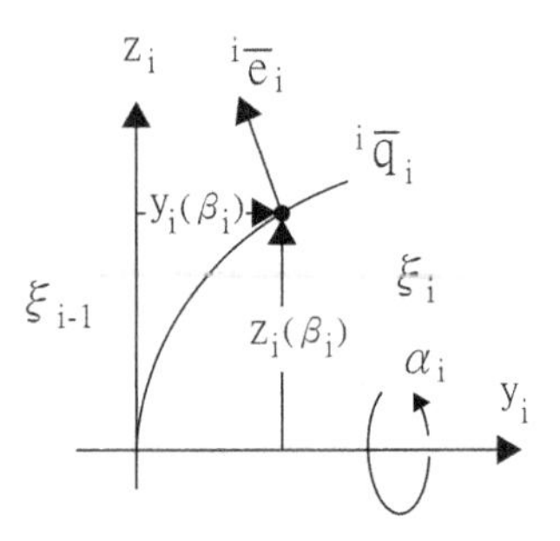

Fig. 2.17 Generating curve for aspherical boundary surface

where the optical axis is presumed to lie in the y_i axis direction and $y_i(\beta_i)$ is the sag (i.e., the y_i component of the displacement from the vertex at distance $z_i(\beta_i)$ from the optical axis) (Fig. 2.17, in which $(xyz)_i$ is the boundary coordinate frame). The two unit normal vectors of the generating curve have the form

$$
\begin{aligned}
{}^i\bar{\eta}_i = \begin{bmatrix} \eta_{ix} \\ \eta_{iy} \\ \eta_{iz} \\ 0 \end{bmatrix} &= \frac{s_i}{\sqrt{[d(y_i(\beta_i))/d\beta_i]^2 + [d(z_i(\beta_i))/d\beta_i]^2}} \begin{bmatrix} 0 \\ -d(z_i(\beta_i))/d\beta_i \\ d(y_i(\beta_i))/d\beta_i \\ 0 \end{bmatrix} \\
&= \frac{s_i}{\sqrt{y_i'^2 + z_i'^2}} \begin{bmatrix} 0 \\ -z_i' \\ y_i' \\ 0 \end{bmatrix},
\end{aligned}
\tag{2.50}
$$

where s_i is again set to either +1 or −1 to indicate the existence of two possible unit normal vectors with directions opposite to one another. The aspherical boundary surface ${}^i\bar{r}_i$ and its two unit normal vectors ${}^i\bar{n}_i$ can be obtained by rotating ${}^i\bar{q}_i$ and ${}^i\bar{\eta}_i$, respectively, about the y_i axis through an angle α_i $(0 \le \alpha_i < 2\pi)$, i.e.,

$$
\begin{aligned}
{}^i\bar{r}_i = \begin{bmatrix} x_i \\ y_i \\ z_i \\ 1 \end{bmatrix} = \text{rot}(\bar{y}, \alpha_i)\,{}^i\bar{q}_i &= \begin{bmatrix} C\alpha_i & 0 & S\alpha_i & 0 \\ 0 & 1 & 0 & 0 \\ -S\alpha_i & 0 & C\alpha_i & 0 \\ 0 & 0 & 0 & 1 \end{bmatrix} \begin{bmatrix} 0 \\ y_i(\beta_i) \\ z_i(\beta_i) \\ 1 \end{bmatrix} \\
&= \begin{bmatrix} z_i(\beta_i)S\alpha_i \\ y_i(\beta_i) \\ z_i(\beta_i)C\alpha_i \\ 1 \end{bmatrix},
\end{aligned}
\tag{2.51}
$$

$$
\begin{aligned}
{}^{i}\bar{n}_i = \mathrm{rot}(\bar{y}, \alpha_i)\,{}^{i}\bar{\eta}_i &= \begin{bmatrix} C\alpha_i & 0 & S\alpha_i & 0 \\ 0 & 1 & 0 & 0 \\ -S\alpha_i & 0 & C\alpha_i & 0 \\ 0 & 0 & 0 & 1 \end{bmatrix} \begin{bmatrix} \eta_{ix} \\ \eta_{iy} \\ \eta_{iz} \\ 0 \end{bmatrix} \\
&= \frac{s_i}{\sqrt{y_i'^2 + z_i'^2}} \begin{bmatrix} y_i' S\alpha_i \\ -z_i' \\ y_i' C\alpha_i \\ 0 \end{bmatrix}.
\end{aligned} \tag{2.52}
$$

Equations (2.51) and (2.52) give ${}^{i}\bar{r}_i$ and ${}^{i}\bar{n}_i$ with respect to the coordinate frame $(xyz)_i$. However, many derivations in this book are referred to the world coordinate frame $(xyz)_o$. Consequently, the pose matrix ${}^{0}\bar{A}_i$ of $(xyz)_i$ with respect to $(xyz)_0$ given in Eq. (2.9) is required. The unit normal vectors of the aspherical surface relative to the world coordinate frame $(xyz)_0$ can then be obtained from the following transformation:

$$
\begin{aligned}
\bar{n}_i = \begin{bmatrix} n_{ix} \\ n_{iy} \\ n_{iz} \\ 0 \end{bmatrix} = {}^{0}\bar{A}_i\,{}^{i}\bar{n}_i &= \frac{s_i}{\sqrt{y_i'^2 + z_i'^2}} \begin{bmatrix} I_{ix} & J_{ix} & K_{ix} & t_{ix} \\ I_{iy} & J_{iy} & K_{iy} & t_{iy} \\ I_{iz} & J_{iz} & K_{iz} & t_{iz} \\ 0 & 0 & 0 & 1 \end{bmatrix} \begin{bmatrix} y_i' S\alpha_i \\ -z_i' \\ y_i' C\alpha_i \\ 0 \end{bmatrix}. \\
&= \frac{s_i}{\sqrt{y_i'^2 + z_i'^2}} \begin{bmatrix} I_{ix} y_i' S\alpha_i - J_{ix} z_i' + K_{ix} y_i' C\alpha_i \\ I_{iy} y_i' S\alpha_i - J_{iy} z_i' + K_{iy} y_i' C\alpha_i \\ I_{iz} y_i' S\alpha_i - J_{iz} z_i' + K_{iz} y_i' C\alpha_i \\ 0 \end{bmatrix}.
\end{aligned} \tag{2.53}
$$

2.4.2 *Incidence Point*

Figure 2.18 shows a typical ray path at an aspherical boundary surface. As shown, the ray originates from point $\bar{P}_{i-1}$ on the previous boundary surface $\bar{r}_{i-1}$ and travels along the unit directional vector $\bar{\ell}_{i-1}$ until it is reflected or refracted at current boundary surface $\bar{r}_i$. Any intermediate point $\bar{P}'_{i-1}$ lying along this ray as it travels from $\bar{P}_{i-1}$ toward $\bar{r}_i$ is given by $\bar{P}'_{i-1} = \bar{P}_{i-1} + \lambda\bar{\ell}_{i-1}$. Let λ_i represent the geometrical path distance from point $\bar{P}_{i-1}$ to the incidence point $\bar{P}_i$ on boundary surface $\bar{r}_i$. The incidence point $\bar{P}_i$ on the boundary surface is given as $\bar{P}_i = \bar{P}_{i-1} + \lambda_i\bar{\ell}_{i-1}$. To solve for λ_i, α_i and β_i, it is necessary to transform $\bar{P}_i$ to coordinate frame $(xyz)_i$ and then equate the result with the boundary surface ${}^{i}\bar{r}_i$ (Eq. (2.51)), i.e.,

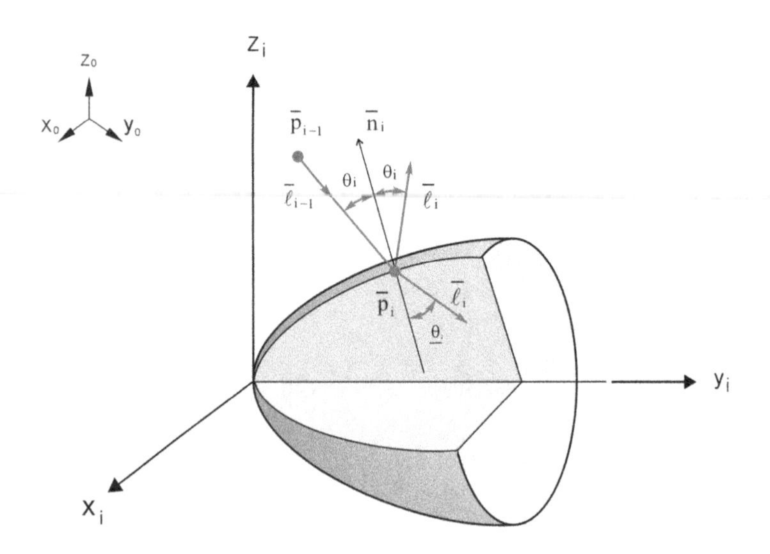

Fig. 2.18 Raytracing at aspherical boundary surface

$$
{}^{i}\bar{P}_{i} = {}^{i}\bar{A}_{0}\bar{P}_{i} = ({}^{0}\bar{A}_{i})^{-1}\bar{P}_{i} = \begin{bmatrix} \sigma_{i} \\ \rho_{i} \\ \tau_{i} \\ 1 \end{bmatrix} = {}^{i}\bar{r}_{i} = \begin{bmatrix} z_{i}(\beta_{i})S\alpha_{i} \\ y_{i}(\beta_{i}) \\ z_{i}(\beta_{i})C\alpha_{i} \\ 1 \end{bmatrix}, \tag{2.54}
$$

where σ_i, ρ_i and τ_i are defined in Eqs. (2.13), (2.14) and (2.15), respectively. The value of parameter α_i at incidence point $\bar{P}_i$ can then be obtained from the first and third components of Eq. (2.54) using the following function:

$$
\begin{cases} \alpha_{i} = \mathrm{atan2}(\sigma_{i},\tau_{i}) & \text{when } z_{i}(\beta_{i}) > 0 \\ \alpha_{i} = \text{any value} & \text{when } z_{i}(\beta_{i}) = 0. \end{cases} \tag{2.55}
$$

Meanwhile, the values of parameters λ_i and β_i at incidence point $\bar{P}_i$ can be determined from the following two independent equations:

$$
z_{i}(\beta_{i})^{2} = \sigma_{i}^{2} + \tau_{i}^{2}, \tag{2.56}
$$

$$
y_{i}(\beta_{i}) = \rho_{i}. \tag{2.57}
$$

The difficulty in tracing a skew-ray at a general aspherical surface lies in determining λ_i from Eqs. (2.56) and (2.57) since the solution cannot usually be determined directly. In other words, some form of numerical method is required (e.g., p. 314 of [34]). However, by adopting a similar procedure to that described in Sect. 2.2, the expressions given in Eqs. (2.26) and (2.29) are still valid for the reflected and refracted unit directional vectors $\bar{\ell}_i$, respectively.

The boundary variable vector $\bar{X}_i$ of the aspherical boundary surface is given as

$$\bar{X}_i = \begin{bmatrix} t_{ix} & t_{iy} & t_{iz} & \omega_{ix} & \omega_{iy} & \omega_{iz} & \xi_{i-1} & \xi_i & \overline{coef_i} \end{bmatrix}^T, \tag{2.58}$$

where $t_{ix}, t_{iy}, t_{iz}, \omega_{ix}, \omega_{iy}$ and ω_{iz} are the six pose variables of the boundary surface; ξ_{i-1} and ξ_i are the refractive indices of media i − 1 and i, respectively; and $\overline{coef_i}$ contains the independent coefficients of $y_i(\beta_i)$ and $z_i(\beta_i)$.

Example 2.6 For an ellipsoidal boundary surface (Fig. 2.19), the generating curve is given as

$$\begin{aligned} {}^i\bar{q}_i &= \begin{bmatrix} 0 & y_i(\beta_i) & z_i(\beta_i) & 1 \end{bmatrix}^T \\ &= \begin{bmatrix} 0 & a_i S\beta_i & b_i C\beta_i & 1 \end{bmatrix}^T \left(0<a_i, 0<b_i, -\frac{\pi}{2} \le \beta_i \le \frac{\pi}{2}\right), \end{aligned}$$

where the geometrical path length λ_i, β_i, and boundary variable vector $\bar{X}_i$ are given respectively as

$$\lambda_i = \frac{-D_i \pm \sqrt{D_i^2 - H_i E_i}}{H_i},$$

$$\beta_i = \operatorname{atan2}\left(\frac{\rho_i}{a_i}, \frac{\sqrt{\sigma_i^2 + \tau_i^2}}{b_i}\right),$$

$$\bar{X}_i = \begin{bmatrix} t_{ix} & t_{iy} & t_{iz} & \omega_{ix} & \omega_{iy} & \omega_{iz} & \xi_{i-1} & \xi_i & a_i & b_i \end{bmatrix}^T,$$

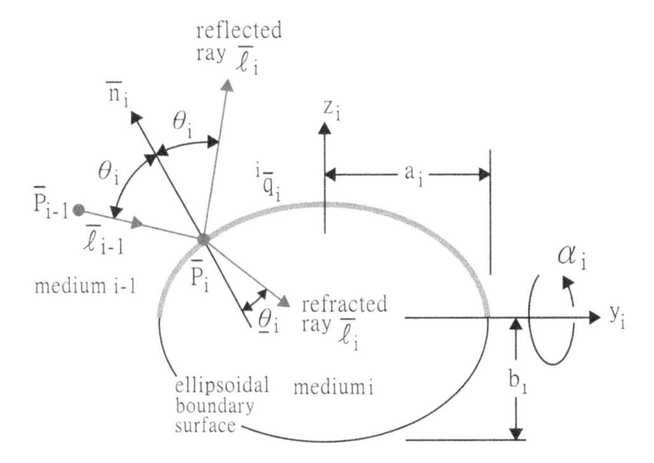

Fig. 2.19 Raytracing at ellipsoidal boundary surface

in which

$$\begin{aligned}
H_i &= 2\left(\frac{1}{a_i^2}-\frac{1}{b_i^2}\right)\left(J_{ix}J_{iy}\ell_{i-1x}\ell_{i-1y}+J_{ix}J_{iz}\ell_{i-1x}\ell_{i-1z}+J_{iy}J_{iz}\ell_{i-1y}\ell_{i-1z}\right)\\
&\quad+\frac{1}{b_i^2}+\left(\frac{1}{a_i^2}-\frac{1}{b_i^2}\right)\left(J_{ix}^2\ell_{i-1x}^2+J_{iy}^2\ell_{i-1y}^2+J_{iz}^2\ell_{i-1z}^2\right),\\
D_i &= \left(\frac{1}{b_i^2}-\frac{J_{ix}^2}{b_i^2}+\frac{J_{ix}^2}{a_i^2}\right)P_{i-1x}\ell_{i-1x}+\left(\frac{1}{b_i^2}-\frac{J_{iy}^2}{b_i^2}+\frac{J_{iy}^2}{a_i^2}\right)P_{i-1y}\ell_{i-1y}+\left(\frac{1}{b_i^2}-\frac{J_{iz}^2}{b_i^2}+\frac{J_{iz}^2}{a_i^2}\right)P_{i-1z}\ell_{i-1z}\\
&\quad+\left(\frac{1}{a_i^2}-\frac{1}{b_i^2}\right)\left[J_{ix}J_{iy}\left(P_{i-1x}\ell_{i-1y}+\ell_{i-1x}P_{i-1y}\right)+J_{ix}J_{iz}\left(P_{i-1x}\ell_{i-1z}+\ell_{i-1x}P_{i-1z}\right)\right.\\
&\quad\left.+J_{iy}J_{iz}\left(P_{i-1y}\ell_{i-1z}+\ell_{i-1y}P_{i-1z}\right)\right]-\frac{\left(t_{ix}\ell_{i-1x}+t_{iy}\ell_{i-1y}+t_{iz}\ell_{i-1z}\right)}{b_i^2}\\
&\quad+\left(\frac{1}{b_i^2}-\frac{1}{a_i^2}\right)\left(J_{ix}t_{ix}+J_{iy}t_{iy}+J_{iz}t_{iz}\right)\left(J_{ix}\ell_{i-1x}+J_{iy}\ell_{i-1y}+J_{iz}\ell_{i-1z}\right),\\
E_i &= \left(\frac{1}{b_i^2}-\frac{J_{ix}^2}{b_i^2}+\frac{J_{ix}^2}{a_i^2}\right)P_{i-1x}^2+\left(\frac{1}{b_i^2}-\frac{J_{iy}^2}{b_i^2}+\frac{J_{iy}^2}{a_i^2}\right)P_{i-1y}^2+\left(\frac{1}{b_i^2}-\frac{J_{iz}^2}{b_i^2}+\frac{J_{iz}^2}{a_i^2}\right)P_{i-1z}^2\\
&\quad+\frac{t_{ix}^2}{b_i^2}+\frac{t_{iy}^2}{a_i^2}+\frac{t_{iz}^2}{b_i^2}-\frac{2}{b_i^2}\left(t_{ix}P_{i-1x}+t_{iy}P_{i-1y}+t_{iz}P_{i-1z}\right)-1\\
&\quad+2\left(\frac{1}{a_i^2}-\frac{1}{b_i^2}\right)\left(J_{ix}J_{iy}P_{i-1x}P_{i-1y}+J_{ix}J_{iz}P_{i-1x}P_{i-1z}+J_{iy}J_{iz}P_{i-1y}P_{i-1z}\right)\\
&\quad-2\left(\frac{1}{a_i^2}-\frac{1}{b_i^2}\right)\left(t_{ix}J_{ix}+t_{iy}J_{iy}+t_{iz}J_{iz}\right)\left(J_{ix}P_{i-1x}+J_{iy}P_{i-1y}+J_{iz}P_{i-1z}\right).
\end{aligned}$$

$\beta_i = \pm\pi/2$ are two pseudo-singular points on the ellipsoidal boundary surface. The cross product $\partial^i\bar{r}_i/\partial\beta_i \times \partial^i\bar{r}_i/\partial\alpha_i$ cannot be obtained at these two points, and thus some difficulties occur in computing the point spread function and modulation transfer function.

Example 2.7 The boundary variable vector $\bar{X}_i$ of a paraboloidal boundary surface defined by generating curve (see Fig. 2.20)

$$^i\bar{q}_i = \begin{bmatrix}0 & y_i(\beta_i) & z_i(\beta_i) & 1\end{bmatrix}^T = \begin{bmatrix}0 & a_i\beta_i^2 & \beta_i & 1\end{bmatrix}^T (0 \le \beta_i, 0 < a_i)$$

is given by

$$\bar{X}_i = \begin{bmatrix}t_{ix} & t_{iy} & t_{iz} & \omega_{ix} & \omega_{iy} & \omega_{iz} & \xi_{i-1} & \xi_i & a_i\end{bmatrix}^T.$$

Its geometrical path length λ_i is given as follows:

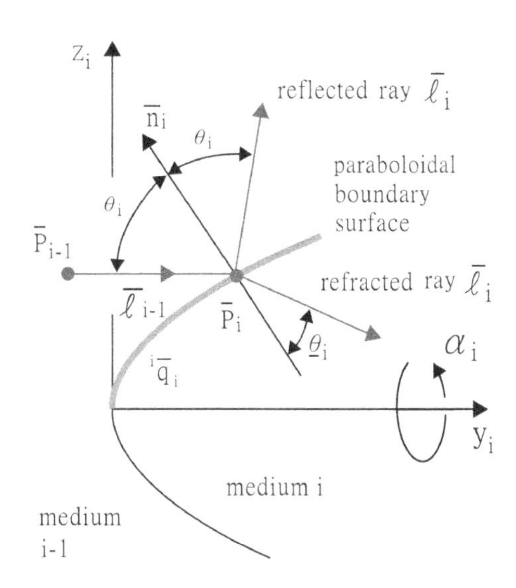

Fig. 2.20 Raytracing at paraboloidal boundary surface

(a) If $J_{ix}\ell_{i-1x} + J_{iy}\ell_{i-1y} + J_{iz}\ell_{i-1z} \neq \pm 1$, then

$$\lambda_i = \frac{-D_i \pm \sqrt{D_i^2 - H_i E_i}}{H_i},$$

where

$$H_i = 1 - \left(J_{ix}\ell_{i-1x} + J_{iy}\ell_{i-1y} + J_{iz}\ell_{i-1z}\right)^2,$$

$$\begin{aligned}
D_i = &\left(1 - J_{ix}^2\right)P_{i-1x}\ell_{i-1x} + \left(1 - J_{iy}^2\right)P_{i-1y}\ell_{i-1y} + \left(1 - J_{iz}^2\right)P_{i-1z}\ell_{i-1z} \\
&- \left(t_{ix}\ell_{i-1x} + t_{iy}\ell_{i-1y} + t_{iz}\ell_{i-1z}\right) - J_{ix}J_{iy}\left(P_{i-1x}\ell_{i-1y} + P_{i-1y}\ell_{i-1x}\right) \\
&- J_{ix}J_{iz}\left(P_{i-1x}\ell_{i-1z} + P_{i-1z}\ell_{i-1x}\right) - J_{iy}J_{iz}\left(P_{i-1y}\ell_{i-1z} + P_{i-1z}\ell_{i-1y}\right) \\
&+ \left(J_{ix}t_{ix} + J_{iy}t_{iy} + J_{iz}t_{iz}\right)\left(J_{ix}\ell_{i-1x} + J_{iy}\ell_{i-1y} + J_{iz}\ell_{i-1z}\right) \\
&- \left(J_{ix}\ell_{i-1x} + J_{iy}\ell_{i-1y} + J_{iz}\ell_{i-1z}\right)/(2a_i),
\end{aligned}$$

$$\begin{aligned}
E_i = &\left(P_{i-1x}^2 + P_{i-1y}^2 + P_{i-1z}^2\right) + t_{ix}^2 + t_{iy}^2 + t_{iz}^2 - 2\left(t_{ix}P_{i-1x} + t_{iy}P_{i-1y} + t_{iz}P_{i-1z}\right) \\
&+ \left[J_{ix}\left(t_{ix} + P_{i-1x}\right) + J_{iy}\left(t_{iy} + P_{i-1y}\right) + J_{iz}\left(t_{iz} + P_{i-1z}\right)\right]^2 \\
&+ \left[J_{ix}\left(t_{ix} - P_{i-1x}\right) + J_{iy}\left(t_{iy} - P_{i-1y}\right) + J_{iz}\left(t_{iz} - P_{i-1z}\right)\right]/a_i.
\end{aligned}$$

(b) If $J_{ix}\ell_{i-1x} + J_{iy}\ell_{i-1y} + J_{iz}\ell_{i-1z} = \pm 1$, then

$$\lambda_i = \frac{D_i}{H_i},$$

where

$$\begin{aligned} D_i &= J_{ix}(t_{ix} - P_{i-1x}) + J_{iy}(t_{iy} - P_{i-1y}) + J_{iz}(t_{iz} - P_{i-1z}) \\ &\quad + a_i\left[(t_{ix} - P_{i-1x})^2 + (t_{iy} - P_{i-1y})^2 + (t_{iz} - P_{i-1z})^2\right] \\ &\quad - a_i\left[J_{ix}(t_{ix} - P_{i-1x}) + J_{iy}(t_{iy} - P_{i-1y}) + J_{iz}(t_{iz} - P_{i-1z})\right]^2, \\ H_i &= J_{ix}\ell_{i-1x} + J_{iy}\ell_{i-1y} + J_{iz}\ell_{i-1z}, \end{aligned}$$

$\beta_i = 0$ is the only pseudo-singular point on the paraboloidal boundary surface.

Example 2.8 The geometrical path length λ_i of a hyperboloidal boundary surface defined by the generating curve (see Fig. 2.21)

$$\begin{aligned} {}^i\bar{q}_i &= \begin{bmatrix} 0 & y_i(\beta_i) & z_i(\beta_i) & 1 \end{bmatrix}^T \\ &= \begin{bmatrix} 0 & a_i/C\beta_i & b_iS\beta_i/C\beta_i & 1 \end{bmatrix}^T (0 \le \beta_i < \pi/2, 0 < a_i, 0 < b_i) \end{aligned}$$

is given by

$$\lambda_i = \frac{-D_i \pm \sqrt{D_i^2 - H_iE_i}}{H_i},$$

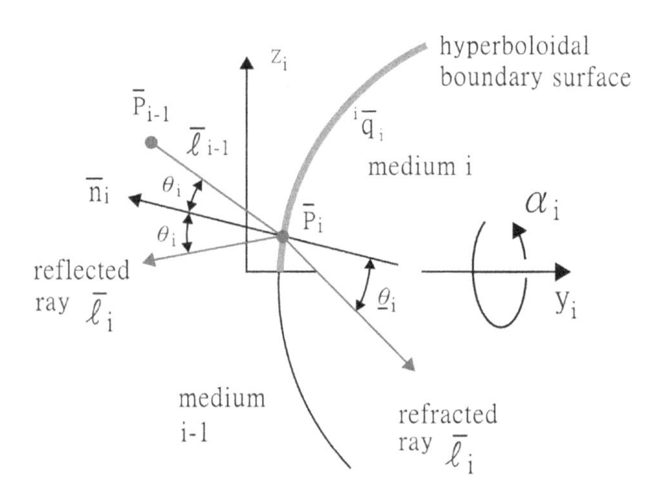

Fig. 2.21 Raytracing at hyperboloidal boundary surface

where

$$
\begin{aligned}
H_i = {} & \left(\frac{1}{b_i^2} - \frac{J_{ix}^2}{b_i^2} - \frac{J_{ix}^2}{a_i^2}\right)\ell_{i-1x}^2 + \left(\frac{1}{b_i^2} - \frac{J_{iy}^2}{b_i^2} - \frac{J_{iy}^2}{a_i^2}\right)\ell_{i-1y}^2 + \left(\frac{1}{b_i^2} - \frac{J_{iz}^2}{b_i^2} - \frac{J_{iz}^2}{a_i^2}\right)\ell_{i-1z}^2 \\
& - 2\left(\frac{1}{a_i^2} + \frac{1}{b_i^2}\right)\left(J_{ix}J_{iy}\ell_{i-1x}\ell_{i-1y} + J_{ix}J_{iz}\ell_{i-1x}\ell_{i-1z} + J_{iy}J_{iz}\ell_{i-1y}\ell_{i-1z}\right), \\
D_i = {} & \left(\frac{1}{b_i^2} - \frac{J_{ix}^2}{b_i^2} - \frac{J_{ix}^2}{a_i^2}\right)P_{i-1x}\ell_{i-1x} + \left(\frac{1}{b_i^2} - \frac{J_{iy}^2}{b_i^2} - \frac{J_{iy}^2}{a_i^2}\right)P_{i-1y}\ell_{i-1y} + \left(\frac{1}{b_i^2} - \frac{J_{iz}^2}{b_i^2} - \frac{J_{iz}^2}{a_i^2}\right)P_{i-1z}\ell_{i-1z} \\
& - \left(\frac{1}{a_i^2} + \frac{1}{b_i^2}\right)\left[J_{ix}J_{iy}(P_{i-1x}\ell_{i-1y} + P_{i-1y}\ell_{i-1x}) + J_{ix}J_{iz}(P_{i-1x}\ell_{i-1z} + P_{i-1z}\ell_{i-1x})\right. \\
& \left. + J_{iy}J_{iz}(P_{i-1y}\ell_{i-1z} + P_{i-1z}\ell_{i-1y})\right] - \left(\ell_{i-1x}t_{ix} + \ell_{i-1y}t_{iy} + \ell_{i-1z}t_{iz}\right)/b_i^2 \\
& + \left(\frac{1}{a_i^2} + \frac{1}{b_i^2}\right)\left(J_{ix}\ell_{i-1x} + J_{iy}\ell_{i-1y} + J_{iz}\ell_{i-1z}\right)\left(J_{ix}t_{ix} + J_{iy}t_{iy} + J_{iz}t_{iz}\right), \\
E_i = {} & \left(\frac{1}{b_i^2} - \frac{J_{ix}^2}{b_i^2} - \frac{J_{ix}^2}{a_i^2}\right)P_{i-1x}^2 + \left(\frac{1}{b_i^2} - \frac{J_{iy}^2}{b_i^2} - \frac{J_{iy}^2}{a_i^2}\right)P_{i-1y}^2 + \left(\frac{1}{b_i^2} - \frac{J_{iz}^2}{b_i^2} - \frac{J_{iz}^2}{a_i^2}\right)P_{i-1z}^2 \\
& - 2\left(\frac{1}{a_i^2} + \frac{1}{b_i^2}\right)\left(J_{ix}J_{iy}P_{i-1x}P_{i-1y} + J_{ix}J_{iz}P_{i-1x}P_{i-1z} + J_{iy}J_{iz}P_{i-1y}P_{i-1z}\right) \\
& + 2\left(\frac{1}{a_i^2} + \frac{1}{b_i^2}\right)(J_{ix}P_{i-1x} + J_{iy}P_{i-1y} + J_{iz}P_{i-1z})(J_{ix}t_{ix} + J_{iy}t_{iy} + J_{iz}t_{iz}) \\
& - 2(P_{i-1x}t_{ix} + P_{i-1y}t_{iy} + P_{i-1z}t_{iz})/b_i^2 + (t_{ix}^2 + t_{iy}^2 + t_{iz}^2)/a_i^2 + 1.
\end{aligned}
$$

Its boundary variable vector $\bar{X}_i$ is

$$\bar{X}_i = \begin{bmatrix} t_{ix} & t_{iy} & t_{iz} & \omega_{ix} & \omega_{iy} & \omega_{iz} & \xi_{i-1} & \xi_i & a_i & b_i \end{bmatrix}^T.$$

$\beta_i = 0$ is the only pseudo-singular point on the hyperboloidal boundary surface.

Example 2.9 The geometrical path length λ_i of a cylindrical boundary surface defined by the generating curve (see Fig. 2.22)

$${}^i\bar{q}_i = \begin{bmatrix} 0 & y_i(\beta_i) & z_i(\beta_i) & 1 \end{bmatrix}^T = \begin{bmatrix} 0 & \beta_i & R_i & 1 \end{bmatrix}^T \; (0 \le \beta_i,\ 0 < R_i)$$

is given by

$$\lambda_i = \frac{-D_i \pm \sqrt{D_i^2 - H_iE_i}}{H_i},$$

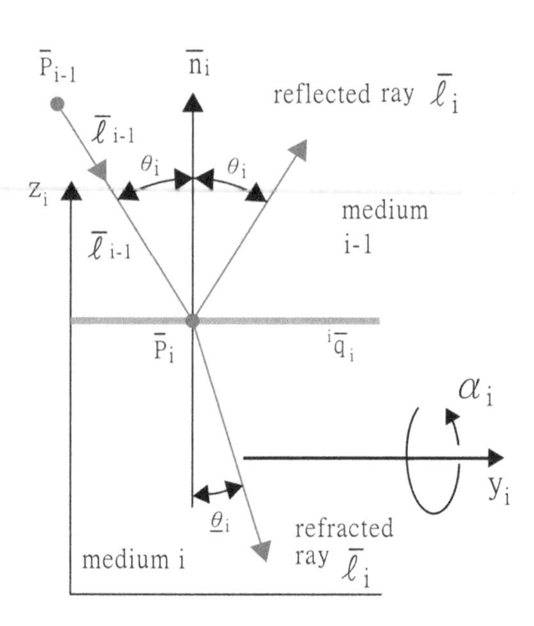

Fig. 2.22 Raytracing at cylindrical boundary surface

where

$$
\begin{aligned}
H_i &= 1 - (J_{ix}\ell_{i-1x} + J_{iy}\ell_{i-1y} + J_{iz}\ell_{i-1z})^2, \\
D_i &= (P_{i-1x} - t_{ix})\ell_{i-1x} + (P_{i-1y} - t_{iy})\ell_{i-1y} + (P_{i-1z} - t_{iz})\ell_{i-1z} \\
&\quad + [J_{ix}(t_{ix} - P_{i-1x}) + J_{iy}(t_{iy} - P_{i-1y}) + J_{iz}(t_{iz} - P_{i-1z})](J_{ix}\ell_{i-1x} + J_{iy}\ell_{i-1y} + J_{iz}\ell_{i-1z}), \\
E_i &= P_{i-1x}^2 + P_{i-1y}^2 + P_{i-1z}^2 + t_{ix}^2 + t_{iy}^2 + t_{iz}^2 - R_i^2 - 2(t_{ix}P_{i-1x} + t_{iy}P_{i-1y} + t_{iz}P_{i-1z}) \\
&\quad - [J_{ix}(t_{ix} - P_{i-1x}) + J_{iy}(t_{iy} - P_{i-1y}) + J_{iz}(t_{iz} - P_{i-1z})]^2.
\end{aligned}
$$

Its boundary variable vector $\bar{X}_i$ is

$$\bar{X}_i = \begin{bmatrix} t_{ix} & t_{iy} & t_{iz} & \omega_{ix} & \omega_{iy} & \omega_{iz} & \xi_{i-1} & \xi_i & R_i \end{bmatrix}^T.$$

There are no pseudo-singular points on the cylindrical boundary surface.

2.5 The Unit Normal Vector of a Boundary Surface for Given Incoming and Outgoing Rays

In the sections above, Snell's law is used to determine the unit directional vectors $\bar{\ell}_i$ of the reflected and refracted rays for an incoming ray impinging on a boundary surface $\bar{r}_i$ between two different isotropic media. In this section, Snell's law is used to derive the active unit normal vector $\bar{n}_i$ at incidence point $\bar{P}_i$ on a boundary surface

$\bar{r}_i$ given the unit directional vectors of the incoming and outgoing rays (i.e., $\bar{\ell}_{i-1}$ and $\bar{\ell}_i$) [38–41]. Note that this problem has important applications in the design and fabrication of aspherical surfaces, since the surface normal vectors determine not only the optical performance of the surface, but also the cutting tool angles required to machine the surface.

2.5.1 Unit Normal Vector of Refractive Boundary Surface

As shown in Fig. 2.6, the design of a refractive surface requires a knowledge of the incidence angle θ_i and refraction angle $\underline{\theta}_i$ corresponding to the unit directional vectors of the incoming and outgoing rays (i.e., $\bar{\ell}_{i-1}$ and $\bar{\ell}_i$). In accordance with Snell's law (Eq. (2.22))

$$S\underline{\theta}_i = \frac{\xi_{i-1}}{\xi_i} S\theta_i = N_i S\theta_i,$$

the refraction phenomenon at the interface between two different isotropic media can be modeled as follows:

$$\bar{\ell}_{i-1} \cdot \bar{\ell}_i = C(\theta_i - \underline{\theta}_i) = C\theta_i C\underline{\theta}_i + S\theta_i S\underline{\theta}_i. \tag{2.59}$$

In geometrical optics, the incidence angle θ_i and refraction angle $\underline{\theta}_i$ lie in the domains of $0° \leq \theta_i \leq 90°$ and $0° \leq \underline{\theta}_i \leq 90°$. Since the trigonometric functions of θ_i and $\underline{\theta}_i$ are positive, the following four equations can be obtained from Eqs. (2.59) and (2.22):

$$S\theta_i = \frac{\sqrt{1 - (\bar{\ell}_{i-1} \cdot \bar{\ell}_i)^2}}{\sqrt{N_i^2 + 1 - 2N_i(\bar{\ell}_{i-1} \cdot \bar{\ell}_i)}}, \tag{2.60}$$

$$C\theta_i = \frac{\left|N_i - (\bar{\ell}_{i-1} \cdot \bar{\ell}_i)\right|}{\sqrt{N_i^2 + 1 - 2N_i(\bar{\ell}_{i-1} \cdot \bar{\ell}_i)}}, \tag{2.61}$$

$$S\underline{\theta}_i = \frac{N_i\sqrt{1 - (\bar{\ell}_{i-1} \cdot \bar{\ell}_i)^2}}{\sqrt{N_i^2 + 1 - 2N_i(\bar{\ell}_{i-1} \cdot \bar{\ell}_i)}}, \tag{2.62}$$

$$C\underline{\theta}_i = \frac{\left|1 - N_i(\bar{\ell}_{i-1} \cdot \bar{\ell}_i)\right|}{\sqrt{N_i^2 + 1 - 2N_i(\bar{\ell}_{i-1} \cdot \bar{\ell}_i)}}. \tag{2.63}$$

Referring to Fig. 2.6, to determine the active unit normal vector $\bar{n}_i$ at any incidence point $\bar{P}_i$ on a refractive surface $\bar{r}_i$, it is first necessary to compute the common unit normal $\bar{m}_i$ of the unit directional vectors $\bar{\ell}_{i-1}$ and $\bar{\ell}_i$ of the incoming and outgoing rays, respectively. Given the assumption $\theta_i \neq \underline{\theta}_i$ when $N_i \neq 1$, $\bar{m}_i$ can be obtained as

$$\bar{m}_i = \begin{bmatrix} m_{ix} & m_{iy} & m_{iz} & 0 \end{bmatrix}^T = \frac{\bar{\ell}_{i-1} \times \bar{\ell}_i}{S(\theta_i - \underline{\theta}_i)}. \tag{2.64}$$

To simplify the expression for the unit normal vector $\bar{n}_i$, it is useful to have the following equation, obtained by taking the post cross product of Eq. (2.64) by $S(\theta_i - \underline{\theta}_i)\bar{\ell}_{i-1}$, i.e.,

$$\begin{aligned} S(\theta_i - \underline{\theta}_i)(\bar{m}_i \times \bar{\ell}_{i-1}) &= (\bar{\ell}_{i-1} \times \bar{\ell}_i) \times \bar{\ell}_{i-1} = \bar{\ell}_i(\bar{\ell}_{i-1} \cdot \bar{\ell}_{i-1}) - \bar{\ell}_{i-1}(\bar{\ell}_i \cdot \bar{\ell}_{i-1}) \\ &= \bar{\ell}_i - \bar{\ell}_{i-1} C(\theta_i - \underline{\theta}_i). \end{aligned} \tag{2.65}$$

Note that $\bar{\ell}_{i-1} \cdot \bar{\ell}_{i-1} = 1$ and $\bar{\ell}_i \cdot \bar{\ell}_{i-1} = C(\theta_i - \underline{\theta}_i)$ are used in deriving Eq. (2.65). As shown in Fig. 2.6, the active unit normal vector $\bar{n}_i$ can be obtained by rotating $-\bar{\ell}_{i-1}$ about $\bar{m}_i$ through an angle θ_i (see Eq. (1.25)). This leads to

$$\begin{aligned} \bar{n}_i &= \begin{bmatrix} n_{ix} & n_{iy} & n_{iz} & 0 \end{bmatrix}^T = \mathrm{rot}(\bar{m}_i, \theta_i)(-\bar{\ell}_{i-1}) \\ &= \begin{bmatrix} m_{ix}^2(1 - C\theta_i) + C\theta_i & m_{iy}m_{ix}(1 - C\theta_i) - m_{iz}S\theta_i & m_{iz}m_{ix}(1 - C\theta_i) + m_{iy}S\theta_i & 0 \\ m_{ix}m_{iy}(1 - C\theta_i) + m_{iz}S\theta_i & m_{iy}^2(1 - C\theta_i) + C\theta_i & m_{iz}m_{iy}(1 - C\theta_i) - m_{ix}S\theta_i & 0 \\ m_{ix}m_{iz}(1 - C\theta_i) - m_{iy}S\theta_i & m_{iy}m_{iz}(1 - C\theta_i) + m_{ix}S\theta_i & m_{iz}^2(1 - C\theta_i) + C\theta_i & 0 \\ 0 & 0 & 0 & 1 \end{bmatrix} \\ &\quad \begin{bmatrix} -\ell_{i-1x} \\ -\ell_{i-1y} \\ -\ell_{i-1z} \\ 0 \end{bmatrix}. \end{aligned} \tag{2.66}$$

Utilizing Eq. (2.65), Eq. (2.66) can be simplified as

$$\bar{n}_i = \left[\frac{(C\theta_i C\underline{\theta}_i + S\theta_i S\underline{\theta}_i)S\theta_i}{S\theta_i C\underline{\theta}_i - C\theta_i S\underline{\theta}_i} - C\theta_i\right]\bar{\ell}_{i-1} - \left[\frac{S\theta_i}{S\theta_i C\underline{\theta}_i - C\theta_i S\underline{\theta}_i}\right]\bar{\ell}_i, \tag{2.67}$$

where $S\theta_i$, $C\theta_i$, $S\underline{\theta}_i$ and $C\underline{\theta}_i$ are given in Eqs. (2.60), (2.61), (2.62) and (2.63), respectively.

2.5.2 *Unit Normal Vector of Reflective Boundary Surface*

Referring to Fig. 2.5, to determine the active unit normal vector $\bar{n}_i$ at any incidence point $\bar{P}_i$ on a reflective surface $\bar{r}_i$, it is first necessary to calculate the common normal vector $\bar{m}_i$ of the unit directional vectors of the incoming and outgoing rays (i.e., $\bar{\ell}_{i-1}$ and $\bar{\ell}_i$). Given the assumption $\theta_i \neq 0$, $\bar{m}_i$ can be obtained as

$$\bar{m}_i = \begin{bmatrix} m_{ix} & m_{iy} & m_{iz} & 0 \end{bmatrix}^T = \frac{\bar{\ell}_i \times \bar{\ell}_{i-1}}{S(2\theta_i)}, \tag{2.68}$$

where the incidence angle θ_i is determined by

$$C(2\theta_i) = \left|\bar{\ell}_{i-1} \cdot \bar{\ell}_i\right|. \tag{2.69}$$

Note that Eq. (2.69) is also applicable for an incidence angle equal to zero.

To simplify the expression for the unit normal vector $\bar{n}_i$, it is useful to have the following equation, obtained by taking the post cross product of Eq. (2.68) by $S(2\theta_i)\bar{\ell}_{i-1}$, i.e.,

$$\begin{aligned} S(2\theta_i)\left(\bar{m}_i \times \bar{\ell}_{i-1}\right) &= (\bar{\ell}_i \times \bar{\ell}_{i-1}) \times \bar{\ell}_{i-1} \\ &= \bar{\ell}_{i-1}(\bar{\ell}_i \cdot \bar{\ell}_{i-1}) - \bar{\ell}_i(\bar{\ell}_{i-1} \cdot \bar{\ell}_{i-1}) = -C(2\theta_i)\bar{\ell}_{i-1} - \bar{\ell}_i. \end{aligned} \tag{2.70}$$

Note that $\bar{\ell}_{i-1} \cdot \bar{\ell}_{i-1} = 1$ and $\bar{\ell}_i \cdot \bar{\ell}_{i-1} = -C(2\theta_i)$ are used in deriving Eq. (2.70). As shown in Fig. 2.5, the active unit normal vector $\bar{n}_i$ can be determined by rotating $-\bar{\ell}_{i-1}$ about $\bar{m}_i$ through an angle θ_i (i.e., Eq. (2.66)). Simplifying Eq. (2.66) using Eq. (2.70), the following expression is obtained for the unit normal vector $\bar{n}_i$ at any incidence point on the reflective surface when $\bar{\ell}_{i-1}$ and $\bar{\ell}_i$ are given:

$$\bar{n}_i = \frac{1}{2C\theta_i}\left(\bar{\ell}_i - \bar{\ell}_{i-1}\right) = \frac{1}{\sqrt{2\left(1 - \bar{\ell}_{i-1} \cdot \bar{\ell}_i\right)}}\left(\bar{\ell}_i - \bar{\ell}_{i-1}\right). \tag{2.71}$$

The fabrication of aspherical surfaces requires the use of high-precision manufacturing techniques in order to achieve the necessary surface accuracy and smoothness. Large aspherical lenses are typically produced using grinding and polishing techniques. Single-point diamond turning [35] is an emerging technique for fabricating large aspherical surfaces, and typically results in a better metallurgical structure than that produced by polishing and lapping. However, in using such a method, precise tool angle settings must be determined in advance in order to obtain the desired surface profile [40, 41]. The setting angles are determined by both the tool geometry and the normal vectors of the aspherical surface. Thus, as described above, Eqs. (2.67) and (2.71) are important not only in predicting the optical performance of the surface, but also in formulating the numerical codes required to machine the surface during the fabrication process.

References

1. Cornbleet S (1983) Geometrical optics reviewed: a new light on an old subject. Proc IEEE 71:471–502
2. Hamilton WR (1830) Supplement to an essay on the theory of systems of rays. Trans R Irish Acad 16:1–61
3. Silverstein L (1918), Simplified method of tracing rays through any optical system of lenses, prism and mirrors. Longmans, Green and Company, New York
4. Spencer CH, Murty MVRK (1962) General ray-tracing Procedure. J Opt Soc Am 52:672–678
5. Stavroudis ON (1951) Ray-tracing formulas for uniaxial crystals. J Opt Soc Am 52:187–191
6. Conrady AE (1929) Applied optics and optical design. Oxford University Press, New York, first edition, Part I, p. 413
7. Luneburg R (1965) Mathematical theory of optics, 3rd edn. Pergamon, New York, p 373
8. Born M, Wolf E (1980) Principles of optics, Chap. 14, 6th edn. Pergamon Press, p 668
9. Malacara D, Malacara Z (2004) Handbook of optical design, Section 3.5, Equation (3.37), 2nd edn. Marcel Dekker Inc. New York
10. Allen WA, Snyder JR (1952) Ray tracing through uncentered and aspherical surfaces. J Opt Soc Am A 42:243–249
11. Ford PW (1960) New ray tracing scheme. J Opt Soc Am 50:528–533
12. Pinto G (1979) A program for ray tracing through tilted and decentered optical surfaces. Opt Acta 65:1321
13. Hanssen JS (1982) Ray-tracing programs for spherical and aspherical surfaces. Appl Opt 21:2184
14. Kasper E (1984) On the numerical determination of optical focusing properties and aberrations. Optik 69:117–125
15. Montagnino L (1968) Ray tracing in inhomogeneous media. J Opt Soc Am 58:1667–1668
16. Buchdahl HA (1973) Rays in gradient index media: separable systems. J Opt Soc Am 63:46
17. Moore DT (1975) Ray tracing in gradient-index media. J Opt Soc Am 65:451–455
18. Moore DT (1980) Gradient-index optics. Appl Opt 19:1035–1038
19. Southwell WH (1982) Ray tracing in gradient-index media. J Opt Soc Am 72:908–911
20. Sharma A, Kumar DV, Ghatak AK (1986), Tracing rays through graded-index media. Appl Opt 21:984–987; 24:4367–4370 (1985); 25:3409–3412
21. Simon MC, Echarri RM (1983) Ray tracing formulas for monoaxial optical components. Appl Opt 22:354–360
22. Simon MC, Echarri RM (1986) Ray tracing formulas for monoaxial optical components: vectorial formulation. Appl Opt 25:1935–1939
23. Simon MC (1987) Refraction in biaxial crystals: a formula for the indices. J Opt Soc Am A 4:2201–2204
24. Yariv A, Yeh P (1984) Optical waves in crystals. Wiley, New York
25. Liang QT (1990) Simple ray-tracing formulas for uniaxial optical crystals. Appl Opt 29:1008–1010
26. Liang QT, Zheng XD (1991) Ray-tracing calculations for uniaxial optical components with curves surfaces. Appl Opt 30:31
27. Trollinger JD, Chipman RA, Wilson DK (1991) Polarization ray tracing in birefringent media. Opt Eng 30:461–466
28. Beyerle G, McDermid IS (1998) Ray-tracing formulas for refraction and internal reflection in uniaxial crystals. Appl Opt 37:7947–7953
29. Zhang WQ (1991) General ray-tracing formulas for crystal. Appl Opt 31:7328–7331
30. Nishidate Y, Nagata T, Morita S, Yamagata Y (2011) Raytracing method for isotropic inhomogeneous refractive-index media from arbitrary discrete input. Appl Opt 50:5192–5199
31. Feder D (1968) Differentiation of ray-tracing equations with respect to construction parameters of rotationally symmetric optics. J Opt Soc Am 58(11):1494–1505

32. Stavroudis O (1976) Simpler derivation of the formulas for generalized ray tracing. J Opt Soc Am 66(12):1330–1333
33. Lu CH, Sung CK (2013) Skew ray tracing and sensitivity analysis of hyperboloid optical boundary surfaces. Optik—Int J Light Electron Optics 124:1159–1169
34. Smith WJ (2001) Modern optical engineering, 3rd edn. Edmund Industrial Optics, Barrington, N.J
35. Haisma J, Hugues E, Babolat C (1979) Realization of a bi-aspherical objective lens for the philips video play system. Opt Lett 4:70–72
36. Lin PD, Tsai CY (2012) Determination of first-order derivatives of skew-ray at aspherical surface. J Opt Soc Am A 29:1141–1153
37. Gutiérrez CE (2013) Aspherical lens design. J Opt Soc Am A 30:1719–1726
38. Lin PD, Tsai CY (2012) Determination of unit normal vectors of aspherical surfaces given unit directional vectors of incoming and outgoing rays. J Opt Soc Am A 29:174–178
39. Lin PD, Tsai CY (2012) Determination of unit normal vectors of aspherical surfaces given unit directional vectors of incoming and outgoing rays: reply. J Opt Soc Am A 29:1358
40. Boothroyd G (1975), Fundamentals of metal machining and machine tools, Chap. 7. McGraw-Hill, New York, NJ
41. Biddut AQ, Rahman M, Neo KS, Rezaur KM, Sawa M, Maeda Y (2007) Performance of single crystal diamond tools with different rake angles during micro-grooving on electroless nickel plated die materials. Int J Adv Manuf Technol 33:891–899

Chapter 3
Geometrical Optical Model

In geometrical optics, the problem of modeling an optical system in an efficient and systematic manner is highly challenging. This chapter addresses recent developments in mathematical modeling in geometrical optics based on a homogeneous coordinate notation approach. Particular emphasis is placed on the determination of the boundary variable vectors and system variable vectors. Section 3.1 introduces a modeling technique for axis-symmetrical optical systems, while Sect. 3.2 extends the modeling technique to the case of optical systems containing prisms. The methods required to determine the spot size, point spread function, and modulation transfer function by means of a raytracing approach are then addressed in the remaining sections.

3.1 Axis-Symmetrical Systems

In general, an optical element (e.g., a dove prism) is simply a block of optical material with a constant refractive index. Optical elements generally possess multiple boundary surfaces; each with its own unique pose matrix. Assuming that an optical element is the jth element in an optical system, and has L_{ej} boundary surfaces, the boundary surfaces can be labeled sequentially as $i = m_{ej} - L_{ej} + 1$ to $i = m_{ej}$ (where m_{ej} is the number assigned to the boundary surface which the exit refracted/reflected by the jth element, see Fig. 3.1). (Note that in this book, the notation "ej" is used to indicate that the associated parameter relates to the jth element. For example, notations ξ_{e3} and q_{e3} represent the refractive index and thickness, respectively, of the third element in an optical system.) In practical applications, several optical elements are arranged together with fixed relative pose matrices in order to form an optical assembly. Furthermore, several optical assemblies are often arranged together to create an optical system designed to perform a particular function. To model such an optical system, it is first necessary

P.D. Lin, *Advanced Geometrical Optics*, Progress in Optical Science and Photonics 4, DOI 10.1007/978-981-10-2299-9_3

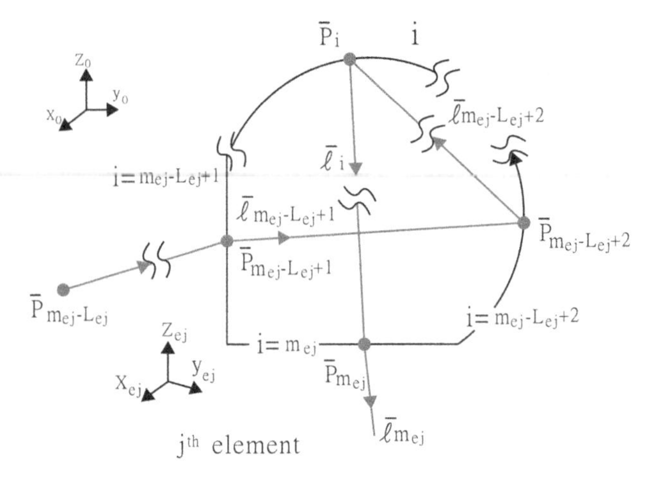

Fig. 3.1 An optical element is a block of optical material with a constant refractive index ξ_{ej}. Each element j contains L_{ej} boundary surfaces labeled from $i = m_{ej} - L_{ej} + 1$ to $i = m_{ej}$

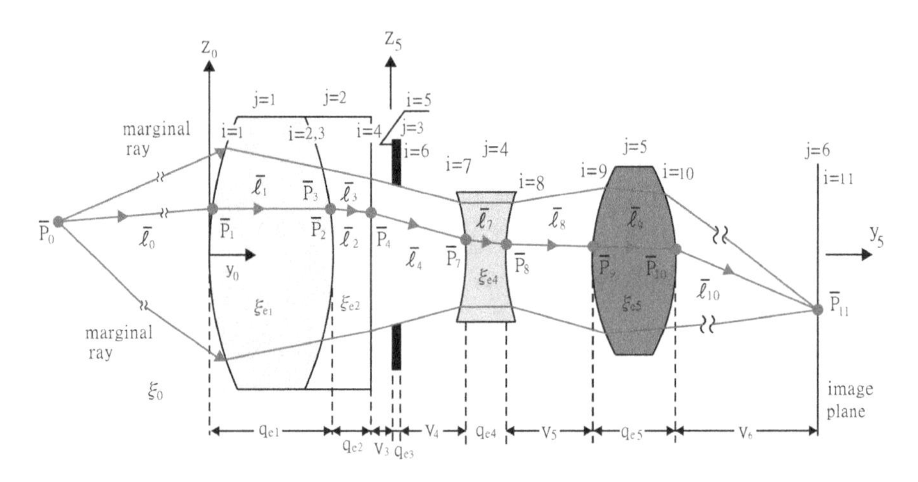

Fig. 3.2 Petzval lens system with n = 11 boundary surfaces [1]

to label its optical elements sequentially from j = 1 to j = k and its boundary surfaces from i = 1 to i = n (see Fig. 3.2, Tables 3.1 and 3.2).

The discussions in this section are limited to axis-symmetrical systems containing a straight-line optical axis. For a system composed of simple lenses, the optical axis passes through the center of curvature of each surface and coincides with the axis of rotational symmetry. Most elements used in axis-symmetrical system are spherical elements; that is, their two boundary surfaces form part of two notional spheres. The radius of the boundary surface may be positive, zero or

Table 3.1 Each element j (j = 1 to j = 5) in the system shown in Fig. 3.2 contains two ($L_{ej} = 2$) boundary surfaces labeled as i = 2j − 1 and i = 2j

Element number	j = 1	j = 2	j = 3	j = 4	j = 5	j = k = 6
L_{ej} (number of boundary surfaces in element j)	2	2	2	2	2	1
The first boundary surface of element j is $i = m_{ej} - L_{ej} + 1$	1	3	5	7	9	11
The last boundary surface of element j is $i = m_{ej}$	2	4	6	8	10	11

Table 3.2 Values of variables for Petzval lens system shown in Fig. 3.2

j	V_j	ξ_{ej}	R_{2j-1}	R_{2j}	q_{ej}
1	0.0000	1.65000	38.2219	−56.0857	15.8496
2	0.0000	1.71736	−56.0857	−590.6820	5.9690
3	3.0226	1.00000			0.0000
4	14.0208	1.52583	−41.7957	29.3446	2.5146
5	7.9248	1.65000	63.5635	−56.8655	6.0960
6	49.6316				

negative, depending on whether the surface is convex, flat or concave, respectively. Nine possible types of spherical element exist depending on the geometries of the two boundary surfaces and their orientation (see Figs. 3.3, 3.4, 3.5, 3.6, 3.7, 3.8, 3.9, 3.10 and 3.11). For each type of element, the two boundary surfaces can be denoted as $\bar{r}_{2j-1}$ and $\bar{r}_{2j}$, respectively, with corresponding boundary coordinate frames $(xyz)_{2j-1}$ and $(xyz)_{2j}$. In performing the 3-D modeling of optical systems, it is first necessary to establish an element coordinate frame $(xyz)_{ej}$ to define the pose of each element j. For simplicity, assume that $(xyz)_{ej}$ always coincides with the first boundary coordinate frame of the element, i.e., $(xyz)_{2j-1}$. In general, the following pose matrix, ${}^0\bar{A}_{ej}$, can then be used to define the position and orientation of this element with respect to the world coordinate frame $(xyz)_0$ in an axis-symmetrical system:

$$ {}^0\bar{A}_{ej} = \text{tran}(0, t_{ejy}, 0) = \begin{bmatrix} I_{ejx} & J_{ejx} & K_{ejx} & t_{ejx} \\ I_{ejy} & J_{ejy} & K_{ejy} & t_{ejy} \\ I_{ejz} & J_{ejz} & K_{ejz} & t_{ejz} \\ 0 & 0 & 0 & 1 \end{bmatrix} = \begin{bmatrix} 1 & 0 & 0 & 0 \\ 0 & 1 & 0 & t_{ejy} \\ 0 & 0 & 1 & 0 \\ 0 & 0 & 0 & 1 \end{bmatrix}. \quad (3.1) $$

It is seen that for reasons of simplicity, Eq. (3.1) is used to define the poses of the elements j directly rather than performing the following more complicated manipulation

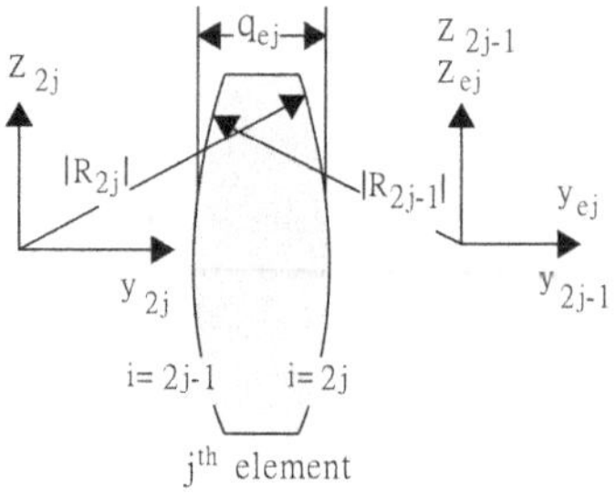

Fig. 3.3 Biconvex element with convex-convex spherical boundary surfaces

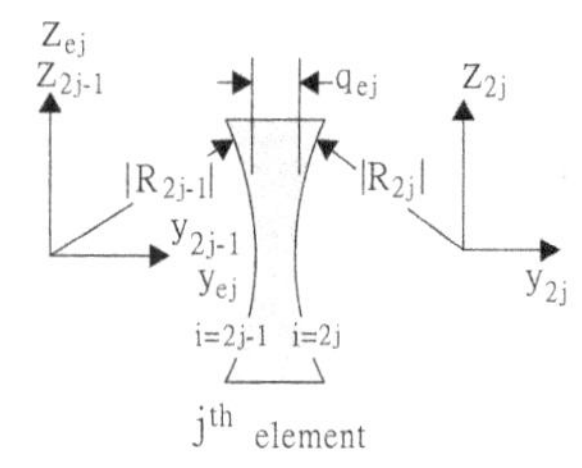

Fig. 3.4 Biconcave element with concave-concave spherical boundary surfaces

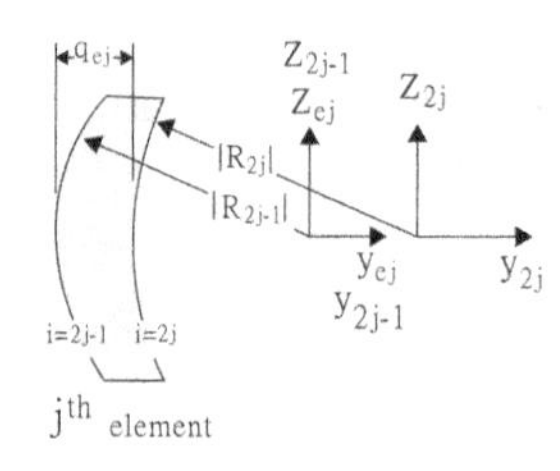

Fig. 3.5 Meniscus element with convex-concave spherical boundary surfaces

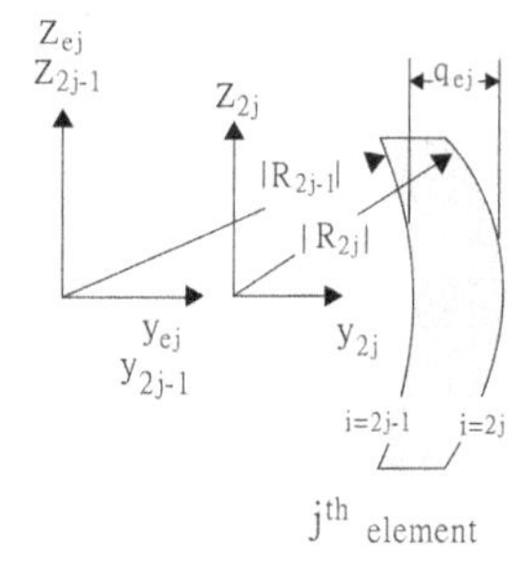

Fig. 3.6 Meniscus element with concave-convex spherical boundary surfaces

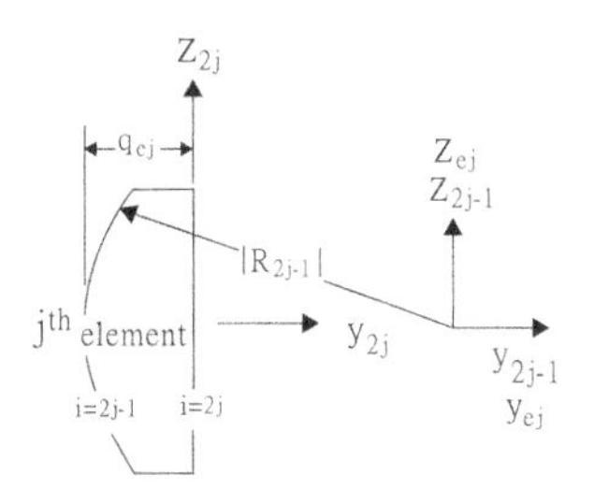

Fig. 3.7 Plano-convex element with convex-flat boundary surfaces

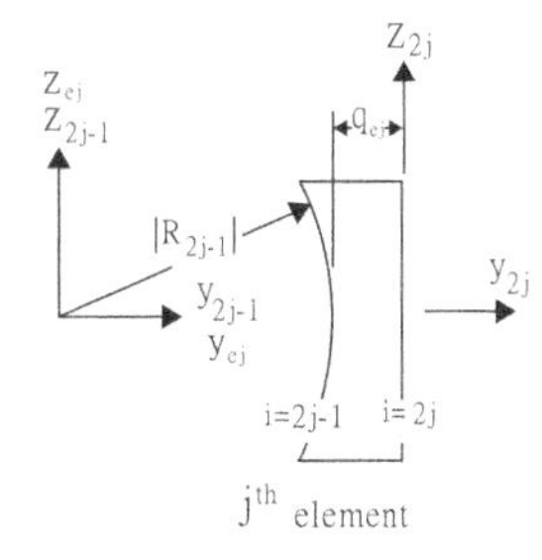

Fig. 3.8 Plano-concave element with concave-flat boundary surfaces

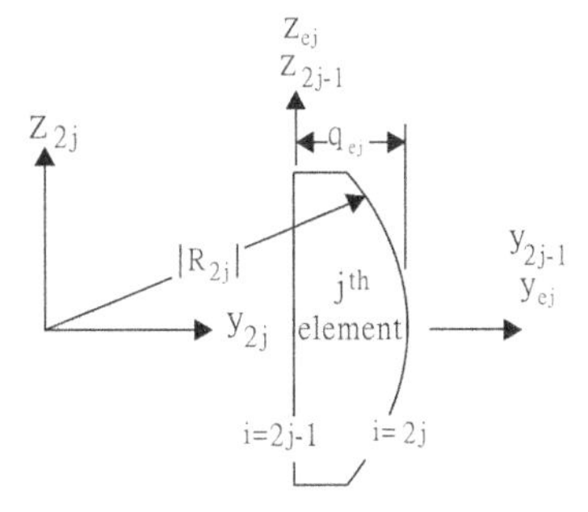

Fig. 3.9 Plano-convex element with flat-convex boundary surfaces

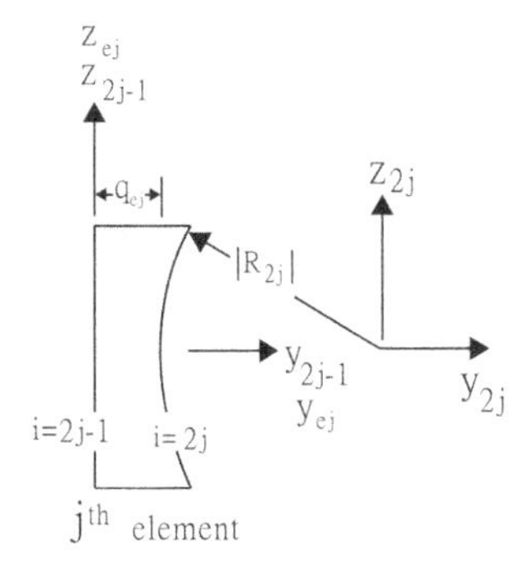

Fig. 3.10 Plano-concave element with flat-concave boundary surfaces

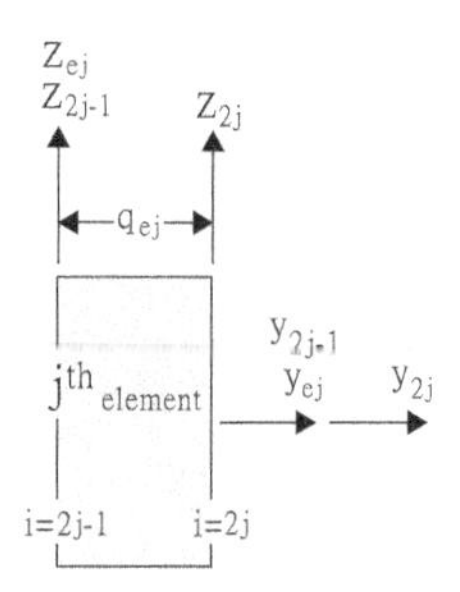

Fig. 3.11 Optical element with two flat boundary surfaces

$$^{0}\bar{A}_{ej} = \text{tran}(t_{ejx}, t_{ejy}, t_{ejz})\text{rot}(\bar{z}, \omega_{ejz})\text{rot}(\bar{y}, \omega_{ejy})\text{rot}(\bar{x}, \omega_{ejx})$$

and then setting $t_{e1x} = t_{e1z} = \omega_{e1x} = \omega_{e1y} = \omega_{e1z} = 0$. It should be noted, however, that by adopting this simplified approach (i.e., Eq. (3.1)), it is impossible to perform error analysis (for t_{ejx}, t_{ejz}, ω_{ejx}, ω_{ejy}, and ω_{ejz}) using Eq. (3.1) since the equation lacks the parameters required for further mathematical manipulations. It only contains the symbol t_{ejy} for analyzing the effect of element positioning error.

As described in Chap. 2, the pose matrix $^{0}\bar{A}_{i}$ (see Eq. (2.9) or Eq. (2.35)) is required to perform raytracing for any optical system. Given the pose $^{ej}\bar{A}_{i}$ of boundary coordinate frame $(xyz)_i$ (i = 2j − 1 and i = 2j for the lenses shown in Figs. 3.3, 3.4, 3.5, 3.6, 3.7, 3.8, 3.9, 3.10 and 3.11) with respect to the element coordinate frame $(xyz)_{ej}$, the pose matrix $^{0}\bar{A}_{i}$ can be computed as

$$^{0}\bar{A}_{i} = {}^{0}\bar{A}_{ej}\,{}^{ej}\bar{A}_{i}. \tag{3.2}$$

$^{ej}\bar{A}_{i}$ can be obtained using the translation and rotation matrices given in Sect. 1.4, and can be represented by the following matrix:

$$^{ej}\bar{A}_{i} = \begin{bmatrix} ^{ej}I_{ix} & ^{ej}J_{ix} & ^{ej}K_{ix} & ^{ej}t_{ix} \\ ^{ej}I_{iy} & ^{ej}J_{iy} & ^{ej}K_{iy} & ^{ej}t_{iy} \\ ^{ej}I_{iz} & ^{ej}J_{iz} & ^{ej}K_{iz} & ^{ej}t_{iz} \\ 0 & 0 & 0 & 1 \end{bmatrix} = \begin{bmatrix} 1 & 0 & 0 & 0 \\ 0 & 1 & 0 & ^{ej}t_{iy} \\ 0 & 0 & 1 & 0 \\ 0 & 0 & 0 & 1 \end{bmatrix}. \tag{3.3}$$

The determinations of $^{ej}\bar{A}_{i}$ for the nine illustrative examples shown in Figs. 3.3, 3.4, 3.5, 3.6, 3.7, 3.8, 3.9, 3.10 and 3.11 are given in the following.

3.1.1 Elements with Spherical Boundary Surfaces

Figure 3.3 shows a biconvex element with thickness q_{ej}. If R_{2j-1} and R_{2j} are the radii of the first and second boundary surfaces, respectively, then the pose matrices of $(xyz)_{2j-1}$ and $(xyz)_{2j}$ with respect to the element coordinate frame $(xyz)_{ej}$ are given respectively by

$$^{ej}\bar{A}_{2j-1} = \bar{I}_{4\times4}, \tag{3.4}$$

$$^{ej}\bar{A}_{2j} = \text{tran}(0, -R_{2j-1} + q_{ej} + R_{2j}, 0). \tag{3.5}$$

The pose matrices $^{0}\bar{A}_{i}$ (i = 2j − 1 and i = 2j) of the two boundary coordinate frames, $(xyz)_{2j-1}$ and $(xyz)_{2j}$, with respect to the world coordinate frame $(xyz)_{0}$ can then be determined from Eq. (3.2) as

$$^{0}\bar{A}_{2j-1} = {}^{0}\bar{A}_{ej}{}^{ej}\bar{A}_{2j-1} = {}^{0}\bar{A}_{ej} = \text{tran}(0, t_{ejy}, 0), \tag{3.6}$$

$$\begin{aligned} ^{0}\bar{A}_{2j} = {}^{0}\bar{A}_{ej}{}^{ej}\bar{A}_{2j} &= \text{tran}(0, t_{ejy}, 0)\text{tran}(0, -R_{2j-1} + q_{ej} + R_{2j}, 0) \\ &= \text{tran}(0, t_{ejy} - R_{2j-1} + q_{ej} + R_{2j}, 0). \end{aligned} \tag{3.7}$$

Equations (3.6) and (3.7) provide the pose matrices, $^{0}\bar{A}_{2j-1}$ and $^{0}\bar{A}_{2j}$, required to perform skew-ray tracing using the methodology presented in Chap. 2. From Eqs. (3.6) and (3.7), it is seen that the element variable vector $\bar{X}_{ej}$ of the biconvex element shown in Fig. 3.3 comprises the pose variable, t_{ejy}; the two radii of the spherical boundary surfaces, R_{2j-1} and R_{2j}; the element thickness, q_{ej}. When the refractive index of air ξ_{air} and element refractive index ξ_{ej} are also taken into account, the variable vector of the biconvex element has the form

$$\bar{X}_{ej} = \begin{bmatrix} t_{ejy} & \xi_{air} & \xi_{ej} & R_{2j-1} & q_{ej} & R_{2j} \end{bmatrix}^{T}. \tag{3.8}$$

Figures 3.4, 3.5and 3.6 show a biconcave element and two meniscus elements, respectively. Equations (3.4), (3.5), (3.6), (3.7) and (3.8) are still valid for these elements since the meniscus, biconcave and biconvex elements are also composed of two spherical boundary surfaces. However, the radii should be assigned appropriate signs in every case.

3.1.2 Elements with Spherical and Flat Boundary Surfaces

Figure 3.7 shows a plano-convex element with i = 2j − 1 and i = 2j being convex and flat boundary surfaces, respectively. The pose matrices, $^{ej}\bar{A}_{2j-1}$ and $^{ej}\bar{A}_{2j}$, of the first and second boundary coordinate frames, $(xyz)_{2j-1}$ and $(xyz)_{2j}$, with respect to the element coordinate frame $(xyz)_{ej}$ are given respectively as

$$^{ej}\bar{A}_{2j-1} = \bar{I}_{4\times4}, \tag{3.9}$$

$$^{ej}\bar{A}_{2j} = \text{tran}(0, -R_{2j-1} + q_{ej}, 0), \tag{3.10}$$

where q_{ej} and R_{2j-1} are the element thickness and the radius of the first boundary surface, respectively. Meanwhile, the pose matrices, ${}^{0}\bar{A}_{2j-1}$ and ${}^{0}\bar{A}_{2j}$, of the first and second boundary coordinate frames, $(xyz)_{2j-1}$ and $(xyz)_{2j}$, with respect to the world coordinate frame $(xyz)_0$ are determined respectively by

$${}^{0}\bar{A}_{2j-1} = {}^{0}\bar{A}_{ej}{}^{ej}\bar{A}_{2j-1} = {}^{0}\bar{A}_{ej} = \text{tran}(0, t_{ejy}, 0), \tag{3.11}$$

$${}^{0}\bar{A}_{2j} = {}^{0}\bar{A}_{ej}{}^{ej}\bar{A}_{2j} = \text{tran}(0, t_{ejy} - R_{2j-1} + q_{ej}, 0). \tag{3.12}$$

Collecting the variables in Eqs. (3.11) and (3.12), and the refractive indices of air and the element, the following variable vector for the element is obtained:

$$\bar{X}_{ej} = \begin{bmatrix} t_{ejy} & \xi_{air} & \xi_{ej} & R_{2j-1} & q_{ej} \end{bmatrix}^T. \tag{3.13}$$

It is noted that compared with a biconvex or biconcave element, the element variable vector in Eq. (3.13) contains only five components since the second boundary surface is flat.

Figure 3.8 shows a plano-concave element with $i = 2j - 1$ and $i = 2j$ representing concave and flat boundary surfaces, respectively. Equations (3.9), (3.10), (3.11), (3.12) and (3.13) are still valid for such an element since the first boundary surfaces of the elements in Figs. 3.7 and 3.8 are both spherical.

3.1.3 Elements with Flat and Spherical Boundary Surfaces

Figure 3.9 shows a plano-convex element with $i = 2j - 1$ and $i = 2j$ representing flat and convex spherical boundary surfaces, respectively. The pose matrices, ${}^{ej}\bar{A}_{2j-1}$ and ${}^{ej}\bar{A}_{2j}$, of the first and second boundary coordinate frames, $(xyz)_{2j-1}$ and $(xyz)_{2j}$, with respect to the element coordinate frame $(xyz)_{ej}$ are given respectively by

$${}^{ej}\bar{A}_{2j-1} = \bar{I}_{4\times4}, \tag{3.14}$$

$${}^{ej}\bar{A}_{2j} = \text{tran}(0, q_{ej} + R_{2j}, 0), \tag{3.15}$$

where q_{ej} and R_{2j} are the element thickness and the radius of the second boundary surface, respectively. The pose matrices, ${}^{0}\bar{A}_{2j-1}$ and ${}^{0}\bar{A}_{2j}$, of $(xyz)_{2j-1}$ and $(xyz)_{2j}$ with respect to the world coordinate frame $(xyz)_0$ are given respectively as

$${}^{0}\bar{A}_{2j-1} = {}^{0}\bar{A}_{ej}{}^{ej}\bar{A}_{2j-1} = {}^{0}\bar{A}_{ej} = \text{tran}(0, t_{ejy}, 0), \tag{3.16}$$

$$^{0}\bar{A}_{2j} = {}^{0}\bar{A}_{ej}{}^{ej}\bar{A}_{2j} = \text{tran}(0, t_{ejy}q_{ej} + R_{2j}, 0). \tag{3.17}$$

Although the element shown in Fig. 3.9 is also a plano-convex element, its variable vector is different from that of the element described in Fig. 3.7, i.e.,

$$\bar{X}_{ej} = \begin{bmatrix} t_{ejy} & \xi_{air} & \xi_{ej} & q_{ej} & R_{2j} \end{bmatrix}^T. \tag{3.18}$$

Figure 3.10 shows another plano-concave element in which $i = 2j - 1$ and $i = 2j$ represent flat and concave spherical boundary surfaces, respectively. Equations (3.14), (3.15), (3.16), (3.17) and (3.18) are still valid for such an element since the first and second boundary surfaces in the elements shown in Figs. 3.9 and 3.10 are flat and spherical, respectively, in both cases. However, an appropriate sign must again be assigned to boundary surface R_{2j}.

3.1.4 Elements with Flat Boundary Surfaces

Figure 3.11 shows an optical element possessing two flat boundary surfaces and a thickness q_{ej}. The pose matrices, ${}^{ej}\bar{A}_{2j-1}$ and ${}^{ej}\bar{A}_{2j}$, of the first and second boundary coordinate frames, $(xyz)_{2j-1}$ and $(xyz)_{2j}$, with respect to the element coordinate frame $(xyz)_{ej}$ are given respectively by

$$^{ej}\bar{A}_{2j-1} = \bar{I}_{4\times4}, \tag{3.19}$$

$$^{ej}\bar{A}_{2j} = \text{tran}(0, q_{ej}, 0). \tag{3.20}$$

In addition, the pose matrices, ${}^{0}\bar{A}_{2j-1}$ and ${}^{0}\bar{A}_{2j}$, of the first and second boundary coordinate frames, $(xyz)_{2j-1}$ and $(xyz)_{2j}$, with respect to the world coordinate frame $(xyz)_0$ are given respectively as

$$^{0}\bar{A}_{2j-1} = {}^{0}\bar{A}_{ej}{}^{ej}\bar{A}_{2j} - 1 = {}^{0}\bar{A}_{ej} = \text{tran}(0, t_{ejy}, 0), \tag{3.21}$$

$$^{0}\bar{A}_{2j} = {}^{0}\bar{A}_{ej}{}^{ej}\bar{A}_{2j} = \text{tran}(0, t_{ejy} + q_{ej}, 0). \tag{3.22}$$

Since both boundary surfaces are flat in this case, the element variable vector has only four components, i.e.,

$$\bar{X}_{ej} = \begin{bmatrix} t_{ejy} & \xi_{air} & \xi_{ej} & q_{ej} \end{bmatrix}^T. \tag{3.23}$$

It is noted that Figs. 3.4, 3.5, 3.6, 3.7, 3.8, 3.9, 3.10 and 3.11 can be deduced from Fig. 3.3 provided that appropriate signs are assigned to the various radii. Consequently, in the design stage of an optical system, the element shown in

Fig. 3.3 can be used to represent any one of the nine possible elements; with the two radii being adjusted as appropriate from a lower bound (which may be a negative value) to an upper bound for optimization purposes. Having obtained the pose matrix ${}^{0}\bar{A}_{i}$ from Eq. (3.2), the skew-ray tracing methodology described in Chap. 2 can be applied successively to trace the rays traveling through the optical system and to obtain the point of incidence of the rays on the nth boundary surface, i.e.,

$$\bar{P}_n = \begin{bmatrix} P_{nx} \\ P_{ny} \\ P_{nz} \\ 1 \end{bmatrix} = \begin{bmatrix} P_{n-1x} + \lambda_n \ell_{n-1x} \\ P_{n-1y} + \lambda_n \ell_{n-1y} \\ P_{n-1z} + \lambda_n \ell_{n-1z} \\ 1 \end{bmatrix} = \bar{P}_{n-1} + \lambda_n \bar{\ell}_{n-1}. \tag{3.24}$$

Example 3.1 The use of multiple elements in an optical system allows more optical aberrations to be corrected than is possible when using only a single element. Referring to the system shown in Fig. 3.2 and Table 3.2, the following pose matrices ${}^{0}\bar{A}_{ej}$ (j = 1 to j = 6) are obtained for the six elements:

$$\begin{aligned}
{}^{0}\bar{A}_{e1} &= \mathrm{tran}(0, v_1 + R_1, 0) = \mathrm{tran}(0, 38.2219, 0), \\
{}^{0}\bar{A}_{e2} &= \mathrm{tran}(0, v_1 + q_{e1} + v_2 + R_3, 0) = \mathrm{tran}(0, -40.2361, 0), \\
{}^{0}\bar{A}_{e3} &= \mathrm{tran}(0, v_1 + q_{e1} + v_2 + q_{e2} + v_3, 0) = \mathrm{tran}(0, 24.8412, 0), \\
{}^{0}\bar{A}_{e4} &= \mathrm{tran}(0, v_1 + q_{e1} + v_2 + q_{e2} + v_3 + q_{e3} + v_4 + R_7, 0) = \mathrm{tran}(0, -2.9337, 0), \\
{}^{0}\bar{A}_{e5} &= \mathrm{tran}(0, v_1 + q_{e1} + v_2 + q_{e2} + v_3 + q_{e3} + v_4 + q_{e4} + v_5 + R_9, 0) = \mathrm{tran}(0, 112.8649, 0), \\
{}^{0}\bar{A}_{e6} &= \mathrm{tran}(0, v_1 + q_{e1} + v_2 + q_{e2} + v_3 + q_{e3} + v_4 + q_{e4} + v_5 + q_{e5} + v_6, 0) = \mathrm{tran}(0, 105.0290, 0).
\end{aligned}$$

Example 3.2 Table 3.3 shows the optical train (i.e., the series of boundary surfaces through which rays pass) of the system shown in Fig. 3.2. It will be recalled that the aperture of an optical system can be treated as a ghost element with refractive index $\xi_{ej} = \xi_{air} = 1$. Let the element thickness be set as $q_{e3} = 0$ in the present example since the thickness of an aperture is usually very small and thus has no effect on the computation result. The pose matrices of the two boundary frames, $(xyz)_{2j-1}$ and $(xyz)_{2j}$, with respect to each element frame $(xyz)_{ej}$ (j = 1 to j = 6) of the system are given by

Table 3.3 Optical train of system shown in Fig. 3.2

Optical train / Element	Boundary surface i	
j = 1	1	2
j = 2	3	4
j = 3	5	6
j = 4	7	8
j = 5	9	10
j = 6	11	

$$
\begin{aligned}
{}^{e1}\bar{A}_1 &= I_{4\times4},\\
{}^{e1}\bar{A}_2 &= \mathrm{tran}(0, -R_1 + q_{e1} + R_2, 0) = \mathrm{tran}(0, -78.4580, 0),\\
{}^{e2}\bar{A}_3 &= I_{4\times4},\\
{}^{e2}\bar{A}_4 &= \mathrm{tran}(0, -R_3 + q_{e2} + R_4, 0) = \mathrm{tran}(0, -528.6273, 0),\\
{}^{e3}\bar{A}_5 &= \bar{I}_{4\times4},
\end{aligned}
$$

$$
\begin{aligned}
{}^{e3}\bar{A}_6 &= \mathrm{tran}(0, q_{e3}, 0) = \mathrm{tran}(0, 0, 0),\\
{}^{e4}\bar{A}_7 &= \bar{I}_{4\times4},\\
{}^{e4}\bar{A}_8 &= \mathrm{tran}(0, -R_7 + q_{e4} + R_8, 0) = \mathrm{tran}(0, 73.6549, 0),\\
{}^{e5}\bar{A}_9 &= \bar{I}_{4\times4},\\
{}^{e5}\bar{A}_{10} &= \mathrm{tran}(0, -R_9 + q_{e5} + R_{10}, 0) = \mathrm{tran}(0, -114.3330, 0),\\
{}^{e6}\bar{A}_{11} &= \bar{I}_{4\times4}
\end{aligned}
$$

Example 3.3 Referring to the element pose matrix ${}^{0}\bar{A}_{ej}$ (j = 1 to j = 6) in Example 3.1 and the boundary pose matrices ${}^{ej}\bar{A}_{2j-1}$ and ${}^{ej}\bar{A}_{2j}$ in Example 3.2, the following pose matrices ${}^{0}\bar{A}_{i}$ (i = 1 to i = 11) are obtained for boundary coordinate frames $(xyz)_i$ with respect to $(xyz)_0$:

$$
\begin{aligned}
{}^{0}\bar{A}_1 &= {}^{0}\bar{A}_{e1}{}^{e1}\bar{A}_1 = \mathrm{tran}(0, v_1 + R_1, 0) = \mathrm{tran}(0, 38.2219, 0),\\
{}^{0}\bar{A}_2 &= {}^{0}\bar{A}_{e1}{}^{e1}\bar{A}_2 = \mathrm{tran}(0, v_1 + q_{e1} + R_2, 0) = \mathrm{tran}(0, -40.2361, 0),\\
{}^{0}\bar{A}_3 &= {}^{0}\bar{A}_{e2}{}^{e2}\bar{A}_3 = \mathrm{tran}(0, v_1 + q_{e1} + v_2 + R_3, 0) = \mathrm{tran}(0, -40.2361, 0),\\
{}^{0}\bar{A}_4 &= {}^{0}\bar{A}_{e2}{}^{e2}\bar{A}_4 = \mathrm{tran}(0, v_1 + q_{e1} + v_2 + q_{e2} + R_4, 0) = \mathrm{tran}(0, -568.8634, 0),\\
{}^{0}\bar{A}_5 &= {}^{0}\bar{A}_{e3}{}^{e3}\bar{A}_5 = \mathrm{tran}(0, v_1 + q_{e1} + v_2 + q_{e2} + v_3, 0) = \mathrm{tran}(0, 24.8412, 0),
\end{aligned}
$$

$$
\begin{aligned}
{}^{0}\bar{A}_6 &= {}^{0}\bar{A}_{e3}{}^{e3}\bar{A}_6 = \mathrm{tran}(0, v_1 + q_{e1} + v_2 + q_{e2} + v_3 + q_{e3}, 0) = \mathrm{tran}(0, 24.8412, 0),\\
{}^{0}\bar{A}_7 &= {}^{0}\bar{A}_{e4}{}^{e4}\bar{A}_7 = \mathrm{tran}(0, v_1 + q_{e1} + v_2 + q_{e2} + v_3 + q_{e3} + v_4 + R_7, 0)\\
&= \mathrm{tran}(0, -2.9337, 0),
\end{aligned}
$$

$$
\begin{aligned}
{}^{0}\bar{A}_8 &= {}^{0}\bar{A}_{e4}{}^{e4}\bar{A}_8 = \mathrm{tran}(0, v_1 + q_{e1} + v_2 + q_{e2} + v_3 + q_{e3} + v_4 + q_{e4} + R_8, 0)\\
&= \mathrm{tran}(0, 70.7212, 0),\\
{}^{0}\bar{A}_9 &= {}^{0}\bar{A}_{e5}{}^{e5}\bar{A}_9 = \mathrm{tran}(0, v_1 + q_{e1} + v_2 + q_{e2} + v_3 + q_{e3} + v_4 + q_{e4} + v_5 + R_9, 0)\\
&= \mathrm{tran}(0, 112.8649, 0),
\end{aligned}
$$

Table 3.4 Values of boundary pose variables for Petzval lens system shown in Fig. 3.2

	t_{ix}	t_{iy}	t_{iz}	ω_{ix}	ω_{iy}	ω_{iz}
i = 1	0	38.2219	0	0°	0°	0°
i = 2	0	−40.2361	0	0°	0°	0°
i = 3	0	−40.2361	0	0°	0°	0°
i − 4	0	−568.8634	0	0°	0°	0°
i = 5	0	24.8412	0	0°	0°	0°
i = 6	0	24.8412	0	0°	0°	0°
i = 7	0	−2.9337	0	0°	0°	0°
i = 8	0	70.7212	0	0°	0°	0°
i = 9	0	112.8649	0	0°	0°	0°
i = 10	0	−1.4681	0	0°	0°	0°
1 = 11	0	105.0290	0	0°	0”	0°

$$\begin{aligned}{}^0\bar{A}_{10} &= {}^0\bar{A}_{e5}{}^{e5}\bar{A}_{10} \\ &= \mathrm{tran}(0, v_1 + q_{e1} + v_2 + q_{e2} + v_3 + q_{e3} + v_4 + q_{e4} + v_5 + q_{e5} + R_{10}, 0) = \mathrm{tran}(0, -1.4681, 0), \\ {}^0\bar{A}_{11} &= {}^0\bar{A}_{e6}{}^{e6}\bar{A}_{11} = \mathrm{tran}(0, v_1 + q_{e1} + v_2 + q_{e2} + v_3 + q_{e3} + v_4 + q_{e4} + v_5 + q_{e5} + v_6, 0) \\ &= \mathrm{tran}(0, 105.0290, 0).\end{aligned}$$

Example 3.4 Recall that when the pose matrix ${}^0\bar{A}_i$ of the ith boundary surface is expressed in terms of Eq. (2.9) or Eq. (2.35), the parameters $t_{ix}, t_{iy}, t_{iz}, \omega_{ix}, \omega_{iy}$ and ω_{iz} represent the pose variables of the ith boundary surface. In this chapter, the values of these pose variables for the system shown in Fig. 3.2 are used to compute various quantities, such as the point spread function and the modulation transfer function. The values of the six variables in pose matrix ${}^0\bar{A}_i$ (i = 1 to i = 11) can be obtained using Eqs. (1.38) to (1.43), where the equations for ${}^0\bar{A}_i$ are formulated in Example 3.3 above. The corresponding results are presented in Table 3.4. It is noted that the values of $t_{ix} = t_{iz} = \omega_{ix} = \omega_{iy} = \omega_{iz} = 0$ shown in Table 3.4 indicate that the optical system is an axis-symmetrical system.

Example 3.5 According to the equations given in Example 3.1 and Eq. (3.8) (or Eqs. (3.13), (3.18), (3.23)), the following element variable vectors can be obtained for the system shown in Fig. 3.2 if the source variable vector $\bar{X}_0$ given in Eq. (2.4) is included:

$$\begin{aligned}\bar{X}_0 &= [P_0x \quad P_0y \quad P_0z \quad \alpha_0 \quad \beta_0]^T, \\ \bar{X}_{e1} &= [v_1 + R_1 \quad \xi_{air} \quad \xi_{e1} \quad R_1 \quad q_{e1} \quad R_2]^T, \\ \bar{X}_{e2} &= [v_1 + q_{e1} + v_2 + R_3 \quad \xi_{air} \quad \xi_{e2} \quad R_3 \quad q_{e2} \quad R_4]^T, \\ \bar{X}_{e3} &= [v_1 + q_{e1} + v_2 + q_{e2} + v_3 \quad \xi_{air} \quad q_{e3}]^T, \\ \bar{X}_{e4} &= [v_1 + q_{e1} + v_2 + q_{e2} + v_3 + q_{e3} + v_4 + R_7 \quad \xi_{air} \quad \xi_{e4} \quad R_7 \quad q_{e4} \quad R_8]^T, \\ \bar{X}_{e5} &= [v_1 + q_{e1} + v_2 + q_{e2} + v_3 + q_{e3} + v_4 + q_{e4} + v_5 + R_9 \quad \xi_{air} \quad \xi_{e5} \quad R_9 \quad q_{e5} \quad R_{10}]^T, \\ \bar{X}_{e6} &= [v_1 + q_{e1} + v_2 + q_{e2} + v_3 + q_{e3} + v_4 + q_{e4} + v_5 + q_{e5} + v_6 \quad \xi_{air}]^T.\end{aligned}$$

Example 3.6 After deleting the repeated variables (e.g., ξ_{air}) in $\bar{X}_{ej}$ (j = 1 to j = 6) in Example 3.5, the following system variable vector $\bar{X}_{sys}$ with dimension $q_{sys} = 29$ is obtained:

$$\bar{X}_{sys} = \begin{bmatrix} \bar{X}_0 & \bar{X}_\xi & \bar{X}_R & \bar{X}_{rest} \end{bmatrix}^T,$$

where $\bar{X}_0$ is the vector with dimension $q_0 = 5$ given in Eq. (2.4), and $\bar{X}_\xi$ is a vector with dimension $q_\xi = 5$ comprising the refractive indices of air and the various elements in the system, i.e.,

$$\bar{X}_\xi = [\xi_{air} \quad \xi_{e1} \quad \xi_{e2} \quad \xi_{e4} \quad \xi_{e5}]^T.$$

In addition, $\bar{X}_R$, with dimension $q_R = 8$, comprises the radii R_i of all the spherical boundary surfaces in the system, i.e.,

$$\bar{X}_R = \begin{bmatrix} R_1 & R_2 & R_3 & R_4 & R_7 & R_8 & R_9 & R_{10} \end{bmatrix}^T.$$

Finally, $\bar{X}_{rest} = \bar{X}_{sys} - \bar{X}_0 - \bar{X}_\xi - \bar{X}_R$, with dimension $q_{rest} = 11$ (where $q_{rest} = q_{sys} - q_0 - q_\xi - q_R$), is a vector containing all of the remaining system variables, i.e.,

$$\bar{X}_{rest} = \begin{bmatrix} v_1 & q_{e1} & v_2 & q_{e2} & v_3 & q_{e3} & v_4 & q_{e4} & v_5 & q_{e5} & v_6 \end{bmatrix}^T.$$

Note that the system variable vector $\bar{X}_{sys}$ is always portioned into four groups in this book for easy computation of the first- and second-order derivatives presented later in Parts Two and Three.

Example 3.7 The cat's eye shown in Fig. 3.12 contains two elements, j = 1 and j = 2, glued together at the interface. As shown in Fig. 3.12c, element 1 is encountered twice by each ray traveling through the system. In sequential geometrical optics, the elements (or boundary surfaces) of an optical system are numbered sequentially in the order in which they are encountered by the ray. Consequently, element 1 in Fig. 3.12a is labeled not only as element j = 1, but also as element j = 3 (see Fig. 3.12c and the optical train in Table 3.5). The pose matrices of element coordinate frames $(xyz)_{e1}$, $(xyz)_{e2}$ and $(xyz)_{e3}$ relative to the world coordinate frame $(xyz)_0$ are given as follows:

$$^0\bar{A}_{e1} = {}^0\bar{A}_{e3} = \bar{I}_{4\times 4},$$
$$^0\bar{A}_{e2} = \mathrm{tran}(t_{e2x}, v_2, t_{e2z}).$$

Note that parameter v_2 in $^0\bar{A}_{e2}$ is a non-negative parameter used to take account of the effect of the separation distance between elements j = 1 and j = 2 caused by the thickness of the glue. Note also that variables t_{e2x} and t_{e2z} account for concentricity errors between the two hemispheres of the cat's eye system.

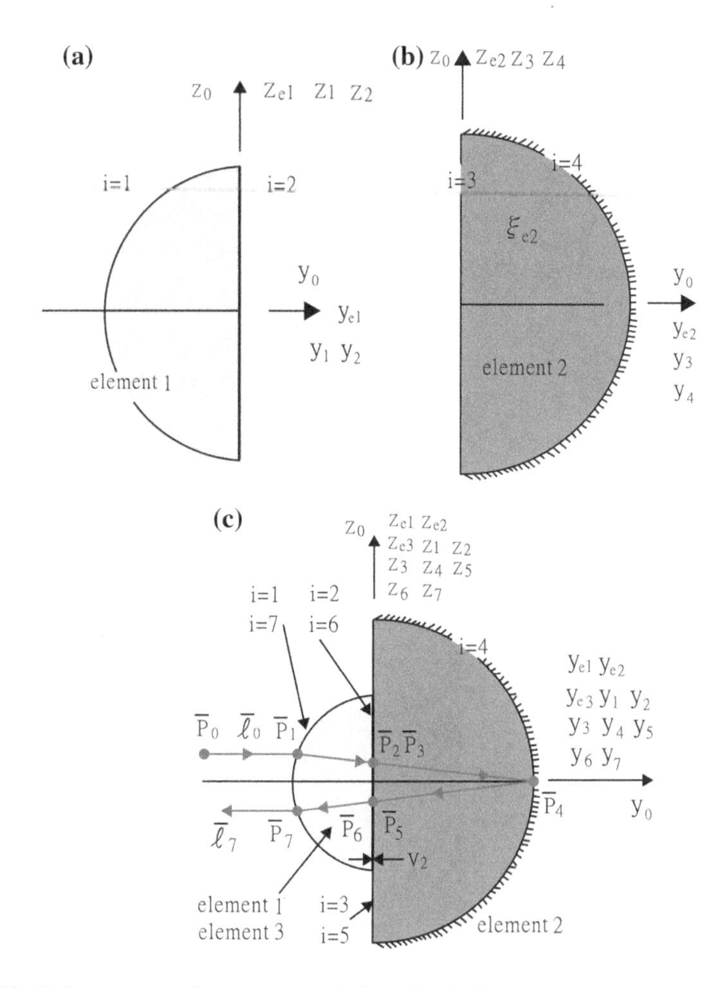

Fig. 3.12 Cat's eye retro-reflector composed of two hemispheres

Table 3.5 Optical train of system shown in Fig. 3.12

Optical train / Element	Boundary surface i		
j = 1	1	2	
j = 2	3	4	5
j = 3	6	7	

Assuming that manufacturing errors are ignored, the pose matrices required to describe the pose of each boundary coordinate frame $(xyz)_i$ ($i = 1$ to $i = 7$) relative to the corresponding element coordinate frame $(xyz)_{ej}$ (refer to Table 3.5) can be expressed as follows:

$$^{e1}\bar{A}_1 = {}^{e1}\bar{A}_2 = {}^{e2}\bar{A}_3 = {}^{e2}\bar{A}_4 = {}^{e2}\bar{A}_5 = {}^{e3}\bar{A}_6 = {}^{e3}\bar{A}_7 = \bar{I}_{4\times 4}.$$

As stated before, the radius R_i of a spherical boundary surface has a positive or negative value depending on whether the surface is convex or concave, respectively. The first boundary surface in Fig. 3.12c is convex for incoming ray $\bar{R}_0$, but concave for ray $\bar{R}_6$. Therefore, R_1 is positive, while $R_7 = -R_1$ is negative. Tracing the path of a ray through the cat's eye system shown in Fig. 3.12c requires prior knowledge of pose matrices $^0\bar{A}_i$ (i = 1 to i = 7). In accordance with the preceding discussions, these matrices can be obtained via the matrix concatenations $^0\bar{A}_i = {}^0\bar{A}_{ej}{}^{ej}\bar{A}_i$. Having collected the variables of interest, the following system variable vector is obtained comprising $m_{sys} = 14$ independent variables:

$$\begin{aligned}\bar{X}_{sys} &= [x_v] = \begin{bmatrix}\bar{X}_0 & \bar{X}_\xi & \bar{X}_R & \bar{X}_{rest}\end{bmatrix}^T \\ &= \begin{bmatrix}P_{0x} & P_{0y} & P_{0z} & \alpha_0 & \beta_0 & \xi_{air} & \xi_{e1} & \xi_{glue} & \xi_{e2} & R_1 & R_4 & t_{e2x} & v_2 & t_{e2z}\end{bmatrix}^T,\end{aligned}$$

where $\bar{X}_0$ is given by Eq. (2.4),

$$\bar{X}_\xi = \begin{bmatrix}\xi_{air} & \xi_{e1} & \xi_{glue} & \xi_{e2}\end{bmatrix}^T,$$
$$\bar{X}_R = \begin{bmatrix}R_1 & R_4\end{bmatrix}^T,$$

and

$$\bar{X}_{rest} = \begin{bmatrix}t_{e2x} & v_2 & t_{e2z}\end{bmatrix}^T.$$

To design and analyze the cat's eye system, the six pose variables (i.e., $t_{ix}, t_{iy}, t_{iz}, \omega_{iz}, \omega_{iy}$ and ω_{ix}, i = 1 to i = 7) of each boundary surface $\bar{r}_i$ must be known. For reasons of simplicity, assume that the system is free of manufacturing errors, i.e., the system is a perfect axis-symmetrical system. The values of the six boundary pose variables for each of the seven boundary surfaces can thus be determined from Eqs. (1.38) to (1.43), and are listed in Table 3.6 given parameter settings of $R_1 = 30$, $R_4 = -60$, $v_2 = 0$, $\xi_{air} = 1$, $\xi_{glue} = 1.2$, $\xi_{e1} = 1.6$ and $\xi_{e2} = 1.7$. It is noted that ω_{iz}, ω_{iy} and ω_{ix} are all equal to zero since the system is an axis-symmetrical system.

Table 3.6 Values of boundary pose variables for cat's eye system shown in Fig. 3.12

i \ Parameter	t_{ix}	t_{iy}	t_{iz}	ω_{ix}	ω_{iy}	ω_{iz}
1, 7	0	0	0	0	0	0
4	0	0	0	0	0	0

i \ Parameter	J_{ix}	J_{iy}	J_{iz}	e_i
2, 6	0	1	0	0
3, 5	0	1	0	0

Note that boundary surfaces $\bar{r}_2$, $\bar{r}_3$, $\bar{r}_5$ and $\bar{r}_6$ have only four pose variables since they are flat boundary surfaces

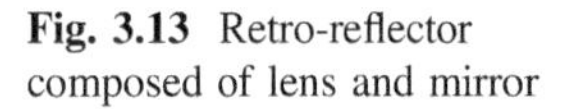

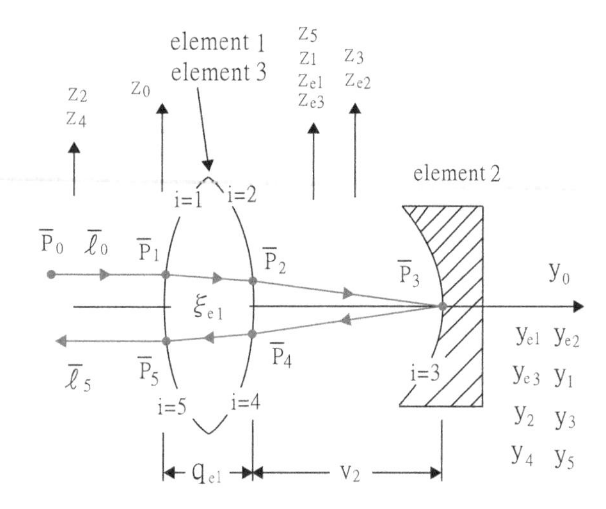

Fig. 3.13 Retro-reflector composed of lens and mirror

Table 3.7 Optical train of system shown in Fig. 3.13

Optical train / Element	Boundary surface i	
j = 1	1	2
j = 2	3	
j = 3	4	5

Example 3.8 Consider the retro-reflector system shown in Fig. 3.13 comprising a bi-convex lens and a concave mirror. As shown, the concave mirror is denoted as element j = 2, while the bi-convex lens is denoted as both j = 1 and j = 3 since it is encountered twice by the rays as they travel through the system. The pose matrices of the element coordinate frames $(xyz)_{ej}$ (j = 1 to j = 3) relative to the world coordinate frame $(xyz)_0$ are given as follows:

$$
{}^{0}\bar{A}_{e1} = {}^{0}\bar{A}_{e3} = \text{tran}(0,\ R_1,\ 0),
$$
$$
{}^{0}\bar{A}_{e2} = \text{tran}(0,\ q_{e1}\ +\ v_2\ +\ R_3,\ 0).
$$

The optical train for the retro-reflector system is shown in Table 3.7. The pose matrices required to describe the pose of each boundary coordinate frame $(xyz)_i$ (i = 1 to i = 5) relative to the corresponding element coordinate frame $(xyz)_{ej}$ (j = 1 to j = 3) have the following forms:

$$
{}^{e1}\bar{A}_1 = {}^{e3}\bar{A}_5 = \bar{I}_{4\times4},
$$
$$
{}^{e2}\bar{A}_3 = \bar{I}_{4\times4},
$$
$$
{}^{e1}\bar{A}_2 = {}^{e3}\bar{A}_4 = \text{trans}(0, -R_1 + q_{e1} + R_2, 0)
$$

Table 3.8 Values of boundary pose variables of retro-reflector system shown in Fig. 3.13

Parameter i	t_{ix}	t_{iy}	t_{iz}	ω_{ix}	ω_{iy}	ω_{iz}
1, 5	0	182.0000	0	0	0	0
2, 4	0	−175.5000	0	0	0	0
3	0	6.5000	0	0	0	0

It is noted in Fig. 3.13 that the first boundary surface of the bi-convex lens is convex for incoming ray $\bar{R}_0$, but concave for ray $\bar{R}_4$. Therefore, R_1 has a positive value, while R_5 has a negative value (i.e., $R_5 = -R_1$). Similarly, R_2 is negative and R_4 is positive (i.e., $R_4 = -R_2$). The required pose matrices ${}^0\bar{A}_i$ (i = 1 to i = 5) can be obtained via the matrix concatenations ${}^0A_i = {}^0A_{e1}{}^{e1}A_i$. The system variable vector $\bar{X}_{sys}$ comprises $q_{sys} = 12$ independent variables, i.e.,

$$\begin{aligned}\bar{X}_{sys} &= [x_v] = \begin{bmatrix}\bar{X}_0 & \bar{X}_\xi & \bar{X}_R & \bar{X}_{rest}\end{bmatrix}^T \\ &= \begin{bmatrix}P_{0x} & P_{0y} & P_{0z} & \alpha_0 & \beta_0 & \xi_{air} & \xi_{e1} & R_1 & R_2 & R_3 & q_{e1} & v_2\end{bmatrix}^T,\end{aligned}$$

where $\bar{X}_0$ is given by Eq. (2.4),

$$\begin{aligned}\bar{X}_\xi &= \begin{bmatrix}\xi_{air} & \xi_{e1}\end{bmatrix}^T, \\ \bar{X}_R &= \begin{bmatrix}R_1 & R_2 & R_3\end{bmatrix}^T,\end{aligned}$$

and

$$\bar{X}_{rest} = \begin{bmatrix}q_{e1} & v_2\end{bmatrix}^T.$$

The values of the six boundary pose variables for each of the five boundary surfaces can be computed from Eqs. (1.38) to (1.43), and are listed in Table 3.8 given parameter settings of $R_1 = 182$, $R_2 = -182$, $R_3 = -200$, $\xi_{air} = 1$, $\xi_{e1} = 1.5$, $q_{e1} = 6.5$ and $v_2 = 200$.

Example 3.9 The variable vector of the system shown in Fig. 2.10 has the form

$$\bar{X}_{sys} = \begin{bmatrix}P_{0x} & P_{0y} & P_{0z} & \alpha_0 & \beta_0 & \xi_{air} & \xi_{e1} & R_1 & R_2 & v_1 & q_{e1}\end{bmatrix}^T.$$

Example 3.10 The variable vector of the system shown in Fig. 2.14 is given by

$$\bar{X}_{sys} = \begin{bmatrix}P_{0x} & P_{0y} & P_{0z} & \alpha_0 & \beta_0 & \xi_{air} & \xi_{e1} & v_1 & q_{e1}\end{bmatrix}^T.$$

3.2 Non-axially Symmetrical Systems

Prisms are commonly used in non-axially symmetrical optical systems due to the ability they provide to output an image with a certain orientation or to relocate the emergent rays in a given manner. To describe the pose of each element of a non-axially symmetrical optical system in 3-D space, it is necessary to establish an element coordinate frame $(xyz)_{ej}$ embedded at a convenient position and orientation in that element. As for axially-symmetrical systems, the following equation can be used to define the pose matrix ${}^{0}\bar{A}_{ej}$ of $(xyz)_{ej}$ with respect to $(xyz)_0$:

$$
\begin{aligned}
{}^{0}\bar{A}_{ej} &= \text{tran}(t_{ejx}, t_{ejy}, t_{ejz})\text{rot}(\bar{z}, \omega_{ejz})\text{rot}(\bar{y}, \omega_{ejy})\text{rot}(\bar{x}, \omega_{ejx}) \\
&= \begin{bmatrix} C\omega_{ejz}C\omega_{ejy} & C\omega_{ejz}S\omega_{ejy}S\omega_{ejx} - S\omega_{ejz}C\omega_{ejx} & C\omega_{ejz}S\omega_{ejy}C\omega_{ejx} + S\omega_{ejz}S\omega_{ejx} & t_{ejx} \\ S\omega_{ejz}C\omega_{ejy} & S\omega_{ejz}S\omega_{ejy}S\omega_{ejx} + C\omega_{ejz}C\omega_{ejx} & S\omega_{ejz}S\omega_{ejy}C\omega_{ejx} - C\omega_{ejz}S\omega_{ejx} & t_{ejy} \\ -S\omega_{ejy} & C\omega_{ejy}S\omega_{ejx} & C\omega_{ejy}C\omega_{ejx} & t_{ejz} \\ 0 & 0 & 0 & 1 \end{bmatrix} \\
&= \begin{bmatrix} I_{ejx} & J_{ejx} & K_{ejx} & t_{ejx} \\ I_{ejy} & J_{ejy} & K_{ejy} & t_{ejy} \\ I_{ejz} & J_{ejz} & K_{ejz} & t_{ejz} \\ 0 & 0 & 0 & 1 \end{bmatrix}.
\end{aligned}
\tag{3.25}
$$

The methodology described in Sect. 1.5 can then be employed to determine the pose matrix ${}^{ej}\bar{A}_i$ ($j = 1$ to $j = k$, $i = m_{ej} - L_{ej} + 1$ to $i = m_{ej}$) of each boundary surface with respect to the corresponding element (see Example 3.12). Finally, Eq. (3.2) (i.e., ${}^{0}\bar{A}_i = {}^{0}\bar{A}_{ej}\,{}^{ej}\bar{A}_i$) can be used to obtain the pose matrices required for successive raytracing at each boundary surface. In general, ${}^{ej}\bar{A}_i$ is more complicated than that for an axis-symmetrical system, and can be given as:

$$
\begin{aligned}
{}^{ej}\bar{A}_i &= \text{tran}({}^{ej}t_{ix}, {}^{ej}t_{iy}, {}^{ej}t_{iz})\text{rot}(\bar{z}, {}^{ej}\omega_{iz})\text{rot}(\bar{y}, {}^{ej}\omega_{iy})\text{rot}(\bar{x}, {}^{ej}\omega_{ix}) \\
&= \begin{bmatrix} {}^{ej}I_{ix} & {}^{ej}J_{ix} & {}^{ej}K_{ix} & {}^{ej}t_{ix} \\ {}^{ej}I_{iy} & {}^{ej}J_{iy} & {}^{ej}K_{iy} & {}^{ej}t_{iy} \\ {}^{ej}I_{iz} & {}^{ej}J_{iz} & {}^{ej}K_{iz} & {}^{ej}t_{iz} \\ 0 & 0 & 0 & 1 \end{bmatrix}.
\end{aligned}
\tag{3.26}
$$

For illustration purposes, consider the non-axially symmetrical optical system shown in Fig. 3.14 containing $k = 4$ optical elements and $n = 13$ boundary surfaces. Table 3.9 lists the labels "i" of the first (i.e., $i = m_{ej} - L_{ej} + 1$) and last (i.e., $i = m_{ej}$) boundary surfaces of each element in the system. Furthermore, Table 3.10 shows the optical train of the system.

Example 3.11 Referring to Figs. 3.14, 3.15, 3.16 and 3.17, the following pose matrices ${}^{0}\bar{A}_{ej}$ ($j = 1$ to $j = 4$) are obtained to describe the position and orientation of each element coordinate frame $(xyz)_{ej}$ with respect to the world coordinate frame $(xyz)_0$:

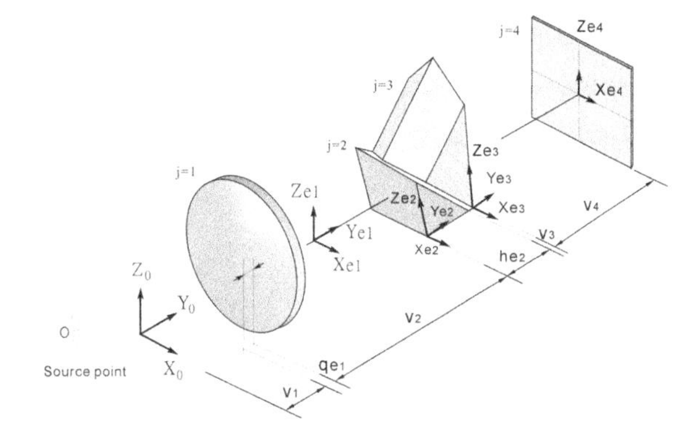

Fig. 3.14 Illustrative non-axially symmetrical optical system consisting of double-convex lens, two prisms and an image plane

Table 3.9 Each element j of system shown in Fig. 3.14 contains L_{ej} boundary surfaces labeled from $i = m_{ej} - L_{ej} + 1$ to $i = m_{ej}$

Element number	j = 1	j = 2	j = 3	j = k = 4
L_{ej} (number of boundary surfaces in element j)	2	4	6	1
The first boundary surface of element j is $i = m_{ej} - L_{ej} + 1$	1	3	7	13
The last boundary surface of element j is $i = m_{ej}$	2	6	12	13

Table 3.10 Optical train of system shown in Fig. 3.14

Optical train / Element	Boundary surface i					
j = 1	1	2				
j = 2	3	4	5	6		
j = 3	7	8	9	10	11	12
j = 4	13					

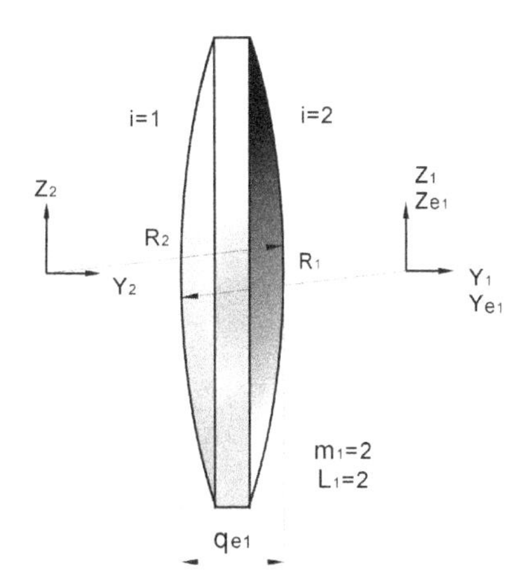

Fig. 3.15 Element j = 1 of system shown in Fig. 3.14 is a double-convex lens

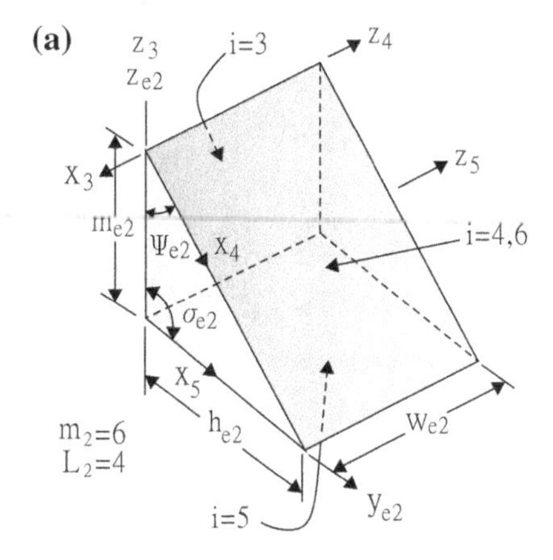

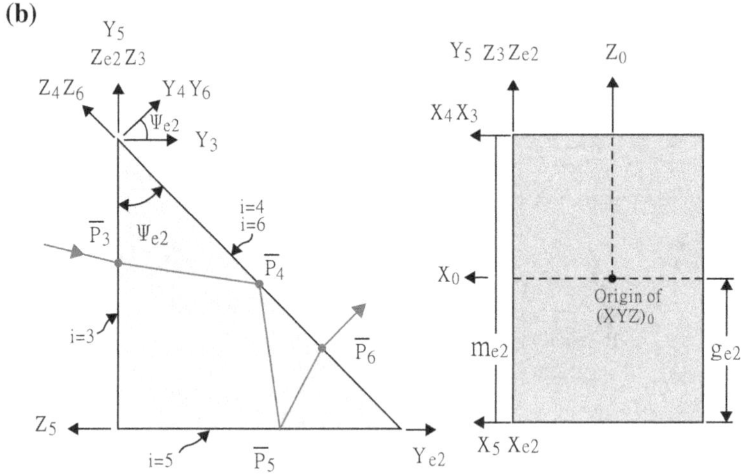

Fig. 3.16 Element j = 2 of system shown in Fig. 3.14 is a right-angle prism if $\sigma_{e2} = 90°$. **a** perspective view of element 2, **b** front and side views of element 2

$$
\begin{aligned}
{}^{0}\bar{A}_{e1} &= \mathrm{tran}(t_{e1x}, v_1 + R_1, t_{e1z})\mathrm{rot}(\bar{z}, \omega_{e1z})\mathrm{rot}(\bar{y}, \omega_{e1y})\mathrm{rot}(\bar{x}, \omega_{e1x}) \\
{}^{0}\bar{A}_{e2} &= \mathrm{tran}(w_{e2}/2, v_1 + q_{e1} + v_2, -g_{e2})\mathrm{rot}(\bar{z}, \omega_{e2z})\mathrm{rot}(\bar{y}, \omega_{e2y})\mathrm{rot}(\bar{x}, \omega_{e2x}) \\
{}^{0}\bar{A}_{e3} &= \mathrm{tran}(w_{e3}/2, v_1 + q_{e1} + v_2 + h_{e2} + v_3, -g_{e3})\mathrm{rot}(\bar{z}, \omega_{e3z})\mathrm{rot}(\bar{y}, \omega_{e3y})\mathrm{rot}(\bar{x}, \omega_{e3x}) \\
{}^{0}\bar{A}_{e4} &= \mathrm{tran}(t_{e4x}, v_1 + q_{e1} + v_2 + h_{e2} + v_3 + v_4, t_{e4z})\mathrm{rot}(\bar{z}, \omega_{e4z})\mathrm{rot}(\bar{y}, \omega_{e4y})\mathrm{rot}(\bar{x}, \omega_{e4x})
\end{aligned}
$$

It is noted that in this particular example, six pose parameters (i.e., $t_{ejx}, t_{ejy}, t_{ejz}, \omega_{ejx}, \omega_{ejy}, \omega_{ejz}$) are considered in each ${}^{0}\bar{A}_{ej}$ to allow an element being posed at any arbitrary position and orientation. Of course, the optical engineer can choose the pose parameters based on his/her problem.

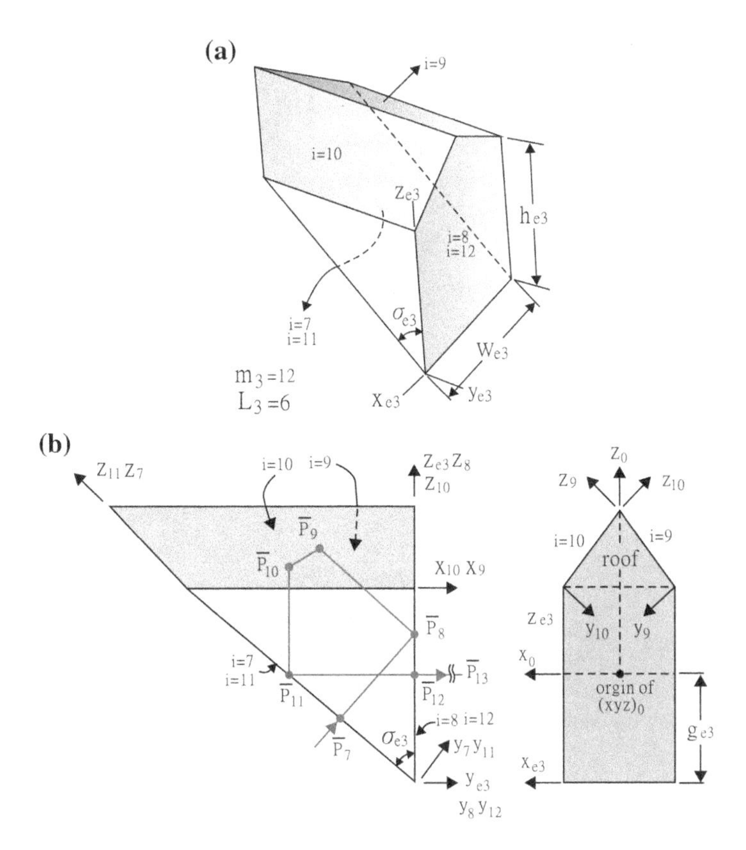

Fig. 3.17 Element j = 3 of system shown in Fig. 3.14 is a prism with a roof. **a** Perspective view of element 3, **b** front and side views of element 3

Example 3.12 Referring to Figs. 3.15, 3.16, 3.17 and 3.18, the following pose matrices ${}^{e1}\bar{A}_i$ (i = 1 to i = 2), ${}^{e2}\bar{A}_i$ (i = 3 to i = 6), ${}^{e3}\bar{A}_i$ (i = 7 to i = 12) and ${}^{e4}\bar{A}_{13}$ are obtained to describe the pose of boundary coordinate frame $(xyz)_i$ with respect to the corresponding element coordinate frame $(xyz)_{ej}$:

$$
\begin{aligned}
{}^{e1}\bar{A}_1 &= \bar{I}_{4\times4}, \\
{}^{e1}\bar{A}_2 &= \text{tran}(0, -R_1 + q_{e1} + R_2, 0), \\
{}^{e2}\bar{A}_3 &= \text{tran}(0, 0, f_{e2}), \\
{}^{e2}\bar{A}_4 &= \text{tran}(0, 0, f_{e2})\text{rot}(\bar{x}, \psi_{e2}), \\
{}^{e2}\bar{A}_5 &= \text{rot}(\bar{x}, -\sigma_{e2} + 180^\circ), \\
{}^{e2}\bar{A}_6 &= {}^{e2}\bar{A}_4,
\end{aligned}
$$

Fig. 3.18 Element j = 4 of system shown in Fig. 3.14 is an image plane

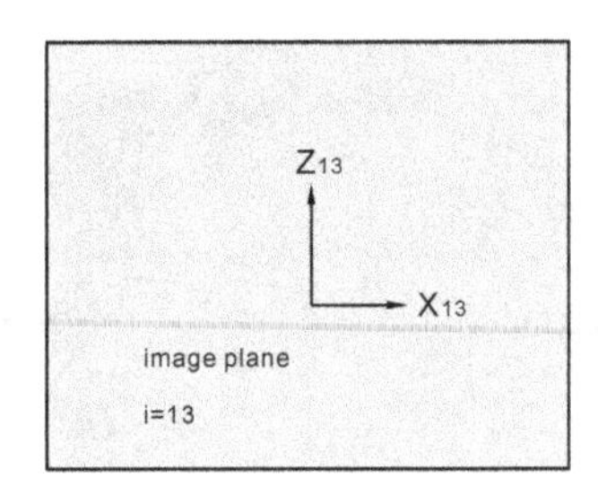

$$
\begin{aligned}
{}^{e3}\bar{A}_7 &= \mathrm{rot}(\bar{x}, \sigma_{e3}),\\
{}^{e3}\bar{A}_8 &= \bar{I}_{4\times4},\\
{}^{e3}\bar{A}_9 &= \mathrm{tran}(-w_{e3},\ 0,\ h_{e3})\mathrm{rot}(\bar{z}, -90^\circ)\mathrm{rot}(\bar{x}, -45^\circ),\\
{}^{e3}\bar{A}_{10} &= \mathrm{tran}(0,\ 0,\ h_{e3})\mathrm{rot}(\bar{z}, 90^\circ)\mathrm{rot}(\bar{x}, -45^\circ),\\
{}^{e3}\bar{A}_{11} &= {}^{e3}\bar{A}_7,\\
{}^{e3}\bar{A}_{12} &= {}^{e3}\bar{A}_8,\\
{}^{e4}\bar{A}_{13} &= \bar{I}_{4\times4}.
\end{aligned}
$$

Example 3.13 As shown in Eqs. (2.9) and (2.35), pose matrices ${}^0\bar{A}_i$ (i = 1 to i = n = 13) are required to perform raytracing. These matrices can be obtained from the following matrix concatenations:

$$
\begin{aligned}
{}^0\bar{A}_1 &= {}^0\bar{A}_{e1}{}^{e1}\bar{A}_1\\
&= \mathrm{tran}(t_{e1x}, v_1 + R_1, t_{e1z})\mathrm{rot}(\bar{z}, \omega_{e1z})\mathrm{rot}(\bar{y}, \omega_{e1y})\mathrm{rot}(\bar{x}, \omega_{e1x}),\\
{}^0\bar{A}_2 &= {}^0\bar{A}_{e1}{}^{e1}\bar{A}_2\\
&= \mathrm{tran}(t_{e1x}, v_1 + R_1, t_{e1z})\mathrm{rot}(\bar{z}, \omega_{e1z})\mathrm{rot}(\bar{y}, \omega_{e1y})\mathrm{rot}(\bar{x}, \omega_{e1x})\mathrm{tran}(0, -R_1 + q_{e1} + R_2, 0),\\
{}^0\bar{A}_3 &= {}^0\bar{A}_{e2}{}^{e2}\bar{A}_3\\
&= \mathrm{tran}(w_{e2}/2, v_1 + q_{e1} + v_2, -g_{e2})\mathrm{rot}(\bar{z}, \omega_{e2z})\mathrm{rot}(\bar{y}, \omega_{e2y})\mathrm{rot}(\bar{x}, \omega_{e2x})\mathrm{tran}(0, 0, f_{e2}),
\end{aligned}
$$

$$
\begin{aligned}
{}^0\bar{A}_4 &= {}^0\bar{A}_{e2}{}^{e2}\bar{A}_4\\
&= \mathrm{tran}(w_{e2}/2, v_1 + q_{e1} + v_2, -g_{e2})\mathrm{rot}(\bar{z}, \omega_{e2z})\mathrm{rot}(\bar{y}, \omega_{e2y})\mathrm{rot}(\bar{x}, \omega_{e2x})\mathrm{tran}(0, 0, f_{e2})\mathrm{rot}(\bar{x}, \psi_{e2}),\\
{}^0\bar{A}_5 &= {}^0\bar{A}_{e2}{}^{e2}\bar{A}_5\\
&= \mathrm{tran}(w_{e2}/2, v_1 + q_{e1} + v_2, -g_{e2})\mathrm{rot}(\bar{z}, \omega_{e2z})\mathrm{rot}(\bar{y}, \omega_{e2y})\mathrm{rot}(\bar{x}, \omega_{e2x})\mathrm{rot}(\bar{x}, -\sigma_{e2} + 180^\circ),
\end{aligned}
$$

$$
\begin{aligned}
{}^0\bar{A}_6 &= {}^0\bar{A}_4\\
{}^0\bar{A}_7 &= {}^0\bar{A}_{e3}{}^{e3}\bar{A}_7\\
&= \mathrm{tran}(w_{e3}/2, v_1 + q_{e1} + v_2 + h_{e2} + v_3, -g_{e3})\mathrm{rot}(\bar{z}, \omega_{e3z})\mathrm{rot}(\bar{y}, \omega_{e3y})\mathrm{rot}(\bar{x}, \omega_{e3x})\mathrm{rot}(\bar{x}, \sigma_{e3}),
\end{aligned}
$$

$$
\begin{aligned}
{}^{0}\bar{A}_{8} &= {}^{0}\bar{A}_{e3}{}^{e3}\bar{A}_{8} \\
&= \mathrm{tran}(w_{e3}/2, v_1 + q_{e1} + v_2 + h_{e2} + v_3, -g_{e3})\mathrm{rot}(\bar{z}, \omega_{e3z})\mathrm{rot}(\bar{y}, \omega_{e3y})\mathrm{rot}(\bar{x}, \omega_{e3x}),
\end{aligned}
$$

$$
\begin{aligned}
{}^{0}\bar{A}_{9} &= {}^{0}\bar{A}_{e3}{}^{e3}\bar{A}_{9} \\
&= \mathrm{tran}(w_{e3}/2, v_1 + q_{e1} + v_2 + h_{e2} + v_3, -g_{e3})\mathrm{rot}(\bar{z}, \omega_{e3z})\mathrm{rot}(\bar{y}, \omega_{e3y})\mathrm{rot}(\bar{x}, \omega_{e3x}) \\
&\quad \mathrm{tran}(-w_{e3}, 0, h_{e3})\mathrm{rot}(\bar{z}, -90^\circ)\mathrm{rot}(\bar{x}, -45^\circ),
\end{aligned}
$$

$$
\begin{aligned}
{}^{0}\bar{A}_{10} &= {}^{0}\bar{A}_{e3}{}^{e3}\bar{A}_{10} \\
&= \mathrm{tran}(w_{e3}/2, v_1 + q_{e1} + v_2 + h_{e2} + v_3, -g_{e3})\mathrm{rot}(\bar{z}, \omega_{e3z})\mathrm{rot}(\bar{y}, \omega_{e3y})\mathrm{rot}(\bar{x}, \omega_{e3x}) \\
&\quad \mathrm{tran}(0, 0, h_{e3})\mathrm{rot}(\bar{z}, 90^\circ)\mathrm{rot}(\bar{x}, -45^\circ),
\end{aligned}
$$

$$
\begin{aligned}
{}^{0}\bar{A}_{11} &= {}^{0}\bar{A}_{7}, \\
{}^{0}\bar{A}_{12} &= {}^{0}\bar{A}_{8},
\end{aligned}
$$

$$
\begin{aligned}
{}^{0}\bar{A}_{13} &= {}^{0}\bar{A}_{e4}{}^{e4}\bar{A}_{13} \\
&= \mathrm{tran}(t_{e4x}, v_1 + q_{e1} + v_2 + h_{e2} + v_3 + v_4, t_{e4z})\mathrm{rot}(\bar{z}, \omega_{e4z})\mathrm{rot}(\bar{y}, \omega_{e4y})\mathrm{rot}(\bar{x}, \omega_{e4x}).
\end{aligned}
$$

Example 3.14 Referring to Fig. 3.15 and Example 3.11, the variable vector $\bar{X}_{e1}$ of the first element in Fig. 3.14 includes six pose variables (i.e., t_{e1x}, $t_{e1y} = v_1 + R_1$, t_{e1z}, ω_{e1x}, ω_{e1y} and ω_{e1z}), the refractive index ξ_{air} of air, the refractive index ξ_{e1} of the element, the two radii, R_1 and R_2, of the spherical boundary surfaces, and the element thickness q_{e1}, i.e.,

$$
\bar{X}_{e1} = \begin{bmatrix} t_{e1x} & v_1 + R_1 & t_{e1z} & \omega_{e1x} & \omega_{e1y} & \omega_{e1z} & \xi_{e1} & \xi_{air} & R_1 & q_{e1} & R_2 \end{bmatrix}^T.
$$

Referring to Fig. 3.16 and Example 3.11, the variable vector $\bar{X}_{e2}$ of the second element in Fig. 3.14 contains six pose variables, the refractive index ξ_{air} of air, the refractive index ξ_{e2} of the element, and the characteristic dimensions h_{e2}, w_{e2}, g_{e2}, f_{e2}, ψ_{e2} and σ_{e2} of the element, i.e.,

$$
\begin{aligned}
\bar{X}_{e2} = \big[& w_{e2}/2 \quad v_1 + q_{e1} + v_2 \quad -g_{e2} \quad \omega_{e2x} \quad \omega_{e2y} \quad \omega_{e2z} \quad \xi_{e2} \quad \xi_{glue} \quad \xi_{air} \quad h_{e2} \quad w_{e2} \\
& g_{e2} \quad f_{e2} \quad \psi_{e2} \quad \sigma_{e2} \big]^T.
\end{aligned}
$$

Referring to Fig. 3.17 and Example 3.11, the variable vector $\bar{X}_{e3}$ of the third element in Fig. 3.14 comprises six pose variables, the gap variable $v_3 = \sqrt{2}\,gap_{glue}$, the refractive index ξ_{air} of air, the refractive index ξ_{glue} of glue, the refractive index ξ_{e3} of the element, and the characteristic dimensions h_{e3}, w_{e3}, g_{e3} and σ_{e3} of the element, i.e.,

$$\bar{X}_{e3} = \left[w_{e3}/2 \quad v_1 + q_{e1} + v_2 + h_{e2} + \sqrt{2}\, gap_{glue} \quad -g_{e3} \quad \psi_{e2} - \sigma_{e3} \quad \omega_{e3y} \quad \omega_{e3z} \quad \xi_{air} \quad \xi_{e3} \right.$$
$$\left. \xi_{glue} \quad h_{e3} \quad w_{e3} \quad g_{e3} \quad \sigma_{e3} \right]^T.$$

Finally, the variable vector $\bar{X}_{e4}$ of the image plane shown in Fig. 3.18 includes six pose variables and the refractive index ξ_{air} of air, i.e.,

$$\bar{X}_{e4} = \left[t_{e4x} \quad v_1 + q_{e1} + v_2 + h_{e2} + \sqrt{2}\, gap_{glue} + v_4 \quad t_{e4z} \quad \omega_{e4x} \quad \omega_{e4y} \quad \omega_{e4z} \quad \xi_{air} \right]^T.$$

Example 3.15 It is seen that some variables (e.g., ξ_{air}) appear repeatedly in $\bar{X}_{ej}$ ($j = 1$ to $j = 4$) in Example 3.14. Furthermore, to simplify the geometry of the system, the characteristic dimension variables can be set as $w_{e3} = w_{e2}$ and $g_{e3} = g_{e2}$. Having deleted these repeated variables, the following variable vector $\bar{X}_{sys}$ with dimension $q_{sys} = 41$ can be obtained as the system variable vector of the system shown in Fig. 3.14:

$$\bar{X}_{sys} = [x_v] = \left[\bar{X}_0 \quad \bar{X}_\xi \quad \bar{X}_R \quad \bar{X}_{rest} \right]^T,$$

where

$$\bar{X}_0 = \left[P_{0x} \quad P_{0y} \quad P_{0z} \quad \alpha_0 \quad \beta_0 \right]^T \text{ with dimension } q_0 = 5,$$

$$\bar{X}_\xi = \left[\xi_{air} \quad \xi_{e1} \quad \xi_{glue} \quad \xi_{e2} \quad \xi_{e3} \right]^T \text{ with dimension } q_\xi = 5,$$

$$\bar{X}_R = \left[R_1 \quad R_2 \right]^T \text{ with dimension } q_R = 2,$$

$$\bar{X}_{rest} = [t_{e1x} \quad v_1 \quad t_{e1z} \quad \omega_{e1x} \quad \omega_{e1y} \quad \omega_{e1z} \quad q_{e1} \quad v_2 \quad \omega_{e2x} \quad \omega_{e2y} \quad \omega_{e2z} \quad h_{e2} \quad w_{e2} \quad g_{e2} \quad f_{e2}$$
$$\psi_{e2} \quad \sigma_{e2} \quad \omega_{e3y} \quad \omega_{e3z} \quad gap_{glue} \quad h_{e3} \quad \sigma_{e3} \quad t_{e4x} \quad v_4 \quad t_{e4z} \quad \omega_{e4x} \quad \omega_{e4y} \quad \omega_{e4z}]^T$$

with dimension $q_{rest} = 28$.

Example 3.16 The six pose variables ${}^0\bar{A}_i$ ($i = 1$ to $i = 13$) of the elements shown in Fig. 3.14 can be obtained using Eqs. (1.38) to (1.43). Assume that the system parameter values are assigned as $R_1 = 38.2219$, $R_2 = -56.0587$, $q_{e1} = 5$, $h_{e2} = f_{e2} = w_{e2} = w_{e3} = 10$, $g_{e2} = 5$, $\psi_{e2} = 45°$, $\sigma_{e2} = 112.5°$, $gap_{glue} = 1.3$, $h_{e3} = 10.\sqrt{2}$, $\sigma_{e3} = 45°$, $v_1 = 10$, $v_2 = 30$, $v_4 = 50$, $\xi_{air} = \xi_{glue} = 1$, $\xi_{e1} = 1.5$, $\xi_{e2} = \xi_{e3} = 1.3$,

$$t_{e1x} = t_{e1z} = \omega_{e1x} = \omega_{e1y} = \omega_{e1z} = \omega_{e2x} = \omega_{e2y} = \omega_{e2z} = \omega_{e3x} = \omega_{e3y} = \omega_{e3z}$$
$$= t_{e4x} = t_{e4z} = \omega_{e4x} = \omega_{e4y} = \omega_{e4z} = 0.$$

The corresponding results are listed in Table 3.11.

Example 3.17 Consider the corner-cube mirror shown in Fig. 3.19 consisting of three mutually perpendicular flat first-surface mirrors. As shown, the source ray $\bar{R}_0$ is reflected three times as it travels through the system (once by each mirror), with the result that its direction is reversed. In analyzing the corner-cube mirror, the element coordinate frame $(xyz)_{e1}$ is required to define the pose of the corresponding element. Referring to Fig. 3.19, the pose of the corner-cube mirror with respect to the world coordinate frame $(xyz)_0$ is defined as follows:

Table 3.11 Boundary pose variables of system shown in Fig. 3.14

	t_{ix}	t_{iy}	t_{iz}	ω_{ix}	ω_{iy}	ω_{iz}
i = 1	0.00000	48.22190	0.00000	0°	0°	0°
i = 2	0.00000	−41.08570	0.00000	0°	0°	0°
i = 3	5.00000	45.00000	5.00000	0°	0°	0°
i = 4	5.00000	45.00000	5.00000	45°	0°	0°
i = 5	5.00000	45.00000	−5.00000	67.5°	0°	0°
i = 6	5.00000	45.00000	5.00000	45°	0°	0°
i = 7	5.00000	56.83848	−5.00000	45°	0°	0°
i = 8	5.00000	56.83848	−5.00000	0°	0°	0°
i = 9	−5.00000	56.83848	9.14214	−45°	0°	−90°
i = 10	5.00000	56.83848	9.14214	−45°	0°	90°
i = 11	5.00000	56.83848	−5.00000	45°	0°	0°
i = 12	5.00000	56.83848	−5.00000	0°	0°	0°
i = 13	0.00000	106.83848	0	0°	0°	0°

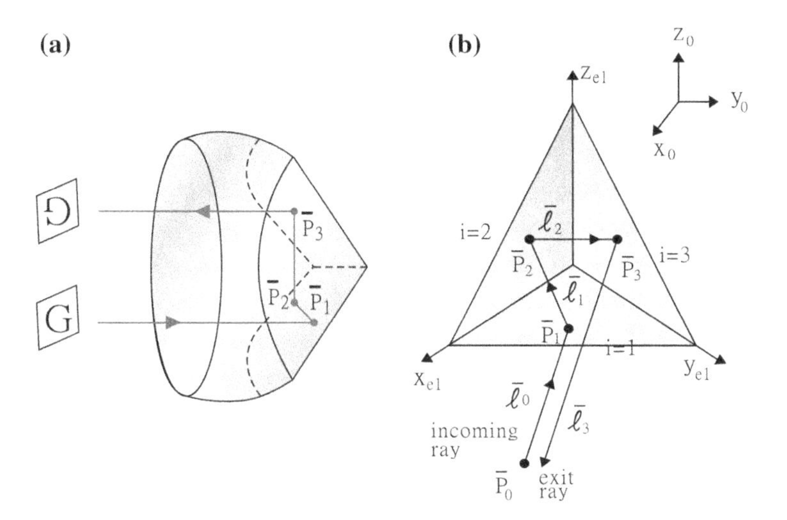

Fig. 3.19 A perfect corner-cube mirror reflects a ray parallel to the incident ray independent of the ray and corner-cube alignment

$$
\begin{aligned}
{}^{0}\bar{A}_{e1} &= \mathrm{tran}(t_{e1x},0,0)\mathrm{tran}(0,t_{e1y},0)\mathrm{tran}(0,0,t_{e1z})\mathrm{rot}(\bar{z},\omega_{e1z})\mathrm{rot}(\bar{y},\omega_{e1y})\mathrm{rot}(\bar{x},\omega_{e1x}) \\
&= \begin{bmatrix} C\omega_{e1z}C\omega_{e1y} & C\omega_{e1z}S\omega_{e1y}S\omega_{e1x}-S\omega_{e1z}C\omega_{e1x} & C\omega_{e1z}S\omega_{e1y}C\omega_{e1x}+S\omega_{e1z}S\omega_{e1x} & t_{e1x} \\ S\omega_{e1z}C\omega_{e1y} & S\omega_{e1z}S\omega_{e1y}S\omega_{e1x}+C\omega_{e1z}C\omega_{e1x} & S\omega_{e1z}S\omega_{e1y}C\omega_{e1x}-C\omega_{e1z}S\omega_{e1x} & t_{e1y} \\ -S\omega_{e1y} & C\omega_{e1y}S\omega_{e1x} & C\omega_{e1y}C\omega_{e1x} & t_{e1z} \\ 0 & 0 & 0 & 1 \end{bmatrix}.
\end{aligned}
$$

When the refractive index ξ_{air} of air is included, the element variable vector $\bar{X}_{e1}$ has the form

$$
\bar{X}_{e1} = \begin{bmatrix} t_{e1x} & t_{e1y} & t_{e1z} & \omega_{e1x} & \omega_{e1y} & \omega_{e1z} & \xi_{air} \end{bmatrix}^T
$$

The corner-cube mirror possesses three flat boundary surfaces (labeled as i = 1, i = 2 and i = 3). Meanwhile, the corresponding boundary variable vectors are given by

$$
\begin{aligned}
\bar{X}_1 &= [J_{1x} \quad J_{1y} \quad J_{1z} \quad e_1 \quad \xi_0 \quad \xi_1]^T, \\
\bar{X}_2 &= [J_2x \quad J_2y \quad J_2z \quad e_2 \quad \xi_1 \quad \xi_2]^T, . \\
\bar{X}_3 &= [J_3x \quad J_3y \quad J_3z \quad e_3 \quad \xi_2 \quad \xi_3]^T
\end{aligned}
$$

Deleting the repeated variables in $\bar{X}_o$, $\bar{X}_{e1}$, $\bar{X}_1$, $\bar{X}_2$ and $\bar{X}_3$ (e.g., $\xi_{air} = \xi_0 = \xi_1 = \xi_2 = \xi_3$), the system vector for the corner-cube mirror system is obtained as

$$
\begin{aligned}
\bar{X}'_{sys} = [&P_{0x} \quad P_{0y} \quad P_{0z} \quad \alpha_0 \quad \beta_0 \quad \xi_{air} \quad t_{e1x} \quad t_{e1y} \quad t_{e1z} \quad \omega_{e1x} \quad \omega_{e1y} \quad \omega_{e1z} \\
&J_{1x} \quad J_{1y} \quad J_{1z} \quad e_1 \quad J_{2x} \quad J_{2y} \quad J_{2z} \quad e_2 \quad J_{3x} \quad J_{3y} \quad J_{3z} \quad e_3]^T.
\end{aligned}
$$

It is noted that $[J_{ix} \quad J_{iy} \quad J_{iz} \quad e_i]^T$ (i = 1 to i = 3) can be computed using the transformation given in Eq. (1.34) with h = 0 and g = e1, i.e., $\bar{r}_i = \left(({}^{0}\bar{A}_{e1})^{-1}\right)^{T} {}^{e1}\bar{r}_i$ (i = 1 to i = 3), where ${}^{e1}\bar{r}_1$, ${}^{e1}\bar{r}_2$, and ${}^{e1}\bar{r}_3$ are given respectively as

$$
\begin{aligned}
{}^{e1}\bar{r}_1 &= \begin{bmatrix} {}^{e1}J_{1x} & {}^{e1}J_{1y} & {}^{e1}J_{1z} & {}^{e1}e_1 \end{bmatrix}^T = [0 \quad 0 \quad 1 \quad 0]^T, \\
{}^{e1}\bar{r}_2 &= \begin{bmatrix} {}^{e1}J_{2x} & {}^{e1}J_{2y} & {}^{e1}J_{2z} & {}^{e1}e_2 \end{bmatrix}^T = [0 \quad 1 \quad 0 \quad 0]^T, \\
{}^{e1}\bar{r}_3 &= \begin{bmatrix} {}^{e1}J_{3x} & {}^{e1}J_{3y} & {}^{e1}J_{3z} & {}^{e1}e_3 \end{bmatrix}^T = [1 \quad 0 \quad 0 \quad 0]^T.
\end{aligned}
$$

In other words, the system variable vector can be rewritten as

$$
\begin{aligned}
\bar{X}''_{sys} = [&P_{0x} \quad P_{0y} \quad P_{0z} \quad \alpha_0 \quad \beta_0 \quad \xi_{air} \quad t_{e1x} \quad t_{e1y} \quad t_{e1z} \quad \omega_{e1x} \quad \omega_{e1y} \quad \omega_{e1z} \\
&{}^{e1}J_{1x} \quad {}^{e1}J_{1y} \quad {}^{e1}J_{1z} \quad {}^{e1}e_1 \quad {}^{e1}J_{2x} \quad {}^{e1}J_{2y} \quad {}^{e1}J_{2z} \quad {}^{e1}e_2 \quad {}^{e1}J_{3x} \quad {}^{e1}J_{3y} \quad {}^{e1}J_{3z} \quad {}^{e1}e_3]^T.
\end{aligned}
$$

Importantly, only two components of the unit normal vector ${}^{e1}\bar{J}_i = \begin{bmatrix} {}^{e1}J_{ix} & {}^{e1}J_{iy} & {}^{e1}J_{iz} & 0 \end{bmatrix}^T$ (i = 1 to i = 3) are independent variables since ${}^{e1}J_{ix}^2 + {}^{e1}J_{iy}^2 + {}^{e1}J_{iz}^2 = 1$. If ${}^{e1}J_{1X}$, ${}^{e1}J_{1y}$, ${}^{e1}J_{2X}$, ${}^{e1}J_{2Z}$, ${}^{e1}J_{3y}$ and ${}^{e1}J_{3Z}$ are taken as independent variables, the system variable vector becomes

$$\bar{X}_{sys} = [P_{0x} \quad P_{0y} \quad P_{0z} \quad \alpha_0 \quad \beta_0 \quad \xi_{air} \quad t_{e1x} \quad t_{e1y} \quad t_{e1z} \quad \omega_{e1x} \quad \omega_{e1y} \quad \omega_{e1z} \\ {}^{e1}J_{1X} \quad {}^{e1}J_{1y} \quad {}^{e1}e_1 \quad {}^{e1}J_{2X} \quad {}^{e1}J_{2Z} \quad {}^{e1}e_2 \quad {}^{e1}J_{3y} \quad {}^{e1}J_{3Z} \quad {}^{e1}e_3]^T.$$

Consider the case where a source ray

$$\bar{R}_0 = \begin{bmatrix} \bar{P}_0 & \bar{\ell}_0 \end{bmatrix}^T = [15 \quad 10 \quad 5 \quad -0.612 \quad -0.354 \quad -0.707]^T$$

with

$$\bar{X}_0 = [15 \quad 10 \quad 5 \quad 210° \quad -45°]^T$$

travels through the corner-cube mirror. The exit ray is given by

$$\bar{R}_3 = [0 \quad -1.34 \quad 12.32 \quad 0.612 \quad 0.354 \quad 0.707]^T.$$

It is noted that the unit directional vector $\bar{\ell}_0$ of the source ray is reversed by the system (i.e., $\bar{\ell}_3 = -\bar{\ell}_0$). In other words, the retro-reflective property of the corner-cube mirror is confirmed numerically.

Example 3.18 Consider the refracting prism shown in Fig. 2.15. If errors in the six pose variables are ignored and the source ray $\bar{R}_0$ is confined to the y_0z_0 plane, the system variable vector is given by

$$\bar{X}_{sys} = \begin{bmatrix} P_{0y} & P_{0Z} & \beta_0 & \xi_{air} & \xi_{e1} & \eta_{e1} \end{bmatrix}^T.$$

3.3 Spot Diagram of Monochromatic Light

Many different image quality evaluation methods are available. Broadly speaking, these methods can be categorized as either geometric image quality metrics or diffractive image quality metrics. Geometric image quality metrics (e.g., spot diagrams, the point spread function (PSF), and the modulation transfer function (MTF)) can be computed using either a ray-counting method or a differential method. The ray-counting method is discussed in the remaining sections of this chapter, while the differential method is addressed in Part Two and Part Three of this book.

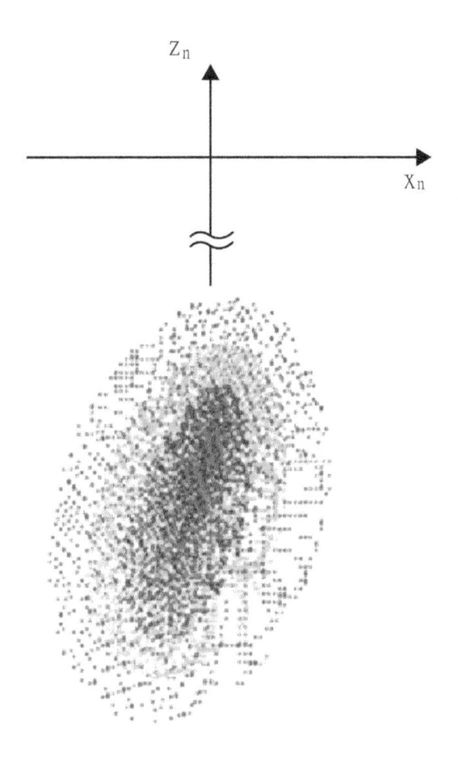

Fig. 3.20 Spot diagram obtained by tracing very large number of skew rays originating from a point source

As shown in Fig. 3.20, a spot diagram indicates with points the ray intersections in the image plane. In other words, a spot diagram provides a geometrical approximation of the aberrated image produced by a point source. Spot diagrams do not give an entirely accurate representation of the images produced by an optical system; particularly for systems with wavefront aberrations much smaller than one wavelength. However, when the system contains several waves of aberrations, spot diagrams nevertheless provide a useful representation of the main characteristics of the image produced by a point source. In general, a spot diagram can be characterized by two main features, namely the spot position and the spot size [2–4]. The spot position is measured by the centroid $\begin{bmatrix} x_{n/\text{centroid}} & y_{n/\text{centroid}} & z_{n/\text{centroid}} & 1 \end{bmatrix}^T$, while the spot size is measured by the root-mean-square (rms) value of the radius from the centroid of the rays at the spot. Prior to calculating the rms value of the radius, it is first necessary to determine the centroid of the imaged spot by averaging the available data obtained from the traces of a finite number (say, G) of rays distributed in the domain of (α_0, β_0) in some regular way (p. 107 of [5]), i.e.,

$$\begin{bmatrix} x_{n/\text{centroid}} & y_{n/\text{centroid}} & z_{n/\text{centroid}} & 1 \end{bmatrix}^T = \begin{bmatrix} \frac{1}{G}\sum_{1}^{G} {}^{n}P_{nx} & 0 & \frac{1}{G}\sum_{1}^{G} {}^{n}P_{nz} & 1 \end{bmatrix}^T, \tag{3.27}$$

where ${}^{n}\bar{P}_{n} = [{}^{n}P_{nx} \quad 0 \quad {}^{n}P_{nz} \quad 1]^{T}$ are the coordinates of the incidence point of a general ray on the image plane computed by ${}^{n}\bar{P}_{n} = {}^{n}\bar{A}_{0}\bar{P}_{n} = ({}^{0}\bar{A}_{n})^{-1}\bar{P}_{n}$. Minimizing the deviations of the spot centroids from their ideal positions at each image spot reduces distortion. Furthermore, minimizing the deviations of the colored spot centroids from one another reduces lateral chromatism.

The rms value of the radius (denoted as rms) of a spot diagram on the image plane can be calculated as

$$\begin{aligned}\mathrm{rms}^2 &= \frac{1}{G}\sum_{1}^{G}\left[({}^{n}P_{nx} - x_{n/centroid})^2 + ({}^{n}P_{nz} - z_{n/centroid})^2\right] \\ &= \frac{1}{G}\sum_{1}^{G}\left[({}^{n}P_{nx})^2 + ({}^{n}P_{nz})^2\right] - \frac{2}{G}\left[x_{n/centroid}\sum_{1}^{G}({}^{n}P_{nx}) + z_{n/centroid}\sum_{1}^{G}({}^{n}P_{nz})\right] + \left(x^2_{n/centroid} + z^2_{n/centroid}\right) \\ &= \frac{1}{G}\sum_{1}^{G}\left[({}^{n}P_{nx})^2 + ({}^{n}P_{nz})^2\right] - \left(x^2_{n/centroid} + z^2_{n/centroid}\right)\end{aligned} \tag{3.28}$$

Minimizing the image spot sizes is beneficial in reducing five main types of aberration, namely spherical, coma, astigmatism, field curvature, and longitudinal color.

Example 3.19 The centroid and rms radius of the image spot diagram for point source $\bar{P}_0 = [0 \quad -507 \quad 170 \quad 1]^T$ in the system shown in Fig. 3.2 are obtained as

$$[x_{11/centroid} \quad y_{11/centroid} \quad z_{11/centroid} \quad 1]^T = [0 \quad 0 \quad -32.59679 \quad 1]^T$$

and 0.11571 mm, respectively, when 27,860 rays are traced. By contrast, the centroid and rms radius are $[0 \quad 0 \quad -32.57903 \quad 1]^T$ and 0.135 mm, respectively, if only 296 rays are traced. In other words, the accuracy of the calculated centroid and rms radius depends fundamentally on the number of rays traced.

3.4 Point Spread Function

The point spread function (PSF) describes the impulse response of an optical system to a point source. Hereafter, the notation $B(x_n, z_n)$ is used to denote the distribution of the PSF, which is a function of the in-plane coordinates (i.e., x_n and z_n) of the image plane. The PSF also represents the distribution of the ray density in a spot diagram. In practical applications, it is important to investigate the PSF since it is the basic "brick" on which an image is "built". However, real point spread functions are rarely (if ever) represented by ordinary analytical functions in the literature (p. 372 of [6]). For example, Smith (p. 372 of [6]) determined the PSF of an optical system by counting the number of rays hitting a system of grids on the image plane (referred to hereafter as the ray-counting method, see Fig. 3.21).

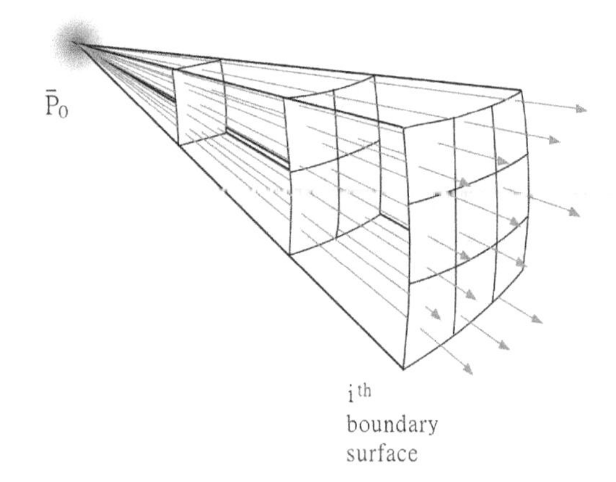

Fig. 3.21 Ray-counting method determines the PSF of an optical system by counting the number of rays hitting a system of grids on the image plane

Table 3.12 Ray density of on-axis point source $\bar{P}_0 = [0 \quad -507 \quad 0 \quad 1]^T$ in system shown in Fig. 3.2 obtained via ray-counting method using 1/30 mm × 1/30 mm grids and 4933 rays

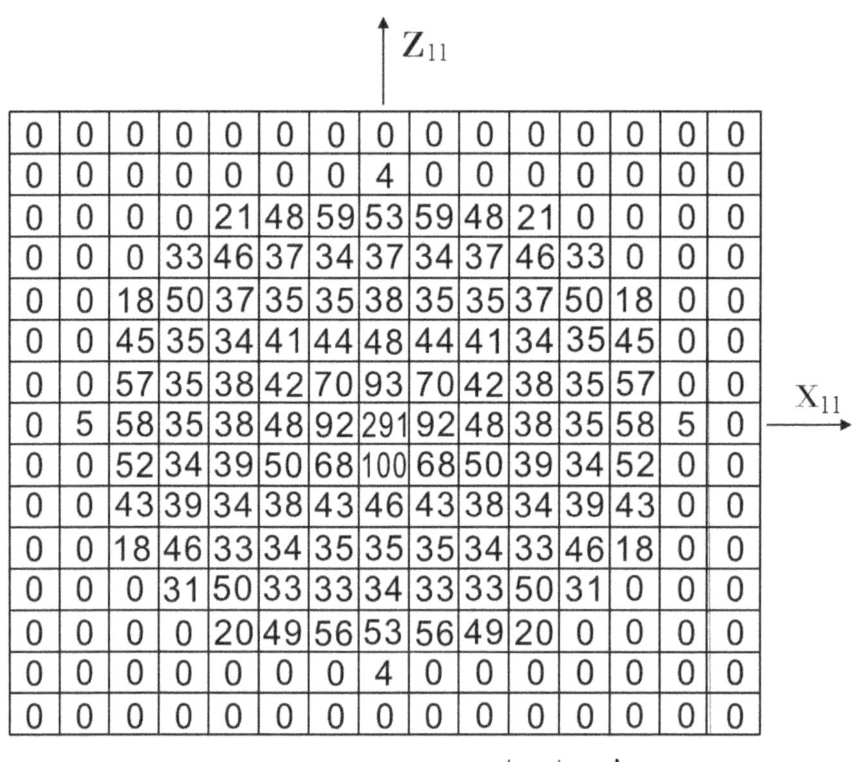

0	0	0	0	0	0	0	0	0	0	0	0	0	0	0
0	0	0	0	0	0	0	4	0	0	0	0	0	0	0
0	0	0	0	21	48	59	53	59	48	21	0	0	0	0
0	0	0	33	46	37	34	37	34	37	46	33	0	0	0
0	0	18	50	37	35	35	38	35	35	37	50	18	0	0
0	0	45	35	34	41	44	48	44	41	34	35	45	0	0
0	0	57	35	38	42	70	93	70	42	38	35	57	0	0
0	5	58	35	38	48	92	291	92	48	38	35	58	5	0
0	0	52	34	39	50	68	100	68	50	39	34	52	0	0
0	0	43	39	34	38	43	46	43	38	34	39	43	0	0
0	0	18	46	33	34	35	35	35	34	33	46	18	0	0
0	0	0	31	50	33	33	34	33	33	50	31	0	0	0
0	0	0	0	20	49	56	53	56	49	20	0	0	0	0
0	0	0	0	0	0	0	4	0	0	0	0	0	0	0
0	0	0	0	0	0	0	0	0	0	0	0	0	0	0

Table 3.13 Normalized PSF computed from Table 3.12

Z_{11} ↑

0	0	0	0	0	0	0	0	0	0	0	0	0	0	0
0	0	0	0	0	0	0	0.01	0	0	0	0	0	0	0
0	0	0	0	0.07	0.16	0.02	0.18	0.2	0.16	0.07	0	0	0	0
0	0	0	0.11	0.16	0.13	0.12	0.13	0.12	0.13	0.16	0	0	0	0
0	0	0.06	0.17	0.13	0.02	0.12	0.13	0.12	0.12	0.13	0.17	0.06	0	0
0	0	0.15	0.12	0.12	0.14	0.15	0.16	0.15	0.14	0.12	0.12	0.15	0	0
0	0	0.20	0.12	0.13	0.14	0.24	0.32	0.24	0.14	0.13	0.12	0.20	0	0
0	0.02	0.20	0.12	0.13	0.16	0.32	1.00	0.32	0.16	0.13	0.12	0.20	0.02	0
0	0	0.18	0.12	0.13	0.17	0.23	0.34	0.23	0.17	0.13	0.12	0.18	0	0
0	0	0.15	0.13	0.12	0.13	0.15	0.16	0.15	0.13	0.12	0.13	0.15	0	0
0	0	0.06	0.16	0.11	0.12	0.12	0.12	0.12	0.12	0.11	0.16	0.06	0	0
0	0	0	0.11	0.17	0.11	0.11	0.12	0.11	0.11	0.17	0.11	0	0	0
0	0	0	0	0.07	0.17	0.19	0.18	0.19	0.17	0.07	0	0	0	0
0	0	0	0	0	0	0	0.01	0	0	0	0	0	0	0
0	0	0	0	0	0	0	0	0	0	0	0	0	0	0

X_{11} →

Example 3.20 Consider again the system shown in Fig. 3.2. Table 3.12 shows the ray density on the image plane as determined using a grid system comprising 1/30 mm × 1/30 mm grids and 4933 traced rays (uniformly distributed over the domain of (α_0, β_0)) originating from on-axis point source $\bar{P}_0 = [0 \quad -507 \quad 0 \quad 1]^T$.

Example 3.21 Table 3.13 and Fig. 3.22 give the variation of the normalized PSF obtained by dividing the number of rays in each grid by 291, i.e., the maximum ray density in Table 3.12. Curve B in Fig. 3.23 shows the normalized PSF of the same point source, but for the case where ray-counting is performed using 1/200 mm × 1/200 mm grids and 122,397 rays. For convenience, Fig. 3.22 is overlaid on Fig. 3.23 for comparison purposes. Note that in this example, the PSF is an axis-symmetrical function since the system is an axis-symmetrical system and the point source lies on the optical axis. Therefore, the two figures, i.e., Figure 3.22 and Fig. 3.23, present only cross-sectional views of the PSF distribution rather than the entire PSF distribution. Comparing the two curves in Fig. 3.23, it is clear that the accuracy of the ray-counting method depends fundamentally on both the number of traced rays and the size of the grids used to mesh the image plane.

Example 3.22 Table 3.14 illustrates the ray density of an off-axis point source $\bar{P}_0 = [0 \quad -507 \quad 170 \quad 1]^T$ obtained by tracing 27,848 rays and using 0.05 mm × 0.05 mm grids. As for the previous example, the normalized PSF can be obtained from Table 3.14 by taking the quotient of the number of rays hitting each grid divided by 1145, i.e., the maximum ray density in Table 3.14. It is

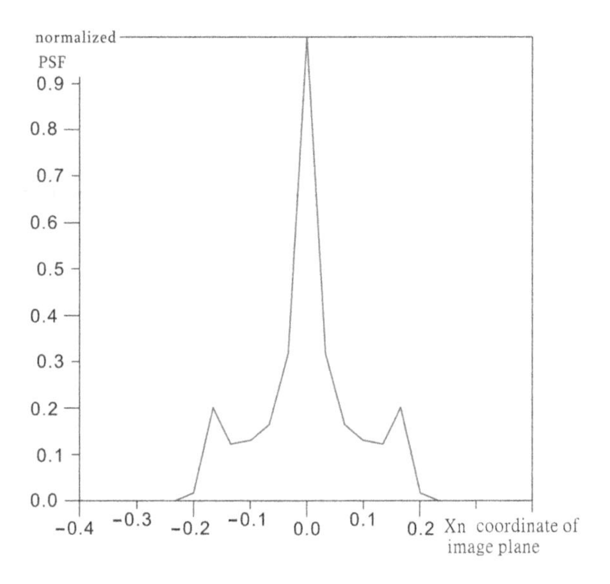

Fig. 3.22 Cross-sectional view of PSF distribution of Table 3.13

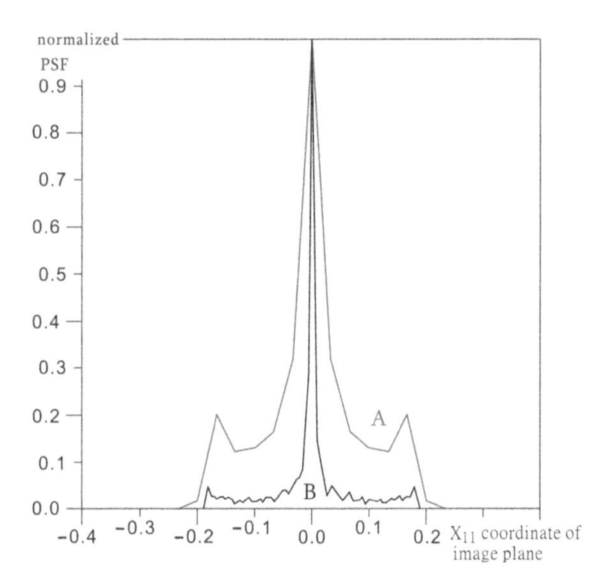

Fig. 3.23 Cross-sectional view of PSF distribution of point source $\bar{P}_0 = [0 \quad -507 \quad 0 \quad 1]^T$ in Fig. 3.2 obtained by tracing different numbers of rays and using different grid sizes. (*A* 1/30 mm × 1/30 mm grids and 4933 rays; *B* 1/200 mm × 1/200 mm grids and 122,397 rays)

observed from Table 3.14 that the energy spread in the sagittal direction is more intense than that in the meridional direction, indicating the existence of coma and astigmatism aberrations.

In general, it should be noted that it is difficult to obtain an accurate estimation of the PSF distribution using the ray-counting method if the PSF distribution contains regions of rapid change. Furthermore, the ray-counting method requires the tracing of many rays in order to estimate the PSF over the entire image plane.

Table 3.14 Ray density of off-axis point source $\bar{P}_0 = [0 \quad -507 \quad 170 \quad 1]^T$ in system shown in Fig. 3.2 obtained via ray-counting method with 0.5mm×0.5mm grids and 27,848 rays

Z_{11}

0	0	0	0	0	0	0	5	0	0	0	0	0	0	0
0	0	0	0	0	0	4	29	4	0	0	0	0	0	0
0	0	0	0	0	16	130	136	130	16	0	0	0	0	0
0	0	0	0	0	104	160	162	160	104	0	0	0	0	0
0	0	0	0	0	159	186	165	186	159	0	0	0	0	0
0	0	0	0	0	190	204	188	204	190	0	0	0	0	0
0	0	0	0	0	196	243	205	243	196	0	0	0	0	0
0	0	0	0	0	99	357	233	357	99	0	0	0	0	0
0	0	0	0	0	0	476	254	476	0	0	0	0	0	0
0	0	0	0	0	0	467	319	467	0	0	0	0	0	0
0	0	0	0	0	0	354	572	354	0	0	0	0	0	0
0	0	0	0	0	0	122	1067	122	0	0	0	0	0	0
0	0	0	0	0	0	83	1145	83	0	0	0	0	0	0
0	0	0	0	0	32	133	1008	133	32	0	0	0	0	0
0	0	0	0	7	87	154	836	154	87	7	0	0	0	0
0	0	0	1	52	100	172	690	172	100	52	1	0	0	0
0	0	0	27	78	106	185	542	185	106	78	27	0	0	0
0	0	9	64	81	111	191	422	191	111	81	64	9	0	0
0	2	49	65	84	113	189	322	189	113	84	65	49	2	0
0	31	56	70	83	108	179	273	179	108	83	70	56	31	0
0	28	59	67	82	112	162	208	162	112	82	67	59	28	0
0	19	58	69	79	107	151	189	151	107	79	69	58	19	0
0	9	51	66	81	104	133	152	133	104	81	66	51	9	0
0	0	37	64	79	92	122	133	122	92	79	64	37	0	0
0	0	20	61	75	89	111	121	111	89	75	61	20	0	0
0	0	0	42	76	82	101	107	101	82	76	42	0	0	0
0	0	0	10	50	81	91	95	91	81	50	10	0	0	0
0	0	0	0	13	54	79	81	79	54	13	0	0	0	0
0	0	0	0	0	7	29	47	29	7	0	0	0	0	0
0	0	0	0	0	0	0	0	0	0	0	0	0	0	0

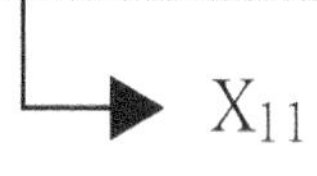

3.5 Modulation Transfer Function

The modulation transfer function (MTF) is the most widely used criterion for evaluating the imaging performance of an optical system ([7–16] and Fig. 3.24). The MTF, which combines both geometric and diffraction metrics, is defined as the ratio of the contrast of the image to the contrast of the object. The MTF is usually computed using the ray-counting method (p. 372 of [6]) described in the previous section. This method is reviewed in the following discussions using an on-axis point source for illustration purposes. Consider an on-axis point source $\bar{P}_0 = [0 \quad P_{0y} \quad 0 \quad 1]^T$ consisting of alternating light and dark bands with a luminous intensity (referred to as the Object Brightness Distribution Function (OBDF)) which varies in accordance with a cosine function. The axis of the OBDF can be either parallel to the direction leading away from the optical axis (i.e., sagittal) or perpendicular to this direction (i.e., meridional) (see Figs. 3.25 and 3.26). The OBDF in the sagittal direction can be expressed as

$$I(x_0) = b_0 + b_1 C(2\pi v x_0), \tag{3.29}$$

where v is the frequency of the brightness variation in cycles per unit length. In addition, $I_{max} = b_0 + b_1$ is the maximum brightness, $I_{min} = b_0 - b_1$ is the minimum

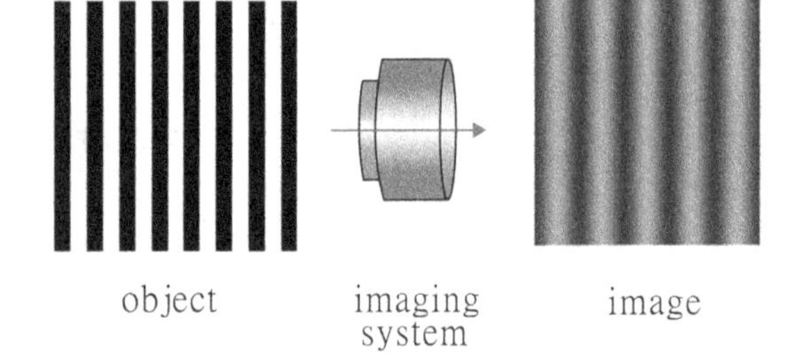

Fig. 3.24 The MTF function specifies the contrast reduction of a periodic pattern after passing through an imaging system. It is noted from this figure that the resulting image is somewhat degraded due to inevitable aberrations and diffraction phenomena

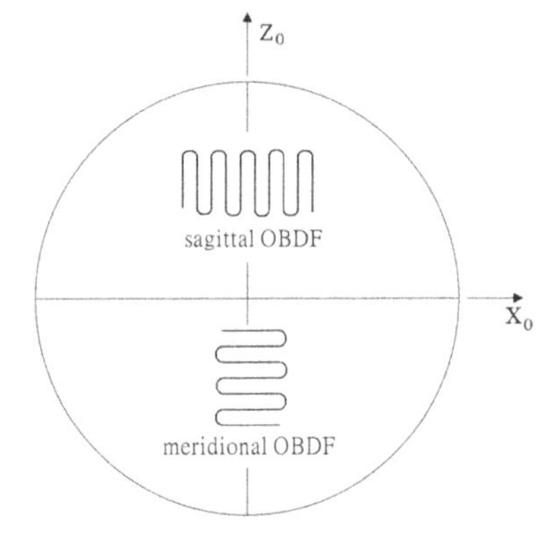

Fig. 3.25 Sagittal and meridional directions of OBDF

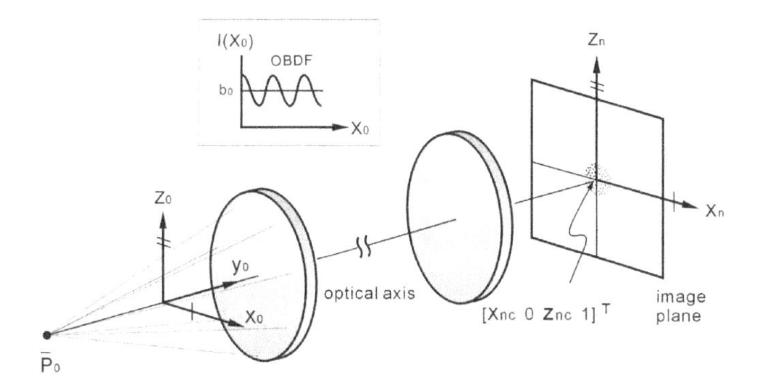

Fig. 3.26 Radiation of light rays from on-axis point source $\bar{P}_0$ onto image plane in axis–symmetrical optical system. Note that the MTF and line spread function are unchanged when the direction of the OBDF is changed

brightness, and x_0 is the coordinate of the world coordinate frame $(xyz)_0$. The modulation of the OBDF is given by

$$M_0 = \frac{I_{max} - I_{min}}{I_{max} + I_{min}} = \frac{(b_0 + b_1) - (b_0 - b_1)}{(b_0 + b_1) + (b_0 - b_1)} = \frac{b_1}{b_0}. \tag{3.30}$$

To compute the MTF, it is necessary to determine the line spread function (LSF) $L(x_n)$. In practice, this can be obtained by integrating the point spread function $B(x_n, z_n)$ along z_n to give $L(x_n) = \int B(x_n, z_n)dz_n$. When the OBDF of $\bar{P}_0$ is imaged by an optical system, the energy intensity at point x_n on the image plane can be determined via the convolution of $I(x_0)$ with the line spread function $L(x_n)$, i.e.,

$$\begin{aligned} I(x_n) &= \int L(\delta)I(x_n - \delta)d\delta \\ &= b_0 \int L(\delta)d\delta + b_1 C(2\pi v x_n)\left(\int L(\delta)C(2\pi v\delta)d\delta\right) + b_1 S(2\pi v x_n)\left(\int L(\delta)S(2\pi v\delta)d\delta\right). \end{aligned} \tag{3.31}$$

The PSF is usually normalized such that $\int L(\delta)d\delta = 1$. In other words,

$$I(x_n) = b_0 + b_1[L_c(v)\, C(2\pi v x_n) + L_s(v)\ S(2\pi v x_n)] = b_0 + b_1 G(v) C(2\pi v x_n - \varpi), \tag{3.32}$$

where

$$L_c(v) = \int L(\delta)\ C(2\pi v\delta)d\delta = \int L(x_n)\ C(2\pi v x_n)dx_n, \tag{3.33}$$

$$L_s(\nu)=\int L(\delta)S(2\pi\nu\delta)d\delta=\int L(x_n)S(2\pi\nu x_n)dx_n, \tag{3.34}$$

$$G(\nu) = \sqrt{L_c^2(\nu)+L_s^2(\nu)}, \tag{3.35}$$

$$\varpi(\nu) = \mathrm{atan2}(L_s(\nu), L_c(\nu)). \tag{3.36}$$

Here, $\varpi(\nu)$ is the phase difference between $I(x_0)$ defined in Eq. (3.29) and the energy distribution function $I(x_n)$ defined in Eq. (3.32). Note that a phase shift of 180° corresponds to a reversal of contrast, i.e., the image pattern is light where it should be dark, and vice versa.

The modulation of the image intensity is given by

$$M_n = b_1/b_0\, G(\nu) = M_0 G(\nu).$$

Therefore, $G(\nu)$ is the MTF, i.e.,

$$\mathrm{MTF}(\nu) = M_n/M_0 = G(\nu) = \sqrt{L_c^2(\nu)+L_s^2(\nu)}. \tag{3.37}$$

Smith [6] obtained the line spread function $L(\delta)$ by counting the number of rays striking the image plane between two parallel lines set a distance of Δx_n apart (e.g., Table 3.12). However, when adopting this approach, the summation of the line spread function is not equal to one (i.e., $\int L(\delta)d\delta = \int L(x_n)dx_n \cong \sum L(x_n)\Delta x_n \neq 1$). Consequently, to obtain the MTF, Smith divided Eq. (3.31) by $\sum L(x_n)\Delta x_n$, resulting in the following energy distribution equation at x_n:

$$\begin{aligned} I^*(x_n) &\cong \frac{I(x_n)}{\sum L(x_n)\Delta x_n} = b_0 + b_1\left[\, L_c^*(\nu)\ C(2\pi\nu x_n) + L_s^*(\nu)\ S(2\pi\nu x_n)\right] \\ &= b_0 + b_1 G^*(\nu)C(2\pi\nu x_n - \varpi^*). \end{aligned} \tag{3.38}$$

Here,

$$L_c^*(\nu) \cong \sum L(x_n)C(2\pi\nu x_n)\Delta x_n/\sum L(x_n)\Delta x_n, \tag{3.39}$$

$$L_s^*(\nu) \cong \sum L(x_n)S(2\pi\nu x_n)\Delta x_n/\sum L(x_n)\Delta x_n, \tag{3.40}$$

$$\varpi^*(\nu) = \mathrm{atan2}(L_s^*(\nu), L_c^*(\nu)). \tag{3.41}$$

It should be noted that $L_s^*(\nu)= 0$ is always true for an on-axis point source in an axis-symmetrical system since $L_s^*(\nu)$ is an odd function with respect to x_n. In [6] image modulation was performed via $M_n^* = (b_1/b_0)G^*(\nu) = M_0G^*(\nu)$. Therefore, $G^*(\nu)$ is the MTF, i.e.,

$$\mathrm{MTF}^*(\nu) = M_n^*/M_0^* = G^*(\nu) = \sqrt{(L_c^*(\nu))^2 + (L_s^*(\nu))^2}. \qquad (3.42)$$

The MTF curve can thus be obtained by taking the frequency ν as the abscissa and $\mathrm{MTF}^*(\nu)$ as the ordinate. The value of the MTF lies in the interval [0, 1] and generally decreases with an increasing frequency ν.

Example 3.23 Figure 3.27 shows the results obtained from the ray-counting method for the line spread function of point source $\bar{P}_0 = [0 \quad -507 \quad 0 \quad 1]^T$ in Fig. 3.2 given different grid sizes on the image plane and a constant number of traced rays (4933). More specifically, the figure shows the normalized line spread function, i.e., the quotient of the line spread function $L(x_n)$ divided by $\sum L(x_n)\Delta x_n$ (i.e. $L(x_n)/\sum L(x_n)\Delta x_n$). It is seen that the grid size has a fundamental effect on the line spread function resolution, which in turn affects the MTF in ray-counting methods. Notably, the irradiance method proposed in Chap. 13 does not need the line spread function. If the mesh is too large (e.g., curves A, B, C and D in

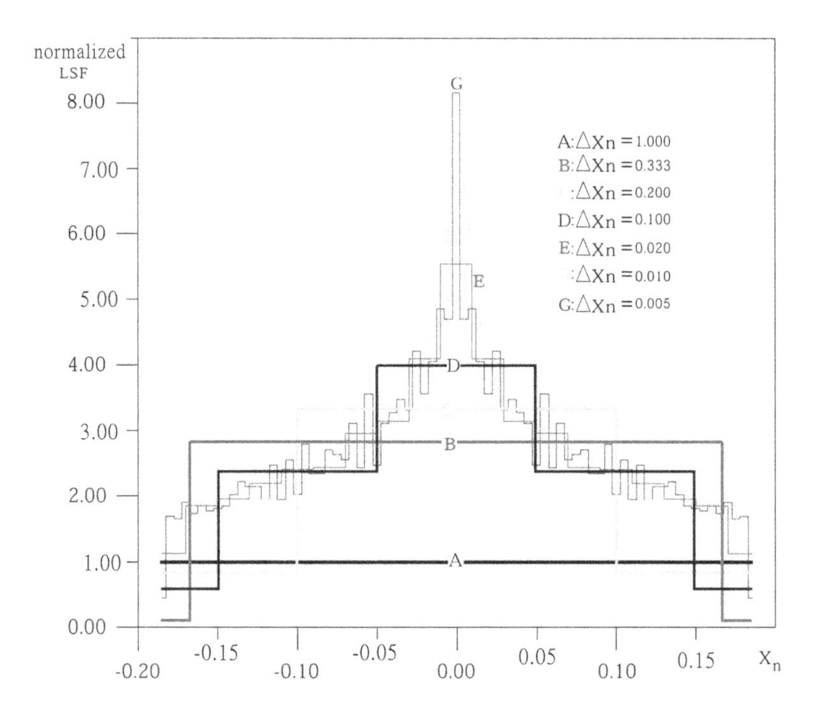

Fig. 3.27 Variation of sagittal line spread function of point source $\bar{P}_0 = [0 \quad -507 \quad 0 \quad 1]^T$ in Fig. 3.2 given different grid sizes on the image plane and 4933 traced rays

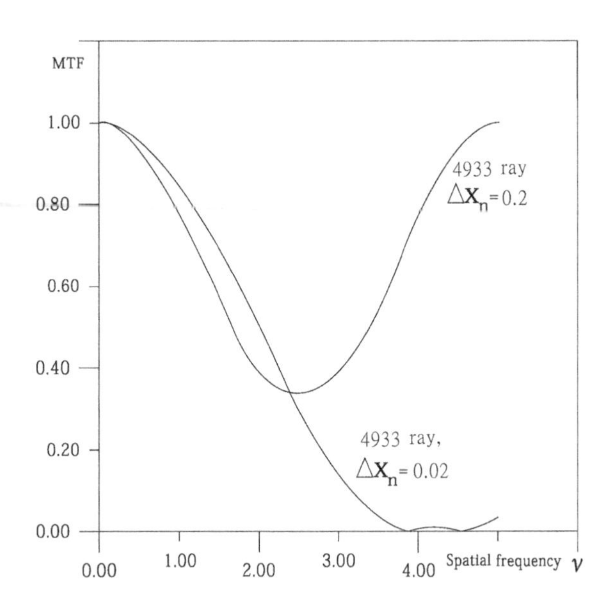

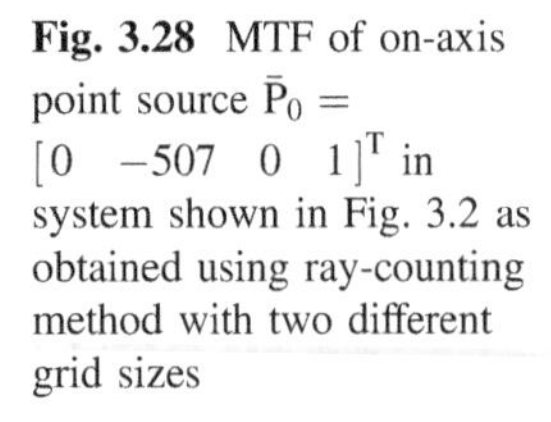
Fig. 3.28 MTF of on-axis point source $\bar{P}_0 = [0 \quad -507 \quad 0 \quad 1]^T$ in system shown in Fig. 3.2 as obtained using ray-counting method with two different grid sizes

Fig. 3.27), the line spread function resolution may be so poor that its value approaches unity (curve A is exactly the same as the aberration-free case), yielding an over-estimated MTF.

Example 3.24 Figure 3.28 presents the MTF for an on-axis point source $\bar{P}_0 = [0 \quad -507 \quad 0 \quad 1]^T$ in the system shown in Fig. 3.2 given two different mesh sizes ($\Delta x_n = 0.2$ and $\Delta x_n = 0.02$). As described above, the ray-counting method requires an estimation of the line spread function, and is thus dependent on both the number of rays traced and the mesh size on the image plane. For an on-axis point source, the sagittal and meridional MTFs are the same, and the phase shift is equal to $\varpi = 0$ for such an axis-symmetrical system. The results show that the MTF is fundamentally dependent on the mesh size (Δx_{11}) of the image plane. Specifically, the MTF may be over-estimated if the mesh is too coarse due to an incorrect line spread function estimation.

Example 3.25 Figure 3.29 shows the variation in the MTF of point source $\bar{P}_0 = [0 \quad -507 \quad 0 \quad 1]^T$ in the system shown in Fig. 3.2 given a various grid size of Δx_{11} and various numbers of traced rays. It is seen that the number of traced rays has a significant effect on the MTF. It is noted that if the mesh size is too fine, the MTF may fluctuate dramatically.

Overall, the discussions above show that the ray-counting method provides a sub-optimal approach for computing the PSF and MTF metrics since its accuracy depends fundamentally on both the number of rays traced and the grid size on the image plane.

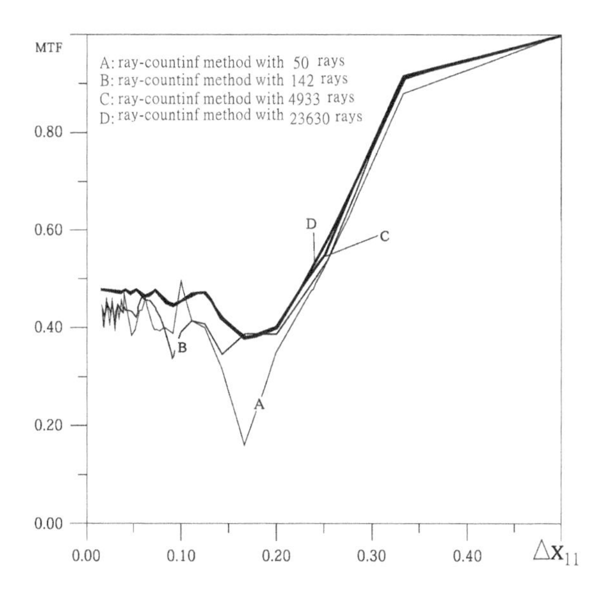

Fig. 3.29 Variation of sagittal MTF of on-axis point source $\bar{P}_0 = [0 \quad -507 \quad 0 \quad 1]^T$ in system shown in Fig. 3.2 given $\nu = 2$(cycle per mm) and different numbers of traced rays

3.6 Motion Measurement Systems

High-accuracy laser-based optoelectronic motion measuring systems typically utilize light rays which travel from one optical boundary surface to another to perform motion measurements [17–23]. Figure 3.30 shows a schematic illustration of the motion measurement system presented by Ni and Wu [17]. As shown, the system consists of two optical assemblies, namely Assembly a = 1, a measuring assembly attached rigidly to a moving table whose position is to be measured; and Assembly

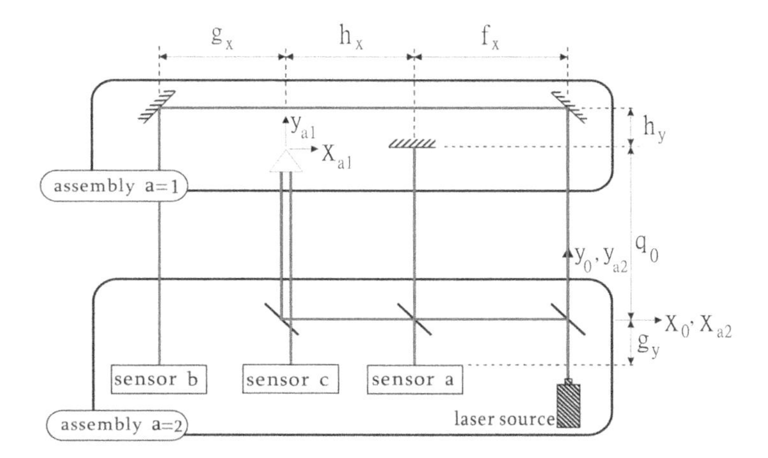

Fig. 3.30 Schematic illustration of motion measurement system proposed in [17]

a = 2, a reference assembly mounted rigidly on the reference base. Assembly a = 2 contains a mixed group of optical elements, two beam splitters and three position sensitive detectors (PSDs) (dual-axis lateral-effect photodetectors) to receive the beams reflected from the measuring assembly. Meanwhile, Assembly a = 1 also contains a mixed group of optical elements, contains a corner-cube and three first-surface mirrors. The unit directional vector $\bar{\ell}_0 = \begin{bmatrix} 0 & 1 & 0 & 0 \end{bmatrix}^T$ of the laser source is aligned with the axis along which the machine tool table travels. Since Assembly a = 1 is attached rigidly to the moving table, it emulates its motion, denoted as

$$\bar{V} = \begin{bmatrix} v_g \end{bmatrix} = \begin{bmatrix} t_{a1x} & t_{a1y} & t_{a1z} & \omega_{a1x} & \omega_{a1y} & y_{a1z} \end{bmatrix}^T, g = 1 \text{ to } g = 6.$$

Two coordinate frames, $(xyz)_{a1}$ and $(xyz)_{a2}$, are embedded into assemblies a = 1 and a = 2, respectively, for modeling purposes. The pose matrix ${}^0\bar{A}_{a2} = \bar{I}_{4\times4}$ of Assembly a = 2 is an identity matrix, since $(xyz)_{a2}$ and $(xyz)_0$ are assumed to coincide for reasons of convenience. For Assembly a = 1, the origin of $(xyz)_{a1}$ is placed at the point where the translational and rotational motions are to be measured. The pose matrix of $(xyz)_{a1}$ with respect to $(xyz)_0$ is defined as

$${}^0\bar{A}_{a1} = \text{tran}(t_{a1x}, t_{a1y}, t_{a1z})\text{rot}(\bar{z}, \omega_{a1z})\text{rot}(\bar{y}, \omega_{a1y})\text{rot}(\bar{x}, \omega_{a1x}).$$

Assuming a zero error motion of the moving table, the parameters of the pose matrix are given as follows: $t_{a1z} = \omega_{a1x} = \omega_{a1y} = \omega_{a1z} = 0$, $t_{a1x} = -(h_x + f_x)$ and $t_{a1y} = q_0$. However, given an error of

$$\Delta\bar{V} = \begin{bmatrix} \Delta v_g \end{bmatrix} = \begin{bmatrix} \Delta t_{a1x} & \Delta t_{a1y} & \Delta t_{a1z} & \Delta\omega_{a1x} & \Delta\omega_{a1y} & \Delta\omega_{a1z} \end{bmatrix}^T$$

in the motion of the table relative to the intended path, the laser beam reflected by the measuring assembly shifts its lateral position or changes its angular orientation, depending on the type of error motion.

Skew-ray tracing provides a convenient means of modeling optical measurement systems designed to measure small changes in a ray's final position (i.e., incidence point $\bar{P}_n$, see Eq. (2.48) with g = n). For such systems, the sensor readings $\begin{bmatrix} X & Z \end{bmatrix}^T$ can be obtained from ${}^n\bar{P}_n$ by means of the following transformation:

$$\begin{bmatrix} X \\ 0 \\ Z \\ 1 \end{bmatrix} = \begin{bmatrix} {}^nP_{nx} \\ 0 \\ {}^nP_{nz} \\ 1 \end{bmatrix} = {}^n\bar{P}_n = {}^n\bar{A}_0\bar{P}_n = ({}^0\bar{A}_n)^{-1}\bar{P}_n$$
$$= \begin{bmatrix} I_{nx} & I_{ny} & I_{nz} & -(I_{nx}t_{nx} + I_{ny}t_{ny} + I_{nz}t_{nz}) \\ J_{nx} & J_{ny} & J_{nz} & -(J_{nx}t_{nx} + J_{ny}t_{ny} + J_{nz}t_{nz}) \\ K_{nx} & K_{ny} & K_{nz} & -(K_{nx}t_{nx} + K_{ny}t_{ny} + K_{nz}t_{nz}) \\ 0 & 0 & 0 & 1 \end{bmatrix} \begin{bmatrix} P_{nx} \\ P_{ny} \\ P_{nz} \\ 1 \end{bmatrix}. \quad (3.43)$$

Since PSD readings can contain both translational and rotational error motions $\Delta\bar{V}$, it is reasonable to express these readings as implicit nonlinear functions of $\bar{V} = [v_g]$, i.e., $[X \quad Z]^T = [X(\bar{V}) \quad Z(\bar{V})]^T$. In practice, the translational and rotational error motions $\Delta\bar{V}$ are very small. As a result, a first-order Taylor series expansion can be used to expand $[X \quad Z]^T$ at $v_g = 0$ (g = 1 to g = 6) to obtain the PSD readings in a linear form, i.e.,

$$\begin{bmatrix} X \\ Z \end{bmatrix} = \begin{bmatrix} X(\bar{0}) \\ Z(\bar{0}) \end{bmatrix} + \begin{bmatrix} \partial X/\partial\bar{V} \\ \partial Z/\partial\bar{V} \end{bmatrix}_{\Delta\bar{v}_g=\bar{0}_{1\times6}} [\Delta v_g]. \tag{3.44}$$

Note that $[\partial X/\partial v_g \quad \partial X/\partial v_g]_{\Delta\bar{V}=\bar{0}_{1\times6}}$ indicates that the matrix is evaluated at $\Delta(\bar{V}) = \bar{0}_{1\times6}$. In the real world, it is impossible to place an element exactly at a specified position and orientation. Consequently, setting errors $[X(\bar{0}) \quad Z(\bar{0})]^T$ inevitably exist. In order to prevent reading errors as a result of these setting errors, the system must first be adjusted such that the setting errors are reduced to the smallest amount possible. The difference between the PSD readings and the setting errors, i.e.,

$$\begin{bmatrix} X \\ Z \end{bmatrix} - \begin{bmatrix} X(\bar{0}) \\ Z(\bar{0}) \end{bmatrix} = \begin{bmatrix} \Delta X \\ \Delta Z \end{bmatrix} \tag{3.45}$$

(designated as the "effective PSD readings"), at any single measurement point can then be used to determine the real positional/angular motion. The effective PSD readings can thus be expressed as

$$\begin{bmatrix} \Delta X \\ \Delta Z \end{bmatrix} = \begin{bmatrix} \partial X/\partial\bar{V} \\ \partial Z/\partial\bar{V} \end{bmatrix} [\Delta v_g]. \tag{3.46}$$

If the translational and rotational motions are very small, the sensor readings can be formulated as

$$\begin{aligned} \begin{bmatrix} \Delta X \\ \Delta Z \end{bmatrix} &= \begin{bmatrix} \partial X/\partial v_g \\ \partial X/\partial v_g \end{bmatrix}_{\Delta\bar{V}=\bar{0}_{1\times6}} \Delta v_g \\ &= \begin{bmatrix} c_{11g}g_y + c_{12g}q_0 + c_{13g}h_y + c_{14g}f_x + c_{15g}h_x + c_{16g}g_x \\ c_{21g}g_y + c_{22g}q_0 + c_{23g}h_y + c_{24g}f_x + c_{25g}h_x + c_{26g}g_x \end{bmatrix} \Delta v_g. \end{aligned} \tag{3.47}$$

The leading coefficient, which represents the weight by which each translational and rotational motion contributes to the sensor readings, describes the sensitivity of the corresponding translational and rotational motions. The coefficients of each translational and rotational motion Δv_g in Eq. (3.47) can be obtained using a finite difference (FD) methodology in accordance with the following step-by-step approach:

(1) Label the boundary surfaces of the system sequentially from i = 1 to i = n.
(2) Assume there exists only small translational or rotational error motions in the measuring assembly, e.g., $\Delta v_2 = \Delta t_{a1y} = 0.00001\,\text{mm}$. Thus, the pose matrix of the measuring assembly can be formulated as ${}^0\bar{A}_{a1} = \text{tran}(-h_x - f_x, q_0 + 0.00001, 0)$.
(3) To determine $\begin{bmatrix} c_{122} & c_{222} \end{bmatrix}^T$, set $q_0 = 1$ while keeping $g_y = h_y = f_x = h_x = g_x = 0$. Calculate the coordinates of incidence point $\bar{P}_n = \begin{bmatrix} P_{nx} & P_{ny} & P_{nz} & 1 \end{bmatrix}^T$ (Eq. (2.48) with g = n) and then determine $\begin{bmatrix} X & 0 & Z & 1 \end{bmatrix}^T$(Eq. (3.43)) on the sensor surface using the skew-ray tracing methodology presented in Chap. 2. From Eq. (3.47), it follows that

$$\begin{bmatrix} c_{122} \\ c_{222} \end{bmatrix} = \begin{bmatrix} \partial X/\partial v_2 \\ \partial Z/\partial v_2 \end{bmatrix} \approx \begin{bmatrix} \Delta X/\Delta v_2 \\ \Delta Z/\Delta v_2 \end{bmatrix}. \tag{3.48}$$

A similar procedure can be applied to obtain the other coefficients.

As shown in Fig. 3.30, the illustrative system contains three separate sensors in Assembly a = 2. Each sensor interacts with a different set of optical elements in Assembly a = 1. Thus, each sensor can be viewed as belonging to a unique sub-system. The following example derives the linear equation of the sub-system containing Sensor b.

Example 3.26 As shown in Fig. 3.31, the sub-system containing Sensor b comprises two first-surface mirrors in Assembly a = 1 and Sensor b in Assembly a = 2.

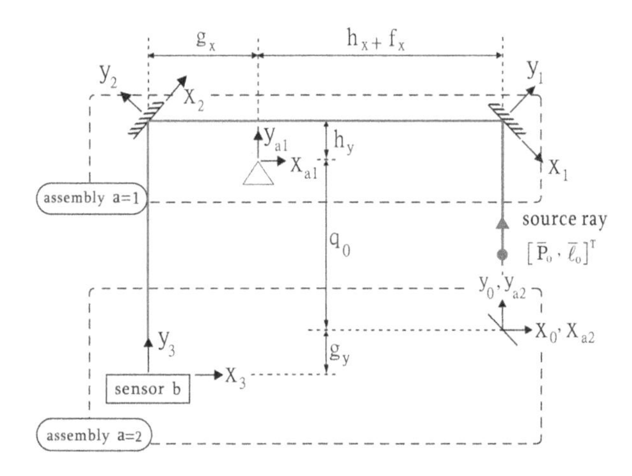

Fig. 3.31 Model of Sensor b sub-system in motion measurement system proposed in [17]

The boundary coordinate frames $(xyz)_1$, $(xyz)_2$, $(xyz)_3$ are defined and the following pose matrices are obtained:

$$^{a1}\bar{A}_1 = \text{tran}(h_x + f_x, h_y, 0)\text{rot}(\bar{z}, -45^\circ),$$
$$^{a1}\bar{A}_2 = \text{tran}(-g_x, h_y, 0)\text{rot}(\bar{z}, 45^\circ),$$
$$^{a2}\bar{A}_3 = \text{tran}(-g_x - h_x - f_x, -g_y, 0).$$

The linear equations for $[\Delta X_b \quad \Delta Z_b]^T$ have the form

$$\begin{bmatrix} \Delta X_b \\ \Delta Z_b \end{bmatrix} = \begin{bmatrix} \Delta^3 P_{3x} \\ \Delta^3 P_{3z} \end{bmatrix} = \begin{bmatrix} -2\Delta t_{a1x} + (2h_y + f_x + h_x + g_x)\Delta\omega_{a1z} \\ [2(g_y + q_0 + h_y) + (f_x + h_x + g_x)]\Delta\omega_{a1x} - (f_x + h_x + g_x)\Delta\omega_{a1y} \end{bmatrix}.$$

In [17], only a single linear equation for Sensor b is given, i.e.,

$$\Delta Z_b = 2q_0\Delta\omega_{a1x} - (f_x + h_x + g_x)\Delta\omega_{a1y}.$$

It is noted that the expressions of ΔZ_b from these two approaches are different. Actually, it was confirmed from simulation that our proposed ΔZ_b is correct. This is example indicates that the work of [17] did not perform a strict skew-ray tracing of their measurement system, leading that the effects of some parameters on the sensor reading were not accounted for.

References

1. Laikin M (1995) Lens design. Marcel Dekker, Inc., pp 71–72
2. Foreman JW (1974) Computation of RMS spot radii by ray tracing. Appl Opt 13:2585–2588
3. Andersen TB (1982) Evaluating RMS spot radii by ray tracing. Appl Opt 21:1241–1248
4. Brixner B (1978) Lens design merit functions: rms image spot size and rms optical path difference. Appl Opt 17:715–716
5. José S (2013) Introduction to aberrations in optical imaging systems. Springer, India
6. Smith WJ (2001) Modern optical engineering, 3rd edn. Edmund Industrial Optics, Barrington, N.J
7. Tseng KH, Kung C, Liao TT, Chang HP (2009) Calculation of modulation transfer function of an optical system by using skew ray tracing, transactions of canadian society for mechanical engineering. The Journal of Mechanical Engineering 33:429–442
8. Inoue S, Tsumura N, Miyake Y (1997) Measuring MTF of paper by sinusoidal test pattern projection. J Imaging Sci Technol 41:657–661
9. Boreman GD, Yang S (1995) Modulation transfer function measurement using three- and four-bar targets. Appl Opt 34:8050–8052
10. Sitter DN, Goddard JS, Ferrell RK (1995) Method for the measurement of the modulation transfer function of sampled imaging systems from bar-target patterns. Appl Opt 34:746–751
11. Barakat R (1965) Determination of the optical transfer function directly from the edge spread function. J Opt Soc Am 55:1217–1221

12. Rogers GL (1998) Measurement of the modulation transfer function of paper. Appl Opt 37:7235–7240
13. Park SK, Schowengerdt R, Kaczynski M (1984) Modulation-transfer-function analysis for sampled image system. Appl Opt 23:2572–2582
14. Inoue S, Tsumura N, Miyake Y (1997) Measuring MTF of paper by sinusoidal test pattern projection. J Imaging Sci Technol 41:657–661
15. Tseng KH, Kung C, Liao TT, Chang HP (2009) Calculation of modulation transfer function of an optical system by using skew ray tracing, transactions of canadian society for mechanical engineering. J Mech Eng 33:429–442
16. Giakoumakis E, Katsarioti MC, Panayiotakis GS (1991) Modulation transfer function of thin transparent foils in radiographic cassettes. Appl Phys A Solids Surf 52:210–212
17. Ni J, Wu SM (1993) An on-line measurement technique for machine volumetric error compensation. ASME J Eng Indus 115:85–92
18. Kim KH, Eman KF, Wu SM (1987), Analysis alignment errors in a laser-based in-process cylindricity measurement system. J Eng Ind-Trans ASME 109:321–329
19. Park CW, Eman KF, Wu SM (1988) An in-process flatness error measurement and compensatory control system. J Eng Ind-Trans ASME 110:263–270
20. Bokelberg EH, Sommer HJ III, Tretheway MW (1994) A six-degree-of-freedom laser vibrometer, part I: Theoretical development. J Sound Vibr 178:643–654
21. Bokelberg EH, Sommer HJ III, Trethewey MW (1994) A six-degree-of-freedom laser vibormeter, part II: experimental validation. J Sound Vibr 178:655–667
22. Lin PD, Ehmann KF (1996) Sensing of motion related errors in multi-axis machines. J Dyn Syst Meas Control-Trans ASME 118:425–433
23. Lee SW, Mayor R, Ni J (2005) Development of a six-degree-of-freedom geometric error measurement system for a meso-scale machine tool. J Manuf Sci Eng-Trans ASME 127:857–865

Chapter 4
Raytracing Equations for Paraxial Optics

Conventional paraxial optics, sometimes known as Gaussian optics, uses 2×2 raytracing matrices as initial estimates in the early design stage of optical systems [1–3]. However, such an approach is suitable only for the analysis of paraxial meridional rays in axis-symmetrical systems. In [4], the present author extended conventional paraxial optics to 3-D optical systems by using a first-order Taylor series expansion to approximate the skew raytracing equations in a 6×6 matrix form. It was shown that the proposed method achieved a good accuracy when limited to the skew rays in the neighborhood of the base line. Section 4.1 of this chapter presents the 6×6 matrices proposed in [4] without rigorous derivation. The matrices are then simplified to conventional 2×2 raytracing matrices in Sect. 4.2. In practice, the derivation procedures used in [4] are rather complex. Thus, the remaining sections in the chapter present a more straightforward geometry-based approach for deriving the conventional 2×2 raytracing matrices. The application of these matrices in paraxial optics to determine the cardinal points for axis-symmetrical systems and to obtain the corresponding image equations is discussed in the following chapter.

4.1 Raytracing Equations of Paraxial Optics for 3-D Optical Systems

In paraxial optics, no distinction is made between a spherical surface and an aspherical surface since, in the paraxial region, spherical and aspherical surfaces are essentially indistinguishable if the vertex radius of curvature of the aspherical surface is used for paraxial raytracing. Furthermore, in paraxial optics, rays are not

P.D. Lin, *Advanced Geometrical Optics*, Progress in Optical Science and Photonics 4, DOI 10.1007/978-981-10-2299-9_4

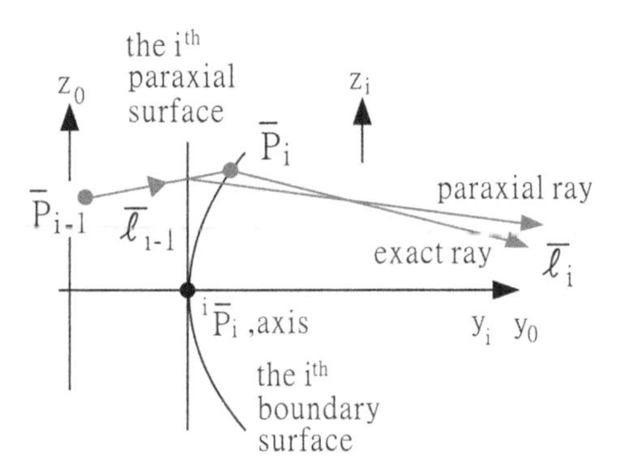

Fig. 4.1 In paraxial optics, rays are reflected (or refracted) at paraxial surfaces, not at real boundary surfaces

reflected (or refracted) at real boundary surfaces, but at imaginary tangent planes (referred to as paraxial surfaces) to real boundary surfaces at their vertices (see Fig. 4.1). Before proceeding, it should be noted that some of the symbols used in this section (e.g., $\bar{M}_i$ in Eq. (4.2)) are also used in other chapters to denote different physical quantities. However, this presents no problem since paraxial optics is an independent topic from the others in geometrical optics.

It will be recalled that the pre-superscript "g" of the leading symbol ${}^g\bar{P}_i$ indicates that the components of this vector are expressed with respect to coordinate frame $(xyz)_g$. The equations presented in this section are referred to boundary coordinate frame $(xyz)_i$. Consequently, the associated rays, incidence points and unit directional vectors are denoted as ${}^i\bar{R}_i$, ${}^i\bar{P}_i$ and ${}^i\bar{\ell}_i$, respectively. In addressing the problem of paraxial optics in 3-D optical systems, it is first necessary to define the paraxial surfaces and differential rays. The ith paraxial surface can be regarded as a virtual flat boundary surface tangent to the real boundary surface at point ${}^i\bar{P}_{i,axis}$, i.e., the vertex of the ith boundary surface (see Figs. 4.1 and 4.2). Note that Fig. 4.2 shows both a skew ray ${}^i\bar{R}_i = \begin{bmatrix} {}^i\bar{P}_i & {}^i\bar{\ell}_i \end{bmatrix}^T$ and an axial ray ${}^i\bar{R}_{i,axis} = \begin{bmatrix} {}^i\bar{P}_{i,axis} & {}^i\bar{\ell}_{i,axis} \end{bmatrix}^T$, where ${}^i\bar{P}_{i,axis}$ is the vertex of the ith boundary surface and ${}^i\bar{\ell}_{i,axis}$ is the unit directional vector of the boundary surface along the optical axis. In general, any skew ray ${}^i\bar{R}_i$ can be expressed as the sum of an axial ray ${}^i\bar{R}_{i,axis}$ oriented along the optical axis and a differential ray $\Delta{}^i\bar{R}_i = \begin{bmatrix} \Delta{}^i\bar{P}_i & \Delta{}^i\bar{\ell}_i \end{bmatrix}^T$. In other words, the differential ray, $\Delta{}^i\bar{R}_i$, is given by the difference between ${}^i\bar{R}_i$ and ${}^i\bar{R}_{i,axis}$. That is,

$$\Delta{}^i\bar{R}_i = \begin{bmatrix} \Delta{}^i\bar{P}_i \\ \Delta{}^i\bar{\ell}_i \end{bmatrix} = \begin{bmatrix} \Delta{}^iP_{ix} \\ \Delta{}^iP_{iy} \\ \Delta{}^iP_{iz} \\ \Delta{}^i\ell_{ix} \\ \Delta{}^i\ell_{iy} \\ \Delta{}^i\ell_{iz} \end{bmatrix} = \begin{bmatrix} {}^i\bar{P}_i \\ {}^i\bar{\ell}_i \end{bmatrix} - \begin{bmatrix} {}^i\bar{P}_{i,axis} \\ {}^i\bar{\ell}_{i,axis} \end{bmatrix} = \begin{bmatrix} {}^iP_{ix} \\ {}^iP_{iy} \\ {}^iP_{iz} \\ {}^i\ell_{ix} \\ {}^i\ell_{iy} \\ {}^i\ell_{iz} \end{bmatrix} - \begin{bmatrix} {}^iP_{ix,axis} \\ {}^iP_{iy,axis} \\ {}^iP_{iz,axis} \\ {}^i\ell_{ix,axis} \\ {}^i\ell_{iy,axis} \\ {}^i\ell_{iz,axis} \end{bmatrix}. \quad (4.1)$$

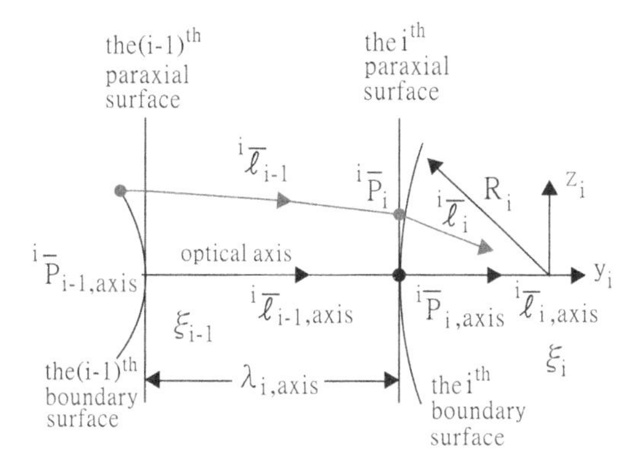

Fig. 4.2 Schematic illustration showing paraxial ray propagating along straight-line path and then reflected or refracted at boundary surface

The main purpose of paraxial optics is to estimate the differential ray, $\Delta^{i}\bar{R}_{i}$, under the condition that ${}^{i}\bar{R}_{i}$ is the neighboring ray of ${}^{i}\bar{R}_{i,axis}$. This section presents the equations of the differential ray, $\Delta^{i}\bar{R}_{i}$, given in a previous study by the current author [4] based on a first-order Taylor series expansion method. As shown in Fig. 4.2, when a skew ray ${}^{i}\bar{R}_{i-1} = \begin{bmatrix} {}^{i}\bar{P}_{i-1} & {}^{i}\bar{\ell}_{i-1} \end{bmatrix}^{T}$ travels from the (i − 1)th surface to the ith surface, it propagates along a straight-line path in medium i − 1 and is then reflected (or refracted) at the ith surface. The differential ray, $\Delta^{i}\bar{R}_{i}$, can be estimated via the following 6 × 6 matrix manipulation:

$$\begin{bmatrix} \Delta^{i}\bar{P}_{i} \\ \Delta^{i}\bar{\ell}_{i} \end{bmatrix} = \bar{M}_{i}\bar{T}_{i}\begin{bmatrix} \Delta^{i}\bar{P}_{i-1} \\ \Delta^{i}\bar{\ell}_{i-1} \end{bmatrix} = \bar{M}_{i}\bar{T}_{i}\left(\begin{bmatrix} {}^{i}\bar{P}_{i-1} \\ {}^{i}\bar{\ell}_{i-1} \end{bmatrix} - \begin{bmatrix} {}^{i}\bar{P}_{i-1,axis} \\ {}^{i}\bar{\ell}_{i-1,axis} \end{bmatrix}\right), \tag{4.2}$$

where $\bar{M}_{i}$ is the reflection or refraction matrix at the ith paraxial surface and $\bar{T}_{i}$ is the transfer matrix when the ray travels in a medium with a refractive index ξ_{i-1}. The vector $\begin{bmatrix} \Delta^{i}\bar{P}_{i-1} & \Delta^{i}\bar{\ell}_{i-1} \end{bmatrix}^{T}$ then gives the differential ray between the skew ray ${}^{i}\bar{R}_{i-1}$ and the axial ray ${}^{i}\bar{R}_{i-1,axis}$.

4.1.1 Transfer Matrix

Denoting the length of the straight-line segment from ${}^{i}\bar{P}_{i-1,axis}$ to ${}^{i}\bar{P}_{i,axis}$ along the optical axis as $\lambda_{i,axis}$, the transfer matrix can be expressed as

$$\bar{T}_i = \begin{bmatrix} 1 & 0 & 0 & \lambda_{i,axis} & 0 & 0 \\ 0 & 1 & 0 & 0 & \lambda_{i,axis} & 0 \\ 0 & 0 & 1 & 0 & 0 & \lambda_{i,axis} \\ 0 & 0 & 0 & 1 & 0 & 0 \\ 0 & 0 & 0 & 0 & 1 & 0 \\ 0 & 0 & 0 & 0 & 0 & 1 \end{bmatrix}. \quad (4.3)$$

4.1.2 *Reflection and Refraction Matrices for Flat Boundary Surface*

The explicit expression of $\bar{M}_i$ depends not only on the boundary surface type (e.g., flat, concave, or convex), but also on the nature of the surface, i.e., reflecting or refracting. Many optical elements (e.g., beam-splitters) have reflecting and/or refracting flat boundary surfaces (Figs. 4.3 and 4.4). Moreover, the boundary coordinate frame $(xyz)_i$ of $\bar{r}_i$ can be oriented in many different ways. If $(xyz)_i$ is oriented in such a way that its $y_i z_i$ plane contains both the optical axis and its y_i axis points from ${}^i\bar{P}_{i-1,axis}$ to ${}^i\bar{P}_{i,axis}$, then the unit directional vector of the incoming ray has the form ${}^i\bar{\ell}_{i-1,axis} = [0 \quad 1 \quad 0 \quad 0]^T$. The 6×6 refraction matrix $\bar{M}_i$ for such a flat boundary surface is given as

$$\bar{M}_i = \begin{bmatrix} 1 & 0 & 0 & 0 & 0 & 0 \\ 0 & 0 & 0 & 0 & 0 & 0 \\ 0 & 0 & 1 & 0 & 0 & 0 \\ 0 & 0 & 0 & N_i & 0 & 0 \\ 0 & 0 & 0 & 0 & N_i^2 & 0 \\ 0 & 0 & 0 & 0 & 0 & N_i \end{bmatrix}, \quad (4.4)$$

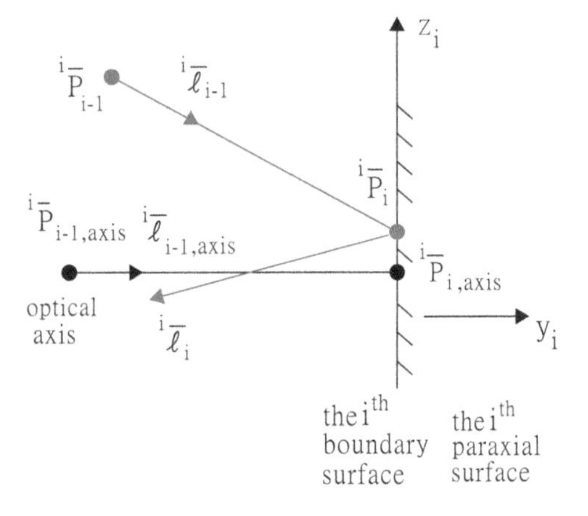

Fig. 4.3 Flat first-surface mirror

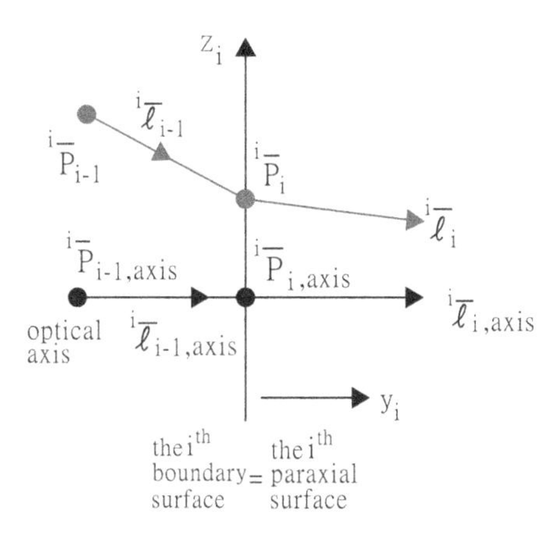

Fig. 4.4 Flat refracting boundary surface

where $N_i = \xi_{i-1}/\xi_i$ is the refractive index of medium $i-1$ relative to that of medium i. The 6×6 reflection matrix $\bar{M}_i$ for the boundary coordinate frame $(xyz)_i$ of a reflecting mirror defined in the same way is given as

$$\bar{M}_i = \begin{bmatrix} 1 & 0 & 0 & 0 & 0 & 0 \\ 0 & 0 & 0 & 0 & 0 & 0 \\ 0 & 0 & 1 & 0 & 0 & 0 \\ 0 & 0 & 0 & 1 & 0 & 0 \\ 0 & 0 & 0 & 0 & -1 & 0 \\ 0 & 0 & 0 & 0 & 0 & 1 \end{bmatrix}. \tag{4.5}$$

4.1.3 *Reflection and Refraction Matrices for Spherical Boundary Surface*

Spherical mirrors can be divided into two classes, namely convex or concave (see Figs. 4.5 and 4.6). A spherical mirror $\bar{r}_i$ is said to be convex if the incoming ray ${}^i\bar{R}_{i-1}$ and center of curvature lie on opposite sides of the mirror. By contrast, the mirror is said to be concave if the incoming ray and center of curvature are located on the same side of the mirror. If the boundary coordinate frame $(xyz)_i$ of $\bar{r}_i$ is defined in such a way that its y_iz_i plane contains the optical axis and its y_i axis points from ${}^i\bar{P}_{i-1,axis}$ to ${}^i\bar{P}_{i,axis}$, then the unit directional vector of the incoming ray is given by ${}^i\bar{\ell}_{i-1,axis} = [0 \quad 1 \quad 0 \quad 0]^T$. Recalling that a convex mirror has a positive value of R_i while a concave mirror has a negative value of R_i, then the following reflection matrix $\bar{M}_i$ holds for both convex and concave mirrors:

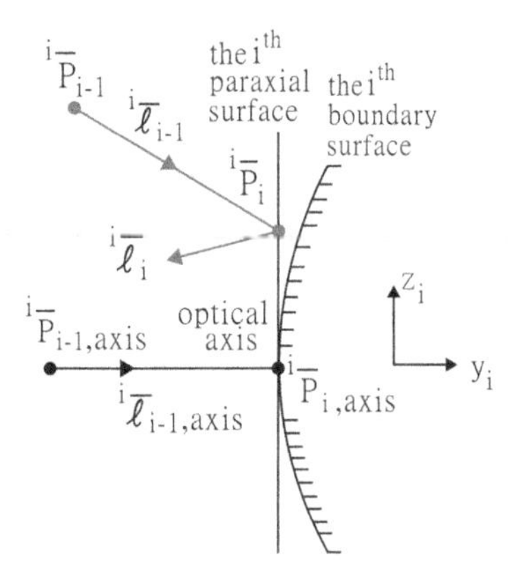

Fig. 4.5 Convex spherical mirror

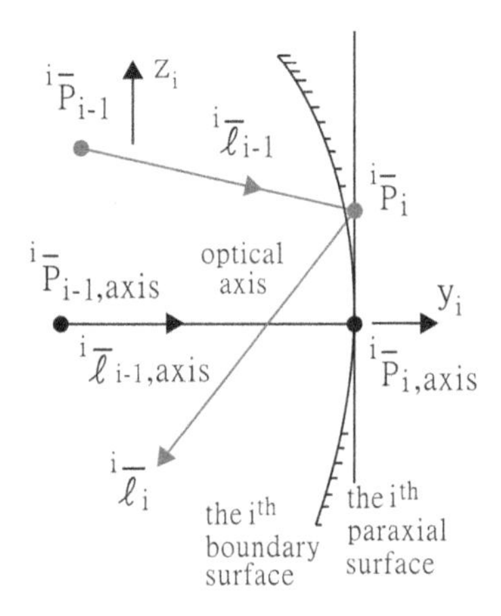

Fig. 4.6 Concave spherical mirror

$$\bar{M}_i = \begin{bmatrix} 1 & 0 & 0 & 0 & 0 & 0 \\ 0 & 0 & 0 & 0 & 0 & 0 \\ 0 & 0 & 1 & 0 & 0 & 0 \\ 2/R_i & 0 & 0 & 1 & 0 & 0 \\ 0 & 0 & 0 & 0 & -1 & 0 \\ 0 & 0 & 2/R_i & 0 & 0 & 1 \end{bmatrix}. \tag{4.6}$$

Spherical refracting boundary surfaces can also be classified as either concave or convex (Figs. 4.7 and 4.8). The optical axis of such a boundary surface is the straight-line segment passing through its geometrical center. Assuming that the y_i axis points from $^i\bar{P}_{i-1,axis}$ to $^i\bar{P}_{i,axis}$, the unit directional vector of the incoming ray has the form $^i\bar{\ell}_{i-1,axis} = [0 \quad 1 \quad 0 \quad 0]^T$. Again, recalling that a convex boundary surface has a positive value of R_i while a concave boundary surface has a negative

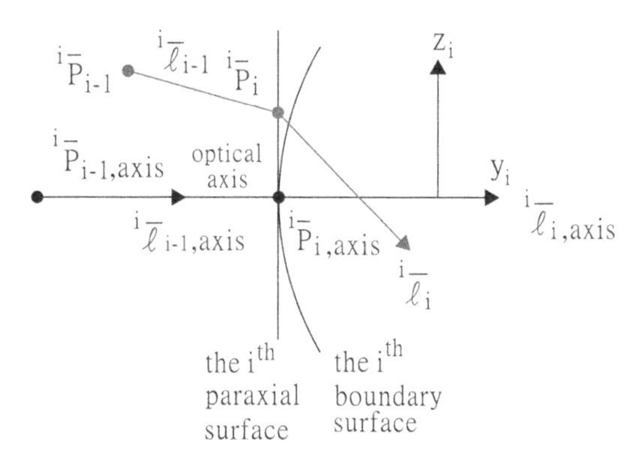

Fig. 4.7 Refracting convex spherical boundary surface

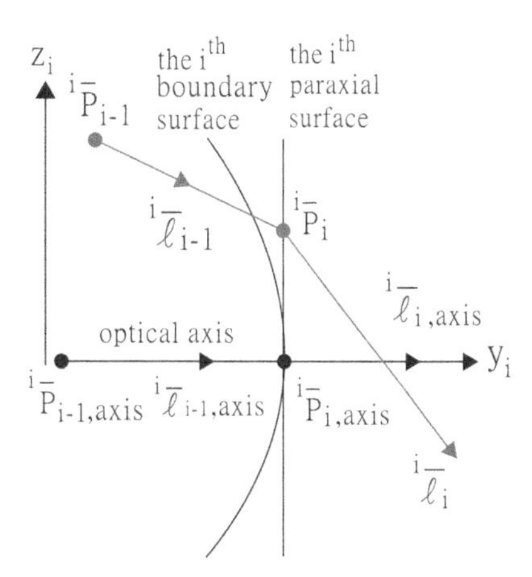

Fig. 4.8 Refracting concave spherical boundary surface

value of R_i, then the following refraction matrix $\bar{M}_i$ is valid for both convex and concave spherical refracting boundary surfaces:

$$\bar{M}_i = \begin{bmatrix} 1 & 0 & 0 & 0 & 0 & 0 \\ 0 & 0 & 0 & 0 & 0 & 0 \\ 0 & 0 & 1 & 0 & 0 & 0 \\ (N_i-1)/R_i & 0 & 0 & N_i & 0 & 0 \\ 0 & 0 & 0 & 0 & N_i^2 & 0 \\ 0 & 0 & (N_i-1)/R_i & 0 & 0 & N_i \end{bmatrix}. \tag{4.7}$$

It is noted that Eqs. (4.4) and (4.5) can be deduced from Eqs. (4.7) and (4.6), respectively, by setting the radius, R_i, to infinity.

To trace a paraxial ray in an optical system possessing n paraxial surfaces, it is first necessary to label the surfaces sequentially from i = 1 to i = n. Having computed the differential ray $\Delta^{i}\bar{R}_i$ at boundary coordinate frame $(xyz)_i$ using Eq. (4.2), $\Delta^{i}\bar{R}_i$ can be transferred to the next boundary coordinate frame $(xyz)_{i+1}$ via

$$\begin{bmatrix} \Delta^{i+1}\bar{P}_i \\ \Delta^{i+1}\bar{\ell}_i \end{bmatrix} = \begin{bmatrix} a_x & b_x & c_x & 0 & 0 & 0 \\ a_y & b_y & c_y & 0 & 0 & 0 \\ a_z & b_z & c_z & 0 & 0 & 0 \\ 0 & 0 & 0 & a_x & b_x & c_x \\ 0 & 0 & 0 & a_y & b_y & c_y \\ 0 & 0 & 0 & a_z & b_z & c_z \end{bmatrix} \begin{bmatrix} \Delta^{i}\bar{P}_i \\ \Delta^{i}\bar{\ell}_i \end{bmatrix} = {}^{i+1}\bar{B}_i \begin{bmatrix} \Delta^{i}\bar{P}_i \\ \Delta^{i}\bar{\ell}_i \end{bmatrix}$$
$$= {}^{i+1}\bar{B}_i\bar{M}_i\bar{T}_i \begin{bmatrix} \Delta^{i}\bar{P}_{i-1} \\ \Delta^{i}\bar{\ell}_{i-1} \end{bmatrix}, \tag{4.8}$$

where the components of ${}^{i+1}\bar{B}_i$ are the orientation components of the transformation matrix ${}^{i+1}\bar{A}_i = ({}^{0}\bar{A}_{i+1})^{-1}\,{}^{0}\bar{A}_i$, i.e., the pose matrix of $(xyz)_i$ with respect to $(xyz)_{i+1}$. That is,

$${}^{i+1}\bar{A}_i = {}^{i+1}\bar{A}_0{}^{0}\bar{A}_i = \left({}^{0}\bar{A}_{i+1}\right)^{-1}{}^{0}\bar{A}_i = \begin{bmatrix} a_x & b_x & c_x & d_x \\ a_y & b_y & c_y & d_y \\ a_z & b_z & c_z & d_z \\ 0 & 0 & 0 & 1 \end{bmatrix}. \tag{4.9}$$

Note that matrices ${}^{0}\bar{A}_i$ and ${}^{0}\bar{A}_{i+1}$ in Eq. (4.9) are given by Eq. (2.9) or Eq. (2.35).

If the boundary surfaces of an optical system with n boundary surfaces are labeled from i = 1 to i = n, then from Eq. (4.2) with i = n, and by successive applications of Eq. (4.8), the differential ray $\Delta^{n}\bar{R}_n$ at the nth boundary surface can be estimated in terms of the differential source ray $\Delta\bar{R}_0$ as follows:

$$\begin{bmatrix} \Delta^{n}\bar{P}_n \\ \Delta^{n}\bar{\ell}_n \end{bmatrix} = \bar{M}_n\bar{T}_n \begin{bmatrix} \Delta^{n}\bar{P}_{n-1} \\ \Delta^{n}\bar{\ell}_{n-1} \end{bmatrix} = (\bar{M}_n\bar{T}_n)({}^{n}\bar{B}_{n-1}\bar{M}_{n-1}\bar{T}_{n-1}) \begin{bmatrix} \Delta^{n-1}\bar{P}_{n-2} \\ \Delta^{n-1}\bar{\ell}_{n-2} \end{bmatrix}$$
$$= (\bar{M}_n\bar{T}_n)({}^{n}\bar{B}_{n-1}\bar{M}_{n-1}\ldots\bar{T}_{i+1})({}^{i+1}\bar{B}_i\bar{M}_i\bar{T}_i)\ldots({}^{2}\bar{B}_1\bar{M}_1\bar{T}_1) \begin{bmatrix} \Delta^{1}\bar{P}_0 \\ \Delta^{1}\bar{\ell}_0 \end{bmatrix} \tag{4.10}$$
$$= (\bar{M}_n\bar{T}_n)({}^{n}\bar{B}_{n-1}\bar{M}_{n-1}\ldots\bar{T}_{i+1})({}^{i+1}\bar{B}_i\bar{M}_i\bar{T}_i)\ldots({}^{2}\bar{B}_1\bar{M}_1\bar{T}_1)({}^{1}\bar{B}_0) \begin{bmatrix} \Delta\bar{P}_0 \\ \Delta\bar{\ell}_0 \end{bmatrix}.$$

Furthermore, the differential source ray $\begin{bmatrix} \Delta\bar{P}_0 & \Delta\bar{\ell}_0 \end{bmatrix}^T$ in Eq. (4.10) can be further estimated by the first-order Taylor series expansion of Eqs. (2.2) and (2.3) with respect to $\bar{X}_0 = \bar{0}$, i.e.,

$$\begin{bmatrix} \Delta\bar{P}_0 \\ \Delta\bar{\ell}_0 \end{bmatrix} = \begin{bmatrix} \Delta P_{0x} \\ \Delta P_{0y} \\ \Delta P_{0z} \\ \Delta\ell_{0x} \\ \Delta\ell_{0y} \\ \Delta\ell_{0z} \end{bmatrix} = \begin{bmatrix} 1 & 0 & 0 & 0 & 0 \\ 0 & 1 & 0 & 0 & 0 \\ 0 & 0 & 1 & 0 & 0 \\ 0 & 0 & 0 & -1 & 0 \\ 0 & 0 & 0 & 0 & 0 \\ 0 & 0 & 0 & 0 & 1 \end{bmatrix} \begin{bmatrix} \Delta P_{0x} \\ \Delta P_{0y} \\ \Delta P_{0z} \\ \Delta\alpha_0 \\ \Delta\beta_0 \end{bmatrix} = \bar{M}_0 \begin{bmatrix} \Delta P_{0x} \\ \Delta P_{0y} \\ \Delta P_{0z} \\ \Delta\alpha_0 \\ \Delta\beta_0 \end{bmatrix}. \quad (4.11)$$

Substituting Eq. (4.11) into Eq. (4.10), the following equation is obtained for the differential ray $\Delta^n\bar{R}_n$:

$$\Delta^n\bar{R}_n = \begin{bmatrix} \Delta^n\bar{P}_n \\ \Delta^n\bar{\ell}_n \end{bmatrix} = (\bar{M}_n\bar{T}_n)(^n\bar{B}_{n-1}\bar{M}_{n-1}\ldots\bar{T}_{i+1})(^{i+1}\bar{B}_i\bar{M}_i\bar{T}_i)\ldots(^2\bar{B}_1\bar{M}_1\bar{T}_1)(^1\bar{B}_0\bar{M}_0)\begin{bmatrix} \Delta P_{0x} \\ \Delta P_{0y} \\ \Delta P_{0z} \\ \Delta\alpha_0 \\ \Delta\beta_0 \end{bmatrix}. \quad (4.12)$$

In addition to the skew-ray tracing methodology presented in Chaps. 2 and 3 of this book, the skew-ray can also be estimated by the sum of the axial ray $^n\bar{R}_{n,axis}$ and the differential ray $\Delta^n\bar{R}_n$ given in Eq. (4.12). It should be noted that the nth boundary surface of an optical system is usually a plane for displaying the image, e.g., a flat screen or a flat photographic sensor. In this case $\Delta^n\bar{\ell}_n$ is of no further interest and can therefore be ignored.

4.2 Conventional 2 × 2 Raytracing Matrices for Paraxial Optics

In addressing paraxial optics for axis-symmetrical systems, it is first necessary to adopt a convention for the algebraic signs given to the various distances involved. In this chapter, the following conventions are applied:

1. In paraxial optics, a distance has both a magnitude (to show how long it is) and a direction. The distance to the left of a reference point is negative; while that to the right is positive. The magnitude of the distance is never negative. However, to show that a particular distance in an illustrative figure is negative, a minus sign, "–", is added to the corresponding symbol.
2. The height above the optical axis is positive, while that below is negative. Again, to show that a particular height in an illustrative figure is negative, a minus sign, "–", is added to the corresponding symbol.

3. The focal length of a converging lens is positive, while that of a diverging lens is negative.
4. The refractive index ξ_i of any medium i is positive.
5. The slope μ_i of a ray (i.e., the angle between the ray and the optical axis, measured in radians) is positive if a counterclockwise rotation turns the ray from the positive direction of the y_0 axis to the direction of the ray, when the ray moves from left to right. Conversely, the slope μ_i of a ray is negative in that case if a clockwise rotation turns the ray from the positive direction of the y_0 axis to the direction of the ray. Again, a negative slope is indicated in the illustrative figures in this book via the addition of a minus sign in front of the corresponding symbol.

It should be noted that the boundary coordinate frames $(xyz)_i$ (i = 1 to i = n) and world coordinate frame $(xyz)_0$ are usually not parallel to one another in non-axially symmetrical optical systems. Thus, a numerical method is required to compute ${}^{i+1}\bar{B}_i$ in Eq. (4.8). However, the paraxial raytracing equations are greatly simplified for the case of axis-symmetrical systems, since ${}^{i+1}\bar{B}_i = \bar{I}_{6\times6}$.

For simplicity, in the following discussions, a flat boundary surface is treated as a special case of a spherical boundary surface with a radius R_i equal to infinity. Consequently, in deriving the equations of paraxial optics, it is necessary to consider only whether the boundary surface is a reflecting surface or a refracting surface.

4.2.1 Refracting Boundary Surfaces

For a refracting boundary surface in an axis-symmetrical system (e.g., Figs. 4.4, 4.7 and 4.8), the x, y and z axes of coordinate frames $(xyz)_0$, $(xyz)_i$ and $(xyz)_{i+1}$, respectively, are parallel. As a result, ${}^{i+1}\bar{B}_i$ in Eq. (4.8) is a unit matrix. Consequently, Eq. (4.8) can be further modified to a 2×2 raytracing equation if rows 1, 2, 4 and 5 and columns 1, 2, 4 and 5 of the refraction matrix, $\bar{M}_i$, and transfer matrix, $\bar{T}_i$, are deleted, i.e.,

$$\begin{bmatrix} \Delta^{i+1}P_{iz} \\ \Delta^{i+1}\ell_{iz} \end{bmatrix} = \begin{bmatrix} 1 & 0 \\ (\xi_{i-1} - \xi_i)/(\xi_i R_i) & \xi_{i-1}/\xi_i \end{bmatrix} \begin{bmatrix} 1 & \lambda_{i,axis} \\ 0 & 1 \end{bmatrix} \begin{bmatrix} \Delta^{i}P_{i-1z} \\ \Delta^{i}\ell_{i-1z} \end{bmatrix}. \quad (4.13)$$

According to the definitions of $\Delta\bar{P}_i$ and $\Delta\bar{\ell}_i$ given in Eq. (4.1), it follows that

$$\Delta^{i+1}P_{iz} = {}^{i+1}P_{iz} - {}^{i+1}P_{iz,axis} = {}^{i+1}P_{iz} - 0 = {}^{i+1}P_{iz} = z_i \quad (4.14)$$

and

$$\Delta^{i+1}\ell_{iz} = {}^{i+1}\ell_{iz} - {}^{i+1}\ell_{iz,axis} = {}^{i+1}\ell_{iz} - 0 = {}^{i+1}\ell_{iz} = S\mu_i \approx \mu_i. \tag{4.15}$$

In paraxial optics, the ray height z_i (Eq. (4.14)) is used rather than ${}^{i+1}P_{iz}$. Furthermore, ${}^{i+1}\ell_{iz}$ is approximated as ${}^{i+1}\ell_{iz} = S\mu_i \approx \mu_i$, where μ_i is the angle (usually small) in radians between the ray and the optical axis (see Table 4.1). Similarly, the expressions $\Delta^i P_{i-1z} = z_{i-1}$ and $\Delta^i \ell_{i-1z} \approx \mu_{i-1}$ also hold. Accordingly, using Eqs. (4.14) and (4.15), Eq. (4.13) can be rewritten as

$$\begin{bmatrix} z_i \\ \mu_i \end{bmatrix} = \begin{bmatrix} 1 & 0 \\ (\xi_{i-1} - \xi_i)/(\xi_i R_i) & \xi_{i-1}/\xi_i \end{bmatrix} \begin{bmatrix} 1 & \lambda_{i,axis} \\ 0 & 1 \end{bmatrix} \begin{bmatrix} z_{i-1} \\ \mu_{i-1} \end{bmatrix}. \tag{4.16}$$

Rearranging Eq. (4.16) yields

$$\begin{bmatrix} z_i \\ \xi_i\mu_i \end{bmatrix} = \begin{bmatrix} 1 & 0 \\ (\xi_{i-1} - \xi_i)/R_i & 1 \end{bmatrix} \begin{bmatrix} 1 & \lambda_{i,axis}/\xi_{i-1} \\ 0 & 1 \end{bmatrix} \begin{bmatrix} z_{i-1} \\ \xi_{i-1}\mu_{i-1} \end{bmatrix} = \bar{M}_i \bar{T}_i \begin{bmatrix} z_{i-1} \\ \xi_{i-1}\mu_{i-1} \end{bmatrix}, \tag{4.17}$$

in which the transfer and refraction matrices are newly defined respectively as

$$\bar{T}_i = \begin{bmatrix} 1 & \lambda_{i,axis}/\xi_{i-1} \\ 0 & 1 \end{bmatrix}, \tag{4.18}$$

$$\bar{M}_i = \begin{bmatrix} 1 & 0 \\ (\xi_{i-1} - \xi_i)/R_i & 1 \end{bmatrix}. \tag{4.19}$$

Note that $\begin{bmatrix} z_i & \xi_i\mu_i \end{bmatrix}^T$ in Eq. (4.17) is referred to as the height-slope matrix at the ith paraxial surface.

4.2.2 Reflecting Boundary Surfaces

For a reflecting boundary surface, the rays are reflected from right to left if the incoming rays move from left to right (see Figs. 4.3, 4.5 and 4.6). Furthermore, ${}^{i+1}\bar{B}_i$ in Eq. (4.8) is again a unit matrix for axis-symmetrical systems. As for a refracting boundary surface, Eq. (4.8) for a reflecting boundary surface can be transformed into a 2 × 2 raytracing equation if rows 1, 2, 4 and 5 and columns 1, 2, 4 and 5 of the transfer matrix, $\bar{T}_i$, and reflection matrix, $\bar{M}_i$, are deleted, i.e.,

$$\begin{bmatrix} \Delta^{i+1}\bar{P}_i \\ \Delta^{i+1}\bar{\ell}_i \end{bmatrix} = \begin{bmatrix} 1 & 0 \\ 2/R_i & 1 \end{bmatrix} \begin{bmatrix} 1 & \lambda_{i,axis} \\ 0 & 1 \end{bmatrix} \begin{bmatrix} \Delta^i\bar{P}_{i-1} \\ \Delta^i\bar{\ell}_{i-1} \end{bmatrix}. \tag{4.20}$$

Table 4.1 Sign notations and formulations used in conventional paraxial optics for axis-symmetrical systems

R_i		μ_i		$\bar{T}_i$	$\bar{M}_i$	
Convex	Concave	Ray moves from left to right	Ray moves from right to left	Transfer matrix	Refraction matrix	Reflection matrix
Positive value	Negative value	ray, μ_i, ray, $-\mu_i$	μ_i, ray, $-\mu_i$, ray	$\begin{bmatrix} 1 & \frac{\lambda_{i,axis}}{\xi_{i-1}} \\ 0 & 1 \end{bmatrix}$	$\begin{bmatrix} 1 & 0 \\ \frac{\xi_{i-1} - \xi_i}{R_i} & 1 \end{bmatrix}$	$\begin{bmatrix} 1 & 0 \\ \frac{2\xi_i}{R_i} & 1 \end{bmatrix}$

Equations (4.14) and (4.15) are still valid with $\Delta^{i+1}P_{iz} = z_i$ and $\Delta^{i+1}\ell_{iz} = {}^{i+1}\ell_{iz} = S(\pi - \mu_i) \approx \mu_i$, where μ_i is the angle (again, usually small) in radians between the ray and the negative direction of the optical axis (see Table 4.1). Similarly, $\Delta^{i}P_{i-1z} = z_{i-1}$ and $\Delta^{i}\ell_{i-1z} \approx \mu_{i-1}$ also hold. Equation (4.20) can thus be rewritten as

$$\begin{bmatrix} z_i \\ \mu_i \end{bmatrix} = \begin{bmatrix} 1 & 0 \\ 2/R_i & 1 \end{bmatrix} \begin{bmatrix} 1 & \lambda_{i,axis} \\ 0 & 1 \end{bmatrix} \begin{bmatrix} z_{i-1} \\ \mu_{i-1} \end{bmatrix}, \tag{4.21}$$

or

$$\begin{bmatrix} z_i \\ \xi_i\mu_i \end{bmatrix} = \begin{bmatrix} 1 & 0 \\ 2\xi_i/R_i & 1 \end{bmatrix} \begin{bmatrix} 1 & \lambda_{i,axis}/\xi_{i-1} \\ 0 & 1 \end{bmatrix} \begin{bmatrix} z_{i-1} \\ \xi_{i-1}\mu_{i-1} \end{bmatrix} = \bar{M}_i\bar{T}_i \begin{bmatrix} z_{i-1} \\ \xi_{i-1}\mu_{i-1} \end{bmatrix}, \tag{4.22}$$

in which the transfer matrix has the form given in Eq. (4.18) and the reflection matrix is given as

$$\bar{M}_i = \begin{bmatrix} 1 & 0 \\ 2\xi_i/R_i & 1 \end{bmatrix}. \tag{4.23}$$

For convenience, the sign notations and formulations of paraxial optics for axis-symmetrical optical systems are summarized in Table 4.1.

It is important to note that the transfer matrix $\bar{T}_i$ (Eq. (4.18)), refraction matrices $\bar{M}_i$ (Eq. (4.19)) and reflection matrices $\bar{M}_i$ (Eq. (4.23)) are also valid for the case where the incoming paraxial ray propagates from right to left provided that the direction of the y_0 axis is reversed while that of the z_0 axis is left unchanged in the upward direction.

Example 4.1 A cat's eye can be optimally achieved using a single transparent sphere provided that the refractive index ξ_{e1} of the cat's eye material is exactly two times the refractive index ξ_{air} of the medium through which the incoming ray travels (see Fig. 4.9). For the case of a sphere with a positive radius R_1, the following matrix equation applies:

$$\begin{bmatrix} z_3 \\ 0 \end{bmatrix} = \begin{bmatrix} 1 & 0 \\ (\xi_{e1} - \xi_{air})/(-R_1) & 1 \end{bmatrix} \begin{bmatrix} 1 & 2R_1/\xi_{e1} \\ 0 & 1 \end{bmatrix} \begin{bmatrix} 1 & 0 \\ 2\xi_{e1}/(-R_1) & 1 \end{bmatrix} \begin{bmatrix} 1 & 2R_1/\xi_{e1} \\ 0 & 1 \end{bmatrix}$$
$$\begin{bmatrix} 1 & 0 \\ (\xi_{air1} - \xi_{e1})/R_1 & 1 \end{bmatrix} \begin{bmatrix} 1 & \lambda_{1,axis}/\xi_{air} \\ 0 & 1 \end{bmatrix} \begin{bmatrix} z_0 \\ 0 \end{bmatrix}.$$

It is noted that the necessary condition $\xi_{e1} = 2\xi_{air}$ is obtained from the second component of the equation.

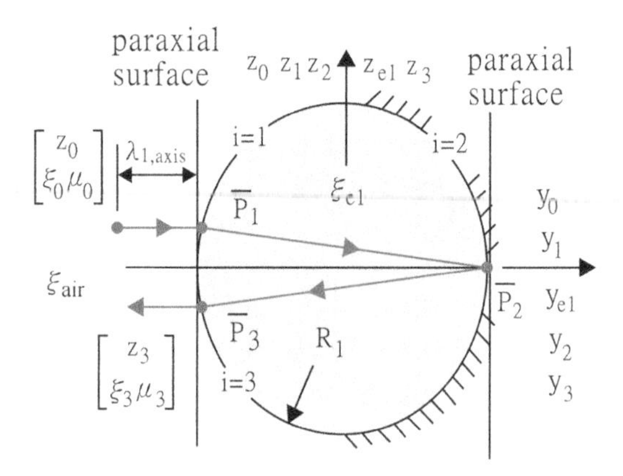

Fig. 4.9 Cat's eye composed of sphere with positive radius R_1

4.3 Conventional Raytracing Matrices for Paraxial Optics Derived from Geometry Relations

As stated at the beginning of this chapter, the derivation procedures described in [4] for 6×6 raytracing matrices for 3-D paraxial optics may be too complex for mathematical treatment. It will also be recalled that the conventional 2×2 raytracing matrices are valid only for paraxial meridional rays traveling in axis-symmetrical systems. However, as discussed in this section, the raytracing matrices required for paraxial optics can be derived using an alternative simpler geometry-based approach. To derive these matrices, it is first necessary to note that a ray at a given point (say, $\bar{Q}'_i$ in Fig. 4.10) in a homogeneous medium confined in the y_0z_0 plane of an axis-symmetrical system can be identified by its height z'_i above the optical axis, the paraxial angle μ'_i the ray makes with the optical axis, and the refractive index ξ'_i of the medium through which it travels. Therefore, the height z'_i

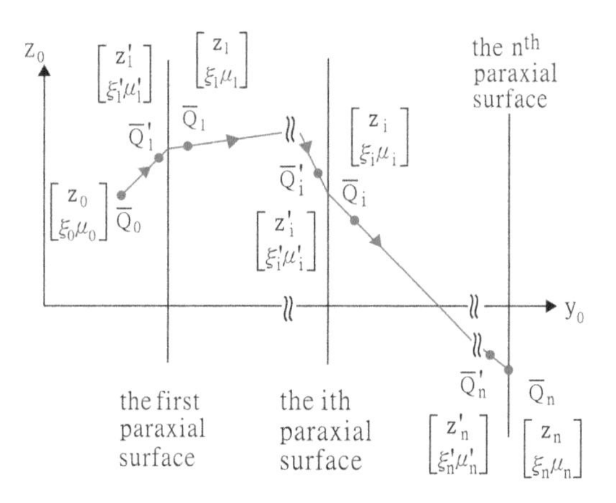

Fig. 4.10 Axis-symmetrical system with multiple paraxial surfaces labeled from i = 1 to i = n

and product $\xi_i' \mu_i'$ are chosen in this chapter to describe the transformation of a ray (denoted as the height-slope matrix $\begin{bmatrix} z_i' & \xi_i' \mu_i' \end{bmatrix}^T$).

Figure 4.10 shows an axis-symmetrical system containing multiple paraxial surfaces labeled sequentially from i = 1 to i = n. It is noted that two points, $\bar{Q}_i'$ and $\bar{Q}_i$, are shown near every real boundary surface. These points indicate the infinitesimal neighboring points along the ray before and after the refraction/reflection occurred at the real boundary surface, respectively. The height-slope matrices of points $\bar{Q}_i'$ and $\bar{Q}_i$ are also given in the figure. In the remaining sections of this chapter, the 2 × 2 raytracing matrices for such an axis-symmetrical system are derived via geometry relations for the following five cases:

(1) the transfer matrix for a ray propagating along a straight-line path;
(2) the refraction matrix at a flat refractive boundary surface;
(3) the reflection matrix at a flat mirror;
(4) the refraction matrix at a spherical refractive boundary surface;
(5) the reflection matrix at a spherical mirror.

4.3.1 Transfer Matrix for Ray Propagating Along Straight-Line Path

When a meridional ray propagates along a straight-line path, e.g., from $\bar{Q}_{i-1}$ to $\bar{Q}_i'$ in Fig. 4.11, the ray height z_i' at $\bar{Q}_i'$ can be described in terms of the ray height z_{i-1} at $\bar{Q}_{i-1}$ as

$$z_i' = z_{i-1} + \lambda_{i,axis} \tan(\mu_{i-1}), \tag{4.24}$$

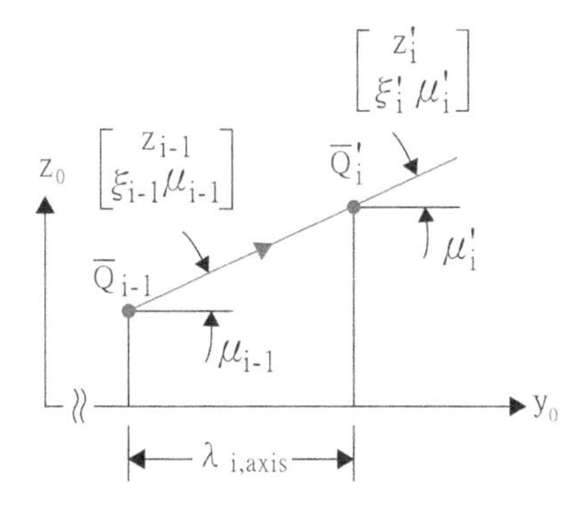

Fig. 4.11 Meridional ray, with positive small angles μ_{i-1} and μ_i', traveling along straight-line path

where μ_{i-1} is the angle in radians between the ray and the optical axis and has a positive value when lying in the counterclockwise direction. If μ_{i-1} is sufficiently small, Eq. (4.24) can be approximated by

$$z_i' = z_{i-1} + \lambda_{i,axis}\,\mu_{i-1}. \tag{4.25}$$

Furthermore, since the ray is assumed to travel in a medium with a constant refractive index $\xi_i' = \xi_{i-1}$ and with a fixed direction $\mu_i' = \mu_{i-1}$, the following equation applies:

$$\xi_i'\,\mu_i' = \xi_{i-1}\,\mu_{i-1}. \tag{4.26}$$

Equations (4.25) and (4.26) can be rewritten in the following matrix form:

$$\begin{bmatrix} z_i' \\ \xi_i'\mu_i' \end{bmatrix} = \begin{bmatrix} 1 & \lambda_{i,axis}/\xi_{i-1} \\ 0 & 1 \end{bmatrix} \begin{bmatrix} z_{i-1} \\ \xi_{i-1}\,\mu_{i-1} \end{bmatrix}. \tag{4.27}$$

The transfer matrix $\bar{T}_i$ representing a translation to the right through a distance $\lambda_{i,axis}$ along the optical axis has the form

$$\bar{T}_i = \begin{bmatrix} 1 & \lambda_{i,axis}/\xi_{i-1} \\ 0 & 1 \end{bmatrix}. \tag{4.28}$$

Note that its determinant, $\det(\bar{T}_i)$, is obviously unity.

It should be noted that Eqs. (4.24) and (4.25) are equally valid for the case shown in Fig. 4.12, in which μ_{i-1} and μ' are both negative (as shown in Table 4.2). Therefore, Eq. (4.27) can again be used to estimate the height-slope matrix at point $\bar{Q}_i'$ if the height-slope matrix at point $\bar{Q}_{i-1}$ is known and the ray propagates along the y_0 axis.

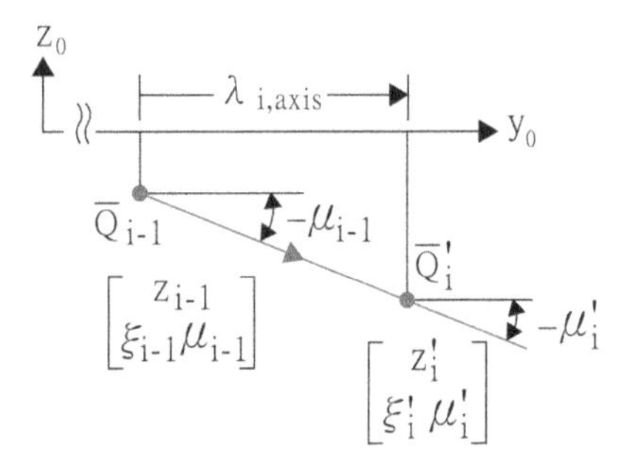

Fig. 4.12 Meridional ray, with negative small angles μ_{i-1} and μ_i', traveling along straight-line path

Table 4.2 Signs of paraxial angles shown in Figs. 4.11 and 4.12

Parameter / Case	μ_{i-1}	μ_i'
Figure 4.11	+	+
Figure 4.12	–	–

4.3.2 *Refraction Matrix at Refractive Flat Boundary Surface*

Figure 4.13 shows a paraxial ray traveling upward from point $\bar{Q}_i'$ to point $\bar{Q}_i$ with greatly exaggerated angles. If the incidence and refraction angles, θ_i and $\underline{\theta}_i$, are sufficiently small, the following simplified equation is obtained for Snell's law $\xi_i' S\theta_i = \xi_i S\underline{\theta}_i$:

$$\xi_i \underline{\theta}_i = \xi_i' \theta_i. \tag{4.29}$$

An observation of Fig. 4.13 shows the existence of the following relations: $\theta_i = \mu_i'$ and $\underline{\theta}_i = \mu_i$ (see also Table 4.3). Thus, Eq. (4.29) can be reformulated as

$$\xi_i \mu_i = \xi_i' \mu_i'. \tag{4.30}$$

If $\bar{Q}_i'$ and $\bar{Q}_i$ are infinitesimal neighboring points of incidence point $\bar{P}_i$ at the real boundary surface $\bar{r}_i$, then their heights, z_i' and z_i, are almost equal. That is

$$z_i = z_i'. \tag{4.31}$$

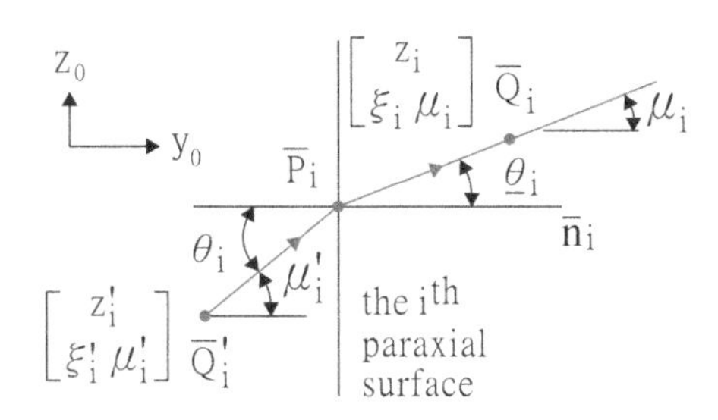

Fig. 4.13 Meridional ray, with positive small angle μ_i', refracted by flat boundary surface perpendicular to y_0 axis

Table 4.3 Relationships among parameters for paraxial ray refracted at refractive flat boundary surface

Parameter / Case	μ_i'	μ_i	Relation of θ_i and μ_i'	Relation of $\underline{\theta}_i$ and μ_i
Figure 4.13	+	+	$\theta_i = \mu_i'$	$\underline{\theta}_i = \mu_i$
Figure 4.14	–	–	$\theta_i = -\mu_i'$	$\underline{\theta}_i = -\mu_i$

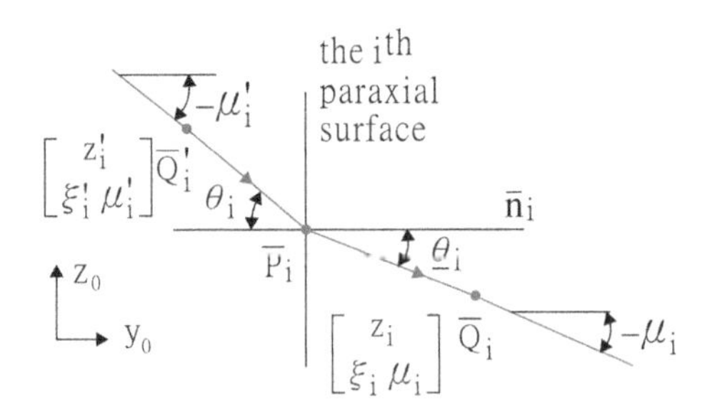

Fig. 4.14 Meridional ray, with negative small angle μ_i', refracted by flat boundary surface perpendicular to y_0 axis

Rearranging Eqs. (4.30) and (4.31) yields the matrix equation shown in Eq. (4.32), which is valid for a paraxial meridional ray refracted by a flat boundary surface provided that the ray travels upward along the y_0 axis:

$$\begin{bmatrix} z_i \\ \xi_i \mu_i \end{bmatrix} = \begin{bmatrix} 1 & 0 \\ 0 & 1 \end{bmatrix} \begin{bmatrix} z_i' \\ \xi_i' \mu_i' \end{bmatrix}. \tag{4.32}$$

The 2 × 2 refraction matrix $\bar{M}_i$ of the paraxial meridional ray is given by

$$\bar{M}_i = \begin{bmatrix} 1 & 0 \\ 0 & 1 \end{bmatrix}, \tag{4.33}$$

with the determinant being $\det(\bar{M}_i) = 1$.

It is noted from Fig. 4.14 and Table 4.3 that the relations $\theta_i = -\mu_i'$ and $\underline{\theta}_i = -\mu_i$ apply when the ray travels in the downward direction and is refracted at flat boundary surface $\bar{r}_i$. Equations (4.30)–(4.33) still hold. Therefore, provided that the ray travels along the y_0 axis, the refraction matrix of the refractive flat boundary surface is once again given by Eq. (4.33).

Example 4.2 Consider the rectangular optical flat shown in Fig. 2.14. (1) Determine the refraction angles μ_1 and μ_2 at boundary surfaces i = 1 and i = 2, respectively, by paraxial optics if $\xi_{air} = 1$. (2) Prove that the exit ray $\bar{R}_2$ is parallel to $\bar{R}_0$. (3) Determine the ray displacement D by paraxial optics.

Solution With no loss of generality, assume that the source ray originates from the origin of frame $(xyz)_0$.

(1) The following paraxial raytracing equation holds for the refracted ray $\bar{R}_1$:

$$\begin{bmatrix} z_1 \\ \xi_{e1} \mu_1 \end{bmatrix} = \begin{bmatrix} 1 & 0 \\ 0 & 1 \end{bmatrix} \begin{bmatrix} 1 & v_1 \\ 0 & 1 \end{bmatrix} \begin{bmatrix} 0 \\ \beta_0 \end{bmatrix} = \begin{bmatrix} v_1 \beta_0 \\ \beta_0 \end{bmatrix}.$$

From the second component of the equation above, the refraction angle at boundary surface i = 1 is obtained as $\mu_1 = \beta_0/\xi_{e1}$.

The height-slope matrix of ray $\bar{R}_2$ is given by

$$\begin{bmatrix} z_2 \\ \xi_2\mu_2 \end{bmatrix} = \begin{bmatrix} z_2 \\ \mu_2 \end{bmatrix} = \begin{bmatrix} 1 & 0 \\ 0 & 1 \end{bmatrix} \begin{bmatrix} 1 & q_{e1}/\xi_{e1} \\ 0 & 1 \end{bmatrix} \begin{bmatrix} v_1\beta_0 \\ \beta_0 \end{bmatrix} = \begin{bmatrix} \beta_0(v_1 + q_{e1}/\xi_{e1}) \\ \beta_0 \end{bmatrix}.$$

By setting $\xi_0 = \xi_2 = \xi_{air} = 1$ in the equation above, the refraction angle at boundary surface i = 2 is obtained as $\mu_2 = \beta_0$.

(2) Since the refraction angle at the second paraxial surface is $\mu_2 = \beta_0$, the exit ray $\bar{R}_2$ is parallel to the source ray $\bar{R}_0$.

(3) From Fig. 2.14, it can be shown that

$$D = [q_{e1}\tan\beta_0 - (z_2 - z_1)]C\beta_0 = q_{e1}(S\beta_0 - \beta_0 C\beta_0/\xi_{e1}).$$

It is noted that the displacement D in the equation above is equal to that given in Example 2.3 when β_0 is small.

4.3.3 *Reflection Matrix at Flat Mirror*

Figure 4.15 shows a paraxial ray traveling upward from $\bar{Q}_i'$ and reflected at a flat mirror to point $\bar{Q}_i$. It is seen that the incidence angle θ_i is equal to the paraxial angle μ_i' before reflection. It is also observed that the reflection angle θ_i is equal to μ_i, i.e., the paraxial angle after reflection (Table 4.3). The reflection event occurs in a medium with constant refractive index (i.e., $\xi_i = \xi_i'$). Thus, the following equation can be derived:

$$\xi_i\,\mu_i = \xi_i'\,\mu_i'. \tag{4.34}$$

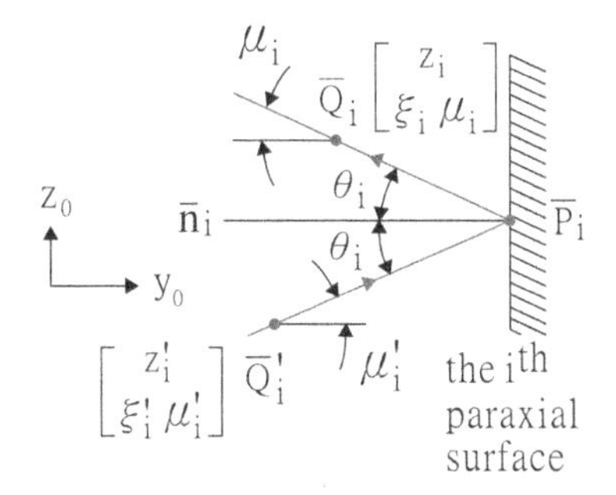

Fig. 4.15 Meridional ray, with positive small angle μ_i', reflected by flat mirror perpendicular to y_0 axis

If $\bar{Q}_i$ and $\bar{Q}'_i$ are infinitesimal neighboring points, then their heights, z_i and z'_i, are almost the same, i.e.,

$$z_i = z'_i. \tag{4.35}$$

Rearranging Eqs. (4.34) and (4.35), the following matrix equation is obtained, which is valid for a paraxial meridional ray reflected by a flat mirror when the ray travels upward along the y_0 axis:

$$\begin{bmatrix} z_i \\ \xi_i \mu_i \end{bmatrix} = \begin{bmatrix} 1 & 0 \\ 0 & 1 \end{bmatrix} \begin{bmatrix} z'_i \\ \xi'_i \mu'_i \end{bmatrix}. \tag{4.36}$$

As shown in Fig. 4.16 and Table 4.4, we have $\theta_i = -\mu'_i$ and $\theta_i = -\mu_i$ when the ray travels in the downward direction and the paraxial angle μ'_i has a small negative value. In other words, Eqs. (4.34)–(4.36) are still valid. Consequently, when a paraxial meridional ray travels from left to right and is reflected by a flat mirror, the following 2×2 reflection matrix $\bar{M}_i$ is obtained:

$$\bar{M}_i = \begin{bmatrix} 1 & 0 \\ 0 & 1 \end{bmatrix}. \tag{4.37}$$

The determinant $\det(\bar{M}_i)$ is obviously unity.

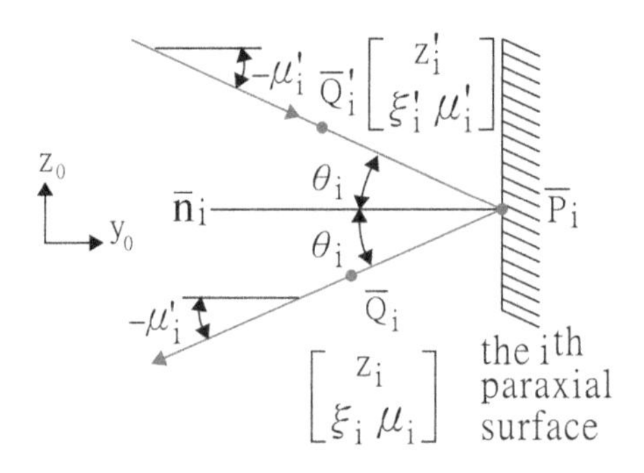

Fig. 4.16 Meridional ray, with negative small angle μ'_i, reflected by flat mirror perpendicular to y_0 axis

Table 4.4 Relationships among parameters for paraxial meridional ray reflected by flat mirror

Parameter / Case	μ'_i	μ_i	Relation of θ_i and μ'_i	Relation of θ_i and μ_i
Figure 4.15	+	+	$\theta_i = \mu'_i$	$\theta_i = \mu_i$
Figure 4.16	–	–	$\theta_i = -\mu'_i$	$\theta_i = -\mu_i$

4.3.4 *Refraction Matrix at Refractive Spherical Boundary Surface*

Figure 4.17 shows a paraxial meridional ray traveling upward from $\bar{Q}'_i$ and refracted at a spherical boundary surface such that it propagates to point $\bar{Q}_i$. If the incidence and refraction angles, θ_i and $\underline{\theta}_i$, are sufficiently small, the following simplified equation for Snell's law ($\xi'_i\ S\theta_i = \xi_i\ S\underline{\theta}_i$) is obtained at incidence point $\bar{P}_i$:

$$\xi'_i\,\theta_i = \xi_i\,\underline{\theta}_i. \tag{4.38}$$

According to the exterior angle theorem, the exterior angle of a triangle is equal to the sum of the two non-adjacent interior angles. Thus, the following two relations are obtained from Fig. 4.17 (see also Table 4.5):

$$\theta_i = \mu'_i + \phi_i. \tag{4.39}$$

and

$$\underline{\theta}_i = \mu_i + \phi_i. \tag{4.40}$$

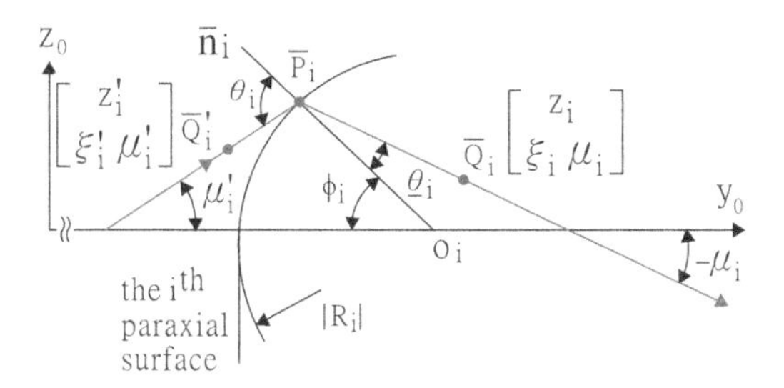

Fig. 4.17 Paraxial meridional ray, with positive angle μ'_i, refracted by convex spherical refractive boundary surface. Note that refracted ray has negative angle μ_i

Table 4.5 Relationships among parameters for paraxial meridional ray refracted by spherical boundary surface

Parameter / Case	θ_i	$\underline{\theta}_i$	μ'_i	μ_i	ϕ_i	Relation of θ_i,ϕ_i, and μ'_i	Relation of $\underline{\theta}_i$, ϕ_i, and μ_i
Figure 4.17	+	+	+	–	Z_i/R_i	$\theta_i = \mu'_i + \phi_i$	$\underline{\theta}_i = \mu_i + \phi_i$
Figure 4.18	+	+	–	+	$-Z_i/R_i$	$\theta_i = -\mu'_i + \phi_i$	$\underline{\theta}_i = -\mu_i + \phi_i$
Figure 4.19	+	+	–	–	Z_i/R_i	$\theta_i = -\mu'_i - \phi_i$	$\underline{\theta}_i = -\mu_i - \phi_i$
Figure 4.20	+	+	+	+	$-Z_i/R_i$	$\theta_i = \mu'_i - \phi_i$	$\underline{\theta}_i = \mu_i - \phi_i$
Figure 4.21	+	+	+	–	$Z_i/(-R_i)$	$\theta_i = -\mu'_i + \phi_i$	$\underline{\theta}_i = -\mu_i + \phi_i$
Figure 4.22	+	+	–	+	$-Z_i/(-R_i)$	$\theta_i = \mu'_i + \phi_i$	$\underline{\theta}_i = \mu_i + \phi_i$
Figure 4.23	+	+	+	+	$Z_i/(-R_i)$	$\theta_i = \mu'_i - \phi_i$	$\underline{\theta}_i = \mu_i - \phi_i$
Figure 4.24	+	+	–	–	$-Z_i/(-R_i)$	$\theta_i = -\mu'_i - \phi_i$	$\underline{\theta}_i = -\mu_i - \phi_i$

The radius of convex spherical boundary surfaces has a positive value. If μ_i' is sufficiently small and $\bar{Q}_i'$ is very near the incidence point $\bar{P}_i$ at the spherical boundary surface $\bar{r}_i$, angle ϕ_i can be approximated as

$$\phi_i \cong z_i'/R_i. \tag{4.41}$$

The following equation can then be obtained after substituting Eqs. (4.39), (4.40) and (4.41) into Eq. (4.38):

$$\xi_i \mu_i = \frac{(\xi_i' - \xi_i) z_i'}{R_i} + \xi_i' \mu_i'. \tag{4.42}$$

Again, if $\bar{Q}_i$ and $\bar{Q}_i'$ are infinitesimal neighboring points, then their heights, z_i and z_i', are almost equal, i.e.,

$$z_i = z_i'. \tag{4.43}$$

Rearranging Eqs. (4.42) and (4.43), the following matrix equation is obtained, which is valid for a paraxial meridional ray refracted by a convex spherical boundary surface provided that the ray travels upward along the y_0 axis:

$$\begin{bmatrix} z_i \\ \xi_i \mu_i \end{bmatrix} = \begin{bmatrix} 1 & 0 \\ (\xi_i' - \xi_i)/R_i & 1 \end{bmatrix} \begin{bmatrix} z_i' \\ \xi_i' \mu_i' \end{bmatrix}. \tag{4.44}$$

The refraction matrix $\bar{M}_i$ of the paraxial meridional ray is given by

$$\bar{M}_i = \begin{bmatrix} 1 & 0 \\ (\xi_i' - \xi_i)/R_i & 1 \end{bmatrix} \tag{4.45}$$

with the determinant being $\det(\bar{M}_i) = 1$. It is noted that Eq. (4.33), i.e., the refraction matrix of a refractive flat boundary surface, can be deduced from Eq. (4.45) by setting the radius R_i to infinity.

Equations (4.38) to (4.45) above are derived for the particular case of the convex spherical boundary surface shown in Fig. 4.17. However, if a thorough analysis is made of the other cases shown in Figs. 4.18, 4.19, 4.20, 4.21, 4.22, 4.23 and 4.24,

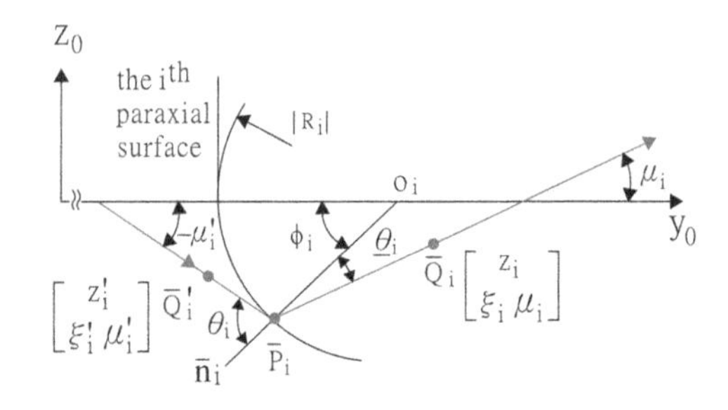

Fig. 4.18 Paraxial meridional ray, with negative angle μ_i', refracted by convex spherical refractive boundary surface. Note that refracted ray has positive angle μ_i

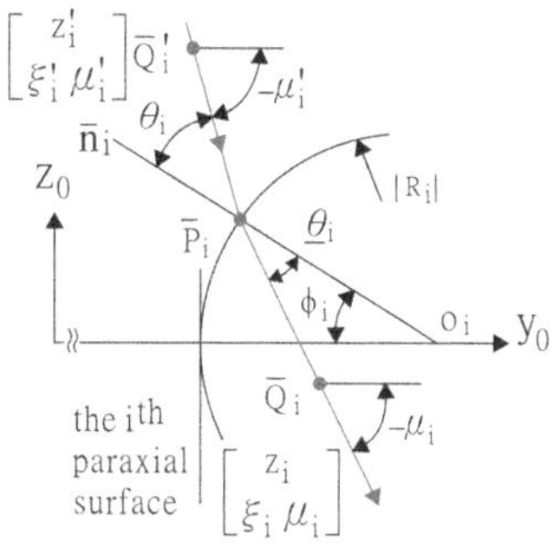

Fig. 4.19 Paraxial meridional ray, with negative angle μ_i', refracted by convex spherical refractive boundary surface. Note that refracted ray has negative angle μ_i

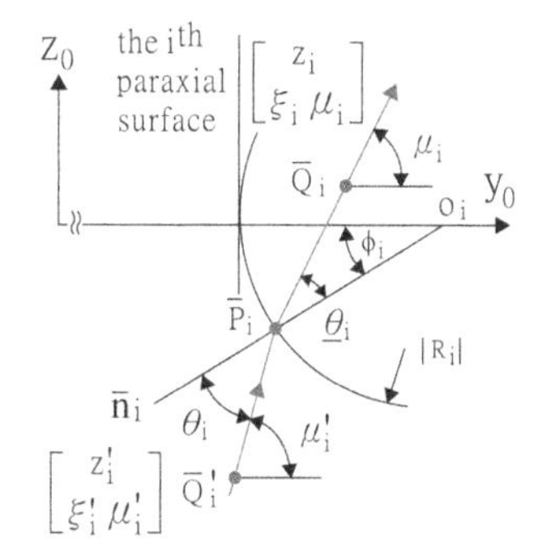

Fig. 4.20 Paraxial meridional ray, with positive angle μ_i', refracted by convex spherical refractive boundary surface. Note that refracted ray has positive angle μ_i

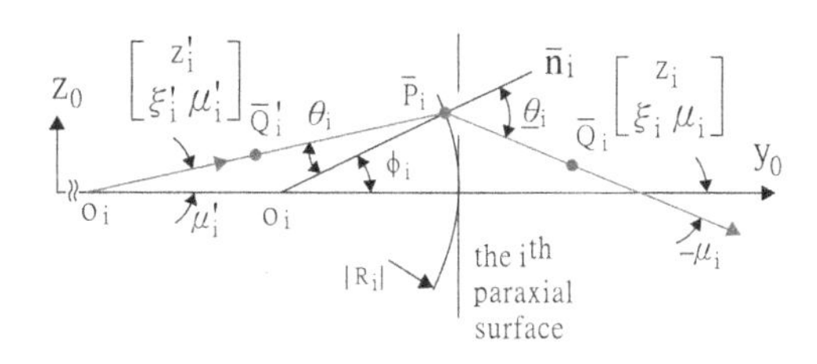

Fig. 4.21 Paraxial meridional ray, with positive angle μ_i', refracted by concave spherical refractive boundary surface. Note that refracted ray has negative angle μ_i

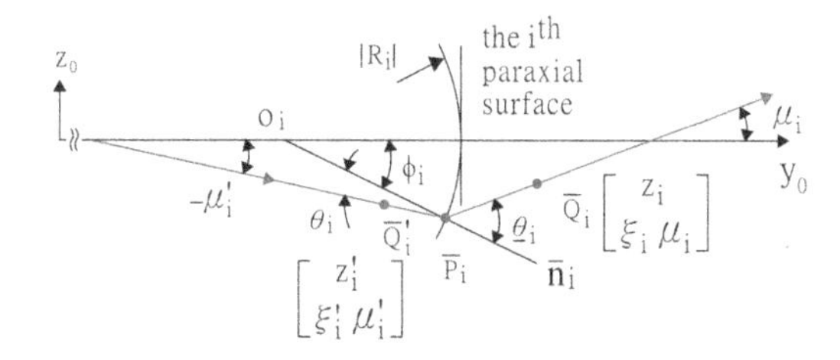

Fig. 4.22 Paraxial meridional ray, with negative angle μ_i', refracted by concave spherical refractive boundary surface. Note that refracted ray has positive angle μ_i

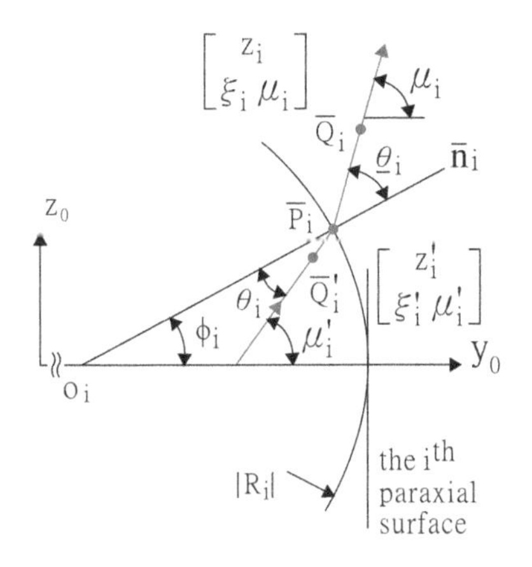

Fig. 4.23 Paraxial meridional ray, with positive angle μ_i', refracted by concave spherical refractive boundary surface. Note that refracted ray has positive angle μ_i

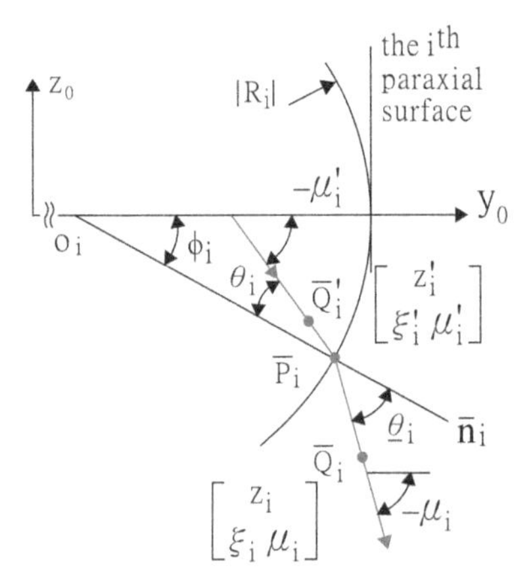

Fig. 4.24 Paraxial meridional ray, with negative angle μ_i', refracted by concave spherical refractive boundary surface. Note that refracted ray has negative angle μ_i

the same refraction matrix $\bar{M}_i$ is obtained. The corresponding parameter relations are listed in Table 4.5. Therefore, Eq. (4.45) is valid for both convex and concave spherical mirrors provided that the incoming ray travels along the y_0 axis.

4.3.5 Reflection Matrix at Spherical Mirror

Figure 4.25 shows a paraxial meridional ray traveling upward and reflected at point $\bar{P}_i$ on a spherical mirror such that it propagates to point $\bar{Q}_i$. In accordance with the basic laws of reflection, the incidence angle is equal to the reflection angle at incidence point $\bar{P}_i$, i.e.,

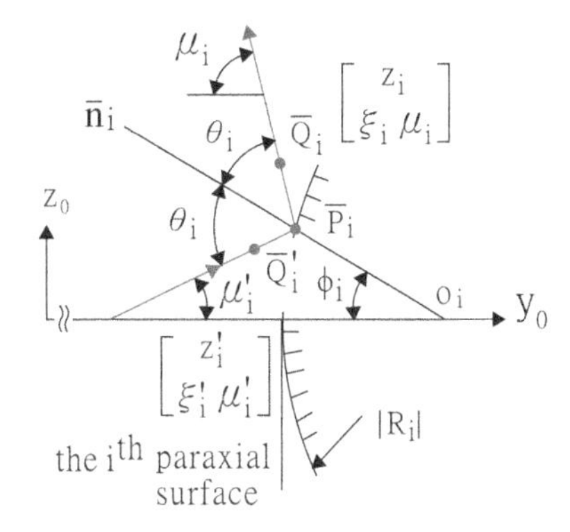

Fig. 4.25 Paraxial meridional ray, with positive angle μ_i', reflected by convex spherical mirror. Note that refracted ray has positive angle μ_i

Table 4.6 Relationships among parameters for paraxial meridional ray reflected by spherical boundary surface

Parameter / Case	θ_i	θ_i	μ_i'	μ_i	ϕ_i	Relation of θ_i, ϕ_i, and μ_i'	Relation of θ_i, ϕ_i, and μ_i
Figure 4.25	+	+	+	+	Z_i/R_i	$\theta_i = \mu_i' + \phi_i$	$\theta_i = \mu_i - \phi_i$
Figure 4.26	+	+	–	–	$-Z_i/R_i$	$\theta_i = -\mu_i' + \phi_i$	$\theta_i = -\mu_i - \phi_i$
Figure 4.27	+	+	–	–	Z_i/R_i	$\theta_i = -\mu_i' - \phi_i$	$\theta_i = -\mu_i + \phi_i$
Figure 4.28	+	+	+	+	$-Z_i/R_i$	$\theta_i = \mu_i' - \phi_i$	$\theta_i = \mu_i + \phi_i$
Figure 4.29	+	+	+	–	$Z_i/(-R_i)$	$\theta_i = -\mu_i' + \phi_i$	$\theta_i = -\mu_i - \phi_i$
Figure 4.30	+	+	–	+	$-Z_i/(-R_i)$	$\theta_i = \mu_i' + \phi_i$	$\theta_i = \mu_i - \phi_i$
Figure 4.31	+	+	+	–	$Z_i/(-R_i)$	$\theta_i = \mu_i' - \phi_i$	$\theta_i = \mu_i + \phi_i$
Figure 4.32	+	+	–	+	$-Z_i/(-R_i)$	$\theta_i = -\mu_i' - \phi_i$	$\theta_i = -\mu_i + \phi_i$

$$\xi_i' \theta_i = \xi_i \theta_i. \tag{4.46}$$

Since θ_i, μ_i', μ_i and ϕ_i are positive acute angles, the following relations are obtained from the exterior angle theorem (see also Table 4.6):

$$\theta_i = \mu_i' + \phi_i, \tag{4.47}$$

$$\mu_i = \theta_i + \phi_i. \tag{4.48}$$

The radius of convex spherical boundary surfaces has a positive value. If μ_i' is sufficiently small and $\bar{Q}_i'$ is very near the incidence point $\bar{P}_i$, the angle ϕ_i can be estimated as

$$\phi_i \cong z_i'/R_i. \tag{4.49}$$

Substituting Eqs. (4.47), (4.48) and (4.49) into Eq. (4.46), it can be shown that

$$\xi_i \mu_i = \frac{2\xi_i z_i'}{R_i} + \xi_i' \mu_i'. \tag{4.50}$$

If $\bar{Q}_i'$ and $\bar{Q}_i$ are infinitesimal neighboring points, their heights are almost equal, i.e.,

$$z_i = z_i'. \tag{4.51}$$

Rearranging Eqs. (4.50) and (4.51), the following matrix equation is obtained, which is valid for a paraxial meridional ray reflected by a convex spherical boundary surface provided that the ray travels upward from left to right:

$$\begin{bmatrix} z_i \\ \xi_i \mu_i \end{bmatrix} = \begin{bmatrix} 1 & 0 \\ 2\xi_i/R_i & 1 \end{bmatrix} \begin{bmatrix} z_i' \\ \xi_i' \mu_i' \end{bmatrix}. \tag{4.52}$$

The refraction matrix $\bar{M}_i$ of the paraxial meridional ray is given by

$$\bar{M}_i = \begin{bmatrix} 1 & 0 \\ 2\xi_i/R_i & 1 \end{bmatrix} \tag{4.53}$$

with its determinant being $\det(\bar{M}_i) = 1$. It is noted that Eq. (4.37), i.e., the reflection matrix of a flat mirror, can be deduced from Eq. (4.53) by setting the radius R_i to infinity.

Equations (4.46)–(4.53) consider the particular case shown in Fig. 4.25, in which the angles (i.e., $\mu_i, \mu_i', \phi_i, \theta_i$), radius R_i, and heights (z_i, z_i') are all positive. However, a thorough analysis of the other cases shown in Figs. 4.26, 4.27, 4.28, 4.29, 4.30, 4.31 and 4.32, where some of the angles, radii and heights are negative, shows that the same refraction matrix $\bar{M}_i$ given in Eq. (4.53) is obtained (Table 4.6). In other words, Eq. (4.53) can be used to estimate the ray height z_i and direction angle μ_i for a paraxial ray refracted by both convex and concave spherical mirrors provided that the ray propagates along the y_0 axis.

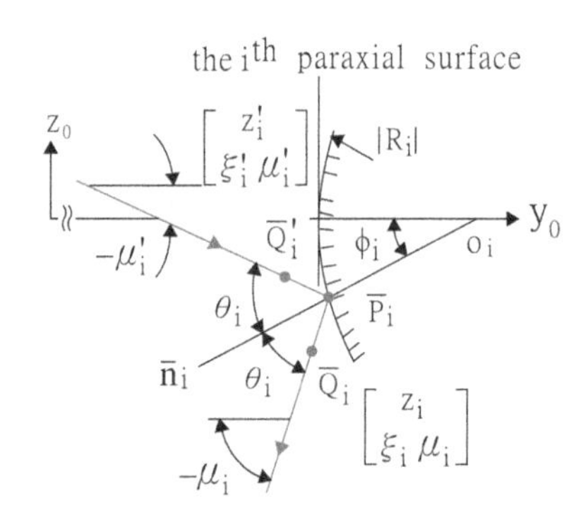

Fig. 4.26 Paraxial meridional ray, with negative angle μ_i', reflected by convex spherical mirror. Note that refracted ray has negative angle μ_i

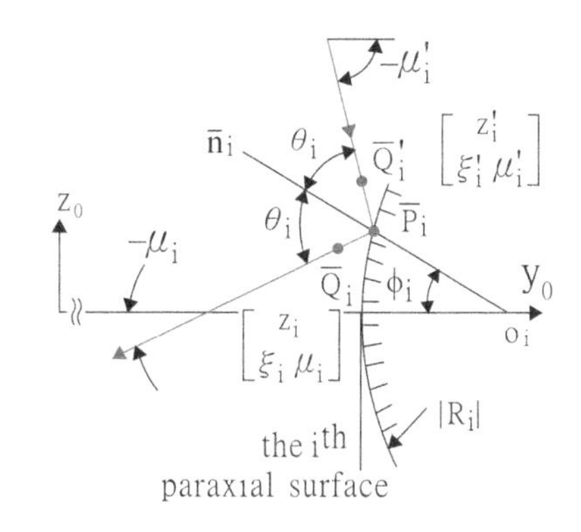

Fig. 4.27 Paraxial meridional ray, with negative angle μ_i', reflected by convex spherical mirror. Note that refracted ray has negative angle μ_i

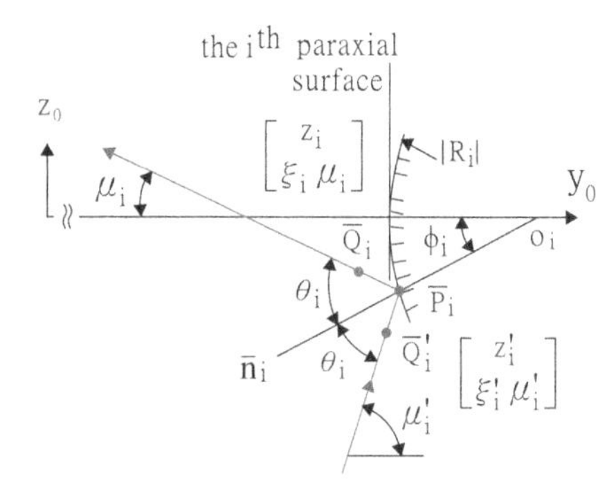

Fig. 4.28 Paraxial meridional ray, with positive angle μ_i', reflected by convex spherical mirror. Note that refracted ray has positive angle μ_i

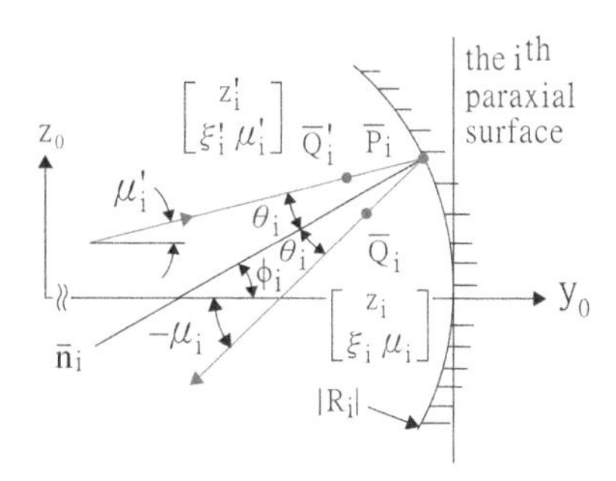

Fig. 4.29 Paraxial meridional ray, with positive angle μ_i', reflected by concave spherical mirror. Note that refracted ray has negative angle μ_i

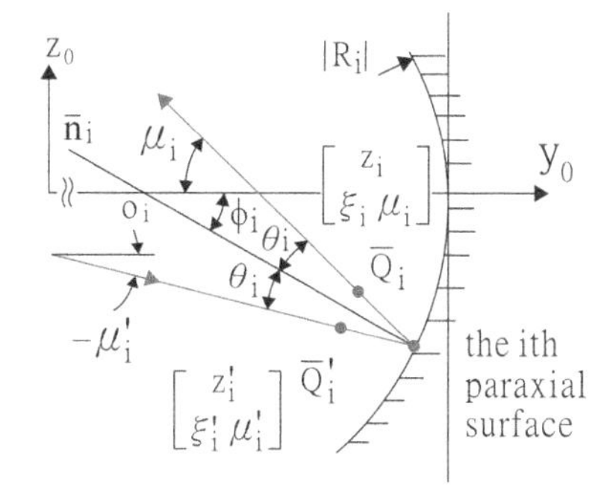

Fig. 4.30 Paraxial meridional ray, with negative angle μ_i', reflected by concave spherical mirror. Note that refracted ray has positive angle μ_i

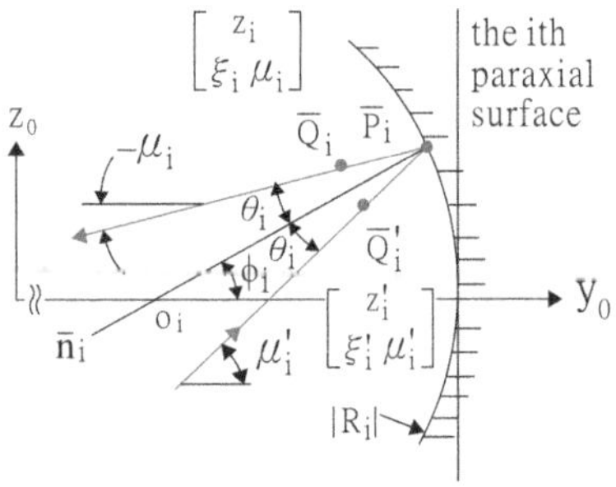

Fig. 4.31 Paraxial meridional ray, with positive angle μ_i', reflected by concave spherical mirror. Note that refracted ray has negative angle μ_i

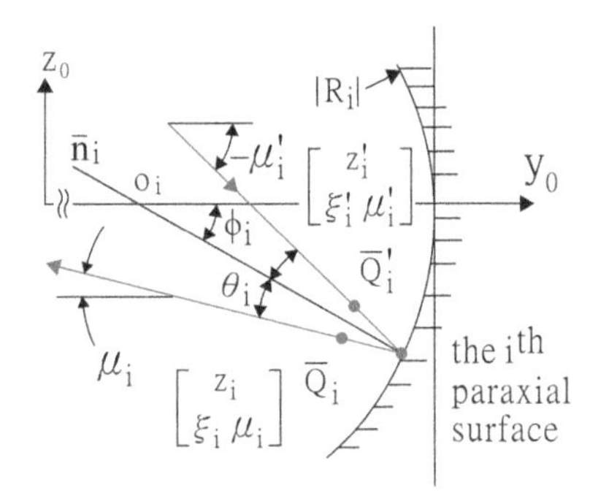

Fig. 4.32 Paraxial meridional ray, with negative angle μ_i', reflected by concave spherical mirror. Note that refracted ray has positive angle μ_i

It should be noted that the transfer matrix $\bar{T}_i$ (Eq. (4.28)), refraction matrices $\bar{M}_i$ (Eqs. (4.33) and (4.45)), and reflection matrices $\bar{M}_i$ (Eqs. (4.37) and (4.53)) are also valid when the incoming paraxial ray propagates from right to left provided that the direction of the y_0 axis is reversed while that of the z_0 axis still points in the upward direction.

References

1. Gerrard A, Burch JM (1975), *Introduction to matrix methods in optics*, John Wiley & Sons, Ltd
2. Attard AE (1984) Matrix optical analysis of skew rays in mixed systems of spherical and orthogonal cylindrical lenses. Appl Opt 23:2706–2709
3. Smith WJ (2001) Modern optical engineering, 3rd edn. Edmund Industrial Optics, Barrington, N.J
4. Lin PD, Sung CK (2006) Matrix-based paraxial skew raytracing in 3D systems with non-coplanar optical axis. OPTIK—Int J Light Electron Opt 117:329–340

Chapter 5
Cardinal Points and Image Equations

2×2 raytracing matrices are extremely useful in geometrical optics. In this chapter, the 2×2 raytracing matrix approach is used to describe the path of a paraxial meridional ray traveling along the y_0z_0 plane through an axis-symmetrical system possessing multiple boundary surfaces. Note that in defining the y_0z_0 plane, it is assumed that the y_0 axis is aligned with the optical axis of the system. It is shown that even complex optical systems can be reduced to a set of only six cardinal points, for the purpose of determining the image close to the optical axis.

5.1 Paraxial Optics

Paraxial optics uses the 2×2 matrix method and the concepts of cardinal planes and cardinal points to trace paraxial rays through axis-symmetrical optical systems. To trace a paraxial ray, it is first necessary to establish and label the paraxial surfaces sequentially from i = 1 to i = n. Two height-slope matrices, $\begin{bmatrix} z_i' & \xi_i'\mu_i' \end{bmatrix}^T$ and $\begin{bmatrix} z_i & \xi_i\mu_i \end{bmatrix}^T$, are assigned to the infinitesimal neighboring points along the ray before and after every paraxial surface i (i = 1 to i = n), as shown in Fig. 4.10. Equations (4.18), (4.19) and (4.23) (or Eqs. (4.28), (4.33), (4.45), (4.37) and (4.53)) are then applied sequentially at each surface in order to determine the height-slope matrix $\begin{bmatrix} z_n & \xi_n\mu_n \end{bmatrix}^T$ at the final paraxial surface, i.e.,

P.D. Lin, *Advanced Geometrical Optics*, Progress in Optical Science and Photonics 4, DOI 10.1007/978-981-10-2299-9_5

$$\begin{bmatrix} z_n \\ \xi_n \mu_n \end{bmatrix} = (\bar{M}_n \bar{T}_n \bar{M}_{n-1} \bar{T}_{n-1} \cdots \bar{M}_i \bar{T}_i \cdots \bar{M}_2 \bar{T}_2 \bar{M}_1 \bar{T}_1) \begin{bmatrix} z_0 \\ \xi_0 \mu_0 \end{bmatrix}$$
$$= \begin{bmatrix} A & B \\ C & D \end{bmatrix} \begin{bmatrix} 1 & \lambda_{1,\text{axis}}/\xi_0 \\ 0 & 1 \end{bmatrix} \begin{bmatrix} z_0 \\ \xi_0 \mu_0 \end{bmatrix} = \begin{bmatrix} A & B \\ C & D \end{bmatrix} \begin{bmatrix} z_1' \\ \xi_1' \mu_1' \end{bmatrix}, \tag{5.1}$$

where ABCD is a matrix describing the transformation between $\begin{bmatrix} z_n & \xi_n \mu_n \end{bmatrix}^T$, i.e., the height-slope matrix of the ray after the final paraxial surface, and $\begin{bmatrix} z_1' & \xi_1' \mu_1' \end{bmatrix}^T$, i.e., the height-slope matrix of the ray at a infinitesimal neighboring point Q_1' along the ray before the first paraxial surface. From Eq. (5.1), matrix ABCD can be obtained as

$$\begin{bmatrix} A & B \\ C & D \end{bmatrix} = \bar{M}_n \bar{T}_n \bar{M}_{n-1} \bar{T}_{n-1} \cdots \bar{M}_i \bar{T}_i \cdots \bar{M}_2 \bar{T}_2 \bar{M}_1. \tag{5.2}$$

It will be recalled that the components of matrix ABCD are such that the determinant, AD – BC, equals unity. Therefore, to trace a ray in the backward direction, the matrix equation should be inverted to yield

$$\begin{bmatrix} z_1' \\ \xi_1' \mu_1' \end{bmatrix} = \begin{bmatrix} D & -B \\ -C & A \end{bmatrix} \begin{bmatrix} z_n \\ \xi_n \mu_n \end{bmatrix}. \tag{5.3}$$

However, it should be noted that even though Eq. (5.3) enables a ray to be traced backward through the system, the ray directions and sign conventions described in the following section for Eq. (5.1) must be retained. Equations (5.1) and (5.3) describe the behavior of the optical system by summarizing the combined effects of the individual optical boundary surfaces within the system.

Example 5.1 Consider the cat's eye system shown in Fig. 3.12. If radii R_1 and R_4 have positive and negative values, respectively, matrix ABCD has the form

$$\begin{bmatrix} A & B \\ C & D \end{bmatrix} = \bar{M}_7 \bar{T}_7 \bar{M}_6 \bar{T}_6 \bar{M}_5 \bar{T}_5 \bar{M}_4 \bar{T}_4 \bar{M}_3 \bar{T}_3 \bar{M}_2 \bar{T}_2 \bar{M}_1$$
$$= \begin{bmatrix} 1 & 0 \\ (\xi_{e1} - \xi_{air})/(-R_1) & 1 \end{bmatrix} \begin{bmatrix} 1 & R_1/\xi_{e1} \\ 0 & 1 \end{bmatrix} \begin{bmatrix} 1 & 0 \\ 0 & 1 \end{bmatrix} \begin{bmatrix} 1 & v_2/\xi_{glue} \\ 0 & 1 \end{bmatrix} \begin{bmatrix} 1 & 0 \\ 0 & 1 \end{bmatrix}$$
$$\begin{bmatrix} 1 & (-R_4)/\xi_{e2} \\ 0 & 1 \end{bmatrix} \begin{bmatrix} 1 & 0 \\ 2\xi_{e2}/R_4 & 1 \end{bmatrix} \begin{bmatrix} 1 & (-R_4)/\xi_{e2} \\ 0 & 1 \end{bmatrix} \begin{bmatrix} 1 & 0 \\ 0 & 1 \end{bmatrix} \begin{bmatrix} 1 & v_2/\xi_{glue} \\ 0 & 1 \end{bmatrix} \begin{bmatrix} 1 & 0 \\ 0 & 1 \end{bmatrix}$$
$$\begin{bmatrix} 1 & R_1/\xi_{e1} \\ 0 & 1 \end{bmatrix} \begin{bmatrix} 1 & 0 \\ (\xi_{air} - \xi_{e1})/R_1 & 1 \end{bmatrix}.$$

5.2 Cardinal Planes and Cardinal Points

To treat a well-corrected optical system (e.g., that shown in Fig. 3.2) as a "black box" in paraxial optics, it is first necessary to define the basic concepts of cardinal planes and cardinal points underpinning the paraxial optics approach (see Figs. 5.1 and 5.2). For any axis-symmetrical optical system, the back principal plane has the property that a ray emerging from the system appears to have crossed this plane at the same distance from the optical axis that it appears to cross the front principal plane. In practice, this means that if the ray comes from the left, the system can be treated as if all of the refraction occurs at the back principal plane. Similarly, for a ray coming from the right, all of the refraction can be considered to occur at the front principal plane. The front and back principal points, designated as H_1 and H_2, respectively, are

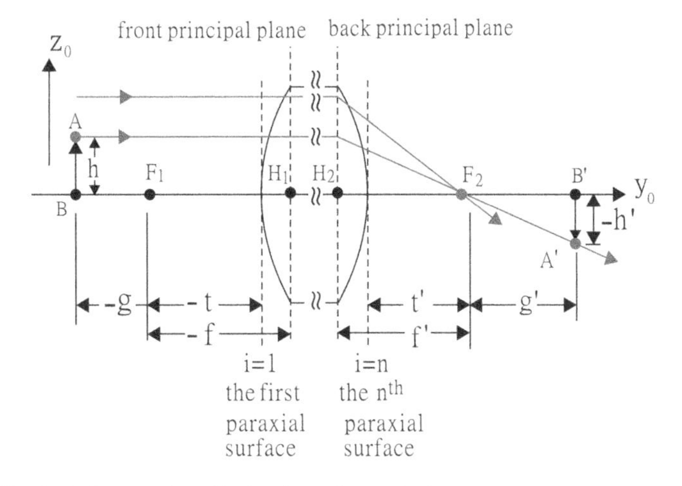

Fig. 5.1 Cardinal planes and cardinal points of an optical system

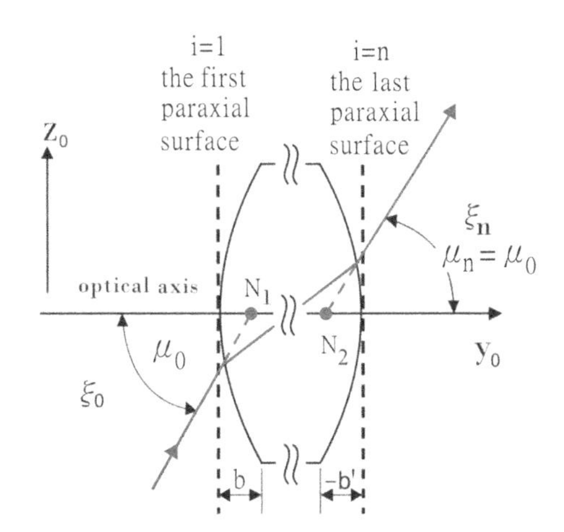

Fig. 5.2 Nodal points of an optical system

the points at which the front and back principal planes cross the optical axis of the system. Similarly, the front and back focal points, designated as F_1 and F_2, respectively, are the points at which rays parallel to the optical axis are brought to a common focus point on the optical axis in front and behind the optical system, respectively. The front and back focal planes are then defined as the planes which pass through the front and back focal points, respectively, in a direction perpendicular to the optical axis. Nodal points N_1 and N_2 are two axial points located such that a ray directed toward the front nodal point N_1 appears (after passing through the system) to emerge from the back nodal point N_2 in a direction parallel to its original direction. (Note that nodal points are also defined as those points having unit angular magnification.)

The effective front focal length, f, of a system is the distance from the front principal point H_1 to the front focal point F_1. Similarly, the effective back focal length, f′, is the distance from the back principal point H_2 to the back focal point F_2. Finally, the front focal length, t, is the distance from the vertex of the first boundary surface (i = 1) to the front focal point F_1, while the back focal length, t′, is the distance from the vertex of the last boundary surface (i = n) to the back focal point, F_2.

5.2.1 *Location of Focal Points*

Assume that matrix ABCD is known for a given system and the aim of this section is to locate the four cardinal points, namely two focal points and two principal points. In general, the focal points F_2 and F_2 of the optical system can be easily determined by tracing the path of a ray as it travels through the system parallel to the optical axis from the first to last paraxial surface using Eq. (5.1) (see Fig. 5.3 and [2]), yielding

$$\begin{bmatrix} z_n \\ \xi_n \mu_n \end{bmatrix} = \begin{bmatrix} A & B \\ C & D \end{bmatrix} \begin{bmatrix} z_1' \\ 0 \end{bmatrix}. \tag{5.4}$$

The two components of Eq. (5.4) yield the equations $z_n = Az_1'$ and $\mu_n = Cz_1'/\xi_n$. The effective back focal length, f′, is then given by the ray height at the back principal plane (i.e., z_1') divided by the slope angle $-\mu_n$ (which is positive value in Fig. 5.3) of the ray emerging from this plane, i.e.,

$$f' = \frac{z_1'}{-\mu_n} = \frac{-\xi_n}{C}. \tag{5.5}$$

Similarly, the back focal length, t′, is given by the ray height at the last paraxial surface (z_n) divided by the slope angle $-\mu_n$ of the ray emerging from this surface, i.e.,

$$t' = \frac{z_n}{-\mu_n} = \frac{-\xi_n A}{C}. \tag{5.6}$$

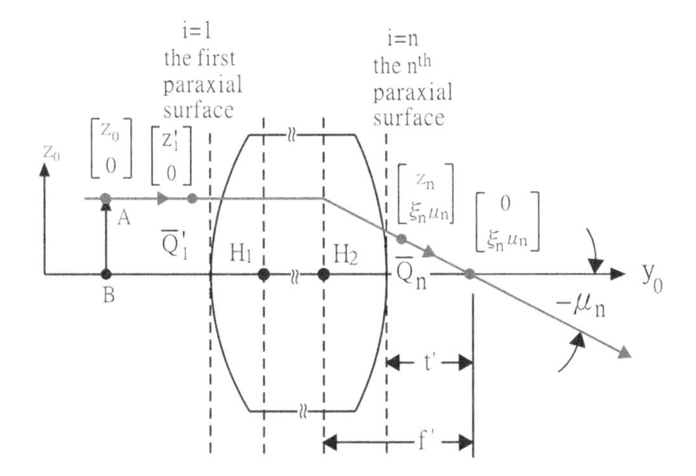

Fig. 5.3 Effective back focal length f' and back focal length t' of an optical system with n boundary surfaces

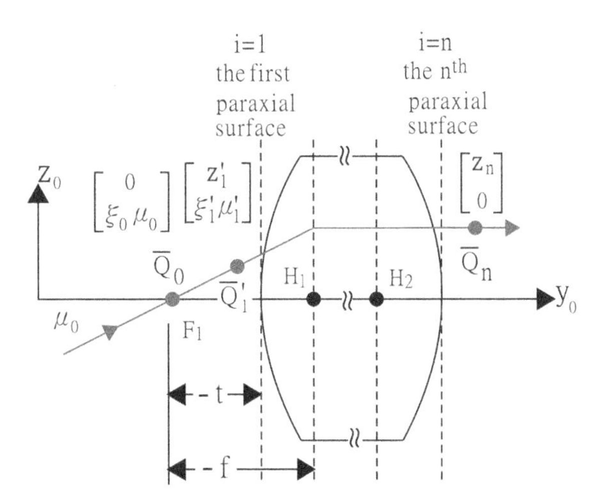

Fig. 5.4 Determination of effective front focal length f and front focal length t of an optical system with n boundary surfaces

The distance from the last paraxial surface to the back principal point, H_2, is obtained simply as the difference between t' and f', i.e., $t' - f' = \xi_n(1 - A)/C$.

The front focal point, F_1, and front principal point, H_1, are found simply by tracing a ray passes through the front focal point F_1 and emerges from the system in a direction parallel to the optical axis (with a height-slope matrix $[z_n \quad 0]^T$) by using Eq. (5.3) (see Fig. 5.4), i.e.,

$$\begin{bmatrix} z_1' \\ \xi_1' \mu_1' \end{bmatrix} = \begin{bmatrix} D & -B \\ -C & A \end{bmatrix} \begin{bmatrix} z_n \\ 0 \end{bmatrix}. \tag{5.7}$$

The two components of Eq. (5.7) yield equations $z'_1 = D\, z_n$ and $\mu'_1 = -C\, z_n/\xi'_1$, respectively. Assuming $\xi'_1 = \xi_0$, (ξ_0 is the refractive index of the medium before the first bounady surface) the following expressions for t and f are obtained:

$$-t = \frac{z'_1}{\mu'_1} = \frac{-\xi'_1 D}{C}$$

or

$$t = \frac{\xi_0 D}{C}, \tag{5.8}$$

and

$$-f = \frac{z_n}{\mu'_1} = -\frac{\xi'_1}{C}$$

or

$$f = \frac{\xi_0}{C}. \tag{5.9}$$

5.2.2 *Location of Nodal Points*

In this section, we wish to locate two nodal points N_1 and N_2 such that any ray entering the system directed towards N_1 appears on emergence as a ray coming from N_2 and making the same angle $\mu_0 = \mu_n$ with the optical axis (see Fig. 5.2). Let the distance of N_1 from the vertex of the first boundary surface (i = 1) be denoted as b, and the distance of N_2 from the vertex of the last boundary surface (i = n) be denoted as b'. It is noted that the ray path in Fig. 5.2 comprises three straight-line paths, namely one path from the first paraxial surface to N_1, a second path from N_1 to N_2, and a third path from N_2 to the last paraxial surface. Therefore, matrix ABCD linking the first paraxial surface and the last paraxial surface can be formulated as (pp. 56-57 of [1])

$$\begin{bmatrix} A & B \\ C & D \end{bmatrix} = \begin{bmatrix} 1 & -b'/\xi_n \\ 0 & 1 \end{bmatrix} \begin{bmatrix} \phi_{11} & \phi_{12} \\ \phi_{21} & \phi_{22} \end{bmatrix} \begin{bmatrix} 1 & b/\xi_0 \\ 0 & 1 \end{bmatrix}. \tag{5.10}$$

The second matrix in Eq. (5.10) links the second nodal point N_2 to the first nodal point N_1. According to the basic definition of nodal points (i.e., $\mu_n = \mu_0$), it follows that

Table 5.1 Values of cardinal points

Cardinal points	Measured from to	Value	Special case $\xi_n = \xi_0 = 1$
f	from H_1 to F_1	$\frac{\xi_0}{C}$	$\frac{1}{C}$
t	from the vertex of the first paraxial surface to F_1	$\frac{\xi_0 D}{C}$	$\frac{D}{C}$
b	from the vertex of the first paraxial surface to N_1	$\frac{D\xi_0 - \xi_n}{C}$	$\frac{D-1}{C}$
f′	from H_2 to F_2	$\frac{-\xi_n}{C}$	$\frac{-1}{C}$
t′	from the vertex of the last paraxial surface to F_2	$\frac{-\xi_n A}{C}$	$\frac{-A}{C}$
b′	from the vertex of the last paraxial surface to N_2	$\frac{A(D\xi_0 - \xi_n)}{C} - \xi_0 B = \frac{\xi_0 - A\xi_n}{C}$	$\frac{1-A}{C}$

$$\begin{bmatrix} 0 \\ \xi_n \mu_n \end{bmatrix} = \begin{bmatrix} 0 \\ \xi_n \mu_0 \end{bmatrix} = \begin{bmatrix} \phi_{11} & \phi_{12} \\ \phi_{21} & \phi_{22} \end{bmatrix} \begin{bmatrix} 0 \\ \xi_0 \mu_0 \end{bmatrix}. \tag{5.11}$$

Note that Eq. (5.11) is true if, and only if, $\phi_{12} = 0$ and $\phi_{22} = \xi_n/\xi_0$. One can therefore obtain b and b′ by substituting ϕ_{12} and ϕ_{22}, respectively, into Eq. (5.10), yielding

$$b = \frac{D\xi_0 - \xi_n}{C}, \tag{5.12}$$

$$b' = \frac{A(D\xi_0 - \xi_n)}{C} - \xi_0 B. \tag{5.13}$$

For convenience, the formulations derived above for the six cardinal points are summarized in Table 5.1. If the medium on both sides of the optical system is the same (e.g., air), then the front and back nodal points, N_1 and N_2, coincide with the front and back principal points, H_1 and H_2, respectively. It should be noted that most textbooks do not consider the effective front focal length of the optical system, and therefore use the notation f to represent the effective back focal length rather than f′, as used in this book.

5.3 Thick and Thin Lenses

Consider the thick double convex lens shown in Fig. 5.5 with thickness q_{el} and refractive index ξ_{el}. The lens has two spherical boundary surfaces with radii R_1 and R_2, respectively, and is surrounded by two different media with refractive indices ξ_0 and ξ_2, respectively. Any incident ray that passes through the lens will be refracted

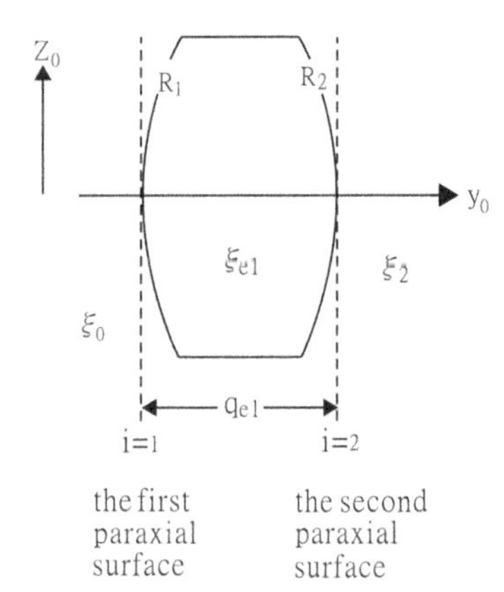

Fig. 5.5 Thick double convex lens surrounded by two different refractive media

twice; one at each surface, and the other a straight-line path between the surfaces. Matrix ABCD is obtained as

$$\begin{bmatrix} A & B \\ C & D \end{bmatrix} = \begin{bmatrix} 1 & 0 \\ (\xi_{e1} - \xi_2)/R_2 & 1 \end{bmatrix} \begin{bmatrix} 1 & q_{e1}/\xi_{e1} \\ 0 & 1 \end{bmatrix} \begin{bmatrix} 1 & 0 \\ (\xi_0 - \xi_{e1})/R_1 & 1 \end{bmatrix}$$
$$= \begin{bmatrix} 1 + q_{e1}(\xi_0 - \xi_{e1})/(\xi_{e1}R_1) & q_{e1}/\xi_{e1} \\ (\xi_0 - \xi_{e1})/R_1 + (\xi_{e1} - \xi_2)/R_2 + q_{e1}(\xi_{e1} - \xi_2)(\xi_0 - \xi_{e1})/(\xi_{e1}R_1R_2) & 1 + q_{e1}(\xi_{e1} - \xi_2)/(R_2\xi_{e1}) \end{bmatrix}. \tag{5.14}$$

Notably, Eq. (5.14) is valid for biconvex, biconcave and meniscus elements provided that the sign notations given in Sect. 4.2 and Table 4.1 are properly assigned to R_1 and R_2. Furthermore, Eq. (5.14) is also applicable for plano-convex and plano-concave elements if R_1 or R_2 is equal to infinity. The cardinal points for the thick lens shown in Fig. 5.5 can be obtained using Eqs. (5.5), (5.6), (5.8), (5.9), (5.12) and (5.13) given in the previous section.

If the thickness q_{e1} of a lens is sufficiently small that it has no effect on the calculation accuracy, the element is referred to as a thin lens. When such a lens is surrounded by a single refractive medium, i.e., $\xi_0 = \xi_2 = \xi$, matrix ABCD can be deduced from Eq. (5.14) (with $q_{e1} = 0$) as

$$\begin{bmatrix} A & B \\ C & D \end{bmatrix} = \begin{bmatrix} 1 & 0 \\ -(\xi_{e1} - \xi)/(1/R_1 - 1/R_2) & 1 \end{bmatrix}. \tag{5.15}$$

The well-known lens maker's formula of the thin lens can be obtained from Eqs. (5.5) and (5.9) as

$$\frac{1}{f} = \frac{-1}{f'} = \frac{-(\xi_{e1} - \xi)}{\xi}\left(\frac{1}{R_1} - \frac{1}{R_2}\right). \tag{5.16}$$

Equation (5.16) is one of the most basic equations in optics, and states that the effective focal length of a thin lens is a function of the radii R_1 and R_2 of the lens surfaces and the refractive index ξ_{e1} of the lens material. The reciprocal of f′ is defined as the optical power ϕ of the lens, and is expressed in diopters (m^{-1}) as

$$\phi = \frac{1}{f'} = \frac{-1}{f} = \frac{(\xi_{e1} - \xi)}{\xi}\left(\frac{1}{R_1} - \frac{1}{R_2}\right). \tag{5.17}$$

In practice, the optical power ϕ evaluates the ability of a thin lens surrounded by a single refractive medium to focus a parallel beam incident on a thin lens. Substituting Eq. (5.17) into Eq. (5.15), the following simplified ABCD matrix (denoted as M_{lens}) is obtained for a thin lens surrounded by a medium with refractive index ξ :

$$M_{lens} = \begin{bmatrix} 1 & 0 \\ -\xi/f' & 1 \end{bmatrix} = \begin{bmatrix} 1 & 0 \\ -\xi\phi & 1 \end{bmatrix}. \tag{5.18}$$

5.4 Curved Mirrors

The discussions above have considered systems in which the rays travel from left to right. However, curved mirrors reverse the direction of the incident rays. Many different types of curved mirrors are used in optical systems; with spherical, parabolic and hyperbolic mirrors being among the most common. Generally speaking, such mirrors are used either singly or in combination to produce the desired optical effect. For convex mirrors, also known as diverging mirrors, the incoming ray and center of curvature lie on opposite sides of the mirror (Fig. 5.6). By contrast, for concave mirrors, also known as converging mirrors, the incoming ray and center of curvature are located on the same side of the mirror (Fig. 5.7). In both cases, the mirror has only one focal point F. Moreover, the line joining the

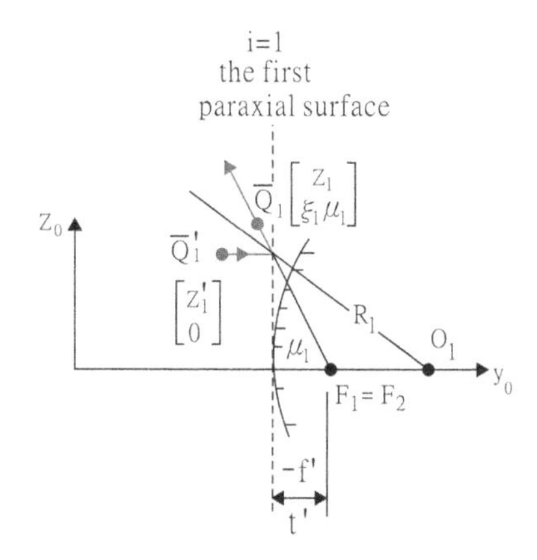

Fig. 5.6 A convex mirror, whose center of curvature is located to the *right* of its surface, forms an erect *virtual image*

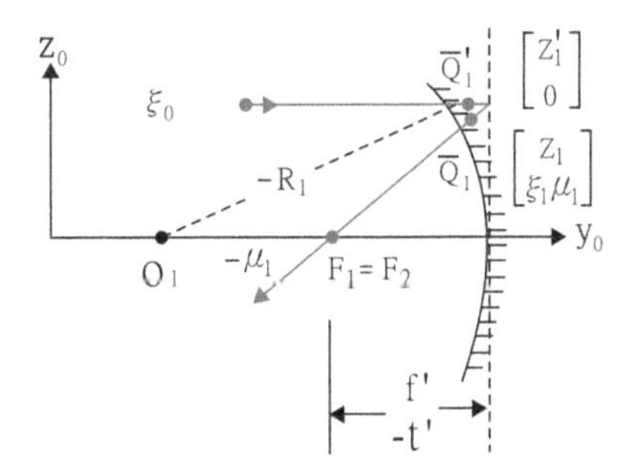

Fig. 5.7 A concave mirror, whose center of curvature is located to the *left* of its surface, forms a *real image*

mirror vertex and its center o_1 represents the optical axis of the system. For both convex and concave mirrors, matrix ABCD has the form

$$\begin{bmatrix} A & B \\ C & D \end{bmatrix} = \begin{bmatrix} 1 & 0 \\ 2\xi_0/R_1 & 1 \end{bmatrix}, \tag{5.19}$$

where R_1 is positive for a convex mirror (Fig. 5.6) and negative for a concave mirror (Fig. 5.7). When we trace a ray parallel to the optical axis through the mirror, the height-slope matrix $[z_1 \quad \xi_1\mu_1]^T$ of the reflected ray is given as

$$\begin{bmatrix} z_1 \\ \xi_1\mu_1 \end{bmatrix} = \begin{bmatrix} z_1 \\ \xi_0\mu_1 \end{bmatrix} = \begin{bmatrix} 1 & 0 \\ 2\xi_0/R_1 & 1 \end{bmatrix} \begin{bmatrix} z_1' \\ 0 \end{bmatrix} = \begin{bmatrix} 1 \\ 2\xi/R_1 \end{bmatrix} z_1'. \tag{5.20}$$

Recall that the focal length of a converging lens is positive while that of a diverging lens is negative (see Sect. 4.2). Therefore, by definition, the focal lengths of a convex mirror and concave mirror are determined from $\mu_1 = z_1'/(-f')$ and $-\mu_1 = z_1'/f'$, respectively. For both mirrors, the effective back focal length is given by

$$f' = \frac{-z_1}{\mu_1} = \frac{-R_1}{2}. \tag{5.21}$$

As described in Sect. 5.2, t' is the back focal length and is measured from the vertex of the boundary surface to focal point F_2. In other words, $t' = z_1/\mu_1 = R_1/2$ and $(-t') = z_1/(-\mu_1) = R_1/2$ for convex and concave mirrors, respectively. Therefore, the principal plane coincides with the paraxial surface since $f' = -t'$ for both convex and concave mirrors. From Eq. (5.21), the effective focal length is equal to half the mirror radius for both concave and convex mirrors. Moreover, the optical power $\phi = 1/f' = -2/R_1$ of a concave mirror is positive since R_1 is negative, whereas the optical power $\phi = -2/R_1$ of a convex mirror is negative since R_1 is positive.

5.5 Determination of Image Position Using Cardinal Points

When the cardinal points of an optical system are known, the location and size of the image formed by the system can be readily determined. Consider the optical system shown in Fig. 5.8, in which A is an off-axis point source, B is an on-axis point source, and A′ and B′ are the imaged points of A and B, respectively. Let the length measured from the front principal point H_1 to B be denoted as s (however, that length in Fig. 5.8 should be −s) and the length measured from the back principal point H_2 to image B′ be denoted as s′. Similarly, let g ($(-g) = (-s) - (-f)$ in Fig. 5.8) be the length measured from the front focal point F_1 to point B and g′ ($g' = s' - f'$) be the length measured from the back focal point F_2 to image point B′. Finally, let h and h′ be the heights of the object and image, respectively. (Note that the preceding algebraic signs follow the sign convention described in Sect. 4.2.)

In practice, imaged point A′ can be determined by tracing any two rays originating from point A with different slope angles. However, most textbooks usually trace three rays, i.e.,

1. A parallel ray with a height-slope matrix $[z_0 \quad \xi_0\mu_0]^T = [h \quad 0]^T$. This ray passes through the back focal point F_2 such that refraction appears to occur at the back principal plane.
2. A ray with a height-slope matrix $[z_0 \quad \xi_0\mu_0]^T = [h \quad \xi_0 h/g]^T$ (where μ_0 is the negative slope angle and is determined from $(-\mu_0) = h/(-g)$). This ray passes through the front focal point F_1 and emerges from the system in a direction parallel to the optical axis.
3. A third ray constructed from point A to the front nodal point N_1. This ray emerges from the back nodal point N_2 in a direction parallel to the entering ray.

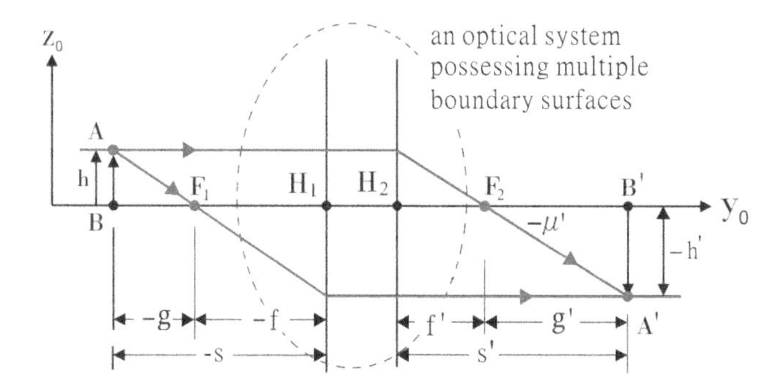

Fig. 5.8 Imaged point A′ can be determined by tracing different rays originating from point source A

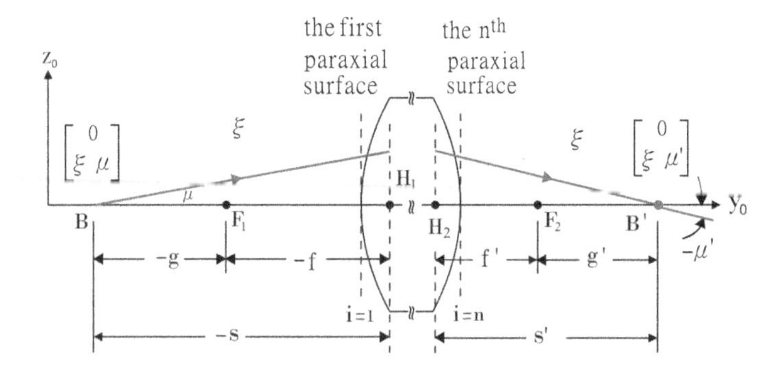

Fig. 5.9 Object and image in an optical system

Figure 5.9 shows an on-axis object point, B, and corresponding imaged point, B′, formed by an optical system (consisting of a thin lens only) with an optical power $\phi = 1/f'$ and surrounded by a medium with refractive index ξ. As shown, a ray originating from B with slope angle μ propagates into the optical system. After refracting, the ray intersects the optical axis with a slope angle μ'. The raytracing equation for point B′ has the form

$$\begin{bmatrix} 0 \\ \xi\mu' \end{bmatrix} = \begin{bmatrix} 1 & s'/\xi \\ 0 & 1 \end{bmatrix} \begin{bmatrix} 1 & 0 \\ -\xi/f' & 1 \end{bmatrix} \begin{bmatrix} 1 & -s/\xi \\ 0 & 1 \end{bmatrix} \begin{bmatrix} 0 \\ \xi\mu \end{bmatrix}$$
$$= \begin{bmatrix} -s/\xi + s'(s/f' + 1)/\xi \\ s/f' + 1 \end{bmatrix} \xi\mu. \quad (5.22)$$

The image location can be determined from the first component of Eq. (5.22) as

$$\frac{1}{f'} = \frac{1}{(-s)} + \frac{1}{s'}, \quad (5.23)$$

where f' is the effective back focal length. It is noted that Eq. (5.23) provides a generic solution for the image location of any simplified optical system surrounded by a single medium with refractive index ξ.

5.6 Equation of Lateral Magnification

Optical magnification includes lateral magnification, longitudinal magnification, angular magnification, and others. Lateral magnification, $m^*_{paraxial}$, is the ratio of some linear dimension h', perpendicular to the optical axis, of an image formed by an optical system, to the corresponding linear dimension h of the object. Figure 5.10 shows a simplified optical system with optical power $\phi = 1/f'$ surrounded by a single medium with refractive index ξ. Also shown in the figure are

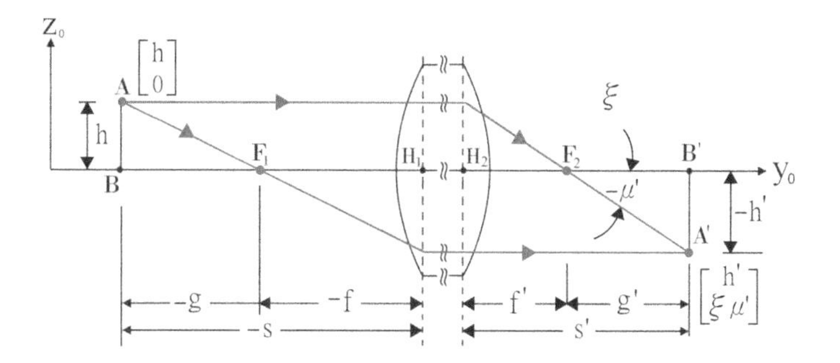

Fig. 5.10 Lateral magnification is the ratio of the image height h′ to the object height h

the ray heights and slope angles at object point A and corresponding imaged point A′. The system equation has the form

$$\begin{bmatrix} h' \\ \xi\mu' \end{bmatrix} = \begin{bmatrix} 1 & s'/\xi \\ 0 & 1 \end{bmatrix} \begin{bmatrix} 1 & 0 \\ -\xi/f' & 1 \end{bmatrix} \begin{bmatrix} 1 & -s/\xi \\ 0 & 1 \end{bmatrix} \begin{bmatrix} h \\ 0 \end{bmatrix} = h \begin{bmatrix} 1 - s'/f' \\ -\xi/f' \end{bmatrix}. \tag{5.24}$$

By definition, the lateral magnification, $m^*_{paraxial}$, can be estimated from the first component of Eq. (5.24) and using Eq. (5.23), to give

$$m^*_{paraxial} = \frac{h'}{h} = 1 - \frac{s'}{f'} = \frac{f'}{f' + s} = \frac{s'}{s}. \tag{5.25}$$

It is noted from Eq. (5.25) that $m^*_{paraxial}$ has a negative value.

5.7 Equation of Longitudinal Magnification

Longitudinal magnification, $\hat{m}_{paraxial}$, is the magnification of the longitudinal thickness or longitudinal motion of an object along the optical axis. For a simplified system with power $\phi = 1/f'$ and surrounded by a single refractive medium ξ, $\hat{m}_{paraxial}$ can be estimated directly from Eq. (5.23) as

$$\hat{m}_{paraxial} = \lim_{\Delta s \to 0} \frac{\Delta s'}{\Delta s} = \frac{ds'}{ds} = \frac{f'^2}{(f' + s)^2} = \left(m^*_{paraxial}\right)^2. \tag{5.26}$$

The longitudinal magnification can also be estimated using the finite difference (FD) method, i.e.,

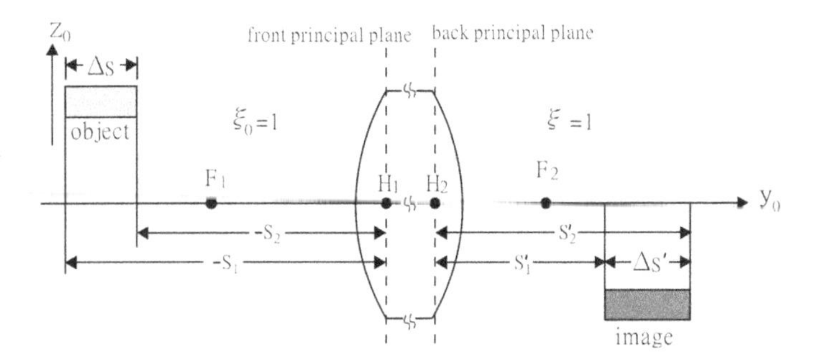

Fig. 5.11 Longitudinal magnification can be estimated using the FD method

$$\hat{m}_{paraxial} = \frac{\Delta s'}{\Delta s} = \frac{s'_2 - s'_1}{(-s_1) - (-s_2)}, \tag{5.27}$$

where s_1 and s_2 denote the distances from the front principal point H_1 to the front and back edges of the object, respectively (These two distances should be $-s_1$ and $-s_2$ in Fig. 5.11), while s'_1 and s'_2 are the distances from the back principal point H_2 to the front and back edges of the image, respectively. From Eq. (5.23), it follows that $s'_1 = s_1 f'/(s_1 + f')$ and $s'_2 = s_2 f'/(s_2 + f')$. Substituting these two equations into Eq. (5.27) yields the same formulation as that given in Eq. (5.26). In contrast to the lateral magnification, $\hat{m}_{paraxial}$ is ordinarily positive; indicating that the object and image move in the same direction. Furthermore, since the lateral magnification $m^*_{paraxial}$ varies with the position y_0, the longitudinal magnification $\hat{m}_{paraxial}$ is also a function of y_0.

5.8 Two-Element Systems

Consider the optical system shown in Fig. 5.12 consisting of two elements surrounded by air with refractive index $\xi_{air} = 1$. Assume that the two elements have a separation distance v_2; optical powers ϕ_a and ϕ_b, respectively; and a back focus length f'. Matrix ABCD has the form

$$\begin{aligned}\begin{bmatrix} A & B \\ C & D \end{bmatrix} &= \begin{bmatrix} 1 & 0 \\ -\phi_b & 1 \end{bmatrix}\begin{bmatrix} 1 & v_2 \\ 0 & 1 \end{bmatrix}\begin{bmatrix} 1 & 0 \\ -\phi_a & 1 \end{bmatrix} \\ &= \begin{bmatrix} 1 - v_2\phi_a & v_2 \\ -(\phi_a + \phi_b - \phi_a\phi_b v_2) & -\phi_b v_2 + 1 \end{bmatrix} \cong \begin{bmatrix} 1 & 0 \\ -\phi_{ab} & 1 \end{bmatrix}. \end{aligned} \tag{5.28}$$

From Eq. (5.28), the optical power ϕ_{ab} and effective back focal length f'_{ab} can be estimated as

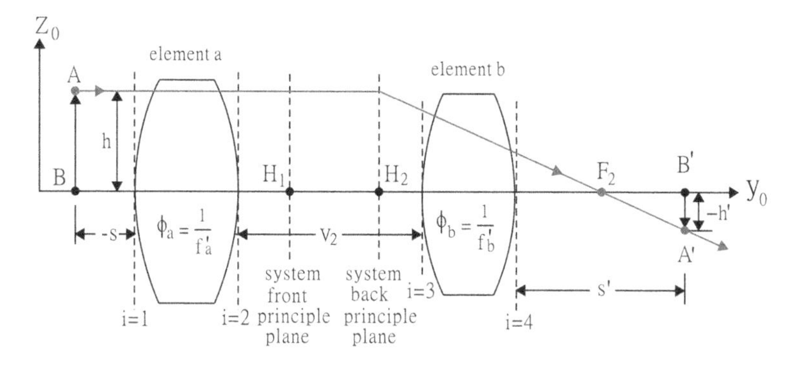

Fig. 5.12 Two-element optical system

$$\phi_{ab} = \frac{1}{f'_{ab}} = \phi_a + \phi_b - \phi_a\phi_b v_2 = \frac{1}{f'_a} + \frac{1}{f'_b} - \frac{v_2}{f'_a f'_b}. \tag{5.29}$$

Moreover, the front and back focal lengths (t_{ab} and t'_{ab}) and effective front and back focal lengths (f_{ab} and f'_{ab}) can be estimated from Eqs. (5.5), (5.6), (5.8) and (5.9) as

$$t'_{ab} = -t_{ab} = \frac{1 - \phi_a v_2}{\phi_a + \phi_b - \phi_a\phi_b v_2} = \frac{f'_b(f'_a - v_2)}{f'_a + f'_b - v_2}, \tag{5.30}$$

$$f'_{ab} = -f_{ab} = \frac{1}{\phi_a + \phi_b - \phi_a\phi_b v_2} = \frac{f'_a f'_b}{f'_a + f'_b - v_2}. \tag{5.31}$$

For general two-element optical systems such as that shown in Fig. 5.13, two main design problems must commonly be solved. In the first problem, most of the system parameters are either given, e.g., the lateral magnification $m^*_{paraxial}$, element spacing v_2, object position $-s$, and image position s', or ignored (e.g., the

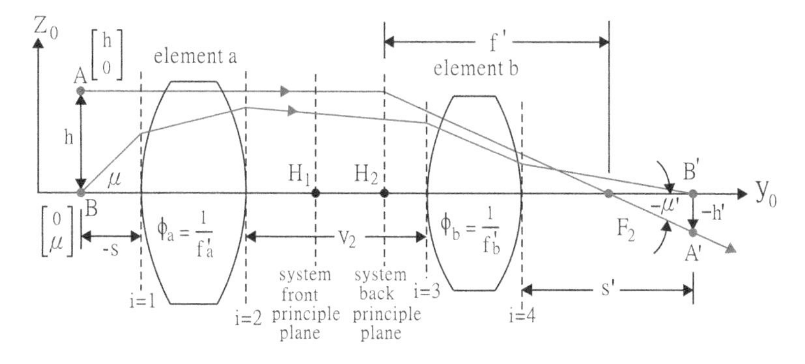

Fig. 5.13 Two-element optical system operating at finite conjugate

thicknesses of the two elements), and the aim is to determine the required optical powers of the two elements. This problem can be solved by tracing two rays originating from A and B with height-slope matrices $[z_0 \quad \xi_0\mu_0]^T = [h \quad 0]^T$ and $[z_0 \quad \xi_0\mu_0]^T = [0 \quad \mu]^T$, respectively. The two rays yield the following two equations:

$$\begin{aligned}\begin{bmatrix} h' \\ \mu' \end{bmatrix} &= \begin{bmatrix} 1 & s' \\ 0 & 1 \end{bmatrix}\begin{bmatrix} 1 & 0 \\ -\phi_b & 1 \end{bmatrix}\begin{bmatrix} 1 & v_2 \\ 0 & 1 \end{bmatrix}\begin{bmatrix} 1 & 0 \\ -\phi_a & 1 \end{bmatrix}\begin{bmatrix} 1 & -s \\ 0 & 1 \end{bmatrix}\begin{bmatrix} h \\ 0 \end{bmatrix} \\ &= \begin{bmatrix} (1-\phi_b s')(1-\phi_a v_2) - \phi_a s' \\ -\phi_b(1-\phi_a v_2) - \phi_a \end{bmatrix} h, \end{aligned} \tag{5.32}$$

$$\begin{aligned}\begin{bmatrix} 0 \\ \mu'' \end{bmatrix} &= \begin{bmatrix} 1 & s' \\ 0 & 1 \end{bmatrix}\begin{bmatrix} 1 & 0 \\ -\phi_b & 1 \end{bmatrix}\begin{bmatrix} 1 & v_2 \\ 0 & 1 \end{bmatrix}\begin{bmatrix} 1 & 0 \\ -\phi_a & 1 \end{bmatrix}\begin{bmatrix} 1 & -s \\ 0 & 1 \end{bmatrix}\begin{bmatrix} 0 \\ \mu \end{bmatrix} \\ &= \begin{bmatrix} -s + v_2 + s v_2 \phi_a + s'[s(\phi_b + \phi_a - \phi_a\phi_b v_2) - \phi_b v_2 + 1] \\ s(\phi_b + \phi_a - \phi_a\phi_b v_2) - \phi_b v_2 + 1 \end{bmatrix} \mu. \end{aligned} \tag{5.33}$$

The required optical powers, ϕ_a and ϕ_b, can then be solved from the first components of Eqs. (5.32) and (5.33) with $m^*_{paraxial} = h'/h$, i.e.,

$$\phi_a = \frac{m^*_{paraxial} s - m^*_{paraxial} v_2 - s'}{m^*_{paraxial} s v_2}, \tag{5.34}$$

$$\phi_b = \frac{v_2 - m^*_{paraxial} s + s'}{v_2 s'}. \tag{5.35}$$

The second problem involves the inverse case, in which the optical powers ϕ_a and ϕ_b (or effective back focal lengths f'_a and f'_b), desired object-to-image distance $L = -s + v_2 + s'$, and lateral magnification $m^*_{paraxial}$ are given, and the aim is to determine the required locations of the two elements. Substituting the first component of Eq. (5.32) into the first component of Eq. (5.33), the following quadratic equation is obtained:

$$0 = v_2^2 - v_2 L + L(f'_a + f'_b) + \frac{\left(m^*_{paraxial} - 1\right)^2 f'_a f'_b}{m^*_{paraxial}}. \tag{5.36}$$

Dimensions s and s′ are then readily determined from the valid solution of Eq. (5.36) as

$$s = \frac{\left(m^*_{paraxial} - 1\right) v_2 + L}{\left(m^*_{paraxial} - 1\right) - m^*_{paraxial} v_2 \phi_a}, \tag{5.37}$$

$$s' = L + s - v_2. \tag{5.38}$$

Two-element systems constitute the vast majority of optical systems, and thus the equations given above are of great practical interest. However, it is noted that a sign change of the lateral magnification $m^*_{paraxial}$ from plus to minus results in two completely different optical systems. For example, while both systems produce the same enlargement (or reduction) of the image, the former system yields an erect image while the latter yields an inverted image.

5.9 Optical Invariant

The linearity of the paraxial raytracing equations results in the existence of an optical invariant (denoted as G in this chapter). Figure 5.14 shows two paraxial rays (a and b) traveling along straight paths in a medium of refractive index ξ_{i-1} and then refracted at a spherical boundary surface with radius R_i. Following refraction, the two rays propagate inside an optical element with a refractive index ξ_i. When the two paraxial rays propagate along their straight-line paths, the optical invariant can be obtained from the following raytracing equations:

$$\begin{bmatrix} z'_{ia} \\ \xi'_i \mu'_{ia} \end{bmatrix} = \begin{bmatrix} 1 & \lambda_{i,axis}/\xi_{i-1} \\ 0 & 1 \end{bmatrix} \begin{bmatrix} z_{i-1a} \\ \xi_{i-1}\mu_{i-1a} \end{bmatrix}, \tag{5.39}$$

$$\begin{bmatrix} z'_{ib} \\ \xi'_i \mu'_{ib} \end{bmatrix} = \begin{bmatrix} 1 & \lambda_{i,axis}/\xi_{i-1} \\ 0 & 1 \end{bmatrix} \begin{bmatrix} z_{i-1b} \\ \xi_{i-1}\mu_{i-1b} \end{bmatrix}. \tag{5.40}$$

From Eqs. (5.39) and (5.40), it follows that

$$\frac{\lambda_{i,axis}}{\xi_{i-1}} = \frac{z'_{ia} - z_{i-1a}}{\xi_{i-1}\mu_{i-1a}} = \frac{z'_{ib} - z_{i-1b}}{\xi_{i-1}\mu_{i-1b}}. \tag{5.41}$$

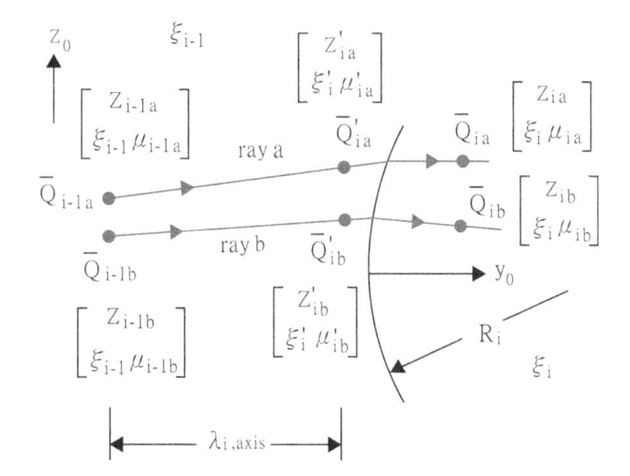

Fig. 5.14 Two paraxial rays refracted at a spherical surface and then travelling in *straight-line* paths can be used to determine the paraxial invariant

Setting $\xi_{i-1}=\xi'_i$, $\mu_{i-1a}=\mu'_{ia}$ and $\mu_{i-1b}=\mu'_{ib}$, the optical invariant can be obtained from Eq. (5.41) as

$$G=\xi_{i-1}\left(\mu_{i-1a}z_{i-1b}-\mu_{i-1b}z_{i-1a}\right)=\xi'_i\left(\mu'_{ia}z'_{ib}-\mu'_{ib}z'_{ia}\right). \quad (5.42)$$

The invariant in a refraction (or reflection) process can also be derived by continuously tracing the paths of two rays at infinitesimal neighboring points before and after refraction, respectively, i.e.,

$$\begin{bmatrix} z_{ia} \\ \xi_i\mu_{ia} \end{bmatrix}=\begin{bmatrix} 1 & 0 \\ (\xi_{i-1}-\xi_i)/R_i & 1 \end{bmatrix}\begin{bmatrix} z'_{ia} \\ \xi'_i\mu'_{ia} \end{bmatrix}, \quad (5.43)$$

$$\begin{bmatrix} z_{ib} \\ \xi_i\mu_{ib} \end{bmatrix}=\begin{bmatrix} 1 & 0 \\ (\xi_{i-1}-\xi_i)/R_i & 1 \end{bmatrix}\begin{bmatrix} z'_{ib} \\ \xi'_i\mu'_{ib} \end{bmatrix}. \quad (5.44)$$

Thus, one has

$$\frac{\xi_{i-1}-\xi_i}{R_i}=\frac{\xi_i\mu_{ia}-\xi'_i\mu'_{ia}}{z'_{ia}}=\frac{\xi_i\mu_{ib}-\xi'_i\mu'_{ib}}{z'_{ib}}. \quad (5.45)$$

From Eq. (5.45), the invariant $G=\xi'_i\left(\mu'_{ia}z'_{ib}-\mu'_{ib}z'_{ia}\right)=\xi_i\left(\mu_{ia}z_{ib}-\mu_{ib}z_{ia}\right)$ defined in Eq. (5.42) is also valid since $z'_{ia}=z_{ia}$ and $z'_{ib}=z_{ib}$. Through a similar series of operations, it can be shown that $G=\xi_i\left(\mu_{ia}z_{ib}-\mu_{ib}z_{ia}\right)$ for a given surface is equal to $G=\xi_{i+1}\left(\mu_{i+1a}z_{i+1b}-\mu_{i+1b}z_{i+1a}\right)$ for the next surface. In other words, G is invariant not only across the surface, but also across the space between two paraxial surfaces. That is, G is invariant both throughout the entire optical system and in any continuous part of the system.

The numerical value of G can be calculated in any one of several ways. Having obtained G, it can be used to determine the value of many other quantities of the optical system without the need for intermediate operations or raytracing calculations which would otherwise be required. The optical invariant is thus one of the most useful tools available to optical engineers in developing optical layouts and system concepts.

5.9.1 Optical Invariant and Lateral Magnification

Having determined the optical invariant G, the lateral magnifications and image height of an optical system can be obtained by means of just two traced rays. Consider the simple optical system shown in Fig. 5.15. From Eq. (5.42), the optical invariant is obtained as

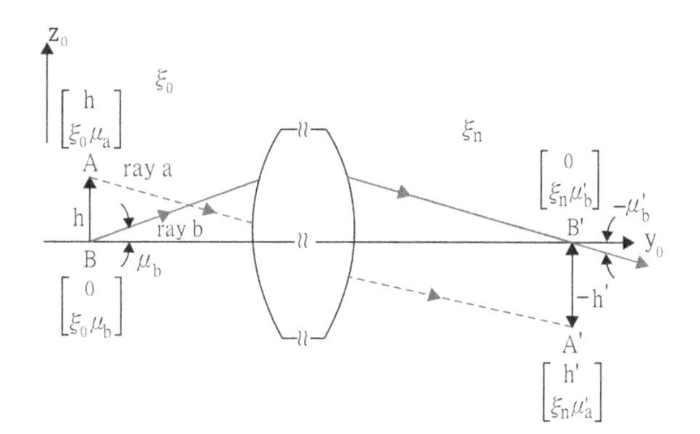

Fig. 5.15 The optical invariant of a system allows the lateral magnification to be estimated by tracing only one on-axis point source

$$G = \xi_0(\mu_a \cdot 0 - \mu_b h) = \xi_n(\mu'_a \cdot 0 - \mu'_b h') = -\xi_0 \mu_b h = -\xi_n \mu'_b h'. \qquad (5.46)$$

Equation (5.46) can be rearranged to give the following generalized expression for the lateral magnification $m^*_{paraxial}$ of the system:

$$m^*_{paraxial} = \frac{h'}{h} = \frac{\xi_0 \mu_b}{\xi_n \mu'_b}. \qquad (5.47)$$

This equation is useful for systems in which the object and image are not in air but in two media with different refractive indices. Equation (5.47) indicates that it is necessary only to trace the ray originating from point B (an on-axis point source) with height-slope matrix $\begin{bmatrix}0 & \xi_0\mu_b\end{bmatrix}^T$ to obtain the height-slope matrix $\begin{bmatrix}0 & \xi_n\mu'_b\end{bmatrix}^T$ at point B′ (the image of B), thereby enabling the lateral magnification of the system to be obtained. If the object and image are both in air, then $m^*_{paraxial} = \mu_b/\mu'_b$. In other words, the lateral magnification is given by the ratio of the slope angles, μ_b and μ'_b, of the incoming ray from B and exiting ray at B′, respectively.

5.9.2 *Image Height for Object at Infinity*

The optical invariant G also enables the estimation of the image height for an object located at infinity in an optical system (see Fig. 5.16). From Eq. (5.42), the optical invariant can be obtained as

$$G = \xi_0(\mu_a \cdot z_b - 0 \cdot z_a) = \xi_n(\mu'_a \cdot 0 - \mu'_b h') = \xi_0 \mu_a z_b = -\xi_n \mu'_b h'. \qquad (5.48)$$

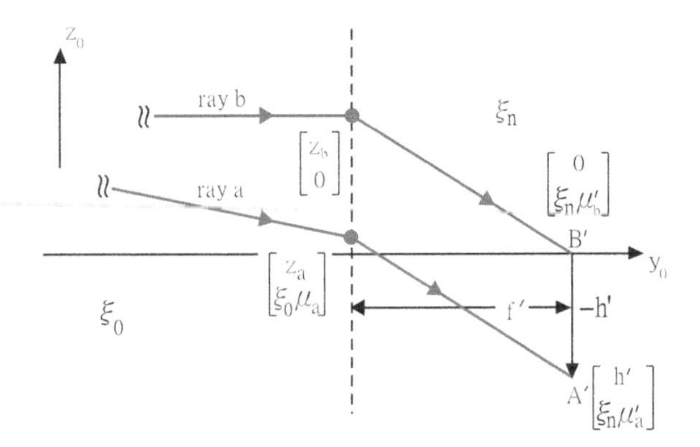

Fig. 5.16 Image height of an optical system with its object located at infinity can be estimated using the optical invariant

The image height is then given as

$$h' = -\frac{\xi_0 \mu_a}{\xi_n \mu_b'} z_b. \tag{5.49}$$

If both the object and the image are in air (i.e., $\xi_0 = \xi_n = \xi_{air}$), then $h' = -z_b\mu_a/\mu_b'$. Note that the back focal length of Fig. 5.16 is given by $t' = z_n/(-\mu_n')$(see Eq. (5.6)). Therefore, Eq. (5.49) can be further simplified as

$$h' = \left(-\frac{z_b}{\mu_b'}\right)\mu_a = \mu_a t'. \tag{5.50}$$

Equation (5.50) shows that for an object located at infinity, the image height is given by the product of the slope angle μ_a and back focal length t' of the optical system.

5.9.3 Data of Third Ray

Figure 5.17 shows three rays propagating from a medium of refractive index ξ to a medium of refractive index ξ'. If the height-slope matrices of rays a and b are known, the optical invariant can be used to determine the ray height-slope matrix of ray c directly. In other words, a paraxial system can be completely described by the ray data of any two unrelated rays. From Eq. (5.42), the following two equations can be obtained:

$$\xi(\mu_c z_a - \mu_a z_c) = (\xi' z_a')\mu_c' - (\xi' \mu_a')z_c', \tag{5.51}$$

$$\xi(\mu_c z_b - \mu_b z_c) = (\xi' z_b')\mu_c' - (\xi' \mu_b')z_c'. \tag{5.52}$$

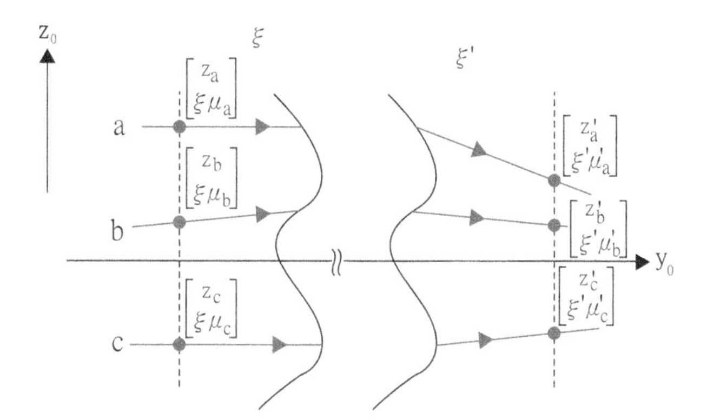

Fig. 5.17 Height-slope matrix of a *third ray* can be determined from the height-slope matrices of the *first* and *second rays*

From Eqs. (5.51) and (5.52), μ_c' and z_c' can be obtained as

$$\mu_c' = \frac{\det\left(\begin{bmatrix} \xi(\mu_c z_a - \mu_a z_c) & -\xi'\mu_a' \\ \xi(\mu_c z_b - \mu_b z_c) & -\xi'\mu_b' \end{bmatrix}\right)}{\det\left(\begin{bmatrix} \xi' z_a' & -\xi'\mu_a' \\ \xi' z_b' & -\xi'\mu_b' \end{bmatrix}\right)} = \frac{\xi(\mu_c z_b - \mu_b z_c)}{\xi'(\mu_a' z_b' - \mu_b' z_a')}\mu_a' + \frac{\xi(\mu_a z_c - \mu_c z_a)}{\xi'(\mu_a' z_b' - \mu_b' z_a')}\mu_b', \tag{5.53}$$

$$z_c' = \frac{\det\left(\begin{bmatrix} \xi' z_a' & \xi(\mu_c z_a - \mu_a z_c) \\ \xi' z_b' & \xi(\mu_c z_b - \mu_b z_c) \end{bmatrix}\right)}{\det\left(\begin{bmatrix} \xi' z_a' & -\xi'\mu_a' \\ \xi' z_b' & -\xi'\mu_b' \end{bmatrix}\right)} = \frac{\xi(\mu_c z_b - \mu_b z_c)}{\xi'(\mu_a' z_b' - \mu_b' z_a')} z_a' + \frac{\xi(\mu_a z_c - \mu_c z_a)}{\xi'(\mu_a' z_b' - \mu_b' z_a')} z_b'. \tag{5.54}$$

Equations (5.53) and (5.54) show that the ray data of a third ray can be determined without further raytracing by means of

$$\mu_c' = E\mu_a' + F\mu_b', \tag{5.55}$$

$$z_c' = Ez_a' + Fz_b', \tag{5.56}$$

where

$$E = \frac{\xi(\mu_c z_b - \mu_b z_c)}{\xi'(\mu_a' z_b' - \mu_b' z_a')}, \tag{5.57}$$

$$F = \frac{\xi(\mu_a z_c - \mu_c z_a)}{\xi'(\mu'_a z'_b - \mu'_b z'_a)}. \tag{5.58}$$

5.9.4 Focal Length Determination

The optical invariant can also be used to determine the back focal length, t', and effective back focal length, f', of an optical system. The height z'_c and angle μ'_c of ray c in Fig. 5.18 (parallel to the optical axis) can be determined from Eqs. (5.59) and (5.60) as

$$\mu'_c = E\mu'_a + F\mu'_b, \tag{5.59}$$

$$z'_c = Ez'_a + Fz'_b, \tag{5.60}$$

where

$$E = \frac{\xi(-\mu_b z_c)}{\xi'(\mu'_a z'_b - \mu'_b z'_a)}, \tag{5.61}$$

$$F = \frac{\xi(\mu_a z_c)}{\xi'(\mu'_a z'_b - \mu'_b z'_a)}. \tag{5.62}$$

The effective back focal length, f', and back focal length, t', can then be determined from Eqs. (5.5) and (5.6) as

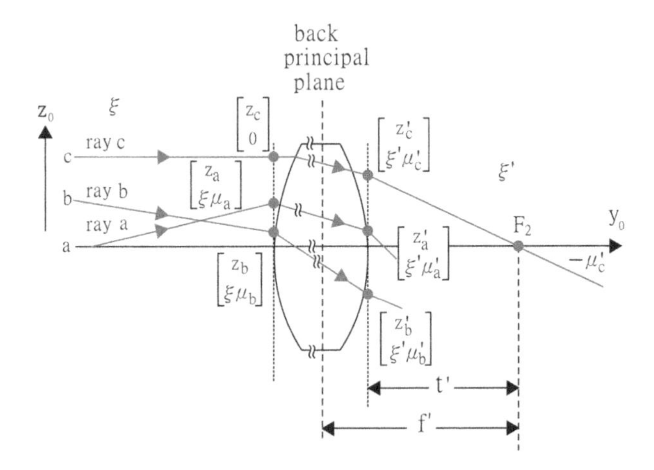

Fig. 5.18 Back focal length and effective back focal length can be determined from the optical invariant

$$\mathrm{f}' = -\frac{\mathrm{z_c}}{\mu_c'} = -\frac{\mathrm{z_c}}{\mathrm{E}\mu_a' + \mathrm{F}\mu_b'} = \frac{\xi'\left(\mu_b' z_a' - \mu_a' z_b'\right)}{\xi\left(\mu_a \mu_b' - \mu_b \mu_a'\right)}, \tag{5.63}$$

$$\mathrm{t}' = -\frac{z_c'}{\mu_c'} = \frac{-\left(\mathrm{E}z_a' + \mathrm{F}z_b'\right)}{\mathrm{E}\mu_a' + \mathrm{F}\mu_b'} = \frac{-\left(\mu_b z_a' - \mu_a z_b'\right)}{\mu_b \mu_a' - \mu_a \mu_b'}. \tag{5.64}$$

Most optical computer programs make use of Eqs. (5.63) and (5.64) to calculate the back focal points of a system. However, such programs usually place the object at a remote, but finite distance, and cannot therefore calculate the focal lengths directly without special calculations.

References

1. Gerrard A, Burch JM (1975) Introduction to matrix methods in optics. John Wiley & Sons, Ltd
2. Kim DH, Shi D, Ilev IK (2011) Alternative method for measuring effective focal length of lenses using the front and back surface reflections from a reference plate. Appl Opt 50:5163–5168

Chapter 6
Ray Aberrations

In Chap. 5, imaging was considered to be ideal. In other words, the rays originating from an object point and passing through an optical system converge to the Gaussian image point in accordance with the principles of paraxial optics. However, as discussed in this chapter, most object points do not in fact form a perfect point image due to aberrations in the optical system. Notably, these aberrations are not necessarily caused by defects in grinding, centering or assembling the lenses, or to material inhomogeneities, but because the lenses inherently refract (or reflect) an incident spherical wavefront into an aberrated wavefront. Aberrations fall into two main classes, namely monochromatic and chromatic. Monochromatic aberrations are caused by the geometry of the boundary surfaces and occur when the ray is reflected (or refracted). They appear even when using monochromatic light, hence their name. By contrast, chromatic aberrations are caused by dispersion due to the variation of the lens's refractive index with the incident wavelength [1–3]. Such aberrations do not appear when monochromatic light is used. This chapter presents a mathematical description of the aberrations in an axis-symmetrical optical system from the viewpoint of ray deviation errors.

6.1 Stops and Aperture

An optical system typically has many openings, or structures, which limit the ray bundles. These structures may be the edge of a lens or mirror, or a ring or some other fixture designed to hold an optical element in place and to limit the amount of light admitted into the system. Such structures are known as stops. The aperture of

P.D. Lin, *Advanced Geometrical Optics*, Progress in Optical Science and Photonics 4, DOI 10.1007/978-981-10-2299-9_6

an optical system is the stop which determines the number of incoming rays which actually enter the optical system. For an imaging system, the aperture thus determines the brightness of the image point. When an aperture has the form of a hole or an opening (such as a diaphragm, e.g., element j = 3 in Fig. 3.2), it is convenient to treat the aperture as a ghost element, i.e., an element with a constant refractive index $\xi_{ej} = \xi_{air} = 1$ (assuming the system is surrounded with air) and a thickness $q_{ej} = 0$.

As described in the following, there are several rays of particular interest when analyzing aberrations.

(a) When a point source $\bar{P}_0$ is confined to the y_0z_0 plane of an axis-symmetrical system (where y_0 points along the optical axis of the system), then a meridional ray (or tangential ray) is a ray lying on that plane. The y_0z_0 plane is thus called the meridional plane of the system (Fig. 3.2). Note that when $\bar{P}_0$ lies on the y_0z_0 plane, any source ray with $\alpha_0 = 0$ (Eq. (2.3)) always travels on the meridional plane in an axis-symmetrical system.

(b) The marginal ray in an axis-symmetrical system is a ray which starts at a point source $\bar{P}_0$ and touches the edge of the aperture of the system as it propagates through the system.

(c) The chief ray $\bar{R}_{i/chief} = \begin{bmatrix} \bar{P}_{i/chief} & \bar{\ell}_{i/chief} \end{bmatrix}^T$ at the ith boundary surface in an axis-symmetrical system is the meridional ray which starts from a point source $\bar{P}_0$ and passes (or has passed) through the center of the aperture (e.g., $x_5 = z_5 = 0$ in Fig. 3.2). The exact spherical coordinates $\alpha_{0/chief}$ and $\beta_{0/chief}$ of the unit directional vector of the chief ray from the point source $\bar{P}_0$, i.e.,

$$\bar{\ell}_{0/chief} = \begin{bmatrix} \ell_{0x/chief} \\ \ell_{0y/chief} \\ \ell_{0z/chief} \\ 0 \end{bmatrix} = \begin{bmatrix} C\beta_{0/chief}C(90^\circ + \alpha_{0/chief}) \\ C\beta_{0/chief}S(90^\circ + \alpha_{0/chief}) \\ S\beta_{0/chief} \\ 0 \end{bmatrix}, \tag{6.1}$$

must be determined using numerical methods. However, $\bar{\ell}_{0/chief}$ can generally be approximated by the straight line connecting $\bar{P}_0$ to the center of the entrance pupil of the system. Notably, this estimation can also be used to construct the marginal rays or a ray passing through a particular point of the aperture. However, these rays can be determined more accurately using the Newton-Raphson method given a knowledge of the Jacobian matrices, i.e., $\partial\bar{\ell}_0/\partial\alpha_0$ and $\partial\bar{\ell}_0/\partial\beta_0$.

(d) A skew-ray is a ray which does not propagate in the plane which contains both the point source and the optical axis. Such rays neither cross, nor are parallel to, the optical axis anywhere.

(e) If the optical axis lies along the y_0 axis direction and the meridional plane is the y_0z_0 plane, sagittal rays intersect the aperture at $z_5 = 0$ in Fig. 3.2, where z_5 is the in-plane coordinate of aperture. In conventional textbooks, a sagittal ray from an off-axis point source $\bar{P}_0$ is defined as a ray which propagates in the "plane" perpendicular to the meridional plane and contains the chief ray.

However, numerical results show that a sagittal plane is not in fact a flat plane through the system, but rather a 3-D curved surface whose shape changes following each refraction (or reflection) process.

(f) A ray which travels along the optical axis of an axis-symmetrical system is called an axial ray.

Figure 3.2 shows an illustrative axis-symmetrical system with symmetry about the optical axis such that every surface is a figure of rotation about the optical axis. Due to the symmetry of the system, it is possible, with no loss of generality, to define the object point $\bar{P}_0$ as lying on the y_0z_0 plane. It will be recalled that the ith ray $\bar{R}_i = \begin{bmatrix} \bar{P}_i & \bar{\ell}_i \end{bmatrix}^T$ is described by its incidence point $\bar{P}_i$ and reflected (or refracted) unit directional vector $\bar{\ell}_i$.

6.2 Ray Aberration Polynomial and Primary Aberrations

In 1856, Philip Ludwig von Seidel published a classical paper in which he extended Gaussian theory to axis-symmetrical systems so as to include all aberrations of the third-order, which had previously been ignored by Gauss. As a result of his work, Seidel proved that the images produced by such systems are subject to five, and only five, distinct forms of aberration. These aberrations are now commonly referred to as the Seidel aberrations (or primary aberrations) and are denoted in order as spherical, coma, astigmatism, field curvature and distortion. Seidel's treatment of optical aberrations was thorough and exhaustive, but every lengthy and complicated. As a consequence, the value of his work has only in recent times been widely recognized. Over the years, his expansion of Gaussian theory has been modified to express imaging errors in terms of either "wavefront aberrations" or "ray aberrations". Notably, these two terminologies do not refer to different things, but to different aspects of the same thing, since an imperfection in the image formed by an optical system can be described in a variety of ways.

Ray aberrations clearly depend on the chosen ray. Consider the axis-symmetrical optical system shown in Fig. 6.1. Due to its symmetry, it is possible, with no loss of generality, to define the object point as $\bar{P}_0 = \begin{bmatrix} 0 & P_{0y} & h & 1 \end{bmatrix}^T$, where h denotes its height above the optical axis. For a ray originating from the object point and passing through the aperture at a point with polar coordinates ρ and θ, the intersection point of the ray on the image plane is given by

$$ {}^{n}\bar{P}_n = \begin{bmatrix} x_n \\ 0 \\ z_n \\ 1 \end{bmatrix} = \begin{bmatrix} \Delta x_n \\ 0 \\ A_2 h + \Delta z_n \\ 1 \end{bmatrix}, \tag{6.2} $$

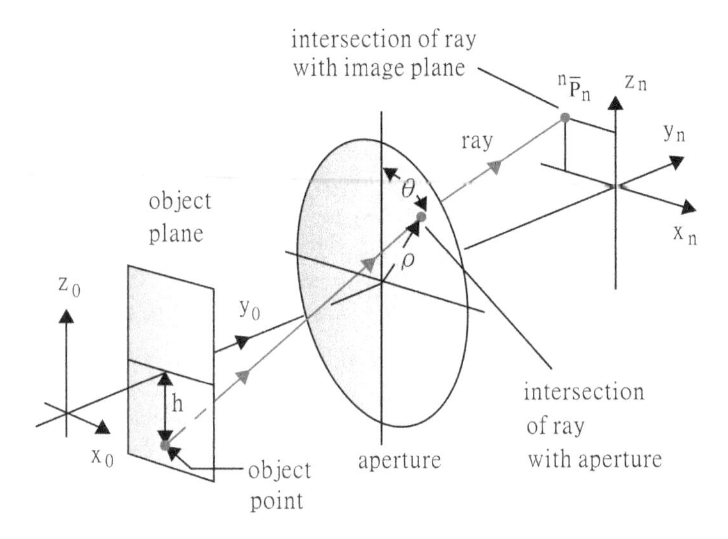

Fig. 6.1 Ray originating from point $\bar{P}_0 = [0 \quad P_{0y} \quad h \quad 1]^T$ in the object plane passes through the aperture at a point defined by polar coordinates (ρ, θ) and then intersects the image plane at ${}^{n}\bar{P}_n = [x_n \quad 0 \quad z_n \quad 1]^T$

where Δx_n and Δz_n denote the transverse aberrations on the image plane. The full seventh-order Taylor series expansions of these two aberrations are given as [4]

$$\begin{aligned}\Delta z_n = {} & A_1 \rho C\theta \\ & + B_1 \rho^3 C\theta + B_2 \rho^2 h(2 + C(2\theta)) + (3B_3 + B_4)\rho h^2 C\theta + B_5 h^3 \\ & + C_1 \rho^5 C\theta + \rho^4 h(C_2 + C_3 C(2\theta)) + \rho^3 h^2 (C_4 + C_6 C^2\theta) C\theta \\ & + \rho^2 h^3 (C_7 + C_8 C(2\theta)) + C_{10} \rho h^4 C\theta + C_{12} h^5 + D_1 \rho^7 C\theta + \ldots,\end{aligned} \tag{6.3}$$

$$\begin{aligned}\Delta x_n = {} & A_1 \rho S\theta \\ & + B_1 \rho^3 S\theta + B_2 \rho^2 h S(2\theta) + (B_3 + B_4)\rho h^2 S\theta \\ & + C_1 \rho^5 S\theta + C_3 \rho^4 h S(2\theta) + \rho^3 h^2 (C_5 + C_6 C^2\theta) S\theta \\ & + C_9 \rho^2 h^3 S(2\theta) + C_{11} \rho h^4 S\theta + D_1 \rho^7 S\theta + \ldots,\end{aligned} \tag{6.4}$$

where A_n, B_n, and so on, are constants. Collectively, all of the terms in Eqs. (6.3) and (6.4) are referred to as transverse aberrations. In other words, they represent the distance by which the ray misses the ideal image point, as defined by the paraxial imaging equations given in Chap. 5. Note that in the A terms, the exponents of ρ and h are unity. Moreover, in the B terms, the exponents are all equal to 3 (i.e., ρ^3, $\rho^2 h$, ρh^2 and h^3), while in the C terms, they are equal to 5, and in the D terms, 7. Thus, the corresponding terms are referred to as first-order, third-order, fifth-order terms, seventh-order terms, and so on. It is noted that the Taylor series expansions contain two first-order terms (when $A_2 h$ is included), five third-order terms, nine fifth-order terms, and (n + 3)(n + 5)/8−1 nth-order terms. In an axis-symmetrical system, there are no even-order terms, i.e., the Taylor series expansions contain only odd-order terms. The A terms relate to the paraxial (or first-order) imagery discussed in Chap. 5. A_1 is a transverse measure of the distance from the paraxial focus point to the defocused image plane (i.e., an image plane which deviates from the paraxial/Gaussian image plane). In other words, in the Gaussian image plane, $A_1 = 0$. Meanwhile, A_2 is simply the lateral magnification (i.e., $A_2 = m^*_{paraxial}$) in the Gaussian image plane. The B terms are referred to as third-order, or Seidel, or primary aberrations. More specifically, B_1 denotes spherical aberration; B_2, coma aberration; B_3, astigmatism aberration; B_4, field curvature aberration; and B_5, distortion aberration. Similarly, the C terms are known as fifth-order or secondary aberrations, where C_1 is the fifth-order spherical aberration; C_2 and C_3 are linear coma aberrations; C_4, C_5 and C_6 are oblique spherical aberrations; C_7, C_8 and C_9 are elliptical coma aberrations; C_{10} and C_{11} are field curvature and astigmatism aberrations, respectively; and C_{12} is the distortion aberration. The fourteen D terms are seventh-order or tertiary aberrations, e.g., D_1 is the seventh-order spherical aberration.

As noted above, the primary aberrations of a system under monochromatic light are classed as spherical, coma, astigmatism, field curvature or distortion. The importance of decomposing the ray aberration in this manner will become evident as the discussions proceed. The following sections discuss each primary aberration and its representation, as well as its general effect on the image appearance. For convenience, each aberration is discussed as if it alone were present. However, it should be noted that, in practical systems, aberrations are more likely to be encountered in combination than singly.

6.3 Spherical Aberration

The components of the primary spherical aberration on the Gaussian image plane are given as $\Delta z_n = B_1 \rho^3 C\theta$ and $\Delta x_n = B_1 \rho^3 S\theta$, where $B_1 > 0$ is referred to as a positive spherical aberration. It is noted from Eqs. (6.3) and (6.4) that spherical aberration is the only third-order term which does not depend on h. In other words, it is the only aberration which can be investigated by an on-axis object point. Furthermore, primary spherical aberration is symmetrical about the optical axis, and hence it is sufficient to consider $\Delta z_n = B_1 \rho^3 C\theta$ by dealing only with meridional rays and setting $\theta = 0$. Figures 6.2 and 6.3 show two somewhat exaggerated sketches of the spherical aberration produced by an on-axis object point. It is noted that the rays close to the optical axis intersect the axis very close to the paraxial focus position. Furthermore, it is seen that as the ray height increases, the position at which the ray intersects the optical axis moves further and further away from the

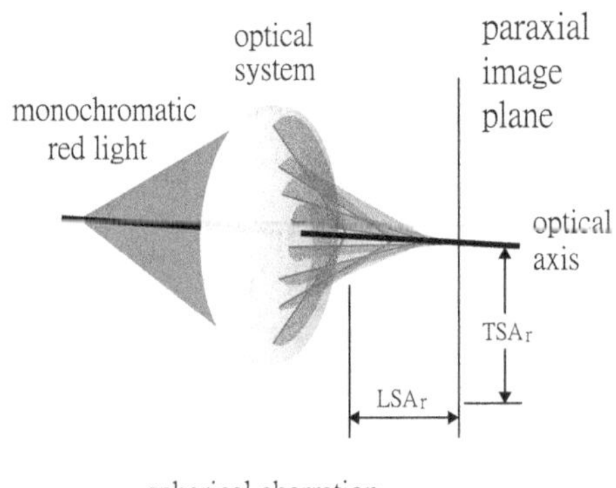

Fig. 6.2 Spherical aberration for on-axis object point. Note that TSA_r denotes transverse spherical aberration, while LSA_r denotes longitudinal spherical aberration

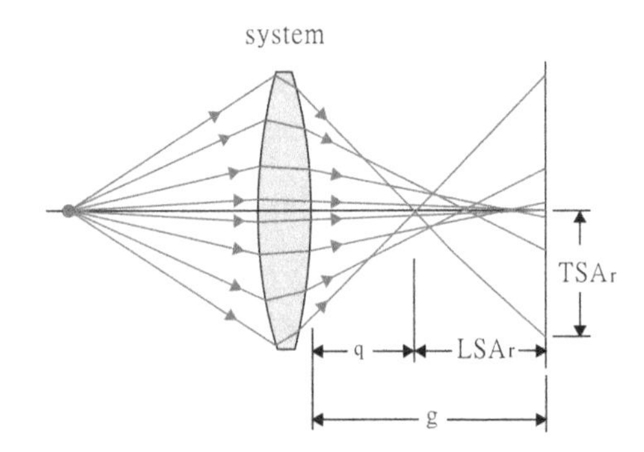

Fig. 6.3 The different rays passing through the system do not meet at one focal point. The further the rays are from the optical axis, the closer to the system they intersect the axis when $B_1 > 0$. ($B_1 > 0$ denotes positive spherical aberration (shown), in which the marginal rays are bent more strongly.)

paraxial focus point. The distance from the focus point to the axial intersection of the ray is known as the longitudinal spherical aberration (LSA). By contrast, the magnitude of the aberration in the vertical direction is referred to as the transverse spherical aberration (TSA). It is noted that the magnitudes of LSA and TSA both depend on the height of the ray. The TSA and LSA for a particular ray (e.g., a ray hitting the aperture at $(\rho, \theta) = (r, 0)$) can be determined by tracing a paraxial ray and meridional ray from the same on-axis object point and determining their final intercept distances g and q, respectively. The LSA of that ray is then obtained as

$$LSA_r = g - q. \tag{6.5}$$

LSA and TSA are not independent of one another since they are related by the included angle μ_r between the ray and the optical axis. More specifically,

$$TSA_r = -LSA_r \tan(|\mu_r|), \tag{6.6}$$

where $|\mu_r|$ is the absolute value of the included angle μ_r. The radius of the blur disc on the Gaussian image plane represents the TSA of the marginal ray. Conventionally, spherical aberration with a negative sign is known as undercorrected spherical aberration since it is usually associated with simple uncorrected

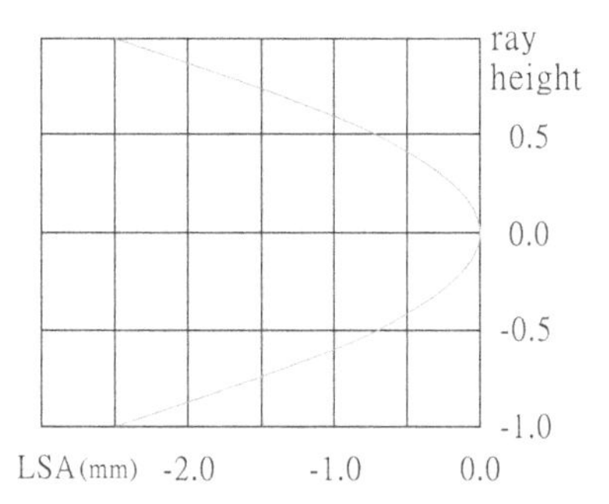

Fig. 6.4 LSA of optical system shown in Fig. 3.2 with collimated rays parallel to optical axis. Note that the ray height is normalized to the range of −1 to +1

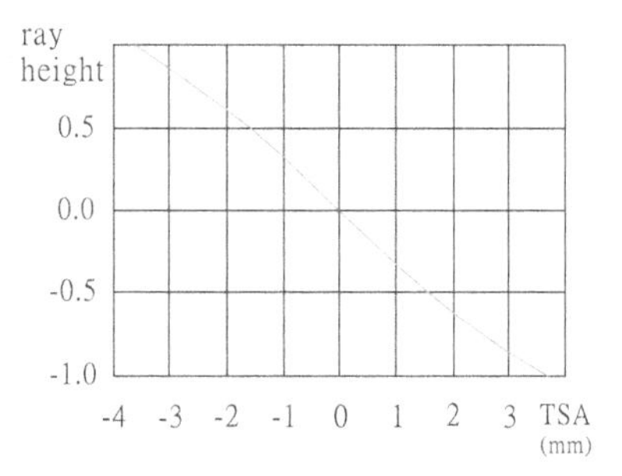

Fig. 6.5 TSA of optical system shown in Fig. 3.2 with collimated rays parallel to optical axis. Note that the ray height is normalized to the range of −1 to +1

positive elements. Similarly, spherical aberration with a positive sign is called overcorrected spherical aberration and is generally associated with diverging elements [4]. The spherical aberration of a system is usually represented graphically, with the magnitudes of the LSA and TSA being plotted against the ray height, as shown in Figs. 6.4 and 6.5, respectively. The image of a point formed by an optical system with spherical aberration usually has the form of a bright dot surrounded by a halo of light. Thus, the effect of spherical aberration on an extended image is to soften the contrast of the image and to blur its details.

6.4 Coma

A coma aberration causes rays from an off-axis object point to create a trailing "comet-like" blur directed away from the optical axis. Assuming that the primary coma exists alone (i.e., no other aberrations are present), the rays from the off-axis object point intersect the image plane at

$$z_n = A_2 h + B_2 \rho^2 h (2 + C(2\theta)) \tag{6.7}$$

and

$$x_n = B_2\rho^2 hS(2\theta), \tag{6.8}$$

where $B_2 > 0$ is referred to as a positive coma, i.e., the ray cone opens away from the optical axis. For a given value of ρ, the locus of the points of intersection of the rays with the image plane is given by

$$(x_n)^2 + \left[z_n - \left(A_2 + 2B_2\rho^2\right)h\right]^2 = \left(B_2\rho^2 h\right)^2. \tag{6.9}$$

Thus, the rays originating from a circle of radius ρ in the aperture lie on a circle of radius $B_2\rho^2 h$ centered at

$$\begin{bmatrix} x_{n/center} & 0 & z_{n/center} & 1 \end{bmatrix}^T = \begin{bmatrix} 0 & 0 & (A_2 + 2B_2\rho^2)h & 1 \end{bmatrix}^T \tag{6.10}$$

on the image plane. As shown in Eq. (6.10), and illustrated in Fig. 6.6, when a bundle of skew rays is incident on an optical system with a coma, the rays passing through the edge zones of the aperture are imaged at a different height than those passing through the center zone. Moreover, the rays which form a circle on the aperture also form a circle on the image plane. The rays from a smaller circle in the aperture form a correspondingly smaller circle on the image plane. Finally, the intersection of the chief ray $\bar{R}_{n/chief}$ with the image plane lies at the apex of the coma patch. Thus, the comatic image can be viewed as being made up of a series of different-sized circles arranged tangent to a 60° angle. In practice, the size of the image circle is proportional to the square of the diameter of the aperture circle. The appearance of a point image formed by a system suffering a coma is shown in Fig. 6.6. Obviously, the aberration is named after the comet shape of the figure.

It is noted that as the rays go around the aperture circle once, they go around the image circle twice in accordance with the B_2 terms in Eqs. (6.7) and (6.8). Figure 6.7 shows the relationship between the position at which a ray passes

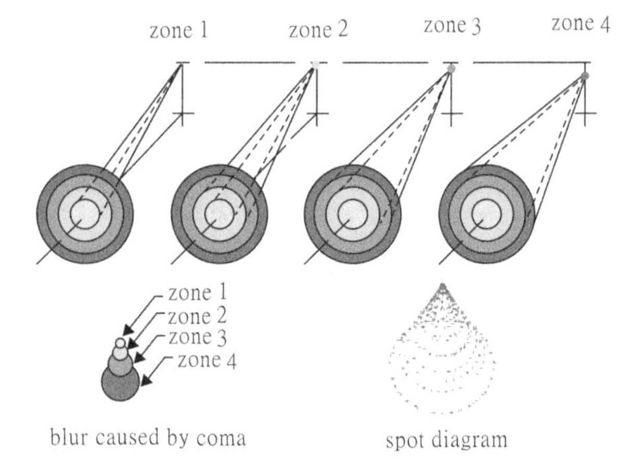

Fig. 6.6 Distinct shape displayed by images with a coma aberration is a result of refraction differences of the rays passing through different regions of the aperture

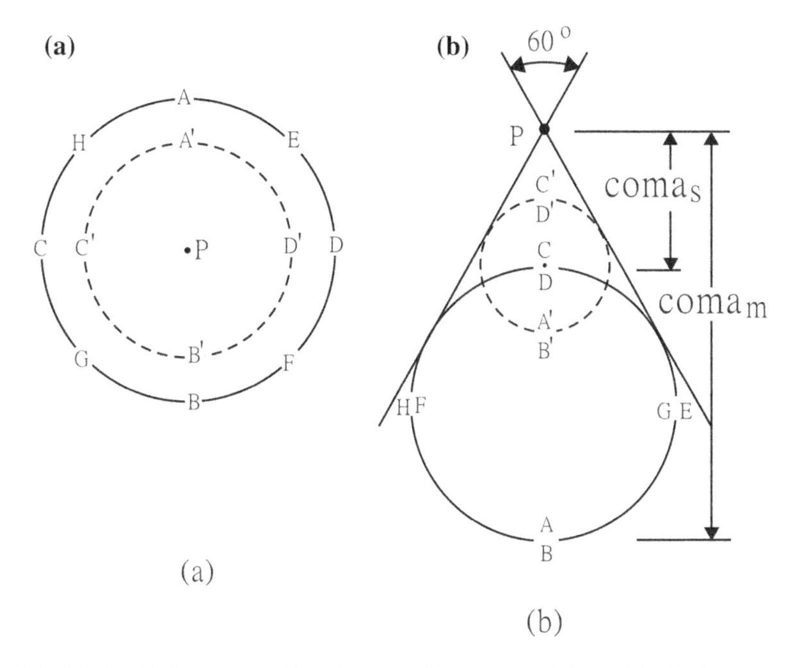

Fig. 6.7 Relationship between position of a ray in the aperture and its position in the coma patch. **a** View of aperture, with individual rays indicated by different letters. **b** Letters indicate positions of corresponding rays in the image

through the aperture and the location which it occupies in the coma patch. Figure 6.7a presents a head-on view of the aperture, with the ray positions indicated by letters A through H and A′ through D′. The resultant coma patch is shown in Fig. 6.7b with the ray locations marked with the corresponding letters.

From Eqs. (6.7) and (6.8), and as shown in Fig. 6.7, the upper and lower marginal rays A and B (striking the aperture at polar coordinates $(\rho_m, 0)$ and (ρ_m, π), respectively) intersect the image plane at

$$ {}^{n}\bar{P}_{n} = \begin{bmatrix} x_{n/marginal} \\ 0 \\ z_{n/marginal} \\ 1 \end{bmatrix} = \begin{bmatrix} 0 \\ 0 \\ A_2 h + 3B_2 \rho_m^2 h \\ 1 \end{bmatrix}. \tag{6.11} $$

The intersection of the chief ray with the image plane (denoted as point P in Fig. 6.7) is expressed as

$$ {}^{n}\bar{P}_{n} = \begin{bmatrix} x_{n/chief} \\ 0 \\ z_{n/chief} \\ 1 \end{bmatrix} = \begin{bmatrix} 0 \\ 0 \\ A_2 h \\ 1 \end{bmatrix}. \tag{6.12} $$

The distance from P to the intersection of AB in Fig. 6.7b is called the meridional coma of the system, and is given by the distance between them, i.e.,

$$\mathrm{coma_m} = z_{n/\mathrm{marginal}} - z_{n/\mathrm{chief}} = 3B_2\rho_m^2 h. \tag{6.13}$$

Furthermore, from Eqs. (6.7) and (6.8), the intersection of sagittal rays C and D (which hit the aperture at $(\rho_m, \pi/2)$ and $(\rho_m, 3\pi/2)$, respectively) with the image plane is given by

$$^{n}\bar{P}_n = \begin{bmatrix} x_n \\ 0 \\ z_n \\ 1 \end{bmatrix} = \begin{bmatrix} 0 \\ 0 \\ A_2 h + B_2\rho_m^2 h \\ 1 \end{bmatrix}. \tag{6.14}$$

The distance from P to CD in Fig. 6.7b is called the sagittal coma and is given as

$$\mathrm{coma_s} = B_2\rho_m^2 h. \tag{6.15}$$

It is noted that the sagittal coma is one-third as large as the meridional coma.

As shown in Fig. 6.8 (and Eqs. (6.7) and (6.8)), the determination of $\mathrm{coma_m}$ and $\mathrm{coma_s}$ involves a particular image plane, on which the upper and lower marginal rays hit at the same point. In other words,

$$\bar{\Phi} = z_{n/\mathrm{upper\ marginal}} - z_{n/\mathrm{lower\ marginal}} = 0. \tag{6.16}$$

For an optical system possessing k elements (e.g., Fig. 3.2), this condition can be satisfied by making an initial guess $v_{k/\mathrm{current}}$ of the separation v_k of the image plane (labeled as the kth element) and the (k − 1)th element. A better approximation of the required separation distance can then be obtained by

$$v_{k/\mathrm{next}} = v_{k/\mathrm{current}} - \frac{\bar{\Phi}}{\partial\bar{\Phi}/\partial v_k}, \tag{6.17}$$

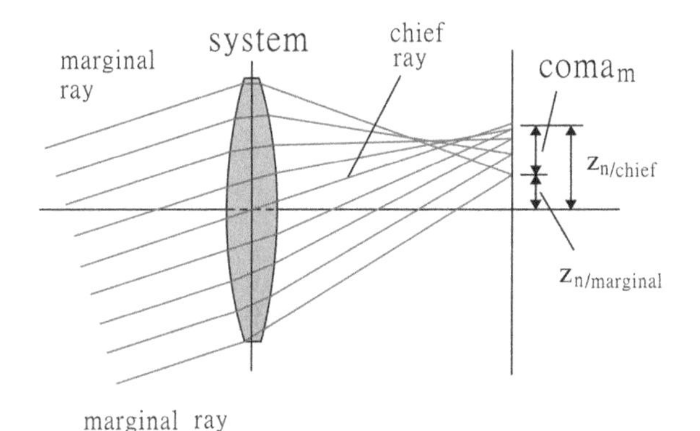

Fig. 6.8 Coma is an aberration which causes rays from an off-axis point of light in the object plane to create a trailing "comet-like" blur. The figure shows the intersections of the meridional rays with the image plane

where the second term is addressed in Part Two of this book. Equation (6.17) is executed iteratively, with the magnitude of $\bar{\Phi}$ checked each time. If the absolute value of $\bar{\Phi}$ satisfies a predefined threshold value, the iteration procedure is terminated and the current value of $v_{k/current}$ is taken as the required image plane.

About half of all the energy in the coma patch is concentrated in the small triangular region between P and CD. Thus, the sagittal coma provides a somewhat better measure of the effective size of the image blur than the meridional coma. Coma aberrations are a particularly disturbing aberration since their flares are nonsymmetrical. Their presence is thus highly detrimental to the accurate determination of the image position since it is much more difficult to locate the "center of gravity" of a coma patch than that of a circular blur such as that produced by spherical aberration. Moreover, the shape of the coma aberration varies both with the shape of the boundary surfaces and with the position of the aperture which determines the bundle of rays forming the image. In an axis-symmetrical system, there is no coma aberration on the optical axis. However, as shown in Eqs. (6.13) and (6.15), the coma patch varies linearly with the distance from the axis.

6.5 Astigmatism

In optical systems with astigmatism, the rays which propagate in the meridional and sagittal planes have different foci. It is noted from Eqs. (6.3) and (6.4) that the components of astigmatism (i.e., $3B_3\rho h^2C\theta$ and $B_3\rho h^2S\theta$) have a square dependence on the ray height, h. The components of the field curvature aberration (i.e., $B_4\rho h^2C\theta$ and $B_4\rho h^2S\theta$) also have an h^2 dependence. However, astigmatism and field curvature represent qualitatively different aberrations despite some similarity between them. Thus, while some books discuss the associated terms together, they are discussed separately in the present book.

The astigmatism effect is best appreciated when considered together with the defocus terms, $A_1\rho C\theta$ and $A_1\rho S\theta$, i.e.,

$$z_n = A_2h + \Delta z_n = A_2h + A_1\rho C\theta + 3B_3\rho h^2C\theta \tag{6.18}$$

and

$$x_n = \Delta x_n = A_1\rho S\theta + B_3\rho h^2S\theta. \tag{6.19}$$

Unless a system is poorly made, no astigmatism occurs when an on-axis point (i.e., $h = 0$) is imaged. However, as the imaged point moves further from the optical axis, the amount of astigmatism gradually increases. For a given value of ρ, the locus of the points of intersection of the rays with the defocused image plane is given by

$$\frac{(z_n - A_2h)^2}{\left(A_1\rho + 3B_3\rho h^2\right)^2} + \frac{(x_n)^2}{\left(A_1\rho + B_3\rho h^2\right)^2} = 1. \tag{6.20}$$

Thus, the rays lying on a circle of radius ρ in the aperture, in general, form an ellipse on the defocused image plane with semi-axes of $\left(A_1 + 3B_3h^2\right)\rho$ and $\left(A_1 + B_3h^2\right)\rho$, respectively. For a given height h, the largest ellipse is obtained from the marginal rays. At the paraxial image plane (i.e.,$A_1 = 0$), the major and minor axes are given as $6B_3\rho h^2$ and $2B_3\rho h^2$, respectively. If the defocused plane is chosen such that $A_1 = -3B_3h^2$, then the transverse aberrations are $\Delta z_n = 0$ and $\Delta x_n = -2B_3\rho h^2 S\theta$, respectively. Consequently, the image has the form of a line with full length $\left|4B_3\rho h^2\right|$ (referred to as the sagittal focal line, as shown in Fig. 6.9). This line, known as the meridional image, is perpendicular to the meridional plane; i.e., it lies on the sagittal plane. Conversely, if the defocused plane is chosen such that $A_1 = -B_3h^2$, the transverse aberrations are given by $\Delta z_n = 2B_3\rho h^2 C\theta$ and $\Delta x_n = 0$, respectively, resulting in a meridional focal line lying on the meridional plane. It is noted that the full length of this focal line is the same as that of the sagittal focal line. Astigmatism occurs when the meridional and sagittal images do not coincide. The distance between the two lines is known as the longitudinal astigmatism.

For the particular case of $A_1 = -2B_3h^2$, Eq. (6.20) becomes

$$(z_n - A_2h)^2 + (x_n)^2 = \left(B_3\rho h^2\right)^2. \tag{6.21}$$

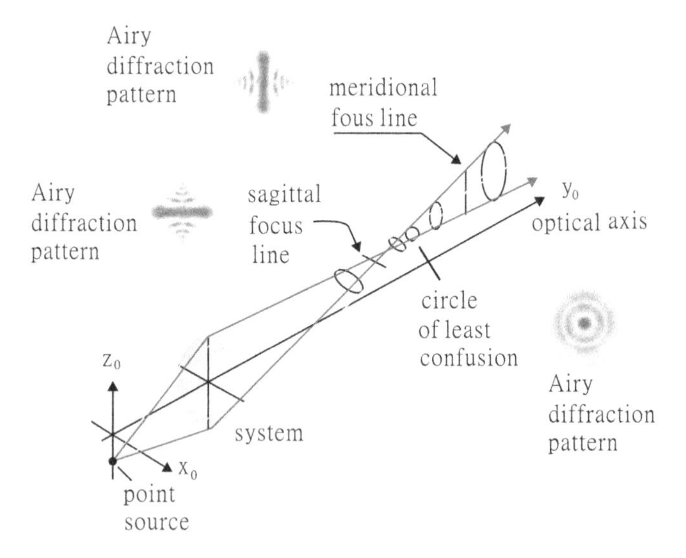

Fig. 6.9 Illustration of astigmatism aberration for off-axis point source

Thus, the imaged spot has the form of a circle with a diameter equal to $\left|2B_3\rho h^2\right|$, i.e., half the length of the meridional (or sagittal) focal line. In other words, it represents the minimum spot size (i.e., the circle of least confusion).

Astigmatism occurs when the optical system is not axis-symmetrical. This may be the result of manufacturing error in the surfaces of the elements or misalignments of the elements. In this case, astigmatism is observed even for rays originating from on-axis object points. Astigmatism is extremely important in vision science and eye care since the human eye often exhibits this form of aberration due to imperfections in the shape of the cornea or lens.

6.6 Field Curvature

Consider an optical system free of all the aberrations thus far considered. For such a system, there exists a one-to-one correspondence between the points on the object plane and those on the focal surface. The field curvature describes the longitudinal departure of the focal surface from the paraxial image plane. Field curvature stems from the fact that the image of a flat object plane becomes curved and not the reciprocal of the radius of the image plane, as shown in Fig. 6.10. As for astigmatism, the field curvature effect is best considered together with the defocus term $A_1\rho C\theta$, i.e.,

$$\Delta z_n = \left(A_1 + B_4 h^2\right)\rho C\theta, \tag{6.22}$$

$$\Delta x_n = \left(A_1 + B_4 h^2\right)\rho S\theta. \tag{6.23}$$

For a given value of ρ, the locus of the intersection points of the rays originating from an off-axis point with the defocused (or paraxial) image plane has the form of a circle. It is noted from Eqs. (6.22) and (6.23) that if the image plane is taken such that $A_1 = -B_4h^2$, the transverse aberration disappears. This implies that the best images of an object plane seldom lie exactly on the paraxial image plane since the focused images actually lie on curved surfaces which are paraboloid in shape

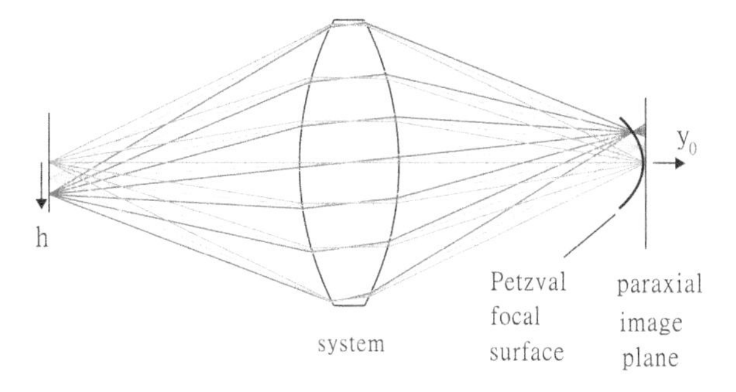

Fig. 6.10 Petzval focal surface deviates from paraxial image plane

(so-called Petzval focal surface, named after the Hungarian mathematician Joseph Max Petzval (1807–1891)). Mathajan presented the following expression for the radii of the focal surfaces (p. 227 of [5]):

$$3R_{sagittal} - R_{meridional} = 2R_{Petzval}, \tag{6.24}$$

or

$$3\left(R_{Petzval} - R_{sagittal}\right) = R_{Petzval} - R_{meridional}, \tag{6.25}$$

where $R_{sagittal}$, $R_{Petzval}$ and $R_{meridional}$ are the radii of the sagittal, meridional and Petzval focal surfaces, respectively. Equation (6.25) implies that the Petzval focal surface is three times as far from the meridional focal surface as it is from the sagittal focal surface. Moreover, the sagittal focal surface always lies between the meridional and Petzval focal surfaces. Finally, in the absence of astigmatism, the sagittal and tangential image surfaces coincide with one another and lie on the Petzval surface.

Field curvature, also referred to as Petzval field curvature, is a function of the refractive indices of elements and the radii of boundary surfaces. When the meridional image lies to the left of the sagittal image (and both are to the left of the Petzval surface), field curvature is referred to as negative, undercorrected, or inward-(toward the lens) curving. By contrast, when the order is reversed, field curvature is referred to as overcorrected, or backward-curving.

6.7 Distortion

One of the requirements for an ideal optical system is that the image which it forms must be geometrically similar to the object. In other words, the image dimensions should be linearly related to those of the object by

$$m^*_{paraxial} = \frac{h'}{h}, \tag{6.26}$$

where the image height h' is determined by the intersection of the chief ray with the paraxial image plane. However, in practice, the geometrical similarity of an image to its object is not linearly related to the object height h. This phenomenon, referred to as image distortion, occurs since the actual lateral magnification m^* of the object varies with the object height h by (see Eq. (6.3))

$$m^* = \frac{h'}{h} = \frac{z_n}{h} = A_2 + B_5 h^2 = m^*_{paraxial} + B_5 h^2, \tag{6.27}$$

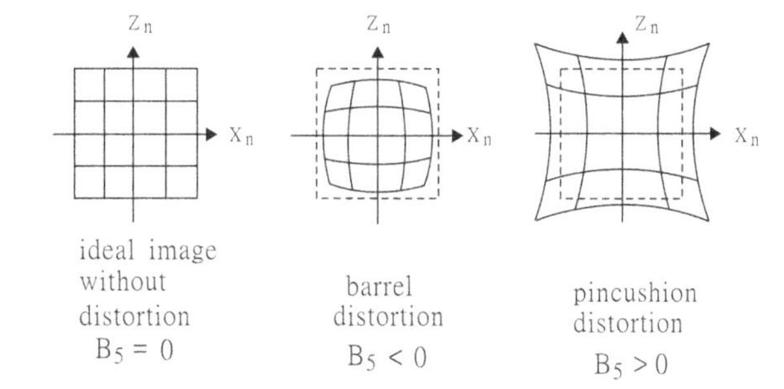

Fig. 6.11 Ideal image without distortion and effects of barrel and pincushion distortion

which is evaluated using the chief ray of the object point. The amount of distortion can be expressed either directly by $m^* - m^*_{paraxial}$ or as a percentage of $m^*_{paraxial}$, i.e.,

$$\bar{\Phi} = 100\left(\frac{m^* - m^*_{paraxial}}{m^*_{paraxial}}\right) = \frac{100B_5h^2}{m^*_{paraxial}}. \tag{6.28}$$

As shown in Eq. (6.27), primary distortion is negative when the actual image is closer to the optical axis than the ideal image, and positive when it lies further from the axis than the ideal image. In practice, this means that the image of a square suffering negative distortion takes on a barrel-like appearance (referred to as barrel distortion), in which the image magnification decreases with an increasing distance from the optical axis (Fig. 6.11). By contrast, in the case of positive distortion, the image takes on a pincushion-like appearance (referred to as pincushion distortion). In such a case, the image magnification increases with an increasing distance from the optical axis. The visible effect is that lines which do not go through the center of the image are bowed inwards, i.e., towards the center of the image.

6.8 Chromatic Aberration

The discussions thus far in this chapter have addressed the five monochromatic aberrations of an axis-symmetrical system. The raytracing equations presented in Eqs. (2.29) and (2.46) are functions of the refractive index $N_i = \xi_{i-1}/\xi_i$, which in turn varies with the wavelength υ. Consequently, different colored rays traverse different paths through the system, and thus so-called chromatic aberration occurs.

In general, the refractive index of optical materials is higher for short wavelengths than for long wavelengths. This dependence can be described, for example, by the following Cauchy equation:

$$\xi_{ej} = G + H/\upsilon^2. \tag{6.29}$$

Table 6.1 The two coefficients of Eq. (6.29) for six common optical materials for incident light in the visible wavelength range

Material	G	H(μm^2)
Fused silica	1.4580	0.00354
Borosilicate glass BK7	1.5046	0.00420
Hard crown glass K5	1.5220	0.00459
Barium crown glass BaK4	1.5690	0.00531
Barium flint glass BaF10	1.6700	0.00743
Dense flint glass SF10	1.7280	0.01342

Table 6.1 lists the two coefficients of Eq. (6.29) for six common optical materials for incident light in the visible wavelength range [1]. Since the paraxial ray-tracing equations presented in Eqs. (4.4), (4.7) and (4.19) are all functions of the refractive index, the cardinal points of an optical system also vary with the wavelength. The term axial chromatic aberration refers to the longitudinal variation of the focus point (or image position) with the wavelength. Figures 6.12 and 6.13 illustrate the chromatic aberration of an optical system. When short-wavelength rays are brought to a focus to the left of long-wavelength rays, the resulting chromatic aberration is termed undercorrected, or negative.

The image of an axial point in the presence of chromatic aberration has the form of a central bright dot surrounded by a halo. The rays of light which are in focus, and those which are nearly in focus, form the bright dot, while the out-of-focus rays form the halo. Thus, in an undercorrected visual instrument, the image has a yellowish dot (formed by the orange, yellow and green rays) and a purplish halo (due

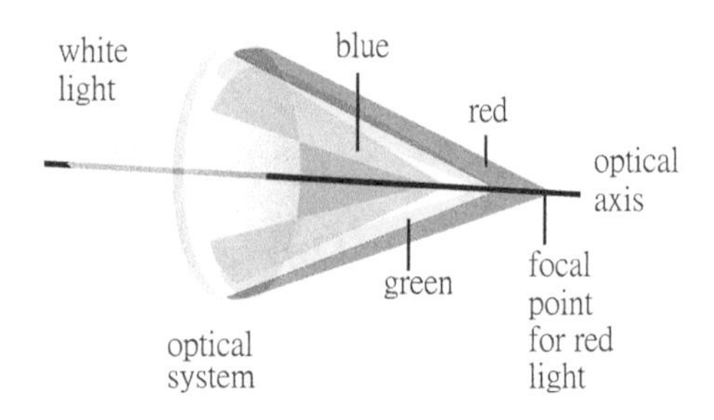

Fig. 6.12 Exaggerated sketch of chromatic aberration in optical system

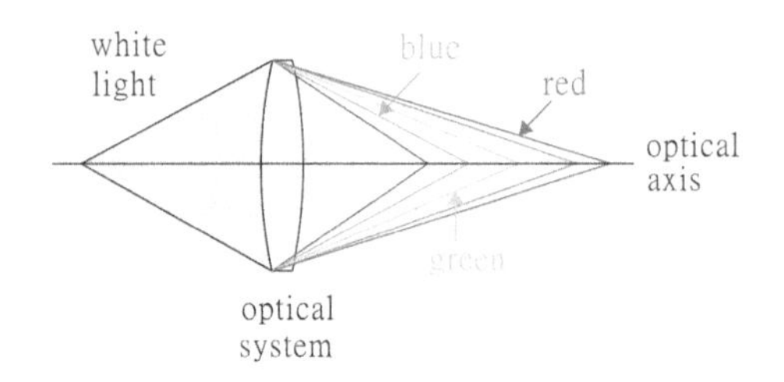

Fig. 6.13 Cross-sectional view of chromatic aberration shown in Fig. 6.12

to the red and blue rays). If the screen on which the image is formed is moved toward the system, the central dot becomes blue. By contrast, if it is moved away, the central dot becomes red. When a system forms images of different sizes for different wavelengths, or spreads the image of an off-axis point into a rainbow, the difference between the image heights for different colors is called lateral color aberration, or the chromatic difference of magnification [4].

The variation of the refractive index of optical materials with the wavelength also produces a variation of the monochromatic aberrations discussed in this chapter. Since each aberration results from the manner in which the rays are bent at the boundary surfaces of the optical system, it is to be expected that, since rays of different wavelength are bent differently, the aberrations will be somewhat different for each wavelength. In general this proves to be the case, and these effects are of practical importance even when the basic aberrations are well corrected.

References

1. Jenkins FA, White HE (1981) Fundamentals of optics, 4th edn. McGraw-Hill, Inc.
2. Kumar V, Singh JK (2010) Model for calculating the refractive index of different materials. Indian J Pure Appl Phys 48:571–574
3. Atchison DA, Simth G (2005) Chromatic dispersions of the ocular media of human eyes. J Opt Soc Am A 22:29–37
4. Smith WJ (2001) Modern optical engineering, 3rd edn. Edmund Industrial Optics, Barrington, N.J
5. Virendra NM (1998) Optical imaging and aberrations, Part I Ray geometrical optics. SPIE Press

Part II
New Tools for Optical Analysis and Design (First-Order Derivative Matrices of a Ray and its OPL)

Optical engineering activities rely heavily on ray-tracing equations. However, the first-order derivative matrices (i.e., the Jacobian matrices) of a ray and its the Optical Path Length (OPL) through the system also provide a useful framework for studying a wide variety of problems in the optical science and engineering fields. While several papers have addressed this issue, they generally fail to take account of all the variables of non-axially symmetrical systems. This may potentially hamper the development of new optical systems; particularly those containing prisms. Accordingly, Chaps. 7 and 8 of this book propose a new computational method for deriving the ray Jacobian matrices of flat and spherical optical boundary surfaces. The validity of the proposed methodology is demonstrated in the remaining chapters of the book. It is shown in Chap. 13 that the ray Jacobian matrices provide a useful means of understanding the Point Spread Function (PSF) and Modulation Transfer Function (MTF) of an optical system, and are also beneficial in predicting the image orientation change produced by prisms and mirrors. The use of the OPL Jacobian matrix to evaluate wavefront aberrations in optical systems is also addressed in Chap. 14. In general, the illustrative examples presented in this part show that the Jacobian matrices of a ray and its OPL, respectively, provide important and highly effective tools for a wide variety of optical systems analysis and design tasks.

"*If I have seen further, it is by standing on the shoulders of giants.*"—Isaac Newton, English physicist and mathematician

Chapter 7
Jacobian Matrices of Ray $\bar{R}_i$ with Respect to Incoming Ray $\bar{R}_{i-1}$ and Boundary Variable Vector $\bar{X}_i$

In automated optical design systems, the Jacobian matrix of an optical quantity with respect to the system variables is generally estimated using the Finite Difference (FD) method [1–15]. In such an approach, each system variable (e.g., ξ_{e1} in Fig. 3.2) is varied by a small amount, $\Delta\xi_{e1}$, and the corresponding changes in the optical quantities of interest (e.g., incidence point $\bar{P}_{11}$ in Fig. 3.2) are calculated using the conventional raytracing method. The differences in the optical quantities of the system, e.g., $\Delta\bar{P}_{11}$, are then divided by the difference in the variable under consideration (e.g., $\Delta\xi_{e1}$ in the present example) in order to estimate the components $\partial\bar{P}_{11}/\partial\xi_{e1}$ of the Jacobian matrix. The FD method is straightforward and easily implemented in computer code, at the cost of additional raytracing steps. However, the accuracy of the results is fundamentally dependent on an appropriate choice of the incremental step size used to adjust each parameter. For example, an excessive step size violates the assumption of local linearity, while an overly-small step size reduces the difference between the original and perturbed solutions and leads to high rounding errors. Many authors have shown that this problem can be avoided by utilizing a differential approach to compute the Jacobian matrix [16–23]. However, the related publications consider only certain system variables and are valid only for axis-symmetrical systems. Consequently, this chapter proposes a more robust differential method for determining the Jacobian matrix $d\bar{R}_i/d\bar{X}_{sys}$ of optical systems containing flat and spherical boundary surfaces. Notably, the proposed methodology is also applicable to systems containing aspherical boundary surfaces if the raytracing algorithm given in [24] is adopted. (Note that the reader may find it helpful to review Sect. 1.10 of this book before reading the remaining discussions in this chapter.)

P.D. Lin, *Advanced Geometrical Optics*, Progress in Optical Science and Photonics 4, DOI 10.1007/978-981-10-2299-9_7

7.1 Jacobian Matrix of Ray

The merit function $\bar{\Phi}$ of an optical system possessing n boundary surfaces may be defined in terms of any ray $\bar{R}_i$, $i \in \{0, 1, 2, \ldots, n\}$. As discussed in Chap. 2, ray $\bar{R}_i$ is a function of both the incoming ray $\bar{R}_{i-1}$ and the boundary variable vector $\bar{X}_i$. Mathematically, $\bar{R}_i$ can be expressed as the following recursive function with the given function $\bar{R}_0$ as:

$$\bar{R}_i = \bar{R}_i(\bar{R}_{i-1}, \bar{X}_i). \tag{7.1}$$

The Jacobian matrix $d\bar{R}_i/d\bar{X}_{sys}$ of ray $\bar{R}_i$ with respect to the system variable vector $\bar{X}_{sys}$ can be determined via the chain rule as (Fig. 7.1)

$$\begin{aligned}
\frac{d\bar{R}_i}{d\bar{X}_{sys}} &= \frac{\partial\bar{R}_i}{\partial\bar{X}_i}\frac{d\bar{X}_i}{d\bar{X}_{sys}} + \frac{\partial\bar{R}_i}{\partial\bar{R}_{i-1}}\frac{d\bar{R}_{i-1}}{d\bar{X}_{sys}} \\
&= \frac{\partial\bar{R}_i}{\partial\bar{X}_i}\frac{d\bar{X}_i}{d\bar{X}_{sys}} + \frac{\partial\bar{R}_i}{\partial\bar{R}_{i-1}}\left(\frac{\partial\bar{R}_{i-1}}{\partial\bar{X}_{i-1}}\frac{d\bar{X}_{i-1}}{d\bar{X}_{sys}} + \frac{\partial\bar{R}_{i-1}}{\partial\bar{R}_{i-2}}\frac{d\bar{R}_{i-2}}{d\bar{X}_{sys}}\right) \\
&= \frac{\partial\bar{R}_i}{\partial\bar{X}_i}\frac{d\bar{X}_i}{d\bar{X}_{sys}} + \frac{\partial\bar{R}_i}{\partial\bar{R}_{i-1}}\frac{\partial\bar{R}_{i-1}}{\partial\bar{X}_{i-1}}\frac{d\bar{X}_{i-1}}{d\bar{X}_{sys}} + \cdots + \frac{\partial\bar{R}_i}{\partial\bar{R}_{i-1}}\frac{\partial\bar{R}_{i-1}}{\partial\bar{R}_{i-2}}\cdots\frac{\partial\bar{R}_2}{\partial\bar{R}_1}\frac{\partial\bar{R}_1}{\partial\bar{X}_1}\frac{d\bar{X}_1}{d\bar{X}_{sys}} \\
&\quad + \frac{\partial\bar{R}_i}{\partial\bar{R}_{i-1}}\frac{\partial\bar{R}_{i-1}}{\partial\bar{R}_{i-2}}\cdots\frac{\partial\bar{R}_1}{\partial\bar{R}_0}\frac{\partial\bar{R}_0}{\partial\bar{X}_0}\frac{d\bar{X}_0}{d\bar{X}_{sys}}.
\end{aligned} \tag{7.2}$$

For a merit function $\bar{\Phi}$ defined in terms of ray $\bar{R}_n$ on the image plane, the Jacobian matrix $d\bar{\Phi}/d\bar{X}_{sys}$ has the form

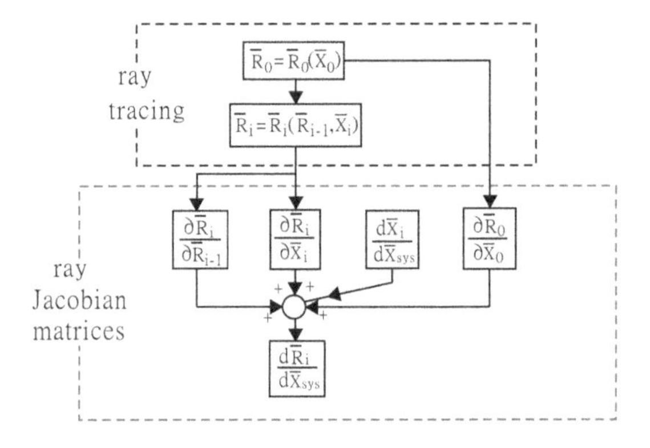

Fig. 7.1 Summary of terms required to determine ray Jacobian matrix $d\bar{R}_i/d\bar{X}_{sys}$

$$
\begin{aligned}
\frac{d\bar{\Phi}}{d\bar{X}_{sys}} &= \frac{\partial\bar{\Phi}}{\partial\bar{R}_n}\frac{d\bar{R}_n}{d\bar{X}_{sys}} \\
&= \frac{\partial\bar{\Phi}}{\partial\bar{R}_n}\frac{\partial\bar{R}_n}{\partial\bar{X}_n}\frac{d\bar{X}_n}{d\bar{X}_{sys}} + \frac{\partial\bar{\Phi}}{\partial\bar{R}_n}\frac{\partial\bar{R}_n}{\partial\bar{R}_{n-1}}\frac{\partial\bar{R}_{n-1}}{\partial\bar{X}_{n-1}}\frac{d\bar{X}_{n-1}}{d\bar{X}_{sys}} + \cdots \\
&+ \frac{\partial\bar{\Phi}}{\partial\bar{R}_n}\frac{\partial\bar{R}_n}{\partial\bar{R}_{n-1}}\frac{\partial\bar{R}_{n-1}}{\partial\bar{R}_{n-2}}\cdots\frac{\partial\bar{R}_2}{\partial\bar{R}_1}\frac{\partial\bar{R}_1}{\partial\bar{X}_1}\frac{d\bar{X}_1}{d\bar{X}_{sys}} \\
&+ \frac{\partial\bar{\Phi}}{\partial\bar{R}_n}\frac{\partial\bar{R}_n}{\partial\bar{R}_{n-1}}\frac{\partial\bar{R}_{n-1}}{\partial\bar{R}_{n-2}}\cdots\frac{\partial\bar{R}_i}{\partial\bar{R}_{i-1}}\cdots\frac{\partial\bar{R}_1}{\partial\bar{R}_0}\frac{\partial\bar{R}_0}{\partial\bar{X}_0}\frac{d\bar{X}_0}{d\bar{X}_{sys}}.
\end{aligned}
\tag{7.3}
$$

Equations (7.2) and (7.3) show that the derivative matrices $\partial\bar{R}_i/\partial\bar{R}_{i-1}$, $\partial\bar{R}_i/\partial\bar{X}_i$ and $d\bar{X}_i/d\bar{X}_{sys}$ are required in advance to determine the Jacobian matrices $d\bar{R}_i/d\bar{X}_{sys}$ and $d\bar{\Phi}/d\bar{X}_{sys}$. Accordingly, Sects. 7.2 and 7.3 derive the Jacobian matrix $\partial\bar{R}_i/\partial\bar{R}_{i-1}$ for flat and spherical boundary surfaces, Sects. 7.4 and 7.5 determine the Jacobian matrix $\partial\bar{R}_i/\partial\bar{X}_i$ for flat and spherical boundary surfaces, and Sect. 7.6 computes the Jacobian matrix $d\bar{R}_g/d\bar{X}_{sys}$ of a general ray $\bar{R}_g$ ($g \in \{0, 1, 2, \ldots, n\}$) with respect to the system variable vector $\bar{X}_{sys}$ of an optical system.

7.2 Jacobian Matrix $\partial\bar{R}_i/\partial\bar{R}_{i-1}$ for Flat Boundary Surface

As shown in Figs. 7.2 and 7.3, any change in the incidence point and unit directional vector of the incoming ray (i.e., $\Delta\bar{P}_{i-1}$ and $\Delta\bar{\ell}_{i-1}$) from the previous boundary surface causes a corresponding change in the reflected (or refracted) ray $\Delta\bar{R}_i$ at the

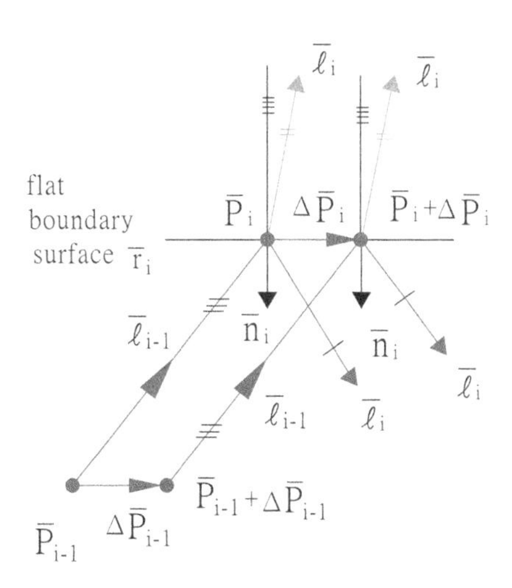

Fig. 7.2 Changes in incidence point and unit directional vector ($\Delta\bar{P}_i$ and $\Delta\bar{\ell}_i$) at flat boundary surface due to changes in incidence point on previous boundary surface ($\Delta\bar{P}_{i-1}$)

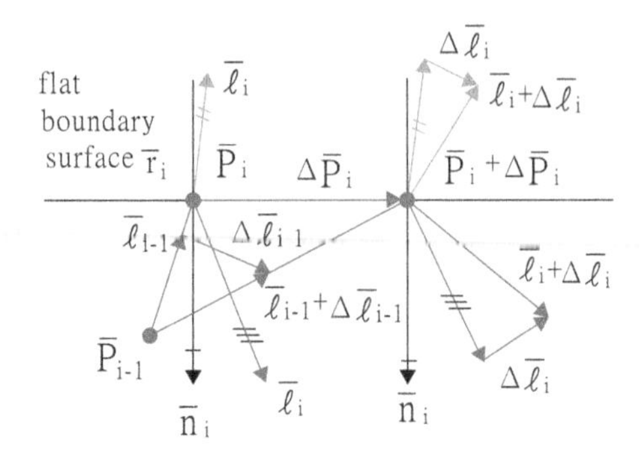

Fig. 7.3 Changes in incidence point and unit directional vector ($\Delta\bar{P}_i$ and $\Delta\bar{\ell}_i$) at flat boundary surface due to changes in unit directional vector at previous boundary surface ($\Delta\bar{\ell}_{i-1}$)

present flat boundary surface $\bar{r}_i$. Therefore, as discussed in the following, the Jacobian matrix $\partial\bar{R}_i/\partial\bar{R}_{i-1}$ comprises two components, namely $\partial\bar{P}_i/\partial\bar{R}_{i-1}$ and $\partial\bar{\ell}_i/\partial\bar{R}_{i-1}$.

7.2.1 Jacobian Matrix of Incidence Point

The Jacobian matrix of incidence point (Eq. (2.37))

$$\bar{P}_i = \bar{P}_{i-1} + \lambda_i \bar{\ell}_{i-1}$$

on a flat boundary surface $\bar{r}_i$ can be obtained by differentiating Eq. (2.37) to give

$$\frac{\partial\bar{P}_i}{\partial\bar{R}_{i-1}} = \begin{bmatrix} \partial P_{ix}/\partial\bar{R}_{i-1} \\ \partial P_{iy}/\partial\bar{R}_{i-1} \\ \partial P_{iz}/\partial\bar{R}_{i-1} \\ \bar{0} \end{bmatrix}_{4\times6} = \frac{\partial\bar{P}_{i-1}}{\partial\bar{R}_{i-1}} + \lambda_i \frac{\partial\bar{\ell}_{i-1}}{\partial\bar{R}_{i-1}} + \frac{\partial\lambda_i}{\partial\bar{R}_{i-1}} \bar{\ell}_{i-1}, \tag{7.4}$$

where

$$\frac{\partial\lambda_i}{\partial\bar{R}_{i-1}} = -\frac{1}{E_i}\frac{\partial D_i}{\partial\bar{R}_{i-1}} + \frac{D_i}{E_i^2}\frac{\partial E_i}{\partial\bar{R}_{i-1}}, \tag{7.5}$$

$$\frac{\partial\bar{P}_{i-1}}{\partial\bar{R}_{i-1}} = \begin{bmatrix} \bar{I}_{3\times3} & \bar{0}_{3\times3} \\ \bar{0}_{1\times3} & \bar{0}_{1\times3} \end{bmatrix}, \tag{7.6}$$

$$\frac{\partial\bar{\ell}_{i-1}}{\partial\bar{R}_{i-1}} = \begin{bmatrix} \bar{0}_{3\times3} & \bar{I}_{3\times3} \\ \bar{0}_{1\times3} & \bar{0}_{1\times3} \end{bmatrix}, \tag{7.7}$$

$$\frac{\partial D_i}{\partial\bar{R}_{i-1}} = [J_{ix} \quad J_{iy} \quad J_{iz} \quad 0 \quad 0 \quad 0], \tag{7.8}$$

$$\frac{\partial E_i}{\partial \bar{R}_{i-1}} = \begin{bmatrix} 0 & 0 & 0 & J_{ix} & J_{iy} & J_{iz} \end{bmatrix}. \tag{7.9}$$

Note that D_i and E_i are defined in Eqs. (2.40) and (2.41), respectively.

7.2.2 *Jacobian Matrix of Unit Directional Vector of Reflected Ray*

Let $\bar{\ell}_i$ be the unit directional vector of a ray reflected at a flat boundary surface $\bar{r}_i$ (Eq. (2.45)), i.e.,

$$\bar{\ell}_i = \bar{\ell}_{i-1} + 2C\theta_i\,\bar{n}_i.$$

The Jacobian matrix of the unit directional vector, i.e., $\partial\bar{\ell}_i/\partial\bar{R}_{i-1}$, can be obtained by differentiating Eq. (2.45). Since $\partial\bar{n}_i/\partial\bar{R}_{i-1} = \bar{0}$, it follows that

$$\frac{\partial\bar{\ell}_i}{\partial\bar{R}_{i-1}} = \begin{bmatrix} \partial\ell_{ix}/\partial\bar{R}_{i-1} \\ \partial\ell_{iy}/\partial\bar{R}_{i-1} \\ \partial\ell_{iz}/\partial\bar{R}_{i-1} \\ \bar{0} \end{bmatrix}_{4\times 6} = \frac{\partial\bar{\ell}_{i-1}}{\partial\bar{R}_{i-1}} + 2\frac{\partial(C\theta_i)}{\partial\bar{R}_{i-1}}\bar{n}_i$$
$$= \begin{bmatrix} 0 & 0 & 0 & 1-2J_{ix}J_{ix} & -2J_{ix}J_{iy} & -2J_{ix}J_{iz} \\ 0 & 0 & 0 & -2J_{iy}J_{ix} & 1-2J_{iy}J_{iy} & -2J_{iy}J_{iz} \\ 0 & 0 & 0 & -2J_{iz}J_{ix} & -2J_{iz}J_{iy} & 1-2J_{iz}J_{iz} \\ 0 & 0 & 0 & 0 & 0 & 0 \end{bmatrix}, \tag{7.10}$$

where $\partial\bar{\ell}_{i-1}/\partial\bar{R}_{i-1}$ is given in Eq. (7.7) and $\partial(C\theta_i)/\partial\bar{R}_{i-1}$ is obtained from Eq. (2.44) as

$$\frac{\partial(C\theta_i)}{\partial\bar{R}_{i-1}} = s_i\frac{\partial E_i}{\partial\bar{R}_{i-1}} = s_i\begin{bmatrix} 0 & 0 & 0 & J_{ix} & J_{iy} & J_{iz} \end{bmatrix}. \tag{7.11}$$

7.2.3 *Jacobian Matrix of Unit Directional Vector of Refracted Ray*

The unit directional vector of a ray refracted at a flat boundary surface $\bar{r}_i$ has the form (Eq. (2.46))

$$\bar{\ell}_i = \left(N_i C\theta_i - \sqrt{1 - N_i^2 + (N_i C\theta_i)^2}\right)\bar{n}_i + N_i \bar{\ell}_{i-1}.$$

The Jacobian matrix of the unit directional vector, i.e., $\partial\bar{\ell}_i/\partial\bar{R}_{i-1}$, can be obtained by differentiating Eq. (2.46). Since $\partial N_i/\partial\bar{R}_{i-1} = \bar{0}$ and $\partial\bar{n}_i/\partial\bar{R}_{i-1} = \bar{0}$, it follows that

$$\frac{\partial\bar{\ell}_i}{\partial\bar{R}_{i-1}} = \begin{bmatrix} \partial\ell_{ix}/\partial\bar{R}_{i-1} \\ \partial\ell_{iy}/\partial\bar{R}_{i-1} \\ \partial\ell_{iz}/\partial\bar{R}_{i-1} \\ \bar{0} \end{bmatrix}_{4\times6} = \left(N_i - \frac{N_i^2 C\theta_i}{\sqrt{1 - N_i^2 + (N_i C\theta_i)^2}}\right)\bar{n}_i \frac{\partial(C\theta_i)}{\partial\bar{R}_{i-1}} + N_i \frac{\partial\bar{\ell}_{i-1}}{\partial\bar{R}_{i-1}}$$

$$= \left(\frac{N_i^2 C\theta_i}{\sqrt{1 - N_i^2 + (N_i C\theta_i)^2}} - N_i\right)\begin{bmatrix} 0 & 0 & 0 & J_{ix}J_{ix} & J_{ix}J_{iy} & J_{ix}J_{iz} \\ 0 & 0 & 0 & J_{iy}J_{ix} & J_{iy}J_{iy} & J_{iy}J_{iz} \\ 0 & 0 & 0 & J_{iz}J_{ix} & J_{iz}J_{iy} & J_{iz}J_{iz} \\ 0 & 0 & 0 & 0 & 0 & 0 \end{bmatrix}$$

$$+ N_i \begin{bmatrix} 0 & 0 & 0 & 1 & 0 & 0 \\ 0 & 0 & 0 & 0 & 1 & 0 \\ 0 & 0 & 0 & 0 & 0 & 1 \\ 0 & 0 & 0 & 0 & 0 & 0 \end{bmatrix}, \tag{7.12}$$

where $\partial\bar{\ell}_{i-1}/\partial\bar{R}_{i-1}$ and $\partial(C\theta_i)/\partial\bar{R}_{i-1}$ are given in Eqs. (7.7) and (7.11), respectively.

Equations (7.10) and (7.12) show that changes in $\bar{P}_{i-1}$ (i.e., the incidence point on the previous boundary surface) have no effect on the unit directional vector $\bar{\ell}_i$ of the rays reflected or refracted at the present flat boundary surface (see Fig. 7.2).

7.2.4 Jacobian Matrix of $\bar{\mathbf{R}}_i$ *with Respect to* $\bar{\mathbf{R}}_{i-1}$ *for Flat Boundary Surface*

Combining Eqs. (7.4) and (7.10) for a reflected ray, or Eqs. (7.4) and (7.12) for a refracted ray, the Jacobian matrix of ray $\bar{R}_i$ with respect to incoming ray $\bar{R}_{i-1}$ is obtained as

$$\frac{\partial\bar{R}_i}{\partial\bar{R}_{i-1}} = \bar{M}_i = \begin{bmatrix} \partial\bar{P}_i/\partial\bar{P}_{i-1} & \partial\bar{P}_i/\partial\bar{\ell}_{i-1} \\ \bar{0} & \partial\bar{\ell}_i/\partial\bar{\ell}_{i-1} \end{bmatrix}. \tag{7.13}$$

Changes in the reflected/refracted ray at the current flat boundary surface, $\Delta\bar{R}_i$, can then be determined directly by taking the product of the Jacobian matrix $\bar{M}_i$ and the corresponding change in the incoming ray, $\Delta\bar{R}_{i-1}$, i.e.,

$$\Delta\bar{R}_i = \bar{M}_i\Delta\bar{R}_{i-1} = \begin{bmatrix} \partial\bar{P}_i/\partial\bar{P}_{i-1} & \partial\bar{P}_i/\partial\bar{\ell}_{i-1} \\ \bar{0} & \partial\bar{\ell}_i/\partial\bar{\ell}_{i-1} \end{bmatrix}\Delta\bar{R}_{i-1}. \tag{7.14}$$

For an optical system in which all of the boundary surfaces are flat, the successive application of Eq. (7.14) enables the change of any intermediate ray, $\Delta\bar{R}_g$, at the gth boundary surface to be expressed in terms of the corresponding change in the source ray, $\Delta\bar{R}_0$, as

$$\begin{aligned}\Delta\bar{R}_g &= \bar{M}_g\bar{M}_{g-1}\ldots\ldots\bar{M}_2\bar{M}_1\Delta\bar{R}_0 = \begin{bmatrix} \partial\bar{P}_g/\partial\bar{P}_0 & \partial\bar{P}_g/\partial\bar{\ell}_0 \\ \partial\bar{\ell}_g/\partial\bar{P}_0 & \partial\bar{\ell}_g/\partial\bar{\ell}_0 \end{bmatrix} \\ &= \begin{bmatrix} \partial\bar{P}_g/\partial\bar{P}_0 & \partial\bar{P}_g/\partial\bar{\ell}_0 \\ \bar{0} & \partial\bar{\ell}_g/\partial\bar{\ell}_0 \end{bmatrix}\Delta\bar{R}_0. \end{aligned} \tag{7.15}$$

Note that the change of the source ray, $\Delta\bar{R}_0$, can be obtained by differentiating the source ray $\bar{R}_0$ (see Eqs. (2.2) and (2.3)), i.e., $\Delta\bar{R}_0 = (\partial\bar{R}_0/\partial\bar{X}_0)\Delta\bar{X}_0 = \bar{S}_0\Delta\bar{X}_0$, where $\Delta\bar{X}_0$ is the change of the variable vector $\bar{X}_0$ (Eq. (2.4)) of the source ray $\bar{R}_0$, and

$$\begin{aligned}\frac{\partial\bar{R}_0}{\partial\bar{X}_0} = \bar{S}_0 &= \begin{bmatrix} \partial P_{0x}/\partial\bar{X}_0 \\ \partial P_{0y}/\partial\bar{X}_0 \\ \partial P_{0z}/\partial\bar{X}_0 \\ \partial\ell_{0x}/\partial\bar{X}_0 \\ \partial\ell_{0y}/\partial\bar{X}_0 \\ \partial\ell_{0z}/\partial\bar{X}_0 \end{bmatrix}_{6\times5} \\ &= \begin{bmatrix} 1 & 0 & 0 & 0 & 0 \\ 0 & 1 & 0 & 0 & 0 \\ 0 & 0 & 1 & 0 & 0 \\ 0 & 0 & 0 & -C\beta_0S(90^\circ+\alpha_0) & -S\beta_0C(90^\circ+\alpha_0) \\ 0 & 0 & 0 & C\beta_0C(90^\circ+\alpha_0) & -S\beta_0S(90^\circ+\alpha_0) \\ 0 & 0 & 0 & 0 & C\beta_0 \end{bmatrix}. \end{aligned} \tag{7.16}$$

The matrix multiplication in Eq. (7.15) yields $\partial\bar{\ell}_g/\partial\bar{P}_0 = \bar{0}_{3\times3}$. Hence, it is inferred that the unit directional vector $\bar{\ell}_g$ of any ray crossing a system possessing only flat boundary surfaces is independent of its point source $\bar{P}_0$. Furthermore, from Eqs. (7.13) and (7.15), it follows that

$$\frac{\partial \bar{\ell}_g}{\partial \bar{\ell}_0} = \frac{\partial \bar{\ell}_g}{\partial \bar{\ell}_{g-1}} \frac{\partial \bar{\ell}_{g-1}}{\partial \bar{\ell}_{g-2}} \cdots \frac{\partial \bar{\ell}_i}{\partial \bar{\ell}_{i-1}} \cdots \frac{\partial \bar{\ell}_2}{\partial \bar{\ell}_1} \frac{\partial \bar{\ell}_1}{\partial \bar{\ell}_0}. \tag{7.17}$$

Applying the definition of the unit normal vector $\bar{n}_i$ given in Eq. (2.36), term $\partial \bar{\ell}_i / \partial \bar{\ell}_{i-1}$ in Eq. (7.17) can be obtained for reflection and refraction processes from Eqs. (7.10) and (7.12), respectively, as

$$\frac{\partial \bar{\ell}_i}{\partial \bar{\ell}_{i-1}} = \frac{\partial \bar{\ell}_i(\bar{n}_i)}{\partial \bar{\ell}_{i-1}} = \begin{bmatrix} 1 - 2n_{ix}n_{ix} & -2n_{ix}n_{iy} & -2n_{ix}n_{iz} \\ -2n_{iy}n_{ix} & 1 - 2n_{iy}n_{iy} & -2n_{iy}n_{iz} \\ -2n_{iz}n_{ix} & -2n_{iz}n_{iy} & 1 - 2n_{iz}n_{iz} \end{bmatrix}, \tag{7.18}$$

and

$$\frac{\partial \bar{\ell}_i}{\partial \bar{\ell}_{i-1}} = \frac{\partial \bar{\ell}_i(\bar{n}_i, N_i)}{\partial \bar{\ell}_{i-1}} = N_i \begin{bmatrix} 1 & 0 & 0 \\ 0 & 1 & 0 \\ 0 & 0 & 1 \end{bmatrix} + \left(\frac{N_i^2 C\theta_i}{\sqrt{1 - N_i^2 + (N_i C\theta_i)^2}} - N_i \right) \begin{bmatrix} n_{ix}n_{ix} & n_{ix}n_{iy} & n_{ix}n_{iz} \\ n_{iy}n_{ix} & n_{iy}n_{iy} & n_{iy}n_{iz} \\ n_{iz}n_{ix} & n_{iz}n_{iy} & n_{iz}n_{iz} \end{bmatrix}. \tag{7.19}$$

Equation (7.18) shows that for a reflected ray, the Jacobian matrix $\partial \bar{\ell}_i / \partial \bar{\ell}_{i-1}$ is a function only of the unit normal vector $\bar{n}_i$ of the ith boundary surface. By contrast, for a refracted ray, $\partial \bar{\ell}_i / \partial \bar{\ell}_{i-1}$ is a function not only of $\bar{n}_i$, but also of refractive index $N_i = \xi_{i-1}/\xi_i$ and parameter E_i.

Example 7.1 Consider the corner-cube mirror shown in Fig. 3.19. The unit normal vectors of the three mirrors have the forms

$$\bar{n}_1 = \begin{bmatrix} 0 & 0 & 1 & 0 \end{bmatrix}^T,$$
$$\bar{n}_2 = \begin{bmatrix} 0 & 1 & 0 & 0 \end{bmatrix}^T,$$
$$\bar{n}_3 = \begin{bmatrix} 1 & 0 & 0 & 0 \end{bmatrix}^T.$$

The retro-reflective property of the corner-cube mirror can be proven directly (i.e., without the need for raytracing) by applying Eq. (7.17) with g = 3, i.e.,

$$\frac{\partial \bar{\ell}_3}{\partial \bar{\ell}_0} = \frac{\partial \bar{\ell}_3}{\partial \bar{\ell}_2} \frac{\partial \bar{\ell}_2}{\partial \bar{\ell}_1} \frac{\partial \bar{\ell}_1}{\partial \bar{\ell}_0} = \begin{bmatrix} -1 & 0 & 0 \\ 0 & -1 & 0 \\ 0 & 0 & -1 \end{bmatrix}.$$

Example 7.2 Consider the optical flat shown in Fig. 2.14. Assume that the flat has a refractive index of $\xi_{e1} = 1.5$ and the source ray is defined as $\bar{P}_0 = \begin{bmatrix} 0 & -5 & 5 & 1 \end{bmatrix}^T$.

In addition, assume that $\alpha_0 = 0°$ and $\beta_0 = 5°$. The Jacobian matrices $\partial \bar{\ell}_1/\partial \bar{\ell}_0$ and $\partial \bar{\ell}_2/\partial \bar{\ell}_1$ are then given by

$$\frac{\partial \bar{\ell}_1}{\partial \bar{\ell}_0} = \begin{bmatrix} 0.6667 & 0 & 0 \\ 0 & 0.4435 & 0 \\ 0 & 0 & 0.6667 \end{bmatrix},$$

$$\frac{\partial \bar{\ell}_2}{\partial \bar{\ell}_1} = \begin{bmatrix} 1.5 & 0 & 0 \\ 0 & 2.2548 & 0 \\ 0 & 0 & 1.5 \end{bmatrix}.$$

The product

$$\frac{\partial \bar{\ell}_2}{\partial \bar{\ell}_0} = \frac{\partial \bar{\ell}_2}{\partial \bar{\ell}_1}\frac{\partial \bar{\ell}_1}{\partial \bar{\ell}_0} = \begin{bmatrix} 1 & 0 & 0 \\ 0 & 1 & 0 \\ 0 & 0 & 1 \end{bmatrix}$$

indicates that the unit directional vector $\bar{\ell}_2$ of the ray exiting the optical flat is parallel to that of the incoming ray $\bar{\ell}_0$. In fact, this property is true for any incoming ray for an optical flat, and is easily proven by Snell's law.

7.3 Jacobian Matrix $\partial \bar{R}_i/\partial \bar{R}_{i-1}$ for Spherical Boundary Surface

As for a flat boundary surface, any change in the incidence point $\Delta\bar{P}_{i-1}$ and/or unit directional vector $\Delta\bar{\ell}_{i-1}$ of the incoming ray on a spherical boundary surface causes a corresponding change in the reflected or refracted ray, $\Delta\bar{R}_i$ (see Figs. 7.4 and 7.5). However, the derivation of the Jacobian matrix $\bar{M}_i = \partial \bar{R}_i/\partial \bar{R}_{i-1}$ for a spherical boundary surface is more cumbersome than that for a flat boundary surface since the raytracing equations for a spherical boundary surface are relatively more

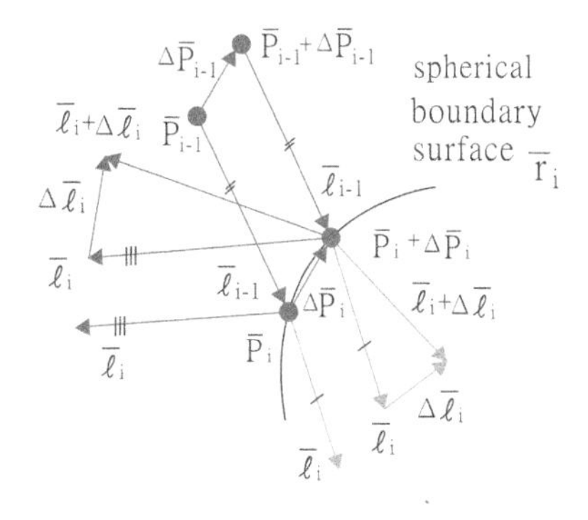

Fig. 7.4 The variations of incidence point and its unit directional vector, $\Delta\bar{P}_i$ and $\Delta\bar{\ell}_i$, due to the change of $\bar{P}_{i-1}$ at a spherical boundary surface

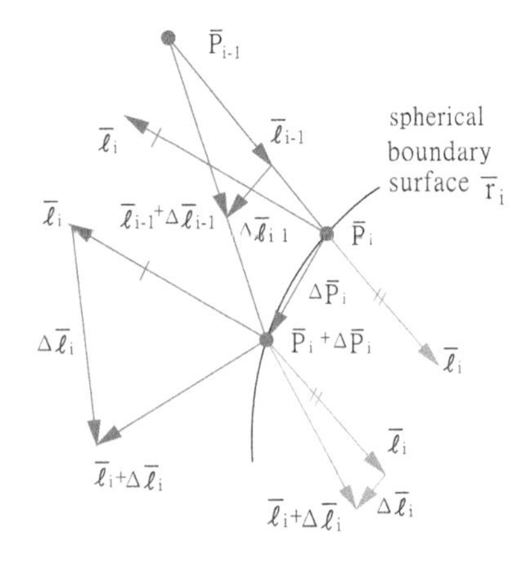

Fig. 7.5 The variations of incidence point and its unit directional vector, $\Delta\bar{P}_i$ and $\Delta\bar{\ell}_i$, due to the change of $\bar{\ell}_{i-1}$ at a spherical boundary surface

complex. The two components of $\bar{M}_i$, namely $\partial\bar{P}_i/\partial\bar{R}_{i-1}$ and $\partial\bar{\ell}_i/\partial\bar{R}_{i-1}$, are derived in the following sections.

7.3.1 *Jacobian Matrix of Incidence Point*

The Jacobian matrix $\partial\bar{P}_i/\partial\bar{R}_{i-1}$ of incidence point (Eq. (2.11))

$$\bar{P}_i = \bar{P}_{i-1} + \lambda_i\bar{\ell}_{i-1}$$

at a spherical boundary surface $\bar{r}_i$ can be obtained by differentiating Eq. (2.11) to give

$$\frac{\partial\bar{P}_i}{\partial\bar{R}_{i-1}} = \begin{bmatrix} \partial P_{ix}/\partial\bar{R}_{i-1} \\ \partial P_{iy}/\partial\bar{R}_{i-1} \\ \partial P_{iz}/\partial\bar{R}_{i-1} \\ \bar{0} \end{bmatrix}_{4\times6} = \frac{\partial\bar{P}_{i-1}}{\partial\bar{R}_{i-1}} + \lambda_i\frac{\partial\bar{\ell}_{i-1}}{\partial\bar{R}_{i-1}} + \frac{\partial\lambda_i}{\partial\bar{R}_{i-1}}\bar{\ell}_{i-1}, \tag{7.20}$$

where λ_i is given in Eq. (2.16). Differentiating λ_i with respect to $\bar{R}_{i-1}$ yields

$$\frac{\partial\lambda_i}{\partial\bar{R}_{i-1}} = -\frac{\partial D_i}{\partial\bar{R}_{i-1}} \pm \frac{1}{2\sqrt{D_i^2 - E_i}}\left(2D_i\frac{\partial D_i}{\partial\bar{R}_{i-1}} - \frac{\partial E_i}{\partial\bar{R}_{i-1}}\right). \tag{7.21}$$

The other two terms in Eq. (7.20), namely $\partial\bar{P}_{i-1}/\partial\bar{R}_{i-1}$ and $\partial\bar{\ell}_{i-1}/\partial\bar{R}_{i-1}$, are given respectively by

$$\frac{\partial\bar{P}_{i-1}}{\partial\bar{R}_{i-1}} = \begin{bmatrix} \bar{I}_{3\times3} & \bar{0}_{3\times3} \\ \bar{0}_{1\times3} & \bar{0}_{1\times3} \end{bmatrix}, \tag{7.22}$$

$$\frac{\partial\bar{\ell}_{i-1}}{\partial\bar{R}_{i-1}} = \begin{bmatrix} \bar{0}_{3\times3} & \bar{I}_{3\times3} \\ \bar{0}_{1\times3} & \bar{0}_{1\times3} \end{bmatrix}. \tag{7.23}$$

Terms $\partial D_i/\partial\bar{R}_{i-1}$ and $\partial E_i/\partial\bar{R}_{i-1}$ in Eq. (7.21) are obtained by differentiating Eqs. (2.17) and (2.18), respectively, to give

$$\frac{\partial D_i}{\partial\bar{R}_{i-1}} = \begin{bmatrix} \ell_{i-1x} & \ell_{i-1y} & \ell_{i-1z} & (P_{i-1x} - t_{ix}) & (P_{i-1y} - t_{iy}) & (P_{i-1z} - t_{iz}) \end{bmatrix}, \tag{7.24}$$

$$\frac{\partial E_i}{\partial\bar{R}_{i-1}} = \begin{bmatrix} 2(P_{i-1x} - t_{ix}) & 2(P_{i-1y} - t_{iy}) & 2(P_{i-1z} - t_{iz}) & 0 & 0 & 0 \end{bmatrix}. \tag{7.25}$$

7.3.2 Jacobian Matrix of Unit Directional Vector of Reflected Ray

Let $\bar{\ell}_i$ be the unit directional vector of a ray reflected at a spherical boundary surface $\bar{r}_i$ (Eq. (2.26)), i.e.,

$$\bar{\ell}_i = \bar{\ell}_{i-1} + 2C\theta_i\bar{n}_i.$$

The Jacobian matrix of the unit directional vector, $\partial\bar{\ell}_i/\partial\bar{R}_{i-1}$, can be obtained by differentiating Eq. (2.26) to give

$$\frac{\partial\bar{\ell}_i}{\partial\bar{R}_{i-1}} = \begin{bmatrix} \partial\ell_{ix}/\partial\bar{R}_{i-1} \\ \partial\ell_{iy}/\partial\bar{R}_{i-1} \\ \partial\ell_{iz}/\partial\bar{R}_{i-1} \\ \bar{0} \end{bmatrix}_{4\times6} = \frac{\partial\bar{\ell}_{i-1}}{\partial\bar{R}_{i-1}} + 2C\theta_i\frac{\partial\bar{n}_i}{\partial\bar{R}_{i-1}} + 2\frac{\partial(C\theta_i)}{\partial\bar{R}_{i-1}}\bar{n}_i, \tag{7.26}$$

where $\bar{n}_i$, $C\theta_i$ and $\partial\bar{\ell}_{i-1}/\partial\bar{R}_{i-1}$ are given in Eqs. (2.10), (2.21) and (7.23), respectively. Note that $\partial\bar{n}_i/\partial\bar{R}_{i-1}$ and $\partial(C\theta_i)/\partial\bar{R}_{i-1}$ are further derived in Eqs. (7. 61) and (7.63) of Appendix 1 in this chapter.

7.3.3 Jacobian Matrix of Unit Directional Vector of Refracted Ray

Let the unit directional vector of a ray refracted at spherical boundary surface $\bar{r}_i$ be given by (Eq. (2.29))

$$\bar{\ell}_i = \left(-\sqrt{1-N_i^2+(N_iC\theta_i)^2}\right)\bar{n}_i + N_i\left(\bar{\ell}_{i-1}+C\theta_i\bar{n}_i\right).$$

The Jacobian matrix of the unit directional vector, $\partial\bar{\ell}_i/\partial\bar{R}_{i-1}$, can be obtained by differentiating Eq. (2.29). Since $\partial N_i/\partial\bar{R}_{i-1}=\bar{0}$, the Jacobian matrix is obtained as

$$\begin{aligned}\frac{\partial\bar{\ell}_i}{\partial\bar{R}_{i-1}} = \begin{bmatrix}\partial\ell_{ix}/\partial\bar{R}_{i-1}\\ \partial\ell_{iy}/\partial\bar{R}_{i-1}\\ \partial\ell_{iz}/\partial\bar{R}_{i-1}\\ \bar{0}\end{bmatrix}_{4\times6} &= N_i\left(\frac{\partial(C\theta_i)}{\partial\bar{R}_{i-1}}\bar{n}_i + C\theta_i\frac{\partial\bar{n}_i}{\partial\bar{R}_{i-1}}\right)\\ &+\left(-\sqrt{1-N_i^2+(N_iC\theta_i)^2}\right)\frac{\partial\bar{n}_i}{\partial\bar{R}_{i-1}}\\ &-\left(\frac{N_i^2C\theta_i}{\sqrt{1-N_i^2+(N_iC\theta_i)^2}}\right)\bar{n}_i\frac{\partial(C\theta_i)}{\partial\bar{R}_{i-1}} + N_i\frac{\partial\bar{\ell}_{i-1}}{\partial\bar{R}_{i-1}},\end{aligned} \tag{7.27}$$

where $\partial\bar{\ell}_{i-1}/\partial\bar{R}_{i-1}$ is given in Eq. (7.23); and $\partial\bar{n}_i/\partial\bar{R}_{i-1}$ and $\partial(C\theta_i)/\partial\bar{R}_{i-1}$ are given in Eqs. (7.61) and (7.63), respectively, of Appendix 1 in this chapter.

7.3.4 Jacobian Matrix of $\bar{\mathbf{R}}_i$ with Respect to $\bar{\mathbf{R}}_{i-1}$ for Spherical Boundary Surface

Combining Eqs. (7.20) and (7.26) for a reflection process, or Eqs. (7.20) and (7.27) for a refraction process, the Jacobian matrix $\partial\bar{R}_i/\partial\bar{R}_{i-1}$ of the reflected / refracted ray $\bar{R}_i$ with respect to the incoming ray $\bar{R}_{i-1}$ at a spherical boundary surface is obtained as

$$\frac{\partial\bar{R}_i}{\partial\bar{R}_{i-1}} = \bar{M}_i = \begin{bmatrix}\partial\bar{P}_i/\partial\bar{P}_{i-1} & \partial\bar{P}_i/\partial\bar{\ell}_{i-1}\\ \partial\bar{\ell}_i/\partial\bar{P}_{i-1} & \partial\bar{\ell}_i/\partial\bar{\ell}_{i-1}\end{bmatrix}. \tag{7.28}$$

Optical systems commonly contain both spherical and flat boundary surfaces. To compute the change in a ray reflected or refracted at the gth boundary surface of

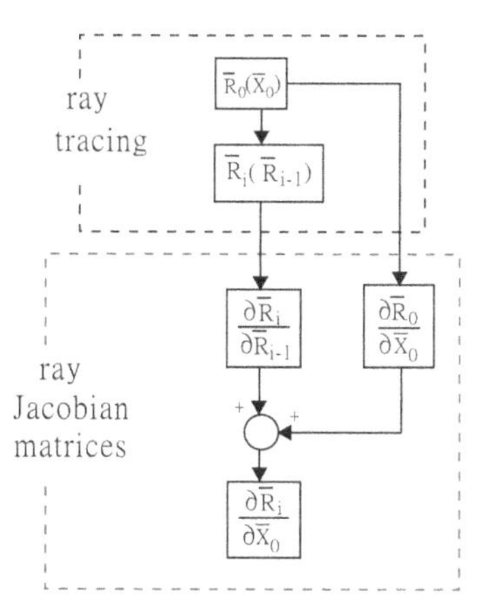

Fig. 7.6 Summary of terms required to determine ray Jacobian matrix $\partial\bar{R}_i/\partial\bar{X}_0$ for an optical system

such an optical system (i.e., $\Delta\bar{R}_g$), it is first necessary to label the boundary surfaces sequentially from i = 0 to i = n (e.g., Figures 3.2 and 3.14). Assuming that the boundary variables in $\bar{X}_i$ (i = 1, 2, …, n) remain constant, $\Delta\bar{R}_g$ can be obtained via the successive application of the Jacobian matrices given in Eqs. (7.13), (7.28) and (7.16) (see Fig. 7.6), i.e.,

$$\Delta\bar{R}_g = \frac{\partial\bar{R}_g}{\partial\bar{X}_0}\Delta\bar{X}_0 = \bar{M}_g\bar{M}_{g-1}\cdots\bar{M}_2\bar{M}_1\bar{S}_0\Delta\bar{X}_0$$

$$= \begin{bmatrix} \partial\bar{P}_g/\partial\bar{P}_0 & \partial\bar{P}_g/\partial(\alpha_0,\beta_0) \\ \partial\bar{\ell}_g/\partial\bar{P}_0 & \partial\bar{\ell}_g/\partial(\alpha_0,\beta_0) \end{bmatrix} \begin{bmatrix} \Delta\bar{P}_0 \\ \Delta\alpha_0 \\ \Delta\beta_0 \end{bmatrix}$$

$$= \begin{bmatrix} \partial P_{gx}/\partial P_{0x} & \partial P_{gx}/\partial P_{0y} & \partial P_{gx}/\partial P_{0z} & \partial P_{gx}/\partial\alpha_0 & \partial P_{gx}/\partial\beta_0 \\ \partial P_{gy}/\partial P_{0x} & \partial P_{gy}/\partial P_{0y} & \partial P_{gy}/\partial P_{0z} & \partial P_{gy}/\partial\alpha_0 & \partial P_{gy}/\partial\beta_0 \\ \partial P_{gz}/\partial P_{0x} & \partial P_{gz}/\partial P_{0y} & \partial P_{gz}/\partial P_{0z} & \partial P_{gz}/\partial\alpha_0 & \partial P_{gz}/\partial\beta_0 \\ \partial\ell_{gx}/\partial P_{0x} & \partial\ell_{gx}/\partial P_{0y} & \partial\ell_{gx}/\partial P_{0z} & \partial\ell_{gx}/\partial\alpha_0 & \partial\ell_{gx}/\partial\beta_0 \\ \partial\ell_{gy}/\partial P_{0x} & \partial\ell_{gy}/\partial P_{0y} & \partial\ell_{gy}/\partial P_{0z} & \partial\ell_{gy}/\partial\alpha_0 & \partial\ell_{gy}/\partial\beta_0 \\ \partial\ell_{gz}/\partial P_{0x} & \partial\ell_{gz}/\partial P_{0y} & \partial\ell_{gz}/\partial P_{0z} & \partial\ell_{gz}/\partial\alpha_0 & \partial\ell_{gz}/\partial\beta_0 \end{bmatrix} \begin{bmatrix} \Delta P_{0x} \\ \Delta P_{0y} \\ \Delta P_{0z} \\ \Delta\alpha_0 \\ \Delta\beta_0 \end{bmatrix}. \tag{7.29}$$

However, it should be noted that Eqs. (7.4) to (7.29) are applicable only to the analysis (not design) of optical systems since, in deriving these equations, all of the boundary variables (i.e., $\bar{X}_i$, i = 1 to i = n) are assumed to be fixed.

When tracing rays through an optical system, sometime it is requested to send a source ray $\bar{R}_0$ from a specific point source $\bar{P}_0$ through a specific point $\bar{P}_g$ on the aperture (where the aperture is labeled as the gth boundary surface in the optical system). The Newton-Raphson method can reach this task. Newton-Raphson method requires a suitable initial estimate (say, $\bar{\ell}_{0/\text{initial}}$) of the unit directional vector. The rate and final accuracy of the convergence process can be improved by adding the initial estimate (or the updated estimate in the previous iteration, i.e., $\bar{\ell}_{0/\text{last}}$) with the product of $\left(\partial \bar{P}_g/\partial(\alpha_0, \beta_0)\right)^{-1}$ and $-\Delta\bar{P}_g$ (i.e., the error of the incidence point). By applying Eq. (7.29) to the Newton-Raphson method, it is possible not only to determine the chief and marginal rays with a greater degree of accuracy, but also to obtain the one-to-one relationship between $\bar{\ell}_0$ (or polar coordinates α_0 and β_0) and the incidence point $\bar{P}_g$ at the aperture for any given point source $\bar{P}_0$.

In Chap. 16, the Jacobian matrix $\partial\bar{R}_i/\partial\bar{X}_0$ of the ith ray with respect to the variable vector $\bar{X}_0$ of source ray $\bar{R}_0$ is required to investigate the wavefront shape of an optical system. This matrix can be obtained by setting g = i in Eq. (7.29) and defining $\Delta\bar{R}_i = (\partial\bar{R}_i/\partial\bar{X}_0)\Delta\bar{X}_0$ (Fig. 7.6), to give

$$\frac{\partial\bar{R}_i}{\partial\bar{X}_0} = \begin{bmatrix} \partial P_{ix}/\partial P_{0x} & \partial P_{ix}/\partial P_{0y} & \partial P_{ix}/\partial P_{0z} & \partial P_{ix}/\partial\alpha_0 & \partial P_{ix}/\partial\beta_0 \\ \partial P_{iy}/\partial P_{0x} & \partial P_{iy}/\partial P_{0y} & \partial P_{iy}/\partial P_{0z} & \partial P_{iy}/\partial\alpha_0 & \partial P_{iy}/\partial\beta_0 \\ \partial P_{iz}/\partial P_{0x} & \partial P_{iz}/\partial P_{0y} & \partial P_{iz}/\partial P_{0z} & \partial P_{iz}/\partial\alpha_0 & \partial P_{iz}/\partial\beta_0 \\ \partial\ell_{ix}/\partial P_{0x} & \partial\ell_{ix}/\partial P_{0y} & \partial\ell_{ix}/\partial P_{0z} & \partial\ell_{ix}/\partial\alpha_0 & \partial\ell_{ix}/\partial\beta_0 \\ \partial\ell_{iy}/\partial P_{0x} & \partial\ell_{iy}/\partial P_{0y} & \partial\ell_{iy}/\partial P_{0z} & \partial\ell_{iy}/\partial\alpha_0 & \partial\ell_{iy}/\partial\beta_0 \\ \partial\ell_{iz}/\partial P_{0x} & \partial\ell_{iz}/\partial_{0y} & \partial\ell_{iz}/\partial P_{0z} & \partial\ell_{iz}/\partial\alpha_0 & \partial\ell_{iz}/\partial\beta_0 \end{bmatrix} = \bar{M}_i\bar{M}_{i-1}\cdots\bar{M}_2\bar{M}_1\bar{S}_0. \tag{7.30}$$

Example 7.3 Consider the cat's eye system shown in Fig. 3.12. The ideal parallelism condition for such an axis-symmetrical system is given as $\partial\ell_{7y}/\partial\ell_{0y} = -1$, and is obtained from the (5,5)th component of $\partial\bar{R}_7/\partial\bar{R}_0 = \bar{M}_7\bar{M}_6\cdots\bar{M}_2\bar{M}_1$. However, numerical results show that $\partial\ell_{7y}/\partial\ell_{0y} = -1$ when $\bar{R}_0 = [0 \quad -40 \quad 0 \quad 0 \quad 1 \quad 0]^T$ and $\partial\ell_{7y}/\partial\ell_{0y} = -1.00041$ when $\bar{R}_0 = [0 \quad -40 \quad 2 \quad 0 \quad 1 \quad 0]^T$, given parameter settings of $R_1 = 30$, $R_4 = -60$, $v_2 = 0$, $\xi_{air} = 1$, $\xi_{glue} = 1.2$, $\xi_{e1} = 1.6$ and $\xi_{e2} = 1.7$. In other words, the cat's eye system achieves the required retro-reflective property only for source rays aligned with the optical axis.

Example 7.4 Assume that for the system shown in Fig. 2.10, $\bar{P}_0 = [0 \quad -5 \quad 5 \quad 1]^T$, $\alpha_0 = 0°$ and $\beta_0 = 5°$. Matrices $\bar{S}_0$, $\bar{M}_1$ and $\bar{M}_2$ are then obtained as

$$\bar{S}_0 = \begin{bmatrix} 1 & 0 & 0 & 0 & 0 \\ 0 & 1 & 0 & 0 & 0 \\ 0 & 0 & 1 & 0 & 0 \\ 0 & 0 & 0 & -0.9962 & 0 \\ 0 & 0 & 0 & 0 & -0.0872 \\ 0 & 0 & 0 & 0 & 0.9962 \end{bmatrix},$$

$$\bar{M}_1 = \begin{bmatrix} 1 & 0 & 0 & 10.3895 & 0 & 0 \\ 0 & -0.0105 & 0.1202 & 0 & -2.2978 & 1.0573 \\ 0 & -0.0884 & 1.0105 & 0 & -1.1100 & 10.4820 \\ -0.0068 & 0 & 0 & 0.5964 & 0 & 0 \\ 0 & 0 & 0.0001 & 0 & 0.4422 & 0.0280 \\ 0 & 0.0006 & -0.0069 & 0 & 0.0361 & 0.5916 \end{bmatrix},$$

$$\bar{M}_2 = \begin{bmatrix} 1 & 0 & 0 & 9.4667 & 0 & 0 \\ 0 & 0.0011 & -0.0608 & 0 & 0.9072 & -0.5595 \\ 0 & -0.0182 & 0.9989 & 0 & -0.1556 & 9.4566 \\ -0.0050 & 0 & 0 & 1.4526 & 0 & 0 \\ 0 & 0 & 0 & 0 & 2.2498 & 0.0455 \\ 0 & 0.0001 & -0.0050 & 0 & 0.0467 & 1.4553 \end{bmatrix}.$$

7.4 Jacobian Matrix $\partial\bar{R}_i/\partial\bar{X}_i$ for Flat Boundary Surface

Sections 7.2 and 7.3 have derived the Jacobian matrix $\bar{M}_i = \partial\bar{R}_i/\partial\bar{R}_{i-1}$ for a reflected or refracted ray $\bar{R}_i$ with respect to the incoming ray $\bar{R}_{i-1}$. However, Eq. (7.1) shows that changes in any of the variables of a flat boundary surface may prompt corresponding changes in the reflected or refracted ray (see Fig. 7.7). Thus, it is necessary to compute the Jacobian matrix of the ray as a function of the boundary variables. Let the Jacobian matrix of the reflected/ refracted ray $\bar{R}_i$ with respect to the boundary variable vector $\bar{X}_i$ be denoted as $\bar{S}_i = \partial\bar{R}_i/\partial\bar{X}_i = \begin{bmatrix} \partial\bar{P}_i/\partial\bar{X}_i & \partial\bar{\ell}_i/\partial\bar{X}_i \end{bmatrix}^T$, where $\bar{X}_i$ is given by Eq. (2.47) for a flat boundary surface as

$$\bar{X}_i = \begin{bmatrix} J_{ix} & J_{iy} & J_{iz} & e_i & \xi_{i-1} & \xi_i \end{bmatrix}^T.$$

The two components of $\bar{S}_i$, namely $\partial\bar{P}_i/\partial\bar{X}_i$ and $\partial\bar{\ell}_i/\partial\bar{X}_i$, are discussed in the following sections.

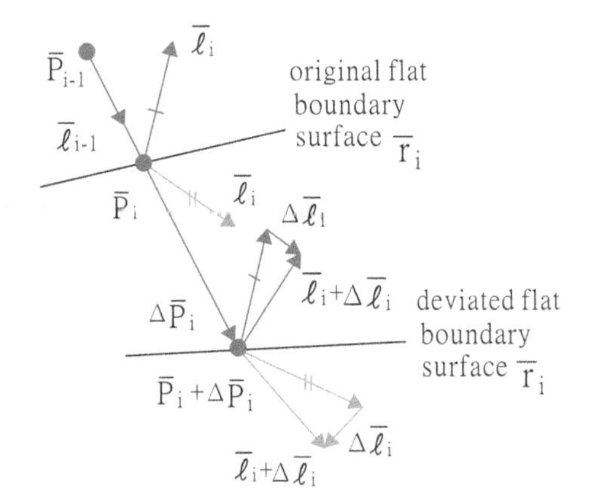

Fig. 7.7 The variations of incidence point and its unit directional vector, $\Delta\bar{P}_i$ and $\Delta\bar{\ell}_i$, due to the change of $\bar{X}_i$ at a flat boundary surface

7.4.1 *Jacobian Matrix of Incidence Point*

The Jacobian matrix of incidence point (Eq. (2.37))

$$\bar{P}_i = \bar{P}_{i-1} + \lambda_i \bar{\ell}_{i-1}$$

on a flat boundary surface $\bar{r}_i$ can be obtained by differentiating Eq. (2.37). Since $\partial\bar{P}_{i-1}/\partial\bar{X}_i = \partial\bar{\ell}_{i-1}/\partial\bar{X}_i = \bar{0}$, it follows that

$$\frac{\partial\bar{P}_i}{\partial\bar{X}_i} = \begin{bmatrix} \partial P_{ix}/\partial\bar{X}_i \\ \partial P_{iy}/\partial\bar{X}_i \\ \partial P_{iz}/\partial\bar{X}_i \\ \bar{0} \end{bmatrix}_{4\times6} = \frac{\partial\lambda_i}{\partial\bar{X}_i}\bar{\ell}_{i-1}, \tag{7.31}$$

where $\partial\lambda_i/\partial\bar{X}_i$ is determined by differentiating Eq. (2.39) to give

$$\frac{\partial\lambda_i}{\partial\bar{X}_i} = -\frac{1}{E_i}\frac{\partial D_i}{\partial\bar{X}_i} + \frac{D_i}{E_i^2}\frac{\partial E_i}{\partial\bar{X}_i}. \tag{7.32}$$

Note that $\partial D_i/\partial\bar{X}_i$ and $\partial E_i/\partial\bar{X}_i$ are obtained from Eqs. (2.40) and (2.41), respectively, as

$$\frac{\partial D_i}{\partial\bar{X}_i} = [P_{i-1x} \quad P_{i-1y} \quad P_{i-1z} \quad 1 \quad 0 \quad 0], \tag{7.33}$$

and

$$\frac{\partial E_i}{\partial\bar{X}_i} = [\ell_{i-1x} \quad \ell_{i-1y} \quad \ell_{i-1z} \quad 0 \quad 0 \quad 0]. \tag{7.34}$$

7.4.2 Jacobian Matrix of Unit Directional Vector of Reflected Ray

Let $\bar{\ell}_i$ be the unit directional vector of a ray reflected at the flat boundary surface $\bar{r}_i$ (Eq. (2.45)), i.e.,

$$\bar{\ell}_i = \bar{\ell}_{i-1} + 2C\theta_i\,\bar{n}_i.$$

The Jacobian matrix of the unit directional vector, i.e., $\partial\bar{\ell}_i/\partial\bar{X}_i$, can be derived by differentiating Eq. (2.45). Since $\partial\bar{\ell}_{i-1}/\partial\bar{X}_i = \bar{0}$, $\partial\bar{\ell}_i/\partial\bar{X}_i$ is obtained as

$$\frac{\partial\bar{\ell}_i}{\partial\bar{X}_i} = \begin{bmatrix} \partial\ell_{ix}/\partial\bar{X}_i \\ \partial\ell_{iy}/\partial\bar{X}_i \\ \partial\ell_{iz}/\partial\bar{X}_i \\ \bar{0} \end{bmatrix}_{4\times 6} = 2C\theta_i\frac{\partial\bar{n}_i}{\partial\bar{X}_i} + 2\frac{\partial(C\theta_i)}{\partial\bar{X}_i}\bar{n}_i, \tag{7.35}$$

where $\bar{n}_i$ and $C\theta_i$ are given in Eqs. (2.36) and (2.44), respectively. Moreover, $\partial\bar{n}_i/\partial\bar{X}_i$ and $\partial(C\theta_i)/\partial\bar{X}_i$ are obtained by differentiating Eqs. (2.36) and (2.44), respectively, to give

$$\frac{\partial\bar{n}_i}{\partial\bar{X}_i} = -s_i\begin{bmatrix} \bar{I}_{3\times 3} & \bar{0}_{3\times 3} \\ \bar{0}_{1\times 3} & \bar{0}_{1\times 3} \end{bmatrix}, \tag{7.36}$$

$$\frac{\partial(C\theta_i)}{\partial\bar{X}_i} = s_i\frac{\partial E_i}{\partial\bar{X}_i} = s_i[\ell_{i-1x} \quad \ell_{i-1y} \quad \ell_{i-1z} \quad 0 \quad 0 \quad 0]. \tag{7.37}$$

7.4.3 Jacobian Matrix of Unit Directional Vector of Refracted Ray

Let the unit directional vector of a ray refracted at flat boundary surface $\bar{r}_i$ be given by (Eq. (2.46))

$$\bar{\ell}_i = \left(N_iC\theta_i - \sqrt{1 - N_i^2 + (N_iC\theta_i)^2}\right)\bar{n}_i + N_i\bar{\ell}_{i-1}$$

The Jacobian matrix of the unit directional vector, i.e., $\partial\bar{\ell}_i/\partial\bar{X}_i$, can be obtained by differentiating Eq. (2.46). Since $\partial\bar{\ell}_{i-1}/\partial\bar{X}_i = \bar{0}$, it follows that

$$\frac{\partial \bar{\ell}_i}{\partial \bar{X}_i} = \begin{bmatrix} \partial \ell_{ix}/\partial \bar{X}_i \\ \partial \ell_{iy}/\partial \bar{X}_i \\ \partial \ell_{iz}/\partial \bar{X}_i \\ \bar{0} \end{bmatrix}_{4\times 6} = \left(N_i C\theta_i - \sqrt{1 - N_i^2 + (N_i C\theta_i)^2} \right) \frac{\partial \bar{n}_i}{\partial \bar{X}_i}$$
$$+ \left(N_i - \frac{N_i^2 C\theta_i}{\sqrt{1 - N_i^2 + (N_i C\theta_i)^2}} \right) \frac{\partial (C\theta_i)}{\partial \bar{X}_i} \bar{n}_i \tag{7.38}$$
$$+ \left(C\theta_i - \frac{N_i\left((C\theta_i)^2 - 1\right)}{\sqrt{1 - N_i^2 + (N_i C\theta_i)^2}} \right) \frac{\partial N_i}{\partial \bar{X}_i} \bar{n}_i + \frac{\partial N_i}{\partial \bar{X}_i} \bar{\ell}_{i-1},$$

where $\partial \bar{n}_i/\partial \bar{X}_i$ and $\partial (C\theta_i)/\partial \bar{X}_i$ are given in Eqs. (7.36) and (7.37), respectively, and $\partial N_i/\partial \bar{X}_i$ is determined by differentiating $N_i = \xi_{i-1}/\xi_i$ with respect to the boundary variable vector $\bar{X}_i$ (Eq. (2.47)) to give

$$\frac{\partial N_i}{\partial \bar{X}_i} = [0 \quad 0 \quad 0 \quad 0 \quad 1/\xi_i \quad -N_i/\xi_i\,]. \tag{7.39}$$

7.4.4 *Jacobian Matrix of* $\bar{\mathbf{R}}_i$ *with Respect to* $\bar{\mathbf{X}}_i$

Combining Eqs. (7.31) and (7.35) for a reflection process, or Eqs. (7.31) and (7.38) for a refraction process, the Jacobian matrix of $\bar{R}_i$ with respect to the boundary variable vector $\bar{X}_i$ for a flat boundary surface is obtained as

$$\bar{S}_i = \frac{\partial \bar{R}_i}{\partial \bar{X}_i} = \begin{bmatrix} \partial \bar{P}_i/\partial \bar{X}_i \\ \partial \bar{\ell}_i/\partial \bar{X}_i \end{bmatrix}. \tag{7.40}$$

The change in the reflected/ refracted ray, $\Delta \bar{R}_i$, at a flat boundary surface caused by a change in the boundary variable vector, $\Delta \bar{X}_i$, can be obtained by taking the product of matrix $\bar{S}_i$ and $\Delta \bar{X}_i$, i.e.,

$$\Delta \bar{R}_i = \bar{S}_i\, \Delta \bar{X}_i = \begin{bmatrix} \partial \bar{P}_i/\partial \bar{X}_i \\ \partial \bar{\ell}_i/\partial \bar{X}_i \end{bmatrix} \Delta \bar{X}_i. \tag{7.41}$$

Summing Eqs. (7.14) and (7.41), the change in the reflected or refracted ray, $\Delta \bar{R}_i$, at a flat boundary surface caused by changes in the incoming ray, $\bar{R}_{i-1}$, and/or boundary variable vector, $\bar{X}_i$, can be obtained as (see Fig. 7.8)

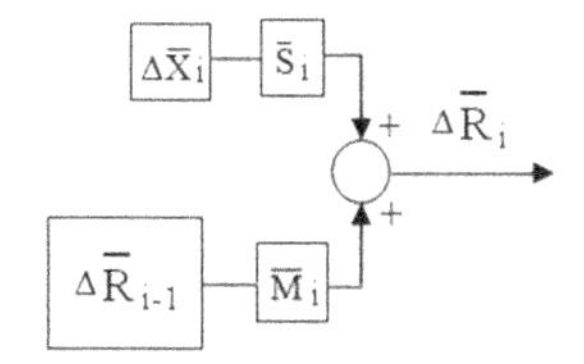

Fig. 7.8 Change in ray, $\Delta\bar{R}_i$, evaluated as sum of $\bar{M}_i\Delta\bar{R}_{i-1}$ and $\bar{S}_i\Delta\bar{X}_i$

$$\Delta\bar{R}_i = \begin{bmatrix} \partial\bar{P}_i/\partial\bar{P}_{i-1} & \partial\bar{P}_i/\partial\bar{\ell}_{i-1} \\ \partial\bar{\ell}_i/\partial\bar{P}_{i-1} & \partial\bar{\ell}_i/\partial\bar{\ell}_{i-1} \end{bmatrix} \Delta\bar{R}_{i-1} + \begin{bmatrix} \partial\bar{P}_i/\partial\bar{X}_i \\ \partial\bar{\ell}_i/\partial\bar{X}_i \end{bmatrix} \Delta\bar{X}_i = \bar{M}_i\Delta\bar{R}_{i-1} + \bar{S}_i\Delta\bar{X}_i. \tag{7.42}$$

The change in a ray at the gth surface of a prism, $\Delta\bar{R}_g$, can be determined by applying Eq. (7.42) successively with $\Delta\bar{R}_0 = \bar{S}_0\Delta\bar{X}_0$ provided that all of the boundary surfaces of the prism are flat. (Note that this issue will be addressed further in Sect. 7.6 for an optical system containing both flat and spherical boundary surfaces.)

Example 7.5 For the optical flat shown in Fig. 2.14, $\Delta\bar{R}_1$ and $\Delta\bar{R}_2$ are given respectively by

$$\Delta\bar{R}_1 = \bar{M}_1\Delta\bar{R}_0 + \bar{S}_1\Delta\bar{X}_1 = \bar{M}_1(\bar{S}_0\Delta\bar{X}_0) + \bar{S}_1\Delta\bar{X}_1 = \bar{M}_1\bar{S}_0\Delta\bar{X}_0 + \bar{S}_1\Delta\bar{X}_1$$

and

$$\begin{aligned} \Delta\bar{R}_2 &= \bar{M}_2\Delta\bar{R}_1 + \bar{S}_2\Delta\bar{X}_2 = \bar{M}_2(\bar{M}_1\Delta\bar{R}_0 + \bar{S}_1\Delta\bar{X}_1) + \bar{S}_2\Delta\bar{X}_2 \\ &= \bar{M}_2\bar{M}_1\bar{S}_0\Delta\bar{X}_0 + \bar{M}_2\bar{S}_1\Delta\bar{X}_1 + \bar{S}_2\Delta\bar{X}_2. \end{aligned}$$

The corresponding flow charts are given in Figs. 7.9 and 7.10, respectively.

Example 7.6 Let the source ray for the optical flat in Fig. 2.14 be defined as $\bar{P}_0 = [0 \quad -5 \quad 5 \quad 1]^T$ with polar coordinates $\alpha_0 = 0°$ and $\beta_0 = 5°$. Matrices $\bar{S}_i$ (i = 1 and i = 2) are then given respectively as

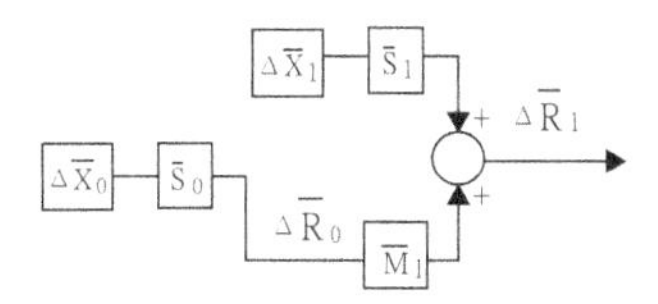

Fig. 7.9 Change in ray $\Delta\bar{R}_1$ evaluated as sum of $\bar{M}_1\bar{S}_0\Delta\bar{X}_0$ (i.e., the change in the source ray) and $\bar{S}_1\Delta\bar{X}_1$ (i.e., the change due to $\Delta\bar{X}_1$)

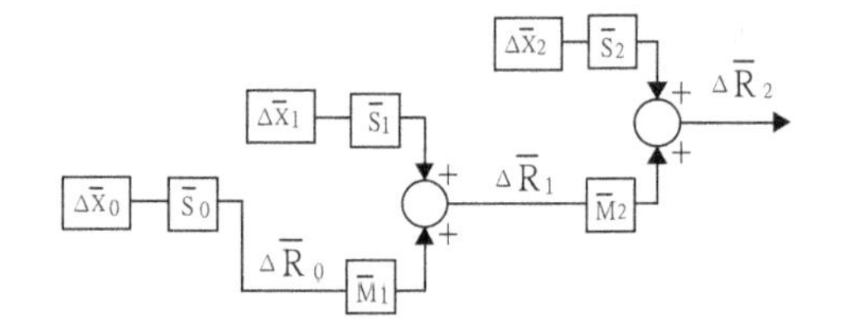

Fig. 7.10 Change in ray $\Delta\bar{R}_2$ evaluated as sum of $\bar{M}_2\bar{M}_1\bar{S}_0\Delta\bar{X}_0$ (i.e., the change in the source ray), $\bar{M}_2\bar{S}_1\Delta\bar{X}_1$(i.e., the change due to $\Delta\bar{X}_1$) and $\bar{S}_2\Delta\bar{X}_2$ (i.e., the change due to $\Delta\bar{X}_2$)

$$\bar{S}_1 = \begin{bmatrix} 0.0000 & 0.0000 & 0.0000 & 0.0000 & 0.0000 & 0.0000 \\ 0.0000 & -5.0000 & -5.8749 & -1.0000 & 0.0000 & 0.0000 \\ 0.0000 & -0.4374 & -0.5140 & -0.0875 & 0.0000 & 0.0000 \\ 0.3342 & 0.0000 & 0.0000 & 0.0000 & 0.0000 & 0.0000 \\ 0.0000 & 0.1119 & -0.0195 & 0.0000 & -0.0034 & 0.0023 \\ 0.0000 & 0.0000 & 0.3342 & 0.0000 & 0.0581 & -0.0387 \end{bmatrix},$$

$$\bar{S}_2 = \begin{bmatrix} 0.0000 & 0.0000 & 0.0000 & 0.0000 & 0.0000 & 0.0000 \\ 0.0000 & -15.0000 & -6.4569 & -1.0000 & 0.0000 & 0.0000 \\ 0.0000 & -0.8730 & -0.3758 & -0.0582 & 0.0000 & 0.0000 \\ -0.5013 & 0.0000 & 0.0000 & 0.0000 & 0.0000 & 0.0000 \\ 0.0000 & 0.2522 & 0.0439 & 0.0000 & -0.0051 & 0.0076 \\ 0.0000 & 0.0000 & -0.5013 & 0.0000 & 0.0581 & -0.0872 \end{bmatrix}.$$

7.5 Jacobian Matrix $\partial\bar{R}_i/\partial\bar{X}_i$ for Spherical Boundary Surface

As for the case of a flat boundary surface, a change in any of the variables associated with a spherical boundary surface may cause a corresponding change in the reflected or refracted ray $\bar{R}_i$ (Fig. 7.11). The objective of this section is to determine the Jacobian matrix $\bar{S}_i = \partial\bar{R}_i/\partial\bar{X}_i = \begin{bmatrix} \partial\bar{P}_i/\partial\bar{X}_i & \partial\bar{\ell}_i/\partial\bar{X}_i \end{bmatrix}^T$ of a reflected/refracted ray $\bar{R}_i$ with respect to the variable vector $\bar{X}_i$ of a spherical boundary surface, given by (Eq. (2.30))

$$\bar{X}_i = \begin{bmatrix} t_{ix} & t_{iy} & t_{iz} & \omega_{ix} & \omega_{iy} & \omega_{iz} & \xi_{i-1} & \xi_i & R_i \end{bmatrix}^T.$$

The two components of $\bar{S}_i$, namely $\partial\bar{P}_i/\partial\bar{X}_i$ and $\partial\bar{\ell}_i/\partial\bar{X}_i$, are discussed in the following.

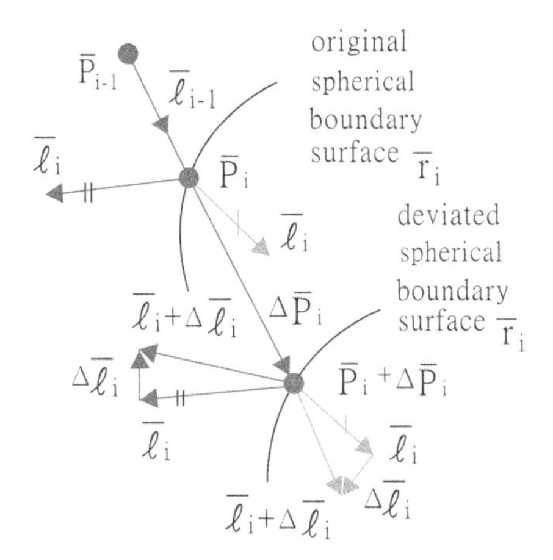

Fig. 7.11 Change in ray $\Delta\bar{R}_i$ as result of change in boundary variable vector, $\Delta\bar{X}_i$

7.5.1 Jacobian Matrix of Incidence Point

The Jacobian matrix of incidence point (Eq. (2.11))

$$\bar{P}_i = \bar{P}_{i-1} + \lambda_i \bar{\ell}_{i-1}$$

for a spherical boundary surface $\bar{r}_i$ is obtained by differentiating $\bar{P}_i = \bar{P}_{i-1} + \bar{\ell}_{i-1}\lambda_i$ given in Eq. (2.11). Since $\partial\bar{P}_{i-1}/\partial\bar{X}_i = \partial\bar{\ell}_{i-1}/\partial\bar{X}_i = \bar{0}$, it follows that

$$\frac{\partial\bar{P}_i}{\partial\bar{X}_i} = \begin{bmatrix} \partial P_{ix}/\partial\bar{X}_i \\ \partial P_{iy}/\partial\bar{X}_i \\ \partial P_{iz}/\partial\bar{X}_i \\ \bar{0} \end{bmatrix}_{4\times 9} = \bar{\ell}_{i-1}\frac{\partial\lambda_i}{\partial\bar{X}_i}. \tag{7.43}$$

The term $\partial\lambda_i/\partial\bar{X}_i$ in Eq. (7.43) is obtained from Eq. (2.16) as

$$\frac{\partial\lambda_i}{\partial\bar{X}_i} = -\frac{\partial D_i}{\partial\bar{X}_i} \pm \frac{D_i}{\sqrt{(D_i^2 - E_i)}}\frac{\partial D_i}{\partial\bar{X}_i} \pm \frac{-1}{2\sqrt{(D_i^2 - E_i)}}\frac{\partial E_i}{\partial\bar{X}_i}, \tag{7.44}$$

in which D_i and E_i are given by Eqs. (2.17) and (2.18), respectively, and

$$\frac{\partial D_i}{\partial\bar{X}_i} = [-\ell_{i-1x} \quad -\ell_{i-1y} \quad -\ell_{i-1z} \quad 0 \quad 0 \quad 0 \quad 0 \quad 0 \quad 0] \tag{7.45}$$

$$\frac{\partial E_i}{\partial\bar{X}_i} = [2(t_{ix} - P_{i-1x}) \quad 2(t_{iy} - P_{i-1y}) \quad 2(t_{iz} - P_{i-1z}) \quad 0 \quad 0 \quad 0 \quad 0 \quad 0 \quad -2R_i]. \tag{7.46}$$

7.5.2 Jacobian Matrix of Unit Directional Vector of Reflected Ray

Let $\bar{\ell}_i$ be the unit directional vector of a ray reflected at a spherical boundary surface $\bar{r}_i$ (Eq. (2.26)), i.e.,

$$\bar{\ell}_i = \bar{\ell}_{i-1} + 2C\theta_i\,\bar{n}_i.$$

The Jacobian matrix of the unit directional vector, i.e., $\partial\bar{\ell}_i/\partial\bar{X}_i$, can be derived by differentiating Eq. (2.26). Since $\partial\bar{\ell}_{i-1}/\partial\bar{X}_i = \bar{0}$, it follows that

$$\frac{\partial\bar{\ell}_i}{\partial\bar{X}_i} = \begin{bmatrix} \partial\ell_{ix}/\partial\bar{X}_i \\ \partial\ell_{iy}/\partial\bar{X}_i \\ \partial\ell_{iz}/\partial\bar{X}_i \\ \bar{0} \end{bmatrix}_{4\times 9} = 2\frac{\partial(C\theta_i)}{\partial\bar{X}_i}\,\bar{n}_i + 2C\theta_i\,\frac{\partial\bar{n}_i}{\partial\bar{X}_i}, \tag{7.47}$$

where $\partial\bar{n}_i/\partial\bar{X}_i$ and $\partial(C\theta_i)/\partial\bar{X}_i$ are given in Eqs. (7.78) and (7.80), respectively, of Appendix 2 in this chapter. In addition, the unit normal vector $\bar{n}_i$ of boundary surface $\bar{r}_i$ and the cosine of the incidence angle, $C\theta_i$, are given in Eqs. (2.10) and (2.21), respectively.

7.5.3 Jacobian Matrix of Unit Directional Vector of Refracted Ray

Let the unit directional vector of a ray refracted at a spherical boundary surface $\bar{r}_i$ be given by (Eq. (2.29))

$$\bar{\ell}_i = \left(-\sqrt{1 - N_i^2 + (N_i C\theta_i)^2}\right)\bar{n}_i + N_i\left(\bar{\ell}_{i-1} + (C\theta_i)\bar{n}_i\right).$$

The Jacobian matrix of the unit directional vector, i.e., $\partial\bar{\ell}_i/\partial\bar{X}_i$, can be obtained by differentiating Eq. (2.29). Since $\partial\bar{\ell}_{i-1}/\partial\bar{X}_i = \bar{0}$, $\partial\bar{\ell}_i/\partial\bar{X}_i$ is obtained as

$$\frac{\partial\bar{\ell}_i}{\partial\bar{X}_i}=\begin{bmatrix}\partial\ell_{ix}/\partial\bar{X}_i\\ \partial\ell_{iy}/\partial\bar{X}_i\\ \partial\ell_{iz}/\partial\bar{X}_i\\ \bar{0}\end{bmatrix}_{4\times 9}=\left(-\sqrt{1-N_i^2+(N_iC\theta_i)^2}\right)\frac{\partial\bar{n}_i}{\partial\bar{X}_i}$$
$$+\left(\frac{-N_i^2C\theta_i}{\sqrt{1-N_i^2+(N_iC\theta_i)^2}}\right)\frac{\partial(C\theta_i)}{\partial\bar{X}_i}\bar{n}_i \tag{7.48}$$
$$+\left(\frac{N_i(1-C^2\theta_i)}{\sqrt{1-N_i^2+(N_iC\theta_i)^2}}\right)\frac{\partial N_i}{\partial\bar{X}_i}\bar{n}_i+\frac{\partial N_i}{\partial\bar{X}_i}\left(\bar{\ell}_{i-1}+\ (C\theta_i)\,\bar{n}_i\right)$$
$$+N_i\left(\frac{\partial(C\theta_i)}{\partial\bar{X}_i}\bar{n}_i+(C\theta_i)\frac{\partial\bar{n}_i}{\partial\bar{X}_i}\right).$$

The terms $\partial n_i/\partial\bar{X}_i$ and $\partial(C\theta_i)/\partial\bar{X}_i$ in Eq. (7.48) are once again given in Eqs. (7.78) and (7.80), respectively, of Appendix 2 in this chapter. $\bar{n}_i$ and $C\theta_i$ are given by Eqs. (2.10) and (2.21), respectively. Finally $\partial N_i/\partial\bar{X}_i$ is determined by differentiating $N_i=\xi_{i-1}/\xi_i$ with respect to the spherical boundary variable vector $\bar{X}_i$ (Eq. (2.30)) to give

$$\frac{\partial N_i}{\partial\bar{X}_i}=\begin{bmatrix}0 & 0 & 0 & 0 & 0 & 0 & 1/\xi_i & -N_i/\xi_i & 0\end{bmatrix}. \tag{7.49}$$

7.5.4 *Jacobian Matrix of* $\bar{\mathbf{R}}_i$ *with Respect to* $\bar{\mathbf{X}}_i$

Combining Eqs. (7.43) and (7.47) for a reflection process, or Eqs. (7.43) and (7.48) for a refraction process, the Jacobian matrix $\bar{S}_i$ of $\bar{R}_i$ with respect to $\bar{X}_i$ at a spherical boundary surface is obtained as

$$\bar{S}_i=\frac{\partial\bar{R}_i}{\partial\bar{X}_i}=\begin{bmatrix}\partial\bar{P}_i/\partial\bar{X}_i\\ \partial\bar{\ell}_i/\partial\bar{X}_i\end{bmatrix}. \tag{7.50}$$

The change in the reflected/ refracted ray $\Delta\bar{R}_i$ at a spherical boundary surface due to the change in the boundary variable vector $\Delta\bar{X}_i$ can be obtained by taking the product of matrix $\bar{S}_i$ and $\Delta\bar{X}_i$, i.e.,

$$\Delta\bar{R}_i=\bar{S}_i\,\Delta\bar{X}_i=\begin{bmatrix}\partial\bar{P}_i/\partial\bar{X}_i\\ \partial\bar{\ell}_i/\partial\bar{X}_i\end{bmatrix}\Delta\bar{X}_i \tag{7.51}$$

As for the case of a flat boundary surface, Eqs. (7.28) and (7.51) yield the following equation for the change in the reflected (or refracted) ray $\Delta\bar{R}_i$ at a spherical boundary surface due to changes in $\Delta\bar{R}_{i-1}$ and $\Delta\bar{X}_i$:

$$\begin{aligned}\Delta\bar{R}_i &= \begin{bmatrix} \partial\bar{P}_i/\partial\bar{P}_{i-1} & \partial\bar{P}_i/\partial\bar{\ell}_{i-1} \\ \partial\bar{\ell}_i/\partial\bar{P}_{i-1} & \partial\bar{\ell}_i/\partial\bar{\ell}_{i-1} \end{bmatrix} \begin{bmatrix} \Delta\bar{P}_{i-1} \\ \Delta\bar{\ell}_{i-1} \end{bmatrix} + \begin{bmatrix} \partial\bar{P}_i/\partial\bar{X}_i \\ \partial\bar{\ell}_i/\partial\bar{X}_i \end{bmatrix} \Delta\bar{X}_i \\ &= \bar{M}_i\Delta\bar{R}_{i-1} + \bar{S}_i\Delta\bar{X}_i. \end{aligned} \tag{7.52}$$

As discussed in the following section, the change of a ray passing through an optical system containing both flat and spherical boundary surfaces can be determined by applying Eqs. (7.42) and (7.52) successively with $\Delta\bar{R}_0 = \bar{S}_0\Delta\bar{X}_0$.

Example 7.7 Consider the system shown in Fig. 2.10. Assume that $\bar{P}_0 = [0 \quad -5 \quad 5 \quad 1]^T$, $\alpha_0 = 0°$ and $\beta_0 = 5°$. Matrices $\bar{S}_i$ (i = 1 and i = 2) are then given as

$$\bar{S}_1 = \begin{bmatrix} 0 & 0 & 0 & 0 & 0 & 0 & 0 & 0 & 0 \\ 0 & 1.0105 & -0.1202 & 0 & 0 & 0 & 0 & 0 & -1.0176 \\ 0 & 0.0884 & -0.0105 & 0 & 0 & 0 & 0 & 0 & -0.0890 \\ 0.0068 & 0 & 0 & 0 & 0 & 0 & 0 & 0 & 0 \\ 0 & 0 & -0.0001 & 0 & 0 & 0 & -0.0025 & 0.0017 & 0 \\ 0 & -0.0006 & 0.0069 & 0 & 0 & 0 & 0.1374 & -0.0916 & 0.0014 \end{bmatrix},$$

$$\bar{S}_2 = \begin{bmatrix} 0 & 0 & 0 & 0 & 0 & 0 & 0 & 0 & 0 \\ 0 & 0.9989 & 0.0608 & 0 & 0 & 0 & 0 & 0 & -1.0007 \\ 0 & 0.0182 & 0.0011 & 0 & 0 & 0 & 0 & 0 & -0.0182 \\ 0.0050 & 0 & 0 & 0 & 0 & 0 & 0 & 0 & 0 \\ 0 & 0 & 0 & 0 & 0 & 0 & -0.0001 & 0.0002 & 0 \\ 0 & -0.0001 & 0.0050 & 0 & 0 & 0 & -0.0427 & 0.0641 & -0.0002 \end{bmatrix}.$$

7.6 Jacobian Matrix of an Arbitrary Ray with Respect to System Variable Vector

This section derives the Jacobian matrix of a ray $\bar{R}_g$ at the gth ($g \in \{0, 1, \ldots, n\}$) boundary surface of an optical system with respect to the system variable vector $\bar{X}_{sys}$. Recall that Eqs. (7.42) and (7.52) express the change of the reflected or refracted ray $\Delta\bar{R}_i$ in terms of $\Delta\bar{R}_{i-1}$ and $\Delta\bar{X}_i$, respectively. By using these two equations successively with $i = g$ to $i = 0$, the change of ray $\bar{R}_g$ at the gth boundary surface can be obtained as (see Fig. 7.12)

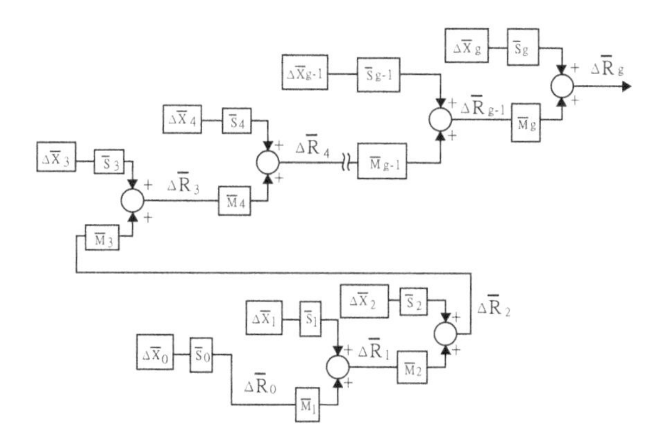

Fig. 7.12 Determination of $\Delta\bar{R}_g$ for an optical system

$$
\begin{aligned}
\Delta\bar{R}_g &= \bar{M}_g\Delta\bar{R}_{g-1} + \bar{S}_g\Delta\bar{X}_g = \bar{M}_g\left(\bar{M}_{g-1}\Delta\bar{R}_{g-2} + \bar{S}_{g-1}\Delta\bar{X}_{g-1}\right) + \bar{S}_g\Delta\bar{X}_g \\
&= \bar{M}_g\bar{M}_{g-1}\Delta\bar{R}_{g-2} + \bar{M}_g\bar{S}_{g-1}\Delta\bar{X}_{g-1} + \bar{S}_g\Delta\bar{X}_g \\
&= \dots \\
&= \bar{M}_g\bar{M}_{g-1}\bar{M}_{g-2}\dots\bar{M}_2\bar{M}_1\bar{S}_0\Delta\bar{X}_0 + \bar{M}_g\bar{M}_{g-1}\bar{M}_{g-2}\dots\bar{M}_3\bar{M}_2\bar{S}_1\Delta\bar{X}_1 + \dots \\
&\quad + \bar{M}_g\bar{M}_{g-1}\bar{S}_{g-2}\Delta\bar{X}_{g-2} + \bar{M}_g\bar{S}_{g-1}\Delta\bar{X}_{g-1} + \bar{S}_g\Delta\bar{X}_g \\
&= \sum_{u=0}^{u=g} \bar{M}_g\bar{M}_{g-1}\dots\bar{M}_{u+2}\bar{M}_{u+1}\bar{S}_u\Delta\bar{X}_u.
\end{aligned}
\tag{7.53}
$$

Equation (7.53) can be expressed in terms of the change of the system variable vector, $\Delta\bar{X}_{sys}$, by means of the following chain rule:

$$
\Delta\bar{X}_i = \frac{d\bar{X}_i}{d\bar{X}_{sys}}\Delta\bar{X}_{sys}. \tag{7.54}
$$

Note that $d\bar{X}_i/d\bar{X}_{sys}$ (i = 0 to i = g), i.e., the Jacobian matrix of the boundary variable vector $\bar{X}_i$ with respect to the system variable vector $\bar{X}_{sys}$, is discussed in more detail in Chap. 8. Meanwhile, defining $\Delta\bar{R}_g = (d\bar{R}_g/d\bar{X}_{sys})\Delta\bar{X}_{sys}$, the Jacobian matrix $d\bar{R}_g/d\bar{X}_{sys}$ can be obtained from Eq. (7.53) as

$$
\begin{aligned}
\frac{d\bar{R}_g}{d\bar{X}_{sys}} &= \bar{M}_g\bar{M}_{g-1}\dots\bar{M}_1\bar{S}_0\frac{d\bar{X}_0}{d\bar{X}_{sys}} + \bar{M}_g\bar{M}_{g-1}\dots\bar{M}_2\bar{S}_1\frac{d\bar{X}_1}{d\bar{X}_{sys}} + \dots \\
&\quad + \bar{M}_g\bar{S}_{g-1}\frac{d\bar{X}_{g-1}}{d\bar{X}_{sys}} + \bar{S}_g\frac{d\bar{X}_g}{d\bar{X}_{sys}}.
\end{aligned}
\tag{7.55}
$$

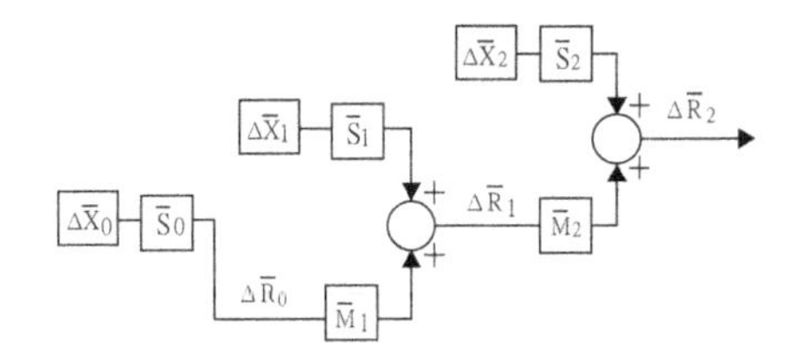

Fig. 7.13 Evaluation of $\Delta\bar{R}_2$ in Fig. 2.14 as sum of $\bar{M}_2\bar{M}_1\bar{S}_0\Delta\bar{X}_0$, $\bar{M}_2\bar{S}_1\Delta\bar{X}_1$ and $\bar{S}_2\Delta\bar{X}_2$

Example 7.8 For the optical flat shown in Fig. 2.14, $\Delta\bar{R}_2$ is given as

$$\begin{aligned}\Delta\bar{R}_2 &= \bar{M}_2\bar{M}_1\bar{S}_0\Delta\bar{X}_0 + \bar{M}_2\bar{S}_1\Delta\bar{X}_1 + \bar{S}_2\Delta\bar{X}_2\\ &= \left(\bar{M}_2\bar{M}_1\bar{S}_0\frac{d\bar{X}_0}{d\bar{X}_{sys}} + \bar{M}_2\bar{S}_1\frac{d\bar{X}_1}{d\bar{X}_{sys}} + \bar{S}_2\frac{d\bar{X}_2}{d\bar{X}_{sys}}\right)\Delta\bar{X}_{sys}\\ &= \begin{bmatrix} 1 & 0 & 0 & -16.6525 & 0 & 0 & 0 & 0 & 0\\ 0 & 0 & 0 & 0 & 0 & 0 & 0 & 1 & 1\\ 0 & -0.0875 & 1 & 0 & 16.7516 & 0.584 & -0.3893 & 0.0875 & 0.0582\\ 0 & 0 & 0 & -0.9962 & 0 & 0 & 0 & 0 & 0\\ 0 & 0 & 0 & 0 & -0.0872 & 0 & 0 & 0 & 0\\ 0 & 0 & 0 & 0 & 0.9962 & 0 & 0 & 0 & 0 \end{bmatrix},\end{aligned}$$

where the system variable vector, $\bar{X}_{sys}$, is given in Example 3.10. The flow chart for $\Delta\bar{R}_2$ is shown in Fig. 7.13. In the equation above, the leading coefficients represent the relative effects of changes in the corresponding variables on the overall value of $\Delta\bar{R}_2$. If the sensitivity of a particular variable is zero, small changes in its value have no effect on $\Delta\bar{R}_2$. The Jacobian matrix $d\bar{R}_2/d\bar{X}_{sys}$ can be obtained from the equation above as

$$\frac{d\bar{R}_2}{d\bar{X}_{sys}} = \bar{M}_2\bar{M}_1\bar{S}_0\frac{d\bar{X}_0}{d\bar{X}_{sys}} + \bar{M}_2\bar{S}_1\frac{d\bar{X}_1}{d\bar{X}_{sys}} + \bar{S}_2\frac{d\bar{X}_2}{d\bar{X}_{sys}}.$$

Example 7.11 As shown in Fig. 7.14, the Jacobian matrix $d\bar{R}_8/d\bar{X}_{sys}$ of the system shown in Fig. 3.2 is given by

$$\begin{aligned}\frac{d\bar{R}_8}{d\bar{X}_{sys}} &= \bar{M}_8\bar{M}_7\bar{M}_6\bar{M}_5\bar{M}_4\bar{M}_3\bar{M}_2\bar{M}_1\bar{S}_0\frac{d\bar{X}_0}{d\bar{X}_{sys}} + \bar{M}_8\bar{M}_7\bar{M}_6\bar{M}_5\bar{M}_4\bar{M}_3\bar{M}_2\bar{S}_1\frac{d\bar{X}_1}{d\bar{X}_{sys}}\\ &+ \bar{M}_8\bar{M}_7\bar{M}_6\bar{M}_5\bar{M}_4\bar{M}_3\bar{S}_2\frac{d\bar{X}_2}{d\bar{X}_{sys}} + \bar{M}_8\bar{M}_7\bar{M}_6\bar{M}_5\bar{M}_4\bar{S}_3\frac{d\bar{X}_3}{d\bar{X}_{sys}} + \bar{M}_8\bar{M}_7\bar{M}_6\bar{M}_5\bar{S}_4\frac{d\bar{X}_4}{d\bar{X}_{sys}}\\ &+ \bar{M}_8\bar{M}_7\bar{M}_6\bar{S}_5\frac{d\bar{X}_5}{d\bar{X}_{sys}} + \bar{M}_8\bar{M}_7\bar{S}_6\frac{d\bar{X}_6}{d\bar{X}_{sys}} + \bar{M}_8\bar{S}_7\frac{d\bar{X}_7}{d\bar{X}_{sys}} + \bar{S}_8\frac{d\bar{X}_8}{d\bar{X}_{sys}}.\end{aligned}$$

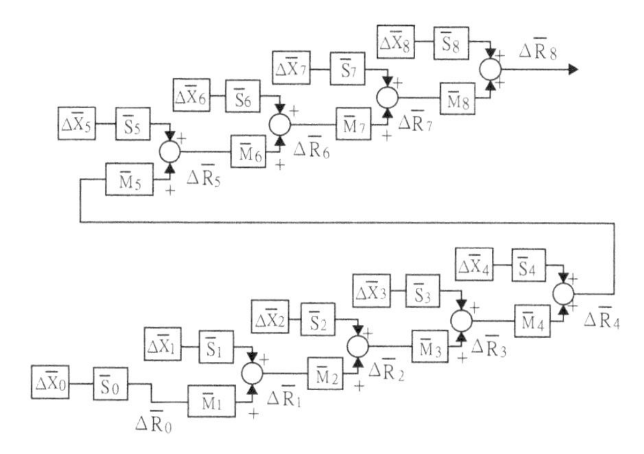

Fig. 7.14 Flow chart for Jacobian matrix $d\bar{R}_8/d\bar{X}_{sys}$ of system shown in Fig. 3.2

Appendix 1

To determine $\partial\bar{R}_i/\partial\bar{R}_{i-1}$ at a spherical boundary surface $\bar{r}_i$, the following terms are required in advance:

(1) $\partial\sigma_i/\partial\bar{R}_{i-1}$, $\partial\rho_i/\partial\bar{R}_{i-1}$ and $\partial\tau_i/\partial\bar{R}_{i-1}$:

From Eq. (2.12), ${}^i\bar{P}_i = {}^i\bar{A}_0\,\bar{P}_i = ({}^0\bar{A}_i)^{-1}\,\bar{P}_i = [\sigma_i \quad \rho_i \quad \tau_i \quad 1]^T$. Thus,

$$\bar{P}_i = {}^0\bar{A}_i \begin{bmatrix} \sigma_i \\ \rho_i \\ \tau_i \\ 1 \end{bmatrix}. \tag{7.56}$$

Differentiating Eq. (7.56) with respect to incoming ray $\bar{R}_{i-1}$ yields

$$\frac{\partial\bar{P}_i}{\partial\bar{R}_{i-1}} = {}^0\bar{A}_i \begin{bmatrix} \partial\sigma_i/\partial\bar{R}_{i-1} \\ \partial\rho_i/\partial\bar{R}_{i-1} \\ \partial\tau_i/\partial\bar{R}_{i-1} \\ \bar{0} \end{bmatrix}, \tag{7.57}$$

where $\partial\bar{P}_i/\partial\bar{R}_{i-1}$ is given in Eq. (7.20). $\partial\sigma_i/\partial\bar{R}_{i-1}$, $\partial\rho_i/\partial\bar{R}_{i-1}$ and $\partial\tau_i/\partial\bar{R}_{i-1}$ can then be determined from

$$\begin{bmatrix} \partial\sigma_i/\partial\bar{R}_{i-1} \\ \partial\rho_i/\partial\bar{R}_{i-1} \\ \partial\tau_i/\partial\bar{R}_{i-1} \\ \bar{0} \end{bmatrix} = \left({}^0\bar{A}_i\right)^{-1} \frac{\partial\bar{P}_i}{\partial\bar{R}_{i-1}}. \tag{7.58}$$

(2) $\partial\alpha_i/\partial\bar{R}_{i-1}$ and $\partial\beta_i/\partial\bar{R}_{i-1}$:

$\partial\alpha_i/\partial\bar{R}_{i-1}$ and $\partial\beta_i/\partial\bar{R}_{i-1}$ can be obtained by differentiating Eqs. (2.19) and (2.20), respectively, with respect to $\bar{R}_{i-1}$, that is,

$$\frac{\partial\alpha_i}{\partial\bar{R}_{i-1}} = \frac{1}{\sigma_i^2+\rho_i^2}\left(\sigma_i\frac{\partial\rho_i}{\partial\bar{R}_{i-1}} - \rho_i\frac{\partial\sigma_i}{\partial\bar{R}_{i-1}}\right), \tag{7.59}$$

$$\begin{aligned}\frac{\partial\beta_i}{\partial\bar{R}_{i-1}} = &\frac{\sqrt{(\sigma_i^2+\rho_i^2)}}{(\sigma_i^2+\rho_i^2+\tau_i^2)}\frac{\partial\tau_i}{\partial\bar{R}_{i-1}}\\ &-\frac{\tau_i}{(\sigma_i^2+\rho_i^2+\tau_i^2)\sqrt{(\sigma_i^2+\rho_i^2)}}\left(\sigma_i\frac{\partial\sigma_i}{\partial\bar{R}_{i-1}}+\rho_i\frac{\partial\rho_i}{\partial\bar{R}_{i-1}}\right).\end{aligned} \tag{7.60}$$

(3) $\partial\bar{n}_i/\partial\bar{X}_i$:

$\partial\bar{n}_i/\partial\bar{R}_{i-1}$ can be obtained by differentiating Eq. (2.10) with respect to $\bar{R}_{i-1}$ to give

$$\frac{\partial\bar{n}_i}{\partial\bar{R}_{i-1}} = \begin{bmatrix}\partial n_{ix}/\partial\bar{R}_{i-1}\\ \partial n_{iy}/\partial\bar{R}_{i-1}\\ \partial n_{iz}/\partial\bar{R}_{i-1}\\ \bar{0}\end{bmatrix} = {}^0\bar{A}_i\frac{\partial({}^i\bar{n}_i)}{\partial\bar{R}_{i-1}}, \tag{7.61}$$

where $\partial({}^i\bar{n}_i)/\partial\bar{R}_{i-1}$ is obtained by differentiating Eq. (2.8) with respect to $\bar{R}_{i-1}$, yielding

$$\frac{\partial({}^i\bar{n}_i)}{\partial\bar{R}_{i-1}} = s_i\begin{bmatrix}-S\beta_iC\alpha_i\\ -S\beta_iS\alpha_i\\ C\beta_i\\ 0\end{bmatrix}\frac{\partial\beta_i}{\partial\bar{R}_{i-1}} + s_i\begin{bmatrix}-C\beta_iS\alpha_i\\ C\beta_iC\alpha_i\\ 0\\ 0\end{bmatrix}\frac{\partial\alpha_i}{\partial\bar{R}_{i-1}}. \tag{7.62}$$

(4) $\partial(C\theta_i)/\partial\bar{R}_{i-1}$:

$\partial(C\theta_i)/\partial\bar{R}_{i-1}$ can be computed directly from Eq. (2.21) as

$$\begin{aligned}\frac{\partial(C\theta_i)}{\partial\bar{R}_{i-1}} &= -\left(\bar{\ell}_{i-1}\cdot\frac{\partial\bar{n}_i}{\partial\bar{R}_{i-1}} + \frac{\partial\bar{\ell}_{i-1}}{\partial\bar{R}_{i-1}}\cdot\bar{n}_i\right),\\ &= -\left(\ell_{i-1x}\frac{\partial n_{ix}}{\partial\bar{R}_{i-1}} + \ell_{i-1y}\frac{\partial n_{iy}}{\partial\bar{R}_{i-1}} + \ell_{i-1z}\frac{\partial n_{iz}}{\partial\bar{R}_{i-1}}\right) - \left(\frac{\partial\ell_{i-1x}}{\partial\bar{R}_{i-1}}n_{ix} + \frac{\partial\ell_{i-1y}}{\partial\bar{R}_{i-1}}n_{iy} + \frac{\partial\ell_{i-1z}}{\partial\bar{R}_{i-1}}n_{iz}\right)\end{aligned} \tag{7.63}$$

where $\partial n_{ix}/\partial\bar{R}_{i-1}$, $\partial n_{iy}/\partial\bar{R}_{i-1}$ and $\partial n_{iz}/\partial\bar{R}_{i-1}$ are given in Eq. (7.61).

Appendix 2

To determine $\partial\bar{R}_i/\partial\bar{X}_i$ at spherical boundary surface $\bar{r}_i$, the following terms are required in advance

(1) $\partial(^0\bar{A}_i)/\partial\bar{X}_i$ (a matrix with dimensions $4 \times 4 \times 9$):

The components of $\partial(^0\bar{A}_i)/\partial\bar{X}_i$ can be determined by differentiating $^0\bar{A}_i$ given in Eq. (2.9) with respect to boundary variable vector (Eq. (2.30))

$$\bar{X}_i = \begin{bmatrix} t_{ix} & t_{iy} & t_{iz} & \omega_{ix} & \omega_{iy} & \omega_{iz} & \xi_{i-1} & \xi_i & R_i \end{bmatrix}^T,$$

to give

$$\frac{\partial(^0\bar{A}_i)}{\partial t_{ix}} = \begin{bmatrix} 0 & 0 & 0 & 1 \\ 0 & 0 & 0 & 0 \\ 0 & 0 & 0 & 0 \\ 0 & 0 & 0 & 0 \end{bmatrix}, \tag{7.64}$$

$$\frac{\partial(^0\bar{A}_i)}{\partial t_{iy}} = \begin{bmatrix} 0 & 0 & 0 & 0 \\ 0 & 0 & 0 & 1 \\ 0 & 0 & 0 & 0 \\ 0 & 0 & 0 & 0 \end{bmatrix}, \tag{7.65}$$

$$\frac{\partial(^0\bar{A}_i)}{\partial t_{iz}} = \begin{bmatrix} 0 & 0 & 0 & 0 \\ 0 & 0 & 0 & 0 \\ 0 & 0 & 0 & 1 \\ 0 & 0 & 0 & 0 \end{bmatrix}, \tag{7.66}$$

$$\frac{\partial(^0\bar{A}_i)}{\partial \omega_{ix}} = \begin{bmatrix} 0 & C\omega_{iz}S\omega_{iy}C\omega_{ix} + S\omega_{iz}S\omega_{ix} & -C\omega_{iz}S\omega_{iy}S\omega_{ix} + S\omega_{iz}C\omega_{ix} & 0 \\ 0 & S\omega_{iz}S\omega_{iy}C\omega_{ix} - C\omega_{iz}S\omega_{ix} & -S\omega_{iz}S\omega_{iy}S\omega_{ix} - C\omega_{iz}C\omega_{ix} & 0 \\ 0 & C\omega_{iy}C\omega_{ix} & -C\omega_{iy}S\omega_{ix} & 0 \\ 0 & 0 & 0 & 0 \end{bmatrix}, \tag{7.67}$$

$$\frac{\partial(^0\bar{A}_i)}{\partial \omega_{iy}} = \begin{bmatrix} -C\omega_{iz}S\omega_{iy} & C\omega_{iz}C\omega_{iy}S\omega_{ix} & C\omega_{iz}C\omega_{iy}C\omega_{ix} & 0 \\ -S\omega_{iz}S\omega_{iy} & S\omega_{iz}C\omega_{iy}S\omega_{ix} & S\omega_{iz}C\omega_{iy}C\omega_{ix} & 0 \\ -C\omega_{iy} & -S\omega_{iy}S\omega_{ix} & -S\omega_{iy}C\omega_{ix} & 0 \\ 0 & 0 & 0 & 0 \end{bmatrix}, \tag{7.68}$$

$$\frac{\partial({}^{0}\bar{A}_{i})}{\partial\omega_{iz}}=\begin{bmatrix}-S\omega_{iz}C\omega_{iy} & -S\omega_{iz}S\omega_{iy}S\omega_{ix}-C\omega_{iz}C\omega_{ix} & -S\omega_{iz}S\omega_{iy}C\omega_{ix}+C\omega_{iz}S\omega_{ix} & 0\\ C\omega_{iz}C\omega_{iy} & C\omega_{iz}S\omega_{iy}S\omega_{ix}-S\omega_{iz}C\omega_{ix} & C\omega_{iz}S\omega_{iy}C\omega_{ix}+S\omega_{iz}S\omega_{ix} & 0\\ 0 & 0 & 0 & 0\\ 0 & 0 & 0 & 0\end{bmatrix}, \tag{7.69}$$

$$\frac{\partial({}^{0}\bar{A}_{i})}{\partial\xi_{i-1}}=\begin{bmatrix}0&0&0&0\\0&0&0&0\\0&0&0&0\\0&0&0&0\end{bmatrix}, \tag{7.70}$$

$$\frac{\partial({}^{0}\bar{A}_{i})}{\partial\xi_{i}}=\begin{bmatrix}0&0&0&0\\0&0&0&0\\0&0&0&0\\0&0&0&0\end{bmatrix}, \tag{7.71}$$

$$\frac{\partial({}^{0}\bar{A}_{i})}{\partial R_{i}}=\begin{bmatrix}0&0&0&0\\0&0&0&0\\0&0&0&0\\0&0&0&0\end{bmatrix}. \tag{7.72}$$

(2) $\partial\sigma_i/\partial\bar{X}_i$, $\partial\rho_i/\partial\bar{X}_i$ and $\partial\tau_i/\partial\bar{X}_i$:

From ${}^{i}\bar{P}_i={}^{i}\bar{A}_0\bar{P}_i=({}^{0}\bar{A}_i)^{-1}\bar{P}_i$ given in Eq. (2.12), it can be shown that

$$\bar{P}_i={}^{0}\bar{A}_i\begin{bmatrix}\sigma_i\\ \rho_i\\ \tau_i\\ 1\end{bmatrix}. \tag{7.73}$$

Differentiating Eq. (7.73) with respect to $\bar{X}_i$ gives

$$\frac{\partial\bar{P}_i}{\partial\bar{X}_i}=\frac{\partial({}^{0}\bar{A}_i)}{\partial\bar{X}_i}\begin{bmatrix}\sigma_i\\ \rho_i\\ \tau_i\\ 1\end{bmatrix}+{}^{0}\bar{A}_i\begin{bmatrix}\partial\sigma_i/\partial\bar{X}_i\\ \partial\rho_i/\partial\bar{X}_i\\ \partial\tau_i/\partial\bar{X}_i\\ \bar{0}\end{bmatrix}, \tag{7.74}$$

where $\partial\bar{P}_i/\partial\bar{X}_i$ is given in Eq. (7.43) and the components of $\partial({}^{0}\bar{A}_i)/\partial\bar{X}_i$ are listed in Eqs. (7.64)–(7.72) of this appendix. $\partial\sigma_i/\partial\bar{X}_i$, $\partial\rho_i/\partial\bar{X}_i$ and $\partial\tau_i/\partial\bar{X}_i$ can then be obtained from

$$\begin{bmatrix} \partial\sigma_i/\partial\bar{X}_i \\ \partial\rho_i/\partial\bar{X}_i \\ \partial\tau_i/\partial\bar{X}_i \\ \bar{0} \end{bmatrix} = \left({}^0\bar{A}_i\right)^{-1}\left(\frac{\partial\bar{P}_i}{\partial\bar{X}_i} - \frac{\partial({}^0\bar{A}_i)}{\partial\bar{X}_i}\begin{bmatrix} \sigma_i \\ \rho_i \\ \tau_i \\ 1 \end{bmatrix}\right). \tag{7.75}$$

(3) $\partial\alpha_i/\partial\bar{X}_i$ and $\partial\beta_i/\partial\bar{X}_i$:

$\partial\alpha_i/\partial\bar{X}_i$ and $\partial\beta_i/\partial\bar{X}_i$ can be obtained by differentiating Eqs. (2.19) and (2.20), respectively, with respect to $\bar{X}_i$ to give

$$\frac{\partial\alpha_i}{\partial\bar{X}_i} = \frac{1}{\sigma_i^2+\rho_i^2}\left(\sigma_i\frac{\partial\rho_i}{\partial\bar{X}_i} - \rho_i\frac{\partial\sigma_i}{\partial\bar{X}_i}\right), \tag{7.76}$$

$$\frac{\partial\beta_i}{\partial\bar{X}_i} = \frac{\sqrt{(\sigma_i^2+\rho_i^2)}}{(\sigma_i^2+\rho_i^2+\tau_i^2)}\frac{\partial\tau_i}{\partial\bar{X}_i} - \frac{\tau_i}{(\sigma_i^2+\rho_i^2+\tau_i^2)\sqrt{(\sigma_i^2+\rho_i^2)}}\left(\sigma_i\frac{\partial\sigma_i}{\partial\bar{X}_i} + \rho_i\frac{\partial\rho_i}{\partial\bar{X}_i}\right). \tag{7.77}$$

(4) $\partial\bar{n}_i/\partial\bar{X}_i$:

$\partial\bar{n}_i/\partial\bar{X}_i$ can be obtained by differentiating Eq. (2.10) with respect to $\bar{X}_i$ to give

$$\frac{\partial\bar{n}_i}{\partial\bar{X}_i} = \begin{bmatrix} \partial n_{ix}/\partial\bar{X}_i \\ \partial n_{iy}/\partial\bar{X}_i \\ \partial n_{iz}/\partial\bar{X}_i \\ \bar{0} \end{bmatrix} = \frac{\partial({}^0\bar{A}_i)}{\partial\bar{X}_i}{}^i\bar{n}_i + {}^0\bar{A}_i\frac{\partial({}^i\bar{n}_i)}{\partial\bar{X}_i}, \tag{7.78}$$

where $\partial({}^i\bar{n}_i)/\partial\bar{X}_i$ is obtained by differentiating Eq. (2.8) to give

$$\frac{\partial({}^i\bar{n}_i)}{\partial\bar{X}_i} = s_i\begin{bmatrix} -S\beta_i C\alpha_i \\ -S\beta_i S\alpha_i \\ C\beta_i \\ 0 \end{bmatrix}\frac{\partial\beta_i}{\partial\bar{X}_i} + s_i\begin{bmatrix} -C\beta_i S\alpha_i \\ C\beta_i C\alpha_i \\ 0 \\ 0 \end{bmatrix}\frac{\partial\alpha_i}{\partial\bar{X}_i}. \tag{7.79}$$

(5) $\partial(C\theta_i)/\partial\bar{X}_i$

$\partial(C\theta_i)/\partial\bar{X}_i$ can be computed directly from Eq. (2.21) as

$$\frac{\partial(C\theta_i)}{\partial\bar{X}_i} = -\left(\ell_{i-1x}\frac{\partial n_{ix}}{\partial\bar{X}_i} + \ell_{i-1y}\frac{\partial n_{iy}}{\partial\bar{X}_i} + \ell_{i-1z}\frac{\partial n_{iz}}{\partial\bar{X}_i}\right), \tag{7.80}$$

where $\partial n_{ix}/\partial\bar{X}_i$, $\partial n_{iy}/\partial\bar{X}_i$ and $\partial n_{iz}/\partial\bar{X}_i$ are given in Eq. (7.78).

References

1. Wynne CG, Wormell P (1963) Lens design by computer. Appl Opt 2:1223–1238
2. Fede DP (1963) Automatic optical design. Appl Opt 2:1209–1226
3. Rimmer M (1970) Analysis of perturbed lens systems. Appl Opt 9:533–537
4. Hopkins HH, Tiziani HJ (1966) A theoretical and experimental study of lens centering errors and their influence on optical image quality. Brit J Appl Phys 17:33–54
5. Andersen TB (1982) Optical aberration functions: chromatic aberrations and derivatives with respect to refractive indices for symmetrical systems. Appl Opt 21:4040–4044
6. Gupta SK, Hradaynath R (1983) Angular tolerance on Dove prisms. Appl Opt 22:3146–3147
7. Lee JF, Leung CY (1989) Method of calculating the alignment tolerance of a Porro prism resonator. Appl Opt 28:3691–3697
8. Stone BD (1997) Perturbations of optical systems. J Opt Soc Am A 14:2837–2849
9. Grey DS (1978) The inclusion of tolerance sensitivities in the merit function for lens optimization. SPIE 147:63–65
10. Herrera EG, Strojnik M (2008) Interferometric tolerance determination for a Dove prism using exact ray trace. Opt Commun 281:897–905
11. Mao W (1995) Adjustment of reflecting prisms. Opt Eng 1(34):79–82
12. Chandler KN (1960) On the effect of small errors in angles of corner-cube reflectors. J Opt Soc Am 50:203–206
13. Lin N (1994) Orientation conjugation of reflecting prism rotation and second-order approximation of image rotation. Opt Eng 33:2400–2407
14. Gutierrez E, Strojnik M, Paez G (2006) Tolerance determination for a Dove prism using exact ray trace. Proc SPIE 6307:63070K
15. Virendra NM (1998) Optical imaging and aberrations. SPIE Press, Part I Ray Geometrical Optics
16. Stone BD (1997) Determination of initial ray configurations for asymmetric systems. J Opt Soc Am A: 14:3415–3429
17. Andersen TB (1982) Optical aberration functions: derivatives with respect to axial distances for symmetrical systems. Appl Opt 21:1817–1823
18. Andersen TB (1985) Optical aberration functions: derivatives with respect to surface parameters for symmetrical systems. Appl Opt 24:1122–1129
19. Feder DP (1957) Calculation of an optical merit function and its derivatives with respect to the system parameters. J Opt Soc Am 47:913–925
20. Feder DP (1968) Differentiation of raytracing equations with respect to constructional parameters of rotationally symmetric systems. J Opt Soc Am 58:1494–1505
21. Stavroudis O (1976) A simpler derivation of the formulas for generalized ray tracing. J Opt Soc Am 66:1330–1333
22. Kross J (1988) Differential ray tracing formulae for optical calculations: principles and applications, SPIE, vol 1013, Optical design method, and large optics, pp 10–18
23. Oertmann W (1988) Differential ray tracing formulae; applications especially to aspheric optical systems, SPIE, vol 1013, Optical design method, and large optics, pp 20–26
24. Lin PD, Tsai CY (2012) Determination of first-order derivatives of skew-ray at aspherical surface. J Opt Soc Am A 29:1141–1153

Chapter 8
Jacobian Matrix of Boundary Variable Vector $\bar{X}_i$ with Respect to System Variable Vector $\bar{X}_{sys}$

The system variable vector $\bar{X}_{sys}$ of an optical system is, nearly always, different from the boundary variable vector $\bar{X}_i$ of a boundary surface. Furthermore, changes in the system variable vector may have a profound effect on the behavior of the rays as they propagate through the system. Therefore, the Jacobian matrix $d\bar{X}_i/d\bar{X}_{sys}$ of the boundary variable vector $\bar{X}_i$ with respect to the system variable vector $\bar{X}_{sys}$ is of crucial concern to optical systems designers. Accordingly, this chapter presents an efficient methodology for evaluating $d\bar{X}_i/d\bar{X}_{sys}$ as the basis for numerical techniques designed to compute the derivatives of the merit function of an optical system [1, 2].

8.1 System Variable Vector

The Jacobian matrix of a merit function is indispensable when designing and analyzing practical optical systems. It is noted from Eqs. (2.11), (2.26) and (2.29) that ray $\bar{R}_i$ in an optical system is a recursive function of the boundary variable vector $\bar{X}_i$. As a result, the Jacobian matrix of a ray $\bar{R}_i$ can only be expressed in terms of $\bar{X}_i$ initially. Consequently, computing the Jacobian matrix of the merit function of an optical system with respect to the system variable vector $\bar{X}_{sys}$ is highly challenging. In this book, the system variable vector $\bar{X}_{sys}$ (with dimension q_{sys}) is partitioned into four sub-matrices, as

$$\bar{X}_{sys} = [x_v] = \begin{bmatrix} \bar{X}_0 & \bar{X}_\xi & \bar{X}_R & \bar{X}_{rest} \end{bmatrix}^T,$$

where the four sub-matrices are defined respectively as

P.D. Lin, *Advanced Geometrical Optics*, Progress in Optical Science and Photonics 4, DOI 10.1007/978-981-10-2299-9_8

(1) $\bar{X}_0 = [P_{0x} \quad P_{0y} \quad P_{0z} \quad \alpha_0 \quad \beta_0]^T$ with dimension $q_0 = 5$: the five independent variables of the source ray $\bar{R}_0$ shown in Eq. (2.4);
(2) $\bar{X}_\xi$ with dimension q_ξ: the refractive indices of air and the elements in the system (i.e., $\bar{X}_\xi = [\xi_{air} \quad \xi_{e1} \quad \xi_{e2} \quad \ldots \quad \xi_{ek}]^T$ for a system with k elements);
(3) $\bar{X}_R$ with dimension q_R: the radii R_i of the spherical boundary surfaces in the system; and
(4) $\bar{X}_{rest} = \bar{X}_{sys} - \bar{X}_0 - \bar{X}_\xi - \bar{X}_R$ with dimension q_{rest} (where $q_{rest} = q_{sys} - q_0 - q_\xi - q_R$): all of the remaining variables of the system.

In evaluating the Jacobian matrix $d\bar{X}_i/d\bar{X}_{sys}$, three different variable vectors should be considered, namely $\bar{X}_0$, i.e., the variable vector of the source ray (given in Eq. (2.4)), and $\bar{X}_i$, i.e., the variable vectors of the spherical and flat boundary surfaces in the optical system (given in Eqs. (2.30) and (2.47), respectively). The determination of $d\bar{X}_i/d\bar{X}_{sys}$ for the three variable vectors is described in the following sections.

8.2 Jacobian Matrix $d\bar{X}_0/d\bar{X}_{sys}$ of Source Ray

Recall that a ray is a recursive function of its source ray $\bar{R}_0$. In other words, the five components of $\bar{R}_0$ are always independent variables of the problem. In general, the Jacobian matrix $d\bar{X}_0/d\bar{X}_{sys} = [J_0(u,v)]$ of the source ray can be partitioned into four sub-matrices, i.e.,

$$\frac{d\bar{X}_0}{d\bar{X}_{sys}} = [J_0(u,v)] = \begin{bmatrix} \frac{\partial \bar{X}_0}{\partial \bar{X}_0} & \frac{\partial \bar{X}_0}{\partial \bar{X}_\xi} & \frac{\partial \bar{X}_0}{\partial \bar{X}_R} & \frac{\partial \bar{X}_0}{\partial \bar{X}_{rest}} \end{bmatrix}_{5 \times q_{sys}}. \tag{8.1}$$

The corresponding solutions are listed in Table 8.1 where $q_{sys} = 5 + q_\xi + q_R + q_{rest}$ and

$$\frac{\partial \bar{X}_0}{\partial \bar{X}_0} = \bar{I}_{5\times 5}, \tag{8.2}$$

$$\frac{\partial \bar{X}_0}{\partial \bar{X}_\xi} = \bar{0}_{5\times q_\xi}, \tag{8.3}$$

$$\frac{\partial \bar{X}_0}{\partial \bar{X}_R} = \bar{0}_{5\times q_R}, \tag{8.4}$$

$$\frac{\partial \bar{X}_0}{\partial \bar{X}_{rest}} = \bar{0}_{5\times q_{rest}}. \tag{8.5}$$

Table 8.1 Jacobian matrix $d\bar{X}_0/d\bar{X}_{sys}$

$\bar{X}_{sys}$ \ $J_i(u,v)$ \ $\bar{X}_i$		i = 0				
		P_{0x} v = 1	P_{0y} v = 2	P_{0z} v = 3	α_0 v = 4	β_0 v = 5
$\bar{X}_0$	P_{0x} u = 1	1	0	0	0	0
	P_{0y} u = 2	0	1	0	0	0
	P_{0z} u = 3	0	0	1	0	0
	α_0 u = 4	0	0	0	1	0
	β_0 u = 5	0	0	0	0	1
$\bar{X}\xi$		$\bar{0}$	$\bar{0}$	$\bar{0}$	$\bar{0}$	$\bar{0}$
$\bar{X}_R$		$\bar{0}$	$\bar{0}$	$\bar{0}$	$\bar{0}$	$\bar{0}$
$\bar{X}_{rest}$		$\bar{0}$	$\bar{0}$	$\bar{0}$	$\bar{0}$	$\bar{0}$

8.3 Jacobian Matrix $d\bar{X}_i/d\bar{X}_{sys}$ of Flat Boundary Surface

It will be recalled from Chap. 2 that the boundary variable vector $\bar{X}_i$ of a flat boundary surface $\bar{r}_i$ is given as (Eq. (2.47))

$$\bar{X}_i = [x_u] = \begin{bmatrix} J_{ix} & J_{iy} & J_{iz} & e_i & \xi_{i-1} & \xi_i \end{bmatrix}^T.$$

The Jacobian matrix $d\bar{X}_i/d\bar{X}_{sys}$ of a flat boundary surface can be partitioned into twelve sub-matrices, i.e.,

$$\frac{d\bar{X}_i}{d\bar{X}_{sys}} = [J_i(u,v)] = \begin{bmatrix} \partial J_{ix}/\partial\bar{X}_0 & \partial J_{ix}/\partial\bar{X}_\xi & \partial J_{ix}/\partial\bar{X}_R & \partial J_{ix}/\partial\bar{X}_{rest} \\ \partial J_{iy}/\partial\bar{X}_0 & \partial J_{iy}/\partial\bar{X}_\xi & \partial J_{iy}/\partial\bar{X}_R & \partial J_{iy}/\partial\bar{X}_{rest} \\ \partial J_{iz}/\partial\bar{X}_0 & \partial J_{iz}/\partial\bar{X}_\xi & \partial J_{iz}/\partial\bar{X}_R & \partial J_{iz}/\partial\bar{X}_{rest} \\ \partial e_i/\partial\bar{X}_0 & \partial e_i/\partial\bar{X}_\xi & \partial e_i/\partial\bar{X}_R & \partial e_i/\partial\bar{X}_{rest} \\ \partial\xi_{i-1}/\partial\bar{X}_0 & \partial\xi_{i-1}/\partial\bar{X}_\xi & \partial\xi_{i-1}/\partial\bar{X}_R & \partial\xi_{i-1}/\partial\bar{X}_{rest} \\ \partial\xi_i/\partial\bar{X}_0 & \partial\xi_i/\partial\bar{X}_\xi & \partial\xi_i/\partial\bar{X}_R & \partial\xi_i/\partial\bar{X}_{rest} \end{bmatrix}_{6\times q_{sys}}$$

$$= \begin{bmatrix} \frac{\partial(J_{ix},J_{iy},J_{iz})}{\partial\bar{X}_0} & \frac{\partial(J_{ix},J_{iy},J_{iz})}{\partial\bar{X}_\xi} & \frac{\partial(J_{ix},J_{iy},J_{iz})}{\partial\bar{X}_R} & \frac{\partial(J_{ix},J_{iy},J_{iz})}{\partial\bar{X}_{rest}} \\ \frac{\partial e_i}{\partial\bar{X}_0} & \frac{\partial e_i}{\partial\bar{X}_\xi} & \frac{\partial e_i}{\partial\bar{X}_R} & \frac{\partial e_i}{\partial\bar{X}_{rest}} \\ \frac{\partial(\xi_{i-1},\xi_i)}{\partial\bar{X}_0} & \frac{\partial(\xi_{i-1},\xi_i)}{\partial\bar{X}_\xi} & \frac{\partial(\xi_{i-1},\xi_i)}{\partial\bar{X}_R} & \frac{\partial(\xi_{i-1},\xi_i)}{\partial\bar{X}_{rest}} \end{bmatrix}, \tag{8.6}$$

where $J_i(u, v)$ specifies the rate of change of the uth boundary variable in $\bar{X}_i$ given an infinitesimal change in the vth variable in $\bar{X}_{sys}$. In the methodology proposed in this book, Eq. (8.6) is solved via the following seven-step procedure:

(1) Initialize the components of $J_i(u, v)$ (u = 1 to u = 6 and v = 1 to v = q_{sys} for a flat boundary surface) to zero.

(2) The five variables in $\bar{X}_0$ of the source ray $\bar{R}_0$ are independent of the variables in $\bar{X}_i$. Therefore, the following results are obtained for the first column of Eq. (8.6):

$$\frac{\partial(J_{ix}, J_{iy}, J_{iz})}{\partial \bar{X}_0} = \bar{0}_{3\times5}, \tag{8.7}$$

$$\frac{\partial e_i}{\partial \bar{X}_0} = \bar{0}_{1\times5}, \tag{8.8}$$

$$\frac{\partial(\xi_{i-1}, \xi_i)}{\partial \bar{X}_0} = \bar{0}_{2\times5}. \tag{8.9}$$

(3) It is noted from $\bar{r}_i = {}^0\bar{A}'_{ej}\,{}^{ej}\bar{r}_i$ (Eq. (8.19)) that no variable in $\bar{X}_\xi$ is used when computing a flat boundary surface $\bar{r}_i$. As a result, the following equations are obtained:

$$\frac{\partial(J_{ix}, J_{iy}, J_{iz})}{\partial \bar{X}_\xi} = \bar{0}_{3\times q_\xi}, \tag{8.10}$$

$$\frac{\partial e_i}{\partial \bar{X}_\xi} = \bar{0}_{1\times q_\xi}. \tag{8.11}$$

(4) Determination of $\partial(\xi_{i-1}, \xi_i)/\partial \bar{X}_\xi$:
Assume that ξ is the vth component of $\bar{X}_{sys}$. The differential matrix $\partial(\xi_{i-1}, \xi_i)/\partial \bar{X}_\xi$ can then be determined from the following functions:

$$J_i(5, v) = \frac{\partial \xi_{i-1}}{\partial \xi} = \begin{cases} 1, & \text{if } \xi_{i-1} = \xi \\ 0, & \text{if } \xi_{i-1} \neq \xi \end{cases}, \tag{8.12}$$

$$J_i(6, v) = \frac{\partial \xi_i}{\partial \xi} = \begin{cases} 1, & \text{if } \xi_i = \xi \\ 0, & \text{if } \xi_i \neq \xi \end{cases}. \tag{8.13}$$

(5) For a flat boundary surface, the variable vector $\bar{X}_i$ does not contain a radius term, R_i. In other words, none of the variables in $\bar{X}_i$ are functions of the components in $\bar{X}_R$. Therefore, it follows that

$$\frac{\partial(J_{ix}, J_{iy}, J_{iz})}{\partial \bar{X}_R} = \bar{0}_{3\times q_R}, \tag{8.14}$$

$$\frac{\partial e_i}{\partial \bar{X}_R} = \bar{0}_{1\times q_R}, \tag{8.15}$$

$$\frac{\partial(\xi_{i-1}, \xi_i)}{\partial \bar{X}_R} = \bar{0}_{2\times q_R}. \tag{8.16}$$

(6) ξ_{i-1} and ξ_i are independent variables. In other words, they are not functions of $\bar{X}_{rest}$. Thus, the following result is obtained:

$$\frac{\partial(\xi_{i-1}, \xi_i)}{\partial \bar{X}_{rest}} = \bar{0}_{2\times q_{rest}}. \tag{8.17}$$

(7) Determination of $\partial(J_{ix}, J_{iy}, J_{iz})/\partial \bar{X}_{rest}$ and $\partial e_i/\partial \bar{X}_{rest}$:

In general, an optical element is a block of optical material possessing multiple boundary surfaces. Defining the pose of the jth element in an optical system in 3-D space requires both the element coordinate frame $(xyz)_{ej}$ and the pose matrix given in Eq. (3.25) (or Eq. (3.1) for an axis-symmetrical system), i.e.,

$$\begin{aligned} {}^0\bar{A}_{ej} &= \mathrm{tran}(t_{ejx}, t_{ejy}, t_{ejz})\mathrm{rot}(\bar{z}, \omega_{ejz})\mathrm{rot}(\bar{y}, \omega_{ejy})\mathrm{rot}(\bar{x}, \omega_{ejx}) \\ &= \begin{bmatrix} I_{ejx} & J_{ejx} & K_{ejx} & t_{ejx} \\ I_{ejy} & J_{ejy} & K_{ejy} & t_{ejy} \\ I_{ejz} & J_{ejz} & K_{ejz} & t_{ejz} \\ 0 & 0 & 0 & 1 \end{bmatrix}. \end{aligned}$$

As stated in Chap. 2, a flat boundary surface can be uniquely defined by its unit normal vector and any point located on its surface. Consequently, the first step in modeling a flat boundary surface (say, $\bar{r}_i$) in an optical element is to express the boundary surface with respect to its element coordinate frame $(xyz)_{ej}$, i.e.,

$${}^{ej}\bar{r}_i = \begin{bmatrix} {}^{ej}n_{ix} & {}^{ej}n_{iy} & {}^{ej}n_{iz} & {}^{ej}e_i \end{bmatrix}^T. \tag{8.18}$$

Equation (8.18) defines the flat boundary surface ${}^{ej}\bar{r}_i$ with respect to the element coordinate frame $(xyz)_{ej}$ (Fig. 8.1). However, many of the derivations presented in this book refer to the world coordinate frame $(xyz)_o$. It is therefore necessary to transform ${}^{ej}\bar{r}_i$ to $\bar{r}_i$ via the following manipulation (Eq. (1.34) with h = 0 and g = ej):

$$\bar{r}_i = {}^0\bar{A}'_{ej}\,{}^{ej}\bar{r}_i = \left(({}^0\bar{A}_{ej})^{-1}\right)^T {}^{ej}\bar{r}_i, \tag{8.19}$$

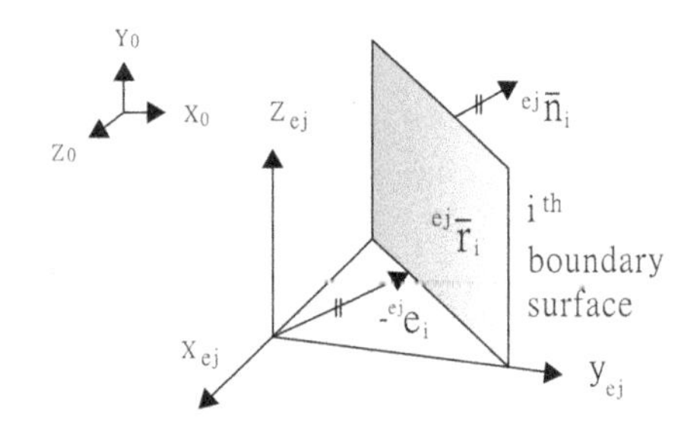

Fig. 8.1 Definition of flat boundary surface in 3-D space by means of four variables

where

$$ {}^{0}\bar{A}'_{ej} = \begin{bmatrix} I_{ejx} & J_{ejx} & K_{ejx} & 0 \\ I_{ejy} & J_{ejy} & K_{ejy} & 0 \\ I_{ejz} & J_{ejz} & K_{ejz} & 0 \\ f_{ejx} & f_{ejy} & f_{ejz} & 1 \end{bmatrix}, \tag{8.20} $$

$$ f_{ejx} = -(I_{ejx}t_{ejx} + I_{ejy}t_{ejy} + I_{ejz}t_{ejz}), \tag{8.21} $$

$$ f_{ejy} = -(J_{ejx}t_{ejx} + J_{ejy}t_{ejy} + J_{ejz}t_{ejz}), \tag{8.22} $$

$$ f_{ejz} = -(K_{ejx}t_{ejx} + K_{ejy}t_{ejy} + K_{ejz}t_{ejz}). \tag{8.23} $$

To obtain these terms, it is first necessary to determine $d\bar{r}_i/d\bar{X}_{sys} = [\partial \bar{r}_i/\partial x_v]$ by differentiating Eq. (8.19) with respect to the vth component of $\bar{X}_{sys}$, i.e.,

$$ \frac{d(\bar{r}_i)}{d\bar{X}_{sys}} = \frac{d({}^{0}\bar{A}'_{ej})}{d\bar{X}_{sys}}({}^{ej}\bar{r}_i) + ({}^{0}\bar{A}'_{ej})\frac{d({}^{ej}\bar{r}_i)}{d\bar{X}_{sys}}. \tag{8.24} $$

It is noted that the computation of Eq. (8.24) presents a significant challenge when $\bar{X}_{sys}$ contains many components. The detailed derivations of $d({}^{0}\bar{A}'_{ej})/d\bar{X}_{sys}$ and $d({}^{ej}\bar{r}_i)/d\bar{X}_{sys}$ are presented in Eqs. (8.77) and (8.83), respectively, in Appendix 3 in this chapter.

Example 8.1 Consider the generic prism shown in Fig. 8.2 comprising n flat boundary surfaces labeled from i = 1 to i = n. When the refractive indices of the element and ambient air (i.e., ξ_{ej} and ξ_{air}, respectively), and the six element pose variables (see Eq. (3.25)) are taken into account, the element variable vector $\bar{X}_{ej}$ of the prism has the form

$$ \bar{X}_{ej} = \begin{bmatrix} t_{ejx} & t_{ejy} & t_{ejz} & \omega_{ejx} & \omega_{ejy} & \omega_{ejz} & \xi_{air} & \xi_{ej} \end{bmatrix}^{T}. $$

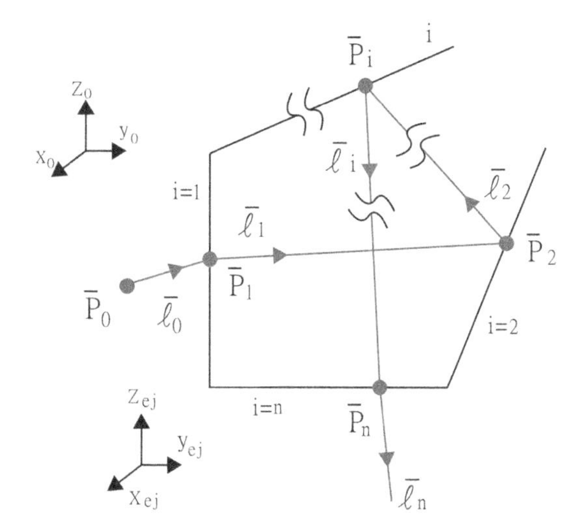

Fig. 8.2 Generic prism possessing n flat boundary surfaces labeled from i = 1 to i = n

When designing and analyzing a single prism, it is possible, with no loss of generality, to simply define ${}^0\bar{A}_{ej} = \bar{I}_{4\times4}$. However, in performing any subsequent alignment analysis, it is preferable to retain the six element pose variables given in Eq. (3.25). Furthermore, in evaluating the manufacturing errors of the boundary surfaces, the system variable vector $\bar{X}_{sys}$ should include the unit normal vectors of each boundary surface. That is, $\bar{X}_{sys}$ should be expressed as

$$\bar{X}_{sys} = \begin{bmatrix} P_{0x} & P_{0y} & P_{0z} & \alpha_0 & \beta_0 & \xi_{air} & \xi_{ej} & t_{ejx} & t_{ejy} & t_{ejz} & \omega_{ejx} & \omega_{ejy} & \omega_{ejz} \\ J_{1x} & J_{1y} & J_{1z} & e_1 & \ldots & J_{nx} & J_{ny} & J_{nz} & e_n \end{bmatrix}^T.$$

For the prism shown in Fig. 8.2, $\bar{X}_{sys}$ has dimensions of $q_{sys} = 13 + 4n$, $q_0 = 5$, $q_\xi = 2$, $q_R = 0$, and $q_{rest} = 4n + 6$. Noting that the boundary variable vector of a flat boundary surface is defined as (Eq. (2.47))

$$\bar{X}_i = [J_{ix} \quad J_{iy} \quad J_{iz} \quad e_i \quad \xi_{i-1} \quad \xi_i]^T,$$

the non-zero values of $J_i(5, v)$ and $J_i(6, v)$ ($i = 1$ to $i = n$ and $v = 1$ to $v = 13 + 4n$) are as follows:

$$J_1(5, 6) = J_1(6, 7) = 1, \tag{a}$$

$$J_i(5, 7) = J_i(6, 7) = 1 \text{ when } i \in \{2, 3, \ldots, n-1\}, \tag{b}$$

$$J_n(5, 7) = J_n(6, 6) = 1. \tag{c}$$

Equation (a) indicates that the 5th and 6th variables ($\xi_0 = \xi_{air}$ and ξ_{e1}) of $\bar{X}_1$ (i.e., the variable vector of the first flat boundary surface) are respectively the 6th and 7th variables of the system variable vector $\bar{X}_{sys}$. Equation (b) shows that the 5th and 6th variables (i.e., the refractive index ξ_{e1} of the prism material) of $\bar{X}_i$ (i = 2 to i = n-1) are the 7th variable of $\bar{X}_{sys}$. A similar explanation can be applied for Eq. (c). Note that the values of $J_i(u, v)$ (i = 1 to i = n, u = 1 to u = 4, v = 1 to v = 13 + 4n) for the prism shown in Fig. 8.2 can be determined numerically from Eq. (8.24).

8.4 Jacobian Matrix $d\bar{X}_i/d\bar{X}_{sys}$ of Spherical Boundary Surface

Assuming that the ith boundary surface ${}^{ej}\bar{r}_i$ of an optical system is a spherical boundary surface (Fig. 2.4), the Jacobian matrix $d\bar{X}_i/d\bar{X}_{sys}$ of the boundary variable vector (Eq. (2.30)

$$\bar{X}_i = \begin{bmatrix} t_{ix} & t_{iy} & t_{iz} & \omega_{ix} & \omega_{iy} & \omega_{iz} & \xi_{i-1} & \xi_i & R_i \end{bmatrix}^T$$

relative to the system variable vector $\bar{X}_{sys}$ is partitioned into the following sub-matrices:

$$\frac{d\bar{X}_i}{d\bar{X}_{sys}} = [J_i(u, v)] = \begin{bmatrix} \partial t_{ix}/\partial\bar{X}_0 & \partial t_{ix}/\partial\bar{X}_\xi & \partial t_{ix}/\partial\bar{X}_R & \partial t_{ix}/\partial\bar{X}_{rest} \\ \partial t_{iy}/\partial\bar{X}_0 & \partial t_{iy}/\partial\bar{X}_\xi & \partial t_{iy}/\partial\bar{X}_R & \partial t_{iy}/\partial\bar{X}_{rest} \\ \partial t_{iz}/\partial\bar{X}_0 & \partial t_{iz}/\partial\bar{X}_\xi & \partial t_{iz}/\partial\bar{X}_R & \partial t_{iz}/\partial\bar{X}_{rest} \\ \partial\omega_{ix}/\partial\bar{X}_0 & \partial\omega_{ix}/\partial\bar{X}_\xi & \partial\omega_{ix}/\partial\bar{X}_R & \partial\omega_{ix}/\partial\bar{X}_{rest} \\ \partial\omega_{iy}/\partial\bar{X}_0 & \partial\omega_{iy}/\partial\bar{X}_\xi & \partial\omega_{iy}/\partial\bar{X}_R & \partial\omega_{iy}/\partial\bar{X}_{rest} \\ \partial\omega_{iz}/\partial\bar{X}_0 & \partial\omega_{iz}/\partial\bar{X}_\xi & \partial\omega_{iz}/\partial\bar{X}_R & \partial\omega_{iz}/\partial\bar{X}_{rest} \\ \partial\xi_{i-1}/\partial\bar{X}_0 & \partial\xi_{i-1}/\partial\bar{X}_\xi & \partial\xi_{i-1}/\partial\bar{X}_R & \partial\xi_{i-1}/\partial\bar{X}_{rest} \\ \partial\xi_i/\partial\bar{X}_0 & \partial\xi_i/\partial\bar{X}_\xi & \partial\xi_i/\partial\bar{X}_R & \partial\xi_i/\partial\bar{X}_{rest} \\ \partial R_i/\partial\bar{X}_0 & \partial R_i/\partial\bar{X}_\xi & \partial R_i/\partial\bar{X}_R & \partial R_i/\partial\bar{X}_{rest} \end{bmatrix}_{9\times q_{sys}}$$

$$= \begin{bmatrix} \partial(t_{ix}, t_{iy}, t_{iz})/\partial\bar{X}_0 & \partial(t_{ix}, t_{iy}, t_{iz})/\partial\bar{X}_\xi & \partial(t_{ix}, t_{iy}, t_{iz})/\partial\bar{X}_R & \partial(t_{ix}, t_{iy}, t_{iz})/\partial\bar{X}_{rest} \\ \partial(\omega_{ix}, \omega_{iy}, \omega_{iz})/\partial\bar{X}_0 & \partial(\omega_{ix}, \omega_{iy}, \omega_{iz})\partial\bar{X}_\xi & \partial(\omega_{ix}, \omega_{iy}, \omega_{iz})/\partial\bar{X}_R & \partial(\omega_{ix}, \omega_{iy}, \omega_{iz})/\partial\bar{X}_{rest} \\ \partial(\xi_{i-1}, \xi_i)/\partial\bar{X}_0 & \partial(\xi_{i-1}, \xi_i)/\partial\bar{X}_\xi & \partial(\xi_{i-1}, \xi_i)/\partial\bar{X}_R & \partial(\xi_{i-1}, \xi_i)/\partial\bar{X}_{rest} \\ \partial R_i/\partial\bar{X}_0 & \partial R_i/\partial\bar{X}_\xi & \partial R_i/\partial\bar{X}_R & \partial R_i/\partial\bar{X}_{rest} \end{bmatrix}, \tag{8.25}$$

where $J_i(u, v)$ (v = 1 to v = q_{sys} and u = 1 to u = 9 for a spherical boundary surface) specifies the rate of change of the uth boundary variable in $\bar{X}_i$ given an infinitesimal change in the vth variable in $\bar{X}_{sys}$. In the methodology proposed in this chapter, the Jacobian matrix $d\bar{X}_i/d\bar{X}_{sys}$ is determined using the following procedure:

(1) Initialize the components of $d\bar{X}_i/d\bar{X}_{sys} = [J_i(u,v)]$ to zero.
(2) The variables in $\bar{X}_0$ (i.e., the five variables of the source ray $\bar{R}_0$) are independent of the variables in $\bar{X}_i$. Therefore, the following results are obtained for the first column of Eq. (8.25):

$$\frac{\partial(t_{ix}, t_{iy}, t_{iz})}{\partial \bar{X}_0} = \bar{0}_{3\times 5}, \tag{8.26}$$

$$\frac{\partial(\omega_{ix}, \omega_{iy}, \omega_{iz})}{\partial \bar{X}_0} = \bar{0}_{3\times 5}, \tag{8.27}$$

$$\frac{\partial(\xi_{i-1}, \xi_i)}{\partial \bar{X}_0} = \bar{0}_{2\times 5}, \tag{8.28}$$

$$\frac{\partial R_i}{\partial \bar{X}_0} = \bar{0}_{1\times 5}. \tag{8.29}$$

(3) ξ_{i-1} and ξ_i are independent variables, i.e., they are not functions of $\bar{X}_{rest}$. Thus, the following result is obtained:

$$\frac{\partial(\xi_{i-1}, \xi_i)}{\partial \bar{X}_{rest}} = \bar{0}_{2\times q_{rest}}. \tag{8.30}$$

Furthermore, no variable in $\bar{X}_\xi$ is used when computing the six boundary pose variables. Consequently, the following equations are obtained:

$$\frac{\partial(t_{ix}, t_{iy}, t_{iz})}{\partial \bar{X}_\xi} = \bar{0}_{3\times q_\xi}, \tag{8.31}$$

$$\frac{\partial(\omega_{ix}, \omega_{iy}, \omega_{iz})}{\partial \bar{X}_\xi} = \bar{0}_{3\times q_\xi}. \tag{8.32}$$

(4) A dimensional analysis reveals that the angular pose variables ω_{ix}, ω_{iy}, and ω_{iz} are independent of the radius R_i of the spherical boundary surface. It therefore follows that

$$\frac{\partial(\omega_{ix}, \omega_{iy}, \omega_{iz})}{\partial \bar{X}_R} = \bar{0}_{3\times q_R}. \tag{8.33}$$

(5) In addition, the radius R_i of the boundary surface is independent of $\bar{X}_\xi$. Consequently, it follows that

$$\frac{\partial R_i}{\partial \bar{X}_\xi} = \bar{0}_{1\times q_\xi}. \tag{8.34}$$

The refractive indices ξ_{i-1} and ξ_i are independent of the radius R_i of the boundary surface. In other words,

$$\frac{\partial(\xi_{i-1}, \xi_i)}{\partial \bar{X}_R} = \bar{0}_{2\times q_R}. \tag{8.35}$$

(6) Similarly, the radius R_i of the boundary surface is independent of the variables in $\bar{X}_{rest}$. Thus,

$$\frac{\partial R_i}{\partial \bar{X}_{rest}} = \bar{0}_{1\times q_{rest}}. \tag{8.36}$$

(7) Determination of $\partial(\xi_{i-1}, \xi_i)/\partial \bar{X}_\xi$:
If ξ, a variable in $\bar{X}_\xi$, is the vth component of $\bar{X}_{sys}$, then $\partial(\xi_{i-1}, \xi_i)/\partial \bar{X}_\xi$ can be determined from the following functions:

$$J_i(7, v) = \frac{\partial \xi_{i-1}}{\partial \xi} = \begin{cases} 1, & \text{if } \xi_{i-1} = \xi \\ 0, & \text{if } \xi_{i-1} \neq \xi \end{cases}, \tag{8.37}$$

$$J_i(8, v) = \frac{\partial \xi_i}{\partial \xi} = \begin{cases} 1, & \text{if } \xi_i = \xi \\ 0, & \text{if } \xi_i \neq \xi \end{cases}. \tag{8.38}$$

(8) Determine $\partial R_i/\partial \bar{X}_R$:
If R, a variable in $\bar{X}_R$, is the vth component of $\bar{X}_{sys}$, then $\partial \bar{R}_i/\partial \bar{X}_R$ can be determined as

$$J_i(9, v) = \frac{\partial R_i}{\partial R} = \begin{cases} 1, & \text{if } R_i = R \\ 0, & \text{if } R_i \neq R \end{cases}. \tag{8.39}$$

(9) Determination of $\partial(t_{ix}, t_{iy}, t_{iz})/\partial \bar{X}_R$, $\partial(t_{ix}, t_{iy}, t_{iz})/\partial \bar{X}_{rest}$ and $\partial(\omega_{ix}, \omega_{iy}, \omega_{iz})/\partial \bar{X}_{rest}$:

Defining the pose of an optical element in 3-D space requires both the element coordinate frame $(xyz)_{ej}$ and the pose matrix given in Eq. (3.25), i.e.,

$$\begin{aligned} {}^0\bar{A}_{ej} &= \text{tran}(t_{ejx}, t_{ejy}, t_{ejz})\text{rot}(\bar{z}, \omega_{ejz})\text{rot}(\bar{y}, \omega_{ejy})\text{rot}(\bar{x}, \omega_{ejx}) \\ &= \begin{bmatrix} I_{ejx} & J_{ejx} & K_{ejx} & t_{ejx} \\ I_{ejy} & J_{ejy} & K_{ejy} & t_{ejy} \\ I_{ejz} & J_{ejz} & K_{ejz} & t_{ejz} \\ 0 & 0 & 0 & 1 \end{bmatrix}. \end{aligned}$$

Furthermore, the pose matrix ${}^{ej}\bar{A}_i$ is required to define the position and orientation of each boundary surface with respect to $(xyz)_{ej}$. Having obtained ${}^0\bar{A}_{ej}$ and

${}^{ej}\bar{A}_i$, the pose matrix ${}^{0}\bar{A}_i$ required for raytracing purposes can be computed via the following matrix multiplication (see Eq. (3.2)):

$$
{}^{0}\bar{A}_i = \text{tran}(t_{ix}, t_{iy}, t_{iz})\text{rot}(\bar{z}, \omega_{iz})\text{rot}(\bar{y}, \omega_{iy})\text{rot}(\bar{x}, \omega_{ix}) = \begin{bmatrix} I_{ix} & J_{ix} & K_{ix} & t_{ix} \\ I_{iy} & J_{iy} & K_{iy} & t_{iy} \\ I_{iz} & J_{iz} & K_{iz} & t_{iz} \\ 0 & 0 & 0 & 1 \end{bmatrix}
$$

$$
= {}^{0}\bar{A}_{ej}{}^{ej}\bar{A}_i = \begin{bmatrix} I_{ejx} & J_{ejx} & K_{ejx} & t_{ejx} \\ I_{ejy} & J_{ejy} & K_{ejy} & t_{ejy} \\ I_{ejz} & J_{ejz} & K_{ejz} & t_{ejz} \\ 0 & 0 & 0 & 1 \end{bmatrix} \begin{bmatrix} {}^{ej}I_{ix} & {}^{ej}J_{ix} & {}^{ej}K_{ix} & {}^{ej}t_{ix} \\ {}^{ej}I_{iy} & {}^{ej}J_{iy} & {}^{ej}K_{iy} & {}^{ej}t_{iy} \\ {}^{ej}I_{iz} & {}^{ej}J_{iz} & {}^{ej}K_{iz} & {}^{ej}t_{iz} \\ 0 & 0 & 0 & 1 \end{bmatrix}, \tag{8.40}
$$

where

$$t_{ix} = I_{ejx}\left({}^{ej}t_{ix}\right) + J_{ejx}\left({}^{ej}t_{iy}\right) + K_{ejx}\left({}^{ej}t_{iz}\right) + t_{ejx}, \tag{8.41}$$

$$t_{iy} = I_{ejy}\left({}^{ej}t_{ix}\right) + J_{ejy}\left({}^{ej}t_{iy}\right) + K_{ejy}\left({}^{ej}t_{iz}\right) + t_{ejy}, \tag{8.42}$$

$$t_{iz} = I_{ejz}\left({}^{ej}t_{ix}\right) + J_{ejz}\left({}^{ej}t_{iy}\right) + K_{ejz}\left({}^{ej}t_{iz}\right) + t_{ejz}. \tag{8.43}$$

To determine $\partial(t_{ix}, t_{iy}, t_{iz})/\partial\bar{X}_R$, $\partial(t_{ix}, t_{iy}, t_{iz})/\partial\bar{X}_{rest}$ and $\partial(\omega_{ix}, \omega_{iy}, \omega_{iz})/\partial\bar{X}_{rest}$, it is first necessary to obtain $d({}^{0}\bar{A}_i)/d\bar{X}_{sys} = [\partial({}^{0}\bar{A}_i)/\partial x_v]$ $(x_v \in \bar{X}_{sys})$ by differentiating Eq. (8.40) with respect to the vth component of $\bar{X}_{sys}$, i.e.,

$$
\frac{d({}^{0}\bar{A}_i)}{d\bar{X}_{sys}} = \begin{bmatrix} \partial I_{ix}/\partial x_v & \partial J_{ix}/\partial x_v & \partial K_{ix}/\partial x_v & \partial t_{ix}/\partial x_v \\ \partial I_{iy}/\partial x_v & \partial J_{iy}/\partial x_v & \partial K_{iy}/\partial x_v & \partial t_{iy}/\partial x_v \\ \partial I_{iz}/\partial x_v & \partial J_{iz}/\partial x_v & \partial K_{iz}/\partial x_v & \partial t_{iz}/\partial x_v \\ 0 & 0 & 0 & 0 \end{bmatrix}
= \frac{d({}^{0}\bar{A}_{ej})}{d\bar{X}_{sys}}{}^{ej}\bar{A}_i + {}^{0}\bar{A}_{ej}\frac{d({}^{ej}\bar{A}_i)}{d\bar{X}_{sys}}, \tag{8.44}
$$

where $d({}^{0}\bar{A}_{ej})/d\bar{X}_{sys}$ and $d({}^{ej}\bar{A}_i)/d\bar{X}_{sys}$ are the first-order derivative matrices of ${}^{0}\bar{A}_{ej}$ and ${}^{ej}\bar{A}_i$, respectively, with respect to $\bar{X}_{sys}$. Computing Eq. (8.44) poses a significant challenge when $\bar{X}_{sys}$ contains many components. In this book, the required terms $d({}^{0}\bar{A}_{ej})/d\bar{X}_{sys}$ and $d({}^{ej}\bar{A}_i)/d\bar{X}_{sys}$ of Eq. (8.44) are obtained using the methods described in Appendices 1 and 2 in this chapter, respectively. The Jacobian matrices of the first six pose variables in $\bar{X}_i$ (i.e., t_{ix}, t_{iy}, t_{iz}, ω_{iz}, ω_{iy} and ω_{ix}) with respect to the system variable vector $\bar{X}_{sys}$ (i.e., $dt_{ix}/d\bar{X}_{sys}$, $dt_{iy}/d\bar{X}_{sys}$, $dt_{iz}/d\bar{X}_{sys}$, $d\omega_{ix}/d\bar{X}_{sys}$, $d\omega_{iy}/d\bar{X}_{sys}$ and $d\omega_{iz}/d\bar{X}_{sys}$) are then determined using the equations given in Appendix 4 in this chapter.

Table 8.2 Jacobian matrix $d(\xi_{i-1}, \xi_i)/d\bar{X}_{sys}$ of cat's eye system shown in Fig. 3.12

$J_i(u,v)$ \ $\bar{X}_i$ / $\bar{X}_{sys}$		i = 1		i = 2		i = 3		i = 4		i = 5		i = 6		i = 7	
		ξ_{i-1} u = 7	ξ_i u = 8	ξ_{i-1} u = 5	ξ_i u = 6	ξ_{i-1} u = 5	ξ_i u = 6	ξ_{i-1} u = 7	ξ_i u = 8	ξ_{i-1} u = 5	ξ_i u = 6	ξ_{i-1} u = 5	ξ_i u = 6	ξ_{i-1} u = 7	ξ_i u = 8
$\bar{X}_0$		$\bar{0}$	$\bar{0}$	$\bar{0}$	$\bar{0}$	$\bar{0}$	$\bar{0}$	$\bar{0}$	$\bar{0}$	$\bar{0}$	$\bar{0}$	$\bar{0}$	$\bar{0}$	$\bar{0}$	$\bar{0}$
$\bar{X}\xi$	ξ_{air} v = 6	1	0	0	0	0	0	0	0	0	0	0	0	0	1
	ξ_{el} v = 7	0	1	1	0	0	0	0	0	0	0	0	1	1	0
	ξ_{glue} v = 8	0	0	0	1	1	0	0	0	0	1	1	0	0	0
	ξ_{e2} v = 9	0	0	0	0	0	1	1	1	1	0	0	0	0	0
$\bar{X}_R$		$\bar{0}$	$\bar{0}$	$\bar{0}$	$\bar{0}$	$\bar{0}$	$\bar{0}$	$\bar{0}$	$\bar{0}$	$\bar{0}$	$\bar{0}$	$\bar{0}$	$\bar{0}$	$\bar{0}$	$\bar{0}$
$\bar{X}_{rest}$		$\bar{0}$	$\bar{0}$	$\bar{0}$	$\bar{0}$	$\bar{0}$	$\bar{0}$	$\bar{0}$	$\bar{0}$	$\bar{0}$	$\bar{0}$	$\bar{0}$	$\bar{0}$	$\bar{0}$	$\bar{0}$

Table 8.3 Jacobian matrix $d(\xi_{i-1}, \xi_i)/d\bar{X}_{sys}$ of retro-reflector system shown in Fig. 3.13

$J_i(u,v)$ \ $\bar{X}_i$ / $\bar{X}_{sys}$		i = 1		i = 2		i = 3		i = 4		i = 5	
		ξ_{i-1} u = 7	ξ_i u = 8	ξ_{i-1} u = 7	ξ_i u = 8	ξ_{i-1} u = 7	ξ_i u = 8	ξ_{i-1} u = 7	ξ_i u = 8	ξ_{i-1} u = 7	ξ_i u = 8
$\bar{X}_0$		$\bar{0}$	$\bar{0}$	$\bar{0}$	$\bar{0}$	$\bar{0}$	$\bar{0}$	$\bar{0}$	$\bar{0}$	$\bar{0}$	$\bar{0}$
$\bar{X}\xi$	ξ_{air} v = 6	1	0	0	1	1	1	1	0	0	1
	ξ_{el} v = 7	0	1	1	0	0	0	0	1	1	0
$\bar{X}_R$		$\bar{0}$	$\bar{0}$	$\bar{0}$	$\bar{0}$	$\bar{0}$	$\bar{0}$	$\bar{0}$	$\bar{0}$	$\bar{0}$	$\bar{0}$
$\bar{X}_{rest}$		$\bar{0}$	$\bar{0}$	$\bar{0}$	$\bar{0}$	$\bar{0}$	$\bar{0}$	$\bar{0}$	$\bar{0}$	$\bar{0}$	$\bar{0}$

Example 8.2 The results of $J_i(7,v) = d\xi_{i-1}/d\bar{X}_{sys}$ and $J_i(8,v) = d\xi_i/d\bar{X}_{sys}$ (i = 1 to i = 7 and v = 1 to v = 17) for the cat's eye system (Fig. 3.12) of Example 3.7 are summarized in Table 8.2. The non-zero values are as follows: $J_1(7,6) = J_1(8,7) = J_4(7,9) = J_4(8,9) = J_7(7,7) = J_7(8,6) = 1$

The result $J_1(7,6) = J_1(8,7) = 1$ indicates that the 7th and 8th variables of $\bar{X}_1$ are respectively the 6th and 7th variables of the system variable vector $\bar{X}_{sys}$. The remaining results can be interpreted in a similar manner.

Example 8.3 The results of $J_i(7,v) = d\xi_{i-1}/d\bar{X}_{sys}$ and $J_i(8,v) = d\xi_i/d\bar{X}_{sys}$ (i = 1 to i = 5 and v = 1 to v = 12) for the retro-reflector system (Fig. 3.13) of Example 3.8 are listed in Table 8.3. The non-zero values are as follows:

$$J_1(7,6) = J_1(8,7) = J_2(7,7) = J_2(8,6) = J_3(7,6) = J_3(8,6) = J_4(7,6) = J_4(8,7) \\ = J_5(7,7) = J_5(8,6) = 1.$$

Example 8.4 The results of $J_i(9,v) = dR_i/d\bar{X}_{sys}$ (i = 1 to i = 7 and v = 1 to $v = q_{sys} = 17$) for the system shown in Fig. 3.12 are given in Table 8.4. The non-zero values are as follows: $J_1(9,10) = 1$, $J_4(9,11) = 1$ and $J_7(9,10) = -1$.

Table 8.4 Jacobian matrix $dR_i/d\bar{X}_{sys}$ of system shown in Fig. 3.12

$J_i(u,v)$ \ $\bar{X}_i$ / $\bar{X}_{sys}$		i = 1	i = 2	i = 3	i = 4	i = 5	i = 6	i = 7
		R_i u = 9			R_i u = 9			R_i u = 9
$\bar{X}_0$		0			0			0
$\bar{X}\xi$		0			0			0
$\bar{X}_R$	R_1 v = 10	1			0			−1
	R_4 v = 11	0			1			0
$\bar{X}_{rest}$		0			0			0

Note that R_i is not a component of boundary variable vector $\bar{X}_i$ for i = 2, 3, 5 and 6 since $\bar{r}_2$, $\bar{r}_3$, $\bar{r}_5$ and $\bar{r}_6$ are flat boundary surfaces

Table 8.5 Jacobian matrix $dR_i/d\bar{X}_{sys}$ of system shown in Fig. 3.13

$\bar{X}_{sys}$	$J_i(u,v)$ $\bar{X}_i$	i = 1	i = 2	i = 3	i = 4	i = 5
		R_i	R_i	R_i	R_i	R_i
		u = 9	u = 9	u = 9	u = 9	u = 9
$\bar{X}_0$		0	0	0	0	0
$\bar{X}_\xi$		0	0	0	0	0
$\bar{X}_R$	R_1 v= 8	1	0	0	0	−1
	R_2 v= 9	0	1	0	−1	0
	R_3 v= 10	0	0	1	0	0
$\bar{X}_{rest}$		0	0	0	0	0

Example 8.5 The results of $J_i(9,v) = dR_i/d\bar{X}_{sys}$ (i = 1 to i = 5 and v = 1 to $v = q_{sys} = 12$) for the system shown in Fig. 3.13 are given in Table 8.5. The non-zero values are as follows: $J_1(9,8) = 1$, $J_2(9,9) = 1$, $J_3(9,10) = 1$, $J_4(9,9) = -1$ and $J_5(9,8) = -1$.

Example 8.6 The non-zero values of $\partial(t_{ix}, t_{iy}, t_{iz})/\partial\bar{X}_R$, $\partial(t_{ix}, t_{iy}, t_{iz})/\partial\bar{X}_{rest}$ and $\partial(\omega_{ix}, \omega_{iy}, \omega_{iz})/\partial\bar{X}_{rest}$ for the system shown in Fig. 3.12 are given as follows:

$$\begin{aligned} J_2(5,7) &= J_2(6,8) = -J_3(1,17) = J_3(3,15) = -J_3(4,13) = J_3(5,8) = J_3(6,9) \\ &= J_4(1,12) = J_4(2,13) = J_4(3,14) = J_4(4,15) = J_4(5,16) = J_4(6,17) \\ &= -J_5(1,17) = J_5(3,15) = -J_5(4,13) = J_5(5,9) = J_5(6,8) = J_6(5,8) \\ &= J_6(6,7) = 1. \end{aligned}$$

Example 8.7 The non-zero values of $\partial(t_{ix}, t_{iy}, t_{iz})/\partial\bar{X}_R$, $\partial(t_{ix}, t_{iy}, t_{iz})/\partial\bar{X}_{rest}$ and $\partial(\omega_{ix}, \omega_{iy}, \omega_{iz})/\partial\bar{X}_{rest}$ for the system shown in Fig. 3.13 are given as follows:

$$\begin{aligned} J_2(2,11) &= J_3(2,11) = J_3(2,12) = J_4(2,11) = 1,\ J_1(2,8) = J_2(2,9) = J_4(2,9) \\ &= J_5(2,8) = -33123.9983 \text{ and } J_3(2,10) = -40000.0018 \end{aligned}$$

Example 8.8 The Jacobian matrix $d\bar{R}_3/d\bar{X}_{sys}$ of the exit ray $\bar{R}_3$ of the corner-cube system shown in Fig. 3.19 can be obtained by using Eq. (7.55) with g = 3. The corresponding results are listed in Table 8.6. It is noted that $\bar{X}_{sys}$ for the system shown in Example 3.18 comprises three components, namely the source ray variable vector $\bar{X}_0$, the element variable vector $\bar{X}_{e1}$ and the boundary variable vectors $\bar{X}_i$ (i = 1 to i = 3). The sub-matrix $\partial\bar{R}_3/\partial\bar{X}_{e1}$ allows the effects of prism misalignments to be explored. For example, $\partial\bar{\ell}_3/\partial\bar{X}_{e1} = \bar{0}_{3\times7}$ proves that the retro-reflective property of the corner-cube mirror is preserved even if the mirror is moved slightly to a new position or orientation. Furthermore, Table 8.6 shows that

Table 8.6 Jacobian matrix of exit ray $\bar{R}_3$ with respect to system variable vector $\bar{X}_{sys}$ for corner-cube mirror shown in Fig. 3.19

		P_{3x}	P_{3y}	P_{3z}	ℓ_{3x}	ℓ_{3y}	ℓ_{3z}
$\bar{X}_0$	P_{0x}	0	1	1	0	0	0
	P_{0y}	0	−1	0	0	0	0
	P_{0z}	0	0	−1	0	0	0
	α_0	0	−51.96	0	0	−1	0
	β_0	0	0	−51.96	0	0	−1
$\bar{X}_{e1}$	ξ_{air}	0	0	0	0	0	0
	t_{e1x}	1	−1	−1	0	0	0
	t_{e1y}	0	2	0	0	0	0
	t_{e1z}	0	0	2	0	0	0
	ω_{e1x}	0	0	0	0	0	0
	ω_{e1y}	20	20	20	0	0	0
	ω_{e1z}	−10	−10	−10	0	0	0
$\bar{X}_i$ i = 1 to i = 3	${}^{e1}J_{1x}$	0	40	40	−1.15	0	1.15
	${}^{e1}J_{1y}$	0	−40	20	0	−1.15	1.15
	${}^{e1}e_1$	0	0	−2	0	0	0
	${}^{e1}J_{2z}$	0	−40	20	0	−1.15	1.15
	${}^{e1}J_{2x}$	0	20	20	−1.15	1.15	0
	${}^{e1}e_2$	0	−2	0	0	0	0
	${}^{e1}J_{3y}$	−10	10	10	−1.15	1.15	0
	${}^{e1}J_{3z}$	−20	20	20	−1.15	0	1.15
	${}^{e1}e_3$	−1	1	1	0	0	0

$\partial\bar{R}_3/\partial\xi_{air} = \bar{0}_{1\times6}$. This result is reasonable since all of the flat boundary surfaces are reflective. The sub-matrix $\partial\bar{R}_3/\partial\bar{X}_i$ (i = 1 to i = 3) indicates the effects of fabrication errors (or tolerances) on the system performance. More specifically, the results presented in Table 8.6 show that the retro-reflective property of the corner-cube mirror is preserved only in the case of no fabrication errors in any of the three reflective flat boundary surfaces (i.e., $\partial\bar{\ell}_3/\partial({}^{e1}J_{1x})$, $\partial\bar{\ell}_3/\partial({}^{e1}J_{1y})$, $\partial\bar{\ell}_3/\partial({}^{e1}J_{2z})$, $\partial\bar{\ell}_3/\partial({}^{e1}J_{2x})$, $\partial\bar{\ell}_3/\partial({}^{e1}J_{3y})$ and $\partial\bar{\ell}_3/\partial({}^{e1}J_{3z})$ are not all zero matrices).

Appendix 1

When using Eqs. (8.24) and (8.44), it is first necessary to determine $d({}^{0}\bar{A}_{ej})/d\bar{X}_{sys}$ in advance, where ${}^{0}\bar{A}_{ej}$ is defined by Eqs. (3.1) and (3.25) as

$$
\begin{aligned}
{}^{0}\bar{A}_{ej} &= \mathrm{tran}(t_{ejx}, t_{ejy}, t_{ejz})\mathrm{rot}(\bar{z}, \omega_{ejz})\mathrm{rot}(\bar{y}, \omega_{ejy})\mathrm{rot}(\bar{x}, \omega_{ejx}) \\
&= \begin{bmatrix} C\omega_{ejy}C\omega_{ejz} & S\omega_{ejx}S\omega_{ejy}C\omega_{ejz} - C\omega_{ejx}S\omega_{ejz} & C\omega_{ejx}S\omega_{ejy}C\omega_{ejz} - S\omega_{ejx}S\omega_{ejz} & t_{ejx} \\ C\omega_{ejy}S\omega_{ejz} & C\omega_{ejx}C\omega_{ejz} + S\omega_{ejx}S\omega_{ejy}S\omega_{ejz} & -S\omega_{ejx}C\omega_{ejz} + C\omega_{ejx}S\omega_{ejy}S\omega_{ejz} & t_{ejy} \\ -S\omega_{ejy} & S\omega_{ejx}C\omega_{ejy} & C\omega_{ejx}C\omega_{ejy} & t_{ejz} \\ 0 & 0 & 0 & 1 \end{bmatrix} \\
&= \begin{bmatrix} I_{ejx} & J_{ejx} & K_{ejx} & t_{ejx} \\ I_{ejy} & J_{ejy} & K_{ejy} & t_{ejy} \\ I_{ejz} & J_{ejz} & K_{ejz} & t_{ejz} \\ 0 & 0 & 0 & 1 \end{bmatrix}.
\end{aligned}
$$

A dimensional analysis reveals that the arguments of the rotation terms in matrix ${}^{0}\bar{A}_{ej}$ can be expressed as linear combinations of the components of $\bar{X}_{sys} = [x_v]$, i.e.,

$$\omega_{ejx} = a_0 + \sum_{v=1}^{q_{sys}} a_v x_v, \tag{8.45}$$

$$\omega_{ejy} = b_0 + \sum_{v=1}^{q_{sys}} b_v x_v, \tag{8.46}$$

$$\omega_{ejz} = c_0 + \sum_{v=1}^{q_{sys}} c_v x_v, \tag{8.47}$$

where $x_v (v \in \{1, 2, \ldots, q_{sys}\})$ is the vth component of $\bar{X}_{sys}$, and a_v, b_v and c_v ($v = 0$ to $v = q_{sys}$) are known constants. Therefore, $d({}^{0}\bar{A}_{ej})/d\bar{X}_{sys}$ can be determined directly by differentiating Eq. (3.1) (or Eq. (3.25)) with respect to x_v to give

$$
\begin{aligned}
\frac{d({}^{0}\bar{A}_{ej})}{d\bar{X}_{sys}} &= \left[\frac{\partial({}^{0}\bar{A}_{ej})}{\partial x_v}\right]_{4\times 4\times q_{sys}} \\
&= \begin{bmatrix} \partial I_{ejx}/\partial x_v & \partial J_{ejx}/\partial x_v & \partial K_{ejx}/\partial x_v & \partial t_{ejx}/\partial x_v \\ \partial I_{ejy}/\partial x_v & \partial J_{ejy}/\partial x_v & \partial K_{ejy}/\partial x_v & \partial t_{ejy}/\partial x_v \\ \partial I_{ejz}/\partial x_v & \partial J_{ejz}/\partial x_v & \partial K_{ejz}/\partial x_v & \partial t_{ejz}/\partial x_v \\ 0 & 0 & 0 & 0 \end{bmatrix},
\end{aligned} \tag{8.48}
$$

where

$$\frac{\partial I_{ejx}}{\partial x_v} = -b_v S\omega_{ejy} C\omega_{ejz} - c_v C\omega_{ejy} S\omega_{ejz}, \tag{8.49}$$

$$\frac{\partial I_{ejy}}{\partial x_v} = -b_v S\omega_{ejy} S\omega_{ejz} + c_v C\omega_{ejy} C\omega_{ejz}, \tag{8.50}$$

$$\frac{\partial I_{ejz}}{\partial x_v} = -b_v C\omega_{ejy}, \tag{8.51}$$

$$\frac{\partial J_{ejx}}{\partial x_v} = a_v C\omega_{ejx} S\omega_{ejy} C\omega_{ejz} + b_v S\omega_{ejx} C\omega_{ejy} C\omega_{ejz} - c_v S\omega_{ejx} S\omega_{ejy} S\omega_{ejz} + a_v S\omega_{ejx} S\omega_{ejz} - c_v C\omega_{ejx} C\omega_{ejz}, \tag{8.52}$$

$$\frac{\partial J_{ejy}}{\partial x_v} = a_v C\omega_{ejx} S\omega_{ejy} S\omega_{ejz} + b_v S\omega_{ejx} C\omega_{ejy} S\omega_{ejz} + c_v S\omega_{ejx} S\omega_{ejy} C\omega_{ejz} - a_v S\omega_{ejx} C\omega_{ejz} - c_v C\omega_{ejx} S\omega_{ejz}, \tag{8.53}$$

$$\frac{\partial J_{ejz}}{\partial x_v} = a_v C\omega_{ejx} C\omega_{ejy} - b_v S\omega_{ejx} S\omega_{ejy}, \tag{8.54}$$

$$\frac{\partial K_{ejx}}{\partial x_v} = -a_v S\omega_{ejx} S\omega_{ejy} C\omega_{ejz} + b_v C\omega_{ejx} C\omega_{ejy} C\omega_{ejz} - c_v C\omega_{ejx} S\omega_{ejy} S\omega_{ejz} - a_v C\omega_{ejx} S\omega_{ejz} - c_v S\omega_{ejx} C\omega_{ejz}, \tag{8.55}$$

$$\frac{\partial K_{ejy}}{\partial x_v} = -a_v S\omega_{ejx} S\omega_{ejy} S\omega_{ejz} + b_v C\omega_{ejx} C\omega_{ejy} S\omega_{ejz} + c_v C\omega_{ejx} S\omega_{ejy} C\omega_{ejz} - a_v C\omega_{ejx} C\omega_{ejz} + c_v S\omega_{ejx} S\omega_{ejz}, \tag{8.56}$$

$$\frac{\partial K_{ejz}}{\partial x_v} = -a_v S\omega_{ejx} C\omega_{ejy} - b_v C\omega_{ejx} S\omega_{ejy}, \tag{8.57}$$

$$\frac{\partial t_{ejx}}{\partial x_v} = \frac{\partial t_{ejx}}{\partial x_v}, \tag{8.58}$$

$$\frac{\partial t_{ejy}}{\partial x_v} = \frac{\partial t_{ejy}}{\partial x_v}, \tag{8.59}$$

$$\frac{\partial t_{ejz}}{\partial x_v} = \frac{\partial t_{ejz}}{\partial x_v}. \tag{8.60}$$

Equations (8.58), (8.59) and (8.60) indicate that $\partial t_{ejx}/\partial x_v$, $\partial t_{ejy}/\partial x_v$ and $\partial t_{ejz}/\partial x_v$ can be obtained simply by differentiating their corresponding expressions since they are always given in explicit form (e.g., $\partial t_{e2x}/\partial x_{25} = 1/2$, $\partial t_{e2y}/\partial x_{20} = 1$ and $\partial t_{e2z}/\partial x_{26} = -1$ for ${}^{0}\bar{A}_{e2}$ in Example 3.11 and $\bar{X}_{sys}$ defined in Example 3.15).

Appendix 2

In using Eqs. (8.24) and (8.44), it is first necessary to determine $d({}^{ej}\bar{A}_i)/d\bar{X}_{sys}$ in advance, where ${}^{ej}\bar{A}_i$ is defined by Eqs. (3.3) and (3.26) as

$$
\begin{aligned}
{}^{ej}\bar{A}_i &= \mathrm{tran}({}^{ej}t_{ix}, {}^{ej}t_{iy}, {}^{ej}t_{iz})\mathrm{rot}(\bar{z}, {}^{ej}\omega_{iz})\mathrm{rot}(\bar{y}, {}^{ej}\omega_{iy})\mathrm{rot}(\bar{x}, {}^{ej}\omega_{ix}) \\
&= \begin{bmatrix} {}^{ej}I_{ix} & {}^{ej}J_{ix} & {}^{ej}K_{ix} & {}^{ej}t_{ix} \\ {}^{ej}I_{iy} & {}^{ej}J_{iy} & {}^{ej}K_{iy} & {}^{ej}t_{iy} \\ {}^{ej}I_{iz} & {}^{ej}J_{iz} & {}^{ej}K_{iz} & {}^{ej}t_{iz} \\ 0 & 0 & 0 & 1 \end{bmatrix},
\end{aligned}
$$

with

$$
\begin{aligned}
{}^{ej}I_{ix} &= C({}^{ej}\omega_{iy})C({}^{ej}\omega_{iz}), \\
{}^{ej}I_{iy} &= C({}^{ej}\omega_{iy})S({}^{ej}\omega_{iz}). \\
{}^{ej}I_{iz} &= -S({}^{ej}\omega_{iy}), \\
{}^{ej}J_{ix} &= S({}^{ej}\omega_{ix})S({}^{ej}\omega_{iy})C({}^{ej}\omega_{iz}) - C({}^{ej}\omega_{ix})S({}^{ej}\omega_{iz}), \\
{}^{ej}J_{iy} &= C({}^{ej}\omega_{ix})C({}^{ej}\omega_{iz}) + S({}^{ej}\omega_{ix})S({}^{ej}\omega_{iy})S({}^{ej}\omega_{iz}), \\
{}^{ej}J_{iz} &= S({}^{ej}\omega_{ix})C({}^{ej}\omega_{iy}), \\
{}^{ej}K_{ix} &= C({}^{ej}\omega_{ix})S({}^{ej}\omega_{iy})C({}^{ej}\omega_{iz}) - S({}^{ej}\omega_{ix})S({}^{ej}\omega_{iz}), \\
{}^{ej}K_{iy} &= -S({}^{ej}\omega_{ix})C({}^{ej}\omega_{iz}) + C({}^{ej}\omega_{ix})S({}^{ej}\omega_{iy})S({}^{ej}\omega_{iz}), \\
{}^{ej}K_{iz} &= C({}^{ej}\omega_{ix})C^{ej}(\omega_{iy}).
\end{aligned}
$$

A dimensional analysis again reveals that the arguments of the rotation terms in matrix ${}^{ej}\bar{A}_i$ can be expressed as linear combinations of the components of $\bar{X}_{sys} = [x_v]$, i.e.,

$$
{}^{ej}\omega_{ix} = \rho_0 + \sum_{v=1}^{q_{sys}} \rho_v x_v, \tag{8.61}
$$

$$ {}^{ej}\omega_{iy} = g_0 + \sum_{v=1}^{q_{sys}} g_v x_v, \tag{8.62} $$

$$ {}^{ej}\omega_{iz} = h_0 + \sum_{v=1}^{q_{sys}} h_v x_v, \tag{8.63} $$

where x_v ($v \in \left\{1, 2, \ldots, q_{sys}\right\}$) is the vth component of $\bar{X}_{sys}$, and ρ_v, g_v and h_v ($v = 0$ to $v = q_{sys}$) are known constants. Therefore, $\partial\left({}^{ej}\bar{A}_i\right)/\partial x_v$ can be determined directly by differentiating Eq. (3.3) (or Eq. (3.26)) with respect to x_v to give

$$ \frac{d\left({}^{ej}\bar{A}_i\right)}{d\bar{X}_{sys}} = \left[\frac{\partial\left({}^{ej}\bar{A}_i\right)}{\partial x_v}\right]_{4\times4\times q_{sys}} = \begin{bmatrix} \partial({}^{ej}I_{ix})/\partial x_v & \partial({}^{ej}J_{ix})/\partial x_v & \partial({}^{ej}K_{ix})/\partial x_v & \partial({}^{ej}t_{ix})/\partial x_v \\ \partial({}^{ej}I_{iy})/\partial x_v & \partial({}^{ej}J_{iy})/\partial x_v & \partial({}^{ej}K_{iy})/\partial x_v & \partial({}^{ej}t_{iy})/\partial x_v \\ \partial({}^{ej}I_{iz})/\partial x_v & \partial({}^{ej}J_{iz})/\partial x_v & \partial({}^{ej}K_{iz})/\partial x_v & \partial({}^{ej}t_{iz})/\partial x_v \\ 0 & 0 & 0 & 0 \end{bmatrix}, \tag{8.64} $$

where

$$ \frac{\partial({}^{ej}I_{ix})}{\partial x_v} = -g_v S({}^{ej}\omega_{iy}) C({}^{ej}\omega_{iz}) - h_v C({}^{ej}\omega_{iy}) S({}^{ej}\omega_{iz}), \tag{8.65} $$

$$ \frac{\partial({}^{ej}I_{iy})}{\partial x_v} = -g_v S({}^{ej}\omega_{iy}) S({}^{ej}\omega_{iz}) + h_v C({}^{ej}\omega_{iy}) C({}^{ej}\omega_{iz}), \tag{8.66} $$

$$ \frac{\partial({}^{ej}I_{iz})}{\partial x_v} = -g_v C({}^{ej}\omega_{iy}), \tag{8.67} $$

$$ \begin{aligned} \frac{\partial({}^{ej}J_{ix})}{\partial x_v} = {} & \rho_v C({}^{ej}\omega_{ix}) S({}^{ej}\omega_{iy}) C({}^{ej}\omega_{iz}) + g_v S({}^{ej}\omega_{ix}) C({}^{ej}\omega_{iy}) C({}^{ej}\omega_{iz}) \\ & - h_v S({}^{ej}\omega_{ix}) S({}^{ej}\omega_{iy}) S({}^{ej}\omega_{iz}) + \rho_v S({}^{ej}\omega_{ix}) S({}^{ej}\omega_{iz}) - h_v C({}^{ej}\omega_{ix}) C({}^{ej}\omega_{iz}), \end{aligned} \tag{8.68} $$

$$ \begin{aligned} \frac{\partial({}^{ej}J_{iy})}{\partial x_v} = {} & \rho_v C({}^{ej}\omega_{ix}) S({}^{ej}\omega_{iy}) S({}^{ej}\omega_{iz}) + g_v S({}^{ej}\omega_{ix}) C({}^{ej}\omega_{iy}) S({}^{ej}\omega_{iz}) \\ & + h_v S({}^{ej}\omega_{ix}) S({}^{ej}\omega_{iy}) C({}^{ej}\omega_{iz}) - \rho_v S({}^{ej}\omega_{ix}) C({}^{ej}\omega_{iz}) - h_v C({}^{ej}\omega_{ix}) S({}^{ej}\omega_{iz}), \end{aligned} \tag{8.69} $$

$$ \frac{\partial({}^{ej}J_{iz})}{\partial x_v} = \rho_v C({}^{ej}\omega_{ix}) C({}^{ej}\omega_{iy}) - g_v S({}^{ej}\omega_{ix}) S({}^{ej}\omega_{iy}), \tag{8.70} $$

$$\frac{\partial({}^{ej}K_{ix})}{\partial x_v} = -\rho_v S({}^{ej}\omega_{ix})S({}^{ej}\omega_{iy})C({}^{ej}\omega_{iz}) + g_v C({}^{ej}\omega_{ix})C({}^{ej}\omega_{iy})C({}^{ej}\omega_{iz}) - h_v C({}^{ej}\omega_{ix})S({}^{ej}\omega_{iy})S({}^{ej}\omega_{iz}) - \rho_v C({}^{ej}\omega_{ix})S({}^{ej}\omega_{iz}) - h_v S({}^{ej}\omega_{ix})C({}^{ej}\omega_{iz}), \quad (8.71)$$

$$\frac{\partial({}^{ej}K_{iy})}{\partial x_v} = -\rho_v S({}^{ej}\omega_{ix})S({}^{ej}\omega_{iy})S({}^{ej}\omega_{iz}) + g_v C({}^{ej}\omega_{ix})C({}^{ej}\omega_{iy})S({}^{ej}\omega_{iz}) + h_v C({}^{ej}\omega_{ix})S({}^{ej}\omega_{iy})C({}^{ej}\omega_{iz}) - \rho_v C({}^{ej}\omega_{ix})C({}^{ej}\omega_{iz}) + h_v S({}^{ej}\omega_{ix})S({}^{ej}\omega_{iz}), \quad (8.72)$$

$$\frac{\partial({}^{ej}K_{iz})}{\partial x_v} = -\rho_v S({}^{ej}\omega_{ix})C({}^{ej}\omega_{iy}) - g_v C({}^{ej}\omega_{ix})S({}^{ej}\omega_{iy}). \quad (8.73)$$

Note that ${}^{ej}t_{ix}$, ${}^{ej}t_{iy}$ and ${}^{ej}t_{iz}$ are always given in explicit form. As a result, $\partial({}^{ej}t_{ix})/\partial x_v$, $\partial({}^{ej}t_{iy})/\partial x_v$ and $\partial({}^{ej}t_{iz})/\partial x_v$ can be obtained simply by differentiating their corresponding expressions, i.e.,

$$\frac{\partial({}^{ej}t_{ix})}{\partial x_v} = \frac{\partial({}^{ej}t_{ix})}{\partial x_v}, \quad (8.74)$$

$$\frac{\partial({}^{ej}t_{iy})}{\partial x_v} = \frac{\partial({}^{ej}t_{iy})}{\partial x_v}, \quad (8.75)$$

$$\frac{\partial({}^{ej}t_{iz})}{\partial x_v} = \frac{\partial({}^{ej}t_{iz})}{\partial x_v}. \quad (8.76)$$

For example, $\partial({}^{e1}t_{2y})/\partial x_{11} = -1$, $\partial({}^{e1}t_{2y})/\partial x_{19} = 1$ and $\partial({}^{e1}t_{2y})/\partial x_{12} = 1$ for ${}^{e1}\bar{A}_2$ in Example 3.12 and $\bar{X}_{sys}$ defined in Example 3.15.

Having obtained numerical values of ${}^{0}\bar{A}_{ej}$, ${}^{ej}\bar{A}_i$, $\partial({}^{0}\bar{A}_{ej})/\partial x_v$ and $\partial({}^{ej}\bar{A}_i)/\partial x_v$, the Jacobian matrices $\partial(\bar{r}_i)/\partial\bar{X}_{sys}$ and $d({}^{0}\bar{A}_i)/d\bar{X}_{sys}$ can be computed from Eqs. (8.24) and (8.44), respectively, using the methods described in Appendices 3 and 4 of this chapter.

Appendix 3

It is noted from Eq. (8.24) that to compute $d(\bar{r}_i)/d\bar{X}_{sys}$, it is first necessary to have the numerical values of ${}^{0}\bar{A}'_{ej}$, ${}^{ej}\bar{r}_i$, $d({}^{0}\bar{A}'_{ej})/d\bar{X}_{sys}$ and $d({}^{ej}\bar{r}_i)/d\bar{X}_{sys}$. The related expressions are presented in the following.

(1) Determination of ${}^{0}\bar{A}'_{ej}$ from ${}^{0}\bar{A}_{ej}$:

If ${}^{0}\bar{A}_{ej}$ is given as Eq. (3.1) (or Eq. (3.25)), then ${}^{0}\bar{A}'_{ej}$ can be determined by Eq. (1.35) with h = 0 and g = ej to give Eqs. (8.20), (8.21), (8.22) and (8.23):

$$
{}^{0}\bar{A}'_{ej} = \begin{bmatrix} I_{ejx} & J_{ejx} & K_{ejx} & 0 \\ I_{ejy} & J_{ejy} & K_{ejy} & 0 \\ I_{ejz} & J_{ejz} & K_{ejz} & 0 \\ f_{ejx} & f_{ejy} & f_{ejz} & 1 \end{bmatrix},
$$

with

$$
f_{ejx} = -(I_{ejx}t_{ejx} + I_{ejy}t_{ejy} + I_{ejz}t_{ejz}),
$$

$$
f_{ejy} = -(J_{ejx}t_{ejx} + J_{ejy}t_{ejy} + J_{ejz}t_{ejz}),
$$

$$
f_{ejz} = -(K_{ejx}t_{ejx} + K_{ejy}t_{ejy} + K_{ejz}t_{ejz}).
$$

(2) Determination of $d({}^{0}\bar{A}'_{ej})/d\bar{X}_{sys}$ from $d({}^{0}\bar{A}_{ej})/d\bar{X}_{sys}$:
Equations (8.20)–(8.23) indicate that if $d({}^{0}\bar{A}_{ej})/d\bar{X}_{sys}$ (see Appendix 1 of this chapter) is known, then $d({}^{0}\bar{A}'_{ej})/d\bar{X}_{sys}$ can be determined as

$$
\frac{d({}^{0}\bar{A}'_{ej})}{d\bar{X}_{sys}} = \left[\frac{\partial({}^{0}\bar{A}'_{ej})}{\partial x_v}\right]_{4\times4\times q_{sys}} = \begin{bmatrix} \partial I_{ejx}/\partial x_v & \partial J_{ejx}/\partial x_v & \partial K_{ejx}/\partial x_v & \bar{0} \\ \partial I_{ejy}/\partial x_v & \partial J_{ejy}/\partial x_v & \partial K_{ejy}/\partial x_v & \bar{0} \\ \partial I_{ejz}/\partial x_v & \partial J_{ejz}/\partial x_v & \partial K_{ejz}/\partial x_v & \bar{0} \\ \partial f_{ejx}/\partial x_v & \partial f_{ejy}/\partial x_v & \partial f_{ejz}/\partial x_v & \bar{0} \end{bmatrix}, \tag{8.77}
$$

with (see Eqs. (8.21), (8.22) and (8.23))

$$
\frac{\partial f_{ejx}}{\partial x_v} = -\left(\frac{\partial I_{ejx}}{\partial x_v}t_{ejx} + \frac{\partial I_{ejy}}{\partial x_v}t_{ejy} + \frac{\partial I_{ejz}}{\partial x_v}t_{ejz} + I_{ejx}\frac{\partial t_{ejx}}{\partial x_v} + I_{ejy}\frac{\partial t_{ejy}}{\partial x_v} + I_{ejz}\frac{\partial t_{ejz}}{\partial x_v}\right), \tag{8.78}
$$

$$
\frac{\partial f_{ejy}}{\partial x_v} = -\left(\frac{\partial J_{ejx}}{\partial x_v}t_{ejx} + \frac{\partial J_{ejy}}{\partial x_v}t_{ejy} + \frac{\partial J_{ejz}}{\partial x_v}t_{ejz} + J_{ejx}\frac{\partial t_{ejx}}{\partial x_v} + J_{ejy}\frac{\partial t_{ejy}}{\partial x_v} + J_{ejz}\frac{\partial t_{ejz}}{\partial x_v}\right), \tag{8.79}
$$

$$\frac{\partial f_{ejz}}{\partial x_v} = -\left(\frac{\partial K_{ejx}}{\partial x_v} t_{ejx} + \frac{\partial K_{ejy}}{\partial x_v} t_{ejy} + \frac{\partial K_{ejz}}{\partial x_v} t_{ejz} + K_{ejx}\frac{\partial t_{ejx}}{\partial x_v} + K_{ejy}\frac{\partial t_{ejy}}{\partial x_v} + K_{ejz}\frac{\partial t_{ejz}}{\partial x_v}\right). \tag{8.80}$$

(3) Determination of ${}^{ej}\bar{r}_i$ from ${}^{ej}\bar{A}_i$:
From Sect. 2.3, it is known that for a flat boundary surface, the boundary surface is expressed as ${}^{i}\bar{r}_i = [0 \quad 1 \quad 0 \quad 0]^T$ when referred to boundary coordinate frame $(xyz)_i$. From Eq. (1.34), the expression ${}^{ej}\bar{r}_i$ for this flat boundary surface is given as

$${}^{ej}\bar{r}_i = {}^{ej}\bar{A}_i' \begin{bmatrix} 0 \\ 1 \\ 0 \\ 0 \end{bmatrix} = \begin{bmatrix} {}^{ej}J_{ix} \\ {}^{ej}J_{iy} \\ {}^{ej}J_{iz} \\ {}^{ej}e_i \end{bmatrix}, \tag{8.81}$$

where ${}^{ej}\bar{A}_i$ is given in Eq. (3.3) (or Eq. (3.26)), and

$${}^{ej}e_i = -\left[({}^{ej}J_{ix})({}^{ej}t_{ix}) + ({}^{ej}J_{iy})({}^{ej}t_{iy}) + ({}^{ej}J_{iz})({}^{ej}t_{iz})\right]. \tag{8.82}$$

(4) Determination of $d({}^{ej}\bar{r}_i)/d\bar{X}_{sys}$ from $d({}^{ej}\bar{A}_i)/d\bar{X}_{sys}$:

Once the numerical values of the components of $d({}^{ej}\bar{A}_i)/d\bar{X}_{sys}$ have been determined (see Appendix 2 of this chapter), $d({}^{ej}\bar{r}_i)/d\bar{X}_{sys}$ can be obtained by differentiating Eq. (8.81) to give

$$\frac{d({}^{ej}\bar{r}_i)}{d\bar{X}_{sys}} = \begin{bmatrix} d({}^{ej}J_{ix})/d\bar{X}_{sys} \\ d({}^{ej}J_{iy})/d\bar{X}_{sys} \\ d({}^{ej}J_{iz})/d\bar{X}_{sys} \\ d({}^{ej}e_i)/d\bar{X}_{sys} \end{bmatrix}, \tag{8.83}$$

with

$$\begin{aligned} \frac{d({}^{ej}e_i)}{d\bar{X}_{sys}} = -&\left[\frac{d({}^{ej}J_{ix})}{d\bar{X}_{sys}}({}^{ej}t_{ix}) + ({}^{ej}J_{ix})\frac{d({}^{ej}t_{ix})}{d\bar{X}_{sys}} + \frac{d({}^{ej}J_{iy})}{d\bar{X}_{sys}}({}^{ej}t_{iy}) + ({}^{ej}J_{iy})\frac{d({}^{ej}t_{iy})}{d\bar{X}_{sys}}\right. \\ &\left. + \frac{d({}^{ej}J_{iz})}{d\bar{X}_{sys}}({}^{ej}t_{iz}) + ({}^{ej}J_{iz})\frac{d({}^{ej}t_{iz})}{d\bar{X}_{sys}}\right]. \end{aligned} \tag{8.84}$$

Having obtained numerical values of ${}^{0}\bar{A}_{ej}$, ${}^{ej}\bar{r}_i$, $d({}^{0}\bar{A}_{ej})/d\bar{X}_{sys}$ and $d({}^{ej}\bar{r}_i)/d\bar{X}_{sys}$, $d(\bar{r}_i)/d\bar{X}_{sys}$ for a flat boundary surface can be computed directly from Eq. (8.24).

Appendix 4

Let the elements of an optical system be labeled from $j = 1$ to $j = k$ sequentially, and the boundary surfaces be marked from $i = 1$ to $i = n$. In addition, let pose matrices ${}^{0}\bar{A}_{ej}$ ($j = 1$ to $j = k$) (Eqs. (3.1) and (3.25)) and ${}^{ej}\bar{A}_{i}$ ($i = 1$ to $i = n$) (Eqs. (3.3) and (3.26)) define the poses of the elements and boundary surfaces, respectively. The matrices ${}^{0}\bar{A}_{i}$ ($i = 1$ to $i = n$) required to perform raytracing can be obtained as (Eqs. (8.40), (8.41), (8.42) and (8.43))

$$
\begin{aligned}
{}^{0}\bar{A}_{i} &= \mathrm{tran}(t_{ix}, t_{iy}, t_{iz})\mathrm{rot}(\bar{z}, \omega_{iz})\mathrm{rot}(\bar{y}, \omega_{iy})\mathrm{rot}(\bar{x}, \omega_{ix}) = {}^{0}\bar{A}_{ej}\,{}^{ej}\bar{A}_{i} \\
&= \begin{bmatrix} I_{ejx} & J_{ejx} & K_{ejx} & t_{ejx} \\ I_{ejy} & J_{ejy} & K_{ejy} & t_{ejy} \\ I_{ejz} & J_{ejz} & K_{ejz} & t_{ejz} \\ 0 & 0 & 0 & 1 \end{bmatrix}
\begin{bmatrix} {}^{ej}I_{ix} & {}^{ej}J_{ix} & {}^{ej}K_{ix} & {}^{ej}t_{ix} \\ {}^{ej}I_{iy} & {}^{ej}J_{iy} & {}^{ej}K_{iy} & {}^{ej}t_{iy} \\ {}^{ej}I_{iz} & {}^{ej}J_{iz} & {}^{ej}K_{iz} & {}^{ej}t_{iz} \\ 0 & 0 & 0 & 1 \end{bmatrix} \\
&= \begin{bmatrix} I_{ix} & J_{ix} & K_{ix} & t_{ix} \\ I_{iy} & J_{iy} & K_{iy} & t_{iy} \\ I_{iz} & J_{iz} & K_{iz} & t_{iz} \\ 0 & 0 & 0 & 1 \end{bmatrix},
\end{aligned}
$$

where

$$t_{ix} = I_{ejx}({}^{ej}t_{ix}) + J_{ejx}({}^{ej}t_{iy}) + K_{ejx}({}^{ej}t_{iz}) + t_{ejx},$$

$$t_{iy} = I_{ejy}({}^{ej}t_{ix}) + J_{ejy}({}^{ej}t_{iy}) + K_{ejy}({}^{ej}t_{iz}) + t_{ejy},$$

$$t_{iz} = I_{ejz}({}^{ej}t_{ix}) + J_{ejz}({}^{ej}t_{iy}) + K_{ejz}({}^{ej}t_{iz}) + t_{ejz}.$$

The numerical values of the six boundary pose variables (t_{ix}, t_{iy}, t_{iz}, ω_{ix}, ω_{iy} and ω_{iz}) for each boundary surface can then be determined from Eqs. (1.38) to (1.43). Furthermore, $d({}^{0}\bar{A}_{i})/d\bar{X}_{sys}$ can be computed from Eq. (8.44) as

$$\frac{d({}^{0}\bar{A}_{i})}{d\bar{X}_{sys}} = \left[\frac{\partial({}^{0}\bar{A}_{i})}{\partial x_{v}}\right]_{4\times 4\times q_{sys}} = \begin{bmatrix} \partial I_{ix}/\partial x_{v} & \partial J_{ix}/\partial x_{v} & \partial K_{ix}/\partial x_{v} & \partial t_{ix}/\partial x_{v} \\ \partial I_{iy}/\partial x_{v} & \partial J_{iy}/\partial x_{v} & \partial K_{iy}/\partial x_{v} & \partial t_{iy}/\partial x_{v} \\ \partial I_{iz}/\partial x_{v} & \partial J_{iz}/\partial x_{v} & \partial K_{iz}/\partial x_{v} & \partial t_{iz}/\partial x_{v} \\ 0 & 0 & 0 & 0 \end{bmatrix}.$$

Now the Jacobian matrix $d\omega_{iz}/d\bar{X}_{sys} = [\partial\omega_{iz}/\partial x_{v}]$ can be computed by differentiating Eq. (1.38) to obtain

$$\frac{\partial\omega_{iz}}{\partial x_{v}} = \frac{E}{F}, \tag{8.85}$$

where

$$F = I_{ix}^2 + I_{iy}^2, \tag{8.86}$$

$$E = I_{ix}\frac{\partial I_{iy}}{\partial x_v} - I_{iy}\frac{\partial I_{ix}}{\partial x_v}. \tag{8.87}$$

Similarly, $d\omega_{iy}/d\bar{X}_{sys} = \left[\partial\omega_{iy}/\partial x_v\right]$ can be obtained by differentiating Eq. (1.39) to give

$$\frac{\partial\omega_{iy}}{\partial x_v} = \frac{G+H}{L}, \tag{8.88}$$

in which

$$L = I_{iz}^2 + (I_{ix}C\omega_{iz} + I_{iy}S\omega_{iz})^2, \tag{8.89}$$

$$G = I_{iz}\left[(-I_{ix}S\omega_{iz} + I_{iy}C\omega_{iz})\frac{\partial\omega_{iz}}{\partial x_v} + \left(\frac{\partial I_{ix}}{\partial x_v}C\omega_{iz} + \frac{\partial I_{iy}}{\partial x_v}S\omega_{iz}\right)\right], \tag{8.90}$$

$$H = -\left(I_{ix}C\omega_{iz} + I_{iy}S\omega_{iz}\right)\frac{\partial I_{iz}}{\partial x_v}. \tag{8.91}$$

Finally, $d\omega_{ix}/d\bar{X}_{sys} = [\partial\omega_{ix}/\partial x_v]$ can be obtained by differentiating Eq. (1.40) to give

$$\frac{\partial\omega_{ix}}{\partial x_v} = \frac{PT - UQ}{M}, \tag{8.92}$$

where

$$M = (K_{ix}S\omega_{iz} - K_{iy}C\omega_{iz})^2 + (-J_{ix}S\omega_{iz} + J_{iy}C\omega_{iz})^2, \tag{8.93}$$

$$P = -J_{ix}S\omega_{iz} + J_{iy}C\omega_{iz}, \tag{8.94}$$

$$T = (K_{ix}C\omega_{iz} + K_{iy}S\omega_{iz})\frac{\partial\omega_{iz}}{\partial x_v} + \left(\frac{\partial K_{ix}}{\partial x_v}S\omega_{iz} - \frac{\partial K_{iy}}{\partial x_v}C\omega_{iz}\right), \tag{8.95}$$

$$U = K_{ix}S\omega_{iz} - K_{iy}C\omega_{iz}, \tag{8.96}$$

$$Q = (-J_{ix}C\omega_{iz} - J_{iy}S\omega_{iz})\frac{\partial\omega_{iz}}{\partial x_v} + \left(-\frac{\partial J_{ix}}{\partial x_v}S\omega_{iz} + \frac{\partial J_{iy}}{\partial x_v}C\omega_{iz}\right). \tag{8.97}$$

It is noted from Eqs. (8.58) to (8.60) of Appendix 1 and Eqs. (8.74)–(8.76) of Appendix 2 that the (1,4)th, (2,4)th and (3,4)th components of matrices ${}^0\bar{A}_{ej}$ and

${}^{ej}\bar{A}_i$ are always given in explicit form. As a result, their Jacobian matrices with respect to $\bar{X}_{sys}$ can be determined simply by direct differentiation. Consequently, $dt_{ix}/d\bar{X}_{sys}$, $dt_{iy}/d\bar{X}_{sys}$, and $dt_{iz}/d\bar{X}_{sys}$ can be obtained from Eqs. (8.41) to (8.43) as

$$\begin{aligned}\frac{\partial t_{ix}}{\partial x_v} &= \frac{\partial I_{ejx}}{\partial x_v}\left({}^{ej}t_{ix}\right) + \frac{\partial J_{ejx}}{\partial x_v}\left({}^{ej}t_{iy}\right) + \frac{\partial K_{ejx}}{\partial x_v}\left({}^{ej}t_{iz}\right) \\ &+ \frac{\partial\left(t_{ejx}\right)}{\partial x_v} + I_{ejx}\frac{\partial\left({}^{ej}t_{ix}\right)}{\partial x_v} + J_{ejx}\frac{\partial\left({}^{ej}t_{iy}\right)}{\partial x_v} + K_{ejx}\frac{\partial\left({}^{ej}t_{iz}\right)}{\partial x_v},\end{aligned} \tag{8.98}$$

$$\begin{aligned}\frac{\partial t_{iy}}{\partial x_v} &= \frac{\partial I_{ejy}}{\partial x_v}\left({}^{ej}t_{ix}\right) + \frac{\partial J_{ejy}}{\partial x_v}\left({}^{ej}t_{iy}\right) + \frac{\partial K_{ejy}}{\partial x_v}\left({}^{ej}t_{iz}\right) \\ &+ \frac{\partial\left(t_{ejy}\right)}{\partial x_v} + I_{ejy}\frac{\partial\left({}^{ej}t_{ix}\right)}{\partial x_v} + J_{ejy}\frac{\partial\left({}^{ej}t_{iy}\right)}{\partial x_v} + K_{ejy}\frac{\partial\left({}^{ej}t_{iz}\right)}{\partial x_v},\end{aligned} \tag{8.99}$$

$$\begin{aligned}\frac{\partial t_{iz}}{\partial x_v} &= \frac{\partial I_{ejz}}{\partial x_v}\left({}^{ej}t_{ix}\right) + \frac{\partial J_{ejz}}{\partial x_v}\left({}^{ej}t_{iy}\right) + \frac{\partial K_{ejz}}{\partial x_v}\left({}^{ej}t_{iz}\right) \\ &+ \frac{\partial\left(t_{ejz}\right)}{\partial x_v} + I_{ejz}\frac{\partial\left({}^{ej}t_{ix}\right)}{\partial x_v} + J_{ejz}\frac{\partial\left({}^{ej}t_{iy}\right)}{\partial x_v} + K_{ejz}\frac{\partial\left({}^{ej}t_{iz}\right)}{\partial x_v}.\end{aligned} \tag{8.100}$$

References

1. Lin PD (2013a) Analysis and design of prisms using the derivatives of a ray, part II: the derivatives of boundary variable vector with respect to system variable vector. Appl Opt 52:4151–4162
2. Lin PD (2013b) Design of optical systems using derivatives of rays: derivatives of variable vector of spherical boundary surfaces with respect to system variable vector. Appl Opt 52:7271–7287

Chapter 9
Prism Analysis

Prisms are common optical elements containing only flat boundary surfaces. In spectral instruments, their function is to disperse light, while in other applications, they are used to displace, deviate or re-orientate an image. Prisms are highly stable systems and are far more robust toward environmental changes of angle than assemblages of mirrors on a metal support block. The propagation of light through prism structures is generally investigated by means of ray tracing. However, in this chapter, the characteristics and performance of common optical prisms are examined using a ray Jacobian matrix approach.

9.1 Retro-reflectors

Retro-reflectors are optical systems designed to reflect a ray back to its source with minimum scattering. Common retro-reflectors include corner-cube mirrors (Fig. 3.19), solid glass corner-cubes (Fig. 9.1), cat's eyes (Fig. 3.12), cat's eyes with primary lenses (Fig. 3.13), cat's eyes with primary mirrors (Fig. 1 of [1]), and transparent spheres with silver linings (Fig. 4.9). However, only corner-cube mirrors and solid glass corner-cubes offer a true retroflection performance.

9.1.1 Corner-Cube Mirror

As shown in Fig. 3.19, a corner-cube mirror consists of three first-surface-mirrors arranged perpendicularly to one another, and juxtaposed so as to form the corner of

P.D. Lin, *Advanced Geometrical Optics*, Progress in Optical Science
and Photonics 4, DOI 10.1007/978-981-10-2299-9_9

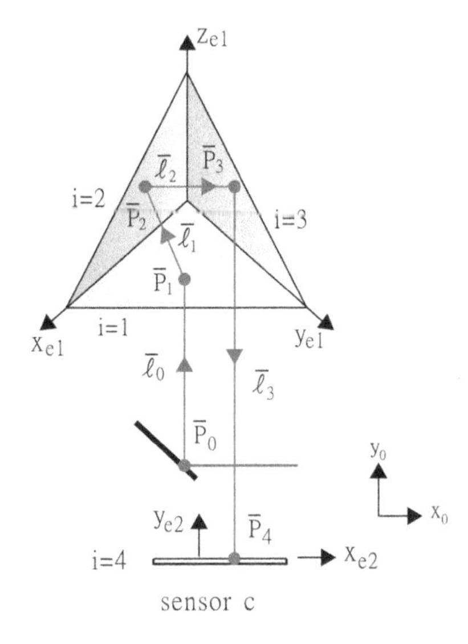

Fig. 9.1 Model of sensor c used in typical motion measurement system

a cube [2]. The retro-reflective property of corner-cube mirrors has been confirmed in Example 7.1. Due to their perfectly retro-reflective properties, corner-cube mirrors are used in many motion measurement systems. A typical example is that shown in Fig. 9.1 (a sub-system of the system shown in Fig. 3.30) comprising a corner-cube mirror and a sensor c. In the following discussions, the equations for the sensor readings of sensor c are determined using a strict skew-ray tracing method. Let the corner-cube mirror and sensor c be denoted as element j = 1 and element j = 2, respectively. From Figs. 3.28 and 9.1, the following pose matrices are obtained:

$$^{0}\bar{A}_{a1} = \text{tran}(-f_x - h_x, q_0, 0)\text{tran}(\Delta t_{a1x}, \Delta t_{a1y}, \Delta t_{a1z})\text{rot}(\bar{z}, \Delta\omega_{a1z})\text{rot}(\bar{y}, \Delta\omega_{a1y})\text{rot}(\bar{x}, \Delta\omega_{a1x}) \tag{9.1}$$

$$^{a1}\bar{A}_{e1} = \begin{bmatrix} -1/\sqrt{2} & 1/\sqrt{2} & 0 & 0 \\ -1/\sqrt{3} & -1/\sqrt{3} & -1/\sqrt{3} & 0 \\ -1/\sqrt{6} & -1/\sqrt{6} & 2/\sqrt{6} & 0 \\ 0 & 0 & 0 & 1 \end{bmatrix}, \tag{9.2}$$

$$^{0}\bar{A}_{e2} = \text{tran}\left(-f_x - h_x, -g_y, 0\right), \tag{9.3}$$

where Δt_{a1x} , Δt_{a1y}, Δt_{a1z}, $\Delta\omega_{a1x}$, $\Delta\omega_{a1y}$ and $\Delta\omega_{a1z}$ are the error motions which are to be measured. Equations (9.1) and (9.3) respectively express the pose matrices of Assembly a = 1 and element 2 with respect to the world coordinate frame $(xyz)_0$, while Eq. (9.2) define the pose of element 1 with the coordinate frame $(xyz)_{a1}$. The reflective boundary surfaces $\bar{r}_i$ (i = 1 to i = 3) required to perform ray tracing can be obtained from $\bar{r}_i = \left(({}^{0}\bar{A}_{a1}{}^{a1}\bar{A}_{e1})^{-1}\right)^{T} {}^{e1}\bar{r}_i$ (Eq. (1.34) with h = 0 and g = e1), where the three boundary surfaces ${}^{e1}\bar{r}_i$ (i = 1 to i = 3) are given in Example 3.17. Setting g = 4 in Eq. (7.55), the following Jacobian matrices are obtained:

$$\begin{bmatrix} \partial P_{4x}/\partial t_{e1x} & \partial P_{4x}/\partial t_{e1y} & \partial P_{4x}/\partial t_{e1z} \\ \partial P_{4y}/\partial t_{e1x} & \partial P_{4y}/\partial t_{e1y} & \partial P_{4y}/\partial t_{e1z} \\ \partial P_{4z}/\partial t_{e1x} & \partial P_{4z}/\partial t_{e1y} & \partial P_{4z}/\partial t_{e1z} \end{bmatrix} = \begin{bmatrix} 2 & 0 & 0 \\ 0 & 0 & 0 \\ 0 & 0 & 2 \end{bmatrix}, \tag{9.4}$$

$$\begin{bmatrix} \partial P_{4x}/\partial \omega_{e1x} & \partial P_{4x}/\partial \omega_{e1y} & \partial P_{4x}/\partial \omega_{e1z} \\ \partial P_{4y}/\partial \omega_{e1x} & \partial P_{4y}/\partial \omega_{e1y} & \partial P_{4y}/\partial \omega_{e1z} \\ \partial P_{4z}/\partial \omega_{e1x} & \partial P_{4z}/\partial \omega_{e1y} & \partial P_{4z}/\partial \omega_{e1z} \end{bmatrix} = \begin{bmatrix} 0 & 0 & 0 \\ 0 & 0 & 0 \\ 0 & 0 & 0 \end{bmatrix}. \tag{9.5}$$

Importantly, Eqs. (9.4) and (9.5) are valid irrespective of the values of the six error motions. Furthermore, the two equations indicate that the readings of sensor c, which are determined by Eq. (3.46), are given as

$$\begin{bmatrix} \Delta X \\ \Delta Z \end{bmatrix} = \begin{bmatrix} 2t_{e1x} \\ 2t_{e1z} \end{bmatrix}. \tag{9.6}$$

Most researchers use a right-angle prism, i.e., a 2-D retro-reflector, to derive the readings of a sensor hit by the reflected ray from a corner-cube mirror. However, their results are identical to those obtained from Eq. (9.6) based on a strict skew-ray tracing approach.

9.1.2 Solid Glass Corner-Cube

The solid glass corner-cube (Fig. 9.2) is a glass-filled half-cube, cut along the diagonal, and designed to provide enhanced dimensional stability under temperature excursions. Mathematically, the corner-cube mirror shown in Fig. 3.19 is a special case of the solid glass corner-cube with $N_1 = \xi_0/\xi_{e1} = N_5 = \xi_{e1}/\xi_0 = 1$ where ξ_{e1} is the refractive index of the corner-cube material. The retro-reflective property of a solid glass corner-cube can be proven numerically by means of Eq. (7.17) with g = 5, i.e.,

$$\frac{\partial \bar{\ell}_5}{\partial \bar{\ell}_0} = \frac{\partial \bar{\ell}_5(\bar{n}_5, N_5)}{\partial \bar{\ell}_4} \frac{\partial \bar{\ell}_4}{\partial \bar{\ell}_3} \frac{\partial \bar{\ell}_3}{\partial \bar{\ell}_2} \frac{\partial \bar{\ell}_2}{\partial \bar{\ell}_1} \frac{\partial \bar{\ell}_1(\bar{n}_1, N_1)}{\partial \bar{\ell}_0} = \begin{bmatrix} -1 & 0 & 0 \\ 0 & -1 & 0 \\ 0 & 0 & -1 \end{bmatrix}, \tag{9.7}$$

in which the unit normal vectors are given by

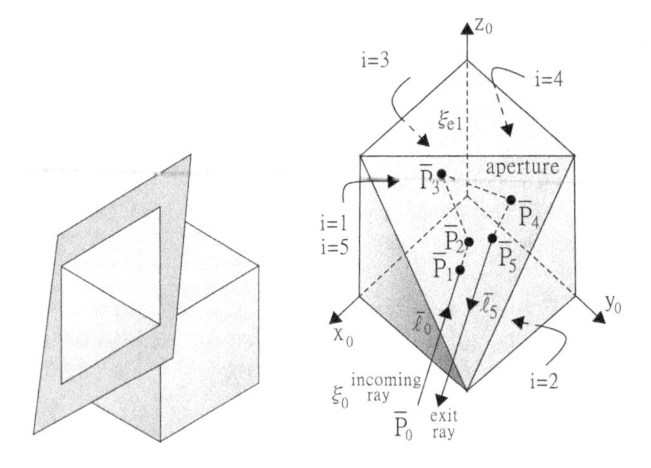

Fig. 9.2 *Solid glass* corner-cube is a glass-filled half-cube, cut along the diagonal

$$\bar{n}_1 = \bar{n}_5 = \pm\left[1/\sqrt{3} \quad 1/\sqrt{3} \quad 1/\sqrt{3} \quad 0\right]^T, \tag{9.8}$$

$$\bar{n}_2 = \pm[0 \quad 0 \quad 1 \quad 0]^T, \tag{9.9}$$

$$\bar{n}_3 = \pm[0 \quad 1 \quad 0 \quad 0]^T, \tag{9.10}$$

$$\bar{n}_4 = \pm[1 \quad 0 \quad 0 \quad 0]^T. \tag{9.11}$$

9.2 Dispersing Prisms

Dispersion is defined as the spreading of white light into its full spectrum of wavelengths through a prism. The performance of a dispersing prism can be characterized by the deviation angle ψ, the dispersion angle Δ, and the spectral dispersion $\partial\psi/\partial\upsilon$, where υ is the wavelength. In general, the deviation angle ψ of a 3-D prism is the angle between the incident ray $\bar{R}_0$ entering the first boundary surface and the exit ray $\bar{R}_n$ exiting the final boundary surface. The deviation angle ψ can thus be computed as

$$\psi = \mathrm{Cos}^{-1}\left(\bar{\ell}_0 \cdot \bar{\ell}_n\right) = \mathrm{Cos}^{-1}\left(\ell_{0x}\ell_{nx} + \ell_{0y}\ell_{ny} + \ell_{0z}\ell_{nz}\right). \tag{9.12}$$

By contrast, the dispersion angle Δ is the difference between the two extreme deviation angles transmitted by the prism, i.e.,

$$\Delta = \psi_{max} - \psi_{min}. \tag{9.13}$$

The variation of the deviation angle ψ with the wavelength υ is called the spectral dispersion, and is determined as $\partial\psi/\partial\upsilon$. The following sections discuss five of the most common dispersive prisms, namely the triangular prism, the Pellin-Broca prism, the dispersive Abbe prism, the achromatic prism, and the direct vision prism.

9.2.1 *Triangular Prism*

The mathematical modeling of a triangular prism has been given in Example 2.4 (Fig. 2.15). This section further discusses the dispersion performance of a triangular prism. As discussed in Example 2.4, the deviation angle ψ of a triangular prism is a function of the refractive index ξ_{e1} of the prism material. Meanwhile, the spectral dispersion is given as

$$\frac{\partial\psi}{\partial\upsilon} = \frac{\xi_{e1} S\eta_{e1}}{\sqrt{\xi_{e1}^2 - (S\theta_1)^2}\sqrt{1 - \left(S\eta_{e1}\sqrt{\xi_{e1}^2 - (S\theta_1)^2} - C\eta_{e1} S\theta_1\right)^2}} \frac{\partial\xi_{e1}}{\partial\upsilon}. \tag{9.14}$$

It is noted that the spectral dispersion is a function of the vertex angle η_{e1}, incidence angle θ_1, refractive index ξ_{e1}, and index of dispersion $\partial\xi_{e1}/\partial\upsilon$ of the prism material. For a triangular prism fabricated of hard crown glass K5 [3], the spectral dispersion is determined numerically to be $\partial\psi/\partial\upsilon = -0.14275$ given a vertex angle of $\eta_{e1} = 60°$, an incidence angle of $\theta_1 = 60°$, a Cauchy equation of $\xi_{e1} = 1.5220 + 0.00459/\upsilon^2$, and a wavelength of $\upsilon = 0.45\ \mu m$.

The spectral dispersion can also be determined via the Jacobian matrix $\partial\bar{\ell}_2/\partial\xi_{e1}$. Assuming the same set of parameters as those described above (i.e., $\eta_{e1} = 60°$, $\upsilon = 0.45\ \mu m$, $\theta_1 = 60°$), and given a point source $\bar{P}_0 = [0 \quad -5 \quad -10 \quad 1]^T$, the numerical values of the exit ray $\bar{R}_2$ and Jacobian matrix $\partial\bar{\ell}_2/\partial\xi_{e1}$ are obtained respectively as

$$\bar{R}_2 = [0 \quad 1.06801 \quad -0.61622 \quad 0 \quad 0.95335 \quad 0.30187]^T \tag{9.15}$$

and

$$\frac{\partial\bar{\ell}_2}{\partial\xi_{e1}} = \begin{bmatrix} \partial\ell_{2x}/\partial\xi_{e1} \\ \partial\ell_{2y}/\partial\xi_{e1} \\ \partial\ell_{2z}/\partial\xi_{e1} \end{bmatrix} = \begin{bmatrix} 0 \\ 0.44917 \\ -1.32303 \end{bmatrix}. \tag{9.16}$$

Taking $\psi = \mathrm{Cos}^{-1}(\bar{\ell}_0 \cdot \bar{\ell}_2)$ (in which $\bar{\ell}_0$ and $\bar{\ell}_2$ are both unit vectors), and applying Cauchy's equation for hard crown glass K5 [3], the spectral dispersion is obtained as

$$\frac{\partial\psi}{\partial\upsilon} = \frac{\partial\psi}{\partial\xi_{e1}}\frac{\partial\xi_{e1}}{\partial\upsilon} = \frac{-1}{\sqrt{1-(\bar{\ell}_0 \cdot \bar{\ell}_2)^2}}\left(\bar{\ell}_0 \cdot \frac{\partial\bar{\ell}_2}{\partial\xi_{e1}}\right)\frac{\partial\xi_{e1}}{\partial\upsilon} = -0.14275. \tag{9.17}$$

It is noted that this value is identical to that obtained from Eq. (9.14).

9.2.2 Pellin-Broca Prism and Dispersive Abbe Prism

For the triangular prism described above, the deviation angle is a function of the refractive index ξ_{e1} of the prism material. However, for certain prisms, the minimum deviation angle is insensitive to the refractive index, and hence to the wavelength. Figures 9.3 and 9.4 show two typical constant-deviation dispersing prisms, namely the Pellin-Broca prism and the Abbe prism. The Pellin-Broca prism consists of a four-sided block of glass shaped as a right prism with 90°, 75°, 135° and 60° angles on the end faces. As shown, a ray enters the prism through face EF, undergoes total internal reflection at face FG, and then exits the prism through face EH. If collimated rays of white light are incident on the prism and observation is performed at a fixed direction of 90° with respect to the incident rays, the observed wavelength depends only on the orientation of the prism and the observation angle represents the minimum deviation angle ψ_{mini} of the wavelength. In practice, ψ_{mini} can be computed as $\partial\psi/\partial\theta_1$, where the deviation angle is given by $\psi = \mathrm{Cos}^{-1}(\bar{\ell}_0 \cdot \bar{\ell}_3)$. From basic geometric principles, a value of $\psi_{mini} = 90°$ is obtained for the prism shown in Fig. 9.3. As the prism is rotated around axis o_{e1}, i.e., the line of intersection of the bisector of ∠FEH and the reflecting face FG, the selected wavelength (which is deviated by 90°) changes without any change in the prism

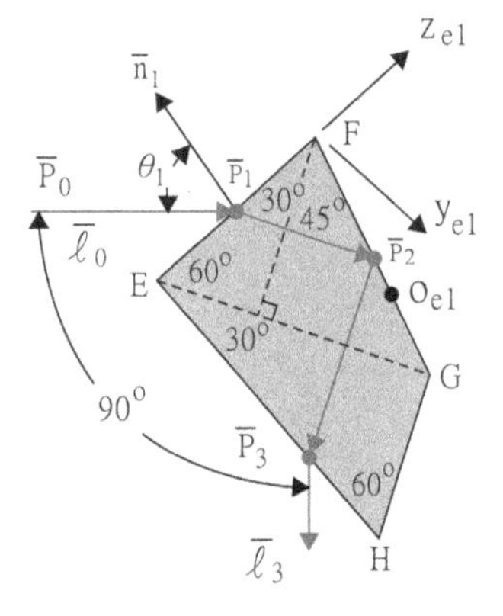

Fig. 9.3 Pellin-Broca prism

Fig. 9.4 Abbe prism

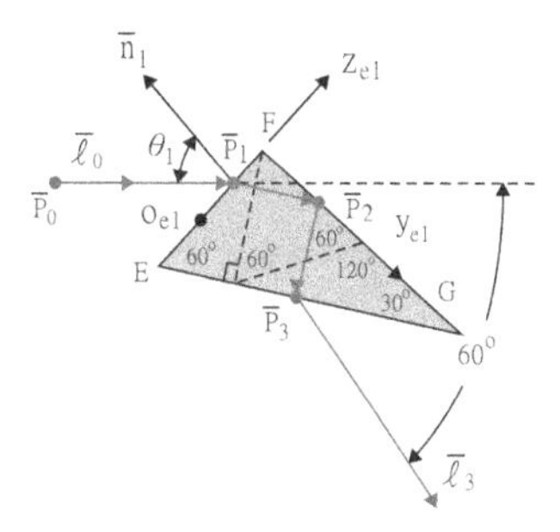

geometry or relative positions of the input and output rays. The Pellin-Broca prism is thus commonly used to separate a single required wavelength from a bundle of rays containing multiple wavelengths.

The Abbe prism shown in Fig. 9.4 consists of a block of glass forming a right prism with 30°–60°–90° triangular faces. When in use, a bundle of collimated white rays is refracted at face EF, undergoes total internal reflection at face FG, and is refracted once again on exiting face EG. The prism shown in Fig. 9.4 is designed in such a way that only one particular wavelength of the incident light exits the prism at a deviation angle of exactly 60°. (Note that 60° is the minimum possible deviation angle ψ_{mini} for the prism, and is determined once again as $\partial\psi/\partial\theta_1$, where $\psi = \mathrm{Cos}^{-1}(\bar{\ell}_0 \cdot \bar{\ell}_3)$.) By rotating the prism (in the plane of the page) around any point o_{e1} on face EF, the wavelength which is deviated by 60° is isolated from the other wavelengths in the bundle and can thus be separately observed.

9.2.3 *Achromatic Prism and Direct Vision Prism*

Achromatic prisms and direct vision prisms are both compound dispersing prisms [4–7]. In other words, they consist of a set of prisms placed in contact with one another and then (generally) cemented together to form a solid assembly. Compound prisms provide the ability to tune the prism dispersion so as to achieve a greater dispersion or higher-order dispersion effects. For example, achromatic prisms produce an angular deviation of the light beam without introducing any chromatic dispersion in the neighborhood of the chosen wavelength. As shown in Fig. 9.5, the exit rays are not coincident, but are almost parallel. In other words, the prism induces a constant angular deviation angle ψ. In practice, an achromatic prism consists of one high-dispersion glass prism and one low-dispersion glass prism. Let the desired deviation be specified as $\psi_{desired} = 20°$ and the desired dispersion be set as zero. Determining the deviation angle ψ from $C\psi = \bar{\ell}_0 \cdot \bar{\ell}_2$ and the dispersion as $d\psi/d\upsilon$ (where υ is the wavelength), these design requirements can be formulated as

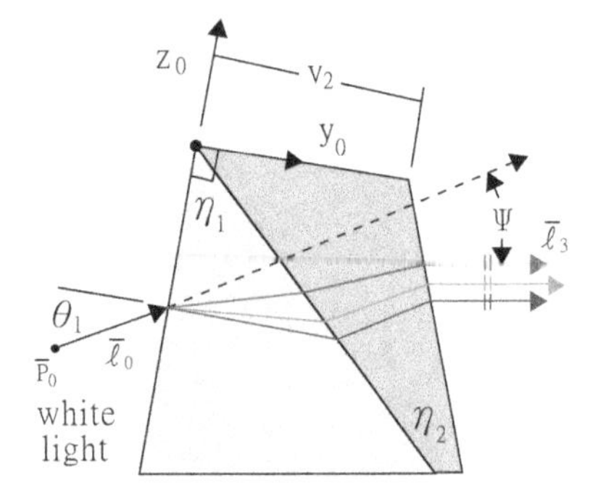

Fig. 9.5 Exit rays of an achromatic prism are not coincident but are parallel, indicating a constant angular deviation angle

$$\psi = \mathrm{Cos}^{-1}(\bar{\ell}_0 \cdot \bar{\ell}_2) = \psi_{\text{desired}}, \tag{9.18}$$

$$\begin{aligned}\frac{\partial \psi}{\partial \upsilon} &= \frac{-1}{\sqrt{1-(\bar{\ell}_0 \cdot \bar{\ell}_3)^2}}\left(\bar{\ell}_0 \cdot \frac{\partial \bar{\ell}_3}{\partial \upsilon}\right) \\ &= \frac{-1}{\sqrt{1-(\bar{\ell}_0 \cdot \bar{\ell}_3)^2}}\left[\bar{\ell}_0 \cdot \left(\frac{\partial \bar{\ell}_3}{\partial \xi_{e1}}\frac{\partial \xi_{e1}}{\partial \upsilon} + \frac{\partial \bar{\ell}_3}{\partial \xi_{e2}}\frac{\partial \xi_{e2}}{\partial \upsilon}\right)\right] = 0.\end{aligned} \tag{9.19}$$

If the first and second elements of the prism (with $v_2 = 15$ mm) are fabricated of dense flint glass SF10 and fused silica, respectively, then the apex angles required to satisfy these design requirements are found to be $\eta_{e1} = 10.6519°$ and $\eta_{e2} = 33.6798°$, respectively, given a chosen wavelength of $\upsilon = 0.55\mu m$ and a source ray $\bar{P}_0 = [0 \quad -5 \quad -10 \quad 1]^T$ with unit directional vector $\bar{\ell}_0 = [\mathrm{C}30°\mathrm{C}90°;\ \mathrm{C}30°\mathrm{S}90° \quad \mathrm{S}30° \quad 0]^T$. The corresponding solutions obtained using the thin prism method (p. 95 of [8] and Appendix 1 of this chapter) are $\eta_{e1} = 10.1589°$ and $\eta_{e2} = 59.2851°$, respectively. These solutions deviate notably from the exact values given above since the Snell's law $\xi_{i-1}\ \mathrm{S}\theta_i = \xi_i\ \mathrm{S}\underline{\theta}_i$ is simplified as $\xi_{i-1}\theta_i = \xi_i\underline{\theta}_i$ in thin prism method.

For a direct vision prism (see Fig. 9.6), the aim is to produce a desired spectral dispersion D_{desired} without any deviation of the ray at the chosen wavelength. In this case, the design task involves determining the apex angles of the two elements which achieve the desired deviation angle ψ_{desired} of zero while preserving the required spectral dispersion value of $(\partial\psi/\partial\upsilon)_{\text{desired}}$ at the chosen wavelength. In other words, the design constraints are formulated as

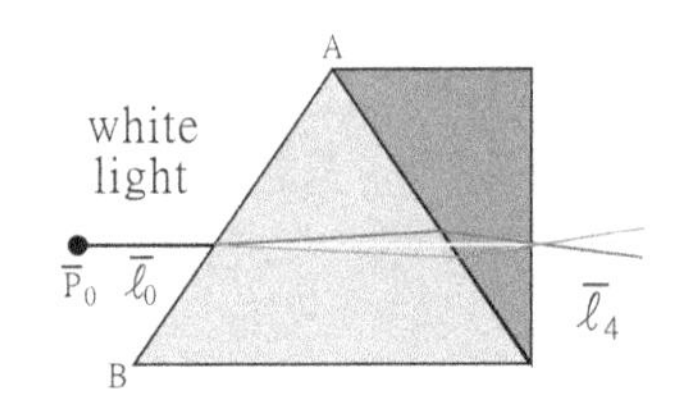

Fig. 9.6 Direct-vision prism disperses incident light into its spectral components without deviating light at the central wavelength

$$\bar{\Phi}_1 = \psi - \psi_{desired} = \mathrm{Cos}^{-1}\left(\bar{\ell}_0 \cdot \bar{\ell}_3\right) = 0 \tag{9.20}$$

$$\bar{\Phi}_2 = \frac{\partial \psi}{\partial \upsilon} - \left(\frac{\partial \psi}{\partial \upsilon}\right)_{desired} = 0. \tag{9.21}$$

Assuming that the first and second elements are made of dense flint glass SF10 and fused silica, respectively, and the chosen wavelength is $\upsilon = 0.55\,\mu\text{m}$, the corresponding apex angles are found to be $\eta_{e1} = 24.5751°$ and $\eta_{e2} = 37.4923°$ for $(\partial\psi/\partial\upsilon)_{desired} = -0.05$ and a source ray $\bar{P}_0 = [0 \quad -5 \quad -10 \quad 1]^T$ with unit directional vector $\bar{\ell}_0 = [C30°C90° \quad C30°S90° \quad S30° \quad 0]^T$. Notably, the thin prism method fails to provide a reasonable outcome in this case.

The following discussions are focused on the analysis of image orientation when objects are imaged by prisms without considering their dispersion. These common prisms include right-angle prisms, double Porro prism, Porro-Abbe prism, Abbe-Koenig prism, roofed Pechan prism, roofed Pechan prism, and Penta prism.

9.3 Right-Angle Prisms

In the right-angle prism shown in Fig. 9.7, the collimated rays enter one of the smaller prism faces at a perpendicular angle and undergo total internal reflection at the hypotenuse face. To determine the image orientation of the object, it is first necessary to establish the world coordinate frame $(xyz)_0$ (referred as object in the study of image orientation) in order to define the orientation of the object before the prism. For simplicity, assume that the object $(xyz)_0$ always emits light rays in the y_0 direction and these rays enter the prism through the first flat boundary surface. The

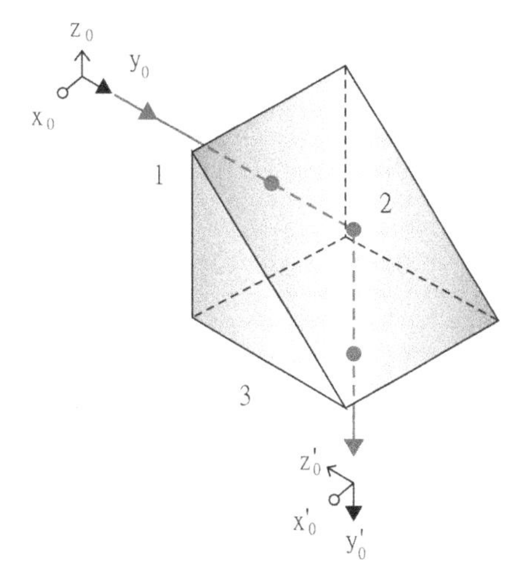

Fig. 9.7 Use of *right-angle* prism to re-direct rays and rotate image

object's image $(xyz)'_0$ will indicate the orientation of the image after the light rays pass through the prism. Let the following unit normal vectors be defined for boundary surfaces i = 1 to i = 3:

$$\bar{n}_1 = \pm[0 \quad -1 \quad 0 \quad 0]^T, \tag{9.22}$$

$$\bar{n}_2 = \pm\left[0 \quad -1/\sqrt{2} \quad -1/\sqrt{2} \quad 0\right]^T, \tag{9.23}$$

$$\bar{n}_3 = \pm[0 \quad 0 \quad -1 \quad 0]^T. \tag{9.24}$$

The image orientation with respect to $(xyz)_0$ can then be obtained by means of Eq. (7.17) with g = 3, i.e.,

$$\frac{\partial \bar{\ell}_3}{\partial \bar{\ell}_0} = \frac{\partial \bar{\ell}_3(\bar{n}_3, N_3)}{\partial \bar{\ell}_2} \frac{\partial \bar{\ell}_2}{\partial \bar{\ell}_1} \frac{\partial \bar{\ell}_1(\bar{n}_1, N_1)}{\partial \bar{\ell}_0} = \begin{bmatrix} 1 & 0 & 0 \\ 0 & 0 & -1 \\ 0 & -1 & 0 \end{bmatrix}. \tag{9.25}$$

The image orientations before and after the prism can be seen from the object and its image, $(xyz)_0$ and $(xyz)'_0$, whose axes have different patterns with arrow, circle, and crossbar.

9.4 Porro Prism

Figure 9.8 shows a parallel bundle of rays passing through a single Porro prism. As shown, the rays enter the prism through the hypotenuse face, undergo reflection at the other two faces, and then exit the prism through the hypotenuse face once again. Since, in this case, the rays exit and enter the prism at normal incidence, the prism is not dispersive. Furthermore, since the incoming rays are perpendicular to the hypotenuse face, the following unit normal vectors are obtained for the four boundary surfaces:

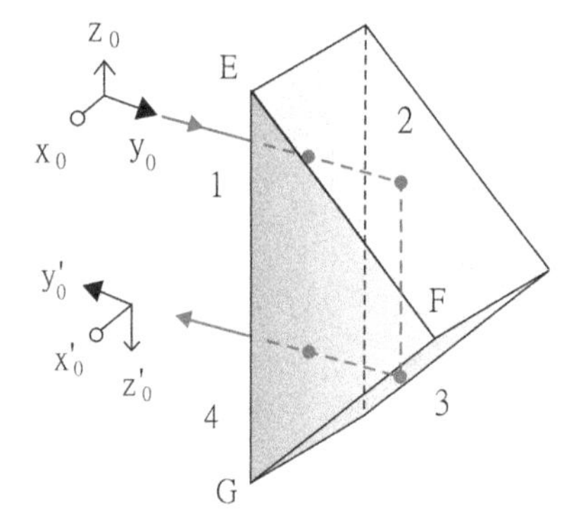

Fig. 9.8 Single Porro prism

$$\bar{n}_1 = \bar{n}_4 = \pm[0 \quad -1 \quad 0 \quad 0]^T, \tag{9.26}$$

$$\bar{n}_2 = \pm\left[0 \quad -1/\sqrt{2} \quad -1/\sqrt{2} \quad 0\right]^T, \tag{9.27}$$

$$\bar{n}_3 = \pm\left[0 \quad -1/\sqrt{2} \quad 1/\sqrt{2} \quad 0\right]^T. \tag{9.28}$$

The image orientation is determined from Eq. (7.17) with g = 4 as

$$\frac{\partial \bar{\ell}_4}{\partial \bar{\ell}_0} = \frac{\partial \bar{\ell}_4(\bar{n}_4, N_4)}{\partial \bar{\ell}_3} \frac{\partial \bar{\ell}_3}{\partial \bar{\ell}_2} \frac{\partial \bar{\ell}_2}{\partial \bar{\ell}_1} \frac{\partial \bar{\ell}_1(\bar{n}_1, N_1)}{\partial \bar{\ell}_0} = \begin{bmatrix} 1 & 0 & 0 \\ 0 & -1 & 0 \\ 0 & 0 & -1 \end{bmatrix}. \tag{9.29}$$

Equation (9.29) shows that a Porro prism inverts the image from top to bottom if the incoming rays enter the prism perpendicularly. Numerical results show that Porro prisms are constant-deviation prisms, in that each incoming ray is reflected through exactly 180°. However, the exit ray is retro-reflective if and only if the incoming ray lies on a plane parallel to face EFG. It is noted from Eq. (9.29) that the first column of Eq. (9.29) has the form $[1 \quad 0 \quad 0]^T$, indicating that the prism is a 2-D retro-reflector. As discussed later in this chapter, Porro prisms are most often used in pairs to erect the image of an object (see Fig. 9.11).

9.5 Dove Prism

A Dove prism (Fig. 9.9) is a truncated version of a right-angle prism and is used almost exclusively for collimated light since for convergent light, it produces substantial astigmatism. The unit normal vectors ${}^{e1}\bar{n}_i$ (i = 1, 2, 3) of the three boundary surfaces of the prism with respect to $(xyz)_{e1}$ are given by

$${}^{e1}\bar{n}_1 = \pm[0 \quad 1 \quad 0 \quad 0]^T, \tag{9.30}$$

$${}^{e1}\bar{n}_2 = \pm\left[0 \quad 1/\sqrt{2} \quad 1/\sqrt{2} \quad 0\right]^T, \tag{9.31}$$

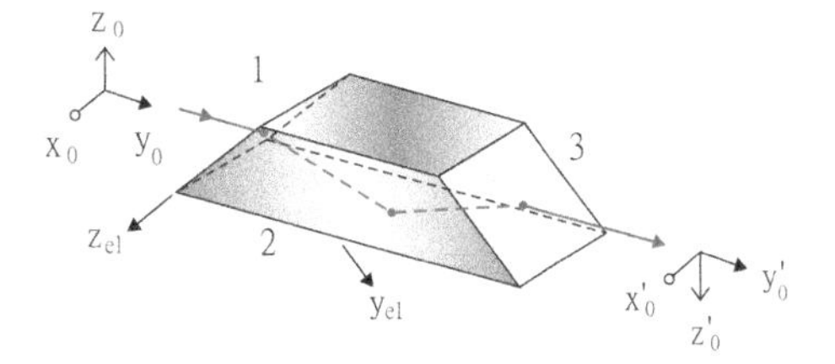

Fig. 9.9 Dove prism

$$ {}^{e1}\bar{n}_3 = \pm[0 \quad 0 \quad 1 \quad 0]^T. \tag{9.32} $$

From Fig. 9.9, the pose matrix ${}^0\bar{A}_{e1}$ of $(xyz)_{e1}$ with respect to $(xyz)_0$ is given by

$$ {}^0\bar{A}_{e1} = tran(0, t_{e1y}, 0)rot(\bar{y}, \omega_{e1y})rot(\bar{z}, 180°)rot(\bar{x}, -135°). $$

when $\omega_{e1y} = 0$, the image orientation with respect to $(xyz)_0$ is given by

$$ \frac{\partial\bar{\ell}_3}{\partial\bar{\ell}_0} = \frac{\partial\bar{\ell}_3(\bar{n}_3, N_3)}{\partial\bar{\ell}_2}\frac{\partial\bar{\ell}_2}{\partial\bar{\ell}_1}\frac{\partial\bar{\ell}_1(\bar{n}_1, N_1)}{\partial\bar{\ell}_0} = \begin{bmatrix} 1 & 0 & 0 \\ 0 & 1 & 0 \\ 0 & 0 & -1 \end{bmatrix}, \tag{9.33} $$

where $\bar{n}_i = {}^0\bar{A}_{e1}{}^{e1}\bar{n}_i$. Equation (9.33) indicates that the object image is inverted from top to bottom but not left to right when $\omega_{e1y} = 0$.

The Dove prism has the interesting property of rotating the image twice as fast as it is itself rotated about the y_0 axis. This can be proven numerically by defining

$$ \begin{bmatrix} \partial\bar{\ell}_3/\partial\bar{\ell}_0 & \bar{0}_{3\times1} \\ \bar{0}_{1\times3} & 1 \end{bmatrix} = rot(\bar{z}, \theta_z)rot(\bar{y}, \theta_y)rot(\bar{x}, \theta_x). \tag{9.34} $$

Equation (9.34) enables the determination of θ_y from Eq. (1.39) and the subsequent proof of $\partial\theta_y/\partial\omega_{e1y} = 2$. It is noted that the second-order derivatives of the system variables are required to complete the proof.

9.6 Roofed Amici Prism

The roofed Amici prism (Fig. 9.10) is essentially a truncated right-angle prism with a roof section added to the hypotenuse face. Such a prism both reverts and inverts the image, and bends the line of sight through 90°. These optical properties of a roofed Amici prism can be proven using the following unit normal vectors of the boundary surfaces:

$$ \bar{n}_1 = \pm[0 \quad -1 \quad 0 \quad 0]^T, \tag{9.35} $$

$$ \bar{n}_2 = \pm[1/\sqrt{2} \quad 1/2 \quad 1/2 \quad 0]^T, \tag{9.36} $$

$$ \bar{n}_3 = \pm[-1/\sqrt{2} \quad 1/2 \quad 1/2 \quad 0]^T, \tag{9.37} $$

$$ \bar{n}_4 = \pm[0 \quad 0 \quad 1 \quad 0]^T. \tag{9.38} $$

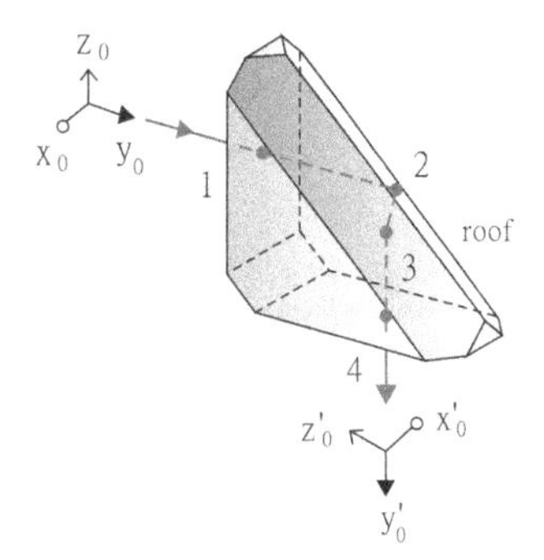

Fig. 9.10 Roofed Amici prism

When the rays enter and exit the faces perpendicularly, the image orientation with respect to $(xyz)_0$ is given by

$$\frac{\partial \bar{\ell}_4}{\partial \bar{\ell}_0} = \frac{\partial \bar{\ell}_4(\bar{n}_4, N_4)}{\partial \bar{\ell}_3} \frac{\partial \bar{\ell}_3}{\partial \bar{\ell}_2} \frac{\partial \bar{\ell}_2}{\partial \bar{\ell}_1} \frac{\partial \bar{\ell}_1(\bar{n}_1, N_1)}{\partial \bar{\ell}_0} = \begin{bmatrix} -1 & 0 & 0 \\ 0 & 0 & -1 \\ 0 & -1 & 0 \end{bmatrix}. \tag{9.39}$$

In other words, the image is both reverted and inverted, as shown in Fig. 9.10.

9.7 Erecting Prisms

In an ordinary telescope, the objective lens forms an inverted image of the object, which is then viewed through the eyepiece. The image seen by the eye is both upside down and reversed from left to right. To avoid the inconvenience of viewing an inverted image, an erecting system is often used to re-invert the image to its proper orientation. This system may have the form of either a lens system or a prism system. The following sections consider four typical prism-based erecting systems, namely the double Porro prism, the Porro-Abbe prism, the Abbe-Koenig prism, and the roofed Pechan prism. (Note that other erecting prisms, such as the Schmidt prism, Leman prism, Goerz prism, roofed Amici prism, and roofed delta prism, are discussed in [8]).

9.7.1 Double Porro Prism

The most commonly used prism-erecting system is the double Porro prism, shown in Fig. 9.11. The system consists of two Porro prisms oriented at 90° to one another. The first prism inverts the image from top to bottom while the second prism reverses the image from left to right. Importantly, the optical axis is displaced laterally, but is not deviated. The unit normal vectors of the boundary surfaces are given by

$$\bar{n}_1 = \pm[0 \quad -1 \quad 0 \quad 0]^T, \tag{9.40}$$

$$\bar{n}_2 = \pm\left[0 \quad -1/\sqrt{2} \quad -1/\sqrt{2} \quad 0\right]^T, \tag{9.41}$$

$$\bar{n}_3 = \pm\left[0 \quad 1/\sqrt{2} \quad 1/\sqrt{2} \quad 0\right]^T, \tag{9.42}$$

$$\bar{n}_4 = \pm[0 \quad 1 \quad 0 \quad 0]^T, \tag{9.43}$$

$$\bar{n}_5 = \pm[0 \quad 1 \quad 0 \quad 0]^T, \tag{9.44}$$

$$\bar{n}_6 = \pm\left[1/\sqrt{2} \quad 1/\sqrt{2} \quad 0 \quad 0\right]^T, \tag{9.45}$$

$$\bar{n}_7 = \pm\left[-1/\sqrt{2} \quad 1/\sqrt{2} \quad 0 \quad 0\right]^T, \tag{9.46}$$

$$\bar{n}_8 = \pm[0 \quad -1 \quad 0 \quad 0]^T. \tag{9.47}$$

Meanwhile, the image orientation with respect to $(xyz)_0$ is given by

$$\frac{\partial \bar{\ell}_8}{\partial \bar{\ell}_0} = \frac{\partial \bar{\ell}_8(\bar{n}_8, N_8)}{\partial \bar{\ell}_7} \frac{\partial \bar{\ell}_7}{\partial \bar{\ell}_6} \frac{\partial \bar{\ell}_6}{\partial \bar{\ell}_5} \frac{\partial \bar{\ell}_5(\bar{n}_5, N_5)}{\partial \bar{\ell}_4} \frac{\partial \bar{\ell}_4(\bar{n}_4, N_4)}{\partial \bar{\ell}_3} \frac{\partial \bar{\ell}_3}{\partial \bar{\ell}_2} \frac{\partial \bar{\ell}_2}{\partial \bar{\ell}_1} \frac{\partial \bar{\ell}_1(\bar{n}_1, N_1)}{\partial \bar{\ell}_0} = \begin{bmatrix} -1 & 0 & 0 \\ 0 & 1 & 0 \\ 0 & 0 & -1 \end{bmatrix}. \tag{9.48}$$

Equation (9.48) indicates that $(xyz)'_0$, i.e., the image of $(xyz)_0$, is inverted from top to bottom and left to right. As shown in Fig. 9.11, for a telescope containing such a system, the final image has the same orientation as the object.

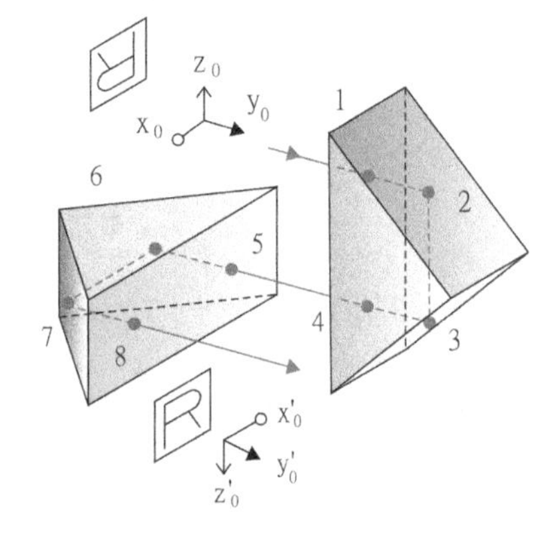

Fig. 9.11 Double Porro prism

9.7.2 Porro-Abbe Prism

The Porro-Abbe prism (Fig. 9.12) is somewhat more difficult to fabricate than the double Porro prism. However, in some applications, its compactness, and the fact that the prisms can be readily cemented together, offer compensating advantages. The unit normal vectors of the boundary surfaces have the forms:

$$\bar{n}_1 = \pm[0 \quad -1 \quad 0 \quad 0]^T, \tag{9.49}$$

$$\bar{n}_2 = \pm\left[0 \quad -1/\sqrt{2} \quad -1/\sqrt{2} \quad 0\right]^T, \tag{9.50}$$

$$\bar{n}_3 = \pm\left[1/\sqrt{2} \quad 0 \quad 1/\sqrt{2} \quad 0\right]^T, \tag{9.51}$$

$$\bar{n}_4 = \pm\left[-1/\sqrt{2} \quad 0 \quad 1/\sqrt{2} \quad 0\right]^T, \tag{9.52}$$

$$\bar{n}_5 = \pm\left[0 \quad 1/\sqrt{2} \quad -1/\sqrt{2} \quad 0\right]^T, \tag{9.53}$$

$$\bar{n}_6 = \pm[0 \quad -1 \quad 0 \quad 0]^T. \tag{9.54}$$

Meanwhile, the image orientation with respect to $(xyz)_0$ is given by

$$\frac{\partial\bar{\ell}_6}{\partial\bar{\ell}_0} = \frac{\partial\bar{\ell}_6(\bar{n}_6, N_6)}{\partial\bar{\ell}_5}\frac{\partial\bar{\ell}_5}{\partial\bar{\ell}_4}\frac{\partial\bar{\ell}_4}{\partial\bar{\ell}_3}\frac{\partial\bar{\ell}_3}{\partial\bar{\ell}_2}\frac{\partial\bar{\ell}_2}{\partial\bar{\ell}_1}\frac{\partial\bar{\ell}_1(\bar{n}_1, N_1)}{\partial\bar{\ell}_0} = \begin{bmatrix} -1 & 0 & 0 \\ 0 & 1 & 0 \\ 0 & 0 & -1 \end{bmatrix}. \tag{9.55}$$

Equation (9.55) confirms that the image is rotated through 180° as it passes through the prism. Due to their relatively small size, Porro-Abbe systems are often used as the image-erecting system in handheld binocular devices.

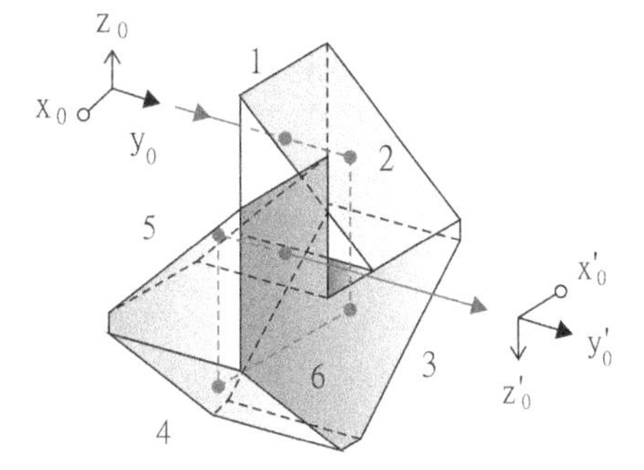

Fig. 9.12 Porro-Abbe prism

9.7.3 Abbe-Koenig Prism

The Abbe-Koenig prism (Fig. 9.13) is not only less bulky than the double Porro and Porro-Abbe prisms, but also has the advantage of erecting the image without displacing the axis. In operation, rays enter one face at normal incidence, are internally reflected from a 30° sloped face, reflected from a roof, reflected from the opposite 30° face, and finally exit the second face, again at normal incidence. The net effect of the internal reflections is to flip the image both vertically and horizontally. This can be proven by using the following unit normal vectors

$$\bar{n}_1 = \pm[0 \quad -1 \quad 0 \quad 0]^T, \tag{9.56}$$

$$\bar{n}_2 = \pm\left[0 \quad -1/2 \quad -\sqrt{3}/2 \quad 0\right]^T, \tag{9.57}$$

$$\bar{n}_3 = \pm\left[-1/\sqrt{2} \quad 0 \quad 1/\sqrt{2} \quad 0\right]^T, \tag{9.58}$$

$$\bar{n}_4 = \pm\left[1/\sqrt{2} \quad 0 \quad 1/\sqrt{2} \quad 0\right]^T, \tag{9.59}$$

$$\bar{n}_5 = \pm\left[0 \quad 1/2 \quad -\sqrt{3}/2 \quad 0\right]^T, \tag{9.60}$$

$$\bar{n}_6 = \pm[0 \quad -1 \quad 0 \quad 0]^T, \tag{9.61}$$

to compute the image orientation change by

$$\frac{\partial \bar{\ell}_6}{\partial \bar{\ell}_0} = \frac{\partial \bar{\ell}_6(\bar{n}_6, N_6)}{\partial \bar{\ell}_5} \frac{\partial \bar{\ell}_5}{\partial \bar{\ell}_4} \frac{\partial \bar{\ell}_4}{\partial \bar{\ell}_3} \frac{\partial \bar{\ell}_3}{\partial \bar{\ell}_2} \frac{\partial \bar{\ell}_2}{\partial \bar{\ell}_1} \frac{\partial \bar{\ell}_1(\bar{n}_1, N_1)}{\partial \bar{\ell}_0} = \begin{bmatrix} -1 & 0 & 0 \\ 0 & 1 & 0 \\ 0 & 0 & -1 \end{bmatrix}. \tag{9.62}$$

Equation (9.62) proves that the Abbe-Koenig prism flips the image both vertically and horizontally. If the prism is made without a roof, it inverts the image in one direction only, just as the Dove prism. However, since its entrance and exit faces are normal to the system axis, it can be placed in a converging beam without introducing astigmatism.

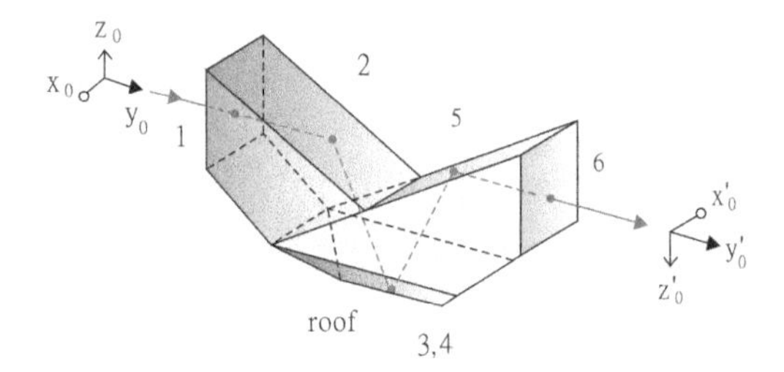

Fig. 9.13 Abbe-Koenig prism

9.7.4 Roofed Pechan Prism

The roofed Pechan prism (Fig. 9.14) comprises two prisms separated by an air gap. The design of the two prisms is such that the entrance rays and exit rays are coaxial, i.e., the prism does not deviate the incoming rays if they are centered on the optical axis. The unit normal vectors of the boundary surfaces are given as

$$\bar{n}_1 = \pm[0 \quad -1 \quad 0 \quad 0]^T, \tag{9.63}$$

$$\bar{n}_2 = \pm\left[0 \quad -1/\sqrt{2} \quad -1/\sqrt{2} \quad 0\right]^T, \tag{9.64}$$

$$\bar{n}_3 = \pm[0 \quad S22.5^\circ \quad C22.5^\circ \quad 0]^T = \pm\left[0 \quad \sqrt{2-\sqrt{2}}/2 \quad \sqrt{2+\sqrt{2}}/2 \quad 0\right]^T, \tag{9.65}$$

$$\bar{n}_4 = \bar{n}_5 = \pm\left[0 \quad -1/\sqrt{2} \quad -1/\sqrt{2} \quad 0\right]^T, \tag{9.66}$$

$$\bar{n}_6 = \pm[0 \quad -1 \quad 0 \quad 0]^T, \tag{9.67}$$

$$\begin{aligned} \bar{n}_7 &= \pm[-C45^\circ \quad S22.5^\circ S45^\circ \quad -C22.5^\circ S45^\circ \quad 0]^T \\ &= \pm\left[-1/\sqrt{2} \quad \sqrt{\sqrt{2}-1}/2 \quad -\sqrt{\sqrt{2}+1}/2 \quad 0\right]^T, \end{aligned} \tag{9.68}$$

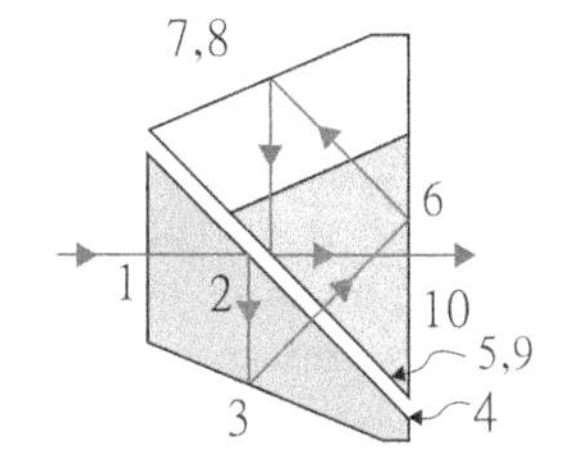

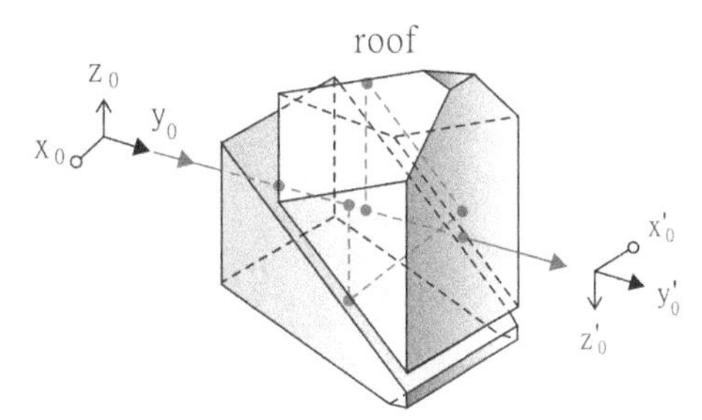

Fig. 9.14 Roofed Pechan prism

$$\begin{aligned}\bar{n}_8 &= \pm[\,C45^\circ \quad S22.5^\circ S45^\circ \quad -C22.5^\circ S45^\circ \quad 0\,]^T \\ &= \pm\left[1/\sqrt{2} \quad \sqrt{\sqrt{2}-1}/2 \quad -\sqrt{\sqrt{2}+1}/2 \quad 0\right]^T,\end{aligned} \tag{9.69}$$

$$\bar{n}_9 = \bar{n}_5, \tag{9.70}$$

$$\bar{n}_{10} = \bar{n}_6. \tag{9.71}$$

Meanwhile, the image orientation with respect to $(xyz)_0$ is given by

$$\begin{aligned}\frac{\partial\bar{\ell}_{10}}{\partial\bar{\ell}_0} &= \frac{\partial\bar{\ell}_{10}(\bar{n}_{10}, N_{10})}{\partial\bar{\ell}_9}\frac{\partial\bar{\ell}_9}{\partial\bar{\ell}_8}\frac{\partial\bar{\ell}_8}{\partial\bar{\ell}_7}\frac{\partial\bar{\ell}_7}{\partial\bar{\ell}_6}\frac{\partial\bar{\ell}_6}{\partial\bar{\ell}_5}\frac{\partial\bar{\ell}_5(\bar{n}_5, N_5)}{\partial\bar{\ell}_4}\frac{\partial\bar{\ell}_4(\bar{n}_4, N_4)}{\partial\bar{\ell}_3}\frac{\partial\bar{\ell}_3}{\partial\bar{\ell}_2}\frac{\partial\bar{\ell}_2}{\partial\bar{\ell}_1}\frac{\partial\bar{\ell}_1(\bar{n}_1, N_1)}{\partial\bar{\ell}_0} \\ &= \begin{bmatrix} -1 & 0 & 0 \\ 0 & 1 & 0 \\ 0 & 0 & -1 \end{bmatrix}.\end{aligned} \tag{9.72}$$

Equation (9.72) proves that the roofed Pechan prism both inverts and reverts the image. By contrast, a roofless Pechan prism inverts or reverts the image (depending on the orientation of the prism), but does not perform both.

9.8 Penta Prism

The Penta prism (Fig. 9.15) is used when it is required to produce an exact 90° deviation of the incoming rays without having to orient the prism too precisely. Penta prisms are often used as the end reflectors in rangefinders or in optical tooling and precise alignment work where it is necessary to establish an exact 90° angle. In investigating the image orientation in a Penta prism, the following unit normal vectors of the boundary surfaces are required:

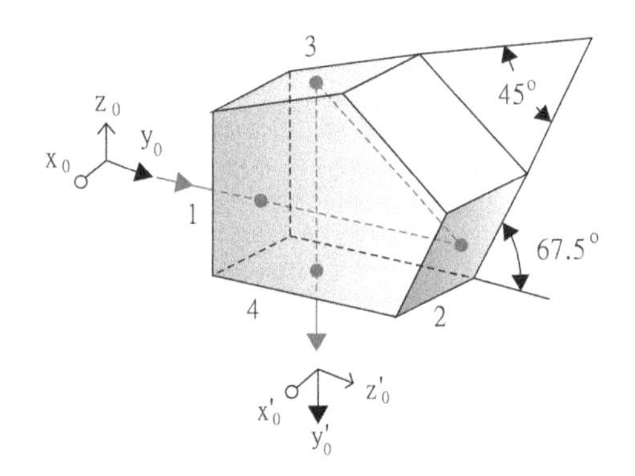

Fig. 9.15 Penta prism

$$\bar{n}_1 = \pm[0 \quad -1 \quad 0 \quad 0]^T, \tag{9.73}$$

$$\begin{aligned}\bar{n}_2 &= \pm[0 \quad -S67.5^\circ \quad C67.5^\circ \quad 0]^T \\ &= \pm\left[0 \quad -\sqrt{(2+\sqrt{2})/2} \quad \sqrt{(2-\sqrt{2})/2} \quad 0\right]^T,\end{aligned} \tag{9.74}$$

$$\begin{aligned}\bar{n}_3 &= \pm[0 \quad C67.5^\circ \quad -S67.5^\circ \quad 0]^T \\ &= \pm\left[0 \quad \sqrt{(2-\sqrt{2})/2} \quad -\sqrt{(2+\sqrt{2})/2} \quad 0\right]^T,\end{aligned} \tag{9.75}$$

$$\bar{n}_4 = \pm[0 \quad 0 \quad 1 \quad 0]^T. \tag{9.76}$$

When the incoming rays are perpendicular to the first flat boundary surface, the image orientation with respect to $(xyz)_0$ is given by

$$\frac{\partial \bar{\ell}_4}{\partial \bar{\ell}_0} = \frac{\partial \bar{\ell}_4(\bar{n}_4, N_4)}{\partial \bar{\ell}_3} \frac{\partial \bar{\ell}_3}{\partial \bar{\ell}_2} \frac{\partial \bar{\ell}_2}{\partial \bar{\ell}_1} \frac{\partial \bar{\ell}_1(\bar{n}_1, N_1)}{\partial \bar{\ell}_0} = \begin{bmatrix} 1 & 0 & 0 \\ 0 & 0 & 1 \\ 0 & -1 & 0 \end{bmatrix}. \tag{9.77}$$

It can be proven numerically that, even if the incoming ray is not orientated at 90° to the prism, it is still bent through a 90° angle as it travels through the prism. In addition, it is noted that one of the reflecting faces of the Penta prism can be replaced with a roof in order to invert the image in a different direction.

Appendix 1

Consider a particular element (labeled as j with vertex angle η_{ej} and refractive index ξ_{ej}, Fig. 2.15) in a compound prism. From an inspection of Fig. 2.15, the deviation angle ψ_j for the element is obtained as

$$\psi_j = \theta_1 - \underline{\theta}_1 + \underline{\theta}_2 - \theta_2. \tag{9.78}$$

Assuming that the incidence angles θ_1 and θ_2 are sufficiently small, the relations $\theta_1 = \xi_{ej}\underline{\theta}_1$ and $\xi_{ej}\theta_2 = \underline{\theta}_2$ can be obtained from the approximated Snell's law. Furthermore, the apex angle η_{ej} of the prism is given by the sum of θ_2 and $\underline{\theta}_1$, i.e., $\eta_{ej} = \theta_2 + \underline{\theta}_1$. Consequently, the deviation angle ψ_j can be approximated as

$$\psi_j = \eta_{ej}(\xi_{ej} - 1). \tag{9.79}$$

The spectral dispersion of the element is estimated as (p. 94 of [8])

$$D_j = \frac{\eta_{ej}(\xi_{ej} - 1)}{\upsilon_{ej}}, \tag{9.80}$$

where υ_{ej} is a basic number used to characterize optical material of the *j*th element, and is called the Abbe V number or simply the V-value.

Equations (9.79) and (9.80) give the deviation angle ψ_j and spectral dispersion D_j of a ray traveling through a single element in a compound prism comprising k elements. The deviation angle ψ and spectral dispersion D of a ray emerging from this prism can be estimated respectively as

$$\psi = \sum_{j=1}^{k} \psi_j = \sum_{j=1}^{k} \eta_{ej}(\xi_{ej} - 1), \tag{9.81}$$

$$D = \sum_{j=1}^{k} D_j = \sum_{j=1}^{k} \frac{\eta_{ej}(\xi_{ej} - 1)}{\upsilon_j}, \tag{9.82}$$

where the apex angle η_{ej} is taken to be positive if the second flat boundary surface $\bar{r}_2$ is rotated counter-clockwise with respect to the first flat boundary surface $\bar{r}_1$, and vice versa.

Let the desired deviation angle and spectral dispersion of a compound prism consisting of two elements be denoted as $\Psi_{desired}$ and $D_{desired}$, respectively. The required apex angles, η_{e1} and η_{e2}, of the two elements can be determined from Eqs. (9.81) and (9.82) with k = 2, to give

$$\eta_{e1} = \frac{\Psi_{desired}\upsilon_{e1} - D_{desired}\upsilon_{e1}\upsilon_{e2}}{(\upsilon_{e1} - \upsilon_{e2})(\xi_{e1} - 1)}, \tag{9.89}$$

$$\eta_{e2} = \frac{D_{desired}\upsilon_{e1}\upsilon_{e2} - \Psi_{desired}\upsilon_{e2}}{(\upsilon_{e1} - \upsilon_{e2})(\xi_{e2} - 1)}, \tag{9.84}$$

where υ_{e1} and υ_{e2} are the Abbe V numbers of element j = 1 and j = 2, respectively.

References

1. Snyder JJ (1975) Paraxial ray analysis of a cat's-eye retroreflector. Appl Opt 14:1825–1828
2. Ni J, Wu S (1993) An on-line measurement technique for machine volumetric error compensation. ASME J Eng Ind 115:85–92
3. Marcuse D (1980) Pulse distortion in single-mode fibers. Appl Opt 19:1653–1860
4. Murty MVRK, Narasimham AL (1970) Some new direct vision dispersion prism systems. Appl Opt 9:859–862
5. Hagen N, Tkaczyk TS (2011) Compound prism design principles, I. Appl Opt 50:4998–5011

6. Hagen N, Tkaczyk TS (2011) Compound prism design priciples, II: triplet and janssen prisms. Appl Opt 50:5012–5022
7. Hagen N, Tkaczyk TS (2011) compound prism design principles, III: linear-wavenumber and optical coherence tomography prisms. Appl Opt 50:5023–5030
8. Smith WJ (2001) Modern optical engineering, 3rd edn. Edmund Industrial Optics, Barrington

Chapter 10
Prism Design Based on Image Orientation

One of the main functions of prisms is that of image reorientation. In designing a prism to produce an image with a particular orientation, most researchers use the trial-and-error method proposed by Smith (p. 100 of [1]), in which a pencil oriented in the pose of the object is used to approach the reflecting surface such that the pencil striking and rebounding from the surface allows the simulation of the image orientation (Fig. 10.1). However, the image orientation following transmission through a prism can be determined more accurately by tracing two rays originating from different points on the object surface (see Fig. 10.2). In addition, Galvez [2] recently proposed a method for analyzing the image orientation using a geometric phase concept. However, these methods are rather awkward. Accordingly, this chapter provides an analytical method based on Eq. (7.17) for solving the prism design problem for a particular image orientation in a more systematic, efficient and accurate manner.

10.1 Reflector Matrix and Image Orientation Function

The discussions in this chapter are valid for all problems involving parallel rays passing through prisms, which comprise only flat boundary surfaces. Assume that the prism shown in Fig. 8.2 has n flat boundary surfaces labeled from i = 1 to i = n, respectively (see also Fig. 10.3). It is noted that for such a prism, the first and last flat boundary surfaces, i.e., the surfaces at which the ray enters and exits the prism, are denoted as i = 1 and i = n, respectively. As shown in Fig. 10.3, to accurately determine the image orientation by tracing a single ray, it is first necessary to

P.D. Lin, *Advanced Geometrical Optics*, Progress in Optical Science and Photonics 4, DOI 10.1007/978-981-10-2299-9_10

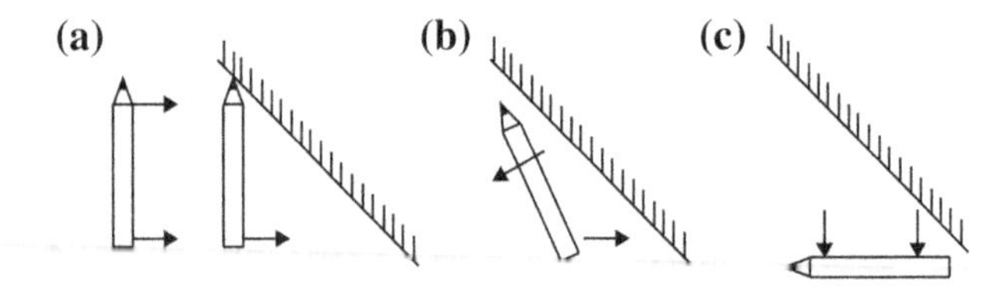

Fig. 10.1 Use of a *pencil* oriented in the pose of the object to approach the reflecting surface and simulate the resulting image orientation

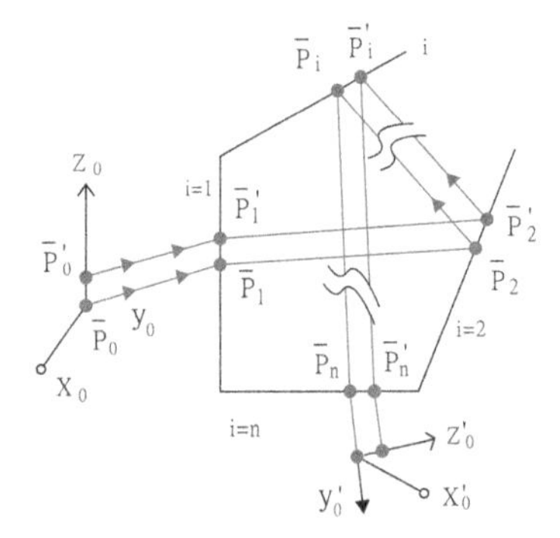

Fig. 10.2 Determination of image orientation by tracing *two rays* originating from different points on the object surface

establish the world coordinate frame $(xyz)_0$ embedded in the object in order to mark its orientation. The image of $(xyz)_0$ can provide an easy reference for any subsequent changes in the image orientation caused by passage through the prism. For simplicity, assume that the object emits parallel source rays $\bar{R}_0$ in the y_0 direction of $(xyz)_0$. In other words, the rays have a unit directional vector of $\bar{\ell}_0 = [0 \quad 1 \quad 0 \quad 0]^T$. The image of $(xyz)_0$ (denoted as $(xyz)'_0$ in consistence with paraxial optics) indicates the orientation of the image formed after the rays pass through the prism (Fig. 10.4). Applying Eq. (7.17) with g = n, and assuming that the rays enter and exit the prism at boundary surfaces i = 1 and i = n, respectively, it is readily shown that

$$\begin{aligned}\bar{\Phi} &= \begin{bmatrix} \bar{a} & \bar{b} & \bar{c} \end{bmatrix} = \begin{bmatrix} a_x & b_x & c_x \\ a_y & b_y & c_y \\ a_z & b_z & c_z \end{bmatrix} \\ &= \frac{\partial \bar{\ell}_n}{\partial \bar{\ell}_0} = \frac{\partial \bar{\ell}_n(\bar{n}_n, N_n)}{\partial \bar{\ell}_{n-1}} \frac{\partial \bar{\ell}_{n-1}(\bar{n}_{n-1})}{\partial \bar{\ell}_{n-2}} \cdots \frac{\partial \bar{\ell}_i(\bar{n}_i)}{\partial \bar{\ell}_{i-1}} \cdots \frac{\partial \bar{\ell}_2(\bar{n}_2)}{\partial \bar{\ell}_1} \frac{\partial \bar{\ell}_1(\bar{n}_1, N_1)}{\partial \bar{\ell}_0},\end{aligned} \tag{10.1}$$

where $\bar{a}$, $\bar{b}$ and $\bar{c}$ are the unit directional vectors of the x'_0, y'_0 and z'_0 axes of $(xyz)'_0$, respectively, referred with respect to $(xyz)_0$. Mathematically, $\bar{\Phi}$ represents the orientation part of matrix ${}^0\bar{A}_{0'}$, i.e., the pose matrix of $(xyz)'_0$ with respect to $(xyz)_0$. That is,

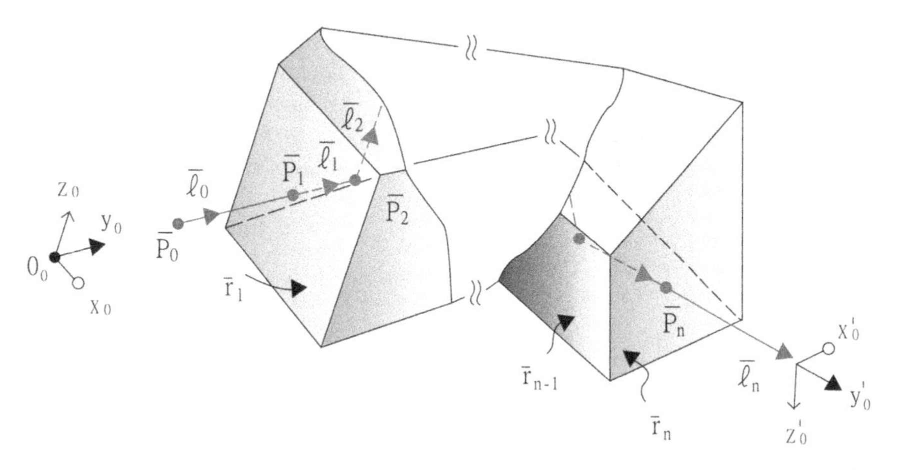

Fig. 10.3 Image orientation of object $(xyz)_0$ imaged by a prism

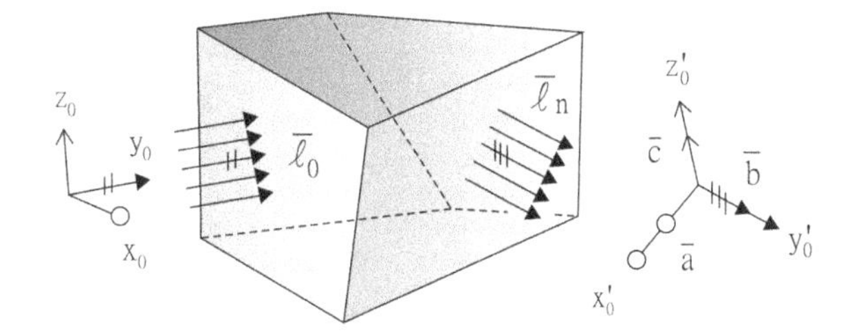

Fig. 10.4 Components $\bar{a}$, $\bar{b}$ and $\bar{c}$ of image orientation function $\bar{\Phi}$ are the unit directional vectors of the x'_0, y'_0 and z'_0 axes, respectively, with respect to $(xyz)_0$

$$
{}^{0}\bar{A}_{0'} = \begin{bmatrix} \bar{\Phi} & \bar{t}_{3\times1} \\ \bar{0}_{1\times3} & 1 \end{bmatrix} = \begin{bmatrix} a_x & b_x & c_x & t_x \\ a_y & b_y & c_y & t_y \\ a_z & b_z & c_z & t_z \\ 0 & 0 & 0 & 1 \end{bmatrix} \tag{10.2}
$$

where the position vector $\bar{t}_{3\times1}$ is of no interest. Let $\bar{\ell}_n$ be the unit directional vector of the exit ray expressed with respect to $(xyz)_0$ and let ${}^{0'}\bar{\ell}_n = [0 \quad 1 \quad 0 \quad 0]^T$ be the unit directional vector of the exit ray with respect to $(xyz)'_0$. Since $\bar{\ell}_0 = [0 \quad 1 \quad 0 \quad 0]^T$, it follows that

$$
\bar{\ell}_n = {}^{0}\bar{A}_{0'}\,{}^{0'}\bar{\ell}_n = {}^{0}\bar{A}_{0'}\,\bar{\ell}_0 = \begin{bmatrix} \bar{\Phi} & \bar{t}_{3\times1} \\ \bar{0}_{1\times3} & 1 \end{bmatrix} \bar{\ell}_0 = \begin{bmatrix} b_x \\ b_y \\ b_z \\ 0 \end{bmatrix} \tag{10.3}
$$

Before proceeding, it is appropriate to formalize the conditions under which Eq. (10.3) is valid: (1) $\bar{\Phi}$ is the orientation component of the pose matrix of $(xyz)'_0$ with respect to $(xyz)_0$; (2) $\bar{\ell}_0 = [0 \quad 1 \quad 0 \quad 0]^T$ lies in the y_0 direction; and

(3) $\bar{\ell}_n = [b_x \quad b_y \quad b_z \quad 0]^T$ is the direction of the y'_0 axis with respect to $(xyz)_0$. The above three conditions are the foundations of the following discussions.

Equation (10.1) can be applied to determine the change in orientation of the image as it passes through a prism. More specifically, for a reflection process, $\partial\bar{\ell}_i/\partial\bar{\ell}_{i-1}$ (i = 2 to i = n − 1) is given by Eq. (7.18) as

$$\frac{\partial\bar{\ell}_i}{\partial\bar{\ell}_{i-1}} = \frac{\partial\bar{\ell}_i(\bar{n}_i)}{\partial\bar{\ell}_{i-1}} = \begin{bmatrix} 1-2n_{ix}n_{ix} & -2n_{ix}n_{iy} & -2n_{ix}n_{iz} \\ -2n_{iy}n_{ix} & 1-2n_{iy}n_{iy} & -2n_{iy}n_{iz} \\ -2n_{iz}n_{ix} & -2n_{iz}n_{iy} & 1-2n_{iz}n_{iz} \end{bmatrix}. \tag{10.4}$$

Equation (10.4) indicates that in a reflection process, $\partial\bar{\ell}_i/\partial\bar{\ell}_{i-1}$ is a function of the unit normal vector $\bar{n}_i$ of the flat boundary surface $\bar{r}_i$ (referred to as reflector $\bar{r}_i$ hereafter). Alternatively, $\partial\bar{\ell}_i/\partial\bar{\ell}_{i-1}$ can be computed from

$$\begin{bmatrix} \partial\bar{\ell}_i/\partial\bar{\ell}_{i-1} & \bar{0}_{3x1} \\ \bar{0}_{1\times3} & 1 \end{bmatrix} = \bar{I}_{4\times4} - 2\bar{n}_i\bar{n}_i^T. \tag{10.5}$$

For a refraction process, $\partial\bar{\ell}_i/\partial\bar{\ell}_{i-1}$ (i = 1 and i = n) is obtained from Eq. (7.19) as

$$\frac{\partial\bar{\ell}_i}{\partial\bar{\ell}_{i-1}} = \frac{\partial\bar{\ell}_i(\bar{n}_i, N_i)}{\partial\bar{\ell}_{i-1}} = N_i\begin{bmatrix} 1 & 0 & 0 \\ 0 & 1 & 0 \\ 0 & 0 & 1 \end{bmatrix} + (H_i - N_i)\begin{bmatrix} n_{ix}n_{ix} & n_{ix}n_{iy} & n_{ix}n_{iz} \\ n_{iy}n_{ix} & n_{iy}n_{iy} & n_{iy}n_{iz} \\ n_{iz}n_{ix} & n_{iz}n_{iy} & n_{iz}n_{iz} \end{bmatrix}, \tag{10.6}$$

with

$$H_i = \frac{N_i^2 C\theta_i}{\sqrt{1 - N_i^2 + N_i^2(C\theta_i)^2}}. \tag{10.7}$$

Alternatively, $\partial\bar{\ell}_i/\partial\bar{\ell}_{i-1}$ of Eq. (10.6) can be computed as

$$\begin{bmatrix} \partial\bar{\ell}_i/\partial\bar{\ell}_{i-1} & \bar{0}_{3\times1} \\ \bar{0}_{1\times3} & N_i \end{bmatrix} = N_i\bar{I}_{4\times4} + (H_i - N_i)\bar{n}_i\bar{n}_i^T. \tag{10.8}$$

In other words, for a refraction process, $\partial\bar{\ell}_i/\partial\bar{\ell}_{i-1}$ is a function not only of $\bar{n}_i$, but also of $N_i = \xi_{i-1}/\xi_i$ and H_i. In general, the refraction processes at $\bar{r}_1$ and $\bar{r}_n$ have a direct effect on the image orientation. However, as described in the following theorem, under two particular conditions, the refraction events at $\bar{r}_1$ and $\bar{r}_n$ have no effect on the image orientation produced by the remaining (n − 2) reflectors in the optical system. In other words,

$$\bar{\Phi} = \frac{\partial \bar{\ell}_n(\bar{n}_n, N_n)}{\partial \bar{\ell}_{n-1}} \bar{\Phi} \frac{\partial \bar{\ell}_1(\bar{n}_1, N_1)}{\partial \bar{\ell}_0} \tag{10.9}$$

Theorem 10.1 *If the incidence angle θ_1 at the first boundary surface $\bar{r}_1$ equal to the refraction angle $\underline{\theta}_n$ at the last boundary surface $\bar{r}_n$, then the following equation gives two sufficient conditions under which Eq. (10.9) is true:*

$$\bar{n}_n = \pm \begin{bmatrix} \bar{\Phi} & \bar{t}_{3\times 1} \\ \bar{0}_{1\times 3} & 1 \end{bmatrix} \bar{n}_1. \tag{10.10}$$

Proof From Eq. (10.7) with i = 1, H_1 is obtained as

$$H_1 = \frac{N_1^2 C\theta_1}{\sqrt{1 - N_1^2 + (N_1 C\theta_1)^2}}.$$

Substituting $N_n = 1/N_1$ and $S\underline{\theta}_n = N_n S\theta_n$ into Eq. (10.7) with i = n, H_n is obtained as

$$H_n = \frac{\sqrt{1 - N_1^2 + N_1^2 (C\underline{\theta}_n)^2}}{N_1^2 C\underline{\theta}_n}.$$

As shown below in Eq. (10.15), the transpose of $\bar{\Phi}$ is equal to its inverse, leading to $\bar{\Phi}\bar{\Phi}^T = \bar{I}_{3\times 3}$. It is also noted that $(\bar{n}_i\bar{n}_i^T)(\bar{n}_i\bar{n}_i^T) = \bar{n}_i\bar{n}_i^T$. Thus, Eq. (10.9) can be proven to be true by substituting Eq. (10.10), $\partial \bar{\ell}_1(\bar{n}_1, N_1)/\partial \bar{\ell}_0$ (Eq. (10.8) with i = 1), and $\partial \bar{\ell}_n(\bar{n}_n, N_n)/\partial \bar{\ell}_{n-1}$ (Eq. (10.8) with i = n) into Eq. (10.9).

Equation (10.10) gives the two sufficient conditions of Eq. (10.9). Consequently, if any one of the two conditions given in Eq. (10.10) is satisfied and $\theta_1 = \underline{\theta}_n$, any prism can be designed through the matrix multiplication of its remaining (n-2) reflector matrices. That is,

$$\bar{\Phi} = \begin{bmatrix} \bar{a} & \bar{b} & \bar{c} \end{bmatrix} = \begin{bmatrix} a_x & b_x & c_x \\ a_y & b_y & c_y \\ a_z & b_z & c_z \end{bmatrix} = \frac{\partial \bar{\ell}_{n-1}}{\partial \bar{\ell}_1} = \frac{\partial \bar{\ell}_{n-1}(\bar{n}_{n-2})}{\partial \bar{\ell}_{n-2}} \cdots \frac{\partial \bar{\ell}_i(\bar{n}_i)}{\partial \bar{\ell}_{i-1}} \cdots \frac{\partial \bar{\ell}_2(\bar{n}_2)}{\partial \bar{\ell}_1}, \tag{10.11}$$

or

$$\begin{bmatrix} \bar{\Phi} & \bar{0}_{3\times1} \\ \bar{0}_{1\times3} & 1 \end{bmatrix} = \begin{bmatrix} \partial\bar{\ell}_{n-1}/\partial\bar{\ell}_{n-2} & \bar{0}_{3\times1} \\ \bar{0}_{1\times3} & 1 \end{bmatrix} \cdots \begin{bmatrix} \partial\bar{\ell}_{i}/\partial\bar{\ell}_{i-1} & \bar{0}_{3\times1} \\ \bar{0}_{1\times3} & 1 \end{bmatrix} \cdots \begin{bmatrix} \partial\bar{\ell}_{2}/\partial\bar{\ell}_{1} & \bar{0}_{3\times1} \\ \bar{0}_{1\times3} & 1 \end{bmatrix}$$
$$= \left[I_{4\times4} - 2\bar{n}_{n-1}\bar{n}_{n-1}^T\right] \cdots \left[I_{4\times4} - 2\bar{n}_{n}\bar{n}_{n}^T\right] \cdots \left[I_{4\times4} - 2\bar{n}_{2}\bar{n}_{2}^T\right]. \tag{10.12}$$

Example 10.1 In most prism applications, the rays both enter $\bar{r}_1$ perpendicularly and exit $\bar{r}_n$ perpendicularly. In other words,

$$\bar{n}_1 = \pm\bar{\ell}_0, \tag{a}$$

$$\bar{n}_n = \pm\bar{\ell}_n. \tag{b}$$

Given Eq. (a) and using (10.3), Eq. (b) yields

$$\bar{n}_n = \pm\bar{\ell}_n = \pm\begin{bmatrix} \bar{\Phi} & \bar{t}_{3\times1} \\ \bar{0}_{1\times3} & 1 \end{bmatrix}\bar{\ell}_0 = \pm\begin{bmatrix} \bar{\Phi} & \bar{t}_{3\times1} \\ \bar{0}_{1\times3} & 1 \end{bmatrix}\bar{n}_1. \tag{c}$$

Equation (c) proves that Eq. (10.10) is satisfied if the rays enter and exit a prism perpendicularly (i.e., $\theta_1 = \underline{\theta}_n = 0°$). In other words, the two refraction processes at $\bar{r}_1$ and $\bar{r}_n$ do not change the image orientation produced by the remaining $(n - 2)$ reflectors. Thus, the following discussions use Eq. (10.11), in which the two refraction processes at $\bar{r}_1$ and $\bar{r}_n$ are ignored, as the basis for solving the prism design problems.

Before proceeding, it is appropriate to review the reflector matrix given in Eq. (10.4). (Note that in mathematics, Eq. (10.4) is also known as the Householder transformation, named for A. S. Householder, who first employed the matrix in matrix computations [3].) The reflector matrix $\partial\bar{\ell}_i/\partial\bar{\ell}_{i-1} = [\bar{u}_i \quad \bar{v}_i \quad \bar{w}_i]$ possesses four important properties:

(1) It is a symmetrical matrix, i.e.,

$$\left(\frac{\partial\bar{\ell}_i}{\partial\bar{\ell}_{i-1}}\right)^T = \frac{\partial\bar{\ell}_i}{\partial\bar{\ell}_{i-1}}. \tag{10.13}$$

(2) It is an orthogonal matrix with a determinant of −1. Its three components (i.e., $\bar{u}_i$, $\bar{v}_i$ and $\bar{w}_i$) form an orthonormal basis. More specifically, the first two column vectors are of unit magnitude and perpendicular, i.e., $|\bar{u}_i| = |\bar{v}_i| = 1$ and $\bar{u}_i \cdot \bar{v}_i = 0$, while the third column vector $\bar{w}_i$ is the vector cross-product of the second and first column vectors $\bar{v}_i \times \bar{u}_i$ (not $\bar{u}_i \times \bar{v}_i$). Therefore, $\partial\bar{\ell}_i/\partial\bar{\ell}_{i-1}$ represents the orientation part of a left-handed transformation matrix.
(3) The inverse of $\partial\bar{\ell}_i/\partial\bar{\ell}_{i-1}$ is itself, i.e.,

$$\left(\frac{\partial \bar{\ell}_i}{\partial \bar{\ell}_{i-1}}\right)^{-1} = \frac{\partial \bar{\ell}_i}{\partial \bar{\ell}_{i-1}}. \tag{10.14}$$

(4) In general, matrix multiplication is not commutative. However, the multiplication of $\partial \bar{\ell}_{i+1}/\partial \bar{\ell}_i$ and $\partial \bar{\ell}_i/\partial \bar{\ell}_{i-1}$ is commutative if $\bar{n}_{i+1}$ and $\bar{n}_i$ are either parallel or perpendicular. (Note that this property is proven in the following theorem.)

Theorem 10.2 *The matrix multiplication of* $\partial \bar{\ell}_{i+1}/\partial \bar{\ell}_i$ and $\partial \bar{\ell}_i/\partial \bar{\ell}_{i-1}$ *is commutative if* $\bar{n}_{i+1}$ *and* $\bar{n}_i$ *are either parallel or perpendicular.*

Proof Since the multiplication of $\partial \bar{\ell}_{i+1}/\partial \bar{\ell}_i$ and $\partial \bar{\ell}_i/\partial \bar{\ell}_{i-1}$ is commutative, the following equation is true:

$$\begin{bmatrix} \partial \bar{\ell}_{i+1}/\partial \bar{\ell}_i & \bar{0}_{3\times1} \\ \bar{0}_{1\times3} & 1 \end{bmatrix} \begin{bmatrix} \partial \bar{\ell}_i/\partial \bar{\ell}_{i-1} & \bar{0}_{3\times1} \\ \bar{0}_{1\times3} & 1 \end{bmatrix} = \begin{bmatrix} \partial \bar{\ell}_i/\partial \bar{\ell}_{i-1} & \bar{0}_{3\times1} \\ \bar{0}_{1\times3} & 1 \end{bmatrix} \begin{bmatrix} \partial \bar{\ell}_{i+1}/\partial \bar{\ell}_i & \bar{0}_{3\times1} \\ \bar{0}_{1\times3} & 1 \end{bmatrix}.$$

From this equation, and using Eq. (10.5), it follows that

$$\left(\bar{I}_{4\times4} - 2\bar{n}_{i+1}\bar{n}_{i+1}^T\right)\left(\bar{I}_{4\times4} - 2\bar{n}_i\bar{n}_i^T\right) = \left(\bar{I}_{4\times4} - 2\bar{n}_i\bar{n}_i^T\right)\left(\bar{I}_{4\times4} - 2\bar{n}_{i+1}\bar{n}_{i+1}^T\right)$$

or

$$\bar{n}_{i+1}\left(\bar{n}_{i+1}^T\bar{n}_i\right)\bar{n}_i^T = \bar{n}_i\left(\bar{n}_i^T\bar{n}_{i+1}\right)\bar{n}_{i+1}^T.$$

It will be recalled that the dot product of two vectors is commutative and a scalar (i.e., $\bar{n}_{i+1}^T\bar{n}_i = \bar{n}_i^T\bar{n}_{i+1}$). Thus, the preceding equation can be rewritten as

$$\left(\bar{n}_{i+1}^T\bar{n}_i\right)\left(\bar{n}_{i+1}\bar{n}_i^T - \bar{n}_i\bar{n}_{i+1}^T\right) = \bar{0}_{3\times3}.$$

The two solutions of this equation are $\bar{n}_{i+1}^T\bar{n}_i = 0$ and $\bar{n}_{i+1}\bar{n}_i^T = \bar{n}_i\bar{n}_{i+1}^T$, respectively. The former solution indicates that $\bar{n}_i + 1$ and $\bar{n}_i$ are perpendicular, while the latter solution indicates that $\bar{n}_{i+1}$ and $\bar{n}_i$ are parallel. Consequently, Theorem 10.2 is proven.

From Eqs. (10.4) and (10.13), it is easily shown that the image orientation function $\bar{\Phi}$ given in Eq. (10.11) has the following four properties:

(1) $\bar{\Phi}$ is an orthogonal matrix (i.e., $|\bar{a}| = |\bar{b}| = |\bar{c}| = 1$, $\bar{a}\cdot\bar{b} = 0$, $\bar{a}\cdot\bar{c} = 0, \bar{b}\cdot\bar{c} = 0$).
(2) $\bar{\Phi}$ can be either right-handed (i.e., $\bar{c} = \bar{a}\times\bar{b}$ when the number of reflectors is even) or left-handed (i.e., $\bar{c} = \bar{b}\times\bar{a}$ when the number of reflectors is odd).

(3) $\bar{\Phi}$ need not be symmetrical.
(4) The transpose of $\bar{\Phi}$ is equal to its inverse, i.e.,

$$\bar{\Phi}^{-1} = \bar{\Phi}^{T}, \tag{10.15}$$

which entails

$$\bar{\Phi}^{T}\bar{\Phi} = \bar{\Phi}\bar{\Phi}^{T} = \bar{I}_{3\times 3}. \tag{10.16}$$

Theorem 10.3 *If the image orientation function $\bar{\Phi}$ is symmetrical (i.e., $\bar{\Phi} = \bar{\Phi}^{T}$), the prism can be used reversely by letting the rays enter and exit the (n − 1)th and 2nd reflectors, respectively, and the required image orientation $\bar{\Phi}$ will still be obtained.*

Proof If $\bar{\Phi}$ is symmetrical we have $\partial\bar{\ell}_{n-1}/\partial\bar{\ell}_1 = (\partial\bar{\ell}_{n-1}/\partial\bar{\ell}_1)^{T}$. From Eqs. (10.11) and (10.13), it can be shown that

$$\begin{aligned}\bar{\Phi} &= \frac{\partial\bar{\ell}_{n-1}}{\partial\bar{\ell}_1} = \frac{\partial\bar{\ell}_{n-1}}{\partial\bar{\ell}_{n-2}}\cdots\frac{\partial\bar{\ell}_i}{\partial\bar{\ell}_{i-1}}\cdots\frac{\partial\bar{\ell}_2}{\partial\bar{\ell}_1} = \left(\frac{\partial\bar{\ell}_{n-1}}{\partial\bar{\ell}_1}\right)^{T} = \left(\frac{\partial\bar{\ell}_2}{\partial\bar{\ell}_1}\right)^{T}\cdots\left(\frac{\partial\bar{\ell}_i}{\partial\bar{\ell}_{i-1}}\right)^{T}\cdots\left(\frac{\partial\bar{\ell}_{n-1}}{\partial\bar{\ell}_{n-2}}\right)^{T} \\ &= \frac{\partial\bar{\ell}_2}{\partial\bar{\ell}_1}\cdots\frac{\partial\bar{\ell}_i}{\partial\bar{\ell}_{i-1}}\cdots\frac{\partial\bar{\ell}_{n-1}}{\partial\bar{\ell}_{n-2}}.\end{aligned} \tag{10.17}$$

Hence, the proof is complete.

10.2 Minimum Number of Reflectors

This section discusses the minimum number (n_{mini}) of reflectors needed by a prism to produce an image with a required orientation function $\bar{\Phi}$. Consider the reflection event at reflector $\bar{r}_i$ shown in Fig. 2.12. From Snell's law: (1) the unit directional vector $\bar{\ell}_{i-1}$ of the incoming ray, the active unit normal vector $\bar{n}_i$, and the reflected unit directional vector $\bar{\ell}_i$ all lie in the same plane with $\bar{m}_i = \bar{n}_i \times \bar{\ell}_{i-1}/S\theta_i$ (Eq. (2.23)) as their common unit normal vector; and (2) the reflection angle is equal to the incidence angle θ_i. Snell's law implies an alternative method of determining the reflected unit directional vector $\bar{\ell}_i$ and active unit normal vector $\bar{n}_i$ of the reflector if $\bar{m}_i$ and $\bar{\ell}_{i-1}$ are known (see Fig. 2.5). Namely:

(1) $\bar{\ell}_i$ can be obtained by rotating $\bar{\ell}_{i-1}$ about $\bar{m}_i$ through an angle $\pi + 2\theta_i$ to give (see Eq. (2.27))

$$\bar{\ell}_i = \text{rot}(\bar{m}_i, \pi + 2\theta_i)\bar{\ell}_{i-1}. \tag{10.18}$$

(2) The active unit normal vector $\bar{n}_i$ can be obtained by rotating $\bar{\ell}_{i-1}$ about $\bar{m}_i$ through an angle $\pi + \theta_i$, i.e.,

$$\bar{n}_i = \text{rot}(\bar{m}_i, \pi + \theta_i)\bar{\ell}_{i-1}. \tag{10.19}$$

The first step in designing a prism to produce a specific image orientation is to construct $\bar{\Phi} = \begin{bmatrix} \bar{a} & \bar{b} & \bar{c} \end{bmatrix}$ in numeric form so as to describe the image orientation of the object after the light rays pass through the (n − 2) reflectors. As described above, $\bar{a}$, $\bar{b}$ and $\bar{c}$ are the unit directional vectors of the x_0', y_0' and z_0' axes of $(xyz)_0'$, respectively, with respect to $(xyz)_0$ (see Fig. 10.4). In discussing the minimum number of reflectors required to produce the desired image orientation function, the two possible types of orientation function, i.e., right-handed and left-handed, must be separately addressed.

10.2.1 Right-Handed Image Orientation Function

It will be recalled that the relative pose matrix of two right-handed coordinate frames, e.g., $(xyz)_0$ and $(xyz)_0'$, can be obtained by rotating $(xyz)_0$ about a unit vector using Eq. (1.25). Therefore, if the image orientation of an object produced by a series of reflectors is described by a right-handed image orientation function $\bar{\Phi}$, then $\bar{\Phi}$ can also be specified by using the orientation part of $\text{rot}(\bar{m}, \mu)$, where $\bar{m} = \begin{bmatrix} m_x & m_y & m_z & 0 \end{bmatrix}^T$ with $m_x^2 + m_y^2 + m_z^2 = 1$, i.e.,

$$\bar{\Phi} = \bar{\Phi}(\bar{m}, \mu) = \begin{bmatrix} m_x^2(1 - C\mu) + C\mu & m_x m_y(1 - C\mu) - m_z S\mu & m_x m_z(1 - C\mu) + m_y S\mu \\ m_x m_y(1 - C\mu) + m_z S\mu & m_y^2(1 - C\mu) + C\mu & m_y m_z(1 - C\mu) - m_x S\mu \\ m_x m_z(1 - C\mu) - m_y S\mu & m_y m_z(1 - C\mu) + m_x S\mu & m_z^2(1 - C\mu) + C\mu \end{bmatrix}. \tag{10.20}$$

It is noted that any unit vector of the form $\bar{m}_2 = \begin{bmatrix} m_{2x} & 0 & m_{2z} & 0 \end{bmatrix}^T$ with $m_{2x}^2 + m_{2z}^2 = 1$ ($m_{2z}m_x \neq m_{2x}m_z$, as explained in the following section) is the common unit normal vector of $\bar{\ell}_1 = \begin{bmatrix} 0 & 1 & 0 & 0 \end{bmatrix}^T$ and $\bar{n}_2$ computed by $\bar{n}_2 = \text{rot}(\bar{m}_2, \pi + \theta_2)\bar{\ell}_1$ (see Fig. 10.5) since $\bar{\ell}_1 \cdot \bar{m}_2 = 0$. Furthermore, any rotation motion

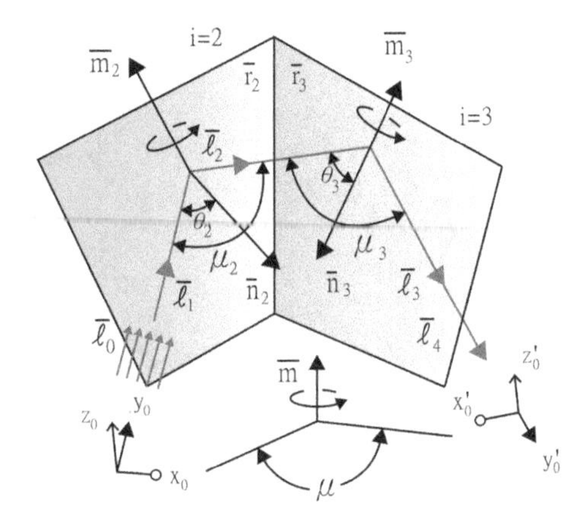

Fig. 10.5 Minimum number of reflectors required to produce a *right-handed* image orientation function $\bar{\Phi}$ is equal to 2

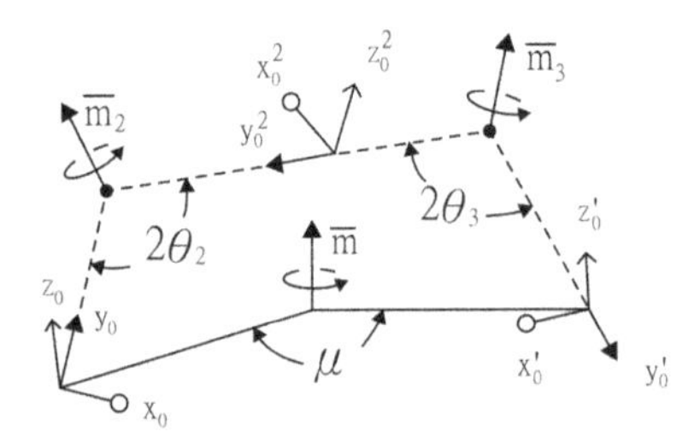

Fig. 10.6 A rotation motion can be regarded as the product of two successive rotation motions. Here, $(xyz)_0^i$ denotes the image of $(xyz)_0$ after the rays are reflected at reflector $\bar{r}_i$

$\text{rot}(\bar{m}, \mu)$ can be regarded as the product of two successive rotation motions, namely an initial rotation $\text{rot}(\bar{m}_2, 2\theta_2)$ followed by a second rotation motion $\text{rot}(\bar{m}_3, 2\theta_3)$ (i.e., $\text{rot}(\bar{m}, \mu) = \text{rot}(\bar{m}_3, 2\theta_3)\text{rot}(\bar{m}_2, 2\theta_2)$, see Fig. 10.6). It will be proven numerically in Sect. 10.4 that $\bar{m}_2$ chosen in this way always gives an angle μ_2 $(0 \le \mu_2 < \pi)$ such that

$$\bar{n}_2 = \text{rot}(\bar{m}_2, \pi + \mu_2/2)\bar{\ell}_1 = \begin{bmatrix} m_{2z}S(\mu_2/2) \\ -C(\mu_2/2) \\ -m_{2x}S(\mu_2/2) \\ 0 \end{bmatrix}$$

(obtained from Eq. (10.19) with i = 2 and then $\theta_2 = \mu_2/2$) is the unit normal vector of reflector $\bar{r}_2$. The corresponding reflector matrix is given by Eq. (10.4) with i = 2 and using the above $\bar{n}_2$, to give

$$\frac{\partial \bar{\ell}_2}{\partial \bar{\ell}_1} = \begin{bmatrix} 1 - m_{2z}^2(1 - C\mu_2) & m_{2z}S\mu_2 & m_{2x}m_{2z}(1 - C\mu_2) \\ m_{2z}S\mu_2 & -C\mu_2 & -m_{2x}S\mu_2 \\ m_{2x}m_{2z}(1 - C\mu_2) & -m_{2x}(1 - C\mu_2) & 1 - m_{2x}^2(1 - C\mu_2) \end{bmatrix}. \tag{10.21}$$

Equation (10.21) guarantees that the matrix $\partial\bar{\ell}_3/\partial\bar{\ell}_2 = \bar{\Phi}(\partial\bar{\ell}_2/\partial\bar{\ell}_1)$ obtained from Eq. (10.11) with n = 4 is always a reflector matrix.

The discussions above show that, at most, two reflectors with unit normal vectors $\bar{n}_2$ and $\bar{n}_3$, respectively, are sufficient to produce the required right-handed image orientation function $\bar{\Phi}$. Importantly, the image orientation function $\bar{\Phi}$ obtained from an even number of reflectors is right-handed, while that obtained from an odd number of reflectors is left-handed. Thus, it can be concluded that in order to produce the image of a right-handed image orientation function $\bar{\Phi}$, the minimum number of reflectors needed is equal to two (e.g., the single Porro prism shown in Fig. 9.8).

10.2.2 Left-Handed Image Orientation Function

It is impossible to obtain a left-handed image $(xyz)_0'$ by rotating a right-handed object $(xyz)_0$ using any of the rotation matrices described in Sect. 1.4 since these matrices are all right-handed. Consequently, if the constructed image orientation function $\bar{\Phi}$ is left-handed, then $\bar{\Phi}$ should first be converted to a right-handed image orientation function, say $\bar{\Phi}'$, by adding a minus sign to the second column of $\bar{\Phi}$. $\bar{\Phi}'$ can then be used to solve for $\bar{m} = [m_x \quad m_y \quad m_z \quad 0]^T$ and μ by $\bar{\Phi}' = \bar{\Phi}'(\bar{m}, \mu)$ using Eq. (10.20). If $m_y = 0$ (i.e., $\bar{m}$ is perpendicular to $\bar{\ell}_1$) and $0 \le \mu < \pi$, it is apparent that a reflector having $\bar{n}_2 = \text{rot}(\bar{m}_2, \pi + \mu/2)\bar{\ell}_1$ as its unit normal vector can produce the desired image orientation function. In this case, only one reflector is necessary to produce the desired image orientation. However, if $\bar{m}$ is not perpendicular to $\bar{\ell}_1$, then (based on the discussions in the previous section), three reflectors are required to output an image with an orientation characterized by a left-handed image orientation function $\bar{\Phi}$.

In summary, the minimum number of reflectors required to produce a left-handed image orientation function is either one (e.g., a second-surface mirror) or three (e.g., a solid glass corner-cube retro-reflector, see Fig. 9.2).

10.3 Prism Design Based on Unit Vectors of Reflectors

This section presents a methodology for determining the unit normal vectors of a prism's reflectors given the need to produce a certain image orientation function $\bar{\Phi}$. Let a recursive function $\bar{G}_i$ be defined as

$$\bar{G}_i = \frac{\partial\bar{\ell}_{n-1}}{\partial\bar{\ell}_{n-2}}\frac{\partial\bar{\ell}_{n-2}}{\partial\bar{\ell}_{n-3}}\cdots\frac{\partial\bar{\ell}_{i+1}}{\partial\bar{\ell}_i}\frac{\partial\bar{\ell}_i}{\partial\bar{\ell}_{i-1}} \quad (i \ge 2 \text{ for a prism}). \tag{10.22}$$

From Eq. (10.22), it follows that

$$\bar{G}_i = \bar{G}_{i+1}\left(\frac{\partial \bar{\ell}_i}{\partial \bar{\ell}_{i-1}}\right). \tag{10.23}$$

Furthermore, from Eqs. (10.23) and (10.14), the following equation is obtained as the basis for the prism design:

$$\bar{G}_{i+1} = \bar{G}_i\left(\frac{\partial \bar{\ell}_i}{\partial \bar{\ell}_{i-1}}\right)^{-1} = \bar{G}_i \frac{\partial \bar{\ell}_i}{\partial \bar{\ell}_{i-1}}. \tag{10.24}$$

The following steps provide a simple and direct procedure for deriving the set of reflectors which produce a desired image orientation function (though not with a guaranteed minimum number of reflectors).

(1) Establish a world coordinate frame $(xyz)_0$ emitting parallel source rays in the y_0 direction and entering the flat boundary surface $\bar{r}_1$ perpendicularly.
(2) Construct the image orientation function $\bar{\Phi} = \begin{bmatrix} \bar{a} & \bar{b} & \bar{c} \end{bmatrix}$ in numeric form (Eq. (10.11)), where $\bar{a}$, $\bar{b}$ and $\bar{c}$ are the unit directional vectors of the x_0', y_0' and z_0' axes of $(xyz)_0'$, respectively, referred with respect to $(xyz)_0$. Set i = 2 and $\bar{G}_2 = \begin{bmatrix} \bar{a}_2 & \bar{b}_2 & \bar{c}_2 \end{bmatrix} = \bar{\Phi} = \begin{bmatrix} \bar{a} & \bar{b} & \bar{c} \end{bmatrix}$.
(3) If $\bar{G}_i$ meets the following conditions: (a) it is symmetric (i.e., $\bar{G}_i = \bar{G}_i^T$); (b) its inverse is itself (i.e., $(\bar{G}_i)^{-1} = \bar{G}_i$); and (c) $\bar{c}_i = \bar{b}_i \times \bar{a}_i$, then $\bar{G}_i$ is a reflector matrix, and the unit normal vector $\bar{n}_i$ can be obtained from Eq. (10.4). Then go to Step (5). Otherwise, choose an arbitrary vector $\bar{n}_i$ as the unit normal vector of the current candidate reflector. Calculate the reflector matrix $\partial \bar{\ell}_i / \partial \bar{\ell}_{i-1}$ from Eq. (10.4).
(4) Calculate the intermediate image orientation function from Eq. (10.24) for the next iteration. Set i = i + 1 and go to Step (3).
(5) Establish the last flat boundary surface $\bar{r}_n$, at which the exit ray is refracted perpendicularly. Check the validity of the prism design by means of 3-D solid modeling software.

Example 10.2 Consider the problem shown in Fig. 10.7 in which an object located at D is to be projected by a projection lens E onto a screen. The screen is parallel to the original projection axis and its center lies above the axis by some distance F. Assume that the problem calls for the design of a reflector system which produces the required image orientation on the screen without the use of any refractive flat boundary surfaces. Let the problem be solved using the notation described in Sect. 10.3.

(1) Start by establishing the world coordinate frame $(xyz)_0$, which is embedded in the projected image after passing through the projection lens. Assume that $(xyz)_0$ emits parallel source rays in the y_0 direction. Furthermore, in order to

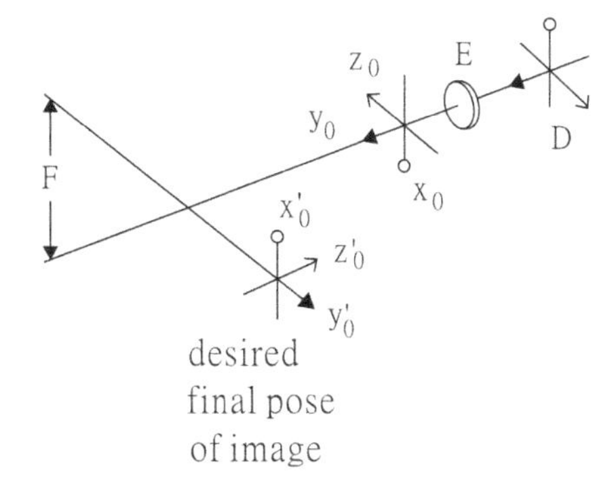

Fig. 10.7 Required poses of object and image

comply with the notation given in Sect. 10.3 (in which $\bar{r}_1$ is a refractive flat boundary surface), let the first reflector in the considered system be labeled as i = 2.

(2) Construct the image orientation function

$$\bar{\Phi} = \begin{bmatrix} \bar{a} & \bar{b} & \bar{c} \end{bmatrix} = \begin{bmatrix} -1 & 0 & 0 \\ 0 & 0 & -1 \\ 0 & -1 & 0 \end{bmatrix}$$

and set $\bar{G}_2 = \begin{bmatrix} \bar{a}_2 & \bar{b}_2 & \bar{c}_2 \end{bmatrix} = \bar{\Phi}$.

(3) It is noted that $\bar{G}_2 = \bar{G}_2^T$ and $(\bar{G}_2)^{-1} = \bar{G}_2$, but with $\bar{c}_i = -\bar{b}_i \times \bar{a}_i$. Thus, choose an arbitrary vector $\bar{n}_2$, say $\bar{n}_2 = \begin{bmatrix} -1/\sqrt{2} & -1/\sqrt{2} & 0 & 0 \end{bmatrix}^T$, as the unit normal vector of the current candidate reflector $\bar{r}_2$. The reflector matrix $\partial\bar{\ell}_2/\partial\bar{\ell}_1$ is given by

$$\frac{\partial\bar{\ell}_2}{\partial\bar{\ell}_1} = \begin{bmatrix} 0 & -1 & 0 \\ -1 & 0 & 0 \\ 0 & 0 & 1 \end{bmatrix}.$$

(4) $\bar{G}_3 = \bar{G}_2 \frac{\partial\bar{\ell}_2}{\partial\bar{\ell}_1} = \begin{bmatrix} 0 & 1 & 0 \\ 0 & 0 & -1 \\ 1 & 0 & 0 \end{bmatrix}$. Set i = 3 and go to Step (3).

(5) It is noted that $\bar{G}_3 \neq \bar{G}_3^T$. Choose an arbitrary vector $\bar{n}_3 = \begin{bmatrix} 0 & 1/\sqrt{2} & -1/\sqrt{2} & 0 \end{bmatrix}^T$ as the unit normal vector of the reflector $\bar{r}_3$. The reflector matrix is given by

$$\frac{\partial \bar{\ell}_3}{\partial \bar{\ell}_2} = \begin{bmatrix} 0 & 1 & 0 \\ 1 & 0 & 0 \\ 0 & 0 & 1 \end{bmatrix}.$$

(4) $\bar{G}_4 = \bar{G}_3 \frac{\partial \bar{\ell}_3}{\partial \bar{\ell}_2} = \begin{bmatrix} 1 & 0 & 0 \\ 0 & 0 & -1 \\ 0 & 1 & 0 \end{bmatrix}$. Set i = 4 and go to Step (3).

(7) It is noted that $\bar{G}_4 \neq \bar{G}_4^T$. Choose an arbitrary vector $\bar{n}_4 = \begin{bmatrix} 0 & 1/\sqrt{2} & 1/\sqrt{2} & 0 \end{bmatrix}^T$ as the unit normal vector of the reflector $\bar{r}_4$. The reflector matrix has the form

$$\frac{\partial \bar{\ell}_4}{\partial \bar{\ell}_3} = \begin{bmatrix} 1 & 0 & 0 \\ 0 & 0 & -1 \\ 0 & -1 & 0 \end{bmatrix}.$$

(4) $\bar{G}_5 = \bar{G}_4 \frac{\partial \bar{\ell}_4}{\partial \bar{\ell}_3} = \begin{bmatrix} 1 & 0 & 0 \\ 0 & 1 & 0 \\ 0 & 0 & -1 \end{bmatrix}$. Set i = 5 and go to Step (3).

(3) $\bar{G}_5$ possesses the required three properties, namely $\bar{G}_5 = \bar{G}_5^T$, $(\bar{G}_5)^{-1} = \bar{G}_5$ and $\bar{c}_5 = \bar{b}_5 \times \bar{a}_5$. Thus, go to Step 5.

(5) Solve the unit normal vector $\bar{n}_5$ of the last reflector $\bar{r}_5$ as $\bar{n}_5 = \begin{bmatrix} 0 & 0 & 1 & 0 \end{bmatrix}^T$.

Figure 10.8 shows the designed system. However, it is noted that the system represents only one of the many possible arrangements of reflector systems which can be utilized to accomplish the same end result.

Example 10.3 In order to further validate the proposed methodology, consider the following image orientation function:

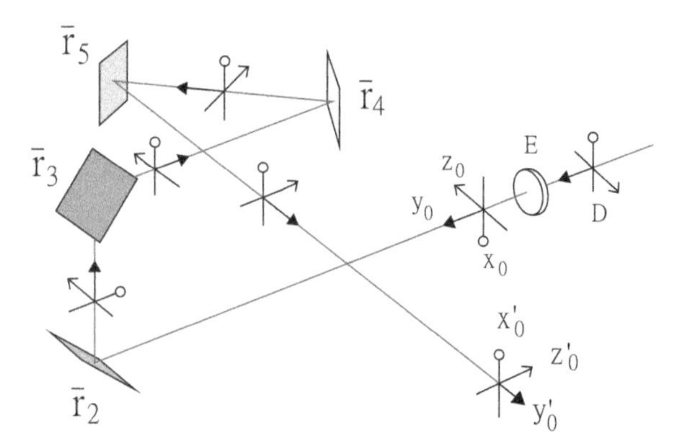

Fig. 10.8 Designed reflector system for Example 10.2

$$\bar{\Phi}_2 = \bar{\Phi} = \begin{bmatrix} -1 & 0 & 0 \\ 0 & -1 & 0 \\ 0 & 0 & -1 \end{bmatrix}.$$

As described in [4], at least eight different reflector configurations can be obtained which produce an identical final image orientation function. Figures 10.9 and 10.10 show two of these configurations. (Note that the corresponding unit normal vectors are listed in Tables 10.1 and 10.2, respectively.) For each of the reflector configurations shown in Figs. 10.9 and 10.10 (and presented in [4]), the number of reflectors is equal to n = 7. However, the prism configuration is different in every case. Thus, the designer should choose the configuration which best meets the particular requirements of the target application. Note that since the unit normal vectors $\bar{n}_4$ and $\bar{n}_5$ in Fig. 10.9 are perpendicular to one another (and hence the multiplication of $\partial\bar{\ell}_4/\partial\bar{\ell}_3$ and $\partial\bar{\ell}_5/\partial\bar{\ell}_4$ is commutative (see Theorem 7.2)), the arrangement shown in Fig. 10.11 (for which the unit normal vectors are listed in Table 10.3) can be obtained from Fig. 10.9 by interchanging reflectors $\bar{r}_4$ and $\bar{r}_5$. (Note that a similar alternative arrangement can be obtained from Fig. 10.10 by exploiting the same commutative property.)

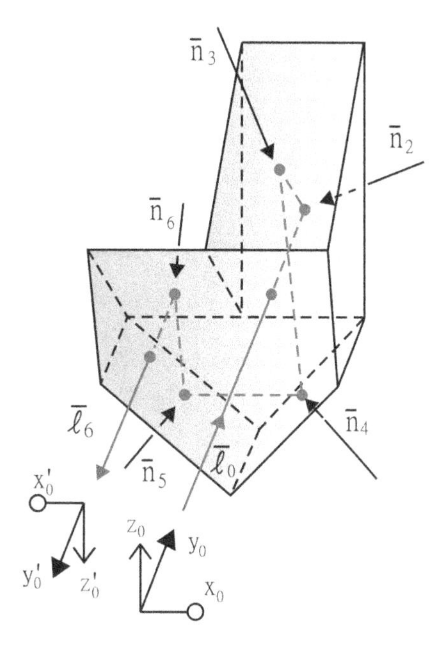

Fig. 10.9 Prism consisting of five reflectors designed to bend incoming rays through exactly 180°

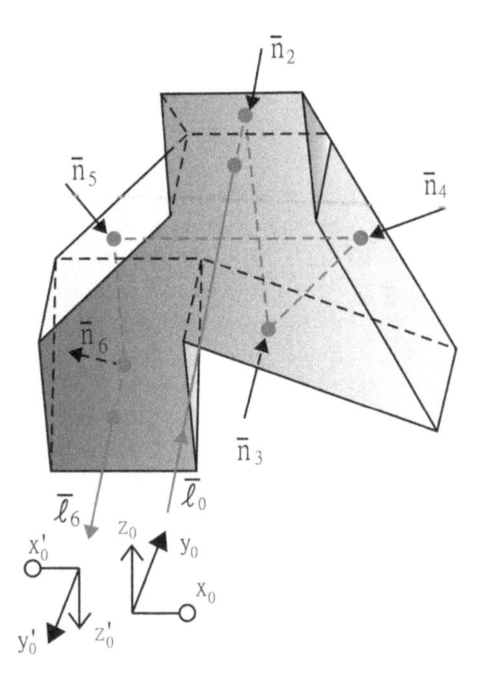

Fig. 10.10 Prism with different configuration from that of Fig. 10.9 which also reflects incoming rays back along their original direction

Table 10.1 Unit normal vectors of reflectors in Fig. 10.9

i	$\bar{n}_i$
2	$[0 \quad -0.9239 \quad 0.3827 \quad 0]^T$
3	$[0 \quad 0.3827 \quad -0.9239 \quad 0]^T$
4	$[-0.7071 \quad 0 \quad 0.7071 \quad 0]^T$
5	$[0.7071 \quad 0 \quad 0.7071 \quad 0]^T$
6	$[0 \quad -0.7071 \quad -0.7071 \quad 0]^T$

Table 10.2 Unit normal vectors of reflectors in Fig. 10.10

i	$\bar{n}_i$
2	$[0 \quad -0.7071 \quad -0.7071 \quad 0]^T$
3	$[0.3827 \quad 0 \quad 0.9239 \quad 0]^T$
4	$[-0.9239 \quad 0 \quad -0.3827 \quad 0]^T$
5	$[0.7071 \quad 0 \quad -0.7071 \quad 0]^T$
6	$[0 \quad -0.7071 \quad 0.7071 \quad 0]^T$

10.4 Exact Analytical Solutions for Single Prism with Minimum Number of Reflectors

This section presents a method for obtaining a prism with the minimum number (n_{mini}) of reflectors required to produce a certain image orientation function $\bar{\Phi}$ based on the auxiliary unit vector $\bar{m}$ and auxiliary angle μ. In general, $\bar{\Phi}$ has nine

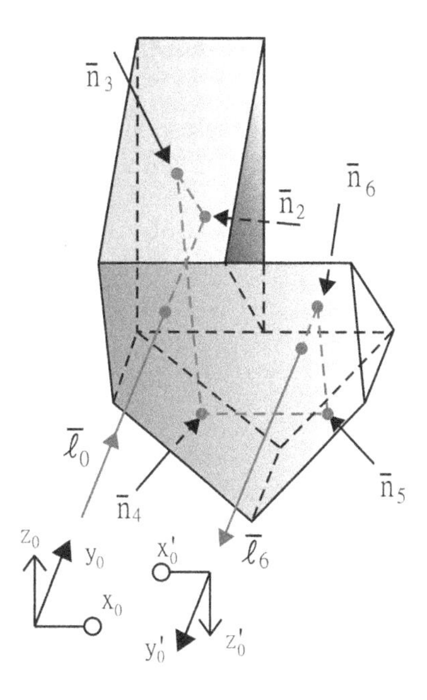

Fig. 10.11 Alternative configuration with same design function as Fig. 10.9

Table 10.3 Unit normal vectors of reflectors in Fig. 10.11

i	$\bar{n}_i$
2	$[0 \quad -0.9239 \quad 0.3827 \quad 0]^T$
3	$[0 \quad 0.3827 \quad -0.9239 \quad 0]^T$
4	$[0.7071 \quad 0 \quad 0.7071 \quad 0]^T$
5	$[-0.7071 \quad 0 \quad 0.7071 \quad 0]^T$
6	$[0 \quad -0.7071 \quad -0.7071 \quad 0]^T$

components. However, of these components, only three are independent. Using $\bar{\Phi}$ with all nine components causes unnecessary mathematical complexity. Thus, to minimize the number of governing equations, $\bar{\Phi}$ should be expressed in terms only of the three independent physical qualities. One set of these physical qualities is the auxiliary unit vector $\bar{m} = [m_x \quad m_y \quad m_z \quad 0]^T$ with $m_x^2 + m_y^2 + m_z^2 = 1$ and auxiliary angle μ $(0 \le \mu \le \pi)$. When seeking to design a prism which outputs an image with an image orientation $\bar{\Phi}$, the first step is to construct the image orientation function $\bar{\Phi} = [\bar{a} \quad \bar{b} \quad \bar{c}]$ in numeric form based on the required image orientation. Note that $\bar{a} = [a_x \quad a_y \quad a_z]^T$, $\bar{b} = [b_x \quad b_y \quad b_z]^T$ and $\bar{c} = [c_x \quad c_y \quad c_z]^T$, are the unit directional vectors of the x'_0, y'_0 and z'_0 axes of $(xyz)'_0$, respectively, referred with respect to $(xyz)_0$ (see Fig. 10.4). In order to obtain the analytical solution of a single prism having the minimum number of reflectors, the auxiliary unit vector $\bar{m}$ and auxiliary angle μ must be determined in accordance with $\bar{\Phi}$, as described in the following sub-sections.

10.4.1 Right-Handed Image Orientation Function

The auxiliary angle μ and auxiliary unit vector $\bar{m}$ can be determined from the components of $\bar{\Phi}$ using the equivalent angle and axis of rotation, respectively, as described in Sect. 1.9. In other words,

$$\mu = \operatorname{atan2}\left(\sqrt{(b_z - c_y)^2 + (c_x - a_z)^2 + (a_y - b_x)^2}, a_x + b_y + c_z - 1\right), \quad (10.25)$$

$$m_x = \frac{0.5(b_z - c_y)}{S\mu}, \quad (10.26)$$

$$m_y = \frac{0.5(c_x - a_z)}{S\mu}, \quad (10.27)$$

$$m_z = \frac{0.5(a_y - b_x)}{S\mu}, \quad (10.28)$$

when the angle μ is very small, the unit vector $\bar{m}$ should be normalized to ensure that $|\bar{m}| = 1$. However, if $\mu \geq 90°$, then another method described in Sect. 1.9 [5] should be adopted to accurately determine $\bar{m}$. The right-handed image orientation function $\bar{\Phi}$ can then be expressed in terms of $\bar{m}$ and μ by using Eq. (10.20).

10.4.2 Left-Handed Image Orientation Function

Again, in order to obtain the minimum number of equations, let $\bar{\Phi}$ be expressed in terms of the auxiliary unit vector $\bar{m}$ and auxiliary angle μ. As stated previously, all of the rotation matrices given in Chap. 1 are right-handed. To overcome this problem, $\bar{\Phi}$ should first be converted to a temporary right-handed image orientation function (denoted as $\bar{\Phi}'$) by adding a minus sign to the second column of $\bar{\Phi}$. Equation (10.20) with μ given by

$$\mu = \operatorname{atan2}\left(\sqrt{(b_z + c_y)^2 + (c_x - a_z)^2 + (a_y + b_x)^2}, a_x - b_y + c_z - 1\right), \quad (10.29)$$

and $\bar{m}$ given as

$$m_x = \frac{-0.5(b_z + c_y)}{S\mu}, \quad (10.30)$$

$$m_y = \frac{0.5(c_x - a_z)}{S\mu}, \quad (10.31)$$

$$m_z = \frac{0.5\left(a_y + b_x\right)}{S\mu}, \tag{10.32}$$

can then be applied once again to obtain an intermediate image orientation function $\bar{\Phi}' = \bar{\Phi}'(\bar{m}, \mu)$. Finally, $\bar{\Phi}'$ can be restored to the required left-handed image orientation function $\bar{\Phi}$ by adding a minus sign to the second column of $\bar{\Phi}'$.

Having obtained the left-handed image orientation function $\bar{\Phi}$ in terms of $\bar{m}$ and μ, the prism can be designed using the methods described in the following sub-sections.

10.4.3 Solution for Right-Handed Image Orientation Function

It was concluded in Sect. 10.2 that two reflectors (i.e., n = 4 if the first and last boundary surfaces are refractive) with unit normal vectors $\bar{n}_2$ and $\bar{n}_3$, respectively, are sufficient to output from a prism an image with a right-handed $\bar{\Phi}$. Consequently, from Eq. (10.11), reflector matrix $\partial\bar{\ell}_3/\partial\bar{\ell}_2$ can be obtained as the product of $\bar{\Phi}$ and the inverse of reflector matrix $\partial\bar{\ell}_2/\partial\bar{\ell}_1$ (note that $\left(\partial\bar{\ell}_2/\partial\bar{\ell}_1\right)^{-1} = \partial\bar{\ell}_2/\partial\bar{\ell}_1$) by setting n = 4, i.e.,

$$\frac{\partial\bar{\ell}_3}{\partial\bar{\ell}_2} = \bar{\Phi}\frac{\partial\bar{\ell}_2}{\partial\bar{\ell}_1}. \tag{10.33}$$

Note that $(xyz)_0$ emits ray $\bar{\ell}_0 = \bar{\ell}_1 = \begin{bmatrix}0 & 1 & 0 & 0\end{bmatrix}^T$ in the y_0 direction. Meanwhile, from Eq. (10.19), $\bar{n}_2$ can be obtained by rotating $\bar{\ell}_1$ about the unit vector $\bar{m}_2 = \begin{bmatrix}m_{2x} & 0 & m_{2z} & 0\end{bmatrix}^T$ (where $\bar{m}_2$ with $m_{2x}^2 + m_{2z}^2 = 1$ is the common unit normal vector of $\bar{\ell}_1$ and $\bar{n}_2$) through an angle $\pi + \theta_2$ (i.e., $\bar{n}_2 = \text{rot}(\bar{m}_2, \pi + \theta_2)\bar{\ell}_1$ with $\bar{\ell}_1 = \begin{bmatrix}0 & 1 & 0 & 0\end{bmatrix}^T$). By substituting

$$\begin{bmatrix} \partial\bar{\ell}_2/\partial\bar{\ell}_1 & \bar{0}_{3\times 1} \\ \bar{0}_{1\times 3} & 1 \end{bmatrix} = \bar{I} - 2\bar{n}_2\bar{n}_2^T$$

(Equation (10.5) with i = 2) (using $\bar{n}_2 = \text{rot}(\bar{m}_2, \pi + \theta_2)\bar{\ell}_1$) and Eq. (10.20) into Eq. (10.33), the following three equations can be obtained based on the fact that $\partial\bar{\ell}_3/\partial\bar{\ell}_2$ (i.e., the reflector matrix with unit normal vector $\bar{n}_3$) should be a symmetrical matrix:

$$\begin{aligned}&\left\{\left[m_x m_z(1-C\mu)+m_y S\mu\right]m_{2x}+\left(m_y^2-m_x^2\right)(1-C\mu)m_{2z}\right\}S(2\theta_2)\\&\quad-\left\{\left[m_y m_x(1-C\mu)+m_z S\mu\right]m_{2x}^2+\left[m_y m_z(1-C\mu)-m_x S\mu\right]m_{2x}m_{2z}\right.\\&\quad\left.-2m_y m_x(1-C\mu)\right\}C(2\theta_2)+\left[m_x m_y(1-C\mu)+m_z S\mu\right]m_{2x}^2\\&\quad+\left[m_y m_z(1-C\mu)-m_x S\mu\right]m_{2x}m_{2z}=0,\end{aligned} \tag{10.34}$$

$$\begin{aligned}&\left\{\left(-m_y^2+m_z^2\right)(1-C\mu)m_{2x}-\left[m_x m_z(1-C\mu)-m_y S\mu\right]m_{2z}\right\}S(2\theta_2)\\&-\left\{\left[m_y m_z(1-C\mu)-m_x S\mu\right]m_{2z}^2+\left[m_x m_y(1-C\mu)+m_z S\mu\right]m_{2x}m_{2z}-2m_y m_z(1-C\mu)\right\}C(2\theta_2)\\&+\left[m_y m_z(1-C\mu)-m_x S\mu\right]m_{2z}^2+\left[m_y m_x(1-C\mu)+m_z S\mu\right]m_{2x}m_{2z}=0,\end{aligned} \tag{10.35}$$

$$\begin{aligned}&\left\{\left[-m_y m_x(1-C\mu)+m_z S\mu\right]m_{2x}-\left[m_y m_z(1-C\mu)+m_x S\mu\right]m_{2z}\right\}S(2\theta_2)\\&\quad+\left[-(m_x m_{2x}+m_z m_{2z})(m_x m_{2z}-m_z m_{2x})(1-C\mu)+m_y S\mu\right]C(2\theta_2)\\&\quad+(m_x m_{2x}+m_z m_{2z})(m_x m_{2z}-m_z m_{2x})(1-C\mu)+m_y S\mu=0.\end{aligned} \tag{10.36}$$

Equations (10.34), (10.35) and (10.36) are all linear dependent equations of $C(2\theta_2)$ and $S(2\theta_2)$. It can be proven numerically that, for an arbitrarily given $\bar{m}_2=[m_{2x}\quad 0\quad m_{2z}\quad 0]^T$ with $m_{2x}^2+m_{2z}^2=1$, the following two equations can be obtained from Eqs. (10.34) and (10.35):

$$S(2\theta_2)=\frac{2m_y(m_x m_{2z}-m_z m_{2x})}{(m_x m_{2z}-m_z m_{2x})^2+m_y^2} \tag{10.37}$$

and

$$C(2\theta_2)=\frac{(m_x m_{2z}-m_z m_{2x})^2-m_y^2}{(m_x m_{2z}-m_z m_{2x})^2+m_y^2}. \tag{10.38}$$

It is noted that these two equations are not functions of μ. Now, θ_2 can be obtained from

$$2\theta_2=\operatorname{atan2}\left(2m_y(m_x m_{2z}-m_z m_{2x}),(m_x m_{2z}-m_z m_{2x})^2-m_y^2\right). \tag{10.39}$$

If $m_x m_{2z}-m_z m_{2x}=0$, then $2\theta_2=\operatorname{atan2}\left(0,-m_y^2\right)$ or $\theta_2=90^\circ$ from Eq. (10.39). It is physically impossible to have such a reflector. Thus, another arbitrarily chosen $\bar{m}_2$ with the form $\bar{m}_2 = [m_{2x} \quad 0 \quad m_{2z} \quad 0]^T$ (where $m_{2x}^2 + m_{2z}^2 = 1$) must be taken.

An alternative method for determining θ_2 and $\bar{m}_2 = [m_{2x} \quad 0 \quad m_{2z} \quad 0]^T$ (where $m_{2x}^2 + m_{2z}^2 = 1$) is to take an arbitrary incidence angle θ_2 (provided that $0 \le \theta_2 < 90°$ and $S^2\theta_2 \ge m_y^2$) and then solve for $\bar{m}_2$ numerically using Eqs. (10.37) and (10.38). Having obtained θ_2 and $\bar{m}_2$, the unit normal vector $\bar{n}_2 = \text{rot}(\bar{m}_2, \pi + \theta_2)\bar{\ell}_1$ with $\bar{\ell}_1 = [0 \quad 1 \quad 0 \quad 0]^T$ and $\bar{m}_2$ of the i = 2 reflector can be determined. Moreover, from Eq. (10.33), the i = 3 reflector matrix $\partial\bar{\ell}_3/\partial\bar{\ell}_2$ can also be obtained. The unit normal vector $\bar{n}_3$ of the i = 3 reflector can then be derived from Eq. (10.4) with i = 3. Having determined the unit normal vectors ($\bar{n}_2$ and $\bar{n}_3$) of the two reflectors, the corresponding prism design can be determined using 3-D solid modeling software.

Example 10.4 Let the validity of the developed methodology be demonstrated using the following right-handed image orientation function:

$$\bar{\Phi} = \begin{bmatrix} 0.229316 & 0.843802 & 0.485192 \\ -0.842691 & 0.421574 & -0.334883 \\ -0.487120 & -0.332073 & 0.807739 \end{bmatrix}.$$

According to the methodology described above, the following results are obtained:

(1) The auxiliary angle μ and auxiliary unit vector $\bar{m}$ are obtained as $\mu = 1.339423$ rad from Eq. (10.25) and $\bar{m} = [0.001443 \quad 0.499466 \quad -0.866333 \quad 0]^T$ from Eqs. (10.26), (10.27) and (10.28).

(2) Choose $\bar{m}_2 = [0.1 \quad 0 \quad 0.994987 \quad 0]^T$ and obtain $\theta_2 = 1.396265$ rad from Eq. (10.39). The unit normal vector of the i = 2 reflector is obtained from $\bar{n}_2 = \text{rot}(\bar{m}_2, \pi + \theta_2)\bar{\ell}_1$ with $\bar{\ell}_1 = [0 \quad 1 \quad 0 \quad 0]^T$ as $\bar{n}_2 = [0.979871 \quad -0.173648 \quad -0.098481 \quad 0]^T$. Consequently, its reflector matrix is determined from Eq. (10.4), as s

$$\frac{\partial\bar{\ell}_2}{\partial\bar{\ell}_1} = \begin{bmatrix} -0.920296 & 0.340305 & 0.192997 \\ 0.340305 & 0.939693 & -0.034202 \\ 0.192997 & -0.034202 & 0.980603 \end{bmatrix}.$$

Finally, from Eq. (10.33), the i = 3 reflector matrix is obtained as

$$\frac{\partial\bar{\ell}_3}{\partial\bar{\ell}_2} = \bar{\Phi}\frac{\partial\bar{\ell}_2}{\partial\bar{\ell}_1} = \begin{bmatrix} 0.169752 & 0.854358 & 0.491179 \\ 0.854358 & 0.120832 & -0.505443 \\ 0.491179 & -0.505443 & 0.709416 \end{bmatrix}.$$

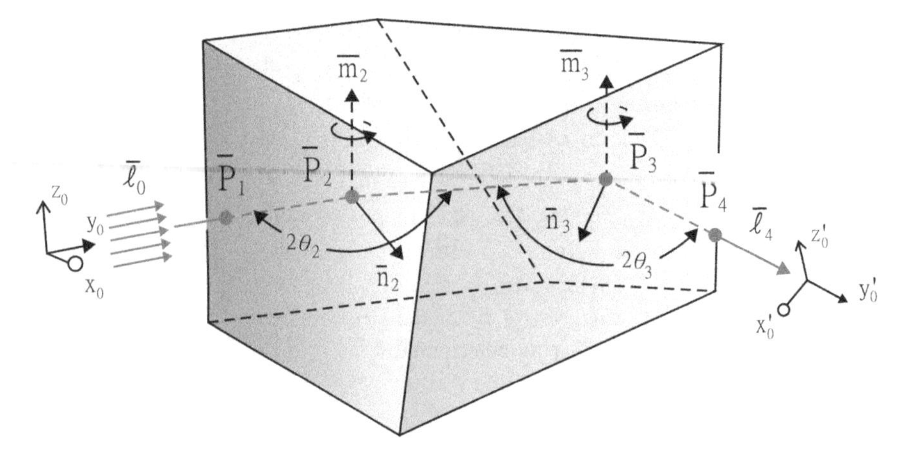

Fig. 10.12 Designed prism in Example 10.4

Table 10.4 Unit normal vectors of reflectors of prism shown in Fig. 10.12

i	$\bar{n}_i$
2	$[0.979871 \quad -0.1736248 \quad -0.098481 \quad 0]^T$
3	$[0.644301 \quad -0.663012 \quad -0.381171 \quad 0]^T$

(3) Substituting $\partial\bar{\ell}_3/\partial\bar{\ell}_2$ into Eq. (10.4) yields $\bar{n}_3 = [0.644301 \quad -0.663012 \quad -0.381171 \quad 0]^T$. The designed prism is shown in Fig. 10.12. (The corresponding unit normal vectors are listed in Table 10.4.)

10.4.4 Solution for Left-Handed Image Orientation Function

A left-handed image orientation function has the form $\bar{\Phi} = [\bar{a} \quad \bar{b} \quad \bar{c}]$ with $\bar{c} = \bar{b} \times \bar{a}$. In order to minimize the number of equations required to complete the design process, let Φ be expressed in terms of the auxiliary angle μ (given in Eq. (10.29)) and auxiliary unit vector $\bar{m}$ (given in Eqs. (10.30), (10.31) and (10.32)). It was shown in Sect. 10.3 that three reflectors (i.e., n = 5 if the first and last flat boundary surfaces are refractive) with unit normal vectors $\bar{n}_2$, $\bar{n}_3$ and $\bar{n}_4$, respectively, are sufficient to output from a prism an image characterized by a left-handed image orientation function $\bar{\Phi}$. After choosing an arbitrary unit vector $\bar{n}_4$ as the unit normal vector of the i = 4 reflector, Eq. (10.11) (with n = 5) yields $(\partial\bar{\ell}_3/\partial\bar{\ell}_2)(\partial\bar{\ell}_2/\partial\bar{\ell}_1)$ as the product of Φ and the reflector matrix $\partial\bar{\ell}_4/\partial\bar{\ell}_3$, i.e.,

$$\frac{\partial\bar{\ell}_3}{\partial\bar{\ell}_2}\frac{\partial\bar{\ell}_2}{\partial\bar{\ell}_1}=\frac{\partial\bar{\ell}_4}{\partial\bar{\ell}_3}\bar{\Phi}. \tag{10.40}$$

It will be recalled that the product of two reflector matrices is a right-handed matrix. Following the procedures described in Sect. 10.4.3, the unit normal vectors $\bar{n}_2$ and $\bar{n}_3$ of the remaining two reflectors can be obtained; thereby completing the design of a prism with three reflectors.

Example 10.5 Let the proposed methodology be demonstrated using the following image orientation function:

$$\bar{\Phi}=\begin{bmatrix}-1&0&0\\0&-1&0\\0&0&-1\end{bmatrix}.$$

Assume that the source, $(xyz)_0$, emits light rays $\bar{\ell}_0=[0\quad 1\quad 0\quad 0]^T$ in the y_0 direction. According to the proposed methodology, the following results are obtained:

(1) Choose $\bar{n}_4=[0.75\quad 0.612372\quad 0.25\quad 0]^T$.
(2) Eq. (10.4) yields

$$\frac{\partial\bar{\ell}_4}{\partial\bar{\ell}_3}=\frac{\partial\bar{\ell}_4(\bar{n}_4)}{\partial\bar{\ell}_3}=\begin{bmatrix}-0.125&-0.918559&-0.375\\-0.918559&0.25&-0.306186\\-0.375&-0.306186&0.875\end{bmatrix}.$$

(3) Eq. (10.40) yields

$$\frac{\partial\bar{\ell}_3}{\partial\bar{\ell}_2}\frac{\partial\bar{\ell}_2}{\partial\bar{\ell}_1}=\frac{\partial\bar{\ell}_4}{\partial\bar{\ell}_3}\bar{\Phi}=\begin{bmatrix}0.125&0.918559&0.375\\0.918559&-0.25&0.306186\\0.375&0.306186&-0.875\end{bmatrix}.$$

(4) The auxiliary angle μ and auxiliary unit vector $\bar{m}$ of the image orientation function are determined to be $\mu=\pi$ from Eq. (10.25) and $\bar{m}=[0.75\quad 0.612372\quad 0.25\quad 0]^T$ from Eqs. (10.26), (10.27) and (10.28).
(5) Choose $\bar{m}_2=[0.6\quad 0\quad 0.8\quad 0]^T$ and obtain $\theta_2=0.9370705\,\text{rad}$ from Eq. (10.39). The unit normal $\bar{n}_2=\text{rot}(\bar{m}_2,\pi+\theta_2)\bar{\ell}_1$ of the i = 2 reflector is obtained as $\bar{n}_2=[-0.612372\quad 0.5\quad 0.612372\quad 0]^T$. Consequently, its reflector matrix is determined from Eq. (10.4) as

$$\frac{\partial \bar{\ell}_2}{\partial \bar{\ell}_1} = \begin{bmatrix} 0.25 & 0.612372 & 0.75 \\ 0.612372 & 0.5 & -0.612372 \\ 0.75 & -0.612372 & 0.25 \end{bmatrix}.$$

Finally, from Eq. (10.40), the i = 3 reflector matrix is obtained as

$$\frac{\partial \bar{\ell}_3}{\partial \bar{\ell}_2} = \frac{\partial \bar{\ell}_4}{\partial \bar{\ell}_3} \bar{\Phi} \frac{\partial \bar{\ell}_2}{\partial \bar{\ell}_1} = \begin{bmatrix} 0.25 & 0.612372 & 0.75 \\ 0.612372 & 0.5 & -0.612372 \\ 0.75 & -0.612372 & 0.25 \end{bmatrix}.$$

(6) $\bar{n}_3 = [0.25 \quad -0.612372 \quad 0.75 \quad 0]^T$ is then obtained by substituting $\partial \bar{\ell}_3 / \partial \bar{\ell}_2$ into Eq. (10.4).

Having obtained the unit normal vectors of the three reflectors, the prism can be designed using 3-D solid modeling software. In this particular example, the prism has the form of a solid glass corner-cube retro-reflector (see Figs. 10.13 and 9.2) with a particular orientation. It is noted that the product of the three matrices $\partial \bar{\ell}_4 / \partial \bar{\ell}_3$, $\partial \bar{\ell}_3 / \partial \bar{\ell}_2$ and $\partial \bar{\ell}_2 / \partial \bar{\ell}_1$ is commutative since $\bar{n}_2$, $\bar{n}_3$ and $\bar{n}_4$ are mutually perpendicular. This confirms a distinguishing feature of solid glass corner-cube retro-reflectors, namely the order of the reflectors encountered by a ray is irrelevant in determining the retro-reflective property.

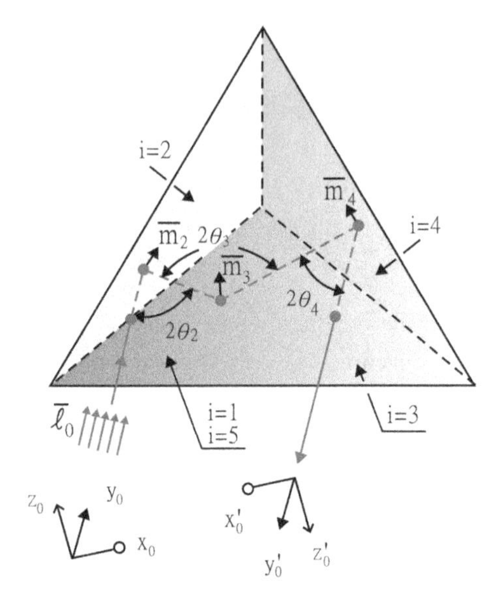

Fig. 10.13 *Solid glass corner-cube* retro-reflector shown in *upper* part of Fig. 9.2a

10.5 Prism Design for Given Image Orientation Using Screw Triangle Method

This section applies the screw triangle method conventionally used in the field of mechanism design to design a prism to produce a specified image orientation. Compared to the methods described in Sects. 10.3 and 10.4, the screw triangle method is more straightforward since its derivations are essentially vector-based calculations.

It will be recalled that an image with a right-handed image orientation function $\bar{\Phi} = [\bar{a} \quad \bar{b} \quad \bar{c}]$ can be expressed in terms of three independent physical qualities, namely the unit vector $\bar{m}$ and auxiliary angle μ, by using the orientation part of $\mathrm{rot}(\bar{m}, \mu)$ to obtain Eq. (10.20). In the screw triangle method, the rotation motion $\mathrm{rot}(\bar{m}, \mu)$ is regarded as a product of two successive rotation motions, namely an initial rotation $\mathrm{rot}(\bar{m}_2, 2\theta_2)$ about the unit vector $\bar{m}_2$ through an angle $2\theta_2$ followed by a second rotation motion $\mathrm{rot}(\bar{m}_3, 2\theta_3)$ about the unit vector $\bar{m}_3$ through an angle $2\theta_3$. From Fig. 10.6, it is seen that the imaged coordinate frame $(xyz)_0^2$ of $(xyz)_0$ reflected by reflector $\bar{r}_2$ changes from a right-handed coordinate frame to a left-handed coordinate frame, and vice versa. Mathematically, $(xyz)_0^2$ can be obtained by rotating the object frame $(xyz)_0$ about the unit directional vector $\bar{m}_2 = [m_{2x} \quad 0 \quad m_{2z} \quad 0]^T$ through a rotation angle $2\theta_2$ $(0 \leq 2\theta_2 < \pi)$ (i.e., $\mathrm{rot}(\bar{m}_2, 2\theta_2)$), and then converting $\mathrm{rot}(\bar{m}_2, 2\theta_2)$ into a left-handed matrix by adding a minus sign to the second column. Here, $\bar{m}_2 = -(\bar{\ell}_1 \times \bar{\ell}_2)/|\bar{\ell}_1 \times \bar{\ell}_2|$. Similarly, the output image, $(xyz)_0'$, can be obtained by rotating $(xyz)_0^2$ about the directional vector $\bar{m}_3 = [m_{3x} \quad m_{3y} \quad m_{3z}]^T$ through a rotation angle $2\theta_3$ $(0 \leq 2\theta_3 < \pi)$, i.e., $\mathrm{rot}(\bar{m}_3, 2\theta_3)$, and then converting the resulting image into the required right-handed coordinate frame $(xyz)_0'$ by adding a minus sign to the second column. Here, $\bar{m}_3 = -(\bar{\ell}_2 \times \bar{\ell}_3)/|\bar{\ell}_2 \times \bar{\ell}_3|$. In other words, the right-handed image orientation function $\bar{\Phi}$ can be regarded as a product of two successive rotation motions, namely a rotation $\mathrm{rot}(\bar{m}_2, 2\theta_2)$ about the unit vector $\bar{m}_2$ through an angle $2\theta_2$ followed by a second rotation $\mathrm{rot}(\bar{m}_3, 2\theta_3)$ about the unit vector $\bar{m}_3$ through an angle $2\theta_3$. Consequently, the image orientation which takes place when an object is reflected by two reflectors can be expressed mathematically as follows:

$$\mathrm{rot}(\bar{m}, \mu) = \mathrm{rot}(\bar{m}_3, 2\theta_3)\mathrm{rot}(\bar{m}_2, 2\theta_2). \tag{10.41}$$

The existence of Eq. (10.41) is in fact already suggested by the result $(\partial\bar{\ell}_3/\partial\bar{\ell}_2)(\partial\bar{\ell}_2/\partial\bar{\ell}_1) = \bar{\Phi}$ obtained from Eq. (10.11) with $n = 4$. However, Eq. (10.41) confirms the feasibility of the screw triangle method as a means of solving the prism design problem for a right-handed image orientation function $\bar{\Phi}$.

The following discussions use the screw triangle geometry method to determine $\bar{m}_3$, given an arbitrarily specified $\bar{m}_2$, in order to design a prism with the minimum number of reflectors (i.e., $n = n_{mini} = 4$). An investigation of the screw triangle

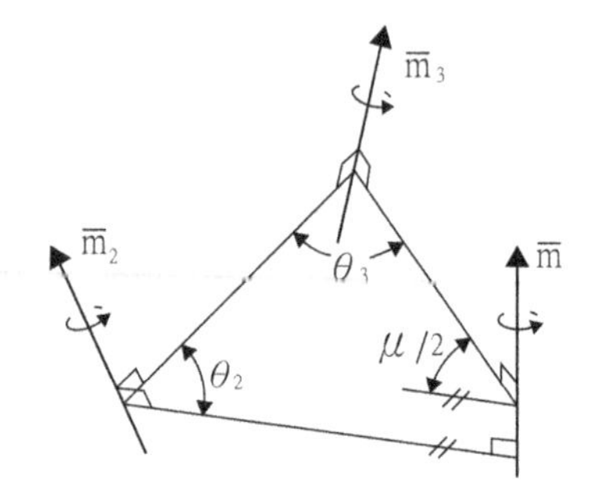

Fig. 10.14 Screw triangle geometry [7, 9]

geometry (see Fig. 10.14) yields the following relations (refer to Eqs. (1), (2) and (3) in [6]):

$$\tan\frac{\mu}{2} = -\frac{\bar{m}_3 \cdot (\bar{m} \times \bar{m}_2)}{(\bar{m}_3 \times \bar{m}) \cdot (\bar{m} \times \bar{m}_2)}, \tag{10.42}$$

$$\tan\theta_2 = \frac{\bar{m}_3 \cdot (\bar{m} \times \bar{m}_2)}{(\bar{m} \times \bar{m}_2) \cdot (\bar{m}_2 \times \bar{m}_3)}, \tag{10.43}$$

$$\tan\theta_3 = \frac{\bar{m}_3 \cdot (\bar{m} \times \bar{m}_2)}{(\bar{m}_2 \times \bar{m}_3) \cdot (\bar{m}_3 \times \bar{m})}. \tag{10.44}$$

It has been proven analytically in [7] that $\lambda = 1$ and $\lambda = e^{\pm i\mu}$ are the eigenvalues of $\bar{\Phi}$ if Eq. (10.20) is used to determine the eigenvalue problem by $\left|\bar{\Phi} - \lambda\bar{I}\right| = 0$. The normalized eigenvector of the first eigenvalue (i.e.,$\lambda = 1$) is the unit directional vector $\bar{m}$. Consequently, the rotation angle μ and directional vector $\bar{m}$ can be determined for a given right-handed image orientation function $\bar{\Phi}$.

As stated above, the aim of the derivations here is to determine $\bar{m}_3$ given an arbitrarily chosen $\bar{m}_2$. According to [8], Eq. (10.42) can be reformulated as

$$\bar{m}_3 \cdot \{(\bar{m} \times \bar{m}_2) + \tan(\mu/2)[\bar{m} \times (\bar{m} \times \bar{m}_2)]\} = 0. \tag{10.45}$$

As described earlier, the y_0 axis of frame $(xyz)_0$ is assumed to be aligned with the direction of the source ray emitted by the object, while the y_0' axis of $(xyz)_0'$ is assumed to be aligned with the output ray. In other words, the unit vector $\bar{b}$ in $\bar{\Phi} = \begin{bmatrix} \bar{a} & \bar{b} & \bar{c} \end{bmatrix}$ describes the direction of the exit ray with respect to $(xyz)_0$ (see Eq. (10.3)). Furthermore, the unit vector $\bar{b}$ is perpendicular to the unit vector $\bar{m}_3$, and thus

$$\bar{m}_3 \cdot \bar{b} = 0. \tag{10.46}$$

Equations (10.45) and (10.46) indicate that $\bar{m}_3$ is the common unit normal vector of $\{(\bar{m} \times \bar{m}_2) + \tan(\mu/2)[\bar{m} \times (\bar{m} \times \bar{m}_2)]\}$ and $\bar{b}$. Consequently, it follows that

$$\bar{m}_3 = \frac{\{(\bar{m} \times \bar{m}_2) + \tan(\mu/2)[\bar{m} \times (\bar{m} \times \bar{m}_2)]\} \times \bar{b}}{|\{(\bar{m} \times \bar{m}_2) + \tan(\mu/2)[\bar{m} \times (\bar{m} \times \bar{m}_2)]\} \times \bar{b}|}, \tag{10.47}$$

where $\bar{m}_2$ is a given arbitrary unit directional vector. θ_2 and θ_3 can then be obtained by substituting $\bar{m}$, $\bar{m}_2$ and $\bar{m}_3$ into Eqs. (10.43) and (10.44), respectively. From Eq. (10.19), $\bar{n}_2$ can be obtained by $\bar{n}_2 = \text{rot}(\bar{m}_2, \pi + \theta_2)\bar{\ell}_1$ with $\bar{\ell}_1 = [0 \quad 1 \quad 0 \quad 0]^T$. Similarly, the unit normal vector $\bar{n}_3$ of the third reflector $\bar{r}_3$ can be obtained from $\bar{n}_3 = \text{rot}(\bar{m}_3, \pi + \theta_3)\bar{\ell}_2$. Having determined the unit normal vectors of all the flat boundary surfaces, i.e., $\bar{n}_1 = \pm[0 \quad 1 \quad 0 \quad 0]^T$, $\bar{n}_2$, $\bar{n}_3$ and $\bar{n}_4 = \pm\bar{b}$, the geometry of the prism is essentially defined.

To design a single prism with the minimum number of reflectors for a particular left-handed image orientation function $\bar{\Phi}$, it is first necessary to choose an arbitrary unit vector $\bar{n}_4$ as the unit normal vector of the fourth reflector. By applying the design procedures described above for the right-handed image orientation function $\bar{\Phi}$, the unit normal vectors $\bar{n}_2$ and $\bar{n}_3$ of the remaining two reflectors can be found; thereby completing the design of a prism with three reflectors.

Example 10.7 Let the right-handed image orientation function given in Example 10.4 be used to demonstrate the validity of the proposed methodology. Recall that object $(xyz)_0$ emits light rays $\bar{\ell}_0 = [0 \quad 1 \quad 0 \quad 0]^T$ in the y_0 direction. Applying the design methodology described above, the following results are obtained:

(1) The eigenvalues of the image orientation function $\bar{\Phi}$ are found to be 1 and $e^{\pm i\mu}$ with $\mu = 1.339423$ rad. The unit directional vector $\bar{m} = [0.001443 \quad 0.499466 \quad -0.866333 \quad 0]^T$ is then obtained from the normalized eigenvector of the eigenvalue equal to 1.
(2) Selecting $\bar{m}_2 = [0.1 \quad 0 \quad 0.994987 \quad 0]^T$ arbitrarily, Eq. (10.47) reveals that the unit directional vector $\bar{m}_3$ is given by $\bar{m}_3 = [0.413752 \quad -0.116977 \quad 0.902843 \quad 0]^T$.
(3) Eqs. (10.43) and (10.44) yield $\theta_2 = 1.396265$ rad and $\theta_3 = 1.16937$ rad, respectively.
(4) Unit normal vectors $\bar{n}_2 = [0.979871 \quad -0.173648 \quad -0.098481 \quad 0]^T$ of the i = 2 reflector and $\bar{n}_3 = [0.644301 \quad -0.663012 \quad -0.381171 \quad 0]^T$ of the i = 3 reflector are obtained from $\bar{n}_2 = \text{rot}(\bar{m}_2, \pi + \theta_2)\bar{\ell}_1$ and $\bar{n}_3 = \text{rot}(\bar{m}_3, \pi + \theta_3)\bar{\ell}_2$, respectively.

The corresponding prism design is shown in Fig. 10.12 with its unit normal vectors listed in Table 10.4.

References

1. Smith WJ (2001) Modern optical engineering, 3rd edn. Edmund Industrial Optics, Barrington
2. Galvez EJ, Holmes CD (1999) Geometric phase of optical rotators. J Opt Soc Am A 16:1981–1985
3. Stewart GW (1973) Introduction to matrix computations. Academic press, Inc., London LTD
4. Tsai CY, Lin PD (2006) Prism design based on image orientation change. Appl Opt 45:3951–3959
5. Paul RP (1982) Robot manipulators-mathematics, programming and control. MIT press, Cambridge
6. Dai JS (2006) An historical review of the theoretical development of rigid body displacements from Rodrigues parameters to the finite twist. Mech Mach Theory 41:41–52
7. Roth B (1967) On the screw axes and other special lines associated with spatial displacements of a rigid body. ASME J Eng Ind 89:102–110
8. Spiegel MR (1989) Schaum's outline of vector analysis. McGraw Hill
9. Tsai LW, Roth B (1972) Desgn of dyads with helical, cylindrical, spherical, revolution and prismatic joints. Mech Mach Theory 7:85–102

Chapter 11
Determination of Prism Reflectors to Produce Required Image Orientation

Optical prisms provide the ability to output an image with a certain orientation and to relocate the exit ray in a given manner. In designing a prism, one of the most important tasks is that of determining the equations of the prism reflectors which produce a given image orientation. Chapter 10 presented a methodology for determining the unit normal vectors $\bar{n}_i$ (i = 2 to i = n − 1) of a prism's reflectors for a given image orientation function $\bar{\Phi}$ without determination the parameters e_i (i = 2 to i = n − 1). In this chapter, an image offset distance is introduced to characterize the length of the common normal segment between the entrance ray of a prism and the exit ray. A methodology is then proposed for determining the equations of the prism reflectors which achieve a given image offset distance [1]. Finally, the conditions under which a bundle of entrance rays can pass fully through a prism without being blocked by any of its boundary surfaces are derived.

11.1 Determination of Reflector Equations

It will be recalled that a reflector can be described completely in terms of its unit normal vector $\bar{n}_i = \begin{bmatrix} n_{ix} & n_{iy} & n_{iz} & 0 \end{bmatrix}^T$ and a parameter e_i. Chapter 10 presented various approaches for determining the unit normal vectors $\bar{n}_i$ (i = 2 to i = n − 1) of a prism's reflectors. However, to fully determine the prism dimensions, parameter e_i (i = 2 to i = n − 1) is also required for each reflective surface. Accordingly, this chapter introduces an image offset vector $\overline{P_2P_{n-1}}$ and image offset distance L in order to facilitate the prism design process.

For most prism applications, the rays enter and exit the first and last surfaces (i.e., $\bar{r}_1$ and $\bar{r}_n$, respectively) perpendicularly. As described in Chap. 10, the refraction processes at these surfaces have no effect on the image orientation produced by the remaining (n − 2) reflectors. Assume that the object emits source rays $\bar{R}_0$ in the y_0 direction and the rays are incident on $\bar{r}_1$ in a normal direction.

P.D. Lin, *Advanced Geometrical Optics*, Progress in Optical Science and Photonics 4, DOI 10.1007/978-981-10-2299-9_11

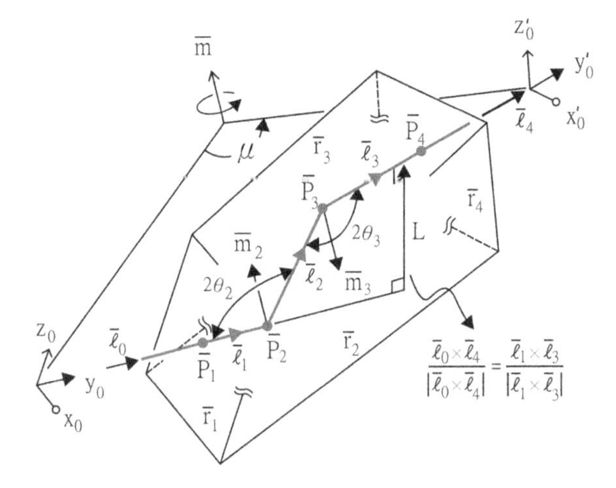

Fig. 11.1 Path of ray traveling through a prism with four flat boundary surfaces

Consequently, the unit normal vector $\bar{n}_1$ of the entry surface and the unit directional vectors $\bar{\ell}_0$ and $\bar{\ell}_1$ of the source and refracted ray, respectively (see Fig. 11.1), are given as

$$\bar{n}_1 = \pm[0 \quad 1 \quad 0 \quad 0]^T, \tag{11.1}$$

$$\bar{\ell}_0 = \bar{\ell}_1 = [0 \quad 1 \quad 0 \quad 0]^T. \tag{11.2}$$

If the image orientation function $\bar{\Phi}$ is given in numeric form as in Eq. (10.11), the unit normal vector $\bar{n}_n$ and unit directional vectors $\bar{\ell}_n$ and $\bar{\ell}_{n-1}$ can be obtained from Eqs. (10.10) and (10.3), respectively, as

$$\bar{n}_n = \pm\begin{bmatrix} \bar{\Phi} & \bar{t}_{3\times 1} \\ \bar{0}_{1\times 3} & 1 \end{bmatrix}\bar{n}_1 = \pm\begin{bmatrix} a_x & b_x & c_x & t_x \\ a_y & b_y & c_y & t_y \\ a_z & b_z & c_z & t_z \\ 0 & 0 & 0 & 1 \end{bmatrix}\begin{bmatrix} 0 \\ 1 \\ 0 \\ 0 \end{bmatrix} = \pm\begin{bmatrix} b_x \\ b_y \\ b_z \\ 0 \end{bmatrix}, \tag{11.3}$$

$$\bar{\ell}_n = \bar{\ell}_{n-1} = \begin{bmatrix} \bar{\Phi} & \bar{t}_{3\times 1} \\ \bar{0}_{1\times 3} & 1 \end{bmatrix}\bar{\ell}_0 = \begin{bmatrix} a_x & b_x & c_x & t_x \\ a_y & b_y & c_y & t_y \\ a_z & b_z & c_z & t_z \\ 0 & 0 & 0 & 1 \end{bmatrix}\begin{bmatrix} 0 \\ 1 \\ 0 \\ 0 \end{bmatrix} = \begin{bmatrix} b_x \\ b_y \\ b_z \\ 0 \end{bmatrix}. \tag{11.4}$$

Note that, with no loss of generality, this chapter takes $\bar{n}_1 = [0 \quad -1 \quad 0 \quad 0]^T$ and $\bar{n}_n = -[b_x \quad b_y \quad b_z \quad 0]^T$ as the unit normal vectors of reflectors $\bar{r}_1$ and $\bar{r}_n$, respectively. It will be recalled that $-e_i$ represents the distance from the origin o_0 of coordinate frame $(xyz)_o$ to the origin o_i of coordinate frame $(xyz)_i$ along the direction of the unit normal vector $\bar{n}_i = [n_{ix} \quad n_{iy} \quad n_{iz} \quad 0]^T$ of reflector $\bar{r}_i$ (see Fig. 1.16 with g = 0). Mathematically, $-e_i$ can be computed as the dot product of vector $\bar{n}_i$ and a vector starting from o_0 and finishing at any point on the reflector. One of the choices for this point is

$$\bar{o}_0\bar{P}_i = \bar{o}_0\bar{P}_0 + \bar{P}_0\bar{P}_1 + \bar{P}_1\bar{P}_2 + \ldots + \bar{P}_{i-1}\bar{P}_i = \bar{o}_0\bar{P}_0 + \lambda_1\bar{\ell}_0 + \lambda_2\bar{\ell}_1 + \ldots + \lambda_i\bar{\ell}_{i-1},$$

i.e., the sum of the vectors of a ray path between neighboring boundary surfaces $\bar{r}_{i-1}$ and $\bar{r}_1$ (i = 1 to i = n) (see Fig. 11.1). In other words, parameter e_i can be determined as

$$e_i = -\bar{n}_i \cdot \bar{o}_0\bar{P}_i = -\bar{n}_i \cdot (\bar{o}_0\bar{P}_0 + \lambda_1\bar{\ell}_0 + \lambda_2\bar{\ell}_1 + \ldots + \lambda_i\bar{\ell}_{i-1}). \tag{11.5}$$

In practice, λ_1, λ_2 and λ_n have arbitrary positive values since, as described above, for a prism, the rays enter $\bar{r}_1$ and exit $\bar{r}_n$ perpendicularly. Thus, parameters e_1, e_2 and e_n of reflectors $\bar{r}_1$, $\bar{r}_2$ and $\bar{r}_n$, respectively, can be determined by assuming $\bar{P}_0 = [0 \quad 0 \quad 0 \quad 1]^T$. Substituting Eqs. (11.2) and (11.4) into Eq. (11.5) with $\bar{n}_1 = [0 \quad -1 \quad 0 \quad 0]^T$ and $\bar{n}_n = -[b_x \quad b_y \quad b_z \quad 0]^T$, e_1, e_2 and e_n are obtained as

$$e_1 = \lambda_1, \tag{11.6}$$

$$e_2 = -n_{2y}(\lambda_1 + \lambda_2), \tag{11.7}$$

$$\begin{aligned} e_n = {} & b_x(\lambda_3\ell_{2x} + \ldots + \lambda_{n-1}\ell_{n-2x} + \lambda_n b_x) + b_y(\lambda_1 + \lambda_2 + \lambda_3\ell_{2y} + \ldots + \lambda_{n-1}\ell_{n-2y} + \lambda_n b_y) \\ & + b_z(\lambda_3\ell_{2z} + \ldots + \lambda_{n-1}\ell_{n-2z} + \lambda_n b_z). \end{aligned} \tag{11.8}$$

Equations (11.5) to (11.8) indicate that with arbitrary positive values of λ_1, λ_2 and λ_n, if $\bar{n}_i$ (i = 2 to i = n − 1), λ_i (i = 3 to i = n) and $\bar{\ell}_i$ (i = 2 to i = n − 2) are known, equations $\bar{r}_i = [n_{ix} \quad n_{iy} \quad n_{iz} \quad e_i]^T$ (i = 1 to i = n) of the prism reflectors are fully defined and hence the prism dimensions can be determined accordingly.

It was concluded in Sect. 10.4 that many different configurations of prisms having the minimum number of reflectors can be found which all produce the required image orientation function. Therefore, in the absence of any other constraint(s), it is impossible to determine the equations of the prism reflectors. To address this problem, let the vector starting from $\bar{P}_2$ and ending at $\bar{P}_{n-1}$ be defined as the image offset vector $\overline{P_2P}_{n-1}$ (see Fig. 10.3) and be obtained by summing up $\bar{P}_{i-1}\bar{P}_i = \lambda_i\bar{\ell}_{i-1}$ (i = 3 to i = n − 1) as follows:

$$\bar{P}_2\bar{P}_{n-1} = \lambda_3\bar{\ell}_2 + \ldots + \lambda_i\bar{\ell}_{i-1} + \ldots + \lambda_{n-1}\bar{\ell}_{n-2} = \sum_{i=3}^{n-1}\lambda_i\bar{\ell}_{i-1}. \tag{11.9}$$

Furthermore, in order to quantitatively describe the offset of the output image, let the distance L of the common perpendicular line segment between ray $\bar{R}_1$ and ray $\bar{R}_{n-1}$ be defined as the image offset distance. Mathematically, L can be obtained as the dot product of the image offset vector $\bar{P}_2\bar{P}_{n-1}$ and $(\bar{\ell}_1 \times \bar{\ell}_{n-1})/|\bar{\ell}_1 \times \bar{\ell}_{n-1}| = (\bar{\ell}_0 \times \bar{\ell}_n)/|\bar{\ell}_0 \times \bar{\ell}_n|$, i.e., the unit vector of the common perpendicular vector of $\bar{\ell}_0$ and $\bar{\ell}_n$. In other words, L is given as

$$L = (\lambda_3\bar{\ell}_2 + \ldots + \lambda_i\bar{\ell}_{i-1} + \ldots + \lambda_{n-1}\bar{\ell}_{n-2}) \cdot \frac{(\bar{\ell}_0 \times \bar{\ell}_n)}{|\bar{\ell}_0 \times \bar{\ell}_n|}. \tag{11.10}$$

It is noted from Eq. (11.10) that L may be either positive or negative. By substituting $\bar{\ell}_0$ and $\bar{\ell}_n$ from Eqs. (11.2) and (11.4), respectively, into Eq. (11.10), the image offset distance is obtained as

$$L = (\lambda_3\bar{\ell}_2 + \ldots + \lambda_i\bar{\ell}_{i-1} + \ldots + \lambda_{n-1}\bar{\ell}_{n-2}) \cdot \left(\begin{bmatrix} b_z & 0 & -b_x & 0 \end{bmatrix}^T \Big/ \sqrt{b_x^2 + b_z^2} \right). \tag{11.11}$$

The magnitude of L (i.e., $|L|$) is the length of the common normal line segment of the entrance ray $\bar{R}_0$ and exit ray $\bar{R}_n$. Therefore, physically, the length of the image offset vector $\bar{P}_2\bar{P}_{n-1}$ must be greater than $|L|$. That is,

$$|\lambda_3\bar{\ell}_2 + \ldots + \lambda_i\bar{\ell}_{i-1} + \ldots + \lambda_{n-1}\bar{\ell}_{n-2}| \geq |L|. \tag{11.12}$$

Notably, the reflector equations $\bar{r}_i$ (i = 1 to i = n) of a prism with the minimum number of reflectors can be expressed in terms of L for a given image orientation function $\bar{\Phi}$ when Eq. (11.12) is satisfied.

11.2 Determination of Prism with n = 4 Boundary Surfaces to Produce Specified Right-Handed Image Orientation

It was concluded in Sect. 10.2 that a prism with two reflectors (i.e., n = 4 if the first and last flat boundary surfaces are refractive; see Fig. 11.1) having unit normal vectors $\bar{n}_2$ and $\bar{n}_3$ is sufficient to produce an image orientation defined by a right-handed function $\bar{\Phi}$. In this case Eq. (11.11) can be simplified as

$$L = (\lambda_3\bar{\ell}_2) \cdot \left(\begin{bmatrix} b_z & 0 & -b_x & 0 \end{bmatrix}^T \Big/ \sqrt{b_x^2 + b_z^2} \right). \tag{11.13}$$

The following discussions present a step-by-step approach for obtaining a prism which produces a given right-handed merit function $\bar{\Phi}$.

(1) Obtain the auxiliary angle μ and auxiliary unit vector $\bar{m} = [m_x \quad m_y \quad m_z \quad 0]^T$ from $\bar{\Phi}$ using Eqs. (10.25) to (10.28).
(2) Arbitrarily select $\bar{m}_2 = [m_{2x} \quad 0 \quad m_{2z} \quad 0]^T$ with $m_{2x}^2 + m_{2z}^2 = 1$ and solve for θ_2 from Eq. (10.39).
(3) Calculate the unit normal vector $\bar{n}_2 = \mathrm{rot}(\bar{m}_2, \pi + \theta_2)\bar{\ell}_1$ from Eq. (10.19) with $\bar{\ell}_1 = [0 \quad 1 \quad 0 \quad 0]^T$.
(4) Calculate the unit directional vector $\bar{\ell}_2$ from Eq. (10.18) to give

$$\bar{\ell}_2 = [m_{2z}S(2\theta_2) \quad -C(2\theta_2) \quad -m_{2x}S(2\theta_2) \quad 0]^T.$$

(5) Substitute the expression for $\bar{\ell}_2$ obtained from step (4) into Eq. (11.13) to give

$$\lambda_3 = \lambda_3(m_{2x}, m_{2z}) = \frac{L\sqrt{b_x^2 + b_z^2}}{2m_y}\left[\frac{(m_x m_{2z} - m_z m_{2x})^2 + m_y^2}{(m_x m_{2z} - m_z m_{2x})(b_x m_{2x} + b_z m_{2z})}\right]. \quad (11.14)$$

It is noted from Eq. (11.14) that λ_3 is a nonlinear function of the auxiliary unit common normal vector $\bar{m}_2 = [m_{2x} \quad 0 \quad m_{2z} \quad 0]^T$, where $m_{2x}^2 + m_{2z}^2 = 1$. The following discussions address the problem of minimizing the prism size by minimizing λ_3 for a given L. Mathematically, this problem can be stated as follows: Minimize the function $\lambda_3 = \lambda_3(m_{2x}, m_{2z})$ subject to the constraint $m_{2x}^2 + m_{2z}^2 = 1$. Applying optimization theory with two independent variables and one constraint on λ_3, the following equation is obtained:

$$\frac{m_{2z}}{m_{2x}} = \frac{b_z m_x(1 - m_x^2) + b_x m_z(1 - m_z^2) \pm \sqrt{m_y^2(m_x^2 + m_z^2)\left[b_x^2 + b_z^2 - (b_x m_z - b_z m_x)^2\right]}}{b_z m_z\left(m_x^2 - m_y^2\right) + b_x m_x(1 - m_z^2)}. \quad (11.15)$$

The ambiguous sign of the root indicates the existence of two possible solutions of $\bar{m}_2 = [m_{2x} \quad 0 \quad m_{2z} \quad 0]^T$. In practice, only one of these solutions is useful, and the appropriate sign must therefore be selected in order to satisfy Eq. (11.12) and obtain a positive value of λ_3.

(6) Calculate $\bar{n}_3$, i.e., the unit normal vector of $\bar{r}_3$, from $\partial\bar{\ell}_3/\partial\bar{\ell}_2 = \bar{\Phi}(\partial\bar{\ell}_2/\partial\bar{\ell}_1)$ obtained from Eq. (10.11) with n = 4.
(7) Obtain the equation of $\bar{r}_1 = [0 \quad -1 \quad 0 \quad e_1]^T$ by using $e_1 = \lambda_1$ with an arbitrary positive value of λ_1.
(8) Obtain the equation of reflector $\bar{r}_2 = [n_{2x} \quad n_{2y} \quad n_{2z} \quad e_2]^T$ from $\bar{n}_2$ in step (3) and Eq. (11.7) using λ_1 obtained in step (7) and an arbitrary positive value of λ_2.

(9) Determine reflector $\bar{r}_3 = [n_{3x} \quad n_{3y} \quad n_{3z} \quad e_3]^T$ with $\bar{n}_3$ from step (6) and e_3 from Eq. (11.5) by using $\bar{\ell}_0 = \bar{\ell}_1 = [0 \quad 1 \quad 0 \quad 0]^T$ and $\bar{\ell}_2$ in step (4) while using Eq. (11.5), i.e.,

$$\begin{aligned} e_3 = & -n_{3x}(\lambda_3 m_{2z} S(2\theta_2)) - n_{3y}(\lambda_1 + \lambda_2 - \lambda_3 C(2\theta_2)) \\ & - n_{3z}(-\lambda_3 m_{2x} S(2\theta_2)). \end{aligned} \tag{11.16}$$

(10) Determine the equation of the last reflector $\bar{r}_4 = [n_{4x} \quad n_{4y} \quad n_{4z} \quad e_4]^T$ with $\bar{n}_4$ from $\bar{n}_4 = -[b_x \quad b_y \quad b_z \quad 0]^T$ and e_4 from Eq. (11.8) with n = 4, i.e.,

$$\begin{aligned} e_4 = & b_x(\lambda_3 m_{2z} S(2\theta_2) + \lambda_4 b_x) + b_y(\lambda_1 + \lambda_2 - \lambda_3 C(2\theta_2) + \lambda_4 b_y) \\ & + b_z(-\lambda_3 m_{2x} S(2\theta_2) + \lambda_4 b_z), \end{aligned} \tag{11.17}$$

by using $\bar{\ell}_2$ in step (4), the chosen values of λ_1 and λ_2, and an arbitrary positive value of λ_4.

Example 11.1 Determine the reflectors of the most compact prism (n = 4) possible to produce the following right-handed merit function Γ given an image offset distance of L = 80 mm:

$$\bar{\Phi} = \begin{bmatrix} 0.607397 & -0.454727 & 0.651377 \\ 0.472274 & 0.866025 & 0.164186 \\ -0.638769 & 0.207902 & 0.740777 \end{bmatrix}.$$

In accordance with the design methodology described above, the following results are obtained:

(1) $\mu = 0.918391$ rad and $\bar{m} = [0.027507 \quad 0.811795 \quad 0.583294 \quad 0]^T$.
(2) $\bar{m}_2 = [-0.803477 \quad 0 \quad 0.595336 \quad 0]^T$ and $\theta_2 = 1.93194$ rad.
(3) $\bar{n}_2 = [0.264569 \quad -0.568619 \quad 0.778894 \quad 0]^T$.
(4) $\bar{\ell}_2 = [0.300878 \quad 0.353345 \quad 0.885787 \quad 0]^T$.
(5) $\lambda_3 = 85.9578$.
(6) $\bar{n}_3 = [-0.664422 \quad 0.450812 \quad -0.596081 \quad 0]^T$.
(7) $e_1 = \lambda_1$ with an arbitrary positive value of λ_1.
(8) $e_2 = 0.568619(\lambda_1 + \lambda_2)$ with an arbitrary positive value of λ_2.
(9) $e_3 = 48.87482 - 0.450812(\lambda_1 + \lambda_2)$.
(10) $e_4 = -30.377488 - 0.866025(\lambda_1 + \lambda_2) - \lambda_4$ with an arbitrary positive value of λ_4.

The designed prism with $\lambda_1 + \lambda_2 = 100$ mm and $\lambda_4 = 40$ mm has the form shown in Fig. 11.2.

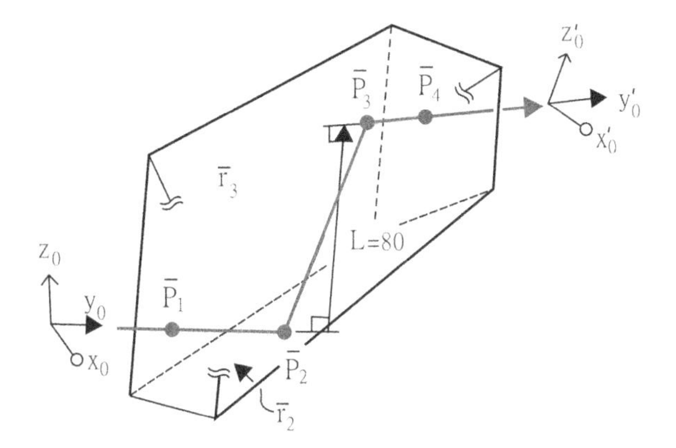

Fig. 11.2 Designed prism for Example 11.1

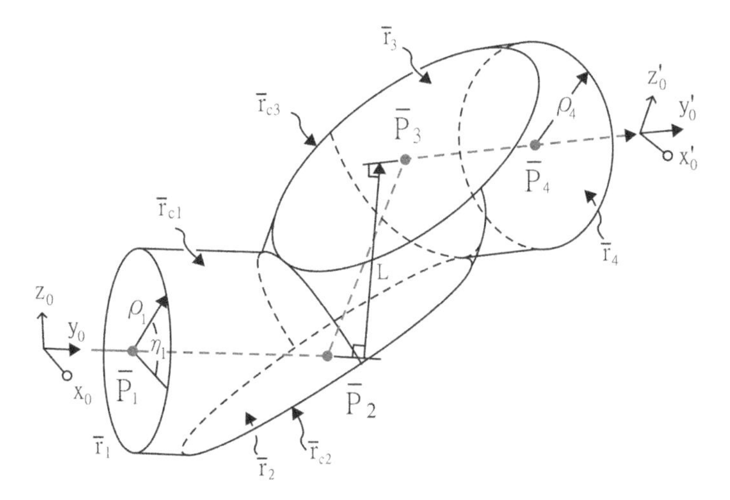

Fig. 11.3 Volume occupied by rays traveling through a prism with n = 4

Example 11.2 The prism obtained in Sect. 11.2 does not guarantee that the entire bundle of entrance rays will pass through the prism. Thus, this example designs the reflectors of a prism which guarantee that a given circular bundle of rays can pass completely through the prism without being blocked by any of the boundary surfaces. Note that if the cross-section of the entrance rays is not circular, say rectangular, then the bundle of entrance rays possessing the cross-section of the minimum circumscribed circle of the rectangular form is used in the current method. Figure 11.3 shows a circular bundle of entrance rays $\bar{R}_0$, which are confined by the surface $\bar{r}_{c1}$. As shown, the rays are reflected first by reflector $\bar{r}_2$ and then by reflector $\bar{r}_3$, and finally pass through the prism. The circular bundle of entrance rays $\bar{R}_0$ are incident on reflector $\bar{r}_2$ in a region bounded by curve $\bar{r}_{c2}$. The rays $\bar{R}_2$

reflected from $\bar{r}_2$ then hit reflector $\bar{r}_3$ in the region bounded by curve $\bar{r}_{c3}$. From the figure, it is seen that for a given L, the entire circular bundle of entrance rays confined by $\bar{r}_{c1} = [\rho_1 C\eta_1 \quad 0 \quad \rho_1 S\eta_1 \quad 1]^T$ (defined with respect to boundary coordinate frame $(xyz)_1$) can pass through the prism without being blocked by $\bar{r}_2$ and $\bar{r}_3$ if one of the following two conditions is met: (1) curves $\bar{r}_{c1}$ and $\bar{r}_{c3}$ intersect at most at one point, or (2) curves $\bar{r}_{c2}$ and $\bar{r}_{c3}$ intersect at most at one point. For a given image offset distance L, these two conditions yield two different solutions for ρ_1. The smaller value of ρ_1 defines the maximum radius of the circular bundle of entrance rays which can pass completely through the prism without being blocked. Taking the right-handed merit function $\bar{\Phi}$ and image offset distance L = 80 considered in Example 11.1 for illustration purposes, the maximum radius of the entrance rays confined by $\bar{r}_{c1}$ is found to be $\rho_1 = 41.6251$.

11.3 Determination of Prism with n = 5 Boundary Surfaces to Produce Specified Left-Handed Image Orientation

A left-handed image orientation function $\bar{\Phi}$ has the form shown in Eq. (10.11) with $\bar{c} = \bar{b} \times \bar{a}$. It was concluded in Sect. 10.2 that three reflectors (i.e., n = 5 if the first and last flat boundary surfaces are refractive) having unit normal vectors $\bar{n}_2$, $\bar{n}_3$ and $\bar{n}_4$, respectively, are sufficient to construct a prism which produces an image orientation defined by a left-handed image orientation function $\bar{\Phi}$ (e.g., Figure 11.4). Consider the prism shown in Fig. 11.4, in which the roof-pair reflectors, $\bar{n}_3$ and $\bar{n}_4$, are perpendicular to one another (denoted as $\bar{r}_3 \perp \bar{r}_4$). For simplicity, let the roof coordinate frame $(xyz)_r$ be used to define the pose of the roof-pair reflectors rather than coordinate frames $(xyz)_3$ and $(xyz)_4$. Furthermore, let the x_r axis be aligned

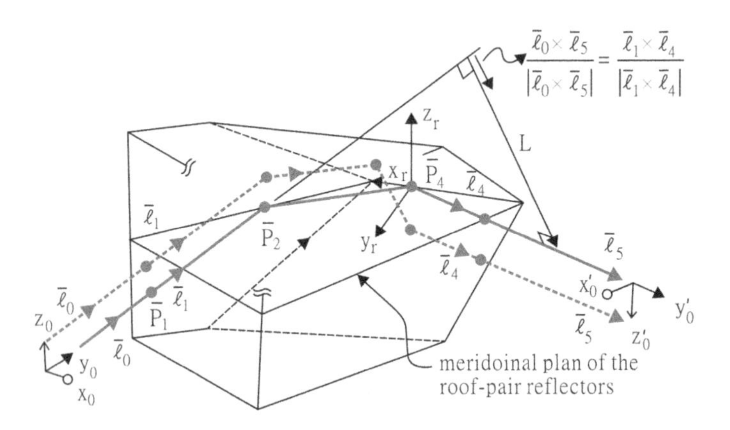

Fig. 11.4 Path of chief ray traveling through a prism of n = 5 with roof-pair reflectors

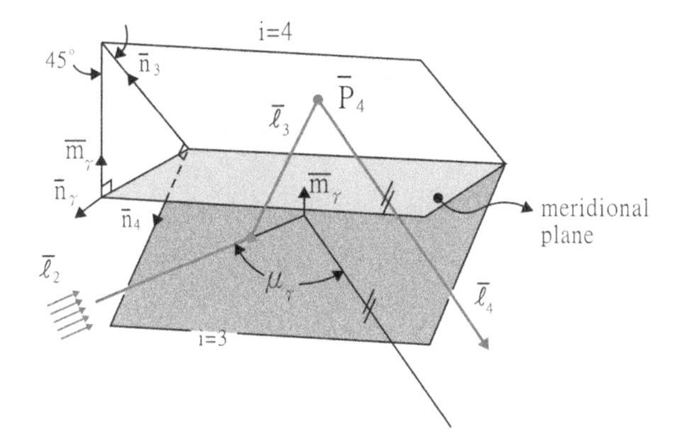

Fig. 11.5 Pose of roof-pair reflectors

with the intersection line of the two reflectors $\bar{r}_3$ and $\bar{r}_4$, and let y_r lie in the meridional plane. From Snell's law, the unit directional vector of the exit ray of the roof-pair reflectors is given by ${}^r\bar{\ell}_4 = [\ell_{2x} \quad -\ell_{2y} \quad -\ell_{2z} \quad 0]^T$, where ${}^r\bar{\ell}_2 = [\ell_{2x} \quad \ell_{2y} \quad \ell_{2z} \quad 0]^T$ is the unit directional vector of the entrance ray. (Note that both rays are expressed with respect to $(xyz)_r$.) Without loss of generality, let ray $\bar{\ell}_2$ be constrained such that it lies on the meridional plane (see Fig. 11.5) of the roof-pair reflectors. Therefore, in the case of a prism with n = 5, Eq. (11.11) can be simplified as

$$\begin{aligned} L &= (\lambda_3\bar{\ell}_2 + \lambda_4\bar{\ell}_3) \cdot \left([b_z \quad 0 \quad -b_x \quad 0]^T \Big/ \sqrt{b_x^2 + b_z^2}\right) \\ &= (\lambda_3\bar{\ell}_2) \cdot \left([b_z \quad 0 \quad -b_x \quad 0]^T \Big/ \sqrt{b_x^2 + b_z^2}\right). \end{aligned} \tag{11.18}$$

The following discussions present a step-by-step approach for obtaining a prism which produces a given left-handed merit function $\bar{\Phi}$.

(1) Obtain the auxiliary angle μ and auxiliary unit vector $\bar{m} = [m_x \quad m_y \quad m_z \quad 0]^T$ with $m_x^2 + m_y^2 + m_z^2 = 1$ from $\bar{\Phi}$ using Eqs. (10.29) to (10.32).
(2) Arbitrarily select $\bar{m}_2 = [m_{2x} \quad 0 \quad m_{2z} \quad 0]^T$ with $m_{2x}^2 + m_{2z}^2 = 1$ and solve for θ_2 from Eq. (10.39).
(3) Calculate the unit normal vector $\bar{n}_2 = \mathrm{rot}(\bar{m}_2, \pi + \theta_2)\bar{\ell}_1$ from Eq. (10.19) with $\bar{\ell}_1 = [0 \quad 1 \quad 0 \quad 0]^T$.
(4). Obtain the equation of $\bar{r}_1 = [0 \quad -1 \quad 0 \quad e_1]^T$ using $e_1 = \lambda_1$ with an arbitrary positive value of λ_1.
(5). Obtain the equation of $\bar{r}_2 = [n_{2x} \quad n_{2y} \quad n_{2z} \quad e_2]^T$ using $\bar{n}_2$ in step (3) and Eq. (11.7) with λ_1 chosen in step (4) and an arbitrary positive value of λ_2

(6) Calculate the unit directional vector $\bar{\ell}_2$ from Eq. (10.18) to give $\bar{\ell}_2 = [m_{2z}S(2\theta_2) \quad -C(2\theta_2) \quad -m_{2x}S(2\theta_2) \quad 0]^T$. Substitute $\bar{\ell}_2$ into Eq. (11.18) to obtain

$$\lambda_3 = \frac{L\sqrt{b_x^2 + b_z^2}}{(b_x m_{2x} + b_z m_{2z})S\mu_2}. \tag{11.19}$$

Note that λ_3 is a function of the image offset distance L and $\bar{m}_2 = [m_{2x} \quad 0 \quad m_{2z} \quad 0]^T$ with $m_{2x}^2 + m_{2z}^2 = 1$. It is possible to determine the most compact prism for a given L by minimizing $\lambda_3 = \lambda_3(m_{2x}, m_{2z})$ subject to the constraint $m_{2x}^2 + m_{2z}^2 = 1$ so as to guarantee $|\lambda_3/L| \geq 1$ (see Eq. (11.12)). Applying optimization theory with two independent variables and one constraint on λ_3 yields

$$\frac{m_{2z}}{m_{2x}} = \frac{-G \pm \sqrt{G^2 - FH}}{F}, \tag{11.20}$$

where

$$F = (m_x^2 - m_z^2)(1 + C\mu) - m_z^2(m_x^2 + m_z^2)(1 - C\mu) + 2m_x m_y m_z S\mu, \tag{11.21}$$

$$G = -2m_x m_z(1 + C\mu) - m_x m_z(m_x^2 + m_z^2)(1 - C\mu) + m_y(m_x^2 - m_z^2)S\mu, \tag{11.22}$$

$$H = -(m_x^2 - m_z^2)(1 + C\mu) - m_x^2(m_x^2 + m_z^2)(1 - C\mu) - 2m_x m_y m_z S\mu. \tag{11.23}$$

Again, the ambiguous sign of the root indicates the existence of two possible solutions of $\bar{m}_2$. As before, only one of these solutions is useful. Hence, the sign should be selected in such a way as to fulfill Eq. (11.12) and obtain a positive value of λ_3.

(7) Having obtained $\bar{\Phi}$ and $\bar{n}_2$, obtain the unit vector $\bar{m}_r$ and angle μ_r from Eq. (10.45).
(8) Determine the unit vector $\bar{n}_r$ of the roof-pair reflectors from $\bar{n}_r = \text{rot}(\bar{m}_r, \pi + \mu_r/2)\bar{\ell}_2$.
(9) Determine e_r of the roof-pair reflectors $\bar{r}_3 \perp \bar{r}_4 = [n_{rx} \quad n_{ry} \quad n_{rz} \quad e_r]^T$ from

$$e_r = -n_{rx}(\lambda_3 m_{2z}S(2\theta_2)) - n_{ry}(d_{2y} - \lambda_3 C(2\theta_2)) - n_{rz}(-\lambda_3 m_{2x}S(2\theta_2)). \tag{11.24}$$

(10) Obtain the equation of the exit reflector $\bar{r}_5 = [n_{5x} \quad n_{5y} \quad n_{5z} \quad e_5]^T$ using e_5 from Eq. (11.8) and an arbitrary positive value of λ_5. Here,

$$e_5 = -b_x(\lambda_3 m_{2z} S\mu_2 + \lambda_5 b_x) - b_y(\lambda_1 + \lambda_2 - \lambda_3 C\mu_2 + \lambda_5 b_y) - b_z(-\lambda_3 m_{2x} S\mu_2 + \lambda_5 b_z). \quad (11.25)$$

Example 11.4 below demonstrates the application of the proposed methodology to the design of the most compact prism with roof-pair reflectors possible for a given left-handed function $\bar{\Phi}$.

Example 11.3 Determine the pose matrices of a prism containing n = 5 boundary surfaces including a roof-pair reflector to produce the following left-handed merit function $\bar{\Phi}$ with an image offset distance of $L = 80\,\text{mm}$:

$$\bar{\Phi} = \begin{bmatrix} -0.607397 & -0.454727 & 0.651377 \\ -0.472274 & 0.866025 & 0.164186 \\ 0.638769 & 0.207902 & 0.740777 \end{bmatrix}.$$

In accordance with the design methodology described above, the following results are obtained:

(1) $\mu = 2.61859\,\text{rad}$ and $\bar{m} = [-0.372472 \quad 0.012621 \quad -0.927957 \quad 0]^T$.
(2) $\bar{m}_2 = [-0.803477 \quad 0 \quad 0.595336 \quad 0]^T$ and $\theta_2 = 0.813145\,\text{rad}$.
(3) $\bar{n}_2 = [0.432484 \quad -0.687216 \quad 0.583688 \quad 0]^T$.
(4) $e_1 = \lambda_1$ with an arbitrary positive value of λ_1.
(5) $e_2 = 0.687216(\lambda_1 + \lambda_2)$ with the chosen λ_1 and an arbitrary positive value of λ_2.
(6) $\bar{\ell}_2 = [0.594420 \quad 0.055468 \quad 0.802239 \quad 0]^T$ and $\lambda_3 = 81.9032$.
(7) $\bar{m}_r = [0.684281 \quad 0.489134 \quad -0.540838 \quad 0]^T$ and $\mu_r = 1.62629\,\text{rad}$.
(8) $\bar{n}_r = [-0.722100 \quad 0.557885 \quad -0.409067 \quad 0]^T$.
(9) $e_r = 59.498995 - 0.557885(\lambda_1 + \lambda_2)$.
(10) $e_5 = -8.00357 - 0.866025(\lambda_1 + \lambda_2) - \lambda_5$ with the chosen values of λ_1 and λ_2, and an arbitrary positive value of λ_5.

The prism designed with $\lambda_1 + \lambda_2 = 100\,\text{mm}$ and $\lambda_5 = 40\,\text{mm}$ is shown in Fig. 11.6.

Example 11.4 Determine the reflector equations of a prism which allows a circular bundle of rays to pass through the prism without being blocked by any of the boundary surfaces (see Fig. 11.7). As shown in Fig. 11.7, let the circular bundle of entrance rays $\bar{R}_0$ be confined by column $\bar{r}_{c1}$ and hit the second boundary surface $\bar{r}_2$ in the area bounded by $\bar{r}_{c2}$. The reflected rays $\bar{R}_2$ hit either $\bar{r}_3$ first and then $\bar{r}_4$, or vice versa. The rays then exit boundary surface $\bar{r}_5$. From Fig. 11.7, it can be seen that, for a given image offset distance L, the circular bundle of entrance rays confined by column $\bar{r}_{c1}$ can pass through the prism without being blocked by $\bar{r}_2$ or $\bar{r}_3 \perp \bar{r}_4$ provided that one of the following four conditions holds: (1) curve $\bar{r}_{c3}$ and column $\bar{r}_{c1}$ intersect one another at most at one point, (2) curve $\bar{r}_{c4}$ and column $\bar{r}_{c1}$ intersect each other at most at one point, (3) curve $\bar{r}_{c3}$ and curve $\bar{r}_{c2}$ intersect each

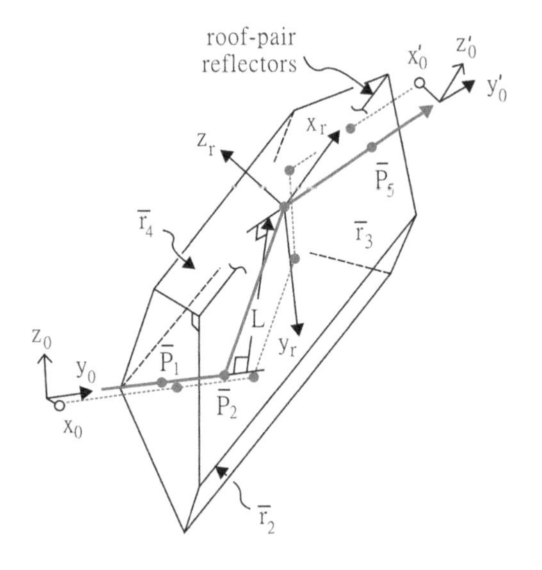

Fig. 11.6 Designed prism for Example 11.3

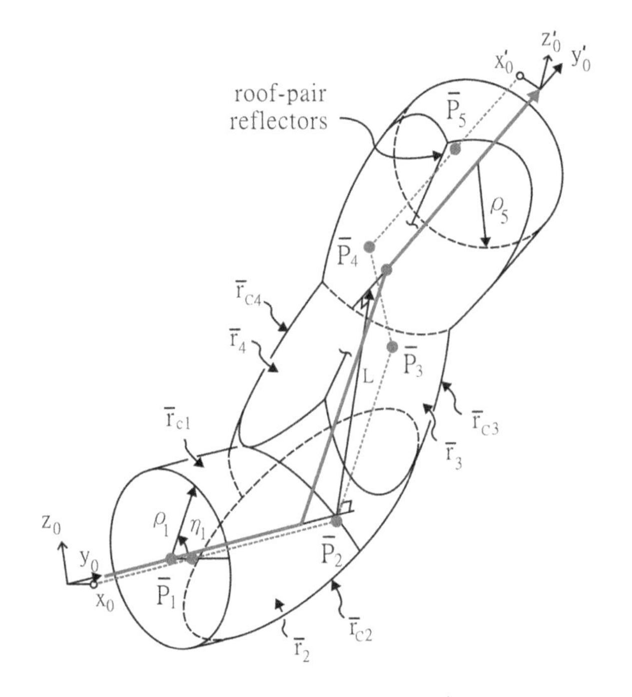

Fig. 11.7 Volume occupied by rays traveling through a prism of n = 5 with roof-pair reflectors

other at most at one point, and (4) curve $\bar{r}_{c4}$ and curve $\bar{r}_{c2}$ intersect each other at most at one point. In other words, for a given L, four different solutions of ρ_1 are obtained if the entrance rays are confined by $\bar{r}_{c1} = [\,\rho_1 C\eta_1 \quad 0 \quad \rho_1 S\eta_1 \quad 1\,]^T$ with respect to the object coordinate frame $(xyz)_0$. The smallest of these four solutions defines the maximum radius of the circular bundle of entrance rays which can pass completely through the prism without being blocked. For the left-handed merit function $\bar{\Phi}$ and image offset distance $L = 80\,\text{mm}$ considered in Example 11.3, the maximum radius of the entrance ray bundle is determined to be $\rho_1 = 29.2197\,\text{mm}$.

Reference

1. Tsai CY, Lin PD (2008) The determination of the position and orientation of a prism's boundary surfaces to produce the required image orientation change. Appl Phys B—Lasers Optics 91:105–114

Chapter 12
Optically Stable Systems

Utilizing an image orientation function, this chapter proves the finding of Schweitzer et al. (Appl Opt 37:5190–5192, 1998) that only two types of optically stable reflector system exist, namely preserving or retro-reflecting. The chapter also presents an analytical method for designing optically stable reflector systems comprising multiple reflectors. It is shown that an infinite number of solutions can be obtained for optically stable systems consisting of more than three reflectors. Furthermore, it is shown that by adding two parallel refracting flat boundary surfaces at the entrance and exit positions of the ray in an optical system with multiple reflectors, an optically stable prism can be obtained.

12.1 Image Orientation Function of Optically Stable Systems

The term "optically stable" is used here to indicate that for a given unit directional vector of the entrance ray, the unit directional vector of the exit ray remains unchanged as the system pose changes, provided that the entrance ray can still pass through the system. Figure 12.1 shows a reflector system containing (n − 2) reflectors and three coordinate frames, $(xyz)_0$, $(xyz)_s$ and $(xyz)'_0$. As discussed in Chap. 10, coordinate frame $(xyz)_0$ represents an object which emits parallel rays in the y_0 direction such that $\bar{\ell}_1 = [0 \quad 1 \quad 0 \quad 0]^T$. Furthermore, $(xyz)_s$ is the coordinate frame imbedded in the system to describe the pose of the system with respect to $(xyz)_0$. Finally, $(xyz)'_0$ indicates the orientation of the image formed by the rays which pass through the system, and can be either right-handed or left-handed. Note that y'_0 is parallel to the unit directional vector $\bar{\ell}_{n-1}$ of the exit ray.

In deriving the necessary and sufficient conditions for an optically stable system, coordinate frames $(xyz)_0$ and $(xyz)_s$ in Fig. 12.1 are assumed to be initially

P.D. Lin, *Advanced Geometrical Optics*, Progress in Optical Science and Photonics 4, DOI 10.1007/978-981-10-2299-9_12

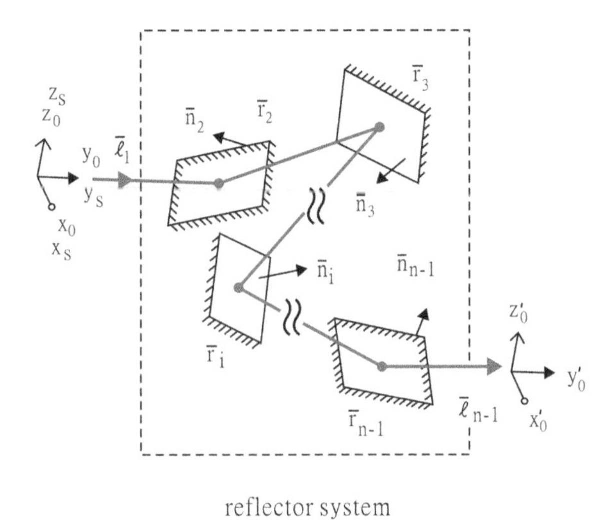

Fig. 12.1 Reflector system comprising (n − 2) reflectors and no refraction process

coincident. Under this condition, $\bar{\ell}_i = {}^s\bar{\ell}_i$ (i = 1 to i = n − 1) and the image orientation function $\bar{\Phi}_s$ is determined from Eq. (10.12) as

$$\begin{bmatrix} \bar{\Phi}_s & \bar{0}_{3\times1} \\ \bar{0}_{1\times3} & 1 \end{bmatrix} = \left[\bar{I}_{4\times4} - 2({}^s\bar{n}_{n-1})({}^s\bar{n}_{n-1})^T\right]\ldots\left[\bar{I}_{4\times4} - 2({}^s\bar{n}_i)({}^s\bar{n}_i)^T\right]\ldots\left[\bar{I}_{4\times4} - 2({}^s\bar{n}_2)({}^s\bar{n}_2)^T\right] \tag{12.1}$$

The unit directional vector $\bar{\ell}_{n-1}$ of the exit ray can be determined from Eq. (10.3) by replacing $\bar{\ell}_n$, $\bar{\Phi}$ and $\bar{\ell}_0$ with ${}^s\bar{\ell}_{n-1}$, $\bar{\Phi}_s$ and ${}^s\bar{\ell}_1$, respectively, yielding

$${}^s\bar{\ell}_{n-1} = \begin{bmatrix} \bar{\Phi}_s & \bar{t}_{3\times1} \\ \bar{0}_{1\times3} & 1 \end{bmatrix} {}^s\bar{\ell}_1 = \bar{\ell}_{n-1} = \begin{bmatrix} \bar{\Phi}_s & \bar{t}_{3\times1} \\ \bar{0}_{1\times3} & 1 \end{bmatrix} \bar{\ell}_1 \tag{12.2}$$

Note that $\bar{\ell}_{n-1} = {}^s\bar{\ell}_{n-1}$ and $\bar{\ell}_1 = {}^s\bar{\ell}_1$ are used in Eq. (12.2) since $(xyz)_0$ and $(xyz)_s$ are assumed to coincide initially.

When the pose of the reflector system changes, the system coordinate frame $(xyz)_s$ deviates from the object coordinate frame $(xyz)_0$. The relative pose of the two frames can be described by a translation motion $\text{tran}(\rho_{sx}, \rho_{sy}, \rho_{sz})$ followed by a rotation motion around a unit vector $\bar{m}_s$ through an angle μ_s, i.e.,

$${}^0\bar{A}_s = \text{tran}(\rho_{sx}, \rho_{sy}, \rho_{sz})\text{rot}(\bar{m}_s, \mu_s). \tag{12.3}$$

As described in the following, the image orientation function $\bar{\Phi}$ of the system in the new pose can be determined using the unit normal vectors $\bar{n}_i$ (i = 1 to i = n − 1) of the system reflectors. Note that $\bar{n}_i$ in Eq. (10.5) is the unit normal vector of the ith reflector expressed with respect to $(xyz)_0$. To compute the reflector matrix given in Eq. (10.5), ${}^s\bar{n}_i$ should first be transformed to $\bar{n}_i$, i.e.,

$$\bar{n}_i = {}^0\bar{A}_s\,{}^s\bar{n}_i = \mathrm{tran}(\rho_{sx}, \rho_{sy}, \rho_{sz})\mathrm{rot}(\bar{m}_s, \mu_s)\,{}^s\bar{n}_i = \mathrm{rot}(\bar{m}_s, \mu_s)\,{}^s\bar{n}_i. \tag{12.4}$$

It is noted from Eq. (12.4) that the translation motion $\mathrm{tran}(\rho_{sx}, \rho_{sy}, \rho_{sz})$ does not affect the computation of $\bar{n}_i$ since its fourth component is zero. Substituting Eq. (12.4) into Eq. (10.5) and noting that

$$\bar{I}_{4\times 4} = \mathrm{rot}(\bar{m}_s, \mu_s)\mathrm{rot}(\bar{m}_s, -\mu_s), \tag{12.5}$$

$$\mathrm{rot}(\bar{m}_s, -\mu_s) = \mathrm{rot}(\bar{m}_s, \mu_s)^T, \tag{12.6}$$

$$\bar{n}_i^T = ({}^s\bar{n}_i)^T\mathrm{rot}(\bar{m}_s, -\mu_s), \tag{12.7}$$

the following equation is obtained:

$$\begin{bmatrix} \partial\bar{\ell}_i/\partial\bar{\ell}_{i-1} & \bar{0}_{3\times 1} \\ \bar{0}_{1\times 3} & 1 \end{bmatrix} = \mathrm{rot}(\bar{m}_s, \mu_s)\left[\bar{I}_{4\times 4} - 2({}^s\bar{n}_i)({}^s\bar{n}_i)^T\right]\mathrm{rot}(\bar{m}_s, -\mu_s). \tag{12.8}$$

Furthermore, substituting Eq. (12.8) into Eq. (10.12), the image orientation function $\bar{\Phi}$ of the reflector system in its new pose is obtained as

$$\begin{bmatrix} \bar{\Phi} & \bar{0}_{3\times 1} \\ \bar{0}_{1\times 3} & 1 \end{bmatrix} = \mathrm{rot}(\bar{m}_s, \mu_s)\begin{bmatrix} \bar{\Phi}_s & \bar{0}_{3\times 1} \\ \bar{0}_{1\times 3} & 1 \end{bmatrix}\mathrm{rot}(\bar{m}_s, -\mu_s), \tag{12.9}$$

where $\bar{\Phi}_s$ is the image orientation function of the system in its initial pose and is defined in Eq. (12.1). Equation (12.9) shows that when $(xyz)_s$ deviates from $(xyz)_0$, a change in the image orientation function $\bar{\Phi}$ inevitably occurs. In the new system pose, the unit directional vector $\bar{\ell}_{n-1}$ of the exit ray can be computed from Eq. (10.3) by replacing $\bar{\ell}_n$ and $\bar{\ell}_0$ with $\bar{\ell}_{n-1}$ and $\bar{\ell}_1$, respectively, to give

$$\bar{\ell}_{n-1} = \begin{bmatrix} \bar{\Phi} & \bar{t}_{3\times 1} \\ \bar{0}_{1\times 3} & 1 \end{bmatrix}\bar{\ell}_1 = \mathrm{rot}(\bar{m}_s, \mu_s)\begin{bmatrix} \bar{\Phi}_s & \bar{t}_{3\times 1} \\ \bar{0}_{1\times 3} & 1 \end{bmatrix}\mathrm{rot}(\bar{m}_s, -\mu_s)\bar{\ell}_1. \tag{12.10}$$

In accordance with the definition of an optically stable system, the orientation of the exit ray $\bar{\ell}_{n-1}$ should remain unchanged when the reflector system moves through $\mathrm{tran}(\rho_{sx}, \rho_{sy}, \rho_{sz})\mathrm{rot}(\bar{m}_s, \mu_s)$. Thus, equating Eqs. (12.10) and (12.2) yields

$$\begin{bmatrix} \bar{\Phi}_s & \bar{t}_{3\times 1} \\ \bar{0}_{1\times 3} & 1 \end{bmatrix} = \mathrm{rot}(\bar{m}_s, \mu_s)\begin{bmatrix} \bar{\Phi}_s & \bar{t}_{3\times 1} \\ \bar{0}_{1\times 3} & 1 \end{bmatrix}\mathrm{rot}(\bar{m}_s, -\mu_s). \tag{12.11}$$

For an optically stable system, Eq. (12.11) must be true for any $\bar{m}_s$ and any angle μ_s. In addition, $\bar{\ell}_0$ should still be able to pass through the prism. The necessary and sufficient condition of Eq. (12.11) for any unit directional vector $\bar{m}_s$ and rotation angle μ_s is $\bar{\Phi}_s = \bar{I}_{3\times3}$ or $\bar{\Phi}_s = -\bar{I}_{3\times3}$. In other words, the finding in [1] that only two types of optically stable system exist, namely direction preserving (i.e., $\bar{\Phi}_s = \bar{I}_{3\times3}$) or retro-reflecting (i.e., $\bar{\Phi}_s = -\bar{I}_{3\times3}$) is confirmed.

12.2 Design of Optically Stable Reflector Systems

Let a reflector system be defined as a system consisting of only reflecting flat boundary surfaces. In addition, in order to comply with the notations given in Sect. 10.3 (in which $\bar{r}_1$ and $\bar{r}_n$ are refractive flat boundary surfaces), let the reflectors in the reflector system be labeled from $i = 2$ to $i = n - 1$. It was shown in Sect. 10.1 that the image orientation function $\bar{\Phi}$ of an optical system with an even number of reflectors is right-handed, while that of a system with an odd number of reflectors is left-handed. Therefore, only optically stable reflector systems comprising an even number of reflectors can produce an image orientation function $\bar{\Phi} = \bar{I}_{3\times3}$ and preserve the ray direction, while only optically stable reflector systems with an odd number of reflectors can generate the image orientation function $\bar{\Phi} = -\bar{I}_{3\times3}$ and retro-reflect the ray. Furthermore, a system containing only one reflector cannot be optically stable (see Eqs. (10.4) and (10.11) with $n = 3$) since a solution for $\bar{n}_2$ cannot be obtained from

$$\bar{\Phi} = -\bar{I}_{3\times3} = \partial\bar{\ell}_2/\partial\bar{\ell}_1 = \begin{bmatrix} 1 - 2n_{2x}n_{2x} & -2n_{2x}n_{2y} & -2n_{2x}n_{2z} \\ -2n_{2y}n_{2x} & 1 - 2n_{2y}n_{2y} & -2n_{2y}n_{2z} \\ -2n_{2z}n_{2x} & -2n_{2z}n_{2y} & 1 - 2n_{2z}n_{2z} \end{bmatrix}. \quad (12.12)$$

Equation (12.12) proves that a minimum of two reflectors are required to produce the image orientation function $\bar{\Phi} = \bar{I}_{3\times3}$, while a minimum of three reflectors are required to generate the image orientation function $\bar{\Phi} = -\bar{I}_{3\times3}$.

In the following discussions, it is proven mathematically that a reflector system with two mutually parallel reflectors and a corner cube (Fig. 3.19) are the only two possible optically stable reflector systems containing two reflectors and three reflectors, respectively. An analytical method is then proposed for designing an optically stable reflector system consisting of more than three reflectors.

12.2.1 Stable Systems Comprising Two Reflectors

For systems comprising just two reflectors, the optical stability condition can only be achieved if the reflectors are positioned such that they are parallel to one another. As stated above, the image orientation function of an optically stable reflector

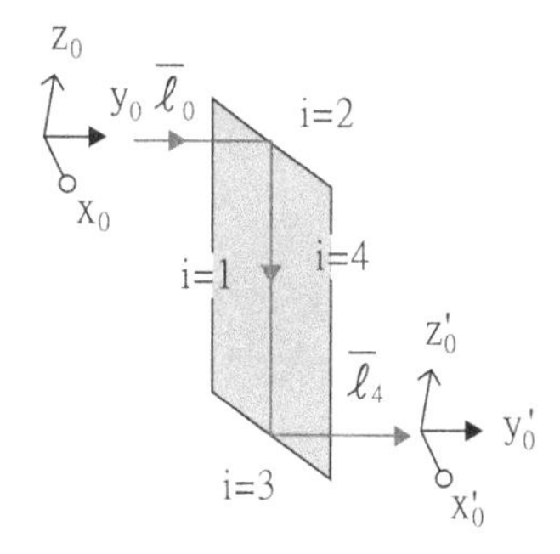

Fig. 12.2 A rhomboid prism is the only optically stable reflector system with two reflectors

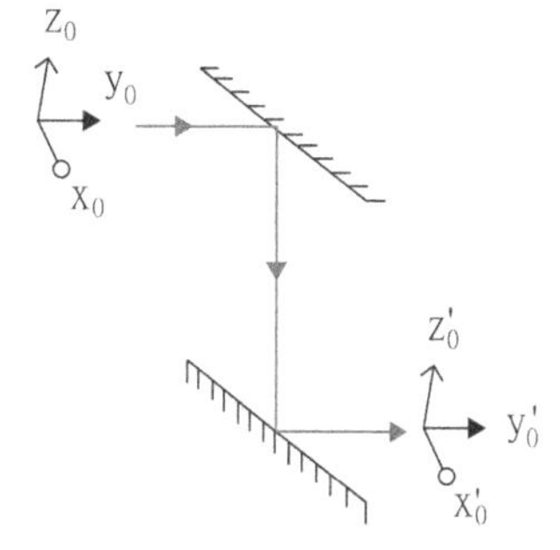

Fig. 12.3 A two-reflector system displaces the optical axis without reorientating the image

system consisting of an even number of reflectors is $\Phi = \bar{I}_{3\times3}$. Substituting $\Phi = \bar{I}_{3\times3}$ into Eq. (10.11) and taking n = 4 yields

$$\bar{I}_{3\times3} = \frac{\partial\bar{\ell}_3(\bar{n}_3)}{\partial\bar{\ell}_2}\frac{\partial\bar{\ell}_2(\bar{n}_2)}{\partial\bar{\ell}_1}. \tag{12.13}$$

Equations (10.13) and (10.14) indicate that the reflector matrix given in Eq. (10.4) is an orthogonal and symmetrical matrix whose inverse matrix is identical to itself (i.e., $\partial\bar{\ell}_i(\bar{n}_i)/\partial\bar{\ell}_{i-1} = \left(\partial\bar{\ell}_i(\bar{n}_i)/\partial\bar{\ell}_{i-1}\right)^{-1} = \left(\partial\bar{\ell}_i(\bar{n}_i)/\partial\bar{\ell}_{i-1}\right)^{T}$). Thus, from Eq. (12.13), it can be shown that $\partial\bar{\ell}_2(\bar{n}_2)/\partial\bar{\ell}_1 = \partial\bar{\ell}_3(\bar{n}_3)/\partial\bar{\ell}_2$, and hence $\bar{n}_2 = \pm\bar{n}_3$. In other words, it is proven that a system with two mutually parallel reflectors (e.g., a rhomboid prism and its equivalent mirror system (shown in Figs. 12.2 and 12.3, respectively)) is the only optically stable reflector system with two reflectors.

12.2.2 *Stable Systems Comprising Three Reflectors*

For reflecting systems comprising three reflectors, the optical stability condition can only be achieved if the system has the form of a corner cube. As described above, the image orientation function of an optically stable reflector system with an odd number of reflectors is $\Phi = -\bar{I}_{3\times3}$. Substituting $\Phi = -\bar{I}_{3\times3}$ into Eq. (10.11) and taking n = 5 yields

$$-\frac{\partial\bar{\ell}_2(\bar{n}_2)}{\partial\bar{\ell}_1} = \frac{\partial\bar{\ell}_4(\bar{n}_4)}{\partial\bar{\ell}_3}\frac{\partial\bar{\ell}_3(\bar{n}_3)}{\partial\bar{\ell}_2}. \tag{12.14}$$

Equation (12.14) indicates that the reflector matrices $\partial\bar{\ell}_3(\bar{n}_3)/\partial\bar{\ell}_2$ and $\partial\bar{\ell}_4(\bar{n}_4)/\partial\bar{\ell}_3$ are commutative since the inverse of the reflector matrix $\partial\bar{\ell}_2(\bar{n}_2)/\partial\bar{\ell}_1$ is identical to itself. It was shown in Theorem 10.2 that the multiplication of $\partial\bar{\ell}_3(\bar{n}_3)/\partial\bar{\ell}_2$ and $\partial\bar{\ell}_4(\bar{n}_4)/\partial\bar{\ell}_3$ is commutative if $\bar{n}_3$ and $\bar{n}_4$ are either parallel or perpendicular to one another. The parallel case is unreasonable since it causes $\bar{n}_2$ to be unsolvable. Thus, $\bar{n}_3$ and $\bar{n}_4$ must be perpendicular. It can be similarly proven that $\bar{n}_4$ and $\bar{n}_2$, and $\bar{n}_2$ and $\bar{n}_3$, must also be perpendicular. Consequently, it can be concluded that an optical system consisting of three reflectors is optically stable if and only if the three reflectors are mutually perpendicular. In other words, in the case of a reflector system with three reflectors, the optical stability condition can only be satisfied if the system has a corner cube configuration (see Fig. 3.19).

12.2.3 Stable Systems Comprising More Than Three Reflectors

The following discussions present a general methodology for the design of optically stable reflector systems consisting of more than three reflectors: The problem can be treated as an image orientation design problem in which the required image orientation function is $\bar{\Phi} = \bar{I}_{3\times3}$ for systems with an even number of reflectors, or $\bar{\Phi} = -\bar{I}_{3\times3}$ for systems with an odd number of reflectors. Substituting $\bar{\Phi} = \bar{I}_{3\times3}$ (or $\bar{\Phi} = -\bar{I}_{3\times3}$) into Eq. (10.11), it can be shown that

$$\pm\left(\frac{\partial\bar{\ell}_2(\bar{n}_2)}{\partial\bar{\ell}_1}\cdots\frac{\partial\bar{\ell}_i(\bar{n}_i)}{\partial\bar{\ell}_{i-1}}\cdots\frac{\partial\bar{\ell}_{n-3}(\bar{n}_{n-3})}{\partial\bar{\ell}_{n-4}}\right) = \frac{\partial\bar{\ell}_{n-1}(\bar{n}_{n-1})}{\partial\bar{\ell}_{n-2}}\frac{\partial\bar{\ell}_{n-2}(\bar{n}_{n-2})}{\partial\bar{\ell}_{n-3}}. \tag{12.15}$$

Note that the left-hand side of Eq. (12.15) is positive (or negative) if the number of reflectors is even (or odd). The unit normal vectors $\bar{n}_i$ $(i = 2, 3, \cdots, n-3)$ on the left-hand side of Eq. (12.15) can be assigned arbitrarily depending on the particular user requirement. Having done so, the two unit normal vectors on the right-hand side of Eq. (12.15), i.e., $\bar{n}_{n-2}$ and $\bar{n}_{n-1}$, can be determined mathematically by defining an auxiliary image orientation function

$$\bar{\Phi}^* = \pm\frac{\partial\bar{\ell}_2(\bar{n}_2)}{\partial\bar{\ell}_1}\cdots\frac{\partial\bar{\ell}_i(\bar{n}_i)}{\partial\bar{\ell}_{i-1}}\cdots\frac{\partial\bar{\ell}_{n-3}(\bar{n}_{n-3})}{\partial\bar{\ell}_{n-4}}$$

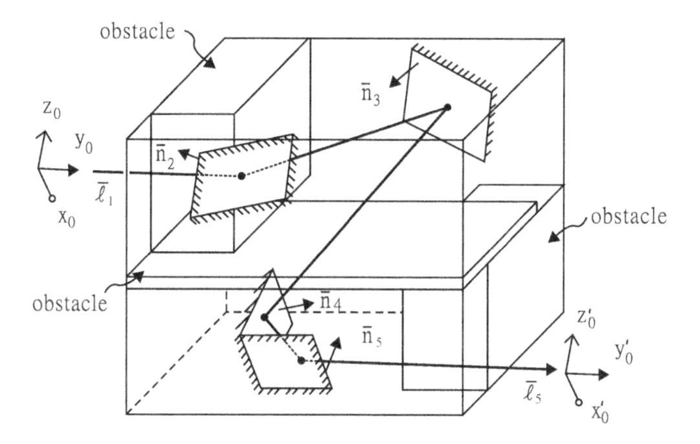

Fig. 12.4 Bypassing obstacles in 3-D optical systems utilizing optically stable system

i.e.,

$$\bar{\Phi}^* = \frac{\partial \bar{\ell}_{n-1}(\bar{n}_{n-1})}{\partial \bar{\ell}_{n-2}} \frac{\partial \bar{\ell}_{n-2}(\bar{n}_{n-2})}{\partial \bar{\ell}_{n-3}}. \tag{12.16}$$

Equation (12.16) can be treated as a simple right-handed image orientation design problem with the given image orientation function of $\bar{\Phi}^*$ comprising two unknown reflectors, i.e., $\partial \bar{\ell}_{n-1}(\bar{n}_{n-1})/\partial \bar{\ell}_{n-2}$ and $\partial \bar{\ell}_{n-2}(\bar{n}_{n-2})/\partial \bar{\ell}_{n-3}$. The unit normal vectors $\bar{n}_{n-1}$ and $\bar{n}_{n-2}$ of the remaining two unknown reflectors can be determined using the image orientation design procedure presented in Sect. 10.4. The fact that the unit normal vectors $\bar{n}_i$ $(i = 2, 3, \cdots, n-3)$ on the left-hand side of Eq. (12.15) can be arbitrarily defined implies that: (1) it is possible to bypass obstacles in a 3-D optical system by using an optically stable reflector system with appropriate unit normal vectors (e.g., Fig. 12.4); and (2) multiple feasible solutions exist for systems comprising more than three reflectors.

Example 12.1 Consider an optically stable reflector system consisting of four reflectors (i.e., n = 6). The image orientation function has the form $\bar{\Phi} = \bar{I}_{3\times3}$ since the system comprises an even number of reflectors. From Eq. (12.15), it follows that

$$\frac{\partial \bar{\ell}_2(\bar{n}_2)}{\partial \bar{\ell}_1} \frac{\partial \bar{\ell}_3(\bar{n}_3)}{\partial \bar{\ell}_2} = \frac{\partial \bar{\ell}_5(\bar{n}_5)}{\partial \bar{\ell}_4} \frac{\partial \bar{\ell}_4(\bar{n}_4)}{\partial \bar{\ell}_3}.$$

Assume that the two arbitrarily assigned unit normal vectors are $\bar{n}_2 = [-0.8809 \quad -0.4134 \quad -0.2303 \quad 0]^T$ and $\bar{n}_3 = [0.5800 \quad -0.7536 \quad -0.3093 \quad 0]^T$, respectively. The auxiliary image orientation function $\bar{\Phi}^*$ is thus given as

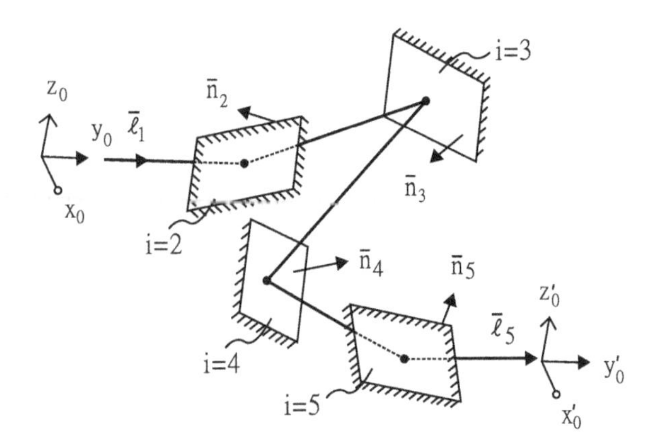

Fig. 12.5 Optically stable system comprising four (n = 6) reflectors

Table 12.1 Unit normal vectors of reflectors shown in Fig. 12.5

i	$\bar{n}_i$
2	$[-0.8809 \quad -0.4134 \quad -0.2303 \quad 0]^T$
3	$[0.5800 \quad -0.7536 \quad -0.3093 \quad 0]^T$
4	$[-0.2943 \quad -0.8661 \quad -0.4040 \quad 0]^T$
5	$[-0.9090 \quad 0.3954 \quad 0.1317 \quad 0]^T$

$$\bar{\Phi}^* = \frac{\partial\bar{\ell}_2(\bar{n}_2)}{\partial\bar{\ell}_1}\frac{\partial\bar{\ell}_3(\bar{n}_3)}{\partial\bar{\ell}_2} = \frac{\partial\bar{\ell}_5(\bar{n}_5)}{\partial\bar{\ell}_4}\frac{\partial\bar{\ell}_4(\bar{n}_4)}{\partial\bar{\ell}_3} = \begin{bmatrix} -0.9630 & 0.2687 & 0.0217 \\ -0.1947 & -0.6373 & -0.7456 \\ -0.1865 & -0.7222 & 0.6660 \end{bmatrix}.$$

The unit normal vectors of the other two unknown reflectors within the system can be easily determined from $\bar{\Phi}^*$ as $\bar{n}_4 = [-0.2943 \quad -0.8661 \quad -0.4040 \quad 0]^T$ and $\bar{n}_5 = [-0.9090 \quad 0.3954 \quad 0.1317 \quad 0]^T$, respectively (see Fig. 12.5 and Table 12.1). Having determined the unit normal vectors $\bar{n}_2$, $\bar{n}_3$, $\bar{n}_4$ and $\bar{n}_5$, the corresponding four-reflector system can be constructed using 3-D solid modeling software.

12.3 Design of Optically Stable Prism

This section presents a method for designing an optically stable prism with n flat boundary surfaces. As with all prisms, refraction occurs at the first (i = 1) and last (i = n) surfaces, while reflection occurs at all of the other surfaces. It will be recalled that if the condition given in Eq. (10.10) is satisfied, Eq. (10.11) expresses the image orientation function $\bar{\Phi}$ produced by the reflectors within the prism. Substituting the image orientation function $\bar{\Phi} = \bar{I}_{3\times3}$ (or $\bar{\Phi} = -\bar{I}_{3\times3}$) for an optically stable prism into Eq. (10.10) yields $\bar{n}_n = \pm\bar{n}_1$. Consequently, a prism is optically stable if it satisfies the following conditions: (1) the image orientation function $\bar{\Phi}$ produced by the reflectors is $\bar{\Phi} = \bar{I}_{3\times3}$ or $\bar{\Phi} = -\bar{I}_{3\times3}$, and (2) the first and last refraction boundaries of the prism are mutually parallel. An optically stable prism with n flat boundary surfaces can therefore be realized by designing its n − 2 reflectors using the methodology presented in Sect. 12.2 and then adding two mutually parallel

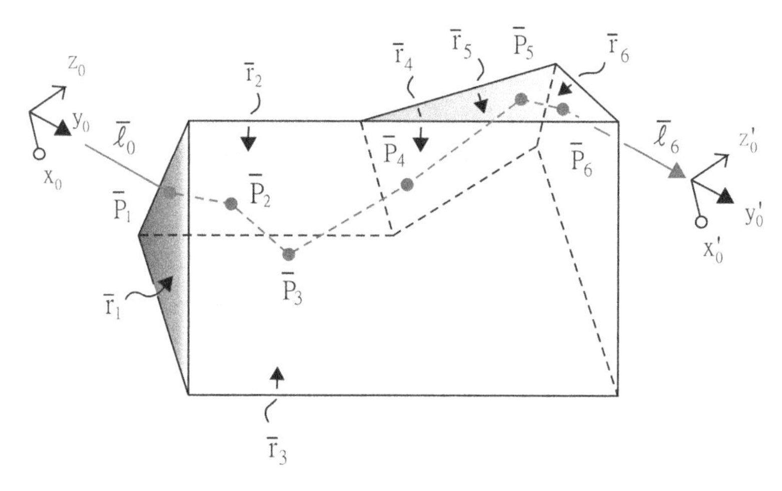

Fig. 12.6 Optically stable prism with n = 6

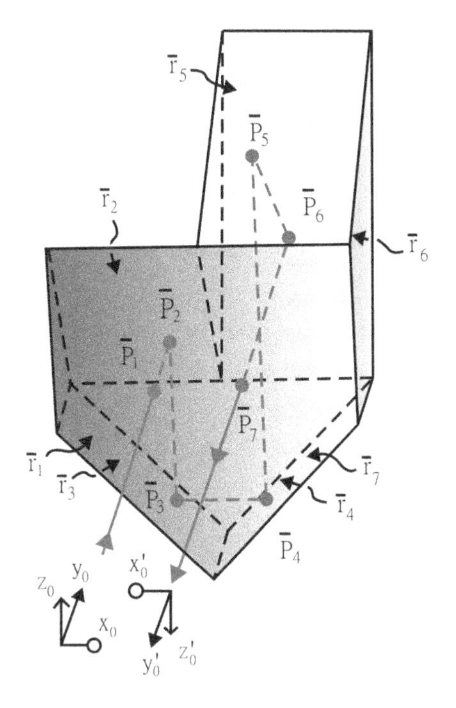

Fig. 12.7 Optically stable prism with n = 7

Table 12.2 Unit normal vectors of reflectors shown in Fig. 12.6

i	$\bar{n}_i$
1	$[0.0222 \quad -0.7710 \quad -0.6364 \quad 0]^T$
2	$[0.9540 \quad -0.1736 \quad -0.2443 \quad 0]^T$
3	$[-0.1193 \quad -0.8509 \quad 0.5115 \quad 0]^T$
4	$[-0.6503 \quad 0.7391 \quad -0.1756 \quad 0]^T$
5	$[0.6367 \quad 0.5565 \quad -0.5337 \quad 0]^T$
6	$[-0.0222 \quad 0.7710 \quad 0.6364 \quad 0]^T$

Table 12.3 Unit normal vectors of reflectors shown in Fig. 12.7

i	$\bar{n}_i$
1	$[0 \quad -1 \quad 0 \quad 0]^T$
2	$[0 \quad -0.7071 \quad -0.7071 \quad 0]^T$
3	$[0.7071 \quad 0 \quad 0.7071 \quad 0]^T$
4	$[-0.7071 \quad 0 \quad 0.7071 \quad 0]^T$
5	$[0 \quad 0.3827 \quad -0.9239 \quad 0]^T$
6	$[0 \quad -0.9239 \quad 0.3827 \quad 0]^T$
7	$[0 \quad 1 \quad 0 \quad 0]^T$

refracting flat boundary surfaces at the entrance ($i = 1$) and exit ($i = n$) positions of the ray path, respectively. Therefore, the rhomboid prism (Fig. 12.2) and solid glass corner-cube (Fig. 9.2) are the only two optically stable prisms containing two and three reflectors, respectively. Figures 12.6 and 12.7 present two illustrative examples of optically stable prisms with $n = 6$ and $n = 7$, respectively. The unit normal vectors of the reflectors in the two prisms are shown in Tables 12.2 and 12.3, respectively.

Reference

1. Schweitzer N, Friedman Y, Skop M (1998) Stability of systems of plane reflecting surfaces. Appl Opt 37:5190–5192

Chapter 13
Point Spread Function, Caustic Surfaces and Modulation Transfer Function

Chapter 3 described the use of a ray-counting method to derive the point spread function (PSF), modulation transfer function (MTF) and spot diagram of an optical system in order to evaluate its image quantity. The PSF plays an important role in image formation theory since it describes the impulse response of an optical system to a point source. However, the literature contains very few proposals for deriving the PSF of an optical system (e.g., [1]). Accordingly, Sects. 13.1 and 13.2 present a differential method based on an irradiance model for computing the PSF. An irradiance-based approach for determining the spot diagram is then presented in Sect. 13.3.

In optical theory, the term "caustic surface" refers to the envelope of a family of light rays. The word "caustic" implies burning, and is appropriate here since the intensity of the light in an optical system increases near a caustic surface. Existing methods for determining caustic surfaces involve computing either the flux density singularity or the center of curvature of the wavefront. However, such methods cannot be applied to complex problems such as determining the 3-D caustic surfaces formed by off-axis point sources or collimated rays skewed with respect to the optical axis of a system with multiple boundary surfaces. Accordingly, Sect. 13.4 proposes a robust numerical method for determining caustic surfaces based on the PSF and the ray Jacobian and Hessian matrices. Notably, the proposed method provides the ability to compute the caustic surface not only for simple optical systems with a single boundary surface, but also for complex (i.e., multi-boundary) systems with 3-D caustic surfaces.

The MTF provides a measure of an optical system's ability to transfer the contrast from a specimen to the image plane at a specific spatial resolution. In general, the MTF can either be computed numerically by geometrical optics or measured experimentally by imaging a knife edge or a bar-target pattern of varying spatial frequency [2–9]. The MTF can also be computed using a ray-counting technique [10]. However, this method is valid only for systems in which the object brightness distribution functions (OBDFs) are oriented along the meridional or sagittal directions. Accordingly, Sects. 13.5 and 13.6 of this chapter propose a

P.D. Lin, *Advanced Geometrical Optics*, Progress in Optical Science and Photonics 4, DOI 10.1007/978-981-10-2299-9_13

method based on an irradiance model and a ray-counting technique for computing the MTF of an optical system with an off-axis point source, for which the OBDF is oriented along any arbitrarily defined direction.

13.1 Infinitesimal Area on Image Plane

Figure 2.1 shows a source ray $\overline{R}_0$ originating from point source $\bar{P}_0 = [P_{0x} \quad P_{0y} \quad P_{0z} \quad 1]^T$ and traveling along the unit directional vector $\bar{\ell}_0 = [C\beta_0 C(90^\circ + \alpha_0) \quad C\beta_0 S(90^\circ + \alpha_0) \quad S\beta_0 \quad 0]^T$. As shown from Eq. (2.32) (with i = n and $y_n = 0$), the in-plane coordinates of the image plane are given by ${}^n\bar{r}_n = [x_n \quad 0 \quad z_n \quad 1]^T$ relative to boundary coordinate frame $(xyz)_n$. To determine the PSF of an optical system using the irradiance method, the Jacobian matrix $\partial(x_n, z_n)/\partial(\alpha_0, \beta_0)$ between the in-plane coordinates (x_n, z_n) and (α_0, β_0) is first required in order to compute the infinitesimal area $d\pi_n = dx_n dz_n$ on the image plane by means of

$$d\pi_n = dx_n dz_n = \left| \det \begin{bmatrix} \partial x_n/\partial\alpha_0 & \partial x_n/\partial\beta_0 \\ \partial z_n/\partial\alpha_0 & \partial z_n/\partial\beta_0 \end{bmatrix} \right| d\alpha_0 d\beta_0 = \left| \det(\frac{\partial(x_n, z_n)}{\partial(\alpha_0, \beta_0)}) \right| d\alpha_0 d\beta_0, \tag{13.1}$$

where || denotes the absolute value of a scalar quantity. The Jacobian matrix $\partial(x_n, z_n)/\partial(\alpha_0, \beta_0)$ describes the extent to which a point (x_n, z_n) is stretched in different directions on the image plane in the neighborhood of (α_0, β_0) (see Fig. 13.1). The literature contains very few proposals for deriving the PSF of optical systems due in large part to a lack of solutions for the Jacobian matrix given in Eq. (13.1). However, in the following discussions, this problem is overcome. Let

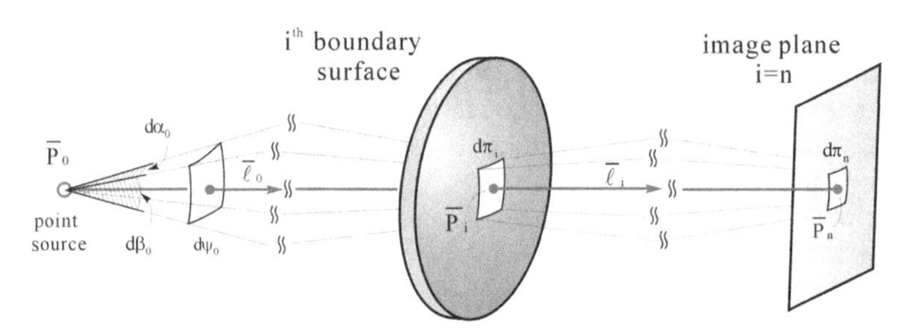

Fig. 13.1 Point source $\bar{P}_0$ radiates light rays over an infinitesimal area $d\pi_n$ centered at $\bar{P}_n$ on the image plane, where $d\psi_0$ is the solid angle

the incidence point $\overline{P}_n$ with respect to the world coordinate frame $(xyz)_0$ be obtained from Eq. (2.48) with g = n. The expression of $\overline{P}_n$ with respect to the coordinate frame $(xyz)_n$ of the image plane can be obtained via the following matrix manipulation:

$$
{}^{n}\overline{P}_n = \begin{bmatrix} {}^{n}P_{nx} \\ {}^{n}P_{ny} \\ {}^{n}P_{nz} \\ 1 \end{bmatrix} = {}^{n}\overline{A}_0 \, \overline{P}_n
$$
$$
= \begin{bmatrix} I_{nx} & I_{ny} & I_{nz} & -(I_{nx}t_{nx} + I_{ny}t_{ny} + I_{nz}t_{nz}) \\ J_{nx} & J_{ny} & J_{nz} & -(J_{nx}t_{nx} + J_{ny}t_{ny} + J_{nz}t_{nz}) \\ K_{nx} & K_{ny} & K_{nz} & -(K_{nx}t_{nx} + K_{ny}t_{ny} + K_{nz}t_{nz}) \\ 0 & 0 & 0 & 1 \end{bmatrix} \begin{bmatrix} P_{nx} \\ P_{ny} \\ P_{nz} \\ 1 \end{bmatrix} = \begin{bmatrix} x_n \\ 0 \\ z_n \\ 1 \end{bmatrix}. \tag{13.2}
$$

Equation (13.2) indicates that if the incidence point $\overline{P}_n$ is expressed with respect to $(xyz)_n$, then its values are equal to the in-plane coordinates of $(xyz)_n$. The required Jacobian matrix $\partial(x_n, z_n)/\partial(\alpha_0, \beta_0)$ in Eq. (13.1) can then be obtained by differentiating Eq. (13.2) with respect to α_0 and β_0, to give

$$
\frac{\partial(x_n, z_n)}{\partial(\alpha_0, \beta_0)} = \begin{bmatrix} I_{nx} & I_{ny} & I_{nz} & -(I_{nx}t_{nx} + I_{ny}t_{ny} + I_{nz}t_{nz}) \\ K_{nx} & K_{ny} & K_{nz} & -(K_{nx}t_{nx} + K_{ny}t_{ny} + K_{nz}t_{nz}) \end{bmatrix} \begin{bmatrix} \partial P_{nx}/\partial\alpha_0 & \partial P_{nx}/\partial\beta_0 \\ \partial P_{ny}/\partial\alpha_0 & \partial P_{ny}/\partial\beta_0 \\ \partial P_{nz}/\partial\alpha_0 & \partial P_{nz}/\partial\beta_0 \\ 0 & 0 \end{bmatrix}
$$
$$
= \begin{bmatrix} I_{nx} & I_{ny} & I_{nz} \\ K_{nx} & K_{ny} & K_{nz} \end{bmatrix} \begin{bmatrix} \partial P_{nx}/\partial\alpha_0 & \partial P_{nx}/\partial\beta_0 \\ \partial P_{ny}/\partial\alpha_0 & \partial P_{ny}/\partial\beta_0 \\ \partial P_{nz}/\partial\alpha_0 & \partial P_{nz}/\partial\beta_0 \end{bmatrix}, \tag{13.3}
$$

where the second matrix on the right-hand side of Eq. (13.3) can be obtained from Eq. (7.30) by setting i = n.

It is noted that an extended source, i.e., a source whose dimensions are significant, must be treated differently than a point source. Specifically, the following term is required in place of $\partial(x_n, z_n)/\partial(\alpha_0, \beta_0)$ to determine the PSF:

$$
\frac{\partial(x_n, z_n)}{\partial(x_0, z_0)} = \begin{bmatrix} I_{nx} & I_{ny} & I_{nz} \\ K_{nx} & K_{ny} & K_{nz} \end{bmatrix} \begin{bmatrix} \partial P_{nx}/\partial x_0 & \partial P_{nx}/\partial z_0 \\ \partial P_{ny}/\partial x_0 & \partial P_{ny}/\partial z_0 \\ \partial P_{nz}/\partial x_0 & \partial P_{nz}/\partial z_0 \end{bmatrix}. \tag{13.4}
$$

13.2 Derivation of Point Spread Function Using Irradiance Method

The PSF of an optical system describes the irradiance distribution on the image plane associated with a point source $\bar{P}_0$. In the irradiance-based method proposed in this chapter, the derivation of the PSF is based on the assumption that $\bar{P}_0$ radiates uniformly with a constant intensity $I_0(\alpha_0, \beta_0) = I_0$ (in watts/steradian) in all directions (see Fig. 13.1). The energy flux dF_0 emitted from $\bar{P}_0$ into a solid angle $d\psi_0 = C\beta_0 d\alpha_0 d\beta_0$ along the ray tube $\bar{R}_0 = [P_{0x} \quad P_{0y} \quad P_{0z} \quad C\beta_0 C(90^\circ + \alpha_0) \quad C\beta_0 S(90^\circ + \alpha_0) \quad S\beta_0]^T$ is given by

$$dF_0 = I_0 d\psi_0 = I_0 C\beta_0 d\alpha_0 d\beta_0. \tag{13.5}$$

Therefore, the total flux F_0 emitted from $\bar{P}_0$ and transmitted into the optical system is given by

$$F_0 = I_0 \iint C\beta_0 d\alpha_0 d\beta_0 = I_0 \psi_0, \tag{13.6}$$

where ψ_0 (see Fig. 13.2) is the solid angle subtended by a ray cone having its apex at $\bar{P}_0$. Notably, all rays within ψ_0 can reach the image plane $\bar{r}_n$ without being blocked by any stops. If $\bar{P}_0$ is an on-axis point, the solid angle ψ_0 can be calculated directly as $\psi_0 = 2\pi(1 - C\eta)$, where 2η is the apex angle between the marginal rays. However, if $\bar{P}_0$ is an off-axis point, numerical integration is required.

In order to normalize the PSF, an assumption is made that the total flux is equal to one (i.e., $F_0 = 1$). Thus, the following equation is obtained from Eq. (13.6):

$$I_0 = 1/\psi_0. \tag{13.7}$$

Defining $B_n(x_n, z_n)$ as the irradiance over an infinitesimal area $d\pi_n$ centered at the incidence point $\bar{P}_n$ on the image plane, the energy flux dF_n received by $d\pi_n$ is given by $dF_n = B_n(x_n, z_n)d\pi_n$, where $d\pi_n$ is defined in Eq. (13.1). Assuming that

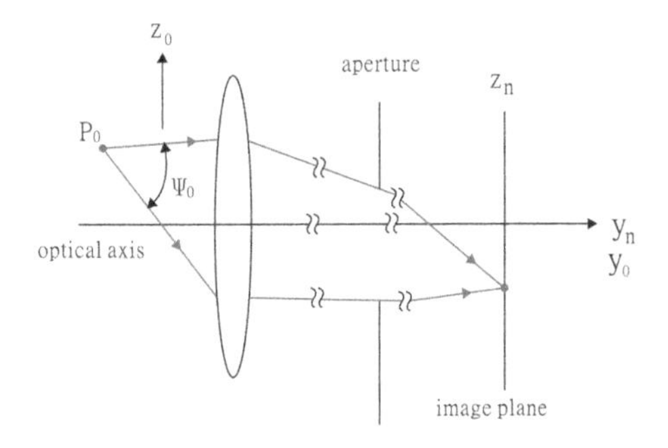

Fig. 13.2 ψ_0 is the solid angle subtended by a ray cone with its apex located at $\bar{P}_0$. All of the rays within ψ_0 reach the image plane without being blocked by any stops

no transmission losses occur (i.e., $dF_0 = dF_n$), the following equation is obtained by applying the principle of energy flux conservation along the ray:

$$dF_0 = I_0 C\beta_0 d\alpha_0 d\beta_0 = dF_n = B_n(x_n, z_n) d\pi_n = B_n(x_n, z_n)\, dx_n dz_n. \tag{13.8}$$

Integrating both sides of Eq. (13.8) yields

$$\int B_n(x_n, z_n) d\pi_n = \iint B_n(x_n, z_n) dx_n dz_n = 1 \tag{13.9}$$

In other words, the total energy flux received at the image plane is normalized to unity.

Substituting Eqs. (13.1) and (13.7) into Eq. (13.8), the following expression is obtained for the irradiance (or PSF) $B_n(x_n, z_n)$ on the image plane:

$$\mathrm{PSF} = B_n(x_n, z_n) = \frac{C\beta_0}{\psi_0 \left| \det\left(\frac{\partial(X_n, Z_n)}{\partial(\alpha_0, \beta_0)}\right) \right|} = \frac{C\beta_0}{\psi_0 \left| \frac{\partial X_n}{\partial \alpha_0}\frac{\partial Z_n}{\partial \beta_0} - \frac{\partial X_n}{\partial \beta_0}\frac{\partial Z_n}{\partial \alpha_0} \right|}. \tag{13.10}$$

Since Eq. (13.10) is a closed-form expression and contains no mathematical integration, a single traced ray $\overline{R}_0$ is sufficient to determine the PSF at the incidence point $\overline{P}_n$ on the image plane. Thus, the proposed irradiance-based PSF computation method is far more computationally efficient than the ray-counting method described in Sect. 3.4 in which a large number of rays must be traced. Furthermore, Eq. (13.10) implies that the PSF is equal to infinity when the Jacobian determinant is zero; yielding the position of the caustic surface.

The line-spread function (LSF) of an optical system, i.e., the impulse response of the system to an infinite number of point sources arranged along a line, can be obtained by integrating $B_n(x_n, z_n)$ along sections parallel to the direction of that line.

A small area of an extended source radiates a certain amount of power per unit of solid angle. Thus, its radiation characteristics can be expressed in terms of the power radiated per unit solid angle per unit area (e.g., the radiance in watts/(steradian · mm^2)). Note that the unit area is measured normal to the direction of radiation, not in the radiating area (referred to as the normal area in this book). If ψ_0 is the normal area of an extended source, in which all of the rays can reach the image plane $\bar{r}_n$ without being blocked, then the PSF is given by

$$\mathrm{PSF} = B_n(x_n, z_n) = \frac{1}{\psi_0 \left| \det\left(\frac{\partial(X_n, Z_n)}{\partial(X_0, Z_0)}\right) \right|} = \frac{1}{\psi_0 \left| \frac{\partial X_n}{\partial X_0}\frac{\partial Z_n}{\partial Z_0} - \frac{\partial X_n}{\partial Z_0}\frac{\partial Z_n}{\partial X_0} \right|}. \tag{13.11}$$

Example 13.1 Figure 13.3 presents the distribution of the PSF of $\overline{P}_0 = \begin{bmatrix} 0 & -507 & 0 & 1 \end{bmatrix}^T$ on the image plane of system shown in Fig. 3.2, as computed by the proposed irradiance method. Note that $\overline{P}_0$ is assumed to have a solid angle of

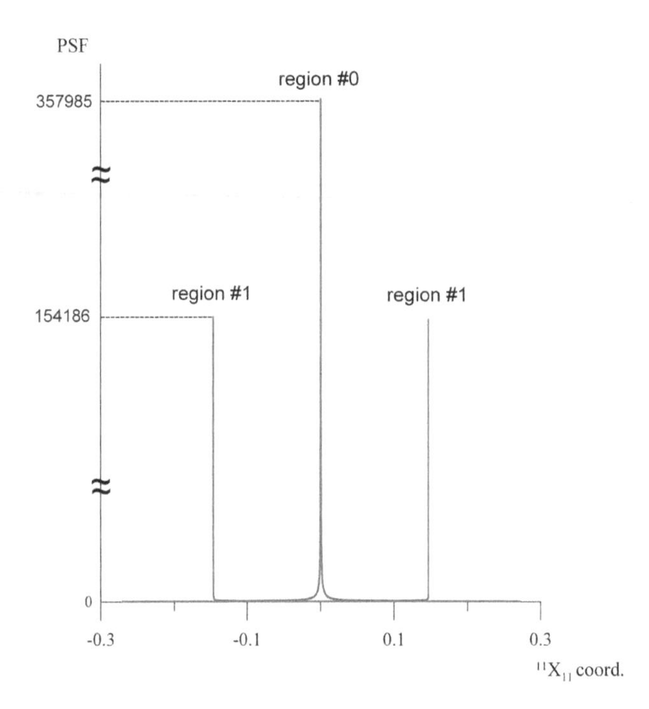

Fig. 13.3 Distribution of PSF on an image plane as computed by the irradiance method

$\psi_0 = 0.005856\,\text{sr}$. Note also that the PSF is an axis-symmetrical function since Fig. 3.2 is an axis-symmetrical system and the point source $\overline{P}_0$ lies on the optical axis. As a result, Fig. 13.3 presents only cross-sectional views rather than the entire PSF distribution. As shown, the distribution contains three regions of high PSF, namely region #0 and regions #1.

Example 13.2 In order to examine the detailed distribution of the PSF in the vicinity of regions #0 and #1 in Fig. 13.3, Table 13.1 lists the PSF values at various points in the three regions. It is seen that the PSF experiences a rapid rate of change in both regions. This is to be expected since all of the rays originating from $\overline{P}_0 = [0 \quad -507 \quad 0 \quad 1]^T$ are well focused on the image plane, which is located at the cusp of the caustic surface. Furthermore, the Jacobian determinant $\det(\partial(x_{11}, z_{11})/\partial(\alpha_0, \beta_0))$ is very small at these points, and hence a ray incident upon an infinitesimal area of the image plane (i = 11) yields a high irradiance.

Example 13.3 For comparison purposes, Fig. 13.4 shows the results obtained when the PSF distribution in Fig. 13.3 is normalized by setting its highest peak equal to unity and is then overlaid on the PSF distribution obtained by the ray-counting method using 192,539 rays and a small grid size of $1/300\,\text{mm} \times 1/300\,\text{mm}$. It is observed that the results obtained from the two methods are similar only when the following conditions are satisfied: (1) the high peak PSF values obtained from the irradiance method in regions #0 and #1 are removed; and (2) the ray-counting method

Table 13.1 PSF values of points in regions #0 and #1 in Fig. 13.3

PSF values of points in region #0			
α_0	$\pm 1.96762408781°$	$\pm 1.9677240878°$	$\pm 1.96792408780°$
β_0	0°	0°	0°
x_{11}	$-3.9343624 \times 10^{-5}$	-7.40454×10^{-7}	7.6483162×10^{-5}
$\left\lvert \det\left(\frac{\partial(X_n,Z_n)}{\partial(\alpha_0,\beta_0)}\right)\right\rvert$	0.025342698582	0.000477001107	0.049280204562
PSF	6738.01	357985.04	3465.07
PSF values of points in region #1			
α_0	$\pm 1.13897583371°$	$\pm 1.113847583372°$	$\pm 1.13857583372°$
β_0	0°	0°	0°
x_{11}	0.146358838756	0.146358863068	0.146358865039
$\left\lvert \det\left(\frac{\partial(X_n,Z_n)}{\partial(\alpha_0,\beta_0)}\right)\right\rvert$	0.056549447153	0.015524786362	0.001107490161
PSF	3019.64	10999.15	154185.9985

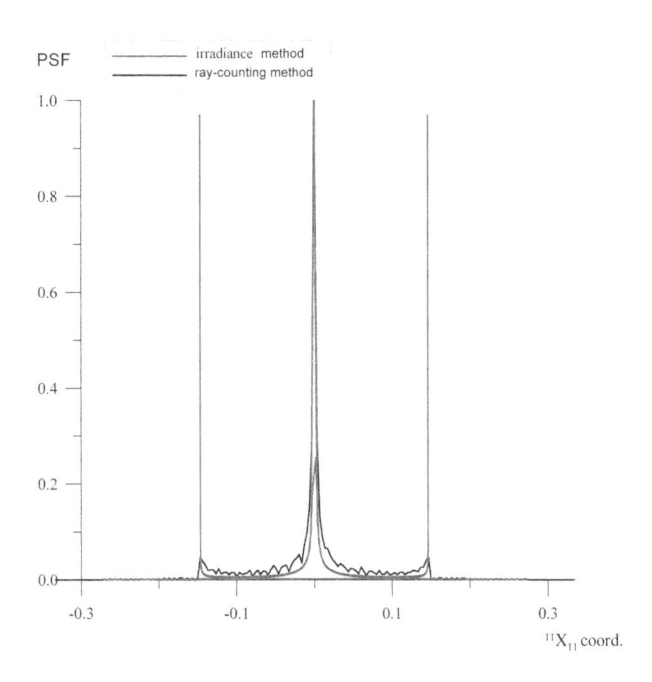

Fig. 13.4 Distribution of PSF on an image plane as computed by the ray-counting method (192,539 rays, grid size: 1/300 mm × 1/300 mm) and the irradiance method, respectively

is implemented using a large number of rays and a very small grid size. In other words, the accuracy of the PSF distribution obtained from the ray-counting method depends significantly on the number of traced rays and the size of the grids used to mesh the image plane. Notably, these problems are avoided in the irradiance method since the PSF is computed using a single ray without the use of a meshed grid.

Overall, the results presented above suggest that it is difficult to obtain an accurate estimation of the PSF distribution using the ray-counting method if the distribution contains regions of rapid change (e.g., as shown in Table 13.1). In addition, the ray-counting method requires the tracing of many rays in order to estimate the PSF over the entire image plane. By contrast, the irradiance method requires just one tracing operation to determine the PSF for a single point on the image plane. In other words, the irradiance method is more computationally efficient than the conventional ray-counting method. However, it should be noted that the irradiance method still requires multiple raytracing operations to construct the entire PSF figure since each figure comprises multiple source rays.

13.3 Derivation of Spot Diagram Using Irradiance Method

Section 3.3 of this book presents a method for evaluating the spot diagram of an optical system using a ray-counting method. In this section, an alternative method is proposed for determining the spot diagram using an irradiance-based approach. To calculate the root mean square (rms) radius of the spot diagram on the image plane, it is first necessary to determine the centroid $\begin{bmatrix} x_{n/centroid} & y_{n/centroid} & z_{n/centroid} & 1 \end{bmatrix}^T$ of the image, i.e.,

$$\begin{bmatrix} x_{n/centroid} \\ y_{n/centroid} \\ z_{n/centroid} \\ 1 \end{bmatrix} = \begin{bmatrix} \iint x_n B_n(x_n, z_n) d\pi_n / \iint B_n(x_n, z_n) d\pi_n \\ 0 \\ \iint z_n B_n(x_n, z_n) d\pi_n / \iint B_n(x_n, z_n) d\pi_n \\ 1 \end{bmatrix} = \begin{bmatrix} \iint x_n B_n(x_n, z_n) d\pi_n \\ 0 \\ \iint z_n B_n(x_n, z_n) d\pi_n \\ 1 \end{bmatrix}. \tag{13.12}$$

It is noted that Eq. (13.12) incorporates Eq. (13.9). Substituting Eqs. (13.1) and (13.10) into Eq. (13.12), the centroid of the image formed on the image plane is obtained as

$$\begin{bmatrix} x_{n/centroid} \\ y_{n/centroid} \\ z_{n/centroid} \\ 1 \end{bmatrix} = \frac{1}{\psi_0} \begin{bmatrix} \iint x_n C\beta_0 d\alpha_0 d\beta_0 \\ 0 \\ \iint z_n C\beta_0 d\alpha_0 d\beta_0 \\ 1 \end{bmatrix}. \tag{13.13}$$

As described in Sect. 3.3, the terms x_n and z_n in Eq. (13.13) for a given source ray $\overline{R}_0$ can be obtained via raytracing; thereby making possible the numerical computation of Eq. (13.13). The rms radius of the spot size on the image plane provides a useful indication of the degree of blurring of the image in an optical system, and can be calculated as

$$\begin{aligned}
\text{rms}^2 &= \iint \left[(x_n - x_{n/\text{centroid}})^2 + (z_n - z_{n/\text{centroid}})^2\right] B_n(x_n, z_n) d\pi_n \\
&= \iint (x_n^2 + z_n^2) B_n(x_n, z_n) d\pi_n + \left(x_{n/\text{centroid}}^2 + z_{n/\text{centroid}}^2\right) \iint B_n(x_n, z_n) d\pi_n \\
&\quad - 2x_{n/\text{centroid}} \iint x_n B_n(x_n, z_n) d\pi_n - 2z_{n/\text{centroid}} \iint z_n B_n(x_n, z_n) d\pi_n \\
&= \iint (x_n^2 + z_n^2) B_n(x_n, z_n) d\pi_n - \left(x_{n/\text{centroid}}^2 + z_{n/\text{centroid}}^2\right) \\
&= \frac{1}{\psi_0} \iint (x_n^2 + z_n^2) C\beta_0 d\alpha_0 d\beta_0 - \left(x_{n/\text{centroid}}^2 + z_{n/\text{centroid}}^2\right).
\end{aligned} \tag{13.14}$$

Conventionally, the image centroid and rms radius of the spot diagram for a given point source $\bar{P}_0$ are estimated using Eqs. (3.27) and (3.28). In practice, the accuracy of the PSF obtained via ray-counting is significantly dependent on the number of rays counted and the size of the grids. However, it can be shown numerically, that the ray-counting method provides sufficiently accurate results for the centroid and rms radius of a spot diagram.

13.4 Caustic Surfaces

The study of caustic surfaces dates back to 1575 and persists to the present (p. 156 of [11]). The literature contains many proposals for determining the caustic surfaces caused by refraction [12–27] or reflection [12–32] in optical systems. Generally speaking, these methods involve determining either the flux density singularity or the center of curvature of the wavefront. However, the analytical expressions for the flux density and wavefront, respectively, are highly complex, and hence traditional methods for determining caustic surfaces are generally applicable only to simple optical systems with a single boundary surface (Sect. 6 of [18]). To overcome this limitation, Burkhard and Shealy [18] proposed a method for evaluating the slope of caustic surfaces using a finite difference (FD) approach. However, the accuracy of FD methods is fundamentally dependent on the step size used in the tuning process. The Jacobian and Hessian matrices provide a convenient means of investigating the caustic surfaces of systems with multiple boundary surfaces. Accordingly, this section proposes an alternative approach for determining caustic surfaces based on the PSF of the virtual image plane of the optical system and the ray Jacobian and Hessian matrices. In developing the proposed approach, the virtual image plane is labeled as the nth boundary surface and is located after the (n − 1)th boundary surface with a separation of v_n. The discussions consider two particular caustic surfaces, namely those formed by point sources and those formed by collimated rays, respectively.

13.4.1 Caustic Surfaces Formed by Point Source

The PSF for a source ray $\overline{R}_0$ originating from point source $\overline{P}_0 = [P_{0x} \quad P_{0y} \quad P_{0z} \quad 1]^T$ has the form shown in Eq. (13.10). Since the PSF of a caustic surface intersected by an image plane is infinite, the points of intersection between the caustic surface and the virtual image plane (if it is denoted as the nth boundary surface) can be determined as

$$\overline{\Phi}(v_n) = \frac{\partial x_n}{\partial \alpha_0}\frac{\partial z_n}{\partial \beta_0} - \frac{\partial x_n}{\partial \beta_0}\frac{\partial z_n}{\partial \alpha_0} = 0. \tag{13.15}$$

The problem then arises as to how to determine the position of the virtual image plane from Eq. (13.15) for a given source ray $\overline{R}_0$. If variable v_n is the separation distance between the virtual image plane and the $(n-1)$th boundary surface (e.g., $v_n = v_6$ in Fig. 3.2), Eq. (13.15) can be solved using the Newton-Raphson method provided that the following Jacobian matrix is available:

$$\frac{\partial \overline{\Phi}(v_n)}{\partial v_n} = \frac{\partial^2 x_n}{\partial v_n \partial \alpha_0}\frac{\partial z_n}{\partial \beta_0} + \frac{\partial x_n}{\partial \alpha_0}\frac{\partial^2 z_n}{\partial v_n \partial \beta_0} - \frac{\partial^2 x_n}{\partial v_n \partial \beta_0}\frac{\partial z_n}{\partial \alpha_0} - \frac{\partial x_n}{\partial \beta_0}\frac{\partial^2 z_n}{\partial v_n \partial \alpha_0}. \tag{13.16}$$

It is noted that Eq. (13.16) involves the Hessian matrices $\partial^2 x_n/\partial v_n \partial \alpha_0$, $\partial^2 z_n/\partial v_n \partial \beta_0$, $\partial^2 x_n/\partial v_n \partial \beta_0$ and $\partial^2 z_n/\partial v_n \partial \alpha_0$, which are derived later in Sect. 17.1 of this book. To determine the root v_n of Eq. (13.15), it is first necessary to make an initial guess, $v_{n/current}$. An improved approximation of the root can then be obtained by

$$v_{n/next} = v_{n/current} - \frac{\overline{\Phi}(v_{n/current})}{\partial \overline{\Phi}(v_{n/current})/\partial v_n}. \tag{13.17}$$

In determining v_n, Eq. (13.17) is executed iteratively, with the magnitude of $\overline{\Phi}(v_{n/next})$ checked each time. If the absolute value of $\overline{\Phi}(v_{n/next})$ fails to meet a predefined threshold value, the iteration procedure continues, i.e., a new value of $v_{n/next}$ is computed using Eq. (13.17). Otherwise, the iteration process is terminated and the current value of $v_{n/next}$ is taken as the root of Eq. (13.15). The incidence point $\overline{P}_n$ on the virtual image plane then defines one point on the caustic surface. It is noted that there generally exist two possible roots of Eq. (13.15) for a given source ray $\overline{R}_0$, namely one root relating to the meridional caustic surface and a second root relating to the sagittal caustic surface. It can be shown numerically that these two points can be determined by using different initial guesses of $v_{n/current}$ (e.g., $v_{n/current} = 0.001$ and $v_{n/current} = 9999$). In both cases, the 3-D surfaces comprising the searched points constitute the corresponding caustic surface.

The derivations above are valid for both axis-symmetrical and non-axially symmetrical systems. However, the caustic surfaces formed by an on-axis point source $\bar{P}_0 = [0 \quad P_{0y} \quad 0 \quad 1]^T$ in an axis-symmetrical system merit further attention. For such a case, the incidence point $[x_n \quad 0 \quad z_n \quad 1]^T$ on the image plane is a function only of parameters α_0 and β_0 for a given point source $\bar{P}_0$, i.e., $x_n = x_n(\alpha_0, \beta_0)$ and $z_n = z_n(\alpha_0, \beta_0)$. Furthermore, it follows that $x_n(\alpha_0, \beta_0) = x_n(\alpha_0, -\beta_0)$ and $z_n(\alpha_0, \beta_0) = z_n(-\alpha_0, \beta_0)$. Finally, it is easily shown that $\partial x_n/\partial \beta_0 = \partial z_n/\partial \alpha_0 = 0$. As a result, Eq. (13.15) can be reduced to the following simplified form:

$$\bar{\Phi}(v_n) = \frac{\partial x_n}{\partial \alpha_0}\frac{\partial z_n}{\partial \beta_0} = 0. \tag{13.18}$$

Equation (13.18) indicates that for the case of an on-axis point source in an axis-symmetrical system, the two possible roots of $\bar{\Phi}(v_n)$ come from either $\partial x_n/\partial \alpha_0 = 0$ or $\partial z_n/\partial \beta_0 = 0$. Furthermore, due to the inherent rotational symmetry of caustic surfaces, only their cross-sections on the y_0z_0 plane need be considered. Note that these cross-sections can be obtained by tracing rays $\bar{R}_0 = [0 \quad P_{0y} \quad 0 \quad 0 \quad C\beta_0 \quad S\beta_0]^T$ originating from a fixed point source P_{0y} with different values of β_0 over the range of $-90° < \beta_0 < 90°$. According to the definition of geometrical optics, $\partial x_n/\partial \alpha_0 = 0$ and $\partial z_n/\partial \beta_0 = 0$ yield the cross-sectional curves of the sagittal caustic surface and meridional caustic surface, respectively, on the y_0z_0 plane.

Example 13.4 Figure 13.5 shows the cross-sectional curves of the caustic surfaces formed by an on-axis point source $\bar{P}_0 = [0 \quad P_{0y} \quad 0 \quad 1]^T$ reflected by a spherical concave mirror with radius $R_1 = -50$ mm. Note that the four curves correspond to four different positions of the point source, namely $P_{0y} = -300$ mm, $P_{0y} = -75$ mm, $P_{0y} = -60$ mm and $P_{0y} = -50$ mm. It can be seen that the sagittal caustic surface has a spike-like characteristic along the optical axis, while the meridional caustic surface has a horn-like appearance. Furthermore, it is observed that the caustic surfaces degenerate and collapse to a point as the location of the point source approaches the center of the spherical mirror. Finally, the presence of negative spherical aberration is noted since the peripheral rays undergo less severe bending than the rays near the optical axis.

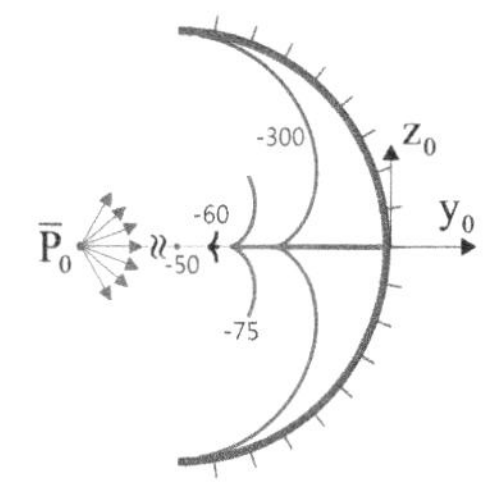

Fig. 13.5 Cross-sectional curves of caustic surfaces on y_0z_0 plane generated by rays originating from point sources located at $P_{0y} = -300$ mm, $P_{0y} = -75$ mm, $P_{0y} = -60$ mm and $P_{0y} = -50$ mm

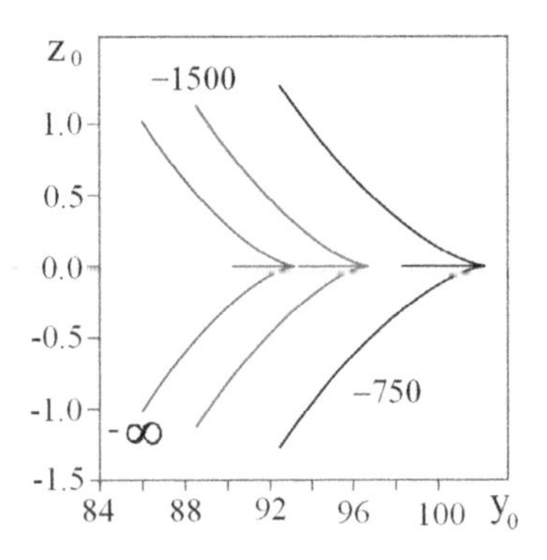

Fig. 13.6 Cross-sectional curves of caustic surfaces on y_0z_0 plane formed by refracted rays originating from on-axis point sources located at $P_{0y} = -\infty$, $P_{0y} = -1500\,\text{mm}$ and $P_{0y} = -750\,\text{mm}$, respectively, in optical system shown in Fig. 3.2 (The unit in figure is mm.)

For the optical system shown in Fig. 13.5, the outer rays (i.e., those rays with a larger value of β_0) may be reflected twice by the mirror. These so-called stray rays originate from the original source, but follow paths other than those intended. The secondary caustic surfaces formed by such rays are considered later in Fig. 13.8.

Note that it is not necessary to discuss the caustic surfaces produced by an off-axis point source $\bar{P}_0 = \begin{bmatrix} 0 & P_{0y} & P_{0z} & 1 \end{bmatrix}^T$ in the system shown in Fig. 13.5 since it is still a system possessing part of a spherical mirror with a new optical axis passing through $\bar{P}_0$.

Example 13.5 Figure 13.6 shows the cross-sectional curves of the caustic surfaces formed on the y_0z_0 plane by on-axis point sources located at $P_{0y} = -\infty$, $P_{0y} = -1500\,\text{mm}$ and $P_{0y} = -750\,\text{mm}$, respectively, in the system shown in Fig. 3.2. It is seen that the curves exhibit positive spherical aberration since the peripheral rays undergo greater bending than those near the optical axis.

Example 13.6 When designing and analyzing axis-symmetrical systems, it is usual to place a point source on the y_0z_0 plane. To illustrate the 3-D shapes of the caustic surfaces formed by an off-axis point source in Fig. 3.2, consider point source $\bar{P}_0 = \begin{bmatrix} 0 & -1500 & 500 & 1 \end{bmatrix}^T$ as an example. Figure 13.7 shows the cross-sectional curves of the caustic surfaces produced by $\bar{P}_0$ on different planes parallel with the y_0z_0 plane. Comparing the results presented in Fig. 13.7 with those presented in Fig. 13.6 for on-axis point sources, it is seen that the sagittal caustic surfaces no longer have a spike-like characteristic.

13.4.2 Caustic Surfaces Formed by Collimated Rays

From Eq. (13.11), the caustic surfaces formed by collimated rays can be determined as

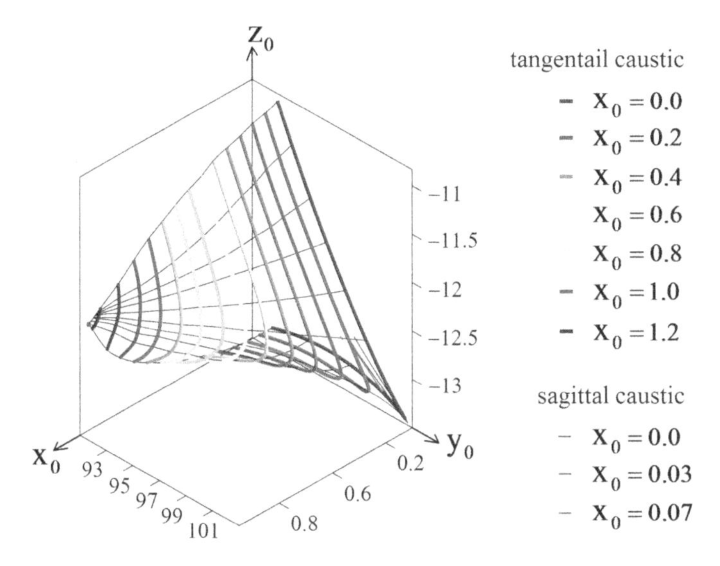

Fig. 13.7 Caustic surfaces formed by off-axis point source $\bar{P}_0 = [0 \quad -1500 \quad 500 \quad 1]^T$ in Fig. 3.2. Note that the caustic surface shapes are illustrated using different cross-sectional planes parallel to the y_0z_0 plane, where unit in figure is mm

$$\bar{\Phi}(v_n) = \frac{\partial x_n}{\partial x_0}\frac{\partial z_n}{\partial z_0} - \frac{\partial x_n}{\partial z_0}\frac{\partial z_n}{\partial x_0} = 0. \tag{13.19}$$

The problem therefore arises as to how to determine the separation v_n of a virtual image plane on which the PSF is infinite for a given ray $\bar{R}_0$. As before, the root v_n of Eq. (13.19) can be solved using the Newton-Raphson method provided that the following Jacobian matrix is available:

$$\frac{\partial\bar{\Phi}}{\partial v_n} = \frac{\partial^2 x_n}{\partial v_n \partial x_0}\frac{\partial z_n}{\partial z_0} + \frac{\partial x_n}{\partial x_0}\frac{\partial^2 z_n}{\partial v_n \partial z_0} - \frac{\partial^2 x_n}{\partial v_n \partial z_0}\frac{\partial z_n}{\partial x_0} - \frac{\partial x_n}{\partial z_0}\frac{\partial^2 z_n}{\partial v_n \partial x_0}. \tag{13.20}$$

It is noted that terms $\partial^2 x_n/\partial v_n \partial x_0$, $\partial^2 z_n/\partial v_n \partial z_0$, $\partial^2 x_n/\partial v_n \partial z_0$ and $\partial^2 z_n/\partial v_n \partial x_0$ are all components of the Hessian matrix $\partial^2 \bar{R}_n \partial \bar{X}_{sys}^2$ discussed in Sect. 17.1 of this book. Using the same iterative procedure as that described in the previous section, two roots can be obtained for each source ray, namely one root for the meridional caustic surface and a second root for the sagittal caustic surface.

When the unit directional vector $\bar{\ell}_0$ of the collimated rays is parallel to the optical axis of an axis-symmetrical system, the caustic surfaces are rotationally symmetric. Consequently, only the cross-sectional curves of the caustic surfaces on the y_0z_0 plane need to be investigated. In practice, these curves can be obtained by considering multiple line sources $\bar{R}_0 = [0 \quad P_{0y} \quad P_{0z} \quad 0 \quad 1 \quad 0]^T$ located at the same point P_{0y} but with different values of P_{0z}. Clearly, the coordinates $[x_n \quad 0 \quad z_n \quad 1]^T$

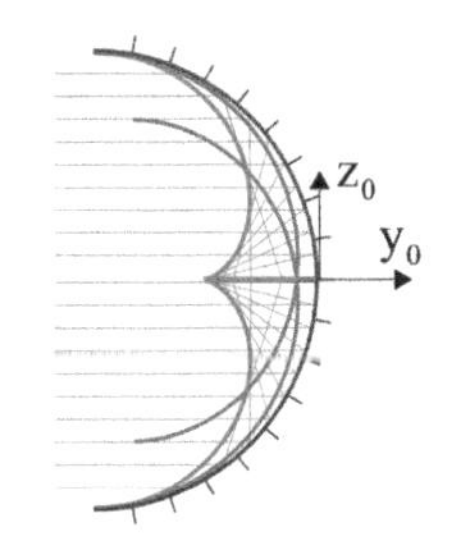

Fig. 13.8 Cross-sectional curves of primary (*red*) and secondary (*blue*) caustic surfaces on y_0z_0 plane generated by collimated rays parallel to the optical axis of a system containing only a single spherical mirror

of the incidence point on the image plane are functions only of x_0 and z_0 when the incoming rays are collimated, i.e., $x_n = x_n(x_0, z_0)$ and $z_n = z_n(x_0, z_0)$. Furthermore, for an axis-symmetrical system, it follows that $x_n(x_0, z_0) = x_n(x_0, -z_0)$ and $z_n(x_0, z_0) = z_n(-x_0, z_0)$, and consequently the condition $\partial x_n/\partial z_0 = \partial z_n/\partial x_0 = 0$ always holds. As a result, for an axis-symmetrical system with collimated rays parallel to the optical axis, Eq. (13.19) can be simplified as

$$\bar{\Phi}(v_n) = \frac{\partial x_n}{\partial x_0}\frac{\partial z_n}{\partial z_0} = 0 \tag{13.21}$$

Equation (13.21) indicates that two roots are obtained for the source ray (from $\partial x_n/\partial x_0 = 0$ and $\partial z_n/\partial z_0 = 0$, respectively). According to the definition of geometrical optics, the former root determines the points of the sagittal caustic surface, while the latter root gives the points of the meridional caustic surface.

Example 13.7 Figure 13.8 shows the caustic surfaces formed by collimated rays parallel to the optical axis of a system containing only a spherical concave mirror (with radius $R_1 = -50\,\text{mm}$). The red curve shows the cross-section of the primary caustic surface formed on the y_0z_0 plane by a line source $\bar{R}_0 = [0 \;\; -100 \;\; P_{0z} \;\; 0 \;\; 1 \;\; 0]^T$ with $-50 \le P_{0z} \le 50$. It is seen that the caustic surface has negative spherical aberration.

It is easily shown that source rays with $P_{0z} \le -35.355$ or $P_{0z} \ge 35.355$ are stray rays, and are thus reflected twice by the mirror. The two blue lines in Fig. 13.8 show the cross-sectional curves of the secondary caustic surfaces formed on the y_0z_0 plane by these stray rays. It is noted that further caustic surfaces may be formed by any stray rays reflected more than twice by the mirror. However, the flux density of such surfaces is usually very low compared with that of the primary caustic surfaces, and hence they can be ignored.

Example 13.8 Figure 13.9 shows the primary caustic surfaces formed by collimated rays oriented at an angle of 10° to the optical axis of the system shown in Fig. 3.2 (i.e., $\alpha_0 = 0°$ and $\beta_0 = 10°$). Note that the shapes of the caustic surfaces are illustrated by means of the corresponding cross-sectional curves on different planes parallel to the y_0z_0 plane. It is seen that the sagittal caustic surface no longer has a spike-like feature.

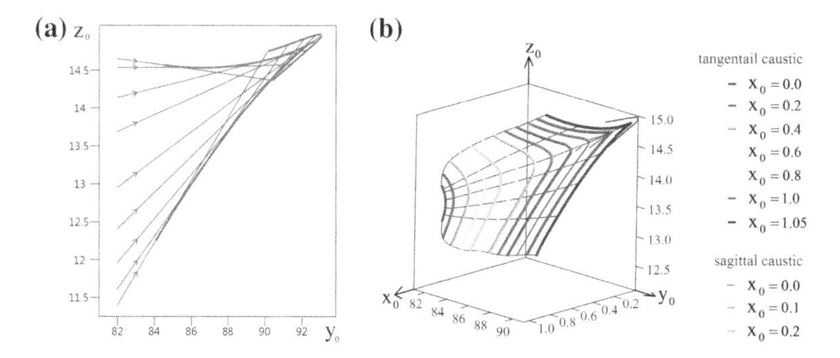

Fig. 13.9 Caustic surfaces formed by collimated rays with $\alpha_0 = 0°$ and $\beta_0 = 10°$ relative to the optical axis of the system shown in Fig. 3.2. Note that the caustic surface shapes are illustrated by cross-sectional curves on different planes parallel to the y_0z_0 plane. **a** Caustic surfaces on y_0z_0 plane. **b** Caustic surfaces on different planes parallel to y_0z_0 plane (The unit in figure is mm.)

13.5 MTF Theory for Any Arbitrary Direction of OBDF

Section 3.5 describes the MTF calculation process for an on-axis point source in an axis-symmetrical system. The imaged spot of such a point source is symmetrical about the optical axis. Therefore, the MTF values of the point source remain unchanged as the object brightness distribution function (OBDF) is rotated. However, for an off-axis point source, the MTF values are not only lower than those of on-axis point sources, but also differ in the sagittal and meridional directions due to astigmatism and coma aberrations. Therefore, in seeking the extreme values of the MTF, it is necessary to calculate the MTF of the off-axis point sources for any arbitrary propagation direction of the OBDF.

Practical optical systems inevitably contain aberrations. Thus, the point sources $\bar{P}_0$ do not form a single image point, but are instead focused over a region of the image plane. Importantly, the degree of blurring of an off-axis point source is not equal in all directions due to the effects of astigmatism and coma aberrations. Figure 13.10 illustrates the same optical system as that shown in Fig. 3.26, but with an off-axis point source $\bar{P}_0 = \begin{bmatrix} P_{0x} & P_{0y} & P_{0z} & 1 \end{bmatrix}^T$ rather than an on-axis point source. As shown, the point source forms a spot on the image plane around the imaged point $\begin{bmatrix} x_{n/chief} & 0 & z_{n/chief} & 1 \end{bmatrix}^T$ of the chief ray. In addition to coordinate frames $(xyz)_0$ and $(xyz)_n$, Fig. 13.10 also shows two parallel coordinate frames, $(x'y'z')_n$ and $(x'y'z')_0$. The origin of frame $(x'y'z')_0$ lies at the point source $\bar{P}_0$ and the x'_0 axis is aligned with the propagation direction of the OBDF. If the point source $\bar{P}_0$ consists of alternating light and dark bands, the OBDF of $\bar{P}_0$ can be expressed as $I(x'_0) = b_0 + b_1\ C(2\pi\nu x'_0)$. Mathematically, the coordinate frame $(x'y'z')_0$ can be obtained by translating the coordinate frame $(xyz)_0$ by the vector $P_{0x}\bar{i} + P_{0y}\bar{j} + P_{0z}\bar{k}$ and then rotating the translated y-axis through an angle $-\mu$. Similarly, $(x'y'z')_n$ can be obtained by translating frame $(xyz)_n$ by a vector

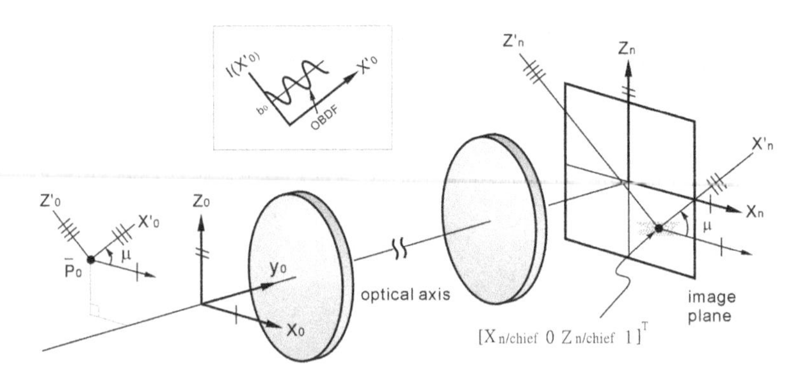

Fig. 13.10 Spot diagram of an off-axis point source $\bar{P}_0$ on the image plane in an axis–symmetrical optical system. Note that the imaged spot is not symmetrical due to coma and astigmatism aberrations

$x_{n/chief}\bar{i} + z_{n/chief}\bar{k}$ and then rotating the translated y-axis through an angle $-\mu$. As a result, the pose matrix of $(x'y'z')_n$ with respect to $(xyz)_n$ is given by ${}^{n}\bar{A}_{n'} = \text{tran}(x_{n/chief}, 0, z_{n/chief})\text{rot}(\bar{y}, -\mu)$, while that of $(xyz)_n$ with respect to $(x'y'z')_n$ is given by ${}^{n'}\bar{A}_n = \text{rot}(\bar{y}, \mu)\text{tran}(-x_{n/chief}, 0, -z_{n/chief})$. The following coordinate transformation exists between $(x'y'z')_n$ and $(xyz)_n$ for coordinates $\begin{bmatrix} x_n & 0 & z_n & 1 \end{bmatrix}^T$ on the image plane:

$$\begin{bmatrix} x'_n \\ 0 \\ z'_n \\ 1 \end{bmatrix} = {}^{n'}\bar{A}_n \begin{bmatrix} x_n \\ 0 \\ z_n \\ 1 \end{bmatrix} = \begin{bmatrix} C\mu & 0 & S\mu & -x_{n/chief}C\mu - z_{n/chief}S\mu \\ 0 & 1 & 0 & 0 \\ -S\mu & 0 & C\mu & x_{n/chief}S\mu - z_{n/chief}C\mu \\ 0 & 0 & 0 & 1 \end{bmatrix} \begin{bmatrix} x_n \\ 0 \\ z_n \\ 1 \end{bmatrix}$$
$$= \begin{bmatrix} (x_n - x_{n/chief})C\mu + (z_n - z_{n/chief})S\mu \\ 0 \\ -(x_n - x_{n/chief})S\mu + (z_n - z_{n/chief})C\mu \\ 1 \end{bmatrix}. \tag{13.22}$$

Since the x'_0 axis of $(x'y'z')_0$ is aligned with the direction of propagation of the OBDF, the equations derived in Sect. 3.5 for the on-axis case are also valid for the off-axis case considered here provided that the OBDF and energy intensity function on the image plane are expressed, respectively, as

$$I(x'_0) = b_0 + b_1 \; C(2\pi v x'_0), \tag{13.23}$$

$$I(x'_n) = \int L(\delta', \mu) I(x'_n - \delta') d\delta', \tag{13.24}$$

where ν is the frequency of the brightness variation in cycles per unit length. Therefore, the following two equations are obtained:

$$L_c(\nu,\mu) = \int L(x'_n,\mu)\, C(2\pi\nu x'_n) dx'_n, \tag{13.25}$$

$$L_s(\nu,\mu) = \int L(x'_n,\mu) S(2\pi\nu x'_n) dx'_n, \tag{13.26}$$

where $x'_n(x_n - x_{n/chief})C\mu + (z_n - z_{n/chief})S\mu$ is obtained from Eq. (13.22). The MTF and phase shift are given by

$$MTF(\nu,\mu) = \sqrt{L_c^2(\nu,\mu) + L_s^2(\nu,\mu)}, \tag{13.27}$$

and

$$\varpi(\nu,\mu) = atan2(L_s(\nu,\mu), L_c(\nu,\mu)), \tag{13.28}$$

respectively, for the propagation direction of the OBDF defined by angle μ. The following two theorems are proposed to clarify the effects of a translation of $(x'y'z')_n$ on $MTF(\nu,\mu)$ and $\varpi(\nu,\mu)$.

Theorem 13.1 *The* MTF *value of a point source* $\bar{P}_0$ *is stationary when the origin of coordinate frame* $(x'y'z')_n$ *is located at* $\begin{bmatrix} x_{n/chief} & 0 & z_{n/chief} & 1 \end{bmatrix}^T$ *without rotation since its gradients, i.e.,* $\partial MTF(\nu,0)/\partial x_{n/chief}$ *and* $\partial MTF(\nu,0)/\partial z_{n/chief}$, *are zero. The proof is provided in Appendix 1 in this chapter. Therefore, the point* $\begin{bmatrix} x_{n/chief} & 0 & z_{n/chief} & 1 \end{bmatrix}^T$ *(i.e., the incident point of the chief ray on the image plane) is taken as the origin of coordinate frame* $(x'y'z')_n$ *in Eq. (13.22) since it usually has the highest irradiance point in the neighborhood.*

Theorem 13.2 *The phase shift* $\varpi(\nu,0)$ *is non-stationary along the* x'_n *axis when the origin of* $(x'y'z')_n$ *is located at* $\begin{bmatrix} x_{n/chief} & 0 & z_{n/chief} & 1 \end{bmatrix}^T$ *without rotation since* $\partial\varpi(\nu,0)/\partial x_{n/chief} = -1$. *However,* $\varpi(\nu,0)$ *is stationary along the* z'_n *axis when the origin of* $(x'y'z')_n$ *is located at* $\begin{bmatrix} x_{n/chief} & 0 & z_{n/chief} & 1 \end{bmatrix}^T$ *without rotation since its gradient with respect to* $z_{n/chief}$ *is zero (i.e.,* $\partial\varpi(\nu,0)/\partial z_{n/chief} = 0$*). The proof is provided in Appendix 2 in this chapter.*

13.6 Determination of MTF for Any Arbitrary Direction of OBDF Using Ray-Counting and Irradiance Methods

This section utilizes two methods, namely the ray-counting method and the irradiance method, to determine the line-spread function (LSF) and MTF for an off-axis point source with an arbitrary propagation direction of the OBDF.

13.6.1 Ray-Counting Method

In the ray-counting method (p. 372 of [10]), the PSF and LSF are expressed in terms of the ray density (i.e., the number of rays intercepted by each grid on the image plane). Figure 13.11 shows a typical example. Thus, the solutions are significantly dependent on both the number of rays traced and the size of the grids used to mesh the image plane. In other words, the ray-counting method provides only a qualitative estimation of the PSF and LSF. In [10], the approximate value of the LSF, i.e., $L(x_n)$, was estimated simply by counting the number of rays in each increment Δx_n (see Fig. 13.11). As a result, the estimated value of the integration of the LSF along the x_n axis was not equal exactly to one (i.e., $\int L(x_n)dx_n \neq 1$). Therefore, an alternative equation, Eq. (3.38), for the energy distribution at x_n was proposed.

To determine the LSF and MTF along any arbitrary direction x'_n shown in Fig. 13.12, it is necessary to transfer all the traced incident points ${}^n\bar{P}_n = [{}^nP_{nx} \quad 0 \quad {}^nP_{nz} \quad 1]^T = [x_n \quad 0 \quad z_n \quad 1]^T$ to coordinate frame $(x'y'z')_n$ by using Eq. (13.22). (Note that ${}^nP_{ny} = y_n = 0$ since the incidence point is expressed with respect to $(xyz)_n$.) The LSF (denoted as $L(x'_n, \mu)$ for a propagation direction defined by angle μ) can then be approximately determined by counting the number of rays in each increment $\Delta x'_n$ (see Fig. 13.12). $L^*_c(\nu)$ and $L^*_s(\nu)$ of Eqs. (3.39) and (3.40),

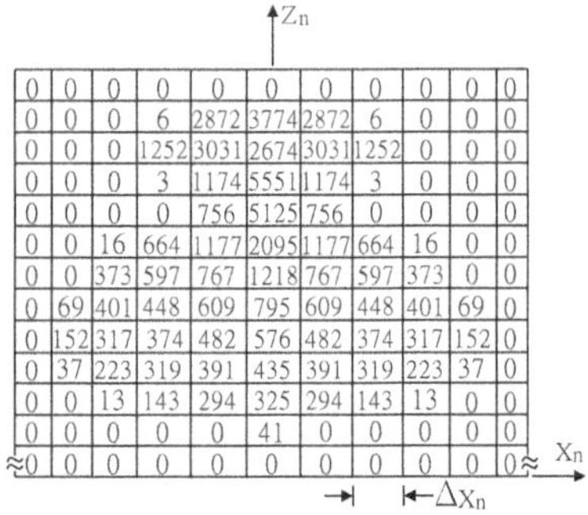

0	0	0	0	0	0	0	0	0	0	0
0	0	0	6	2872	3774	2872	6	0	0	0
0	0	0	1252	3031	2674	3031	1252	0	0	0
0	0	0	3	1174	5551	1174	3	0	0	0
0	0	0	0	756	5125	756	0	0	0	0
0	0	16	664	1177	2095	1177	664	16	0	0
0	0	373	597	767	1218	767	597	373	0	0
0	69	401	448	609	795	609	448	401	69	0
0	152	317	374	482	576	482	374	317	152	0
0	37	223	319	391	435	391	319	223	37	0
0	0	13	143	294	325	294	143	13	0	0
0	0	0	0	0	41	0	0	0	0	0
≈0	0	0	0	0	0	0	0	0	0	0≈

Fig. 13.11 Ray density of point source $\bar{P}_0 = [0 \quad -507 \quad 150 \quad 1]^T$ on the image plane as obtained from the ray-counting method with a grid size of 0.1 mm and 56,529 counted rays distributed uniformly over ψ_0

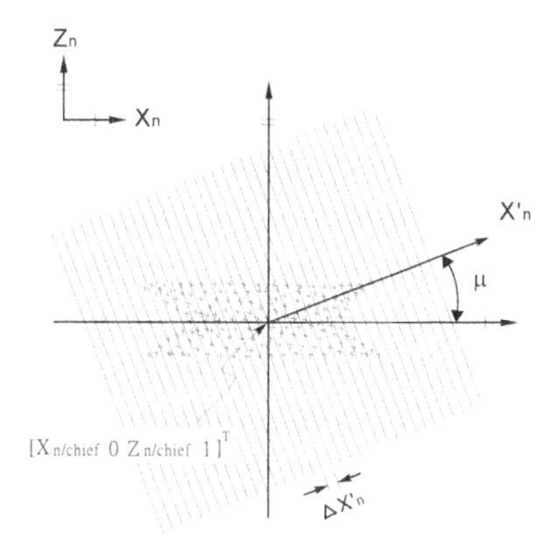

Fig. 13.12 Ray-counting method in which the ray density (i.e., the number of rays intercepted by each grid on the image plane) is taken as a measure of the PSF and LSF

now denoted as $L_c^*(\nu, \mu)$ and $L_s^*(\nu, \mu)$ respectively, can then be approximated via the following summation equations:

$$L_c^*(\nu, \mu) = \frac{\sum L(x_n', \mu) C(2\pi\nu x_n') \Delta x_n'}{\sum L(x_n', \mu) \Delta x_n'}, \tag{13.29}$$

$$L_s^*(\nu, \mu) = \frac{\sum L(x_n', \mu) S(2\pi\nu x_n') \Delta x_n'}{\sum L(x_n', \mu) \Delta x_n'}. \tag{13.30}$$

The phase shift $\varpi^*(\nu, \mu)$ and $\mathrm{MTF}^*(\nu, \mu)$ along the direction defined by angle μ can then be obtained, respectively, as

$$\varpi^*(\nu, \mu) = \mathrm{atan2}(L_s^*(\nu, \mu), L_c^*(\nu, \mu)) \tag{13.31}$$

and

$$\mathrm{MTF}^*(\nu, \mu) = \sqrt{[L_c^*(\nu, \mu)]^2 + [L_s^*(\nu, \mu)]^2}. \tag{13.32}$$

13.6.2 Irradiance Method

Consider the optical system shown in Fig. 13.13, in which a general source ray $\bar{R}_0$ originates from the off-axis point source $\bar{P}_0 = [P_{0x} \quad P_{0y} \quad P_{0z} \quad 1]^T$ and travels along the unit directional vector $\bar{\ell}_0 = [C\beta_0 C(90^\circ + \alpha_0) \quad C\beta_0 S(90^\circ + \alpha_0) \quad S\beta_0 \quad 0]^T$. The refracted ray of $\bar{R}_0$ intersects the image plane at coordinates $[x_n \quad 0 \quad z_n \quad 1]^T$, where $x_n = x_n(\alpha_0, \beta_0)$ and $z_n = z_n(\alpha_0, \beta_0)$. Defining $B_n(x_n', z_n')$ as the irradiance over an infinitesimal area $d\pi_n' = dx_n' dz_n'$ centered at the incident point on the image plane, the energy flux dF_n' received by $d\pi_n'$ is equal to $dF_n' = B(x_n', z_n') d\pi_n'$. Assuming no transmission losses (i.e. $dF_0 = dF_n'$), and applying the principle of energy flux

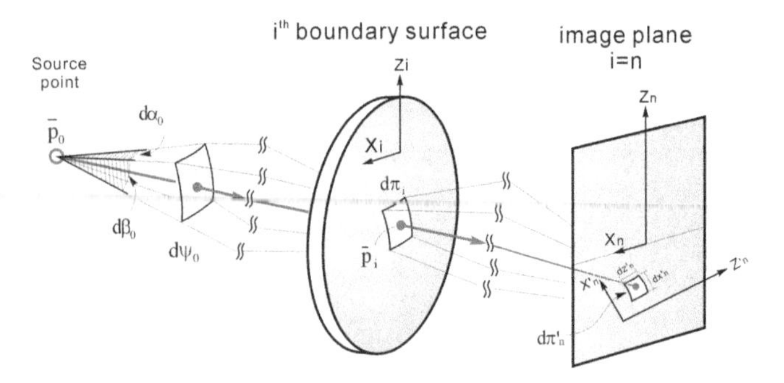

Fig. 13.13 Infinitesimal area $d\pi_n' = dx_n'dz_n'$ centered at the incident point on the image plane receives energy flux from point source $\bar{P}_0$

conservation along the ray path, the energy flux emitted over the infinitesimal solid angle $d\psi_0$ is given by

$$dF_0 = I_0 d\psi_0 = \frac{C\beta_0}{\psi_0} d\alpha_0 d\beta_0 = dF_n' = B(x_n', z_n') dx_n' dz_n'. \tag{13.33}$$

It will be recalled that the Jacobian matrix $\partial(x_n', z_n')/\partial(\alpha_0, \beta_0)$ describes the extent to which a point (x_n', z_n') on the image plane is stretched in different directions in the neighborhood of (α_0, β_0). Given this matrix, the infinitesimal area $d\pi_n' = dx_n'dz_n'$ centered at the point of incidence on the image plane can be determined as

$$d\pi_n' = dx_n'dz_n' = \left| \det\left(\frac{\partial(x_n', z_n')}{\partial(\alpha_0, \beta_0)}\right) \right| d\alpha_0 d\beta_0. \tag{13.34}$$

Substituting Eq. (13.34) into Eq. (13.33), the expression for the PSF $B(x_n', z_n')$ on the image plane is obtained as

$$B(x_n', z_n') = \frac{C\beta_0}{\psi_0 \left|\det(\partial(x_n', z_n')/\partial(\alpha_0, \beta_0))\right|}. \tag{13.35}$$

The LSF, $L(x_n', \mu)$, of the point source $\bar{P}_0$ on the image plane can then be obtained by integrating $B(x_n', z_n')$ along z_n' to give

$$L(x_n', \mu) = \int B(x_n', z_n') dz_n' = \frac{1}{\psi_0} \int \frac{C\beta_0}{\left|\det(\partial(x_n', z_n')/\partial(\alpha_0, \beta_0))\right|} dz_n'. \tag{13.36}$$

Note that in Eq. (13.36), an integration problem may arise as the Jacobian determinant approaches zero. A typical example of this problem occurs when calculating the LSF $L(x'_n, \mu)$ of an optical system in which the image plane is located at the cusp of a caustic surface [11–14, 33–39]. In determining the MTF, this problem can be avoided by substituting Eq. (13.36) into Eqs. (13.25) and (13.26) and then using Eq. (13.34) to obtain the following equations:

$$\begin{aligned} L_c(\nu, \mu) &= \frac{1}{\psi_0}\iint \frac{C\beta_0 C(2\pi\nu x'_n)}{\left|\det(\partial(x'_n, z'_n)/\partial(\alpha_0, \beta_0))\right|} dz'_n dx'_n = \frac{1}{\psi_0}\iint C(2\pi\nu x'_n) C\beta_0 d\alpha_0 d\beta_0 \\ &= \frac{1}{\psi_0}\iint C(2\pi\nu\left[(x_n - x_{n/chief})C\mu + (z_n - z_{n/chief})S\mu\right]) C\beta_0 d\alpha_0 d\beta_0, \end{aligned} \tag{13.37}$$

$$\begin{aligned} L_s(\nu, \mu) &= \frac{1}{\psi_0}\iint \frac{C\beta_0 S(2\pi\nu x'_n)}{\left|\det(\partial(x'_n, z'_n)/\partial(\alpha_0, \beta_0))\right|} dz'_n dx'_n = \frac{1}{\psi_0}\iint S(2\pi\nu x'_n) C\beta_0 d\alpha_0 d\beta_0 \\ &= \frac{1}{\psi_0}\iint S(2\pi\nu\left[(x_n - x_{n/chief})C\mu + (z_n - z_{n/chief})S\mu)\right]) C\beta_0 d\alpha_0 d\beta_0. \end{aligned} \tag{13.38}$$

The MTF and phase shift can then be computed directly as

$$MTF(\nu, \mu) = \sqrt{[L_c(\nu, \mu)]^2 + [L_s(\nu, \mu)]^2} \tag{13.39}$$

and

$$\varpi(\nu, \mu) = \operatorname{atan2}(L_s(\nu, \mu), L_c(\nu, \mu)), \tag{13.40}$$

respectively. In other words, the LSF is not required, and hence the potential integration problem in Eq. (13.36) is avoided. The following two theorems are provided to clarify the variations of $MTF(\nu, \mu)$ and $\varpi(\nu, \mu)$ over the range of $0° \le \mu \le 360°$.

Theorem 13.3 *For the axis-symmetrical system shown in Fig. 13.10,* $L_c(\nu, \mu + 180°) = L_c(\nu, \mu)$ *and* $L_s(\nu, \mu + 180°) = -L_s(\nu, \mu)$, *for which* $MTF(\nu, \mu + 180°) = MTF(\nu, \mu)$ *and* $\varpi(\nu, \mu + 180°) = -\varpi(\nu, \mu)$. *The proof is provided in Appendix 3 in this chapter.*

Theorem 13.4 *If the point source* $\bar{P}_0$ *in Fig. 13.10 lies on the* y_0z_0 *plane (i.e.,* $\bar{P}_0 = [0 \quad P_{0y} \quad P_{0z} \quad 1]^T$*), it follows that* $L_c(\nu, 90° + \mu) = L_c(\nu, 90° - \mu)$ *and* $L_s(\nu, 90° + \mu) = L_s(\nu, 90° - \mu)$. *As a result,* $MTF(\nu, 90° + \mu) = MTF(\nu, 90° - \mu)$ *and* $\varpi(\nu, 90° + \mu) = \varpi(\nu, 90° - \mu)$. *The proof is presented in Appendix 4 in this chapter.*

Due to the symmetry of an axis-symmetrical optical system about the optical axis (i.e., the y_0 axis in Fig. 13.10), it is possible with no loss of generality to define a point source $\bar{P}_0$ as lying on the y_0z_0 plane. In other words, Theorem 13.4 is always applicable for an axis-symmetrical optical system provided that the point sources are located on the y_0z_0 plane. Significantly, Theorem 13.4 indicates that the MTF and phase shift curves are both symmetrical with respect to $\mu = 90^\circ$ for a given frequency ν. Meanwhile, Theorem 13.3 shows that the MTF and phase shift curves are anti-symmetrical with respect to $\mu = 180^\circ$ for a given ν. Therefore, provided that the point source $\bar{P}_0$ is located on the y_0z_0 plane of $(xyz)_0$, a complete understanding of the MTF and phase shift can be obtained by computing the variations of $MTF(\nu, \mu)$ and $\varpi(\nu, \mu)$ over the range of $0^\circ \leq \mu \leq 90^\circ$. In other words, the extreme values of the MTF and phase shift can be determined without the need to explore the full range of $0^\circ \leq \mu \leq 360^\circ$.

The following examples demonstrate the validity of the MTF computation methods presented in this section using the optical system shown in Fig. 3.2 for illustration purposes, in which the 5th boundary surface is the aperture and n = 11 is the image plane. Note that the integrations in Eqs. (13.37) and (13.38) are performed using Simpson's rule. In addition, the off-axis point sources are assumed to be located on the y_0z_0 plane such that Theorem 13.4 can be applied.

Example 13.9 Figures 13.14 and 13.15 present the variations of the LSF, MTF and phase shift of a point source $\bar{P}_0 = [0 \quad -507 \quad 150 \quad 1]^T$ in the system shown in Fig. 3.2. In general, the results show that when using the ray-counting method, the LSF, MTF and phase shift are all dependent on the number of rays traced and the grid size meshed on the image plane. In other words, if the chosen grid mesh is insufficiently fine, the LSF will have poor resolution and the accuracy of the MTF will be degraded. However, Fig. 13.15 shows that the MTF computed using the

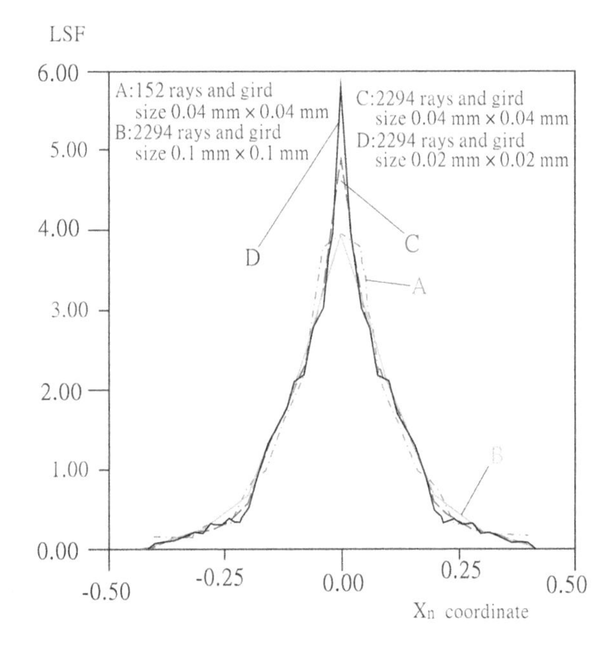

Fig. 13.14 LSF of $\bar{P}_0 = [0 \quad -507 \quad 150 \quad 1]^T$ as determined using the ray-counting method. It is seen that the LSF depends on both the number of rays traced and the grid size on the image plane

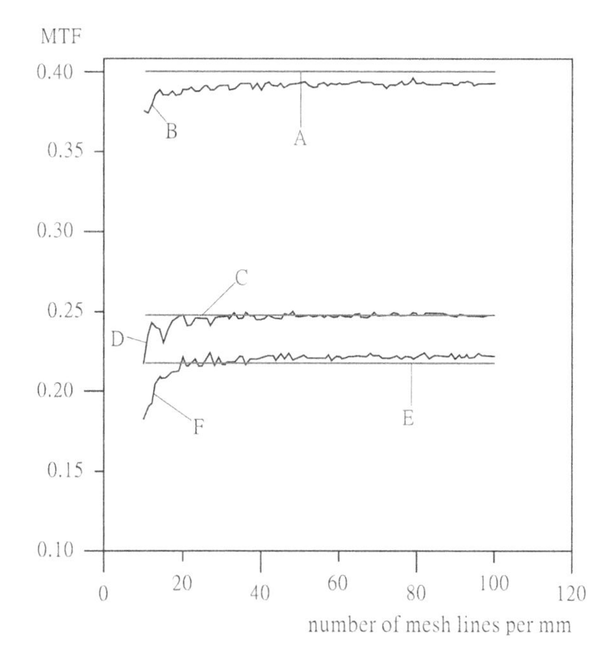

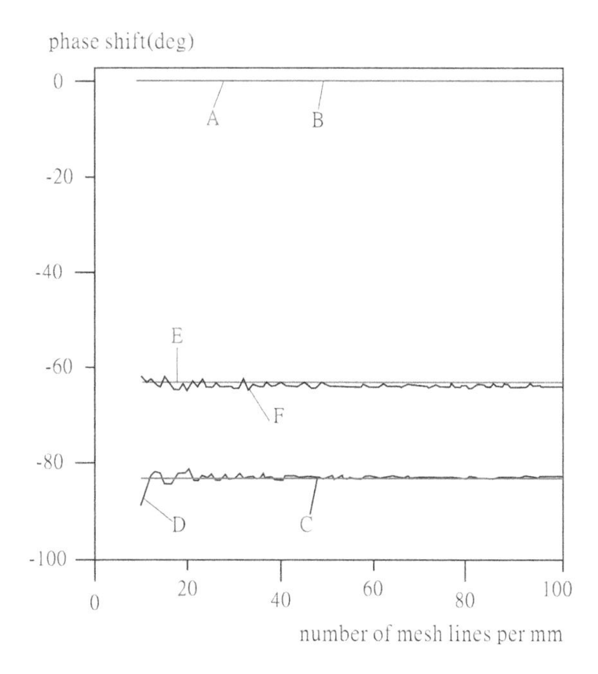

Fig. 13.15 Variations of the MTF and phase shift of a point source $\bar{P}_0 = [0 \quad -507 \quad 150 \quad 1]^T$ when tracing 2294 rays and using $v = 2$. It is shown that the MTF computed using the ray-counting method $(B : \mu = 0°, D : \mu = 45°, F : \mu = 90°)$ is sensitive to the grid size meshed on the image plane. However, the MTF computed using the irradiance method $(A : \mu = 0°, C : \mu = 45°, E : \mu = 90°)$ does not need grids

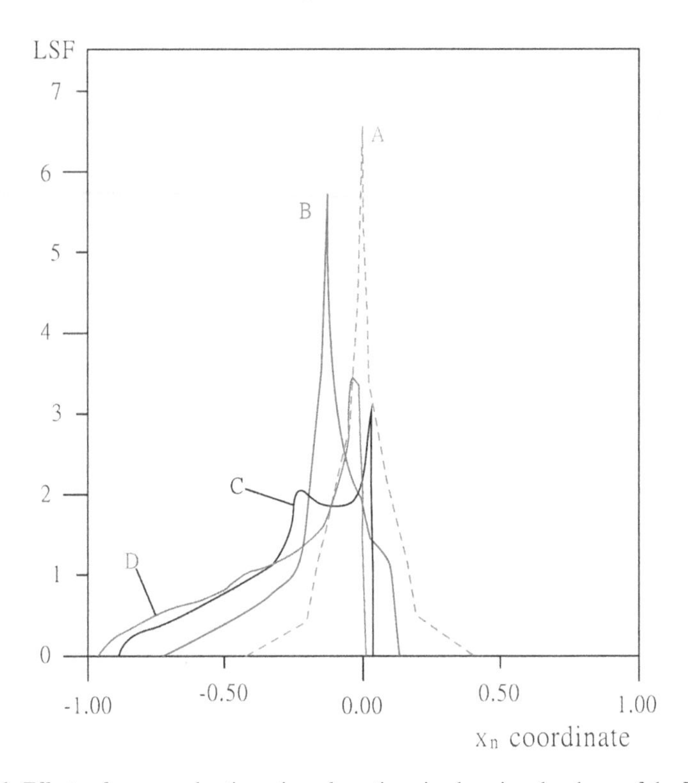

Fig. 13.16 Effects of coma and astigmatism aberrations in changing the shape of the LSF when the OBDF is rotated through different angles ($\bar{P}_0 = [0 \quad -507 \quad 150 \quad 1]^T$, 25, 781rays, $A:\mu = 0°, B:\mu = 30°, C:\mu = 60°, D:\mu = 90°$)

irradiance method has a constant value since it is calculated solely on the basis of irradiance, i.e., it has no need for meshing and/or prior knowledge of the LSF.

Example 13.10 Figure 13.16 shows the variation of the LSF when the OBDF of $\bar{P}_0$ in Example 13.4 is rotated through different angles. It is seen that the shape of the LSF for $\mu = 0°$ is significantly different from that for $\mu = 90°$. Thus, it is inferred that coma and astigmatism aberrations exist in the system shown in Fig. 3.2 for this off-axis point source. Consequently, the ray-counting method requires an appropriate choice of grid mesh size to improve the LSF resolution and MTF accuracy.

Example 13.11 Figure 13.17 illustrates the variation of the MTF of the off-axis point source $\bar{P}_0 = [0 \quad -507 \quad 150 \quad 1]^T$ when the OBDF is rotated from $\mu = 0°$ to $\mu = 90°$. The results show that the MTF varies with the rotation angle μ of the OBDF. Significantly, the MTF neither increases nor decreases monotonically in the domain $0° \le \mu \le 90°$. Therefore, it is impossible to determine the extreme values of the MTF and phase shift by considering only the sagittal and meridional directions.

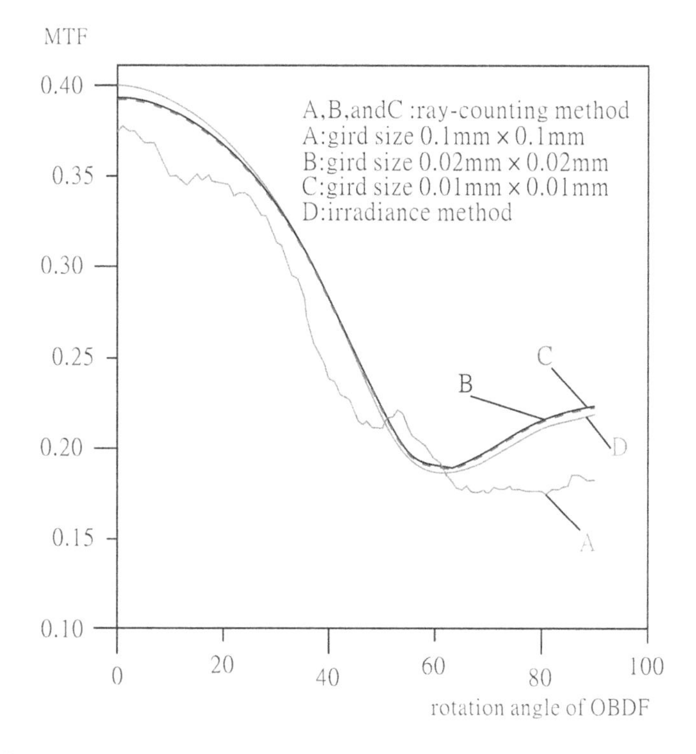

Fig. 13.17 Non-monotonic variations of MTF in the domain $0° \leq \mu \leq 90°$ ($\bar{\mathbf{P}}_0 = [0 \quad -507 \quad 150 \quad 1]^T$, 2294 rays, $\nu = 2$). It is noted that the extreme MTF values occur at intermediate directions between the sagittal and meridional directions

Example 13.12 Figure 13.18 presents the MTF values of $\bar{P}_0$ in Example 13.8 as obtained by the ray-counting method and irradiance method, respectively, for different numbers of traced rays. It is observed that for both methods, the MTF value oscillates if the number of traced rays is less than a certain threshold value ($\sim$500 rays in the present case). For the irradiance method, this problem arises since Eqs. (13.37) and (13.38) involve numerical integration and the accuracy of any numerical integration method is significantly dependent on the step size. In addition, it is seen that in the ray-counting method, even when a sufficient number of rays are traced to ensure a constant MTF, the estimated value of the MTF still depends on the chosen grid size.

Overall, the results presented in Figs. 13.14, 13.15, 13.16, 13.17 and 13.18 confirm that the ray-counting and irradiance methods both provide a feasible means of computing the MTF and phase shift of an off-axis point source for any arbitrary propagation direction of the OBDF. However, for both methods, a sufficient number of rays must be traced in order to ensure the accuracy of the estimated MTF. Moreover, the ray-counting method has two drawbacks: (1) the LSF must be

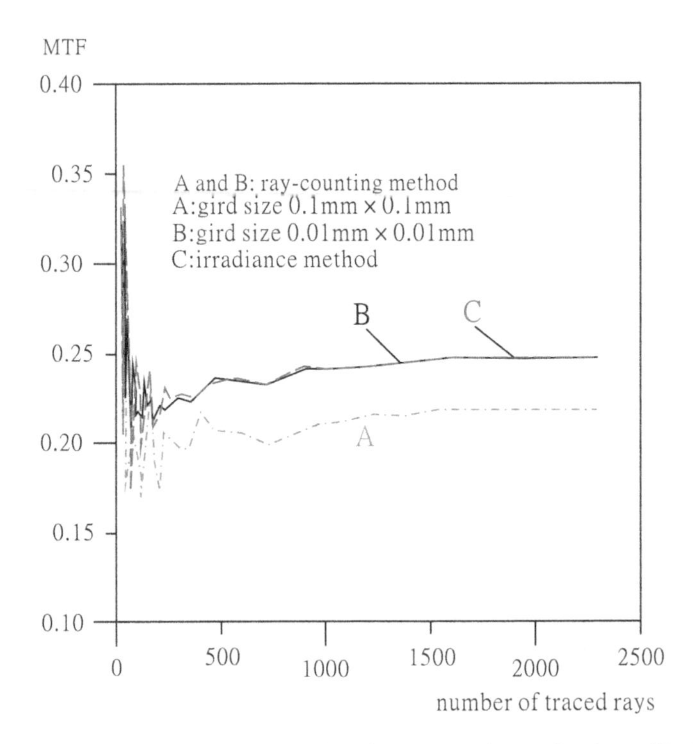

Fig. 13.18 Variations of MTF with the number of traced rays for $v = 2$ and $\mu = 45°$. It is noted that for both the ray-counting method and the irradiance method, the MTF values oscillate significantly if a sufficient number of rays are not traced

computed before the MTF can be obtained; and (2) the accuracy of the MTF and phase shift estimates is dependent on the size of the grids used to mesh the image plane. Notably, the irradiance method for MTF computation (Cancel to improve the sentence flow) does not count the number of rays intercepting a grid. In other words, it is immune to the choice of grid size and thus tends to have a better accuracy than the ray-counting method.

Appendix 1

For the case of zero rotation, the coordinate transformation between coordinate frames $(x'y'z')_n$ and $(xyz)_n$ can be obtained simply by setting $\mu = 0$ in Eq. (13.27). The MTF is therefore denoted as $\mathrm{MTF}(\nu, 0)$. Since $\mathrm{MTF}^2(\nu, 0) = L_c^2(\nu, 0) + L_s^2(\nu, 0)$, the gradient of $\mathrm{MTF}^2(\nu, 0)$ with respect to $x_{n/\mathrm{chief}}$ can be obtained from

$$\begin{aligned}
\frac{\partial \mathrm{MTF}^2(\nu,0)}{\partial x_{n/chief}} &= 2\mathrm{MTF}(\nu,0)\frac{\partial \mathrm{MTF}(\nu,0)}{\partial x_{n/chief}} = \frac{\partial \mathrm{MTF}^2(\nu,0)}{\partial x'_n}\frac{\partial x'_n}{\partial x_{n/chief}} + \frac{\partial \mathrm{MTF}^2(\nu,0)}{\partial z'_n}\frac{\partial z'_n}{\partial x_{n/chief}} \\
&= 2\left(L_c(\nu,0)\frac{\partial L_c(\nu,0)}{\partial x'_n} + L_s(\nu,0)\frac{\partial L_s(\nu,0)}{\partial x'_n}\right)\frac{\partial x'_n}{\partial x_{n/chief}} \\
&\quad + 2\left(L_c(\nu,0)\frac{\partial L_c(\nu,0)}{\partial z'_n} + L_s(\nu,0)\frac{\partial L_s(\nu,0)}{\partial z'_n}\right)\frac{\partial z'_n}{\partial x_{n/chief}} \\
&= 0.
\end{aligned} \tag{13.41}$$

Similarly, one has

$$\frac{\partial \mathrm{MTF}^2(\nu,0)}{\partial z_{n/chief}} = 2\mathrm{MTF}(\nu,0)\frac{\partial \mathrm{MTF}(\nu,0)}{\partial z_{n/chief}} = \frac{\partial \mathrm{MTF}^2(\nu,0)}{\partial x'_n}\frac{\partial x'_n}{\partial z_{n/chief}} + \frac{\partial \mathrm{MTF}^2(\nu,0)}{\partial z'_n}\frac{\partial z'_n}{\partial z_{n/chief}} = 0. \tag{13.42}$$

Equations (13.41) and (13.42) show that $\partial \mathrm{MTF}(\nu,0)\partial x_{n/chief} = \partial \mathrm{MTF}(\nu,0)/\partial z_{n/chief} = 0$. In other words, $\mathrm{MTF}(\nu,0)$ is stationary in the neighborhood of $\begin{bmatrix} x_{n/chief} & 0 & z_{n/chief} & 1 \end{bmatrix}^T$, and hence Theorem 13.1 is proven.

Appendix 2

The gradients of phase shift $\varpi(\nu,0)$ with respect to $x_{n/chief}$ and $z_{n/chief}$ can be obtained respectively as

$$\begin{aligned}
\frac{\partial \varpi(\nu,0)}{\partial x_{n/chief}} &= \frac{\partial \varpi(\nu,0)}{\partial x'_n}\frac{\partial x'_n}{\partial x_{n/chief}} + \frac{\partial \varpi(\nu,0)}{\partial z'_n}\frac{\partial z'_n}{\partial x_{n/chief}} \\
&= \frac{L_c(\nu,0)}{L_c(\nu,0)^2 + L_s(\nu,0)^2}\frac{\partial L_s(\nu,0)}{\partial x'_n}\frac{\partial x'_n}{\partial x_{n/chief}} - \frac{L_s(\nu,0)}{L_c(\nu,0)^2 + L_s(\nu,0)^2}\frac{\partial L_c(\nu,0)}{\partial z'_n}\frac{\partial z'_n}{\partial x_{n/chief}} \\
&= \frac{\partial x'_n}{\partial x_{n/chief}} = -1,
\end{aligned} \tag{13.43}$$

$$\begin{aligned}
\frac{\partial \varpi(\nu,0)}{\partial z_{n/chief}} &= \frac{\partial \varpi(\nu,0)}{\partial x'_n}\frac{\partial x'_n}{\partial z_{n/chief}} + \frac{\partial \varpi(\nu,0)}{\partial z'_n}\frac{\partial z'_n}{\partial z_{n/chief}} \\
&= \frac{L_c(\nu,0)}{L_c(\nu,0)^2 + L_s(\nu,0)^2}\frac{\partial L_s(\nu,0)}{\partial x'_n}\frac{\partial x'_n}{\partial z_{n/chief}} - \frac{L_s(\nu,0)}{L_c(\nu,0)^2 + L_s(\nu,0)^2}\frac{\partial L_c(\nu,0)}{\partial z'_n}\frac{\partial z'_n}{\partial z_{n/chief}} \\
&= \frac{\partial x'_n}{\partial z_{n/chief}} = 0.
\end{aligned} \tag{13.44}$$

Equation (13.43) shows that $\varpi(\nu, 0)$ changes with a change of $x_{n/chief}$. However, Eq. (13.44) shows that $\varpi(\nu, 0)$ is unchanged with small changes of $z_{n/chief}$. Thus, as discussed in Sect. 3.5, a phase shift of $\varpi = 180^\circ$ yields a reversal of contrast.

Appendix 3

From Eqs. (13.37) and (13.38), it follows that

$$\begin{aligned}L_c(\nu, \mu + 180^\circ) &= \frac{1}{\psi_0}\iint C\left[2\pi\nu(x_n - x_{n/chief})C(\mu + 180^\circ) + 2\pi\nu(z_n - z_{n/chief})S(\mu + 180^\circ)\right]C\beta_0 d\alpha_0 d\beta_0 \\ &= \frac{1}{\psi_0}\iint C\left[2\pi\nu(x_n - x_{n/chief})C\mu + 2\pi\nu(z_n - z_{n/chief})S\mu\right]C\beta_0 d\alpha_0 d\beta_0 = L_c(\nu, \mu).\end{aligned} \tag{13.45}$$

$$\begin{aligned}L_s(\nu, \mu + 180^\circ) &= \frac{1}{\psi_0}\iint S\left[2\pi\nu(x_n - x_{n/chief})C(\mu + 180^\circ) + 2\pi\nu(z_n - z_{n/chief})S(\mu + 180^\circ)\right]C\beta_0 d\alpha_0 d\beta_0 \\ &= \frac{-1}{\psi_0}\iint S\left[2\pi\nu(x_n - x_{n/chief})C(\mu) + 2\pi\nu(z_n - z_{n/chief})S\mu\right]C\beta_0 d\alpha_0 d\beta_0 = -L_s(\nu, \mu).\end{aligned} \tag{13.46}$$

As a result, the following two equations are obtained from Eqs. (13.45) and (13.46):

$$\begin{aligned}MTF(\nu, \mu + 180^\circ) &= \sqrt{[L_c(\nu, \mu + 180^\circ)]^2 + [L_s(\nu, \mu + 180^\circ)]^2} \\ &= \sqrt{[L_c(\nu, \mu)]^2 + [L_s(\nu, \mu)]^2} = MTF(\nu, \mu),\end{aligned} \tag{13.47}$$

$$\begin{aligned}\varpi(\nu, \mu + 180^\circ) &= \operatorname{atan2}(L_s(\nu, \mu + 180^\circ), L_c(\nu, \mu + 180^\circ)) \\ &= \operatorname{atan2}(-L_s(\nu, \mu), L_c(\nu, \mu)) = -\varpi(\nu, \mu).\end{aligned} \tag{13.48}$$

Appendix 4

As discussed in Sect. 13.1, a general source ray $\bar{R}_0$ intersects the image plane at coordinates $\begin{bmatrix} x_n & 0 & z_n & 1 \end{bmatrix}^T$, where $x_n = x_n(\alpha_0, \beta_0)$ and $z_n = z_n(\alpha_0, \beta_0)$. As shown in Fig. 13.10, and with no loss in generality, $\bar{R}_0$ can always be defined as lying on the y_0z_0 plane and expressed as $\bar{P}_0 = \begin{bmatrix} 0 & P_{0y} & P_{0z} & 1 \end{bmatrix}^T$ due to the symmetry of the optical system. As a result, its chief ray intersects the z_n axis of the image coordinate frame $(xyz)_n$, and hence

$$x_{n/chief} = y_{n/chief} = 0. \tag{13.49}$$

Furthermore, two general rays $\begin{bmatrix} \bar{P}_0 & \bar{\ell}_{0/1} \end{bmatrix}^T$ and $\begin{bmatrix} \bar{P}_0 & \bar{\ell}_{0/2} \end{bmatrix}^T$, with

$$\bar{\ell}_{0/1} = \begin{bmatrix} C\beta_0 C(90^\circ - \alpha_0) & C\beta_0 S(90^\circ - \alpha_0) & S\beta_0 & 0 \end{bmatrix}^T \tag{13.50}$$

and

$$\bar{\ell}_{0/2} = \begin{bmatrix} C\beta_0 C(90^\circ + \alpha_0) & C\beta_0 S(90^\circ + \alpha_0) & S\beta_0 & 0 \end{bmatrix}^T, \tag{13.51}$$

respectively, always have identical z_n intersections and oppositely-signed x_n values in an axis-symmetrical optical system, i.e.,

$$x_n(90^\circ - \alpha_0, \beta_0) = -x_n(90^\circ + \alpha_0, \beta_0), \tag{13.52}$$

$$z_n(90^\circ - \alpha_0, \beta_0) = z_n(90^\circ + \alpha_0, \beta_0). \tag{13.53}$$

Let $L_c(\nu, 90^\circ + \mu)$ be obtained by substituting Eq. (13.49) into Eq. (13.37), yielding

$$\begin{aligned}
L_c(\nu, 90^\circ + \mu) &= \frac{1}{\psi_0} \iint C(2\pi\nu[x_n C(90^\circ + \mu) + (z_n - z_{n/chief}) S(90^\circ + \mu)]) C\beta_0 d\alpha_0 d\beta_0 \\
&= \frac{C(2\pi\nu z_{n/chief} C\mu)}{\psi_0} \iint C(2\pi\nu(x_n S\mu - z_n C\mu)) C\beta_0 d\alpha_0 d\beta_0 \\
&\quad - \frac{S(2\pi\nu z_{n/chief} C\mu)}{\psi_0} \iint S(2\pi\nu(x_n S\mu - z_n C\mu)) C\beta_0 d\alpha_0 d\beta_0 \\
&= \frac{C(2\pi\nu z_{n/chief} C\mu)}{\psi_0} \left[\iint C(2\pi\nu x_n S\mu) C(2\pi\nu z_n C\mu) C\beta_0 d\alpha_0 d\beta_0 + \iint S(2\pi\nu x_n S\mu) S(2\pi\nu z_n C\mu) C\beta_0 d\alpha_0 d\beta_0 \right] \\
&\quad - \frac{S(2\pi\nu z_{n/chief} C\mu)}{\psi_0} \left[\iint S(2\pi\nu x_n S\mu) C(2\pi\nu z_n C\mu) C\beta_0 d\alpha_0 d\beta_0 - \iint C(2\pi\nu x_n S\mu) S(2\pi\nu z_n C\mu) C\beta_0 d\alpha_0 d\beta_0 \right].
\end{aligned} \tag{13.54}$$

If the upper and lower integration limits of α_0 are given by $90^\circ + \alpha_{0/limit}$ and $90^\circ - \alpha_{0/limit}$, respectively, the integrand of the first term in Eq. (13.54) can be written in the following form when Eqs. (13.52) and (13.53) are used:

$$\begin{aligned}
&\iint C(2\pi v x_n S\mu)C(2\pi v z_n C\mu)C\beta_0 d\alpha_0 d\beta_0 \\
&= \int \left[\int_{90^\circ - \alpha_{0/limit}}^{90^\circ + \alpha_{0/limit}} C(2\pi v x_n S\mu)C(2\pi v z_n C\mu)d(\alpha_0 - 90^\circ) \right] C\beta_0 d\beta_0 \\
&= \int \left[\int_{-\alpha_{0/limit}}^{\alpha_{0/limit}} C(2\pi v x_n S\mu)C(2\pi v z_n C\mu)d\alpha_0 \right] C\beta_0 d\beta_0 \\
&= \int \left[\int_{-\alpha_{0/limit}}^{0} C(2\pi v x_n S\mu)C(2\pi v z_n C\mu)d\alpha_0 + \int_{0}^{\alpha_{0/limit}} C(2\pi v x_n S\mu)C(2\pi v z_n C\mu)d\alpha_0 \right] C\beta_0 d\beta_0 \\
&= 2\int \left[\int_{0}^{\alpha_{0/limit}} C(2\pi v x_n S\mu)C(2\pi v z_n C\mu)d\alpha_0 \right] C\beta_0 d\beta_0 .
\end{aligned} \tag{13.55}$$

Similarly, the second, third and fourth terms in Eq. (13.54) can be rewritten in the forms shown in Eqs. (13.56), (13.57) and (13.58), respectively, i.e.,

$$\iint S(2\pi v x_n S\mu)S(2v\pi z_n C\mu)C\beta_0 d\alpha_0 d\beta_0 = 0, \tag{13.56}$$

$$\iint S(2\pi v x_n S\mu)C(2\pi v z_n C\mu)C\beta_0 d\alpha_0 d\beta_0 = 0, \tag{13.57}$$

$$\iint C(2\pi v x_n S\mu)S(2\pi v z_n C\mu)C\beta_0 d\alpha_0 d\beta_0 = 2\int \left[\int_{0}^{\alpha_{0/limit}} C(2\pi v x_n S\mu)S(2\pi v z_n C\mu)d\alpha_0 \right] C\beta_0 d\beta_0 . \tag{13.58}$$

Substituting Eqs. (13.55), (13.56), (13.57) and (13.58) into Eq. (13.54) gives

$$\begin{aligned}
L_c(v, 90^\circ + \mu) = &\frac{2C(2\pi v z_{n/chief} C\mu)}{\psi_0} \int \left[\int_{0}^{\alpha_{0limit}} C(2\pi v x_n S\mu)C(2\pi v z_n C\mu)d\alpha_0 \right] C\beta_0 d\beta_0 \\
&+ \frac{2S(2\pi v z_{n/chief} C\mu)}{\psi_0} \int \left[\int_{0}^{\alpha_{0/limit}} C(2\pi v x_n S\mu)S(2\pi v z_n C\mu)d\alpha_0 \right] C\beta_0 d\beta_0 .
\end{aligned} \tag{13.59}$$

Equation (13.59) indicates that $L_c(v, 90^\circ + \mu)$ is an even function with respect to μ, i.e.,

$$L_c(v, 90^\circ + \mu) = L_c(v, 90^\circ - \mu). \tag{13.60}$$

Similarly, by substituting $\mu = 90^\circ + \mu$ and Eq. (13.49) into Eq. (13.38), $L_s(v, 90^\circ + \mu)$ can be rewritten as

$$
\begin{aligned}
L_s(\nu, 90^\circ+\mu) &= \frac{1}{\psi_0}\iint S\left(2\pi\nu\left[x_n C(90^\circ+\mu) + (z_n - z_{n/chief})C(90^\circ+\mu)\right]\right)C\beta_0 d\alpha_0 d\beta_0 \\
&= \frac{-S(2\pi\nu C\mu z_{n/chief})}{\psi_0}\iint C(2\pi\nu(x_n S\mu - z_n C\mu))C\beta_0 d\alpha_0 d\beta_0 \\
&\quad - \frac{C(2\pi\nu z_{n/chief} C\mu)}{\psi_0}\iint S(2\pi\nu(x_n S\mu - z_n C\mu))C\beta_0 d\alpha_0 d\beta_0 \\
&= \frac{-S(2\pi\nu z_{n/chief} C\mu)}{\psi_0}\left[\iint C(2\pi\nu x_n S\mu)C(2\pi\nu z_n C\mu)C\beta_0 d\alpha_0 d\beta_0 + \iint S(2\pi\nu x_n S\mu)S(2\pi\nu z_n C\mu)C\beta_0 d\alpha_0 d\beta_0\right] \\
&\quad - \frac{C(2\pi\nu z_{n/chief} C\mu)}{\psi_0}\left[\iint S(2\pi\nu x_n S\mu)C(2\pi\nu z_n C\mu)C\beta_0 d\alpha_0 d\beta_0 - \iint C(2\pi\nu x_n S\mu)S(2\pi\nu z_n C\mu)C\beta_0 d\alpha_0 d\beta_0\right] \\
&= \frac{-2S(2\pi\nu z_{n/chief} C\mu)}{\psi_0}\int\left[\int_0^{\alpha_{0/limit}} C(2\pi\nu x_n S\mu)C(2\pi\nu z_n C\mu)d\alpha_0\right]C\beta_0 d\beta_0 \\
&\quad + \frac{2C(2\pi\nu z_{n/chief} C\mu)}{\psi_0}\int\left[\int_0^{\alpha_{0/limit}} C(2\pi\nu x_n S\mu)S(2\pi\nu z_n C\mu)d\alpha_0\right]C\beta_0 d\beta_0.
\end{aligned}
\tag{13.61}
$$

$L_s(\nu, 90^\circ + \mu)$ is an even function with respect to μ, i.e.,

$$
L_s(\nu, 90^\circ + \mu) = L_s(\nu, 90^\circ - \mu). \tag{13.62}
$$

Thus, from Eqs. (13.60) and (13.62), it follows that

$$
\begin{aligned}
\mathrm{MTF}(\nu, 90^\circ + \mu) &= \sqrt{[L_c(\nu, 90^\circ + \mu)]^2 + [L_s(\nu, 90^\circ + \mu)]^2} \\
&= \sqrt{[L_c(\nu, 90^\circ - \mu)]^2 + [L_s(\nu, 90^\circ - \mu)]^2} = \mathrm{MTF}(\nu, 90^\circ - \mu)
\end{aligned}
\tag{13.63}
$$

and

$$
\begin{aligned}
\varpi(\nu, 90^\circ + \mu) &= \mathrm{atan2}(L_s(\nu, 90^\circ + \mu), L_c(\nu, 90^\circ + \mu)) \\
&= \mathrm{atan2}(L_s(\nu, 90^\circ - \mu), L_c(\nu, 90^\circ - \mu)) = \varpi(\nu, 90^\circ - \mu).
\end{aligned}
\tag{13.64}
$$

References

1. Mahajan VN (1998) Optical imaging and aberrations part I ray geometrical optics. SPIE-The International Society for Optical Engineering
2. Tseng KH, Kung C, Liao TT, Chang HP (2009) Calculation of modulation transfer function of an optical system by using skew ray tracing, transactions of canadian society for mechanical engineering. J Mech Eng 33:429–442
3. Inoue S, Tsumura N, Miyake Y (1997) Measuring MTF of paper by sinusoidal test pattern projection. J Imaging Sci Tech 41:657–661
4. Boreman GD, Yang S (1995) Modulation transfer function measurement using three- and four-bar targets. Appl Opt 34:8050–8052
5. Sitter DN, Goddard JS, Ferrell RK (1995) Method for the measurement of the modulation transfer function of sampled imaging systems from bar-target patterns. Appl Opt 34:746–751
6. Barakat R (1965) Determination of the optical transfer function directly from the edge spread function. J Opt Soc Am 55:1217–1221

7. Rogers GL (1998) Measurement of the modulation transfer function of paper. Appl Opt 37:7235–7240
8. Park SK, Schowengerdt R, Kaczynski M (1984) Modulation-transfer-function analysis for sampled image system. Appl Opt 23:2572–2582
9. Giakoumakis E, Katsarioti MC, Panayiotakis GS (1991) Modulation transfer function of thin transparent foils in radiographic cassettes. Appl Phys Solid Surf 52:210–212
10. Smith WJ (2001) Modern optical engineering, 3rd edn. Edmund Industrial Optics, Barrington, N.J.
11. Herzberger M (1958) Modern geometrical optics. Interscience Publishers, Inc
12. Hoffnagle JA, Shealy DL (2011) Refracting the k-function: Stavroudis's solution to the eikonal equation for multielement optical systems. J Opt Soc Am A: 28:1312–1321
13. Shealy DL, Burkhard DG (1973) Caustic surfaces and irradiance for reflection and refraction from an ellipsoid, elliptic parabolid and elliptic cone. Appl Opt 12:2955–2959
14. Avendaño-Alejo M, Castañeda L, Moreno I (2010) Properties of caustics produced by a positive lens: meridional rays. J Opt Soc Am 27:2252–2260
15. Avendaño-Alejo M, González-Utrera D, Castañeda L (2011) Caustics in a meridional plane produced by plano-convex conic lenses. J Opt Soc Am A: 28:2619–2628
16. Avendaño-Alejo M, Daz-Uribe R, Moreno I (2008) Caustics caused by refraction in the interface between an isotropic medium and a uniaxial crystal. J Opt Soc Am A: 25:1586–1593
17. Lock JA, Adler CL, Hovenac EA (2000) Exterior caustics produced in scattering of a diagonally incident plane wave by a circular cylinder: semiclassical scattering theory analysis. J Opt Soc Am A: 17:1846–1856
18. Burkhard DG, Shealy DL (1981) Simplified formula for the illuminance in an optical system. Appl Opt 20:897–909
19. Avendaño-Alejo M (2013) Caustics in a meridional plane produced by plano-convex aspherical lenses. J Opt Soc Am A: 30:501–508
20. Stavroudis ON, Fronczek RC (1976) Caustic surfaces and the structure of the geometrical image. J Opt Soc Am 66:795–800
21. Stavroudis ON (1995) The k function in geometrical optics and its relationship to the archetypal wave front and the caustic surface. J Opt Soc Am A: 12:1010–1016
22. Burkhard DG, Shealy DL (1973) Flux density for ray propagation in geometrical optics. J Optical Soc Am 63
23. Shealy DL, Burkhard DG (1973) Flux density for ray propagation in discrete index media expressed in terms of the intrinsic geometry of the deflecting surface. Optica Acta 20
24. Theocaris PS (1977) Properties of caustics from conic reflectors. 1: meridional rays. Appl Opt 16:1705–1716
25. Parke Mathematical Laboratories, Inc. (1952) Calculation of the caustic (focal) surface when the reflecting surface is a paraboloid of revolution and the incoming rays are parallel. Contract no. AF 19(122), 484
26. Maca-García S, Avendaño-Alejo M, Castañeda L (2012) Caustics in a meridional plane produced by concave conic mirrors. J Opt Soc Am A: 29:1977–1985
27. Avendaño-Alejo M, Moreno I, Castañeda L (2010) Caustics caused by multiple reflections on a circular surface. Am J Phys 78:1195–1198
28. Hosken RW (2007) Circle of least confusion of a spherical reflector. Appl Opt 46:3107–3117
29. Castro-Ramos J, Prieto O, Silva-Ortigoza G (2004) Computation of the disk of least confusion for conic mirrors. Appl Opt 43:6080–6089
30. Silva-Ortigoza G, Castro-Ramos J, Cordero-Dávila A (2001) Exact calculation of the circle of least confusion of a rotationally symmetric mirror. II Appl Opt 40:1021–1028
31. Cordero-Dávila A, Castro-Ramos J (1998) Exact calculation of the circle of least confusion of a rotationally symmetric mirror. Appl Opt 37:6774–6778
32. Burkhard DG, Shealy DL (1982) Formula for the density of tangent rays over a caustic surface. Appl Opt 21:3299–3306

33. Shealy DL, Burkhard DG (1976) Caustic surface merit functions in optical design. J Opt Soc Am 66:1122
34. Shealy DL (1976) Analytical illuminance and caustic surface calculations in geometrical optics. Appl Opt 15:2588–2596
35. Andersen TB (1981) Optical aberration functions: computation of caustic surfaces and illuminance in symmetrical systems. Appl Opt 20:3723–3728
36. Kassim AM, Shealy DL (1988) Wave front equation, caustics, and wave aberration function of simple lenses and mirrors. Appl Opt 21:516–522
37. Kassim AM, Shealy DL, Burkhard DG (1989) Caustic merit function for optical design. Appl Opt 28:601–606
38. Silva-Orthigoza G, Marciano-Melchor M, Carvente-Munoz O, Silva-Ortigoza R (2002) Exact computation of the caustic associated with the evolution of an aberrated wavefront. J Opt A: Pure Appl Opt 4:358–365
39. Shealy DL, Hoffnagle JA (2008) Wavefront and caustics of a plane wave refracted by an arbitrary surface. J Opt Soc Am A: 25:2370–2382

Chapter 14
Optical Path Length and Its Jacobian Matrix

The optical path length (OPL) is an essential property of optical systems since it determines the phase of the light and governs the interference and diffraction of the rays as they propagate. In fine-tuning the performance of any system, it is essential that the wavefront aberrations be clearly understood such that they can be properly controlled. Accordingly, this chapter presents a mathematical approach for determining the Jacobian matrix of the wavefront aberrations in a system with respect to the system variables by using the OPL. The proposed method not only resolves the error inherent in finite difference (FD) methods caused by the denominator being far smaller than the numerator, but also avoids the need for multiple raytracing operations. In addition, the proposed method provides a powerful tool for the optimization of optical systems using a merit function based on wavefront aberrations.

14.1 Jacobian Matrix of $\mathrm{OPL_i}$ Between (i−1)th and ith Boundary Surfaces

As illustrated in Fig. 14.1, in a medium of constant refractive index ξ_{i-1}, the OPL between points $\bar{P}_{i-1}$ and $\bar{P}_i$ (denoted as $\mathrm{OPL_i}$) is defined as the product of ξ_{i-1} and the geometric length λ_i between the two points. That is

$$\mathrm{OPL_i} = \mathrm{OPL}(\bar{P}_{i-1}, \bar{P}_i) = \xi_{i-1}\lambda_i. \tag{14.1}$$

As described in Chap. 2, λ_i is a recursive function. Consequently, $\mathrm{OPL_i}$ is also a recursive function. Furthermore, as noted from Figs. 7.2, 7.3, 7.4, 7.5, 7.7 and 7.11, λ_i is a function not only of the incoming ray $\bar{R}_{i-1}$, but also of the boundary variable vector $\bar{X}_i$. Therefore, this section formulates the Jacobian matrix of $\mathrm{OPL_i}$ separately with respect to $\bar{R}_{i-1}$ and $\bar{X}_i$, respectively.

P.D. Lin, *Advanced Geometrical Optics*, Progress in Optical Science and Photonics 4, DOI 10.1007/978-981-10-2299-9_14

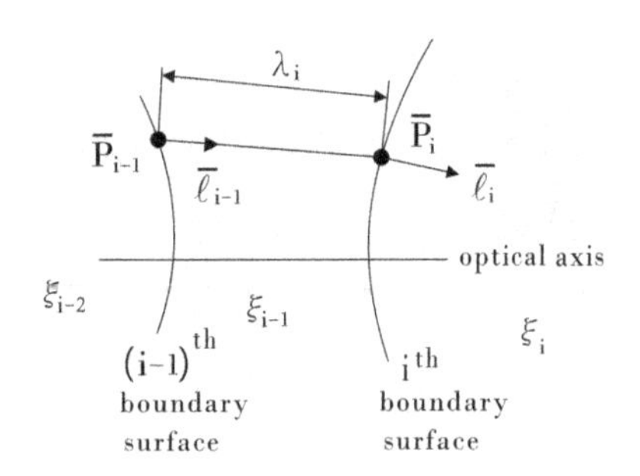

Fig. 14.1 OPL_i between points $\bar{P}_{i-1}$ and $\bar{P}_i$ is defined as the product of the geometric length λ_i between the two points and ξ_{i-1}

14.1.1 Jacobian Matrix of OPL_i with Respect to Incoming Ray $\bar{R}_{i-1}$

Equations (2.16) and (2.39) show that λ_i varies as a function of the incoming ray $\bar{R}_{i-1}$ for both spherical and flat boundary surfaces. Therefore, $OPL_i = \xi_{i-1}\lambda_i$ is also governed by $\bar{R}_{i-1}$. The Jacobian matrix of OPL_i with respect to $\bar{R}_{i-1}$ (i.e., $\partial OPL_i/\partial\bar{R}_{i-1}$) can be obtained directly by differentiating $OPL_i = \xi_{i-1}\lambda_i$ with respect to $\bar{R}_{i-1}$ to give

$$\frac{\partial OPL_i}{\partial \bar{R}_{i-1}} = \xi_{i-1}\frac{\partial \lambda_i}{\partial \bar{R}_{i-1}}. \tag{14.2}$$

The term $\partial\lambda_i/\partial\bar{R}_{i-1}$ in Eq. (14.2) is given by Eqs. (7.5) and (7.21) for flat and spherical boundary surfaces, respectively. The change in OPL_i, ΔOPL_i, induced by a change in the incoming ray, $\Delta\bar{R}_{i-1}$, can then be computed as the matrix product of the Jacobian matrix of OPL_i and $\Delta\bar{R}_{i-1}$, i.e.,

$$\Delta OPL_i = \frac{\partial OPL_i}{\partial \bar{R}_{i-1}}\Delta\bar{R}_{i-1}. \tag{14.3}$$

Chapter 15 investigates the wavefront shape, irradiance and caustic surface along the ray path in a given optical system using $\partial OPL_i/\partial\bar{X}_0$, i.e., the Jacobian matrix of OPL_i with respect to the source variable vector $\bar{X}_0$. This matrix can be obtained by substituting $\Delta\bar{R}_{i-1}=(\partial\bar{R}_{i-1}/\partial\bar{X}_0)\Delta\bar{X}_0$ (where $\partial\bar{R}_{i-1}/\partial\bar{X}_0$ is obtained from Eq. (7.30) by replacing i with i − 1) into Eq. (14.3) to give

$$\Delta OPL_i = \frac{\partial OPL_i}{\partial \bar{R}_{i-1}}\bar{M}_{i-1}\bar{M}_{i-2}\ldots\bar{M}_2\bar{M}_1\bar{S}_0\Delta\bar{X}_0. \tag{14.4}$$

Defining ΔOPL_i as $\Delta OPL_i = (\partial OPL_i/\partial\bar{X}_0)\Delta\bar{X}_0$, $\partial OPL_i/\partial\bar{X}_0$ can be obtained as

$$\frac{\partial OPL_i}{\partial \bar{X}_0} = \frac{\partial OPL_i}{\partial \bar{R}_{i-1}} \bar{M}_{i-1}\bar{M}_{i-2}\ldots\bar{M}_2\bar{M}_1\bar{S}_0. \tag{14.5}$$

Notably, Eq. (14.5) provides the means to investigate the change in the wavefront caused by a change in any of the source ray variables, as defined in Eq. (2.4).

14.1.2 Jacobian Matrix of OPL_i with Respect to Boundary Variable Vector $\bar{X}_i$

Changes in the boundary variables of $\bar{X}_i$ may also cause changes in OPL_i. The Jacobian matrix of OPL_i with respect to boundary variable vector $\bar{X}_i$ can be obtained for both spherical and flat boundary surfaces by taking the first-order differentiation of Eq. (14.1) to give

$$\frac{\partial OPL_i}{\partial \bar{X}_i} = \lambda_i \frac{\partial \xi_{i-1}}{\partial \bar{X}_i} + \xi_{i-1}\frac{\partial \lambda_i}{\partial \bar{X}_i}, \tag{14.6}$$

where $\bar{X}_i$ for spherical and flat boundary surfaces is given by (Eqs. 2.30 and 2.47)

$$\bar{X}_i = [t_{ix} \quad t_{iy} \quad t_{iz} \quad \omega_{ix} \quad \omega_{iy} \quad \omega_{iz} \quad \xi_{i-1} \quad \xi_i \quad R_i]^T$$

and

$$\bar{X}_i = [J_{ix} \quad J_{iy} \quad J_{iz} \quad e_i \quad \xi_{i-1} \quad \xi_i]^T,$$

respectively. Note that $\partial\lambda_i/\partial\bar{X}_i$ in Eq. (14.6) is given by Eqs. (7.32) and (7.44) for flat and spherical boundary surfaces, respectively. In addition, $\partial\xi_{i-1}\partial\bar{X}_i$ is given for spherical and flat boundary surfaces as

$$\frac{\partial \xi_{i-1}}{\partial \bar{X}_i} = [0 \quad 0 \quad 0 \quad 0 \quad 0 \quad 0 \quad 1 \quad 0 \quad 0], \tag{14.7}$$

and

$$\frac{\partial \xi_{i-1}}{\partial \bar{X}_i} = [0 \quad 0 \quad 0 \quad 0 \quad 1 \quad 0], \tag{14.8}$$

respectively. From Eq. (14.6), the change in OPL_i induced by any change in the boundary variables of $\bar{X}_i$ can be determined as

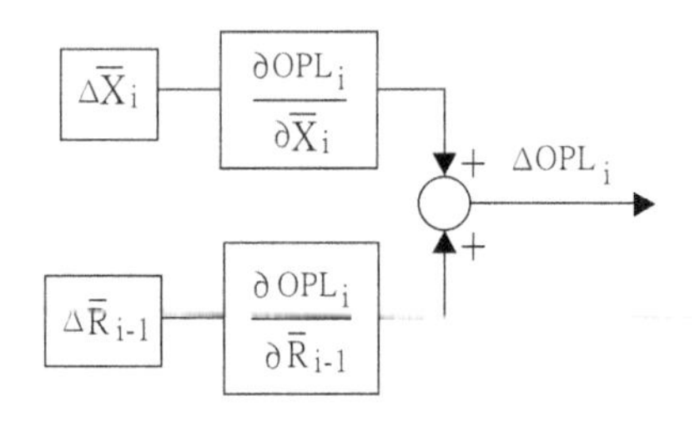

Fig. 14.2 ΔOPL_i is the sum of the changes in OPL_i produced by $\Delta\bar{R}_{i-1}$ and $\Delta\bar{X}_i$

$$\Delta OPL_i = \frac{\partial OPL_i}{\partial \bar{X}_i}\Delta\bar{X}_i. \tag{14.9}$$

As shown in Fig. 14.2, the total change in OPL_i caused by changes in both the incoming ray $\bar{R}_{i-1}$ and the boundary variable vector $\bar{X}_i$ is obtained simply as the sum of Eqs. (14.3) and (14.9), i.e.,

$$\Delta OPL_i = \frac{\partial OPL_i}{\partial \bar{X}_i}\Delta\bar{X}_i + \frac{\partial OPL_i}{\partial \bar{R}_{i-1}}\Delta\bar{R}_{i-1}. \tag{14.10}$$

As stated previously in Chap. 8, not all of the variables in $\bar{X}_i$ (i = 0 to i = n) are independent variables since an element may contain multiple boundary surfaces and a boundary variable may appear repeatedly in different $\bar{X}_i$. Therefore, it is appropriate to transfer $\Delta\bar{X}_i$ in Eq. (14.10) to the system variable vector $\Delta\bar{X}_{sys}$ in accordance with $\Delta\bar{X}_i = (\partial\bar{X}_i/\partial\bar{X}_{sys})\Delta\bar{X}_{sys}$, where $\partial\bar{X}_i/\partial\bar{X}_{sys}$ is computed by Eqs. (8.1), (8.6) and (8.25) for a source variable vector $\bar{X}_0$, flat boundary variable vector $\bar{X}_i$, and spherical boundary variable vector $\bar{X}_i$, respectively. The following equation is then obtained:

$$\Delta OPL_i = \frac{\partial OPL_i}{\partial \bar{X}_i}\frac{\partial \bar{X}_i}{\partial \bar{X}_{sys}}\Delta\bar{X}_{sys} + \frac{\partial OPL_i}{\partial \bar{R}_{i-1}}\Delta\bar{R}_{i-1}, \tag{14.11}$$

where $\Delta\bar{R}_{i-1}$ can be obtained from Eq. (7.53) with $g = i - 1$. Substituting $\Delta\bar{R}_{i-1}$ into Eq. (14.11), and defining $\Delta OPL_i = (dOPL_i/d\bar{X}_{sys})\Delta\bar{X}_{sys}$, $dOPL_i/d\bar{X}_{sys}$ can be formulated as

$$\begin{aligned}\frac{dOPL_i}{d\bar{X}_{sys}} = {} & \frac{\partial OPL_i}{\partial \bar{X}_i}\frac{d\bar{X}_i}{d\bar{X}_{sys}} + \frac{\partial OPL_i}{\partial \bar{R}_{i-1}}\frac{\partial \bar{R}_{i-1}}{\partial \bar{X}_{i-1}}\frac{d\bar{X}_{i-1}}{d\bar{X}_{sys}} + \ldots + \frac{\partial OPL_i}{\partial \bar{R}_{i-1}}\frac{\partial \bar{R}_{i-1}}{\partial \bar{R}_{i-2}}\cdots\frac{\partial \bar{R}_2}{\partial \bar{R}_1}\frac{\partial \bar{R}_1}{\partial \bar{X}_1}\frac{d\bar{X}_1}{d\bar{X}_{sys}} \\ & + \frac{\partial OPL_i}{\partial \bar{R}_{i-1}}\frac{\partial \bar{R}_{i-1}}{\partial \bar{R}_{i-2}}\cdots\frac{\partial \bar{R}_1}{\partial \bar{R}_0}\frac{\partial \bar{R}_0}{\partial \bar{X}_0}\frac{d\bar{X}_0}{d\bar{X}_{sys}}.\end{aligned} \tag{14.12}$$

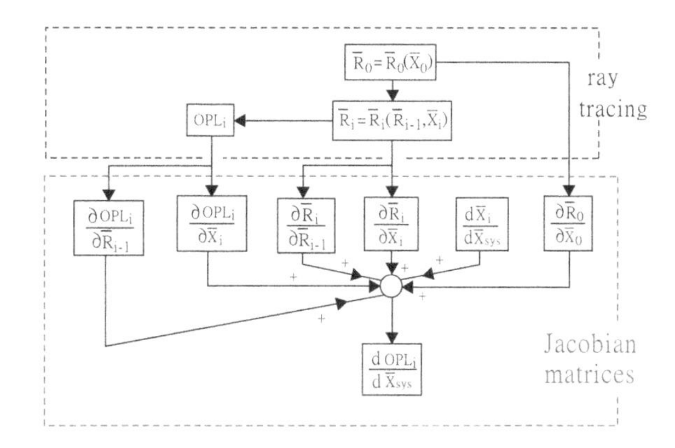

Fig. 14.3 Flowchart showing the determination of OPL_i and its Jacobian matrix

Figure 14.3 presents a hierarchical flowchart summarizing the raytracing and Jacobian matrix operations required to determine the Jacobian matrix of OPL_i relative to the system variable vector $\bar{X}_{sys}$.

14.2 Jacobian Matrix of OPL Between Two Incidence Points

Let the OPL between two incidence points $\bar{P}_g$ and $\bar{P}_h$ of a ray on the gth and hth boundary surfaces, respectively, be denoted as $OPL(\bar{P}_g, \bar{P}_h)$, where g and h are two positive integers and satisfy $0 \leq g < h \leq n$. This section derives the Jacobian matrix of $OPL(\bar{P}_g, \bar{P}_h)$ with respect to the system variable vector $\bar{X}_{sys}$ of the optical system. $OPL(\bar{P}_g, \bar{P}_h)$ can be obtained by summing OPL_i between the two boundary surfaces, i.e.,

$$OPL(\bar{P}_g, \bar{P}_h) = \sum_{i=g+1}^{i=h} OPL_i \tag{14.13}$$

From Eqs. (14.13) and (14.10), the following expression is obtained for $\Delta OPL(\bar{P}_g, \bar{P}_h)$:

$$\Delta OPL(\bar{P}_g, \bar{P}_h) = \sum_{i=g+1}^{i=h} \Delta OPL_i = \sum_{i=g+1}^{i=h} \frac{\partial OPL_i}{\partial \bar{X}_i} \Delta \bar{X}_i + \sum_{i=g+1}^{i=h} \frac{\partial OPL_i}{\partial \bar{R}_{i-1}} \Delta \bar{R}_{i-1}. \tag{14.14}$$

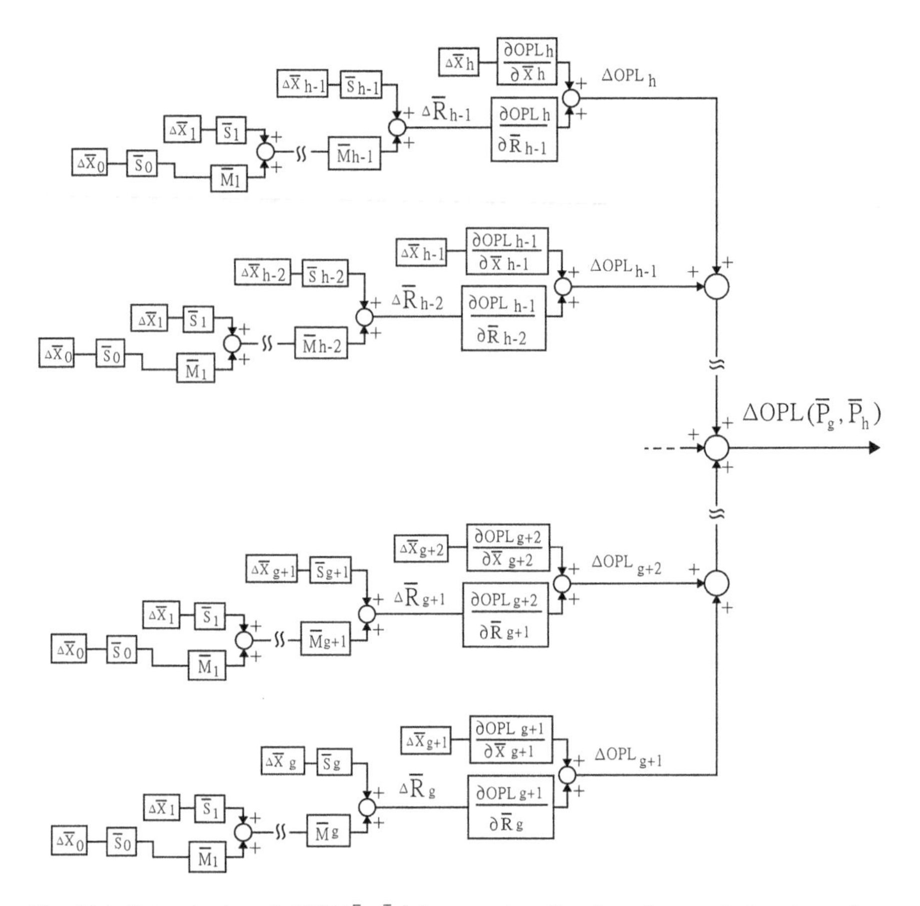

Fig. 14.4 Determination of $\Delta OPL(\bar{P}_g, \bar{P}_h)$ for a ray traveling through an optical system, where $OPL(\bar{P}_g, \bar{P}_h)$ is the optical path length from point $\bar{P}_g$ to point $\bar{P}_h$

The first term in Eq. (14.14) gives the effect of changes in the boundary variables of $\bar{X}_i$ (i = g + 1 to i = h) on $\Delta OPL(\bar{P}_g, \bar{P}_h)$. Changes in the incoming ray $\bar{R}_{i-1}$, i.e., term $\Delta\bar{R}_{i-1}$ in Eq. (14.14), can be obtained from Eq. (7.53) with $g = i - 1$. Equation (14.14) can then be re-written as (see Fig. 14.4)

$$
\begin{aligned}
\Delta \mathrm{OPL}(\bar{P}_g,\bar{P}_h) &= \sum_{i=g+1}^{i=h}\frac{\partial \mathrm{OPL}_i}{\partial \bar{X}_i}\Delta\bar{X}_i + \sum_{i=g+1}^{i=h}\frac{\partial \mathrm{OPL}_i}{\partial \bar{R}_{i-1}}\left(\sum_{u=0}^{u=i-1}\bar{M}_{i-1}\bar{M}_{i-2}\ldots\bar{M}_{u+1}\bar{S}_u\Delta\bar{X}_u\right)\\
&= \frac{\partial \mathrm{OPL}_{g+1}}{\partial \bar{X}_{g+1}}\Delta\bar{X}_{g+1} + \frac{\partial \mathrm{OPL}_{g+1}}{\partial \bar{R}_g}\left(\sum_{u=0}^{u=g}\bar{M}_g\bar{M}_{g-1}\ldots\bar{M}_{u+1}\bar{S}_u\Delta\bar{X}_u\right)\\
&+ \frac{\partial \mathrm{OPL}_{g+2}}{\partial \bar{X}_{g+2}}\Delta\bar{X}_{g+2} + \frac{\partial \mathrm{OPL}_{g+2}}{\partial \bar{R}_{g+1}}\left(\sum_{u=0}^{u=g+1}\bar{M}_{g+1}\bar{M}_{g+2}\ldots\bar{M}_{u+1}\bar{S}_u\Delta\bar{X}_u\right)\\
&+\ldots\\
&+ \frac{\partial \mathrm{OPL}_{h-1}}{\partial \bar{X}_{h-1}}\Delta\bar{X}_{h-1} + \frac{\partial \mathrm{OPL}_{h-1}}{\partial \bar{R}_{h-2}}\left(\sum_{u=0}^{u=h-2}\bar{M}_{h-2}\bar{M}_{h-3}\ldots\bar{M}_{u+1}\bar{S}_u\Delta\bar{X}_u\right)\\
&+ \frac{\partial \mathrm{OPL}_{h}}{\partial \bar{X}_{h}}\Delta\bar{X}_{h} + \frac{\partial \mathrm{OPL}_{h}}{\partial \bar{R}_{h-1}}\left(\sum_{u=0}^{u=h-1}\bar{M}_{h-1}\bar{M}_{h-2}\ldots\bar{M}_{u+1}\bar{S}_u\Delta\bar{X}_u\right).
\end{aligned}
\tag{14.15}
$$

Note that for reasons of simplicity, some terms are deliberately omitted here. The expression given in Eq. (14.15) is sufficient to describe the total OPL change, $\Delta \mathrm{OPL}(\bar{P}_g,\bar{P}_h)$, with respect to the system variable vector $\bar{X}_{sys}$ if $\Delta\bar{X}_i$ (i = 0 to i = h) is transferred to $\Delta\bar{X}_{sys}$ by means of $\Delta\bar{X}_i = (\partial\bar{X}_i/\partial\bar{X}_{sys})\Delta\bar{X}_{sys}$. Consequently, defining $\Delta \mathrm{OPL}(\bar{P}_g,\bar{P}_h) = (\partial \mathrm{OPL}(\bar{P}_g,\bar{P}_h)/\partial\bar{X}_{sys})\Delta\bar{X}_{sys}$, the Jacobian matrix $\partial \mathrm{OPL}(\bar{P}_g,\bar{P}_h)/\partial\bar{X}_{sys}$ can be obtained from Eq. (14.15) as

$$
\begin{aligned}
\frac{d\mathrm{OPL}(\bar{P}_g,\bar{P}_h)}{d\bar{X}_{sys}} &= \frac{\partial \mathrm{OPL}_{g+1}}{\partial \bar{X}_{g+1}}\frac{d\bar{X}_{g+1}}{d\bar{X}_{sys}} + \frac{\partial \mathrm{OPL}_{g+1}}{\partial \bar{R}_g}\left(\sum_{u=0}^{u=g}\bar{M}_g\bar{M}_{g-1}\ldots\bar{M}_{u+1}\bar{S}_u\frac{d\bar{X}_u}{d\bar{X}_{sys}}\right)\\
&+ \frac{\partial \mathrm{OPL}_{g+2}}{\partial \bar{X}_{g+2}}\frac{d\bar{X}_{g+2}}{d\bar{X}_{sys}} + \frac{\partial \mathrm{OPL}_{g+2}}{\partial \bar{R}_{g+1}}\left(\sum_{u=0}^{u=g+1}\bar{M}_{g+1}\bar{M}_{g+2}\ldots\bar{M}_{u+1}\bar{S}_u\frac{d\bar{X}_u}{d\bar{X}_{sys}}\right)\\
&+\ldots\\
&+ \frac{\partial \mathrm{OPL}_{h-1}}{\partial \bar{X}_{h-1}}\frac{d\bar{X}_{h-1}}{d\bar{X}_{sys}} + \frac{\partial \mathrm{OPL}_{h-1}}{\partial \bar{R}_{h-2}}\left(\sum_{u=0}^{u=h-2}\bar{M}_{h-2}\bar{M}_{h-3}\ldots\bar{M}_{u+1}\bar{S}_u\frac{d\bar{X}_u}{d\bar{X}_{sys}}\right)\\
&+ \frac{\partial \mathrm{OPL}_{h}}{\partial \bar{X}_{h}}\frac{d\bar{X}_{h}}{d\bar{X}_{sys}} + \frac{\partial \mathrm{OPL}_{h}}{\partial \bar{R}_{h-1}}\left(\sum_{u=0}^{u=h-1}\bar{M}_{h-1}\bar{M}_{h-2}\ldots\bar{M}_{u+1}\bar{S}_u\frac{d\bar{X}_u}{d\bar{X}_{sys}}\right).
\end{aligned}
\tag{14.16}
$$

It should be noted that $\mathrm{OPL}(\bar{P}_0,\bar{P}_n)$ is the total OPL traversed by a ray through an optical system with n boundary surfaces. Therefore, the change in the total OPL (i.e., $\Delta \mathrm{OPL}(\bar{P}_0,\bar{P}_n)$) due to a change in the boundary variable vector $\Delta\bar{X}_i$ (i = 0 to i = n) can be obtained from Eq. (14.15) by setting g = 0 and h = n. That is,

$$\begin{aligned}\Delta \mathrm{OPL}(\bar{P}_0,\bar{P}_n) = {} & \frac{\partial \mathrm{OPL}_1}{\partial \bar{X}_1}\Delta\bar{X}_1 + \frac{\partial \mathrm{OPL}_1}{\partial \bar{R}_0}(\bar{S}_0\Delta\bar{X}_0) \\ & + \frac{\partial \mathrm{OPL}_2}{\partial \bar{X}_2}\Delta\bar{X}_2 + \frac{\partial \mathrm{OPL}_2}{\partial \bar{R}_1}(\bar{M}_1\bar{S}_0\Delta\bar{X}_0 + \bar{S}_1\Delta\bar{X}_1) \\ & + \ldots \\ & + \frac{\partial \mathrm{OPL}_{n-1}}{\partial \bar{X}_{n-1}}\Delta\bar{X}_{n-1} + \frac{\partial \mathrm{OPL}_{n-1}}{\partial \bar{R}_{n-2}}(\bar{M}_{n-2}\ldots\bar{M}_1\bar{S}_0\Delta\bar{X}_0 + \quad + \bar{M}_{n-2}\bar{S}_{n-3}\Delta\bar{X}_{n-3} + \bar{S}_{n-2}\Delta\bar{X}_{n-2}) \\ & + \frac{\partial \mathrm{OPL}_n}{\partial \bar{X}_n}\Delta\bar{X}_n + \frac{\partial \mathrm{OPL}_n}{\partial \bar{R}_{n-1}}(\bar{M}_{n-1}\ldots\bar{M}_1\bar{S}_0\Delta\bar{X}_0 + \ldots + \bar{M}_{n-1}\bar{S}_{n-2}\Delta\bar{X}_{n-2} + \bar{S}_{n-1}\Delta\bar{X}_{n-1}).\end{aligned} \tag{14.17}$$

The Jacobian matrix $\mathrm{dOPL}(\bar{P}_0,\bar{P}_n)/\mathrm{d}\bar{X}_{sys}$ can then be obtained from Eq. (14.17) as

$$\begin{aligned}\frac{\mathrm{dOPL}(\bar{P}_0,\bar{P}_n)}{\mathrm{d}\bar{X}_{sys}} = {} & \frac{\partial \mathrm{OPL}_1}{\partial \bar{X}_1}\frac{\mathrm{d}\bar{X}_1}{\mathrm{d}\bar{X}_{sys}} + \frac{\partial \mathrm{OPL}_1}{\partial \bar{R}_0}\left(\bar{S}_0\frac{\mathrm{d}\bar{X}_0}{\mathrm{d}\bar{X}_{sys}}\right) \\ & + \frac{\partial \mathrm{OPL}_2}{\partial \bar{X}_2}\frac{\mathrm{d}\bar{X}_2}{\mathrm{d}\bar{X}_{sys}} + \frac{\partial \mathrm{OPL}_2}{\partial \bar{R}_1}\left(\bar{M}_1\bar{S}_0\frac{\mathrm{d}\bar{X}_0}{\mathrm{d}\bar{X}_{sys}} + \bar{S}_1\frac{\mathrm{d}\bar{X}_1}{\mathrm{d}\bar{X}_{sys}}\right) \\ & + \frac{\partial \mathrm{OPL}_{n-1}}{\partial \bar{X}_{n-1}}\frac{\mathrm{d}\bar{X}_{n-1}}{\mathrm{d}\bar{X}_{sys}} + \ldots + \frac{\partial \mathrm{OPL}_{n-1}}{\partial \bar{R}_{n-2}}\left(\bar{M}_{n-2}\ldots\bar{M}_1\bar{S}_0\frac{\mathrm{d}\bar{X}_0}{\mathrm{d}\bar{X}_{sys}} + \ldots + \bar{M}_{n-2}\bar{S}_{n-3}\frac{\mathrm{d}\bar{X}_{n-3}}{\mathrm{d}\bar{X}_{sys}} + \bar{S}_{n-2}\frac{\mathrm{d}\bar{X}_{n-2}}{\mathrm{d}\bar{X}_{sys}}\right) \\ & + \frac{\partial \mathrm{OPL}_n}{\partial \bar{X}_n}\frac{\mathrm{d}\bar{X}_n}{\mathrm{d}\bar{X}_{sys}} + \frac{\partial \mathrm{OPL}_n}{\partial \bar{R}_{n-1}}\left(\bar{M}_{n-1}\ldots\bar{M}_1\bar{S}_0\frac{\mathrm{d}\bar{X}_0}{\mathrm{d}\bar{X}_{sys}} + \ldots + \bar{M}_{n-1}\bar{S}_{n-2}\frac{\mathrm{d}\bar{X}_{n-2}}{\mathrm{d}\bar{X}_{sys}} + \bar{S}_{n-1}\frac{\mathrm{d}\bar{X}_{n-1}}{\mathrm{d}\bar{X}_{sys}}\right).\end{aligned} \tag{14.18}$$

In analyzing an existing optical system, it is sometimes necessary to compute the change of $\mathrm{OPL}(\bar{P}_0,\bar{P}_n)$ after a ray has passed completely through the system in order to analyze the wavefront aberration. This can be achieved using Eq. (14.17) with $\Delta\bar{X}_i = \bar{0}$ (i = 1 to i = n), to give

$$\begin{aligned}& \Delta \mathrm{OPL}(\bar{P}_0,\bar{P}_n) \\ & = \left(\frac{\partial \mathrm{OPL}_1}{\partial \bar{R}_0} + \frac{\partial \mathrm{OPL}_2}{\partial \bar{R}_1}\bar{M}_1 + \ldots + \frac{\partial \mathrm{OPL}_{n-1}}{\partial \bar{R}_{n-2}}\bar{M}_{n-2}\ldots\bar{M}_1 + \frac{\partial \mathrm{OPL}_n}{\partial \bar{R}_{n-1}}\bar{M}_{n-1}\ldots\bar{M}_1\right)\bar{S}_0\Delta\bar{X}_0.\end{aligned} \tag{14.19}$$

The Jacobian matrix of $\mathrm{OPL}(\bar{P}_0,\bar{P}_n)$ with respect to $\bar{X}_0$ can then be obtained from Eq. (14.19) as

$$\frac{\partial \mathrm{OPL}(\bar{P}_0,\bar{P}_n)}{\partial \bar{X}_0} = \left(\frac{\partial \mathrm{OPL}_1}{\partial \bar{R}_0} + \frac{\partial \mathrm{OPL}_2}{\partial \bar{R}_1}\bar{M}_1 + \ldots + \frac{\partial \mathrm{OPL}_{n-1}}{\partial \bar{R}_{n-2}}\bar{M}_{n-2}\ldots\bar{M}_1 + \frac{\partial \mathrm{OPL}_n}{\partial \bar{R}_{n-1}}\bar{M}_{n-1}\ldots\bar{M}_1\right)\bar{S}_0. \tag{14.20}$$

Example 14.1 Referring to the system shown in Fig. 3.2, $\Delta\mathrm{OPL}(\bar{P}_1,\bar{P}_2)$ (i.e., the change in the OPL of a ray traveling through the first element) due to $\Delta\bar{X}_{sys}$ is given by (see Fig. 14.5)

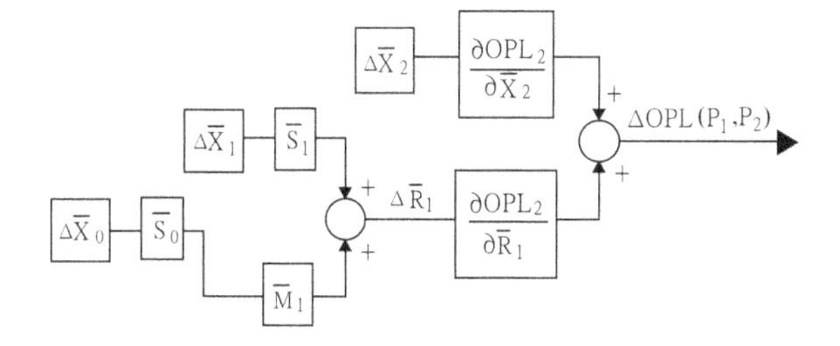

Fig. 14.5 Determination of $\Delta OPL(\bar{P}_1, \bar{P}_2)$ for the system in Fig. 3.2, where $OPL(\bar{P}_1, \bar{P}_2)$ is the OPL measured from incidence point $\bar{P}_1$ to exit point $\bar{P}_2$ of the first element

$$\begin{aligned}\Delta OPL_2 &= \frac{\partial OPL_2}{\partial \bar{X}_2}\Delta\bar{X}_2 + \frac{\partial OPL_2}{\partial \bar{R}_1}(\bar{M}_1\bar{S}_0\Delta\bar{X}_0 + \bar{S}_1\Delta\bar{X}_1)\\ &= \frac{\partial OPL_2}{\partial \bar{X}_2}\frac{d\bar{X}_2}{d\bar{X}_{sys}}\Delta\bar{X}_{sys} + \frac{\partial OPL_2}{\partial \bar{R}_1}\left(\bar{M}_1\bar{S}_0\frac{d\bar{X}_0}{d\bar{X}_{sys}}\Delta\bar{X}_{sys} + \bar{S}_1\frac{d\bar{X}_1}{d\bar{X}_{sys}}\Delta\bar{X}_{sys}\right).\end{aligned}$$

The Jacobian matrix $dOPL_2/d\bar{X}_{sys}$ can then be obtained as

$$\frac{dOPL_2}{d\bar{X}_{sys}} = \frac{\partial OPL_2}{\partial \bar{X}_2}\frac{d\bar{X}_2}{d\bar{X}_{sys}} + \frac{\partial OPL_2}{\partial \bar{R}_1}\left(\bar{M}_1\bar{S}_0\frac{d\bar{X}_0}{d\bar{X}_{sys}} + \bar{S}_1\frac{d\bar{X}_1}{d\bar{X}_{sys}}\right).$$

When ΔOPL_2 is evaluated for the chief ray originating from $\bar{P}_0 = [0 \quad -507 \quad 170 \quad 1]^T$, the numerical result is given by

$$\begin{aligned}\Delta OPL_2 = &-0.098205\Delta P_{0y} - 0.305797\Delta P_{0z} - 176.286003\Delta\beta_0 + 0.098205\Delta v_1\\ &+ 0.305797\Delta t_{e1z} - 7.135859\Delta\omega_{e1x} + 0.644910\Delta\xi_{air} + 15.268118\Delta\xi_{e1}\\ &+ 1.731846\Delta q_{e1} + 0.011776\Delta R_1 + 0.002387\Delta R_2.\end{aligned}$$

Example 14.2 Referring to the system shown in Fig. 3.2, the change in $OPL(\bar{P}_8, \bar{P}_{10})$ is given by (see Fig. 14.6):

$$\begin{aligned}\Delta OPL(\bar{P}_8, \bar{P}_{10}) &= \Delta OPL_9 + \Delta OPL_{10}\\ &= \frac{\partial OPL_9}{\partial \bar{R}_8}\Delta\bar{R}_8 + \frac{\partial OPL_9}{\partial \bar{X}_9}\Delta\bar{X}_9 + \frac{\partial OPL_{10}}{\partial \bar{R}_9}\Delta\bar{R}_9 + \frac{\partial OPL_{10}}{\partial \bar{X}_{10}}\Delta\bar{X}_{10}\\ &= \frac{\partial OPL_9}{\partial \bar{R}_8}\Delta\bar{R}_8 + \frac{\partial OPL_9}{\partial \bar{X}_9}\Delta\bar{X}_9 + \frac{\partial OPL_{10}}{\partial \bar{R}_9}(\bar{M}_9\Delta\bar{R}_8 + \bar{S}_9\Delta\bar{X}_9) + \frac{\partial OPL_{10}}{\partial \bar{X}_{10}}\Delta\bar{X}_{10}.\end{aligned}$$

Note that this equation is obtained from Eq. (14.14) with g = 8 and h = 10. The term $\Delta\bar{R}_8$ is given from Eq. (7.53) with g = 8 as

$$\begin{aligned}\Delta\bar{R}_8 = &\bar{M}_8\bar{M}_7\bar{M}_6\bar{M}_5\bar{M}_4\bar{M}_3\bar{M}_2\bar{M}_1\bar{S}_0\Delta\bar{X}_0 + \bar{M}_8\bar{M}_7\bar{M}_6\bar{M}_5\bar{M}_4\bar{M}_3\bar{M}_2\bar{S}_1\Delta\bar{X}_1\\ &+ \bar{M}_8\bar{M}_7\bar{M}_6\bar{M}_5\bar{M}_4\bar{M}_3\bar{S}_2\Delta\bar{X}_2 + \bar{M}_8\bar{M}_7\bar{M}_6\bar{M}_5\bar{M}_4\bar{S}_3\Delta\bar{X}_3 + \bar{M}_8\bar{M}_7\bar{M}_6\bar{M}_5\bar{S}_4\Delta\bar{X}_4\\ &+ \bar{M}_8\bar{M}_7\bar{M}_6\bar{S}_5\Delta\bar{X}_5 + \bar{M}_8\bar{M}_7\bar{S}_6\Delta\bar{X}_6 + \bar{M}_8\bar{S}_7\Delta\bar{X}_7 + \bar{S}_8\Delta\bar{X}_8.\end{aligned}$$

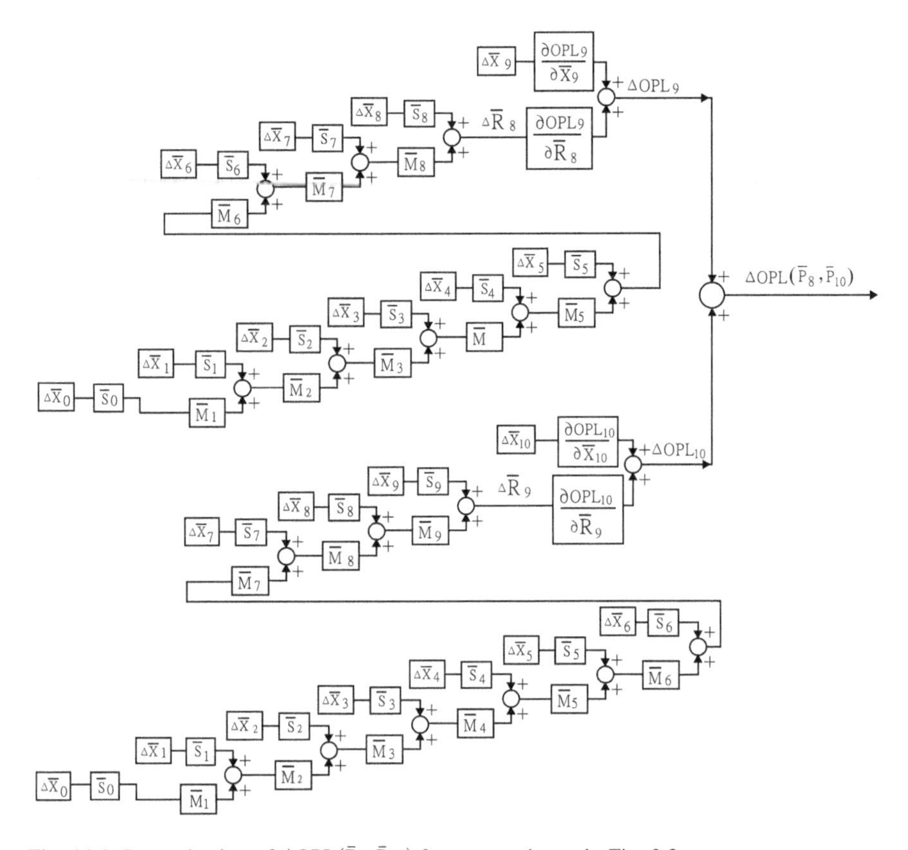

Fig. 14.6 Determination of $\Delta\mathrm{OPL}(\bar{\mathrm{P}}_8, \bar{\mathrm{P}}_{10})$ for system shown in Fig. 3.2

The Jacobian matrix $\partial\mathrm{OPL}(\bar{\mathrm{P}}_8, \bar{\mathrm{P}}_{10})/\partial\bar{\mathrm{X}}_{\mathrm{sys}}$ can be obtained from the above two equations by using $\Delta\bar{\mathrm{X}}_{\mathrm{i}} = (\partial\bar{\mathrm{X}}_{\mathrm{i}}/\partial\bar{\mathrm{X}}_{\mathrm{sys}})\Delta\bar{\mathrm{X}}_{\mathrm{sys}}$ (i = 0 to i = 10).

14.3 Computation of Wavefront Aberrations

In geometrical optics, a wavefront $\bar{\Omega}$ is defined by the locus of points in space having the same OPL from a point source $\bar{\mathrm{P}}_0$. In paraxial optics, the emerging wavefront is essentially spherical and forms a perfect image at its center. However, in practical optical systems, the wavefront propagating through the system is often no longer perfectly spherical by the time it exits the final boundary surface. Therefore, it is important to determine the departure of the wavefront shape traveling along a ray from a perfect spherical form (which denoted as reference sphere $\bar{\mathrm{r}}_{\mathrm{ref}}$). Conventionally, the reference sphere $\bar{\mathrm{r}}_{\mathrm{ref}}$ is the sphere centered at $\bar{\mathrm{P}}_{\mathrm{n/chief}}$, the

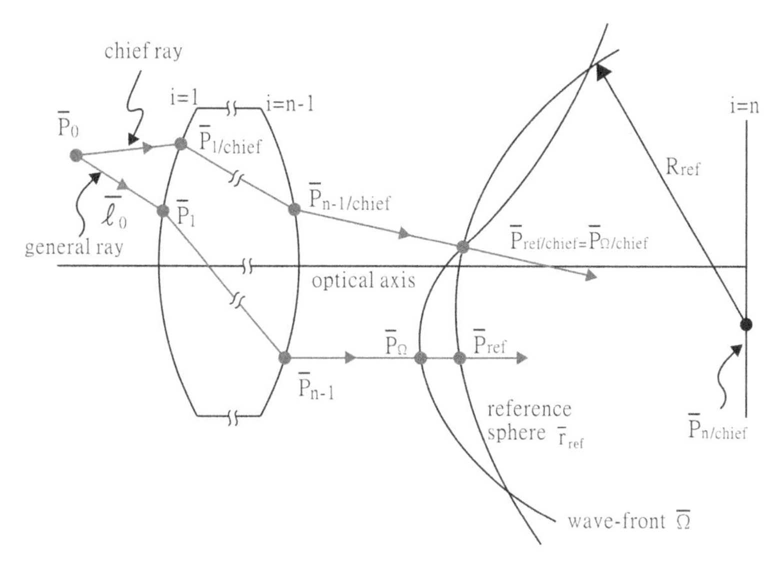

Fig. 14.7 Determination of the wavefront aberration of an axis-symmetrical optical system with n boundary surfaces

incidence point of the chief ray at the last boundary surface i = n, and passing through the center of the exit pupil (Fig. 14.7). The OPL between actual wavefront $\bar{\Omega}$ and the reference sphere $\bar{r}_{ref}$ is the wavefront aberration. However, Mikŝ [1] suggested that the wavefront $\bar{\Omega}$ and reference sphere $\bar{r}_{ref}$ may not pass through the center of the exit pupil. It is possible to define the wavefront aberration by using another reference sphere with a different radius, provided that the chief ray has been refracted by the final boundary surface of the system. Figure 14.7 illustrates the path of a general ray $\bar{R}_0$ originating from a point source $\bar{P}_0 = [P_{0x} \quad P_{0y} \quad P_{0z} \quad 1]^T$ and having a unit directional vector (Eq. 2.3)

$$\bar{\ell}_0 = [\ell_{0x} \quad \ell_{0y} \quad \ell_{0z} \quad 0]^T = [C\beta_0 C(90^\circ + \alpha_0) \quad C\beta_0 S(90^\circ + \alpha_0) \quad S\beta_0 \quad 0]^T$$

as it propagates through an axis-symmetrical optical system containing n boundary surfaces. Figure 14.7 also shows an arbitrary reference sphere $\bar{r}_{ref}$ of radius R_{ref} centered at $\bar{P}_{n/chief}$ used to define the wavefront aberration. As shown, the chief ray originating from the point source $\bar{P}_0$ is incident on the reference sphere at point $\bar{P}_{ref/chief}$. Let $\bar{P}_\Omega$ denote the intersection point of the path of the general source ray $\bar{R}_0$ and the wavefront $\bar{\Omega}$. The wavefront aberration for $\bar{R}_0$ is then determined as

$$W(\bar{X}_0) = OPL(\bar{P}_0, \bar{P}_{ref}) - OPL(\bar{P}_0, \bar{P}_\Omega). \tag{14.21}$$

The wavefront aberration $W(\bar{X}_0)$ is positive if the reference sphere $\bar{r}_{ref}$ leads the aberrated wavefront $\bar{\Omega}$, as shown in Fig. 14.7. Equation (14.21) indicates that for a given system, the wavefront aberration is a function of the variable vector $\bar{X}_0 = [P_{0x} \quad P_{0y} \quad P_{0z} \quad \alpha_0 \quad \beta_0]^T$ of the source ray $\bar{R}_0$. For the wavefront aberration of a fixed point source (e.g., $\bar{P}_0$), $W(\bar{X}_0)$ can be further simplified as a function of α_0 and β_0 only. In other words, the wavefront aberration can be formulated as

$$W(\alpha_0, \beta_0) = OPL(\bar{P}_0, \bar{P}_{ref}) - OPL(\bar{P}_0, \bar{P}_\Omega). \tag{14.22}$$

Significantly, Eq. (14.22) expresses the wavefront aberration in terms of the two polar coordinates α_0 and β_0 of the unit directional vector $\bar{\ell}_0$ of the general ray $\bar{R}_0$. By contrast, most wavefront aberration calculation methods take the entrance or exit pupil coordinates (e.g., x_5 and z_5 of Fig. 3.2) as the variables. Thus, the wavefront aberration function becomes a composite function, i.e., $W(x_5(\alpha_0, \beta_0), z_5(\alpha_0, \beta_0))$. The computation of composite functions is generally extremely tedious. Consequently, Eq. (14.22) has two practical advantages: (a) the shape of the wavefront aberration of a ray following reflection/or refraction can be numerically determined without the need to compute the corresponding coordinates of the entrance or exit pupil (note that these two pupils are virtual surfaces obtained from the images of the aperture); and (b) the local principal curvatures of the wavefront aberration can be investigated simply by differentiating Eq. (14.22) with respect to α_0 and β_0.

Points $\bar{P}_\Omega$ and $\bar{P}_{ref/chief}$ both lie on the wavefront $\bar{\Omega}$. Consequently, Eq. (14.21) can be rewritten as

$$W(\bar{X}_0) = OPL(\bar{P}_0, \bar{P}_{ref}) - OPL(\bar{P}_0, \bar{P}_{ref/chief}), \tag{14.23}$$

leading to $OPL(\bar{P}_0, \bar{P}_\Omega) = OPL(\bar{P}_0, \bar{P}_{ref/chief})$. Equation (14.23) indicates that the wavefront aberration $W(\bar{X}_0)$ of a general source ray $\bar{R}_0$ is equal to the difference between $OPL(\bar{P}_0, \bar{P}_{ref})$ and $OPL(\bar{P}_0, \bar{P}_{ref/chief})$, where $\bar{P}_{ref}$ and $\bar{P}_{ref/chief}$ are the incident points of the source ray $\bar{R}_0$ and chief ray $\bar{R}_0/chief$, respectively, on the reference sphere $\bar{r}_{ref}$.

It should be noted that the reference sphere $\bar{r}_{ref}$ is centered at the imaging point $\bar{P}_{n/chief}$ of the chief ray originating from point source $\bar{P}_0$ (see Fig. 14.7). However, $\bar{P}_{n/chief}$ does not lie on the optical axis when $\bar{P}_0$ is an off-axis point. In other words, it is necessary to compute the wavefront aberration for a non-axially symmetrical system, even though the system is in fact axis-symmetrical. In Sect. 14.2, a methodology was presented for determining the Jacobian matrix of the OPL with respect to the system variable vector $\bar{X}_{sys}$ for non-axially symmetrical optical systems. As stated in Chap. 3, when analyzing a system possessing n optical boundary surfaces, it is first necessary to label the surfaces sequentially from $i = 0$ to $i = n$ (e.g. , Figure 3.2). Figure 14.8 shows the non-axially symmetrical system obtained by treating the reference sphere $\bar{r}_{ref}$ as an additional spherical boundary surface in the system shown in Fig. 14.7. Note that in applying this approach, $\bar{r}_{ref}$ in Fig. 14.7 is labeled as the nth boundary surface. Thus, the image plane, which was originally

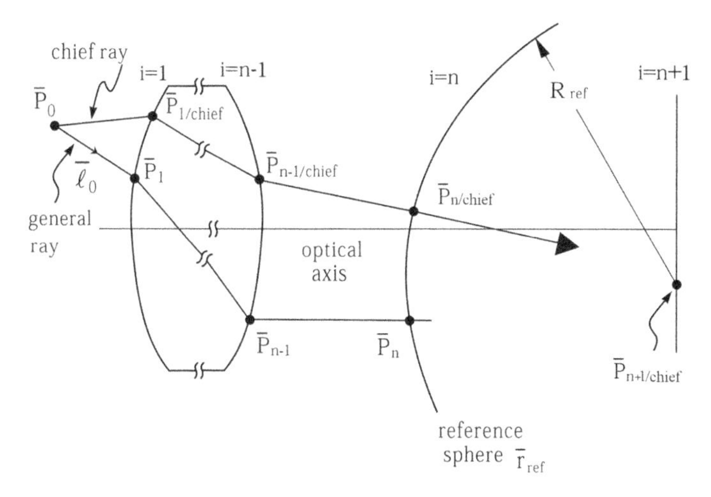

Fig. 14.8 The system shown in Fig. 14.7 becomes a non-axially symmetrical system when the reference sphere $\bar{r}_{ref}$ is treated as a virtual boundary surface with radius R_{ref} centered at the incidence point of the chief ray on the imaging plane

labeled as i = n, is re-labeled as the (n + 1)th boundary surface. Using the modified system shown in Fig. 14.8, the wavefront aberration given in Eq. (14.23) becomes

$$W(\bar{X}_0) = OPL(\bar{P}_0, \bar{P}_n) - OPL(\bar{P}_0, \bar{P}_{n/chief}). \quad (14.24)$$

The first and second terms of Eq. (14.24) are given respectively as

$$OPL(\bar{P}_0, \bar{P}_n) = \sum_{i=1}^{i=n} OPL_i, \quad (14.25)$$

$$OPL(\bar{P}_0, \bar{P}_{n/chief}) = \sum_{i=1}^{i=n} OPL_{i/chief}. \quad (14.26)$$

OPL_i in Eq. (14.25) and $OPL_{i/chief}$ in Eq. (14.26) are the OPLs of the general ray $\bar{R}_0 = \begin{bmatrix} \bar{P}_0 & \bar{\ell}_0 \end{bmatrix}^T$ and the chief ray $\bar{R}_0/chief = \begin{bmatrix} \bar{P}_0 & \bar{\ell}_{0/chief} \end{bmatrix}^T$, respectively, between the (i − 1)th and ith boundary surfaces for a given point source $\bar{P}_0$. Substituting Eqs. (14.25) and (14.26) into Eq. (14.24), the following equation for the wavefront aberration is obtained:

$$W(\bar{X}_0) = \sum_{i=1}^{i=n} OPL_i - \sum_{i=1}^{i=n} OPL_{i/chief} = \sum_{i=1}^{i=n} \left(OPL_i - OPL_{i/chief}\right). \quad (14.27)$$

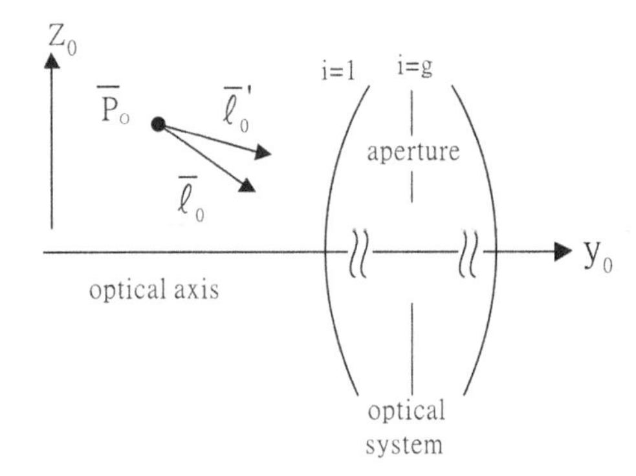

Fig. 14.9 Feasibility of estimating the wavefront aberration of neighboring ray $\bar{R}_0'$ via a Taylor series expansion if the wavefront aberration of reference ray $\bar{R}_0$ is given

Equation (14.27) shows that the wavefront aberration of a system with multiple boundary surfaces is additive, i.e., the wavefront aberration of a ray for the entire system is equal to the sum of the wavefront aberrations at each of the boundary surfaces encountered by the ray as it travels through the system. (Note that this property does not hold true for ray aberrations.)

Using the modified system shown in Fig. 14.8, the Jacobian matrix of the wavefront aberration given in Eq. (14.27) becomes

$$\frac{\partial W(\bar{X}_0)}{\partial \bar{X}_{sys}} = \frac{\partial OPL(\bar{P}_0, \bar{P}_n)}{\partial \bar{X}_{sys}} - \frac{\partial OPL(\bar{P}_0, \bar{P}_{n/chief})}{\partial \bar{X}_{sys}} = \sum_{i=1}^{i=n} \frac{\partial OPL_i}{\partial \bar{X}_{sys}} - \sum_{i=1}^{i=n} \frac{\partial OPL_{i/chief}}{\partial \bar{X}_{sys}}. \tag{14.28}$$

Note that $\bar{P}_n$ and $\bar{P}_{n/chief}$ in Eq. (14.28) are the incident points of the general ray $\bar{R}_0$ and chief ray, respectively, on the reference sphere $\bar{r}_{ref}$ (i.e., the nth boundary surface of the modified system in Fig. 14.8).

Figure 14.9 shows two rays originating from a point source $\bar{P}_0$, namely a reference source ray $\bar{R}_0 = \begin{bmatrix} \bar{P}_0 & \bar{\ell}_0 \end{bmatrix}^T$ with variable vector $\bar{X}_0 = \begin{bmatrix} P_{0x} & P_{0y} & P_{0z} & \alpha_0 & \beta_0 \end{bmatrix}^T$ and a neighboring source ray $\bar{R}_0' = \begin{bmatrix} \bar{P}_0 & \bar{\ell}_0' \end{bmatrix}^T$ with variable vector $\bar{X}_0' = \begin{bmatrix} P_{0x} & P_{0y} & P_{0z} & \alpha_0' & \beta_0' \end{bmatrix}^T$. For an existing optical system, it is possible to estimate the wavefront aberration $W(\bar{X}_0') = W(\bar{X}_0) + \Delta W(\bar{X}_0)$ of the ray $\bar{R}_0'$ by a first-order Taylor series expansion if the wavefront aberration $W(\bar{X}_0)$ of the reference ray $\bar{R}_0$ is given. The change in the wavefront aberration $\Delta W(\bar{X}_0)$ is then given by

$$\Delta W(\bar{X}_0) = \left(\sum_{i=1}^{i=n} \frac{\partial OPL_i}{\partial \bar{X}_0} \right) \Delta \bar{X}_0, \tag{14.29}$$

where $\Delta \bar{X}_0 = \bar{X}_0' - \bar{X}_0 = \begin{bmatrix} 0 & 0 & 0 & \alpha_0' - \alpha_0 & \beta_0' - \beta_0 \end{bmatrix}^T$. The Jacobian matrix of Eq. (14.29) is evaluated by the reference ray $\bar{R}_0$. The wavefront aberration $W(\bar{X}_0')$ of the neighboring ray $\bar{R}_0'$ is then given by the sum of $W(\bar{X}_0)$ and $\Delta W(\bar{X}_0)$.

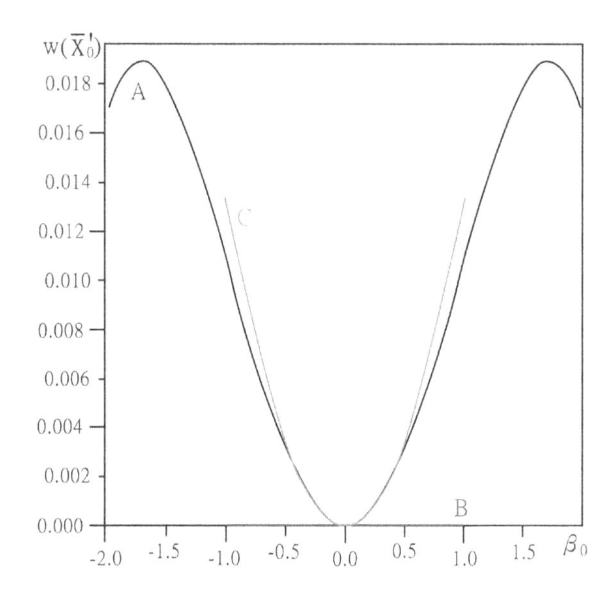

Fig. 14.10 *Curve A* shows the wavefront aberration of an on-axis point source $\bar{P}_0 = [0 \quad -507 \quad 0 \quad 1]^T$ in the system shown in Fig. 3.2 as computed by the raytracing method (Note that *Curves B* and *C* are discussed later in Example 16.1.)

Example 14.3 As shown by Eq. (14.23), it is possible, in principle, to trace a general ray $\bar{R}_0 = [\bar{P}_0 \quad \bar{\ell}_0]^T$ and chief ray $\bar{R}_{0/\text{chief}} = [\bar{P}_0 \quad \bar{\ell}_{0/\text{chief}}]^T$, and then subtract their respective optical path lengths in order to obtain the wavefront aberration $W(\bar{X}_0)$ of $\bar{R}_0$. Curve A in Fig. 14.10 shows the wavefront aberration $W(\bar{X}_0)$ obtained via raytracing for an on-axis point source $\bar{P}_0 = [0 \quad -507 \quad 0 \quad 1]^T$ in the system shown in Fig. 3.2 given a reference sphere $\bar{r}_{\text{ref}}$ with a radius of $R_{\text{ref}} = 45$ mm (see Fig. 14.11). Note that in this particular

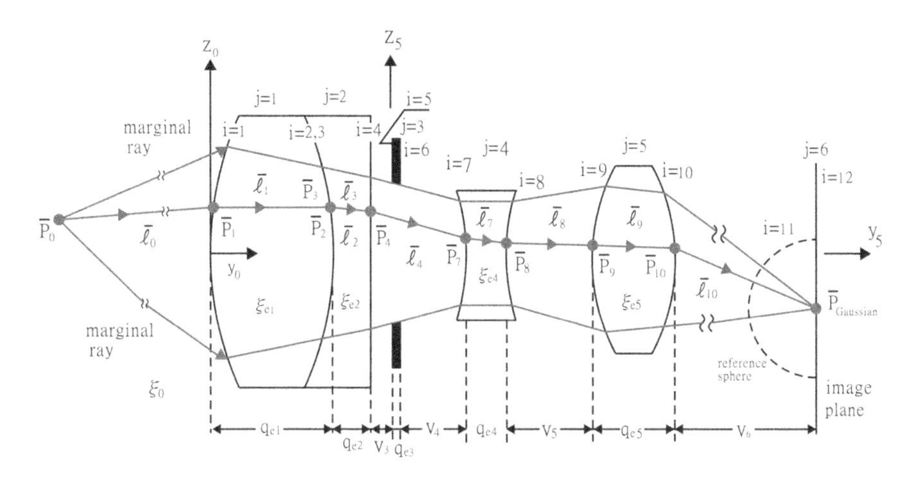

Fig. 14.11 Modified system of Fig. 3.2 in which the reference sphere $\bar{r}_{\text{ref}}$ is considered as a virtual boundary surface (labeled as i = 11) with radius R_{11}

example, $W(\bar{X}_0)$ is an axis-symmetric function since the system shown in Fig. 3.2 (or Fig. 14.11) is axis-symmetric and $\bar{P}_0$ lies on the optical axis. As a result, Curve A in Fig. 14.10 shows only a cross-sectional view of $W(\bar{X}_0)$ rather than its entire distribution. Observing curve A, it is inferred that $\bar{P}_0$ is well focused and has only a small wavefront aberration since it lies within the depth of field of the system. It can be shown numerically that the root mean square, mean value and variance [2] of the wavefront aberration are $\mathrm{rms}(W(\bar{X}_0)) = 0.0002198$, $\mathrm{mean}(W(\bar{X}_0)) = 0.0135739$, and $\mathrm{variance}(W(\bar{X}_0)) = 0.0000355$, respectively.

Example 14.4 The maximum positive and negative wavefront aberrations determine the peak-to-valley (P-V) wavefront aberration and represent the maximum departures of the actual wavefront from the desired wavefront in the positive and negative directions, respectively. When utilizing raytracing methods, it is generally necessary to trace many rays in order to accurately determine the maximum positive and negative values of $W(\bar{X}_0)$. (Note that a smaller number of rays can be traced if an estimated P-V value with lower accuracy is sufficient.) An alternative approach is to use some form of numerical method to obtain the roots of $\left(\sum_{i=1}^{i=n} \partial OPL_i / \partial \bar{X}_0\right) = 0$. Clearly, however, it is also necessary to compute the wavefront aberration $W(\bar{X}_0)$ for all the marginal rays since it is possible that local maximum and minimum values of the wavefront aberration may exist. In the case of Example 14.3, the local minimum value is $W(\bar{X}_0) = 0$ and is located at $\beta_0 = 0°$, while the local maximum value is $W(\bar{X}_0) = 0.0188926$ and is located at $\beta_0 = \pm 1.7053°$. The wavefront aberration of the marginal ray is $W(\bar{X}_0) = 0.0169624$. The P-V wavefront aberration is therefore equal to 0.0188926. However, the P-V property simply states the maximum deviation of the wavefront aberration. In other words, it provides no indication of the area over which the aberration occurs. Moreover, in practice, an optical system with a large P-V error may actually perform better than a system with a small P-V error. As a result, it is generally more meaningful to specify the wavefront quality using the root mean square, mean value and variance of the wavefront aberration rather than its P-V value (see Example 14.3).

14.4 Merit Function Based on Wavefront Aberration

The Jacobian matrix of the wavefront aberration presented in Sect. 14.3 provides an ideal basis for the optimization process performed in automatic optical systems design applications. The two merit functions proposed by Meiron [2] (i.e., the mean square value of the OPD and the variance of the OPD) both require the tracing of a large number of rays since they involve the integration of Eq. (14.27) over all portions of the reference sphere filled by the wavefront. To resolve this problem, let the sum of the squared wavefront aberrations be taken as a merit function. Assume that $\bar{P}_0$ is a single point source which emits a large number of rays into space. To calculate the wavefront aberration, it is necessary to trace a subset of these rays (say, f) through the system. In general, most optimization systems evaluate the merit function for several point sources (say, q) in the field of view. Thus, the following merit function is introduced to evaluate the image quality:

$$\bar{\Phi} = \sum_{1}^{q}\sum_{1}^{f} W^2(\bar{X}_0) = \sum_{1}^{q}\sum_{1}^{f}\left[OPL(\bar{P}_0, \bar{P}_n) - OPL(\bar{P}_0, \bar{P}_{n/chief})\right]^2. \quad (14.30)$$

The Jacobian matrix $\partial\bar{\Phi}/\partial\bar{X}_{sys}$, which is needed to determine the appropriate search direction for optimization programs, can then be obtained by differentiating Eq. (14.30) to give

$$\begin{aligned}\frac{\partial\bar{\Phi}}{\partial\bar{X}_{sys}} &= 2\sum_{1}^{q}\sum_{1}^{f} W(\bar{X}_0)\frac{\partial W(\bar{X}_0)}{\partial\bar{X}_{sys}} \\ &= 2\sum_{1}^{q}\sum_{1}^{f}\left[OPL(\bar{P}_0, \bar{P}_n) - OPL(\bar{P}_0, \bar{P}_{n/chief})\right]\left[\frac{\partial OPL(\bar{P}_0, \bar{P}_n)}{\partial\bar{X}_{sys}} - \frac{\partial OPL(\bar{P}_0, \bar{P}_{n/chief})}{\partial\bar{X}_{sys}}\right].\end{aligned} \quad (14.31)$$

Note that Eq. (14.31) has to be evaluated using Eqs. (14.27) and (14.28). In the simplest sense, optimizing an optical system involves minimizing $\bar{\Phi}$, where the numerical value of $\bar{\Phi}$ depends on the system variable vector $\bar{X}_{sys}$.

References

1. Mikŝ A (2002) Dependence of the wave-front aberration on the radius of the reference sphere. J Opt Soc Am 19:1187–1190
2. Meiron J (1968) The use of merit functions on wave-front aberrations in automatic lens design. Appl Opt 7:667–672

Part III
A Bright Light for Geometrical Optics (Second-Order Derivative Matrices of a Ray and its OPL)

In engineering systems design, iterative optimization methods play a key role in tuning the system variables in such a way as to maximize the system performance. In practice, the performance of such methods is fundamentally dependent on the choice of search direction. Most existing optimization methods for engineering problems use the first- or second-order derivative matrices of the merit function to determine the search direction. However, the rays in an optical system have the form of recursive functions, and thus deriving analytical expressions for the derivative matrices of a merit function is not easily achieved. Consequently, pseudo-derivative matrices based on the Finite Difference (FD) approach are commonly used to approximate the real derivative matrices. However, the accuracy of the results obtained using FD methods is critically dependent on the incremental step size used in the tuning stage. Moreover, the FD method is inefficient since, in attempting to optimize the system performance, the effects of the system parameters must be individually examined. To overcome these limitations, the third part of this book proposes a straightforward computational scheme for deriving the Hessian matrices of a ray and its optical path length, respectively. It is shown that the Hessian matrix provides an effective means of determining the search direction when optimizing the variables of an optical system during the system design process. Moreover, the Hessian matrix provides a useful insight into the system dependencies and is therefore invaluable in evaluating the effects of system tolerances and misalignments.

"*I admire the elegance of your method of computation; it must be nice to ride through these fields upon the horse of true mathematics while the like of us have to make our way laboriously on foot.*."—Albert Einstein, the Italian Mathematicians of Relativity

Chapter 15
Wavefront Aberration and Wavefront Shape

The study of optical systems may be considered from two different perspectives, namely systems analysis and systems design, respectively. Systems analysis involves the determination of the performance inherent in a given system. More specifically, systems analysis involves evaluating the changes which occur in the system when changing one (or more) of the components of the source variable vector $\bar{X}_0$, while keeping the other parameters of $\bar{X}_i$ ($i = 1$ to $i = n$) at their designed values. Figure 15.1 presents a three-level hierarchical chart (i.e., ray tracing, ray Jacobian matrices and ray Hessian matrices) showing the potential steps in a systems analysis procedure. Traditionally, optical systems are analyzed using only raytracing equations (i.e., the upper level of the three-level hierarchy). However, as shown in Part Two of this book, the Jacobian matrices of a ray and its OPL of an optical system also provide a useful framework for studying a wide variety of optical problems. In this chapter, it is shown that the Hessian matrices of the ray and its OPL are also of benefit in analyzing a system; particularly in analyzing the wavefront aberration and waveform shape.

The wavefront shape along a ray path can be computed using either the k-function method [1, 2] or the differential geometry-based method [3]. Shealy et al. [4–8] extended the former method to examine a plane wavefront propagating through a system containing multiple optical elements along a path parallel to the optical axis. However, this chapter shows that a differential geometry-based approach based on the first and second fundamental forms of the wavefront provides a more general approach for discussing the shape of the wavefront. Implementing the proposed method requires: (1) equations which describe the reflection/or refraction process at each boundary surface encountered by the ray (addressed in Chap. 3); (2) expressions for the Jacobian matrices of the ray $\bar{R}_i$ and OPL_i with respect to the variable vector $\bar{X}_0$ of the source ray (i.e., $\partial\bar{R}_i/\partial\bar{X}_0$ and $\partial OPL_i/\partial\bar{X}_0$, given in Chaps. 7 and 14, respectively); and (3) expressions for the Hessian matrices of the ray $\bar{R}_i$ and OPL_i with respect to $\bar{X}_0$ (i.e., $\partial^2\bar{R}_i/\partial\bar{X}_0^2$ and $\partial^2 OPL_i/\partial\bar{X}_0^2$). The literature lacks any formal method for computing the required

P.D. Lin, *Advanced Geometrical Optics*, Progress in Optical Science and Photonics 4, DOI 10.1007/978-981-10-2299-9_15

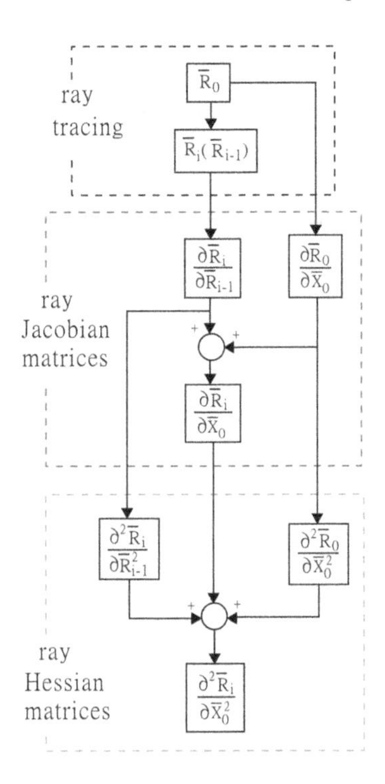

Fig. 15.1 Three-level hierarchical chart showing systems analysis procedure based on raytracing, Jacobian matrix and Hessian matrix

Hessian matrices. Thus, to address this problem, this chapter proposes a method for determining $\partial^2\bar{R}_i/\partial\bar{X}_0^2$ and $\partial^2\mathrm{OPL}_i/\partial\bar{X}_0^2$ using a differential geometry-based approach. Before determining these two Hessian matrices, however, it is first necessary to compute matrix $\partial^2\bar{R}_i/\partial\bar{R}_{i-1}^2$, as described in Sects. 15.1 and 15.2 for flat and spherical boundary surfaces, respectively. (Note that readers may find it useful to review the basic notations associated with the Hessian matrix presented in Sect. 1.10 before reading this chapter.)

15.1 Hessian Matrix $\partial^2\bar{R}_i/\partial\bar{R}_{i-1}^2$ for Flat Boundary Surface

As discussed in Sect. 7.2 and illustrated in Figs. 7.2 and 7.3, any change in the incoming ray $\Delta\bar{R}_{i-1}$ from the previous boundary surface $\bar{r}_{i-1}$ may cause a corresponding change in the refracted/reflected ray $\Delta\bar{R}_i$ at the current boundary surface $\bar{r}_i$. The ith ray $\bar{R}_i$ includes the incidence point $\bar{P}_i$ and unit directional vector $\bar{\ell}_i$. Therefore, the Hessian matrix $\partial^2\bar{R}_i/\partial\bar{R}_{i-1}^2$ comprises two components, namely $\partial^2\bar{P}_i/\partial\bar{R}_{i-1}^2$ and $\partial^2\bar{\ell}_i/\partial\bar{R}_{i-1}^2$, as discussed in the following sub-sections.

15.1.1 Hessian Matrix of Incidence Point $\bar{\mathbf{P}}_i$

The Hessian matrix $\partial^2\bar{P}_i/\partial\bar{R}_{i-1}^2$ of incidence point $\bar{P}_i$ on a flat boundary surface can be obtained directly by differentiating Eq. (7.4). Since $\partial^2\bar{P}_{i-1}/\partial\bar{R}_{i-1}^2 = \partial^2\bar{\ell}_{i-1}/\partial\bar{R}_{i-1}^2 = \bar{0}$, it follows that

$$\frac{\partial^2\bar{P}_i}{\partial\bar{R}_{i-1}^2} = \frac{\partial\lambda_i}{\partial\bar{R}_{\underline{i-1}}}\frac{\partial\bar{\ell}_{i-1}}{\partial\bar{R}_{i-1}} + \frac{\partial\lambda_i}{\partial\bar{R}_{i-1}}\frac{\partial\bar{\ell}_{i-1}}{\partial\bar{R}_{\underline{i-1}}} + \frac{\partial^2\lambda_i}{\partial\bar{R}_{i-1}^2}\bar{\ell}_{i-1}, \tag{15.1}$$

where $\partial\lambda_i/\partial\bar{R}_{i-1}$ and $\partial\bar{\ell}_{i-1}/\partial\bar{R}_{i-1}$ are given by Eqs. (7.5) and (7.7), respectively. Since $\partial^2 D_i/\partial\bar{R}_{i-1}^2 = \partial^2 E_i/\partial\bar{R}_{i-1}^2 = \bar{0}$, the following expression is obtained for $\partial^2\lambda_i/\partial\bar{R}_{i-1}^2$ from Eq. (7.5):

$$\frac{\partial^2\lambda_i}{\partial\bar{R}_{i-1}^2} = \frac{1}{E_i^2}\frac{\partial E_i}{\partial\bar{R}_{\underline{i-1}}}\frac{\partial D_i}{\partial\bar{R}_{i-1}} + \frac{1}{E_i^2}\frac{\partial D_i}{\partial\bar{R}_{\underline{i-1}}}\frac{\partial E_i}{\partial\bar{R}_{i-1}} - \frac{2D_i}{E_i^3}\frac{\partial E_i}{\partial\bar{R}_{\underline{i-1}}}\frac{\partial E_i}{\partial\bar{R}_{i-1}}. \tag{15.2}$$

The terms $\partial D_i/\partial\bar{R}_{i-1}$ and $\partial E_i/\partial\bar{R}_{i-1}$ in Eq. (15.2) are given by Eqs. (7.8) and (7.9), respectively. The explicit expression of $\partial^2\lambda_i/\partial\bar{R}_{i-1}^2$, which is a symmetrical matrix, is given in Eq. (15.60) of Appendix 1 in this chapter.

15.1.2 Hessian Matrix of Unit Directional Vector $\bar{\boldsymbol{\ell}}_i$ of Reflected Ray

For a ray reflected with a unit directional vector $\bar{\ell}_i$ at a flat boundary surface, the Hessian matrix $\partial^2\bar{\ell}_i/\partial\bar{R}_{i-1}^2$ can be obtained directly by differentiating Eq. (7.10). Since $\partial\bar{\ell}_{\mathbf{i}}/\partial\bar{R}_{i-1}$ in Eq. (7.10) is not a function of $\bar{R}_{i-1}$, it follows that

$$\frac{\partial^2\bar{\ell}_i}{\partial\bar{R}_{i-1}^2} = \bar{0}_{4\times6\times6}. \tag{15.3}$$

15.1.3 Hessian Matrix of Unit Directional Vector $\bar{\boldsymbol{\ell}}_i$ of Refracted Ray

For a ray refracted with a unit directional vector $\bar{\ell}_i$ at a flat boundary surface, the Hessian matrix $\partial^2\bar{\ell}_i/\partial\bar{R}_{i-1}^2$ can be obtained directly by differentiating Eq. (7.12). Noting that $N_i = \xi_{i-1}/\xi_i$, $\partial\bar{\ell}_{i-1}/\partial\bar{R}_{i-1}$ and $\partial(C\theta_i)/\partial\bar{R}_{i-1}$ (see Eqs. (7.7) and (7.11)) are not functions of $\bar{R}_{i-1}$, it follows that

$$\frac{\partial^2 \bar{\ell}_i}{\partial \bar{R}_{i-1}^2} = N_i^2 \, \bar{n}_i \left(\frac{-1}{\sqrt{1 - N_i^2 + (N_i C\theta_i)^2}} + \frac{(N_i C\theta_i)^2}{\left[1 - N_i^2 + (N_i C\theta_i)^2\right]^{3/2}} \right) \frac{\partial (C\theta_i)}{\partial \underline{\bar{R}_{i-1}}} \frac{\partial (C\theta_i)}{\partial \bar{R}_{i-1}}. \tag{15.4}$$

The term $\partial(C\theta_i)/\partial\bar{R}_{i-1}$ in Eq. (15.4) is given in Eq. (7.11). The explicit expression of $\partial^2\bar{\ell}_i/\partial\bar{R}_{i-1}^2$ is given in Eq. (15.62) of Appendix 1 in this chapter.

If the ith boundary surface $\bar{r}_i$ of the optical system is flat, then by combining Eqs. (15.1) and (15.3) (when the ray is reflected), or Eqs. (15.1) and (15.4) (when the ray is refracted), the Hessian matrix $\partial^2\bar{R}_i/\partial\bar{R}_{i-1}^2$ is obtained as

$$\frac{\partial^2 \bar{R}_i}{\partial \bar{R}_{i-1}^2} = \begin{bmatrix} \partial^2 \bar{P}_i / \partial \bar{R}_{i-1}^2 \\ \partial^2 \bar{\ell}_i / \partial \bar{R}_{i-1}^2 \end{bmatrix}_{6\times6\times6}. \tag{15.5}$$

15.2 Hessian Matrix $\partial^2\bar{R}_i/\partial\bar{R}_{i-1}^2$ for Spherical Boundary Surface

Equations (7.20), (7.26) and (7.27) give the Jacobian matrices of the incidence point $\bar{P}_i$ and unit directional vectors $\bar{\ell}_i$ of rays reflected or refracted at a spherical boundary surface, respectively, with respect to the incoming ray $\bar{R}_{i-1}$. As described in the following sub-sections, the Hessian matrix $\partial^2\bar{R}_i/\partial\bar{R}_{i-1}^2$ can be determined by further differentiating these equations with respect to the incoming ray $\bar{R}_{i-1}$.

15.2.1 Hessian Matrix of Incidence Point $\bar{P}_i$

The Hessian matrix $\partial^2\bar{P}_i/\partial\bar{R}_{i-1}^2$ of incidence point $\bar{P}_i$ on a spherical boundary surface can be obtained by differentiating Eq. (7.20). Since $\partial^2\bar{P}_{i-1}/\partial\bar{R}_{i-1}^2 = \partial^2\bar{\ell}_{i-1}/\partial\bar{R}_{i-1}^2 = \bar{0}$ (from Eqs. (7.22) and (7.23)), it follows that

$$\frac{\partial^2 \bar{P}_i}{\partial \bar{R}_{i-1}^2} = \frac{\partial \lambda_i}{\partial \underline{\bar{R}_{i-1}}} \frac{\partial \bar{\ell}_{i-1}}{\partial \bar{R}_{i-1}} + \frac{\partial \lambda_i}{\partial \bar{R}_{i-1}} \frac{\partial \bar{\ell}_{i-1}}{\partial \underline{\bar{R}_{i-1}}} + \frac{\partial^2 \lambda_i}{\partial \bar{R}_{i-1}^2} \bar{\ell}_{i-1}, \tag{15.6}$$

where $\partial\lambda_i/\partial\bar{R}_{i-1}$ and $\partial\bar{\ell}_{i-1}/\partial\bar{R}_{i-1}$ are given by Eqs. (7.21) and (7.23), respectively. The following expression for $\partial^2\lambda_i/\partial\bar{R}_{i-1}^2$ is obtained from Eq. (7.21):

$$\frac{\partial^2\lambda_i}{\partial\bar{R}_{i-1}^2} = -\frac{\partial^2 D_i}{\partial\bar{R}_{i-1}^2} \pm \frac{1}{2\sqrt{D_i^2 - E_i}}\left(2\frac{\partial D_i}{\partial\bar{R}_{\underline{i-1}}}\frac{\partial D_i}{\partial\bar{R}_{i-1}} + 2D_i\frac{\partial^2 D_i}{\partial\bar{R}_{i-1}^2} - \frac{\partial^2 E_i}{\partial\bar{R}_{i-1}^2}\right)$$
$$\pm \frac{-1}{4(D_i^2 - E_i)^{3/2}}\left(2D_i\frac{\partial D_i}{\partial\bar{R}_{\underline{i-1}}} - \frac{\partial E_i}{\partial\bar{R}_{\underline{i-1}}}\right)\left(2D_i\frac{\partial D_i}{\partial\bar{R}_{i-1}} - \frac{\partial E_i}{\partial\bar{R}_{i-1}}\right), \tag{15.7}$$

where $D_i, E_i, \partial D_i/\partial\bar{R}_{i-1}$ and $\partial E_i/\partial\bar{R}_{i-1}$ are given by Eqs. (2.17), (2.18), (7.24) and (7.25), respectively. The other two terms in Eq. (15.7), i.e., $\partial^2 D_i/\partial\bar{R}_{i-1}^2$ and $\partial^2 E_i/\partial\bar{R}_{i-1}^2$, are obtained from Eqs. (7.24) and (7.25), respectively, as

$$\frac{\partial^2 D_i}{\partial\bar{R}_{i-1}^2} = \begin{bmatrix} \bar{0}_{1\times3} & \bar{I}_{3\times3} \\ \bar{I}_{3\times3} & \bar{0}_{1\times3} \end{bmatrix}, \tag{15.8}$$

$$\frac{\partial^2 E_i}{\partial\bar{R}_{i-1}^2} = 2\begin{bmatrix} \bar{I}_{3\times3} & \bar{0}_{3\times3} \\ \bar{0}_{3\times3} & \bar{0}_{3\times3} \end{bmatrix}. \tag{15.9}$$

15.2.2 Hessian Matrix of Unit Directional Vector $\bar{\ell}_i$ of Reflected Ray

For a ray reflected with a unit directional vector $\bar{\ell}_i$ at a spherical boundary surface, the Hessian matrix $\partial^2\bar{\ell}_i/\partial\bar{R}_{i-1}^2$ can be obtained by differentiating Eq. (7.26). Since $\partial\bar{\ell}_{i-1}/\partial\bar{R}_{i-1}$ given in Eq. (7.23) is not a function of $\bar{R}_{i-1}$, it follows that

$$\frac{\partial^2\bar{\ell}_i}{\partial\bar{R}_{i-1}^2} = 2\left(\frac{\partial(C\theta_i)}{\partial\bar{R}_{\underline{i-1}}}\frac{\partial\bar{n}_i}{\partial\bar{R}_{i-1}} + C\theta_i\frac{\partial^2\bar{n}_i}{\partial\bar{R}_{i-1}^2} + \frac{\partial^2(C\theta_i)}{\partial\bar{R}_{i-1}^2}\bar{n}_i + \frac{\partial(C\theta_i)}{\partial\bar{R}_{i-1}}\frac{\partial\bar{n}_i}{\partial\bar{R}_{\underline{i-1}}}\right). \tag{15.10}$$

The terms $\partial\bar{n}_i/\partial\bar{R}_{i-1}$ and $\partial(C\theta_i)/\partial\bar{R}_{i-1}$ in Eq. (15.10) are given by Eqs. (7.61) and (7.63), respectively, in Appendix 1 of Chap. 7. In addition, the expressions for $\partial^2\bar{n}_i/\partial\bar{R}_{i-1}^2$ and $\partial^2(C\theta_i)/\partial\bar{R}_{i-1}^2$ are give in Eqs. (15.70) and (15.72), respectively, of Appendix 2 in this chapter.

15.2.3 Hessian Matrix of Unit Directional Vector $\bar{\ell}_i$ of Refracted Ray

For a ray refracted with a unit directional vector $\bar{\ell}_i$ at a spherical boundary surface, the Hessian matrix $\partial^2\bar{\ell}_i/\partial\bar{R}_{i-1}^2$ can be obtained by differentiating Eq. (7.27). Noting that $\partial N_i/\partial\bar{R}_{i-1} = \bar{0}$ and $\partial^2\bar{\ell}_{i-1}/\partial\bar{R}_{i-1}^2 = \bar{0}, \partial^2\bar{\ell}_i/\partial\bar{R}_{i-1}^2$ is obtained as

$$\frac{\partial^2 \bar{\ell}_i}{\partial \bar{R}_{i-1}^2} = N_i \left(\frac{\partial^2 (C\theta_i)}{\partial \bar{R}_{i-1}^2} \bar{n}_i + \frac{\partial (C\theta_i)}{\partial \bar{R}_{i-1}} \frac{\partial \bar{n}_i}{\partial \underline{\bar{R}_{i-1}}} + \frac{\partial (C\theta_i)}{\partial \underline{\bar{R}_{i-1}}} \frac{\partial \bar{n}_i}{\partial \bar{R}_{i-1}} + C\theta_i \frac{\partial^2 \bar{n}_i}{\partial \bar{R}_{i-1}^2} \right)$$
$$+ \left(-\sqrt{1 - N_i^2 + (N_i C\theta_i)^2} \right) \frac{\partial^2 \bar{n}_i}{\partial \bar{R}_{i-1}^2} - \left(\frac{N_i^2 C\theta_i}{\sqrt{1 - N_i^2 + (N_i C\theta_i)^2}} \right) \frac{\partial (C\theta_i)}{\partial \underline{\bar{R}_{i-1}}} \frac{\partial \bar{n}_i}{\partial \bar{R}_{i-1}}$$
$$- \left(\frac{N_i^2 C\theta_i}{\sqrt{1 - N_i^2 + (N_i C\theta_i)^2}} \right) \left(\frac{\partial \bar{n}_i}{\partial \underline{\bar{R}_{i-1}}} \frac{\partial (C\theta_i)}{\partial \bar{R}_{i-1}} + \bar{n}_i \frac{\partial^2 (C\theta_i)}{\partial \bar{R}_{i-1}^2} \right)$$
$$- \left(\frac{N_i^2}{\sqrt{1 - N_i^2 + (N_i C\theta_i)^2}} + \frac{N_i^2 (N_i C\theta_i)^2}{\left(1 - N_i^2 + (N_i C\theta_i)^2\right)^{3/2}} \right) \frac{\partial (C\theta_i)}{\partial \underline{\bar{R}_{i-1}}} \bar{n}_i \frac{\partial (C\theta_i)}{\partial \bar{R}_{i-1}}. \tag{15.11}$$

Terms $\partial^2 \bar{n}_i / \partial \bar{R}_{i-1}^2$ and $\partial^2 (C\theta_i) / \partial \bar{R}_{i-1}^2$ in Eq. (15.11) are given in Eqs. (15.70) and (15.72), respectively, of Appendix 2 in this chapter, while terms $\partial \bar{n}_i / \partial \bar{R}_{i-1}$ and $\partial (C\theta_i) / \partial \bar{R}_{i-1}$ are given in Eqs. (7.61) and (7.63), respectively, of Appendix 1 in Chapter 7.

If the ith boundary surface of the optical system is spherical, then by combining Eqs. (15.6) and (15.10) (when ray is reflected), or Eqs. (15.6) and (15.11) (when ray is refracted), the Hessian matrix $\partial^2 \bar{R}_i / \partial \bar{R}_{i-1}^2$ is obtained as

$$\frac{\partial^2 \bar{R}_i}{\partial \bar{R}_{i-1}^2} = \begin{bmatrix} \partial^2 \bar{P}_i / \partial \bar{R}_{i-1}^2 \\ \partial^2 \bar{\ell}_i / \partial \bar{R}_{i-1}^2 \end{bmatrix}_{6 \times 6 \times 6}. \tag{15.12}$$

As shown in Eqs. (15.1) to (15.12), $\partial^2 \bar{P}_i / \partial \bar{R}_{i-1}^2$ and $\partial^2 \bar{\ell}_i / \partial \bar{R}_{i-1}^2$ are symmetrical matrices. Thus, the Hessian matrix $\partial^2 \bar{R}_i / \partial X_0^2$ computed in the following section is also a symmetrical matrix.

15.3 Hessian Matrix of $\bar{R}_i$ with Respect to Variable Vector $\bar{X}_0$ of Source Ray

The Hessian matrix $\partial^2 \bar{R}_i / \partial \bar{X}_0^2$ of a ray $\bar{R}_i$ with respect to the variable vector (Eq. (2.4))

$$\bar{X}_0 = \begin{bmatrix} P_{0x} & P_{0y} & P_{0z} & \alpha_0 & \beta_0 \end{bmatrix}^T$$

of a source ray $\bar{R}_0$ has the form

$$\frac{\partial^2\bar{R}_i}{\partial\bar{X}_0^2}=\begin{bmatrix}\frac{\partial^2\bar{R}_i}{\partial P_{0x}\partial P_{0x}} & \frac{\partial^2\bar{R}_i}{\partial P_{0x}\partial P_{0y}} & \frac{\partial^2\bar{R}_i}{\partial P_{0x}\partial P_{0z}} & \frac{\partial^2\bar{R}_i}{\partial P_{0x}\partial\alpha_0} & \frac{\partial^2\bar{R}_i}{\partial P_{0x}\partial\beta_0}\\ & \frac{\partial^2\bar{R}_i}{\partial P_{0y}\partial P_{0y}} & \frac{\partial^2\bar{R}_i}{\partial P_{0y}\partial P_{0z}} & \frac{\partial^2\bar{R}_i}{\partial P_{0y}\partial\alpha_0} & \frac{\partial^2\bar{R}_i}{\partial P_{0y}\partial\beta_0}\\ & & \frac{\partial^2\bar{R}_i}{\partial P_{0z}\partial P_{0z}} & \frac{\partial^2\bar{R}_i}{\partial P_{0z}\partial\alpha_0} & \frac{\partial^2\bar{R}_i}{\partial P_{0z}\partial\beta_0}\\ & \text{symm.} & & \frac{\partial^2\bar{R}_i}{\partial\alpha_0\partial\alpha_0} & \frac{\partial^2\bar{R}_i}{\partial\alpha_0\partial\beta_0}\\ & & & & \frac{\partial^2\bar{R}_i}{\partial\beta_0\partial\beta_0}\end{bmatrix}. \tag{15.13}$$

Note that the annotation "symm." in Eq. (15.13) indicates that $\partial^2\bar{R}_i/\partial\bar{X}_0^2$ is a symmetrical matrix. Equation (15.13) can be determined by

$$\frac{\partial^2\bar{R}_i}{\partial\bar{X}_0^2}=\left(\frac{\partial\bar{R}_{i-1}}{\partial\bar{X}_0}\right)^T\frac{\partial^2\bar{R}_i}{\partial\bar{R}_{i-1}^2}\frac{\partial\bar{R}_{i-1}}{\partial\bar{X}_0}+\frac{\partial\bar{R}_i}{\partial\bar{R}_{i-1}}\frac{\partial^2\bar{R}_{i-1}}{\partial\bar{X}_0^2}. \tag{15.14}$$

$\partial\bar{R}_{i-1}/\partial\bar{X}_0$ in Eq. (15.14) can be obtained from Eq. (7.30) by replacing i with $i-1$. Moreover, $\partial\bar{R}_i/\partial\bar{R}_{i-1}$ and $\partial^2\bar{R}_i/\partial\bar{R}_{i-1}^2$ have already been presented in Sects. 7.2, 7.3, 15.1 and 15.2 for flat and spherical boundary surfaces. Therefore, it is possible to compute matrix $\partial^2\bar{R}_i/\partial\bar{X}_0^2$ sequentially from $i=1$ to $i=n$ given $\partial^2\bar{R}_0/\partial\bar{X}_0^2$ which is obtained by directly differentiating Eq. (7.16), to give

$$\frac{\partial^2\bar{R}_0}{\partial\bar{X}_0^2}=\begin{bmatrix}\partial^2P_{0x}/\partial\bar{X}_0^2\\ \partial^2P_{0y}/\partial\bar{X}_0^2\\ \partial^2P_{0z}/\partial\bar{X}_0^2\\ \partial^2\ell_{0x}/\partial\bar{X}_0^2\\ \partial^2\ell_{0y}/\partial\bar{X}_0^2\\ \partial^2\ell_{0z}/\partial\bar{X}_0^2\end{bmatrix}_{6\times5\times5} \tag{15.15}$$

where

$$\frac{\partial^2P_{0x}}{\partial\bar{X}_0^2}=\frac{\partial^2P_{0y}}{\partial\bar{X}_0^2}=\frac{\partial^2P_{0z}}{\partial\bar{X}_0^2}=\bar{0}_{5\times5}, \tag{15.16}$$

$$\frac{\partial^2\ell_{0x}}{\partial\bar{X}_0^2}=\begin{bmatrix}0&0&0&0&0\\0&0&0&0&0\\0&0&0&0&0\\0&0&0&-C\beta_0C(90^\circ+\alpha_0)&S\beta_0S(90^\circ+\alpha_0)\\0&0&0&S\beta_0S(90^\circ+\alpha_0)&-C\beta_0C(90^\circ+\alpha_0)\end{bmatrix}, \tag{15.17}$$

$$\frac{\partial^2 \ell_{0y}}{\partial \bar{X}_0^2} = \begin{bmatrix} 0 & 0 & 0 & 0 & 0 \\ 0 & 0 & 0 & 0 & 0 \\ 0 & 0 & 0 & 0 & 0 \\ 0 & 0 & 0 & -C\beta_0 S(90^\circ + \alpha_0) & -S\beta_0 C(90^\circ + \alpha_0) \\ 0 & 0 & 0 & -S\beta_0 C(90^\circ + \alpha_0) & -C\beta_0 S(90^\circ + \alpha_0) \end{bmatrix}, \tag{15.18}$$

$$\frac{\partial^2 \ell_{0z}}{\partial \bar{X}_0^2} = \begin{bmatrix} 0 & 0 & 0 & 0 & 0 \\ 0 & 0 & 0 & 0 & 0 \\ 0 & 0 & 0 & 0 & 0 \\ 0 & 0 & 0 & 0 & 0 \\ 0 & 0 & 0 & 0 & -S\beta_0 \end{bmatrix}. \tag{15.19}$$

15.4 Hessian Matrix of OPL_i with Respect to Variable Vector $\bar{X}_0$ of Source Ray

The Hessian matrix $\partial^2 OPL_i/\partial \bar{X}_0^2$ of OPL_i with respect to the variable vector (Eq. 2.4)

$$\bar{X}_0 = \begin{bmatrix} P_{0x} & P_{0y} & P_{0z} & \alpha_0 & \beta_0 \end{bmatrix}^T$$

of source ray $\bar{R}_0$ has the form

$$\frac{\partial^2 OPL_i}{\partial \bar{X}_0^2} = \begin{bmatrix} \frac{\partial^2 OPL_i}{\partial P_{0x}\partial P_{0x}} & \frac{\partial^2 OPL_i}{\partial P_{0x}\partial P_{0y}} & \frac{\partial^2 OPL_i}{\partial P_{0x}\partial P_{0z}} & \frac{\partial^2 OPL_i}{\partial P_{0x}\partial \alpha_0} & \frac{\partial^2 OPL_i}{\partial P_{0x}\partial \beta_0} \\ & \frac{\partial^2 OPL_i}{\partial P_{0y}\partial P_{0y}} & \frac{\partial^2 OPL_i}{\partial P_{0y}\partial P_{0z}} & \frac{\partial^2 OPL_i}{\partial P_{0y}\partial \alpha_0} & \frac{\partial^2 OPL_i}{\partial P_{0y}\partial \beta_0} \\ & & \frac{\partial^2 OPL_i}{\partial P_{0z}\partial P_{0z}} & \frac{\partial^2 OPL_i}{\partial P_{0z}\partial \alpha_0} & \frac{\partial^2 OPL_i}{\partial P_{0z}\partial \beta_0} \\ & \text{symm.} & & \frac{\partial^2 OPL_i}{\partial \alpha_0\partial \alpha_0} & \frac{\partial^2 OPL_i}{\partial \alpha_0\partial \beta_0} \\ & & & & \frac{\partial^2 OPL_i}{\partial \beta_0\partial \beta_0} \end{bmatrix}. \tag{15.20}$$

Equation (15.20) can be determined directly from Eq. (15.14) by replacing $\bar{R}_i$ with OPL_i, i.e.,

$$\frac{\partial^2 OPL_i}{\partial \bar{X}_0^2} = \left(\frac{\partial \bar{R}_{i-1}}{\partial \bar{X}_0}\right)^T \frac{\partial^2 OPL_i}{\partial \bar{R}_{i-1}^2} \frac{\partial \bar{R}_{i-1}}{\partial \bar{X}_0} + \frac{\partial OPL_i}{\partial \bar{R}_{i-1}} \frac{\partial^2 \bar{R}_{i-1}}{\partial \bar{X}_0^2}. \tag{15.21}$$

Again, $\partial \bar{R}_{i-1}/\partial \bar{X}_0$ in Eq. (15.21) can be obtained from Eq. (7.30) by replacing i with i − 1. Moreover, $\partial OPL_i/\partial \bar{R}_{i-1}$ is given in Eq. (14.2) and $\partial^2 \bar{R}_{i-1}/\partial \bar{X}_0^2$ is obtained from Eq. (15.13) by replacing i with i − 1. Thus, $\partial^2 OPL_i/\partial \bar{X}_0^2$ can be determined given $\partial^2 \mathrm{OPL_i}/\partial \bar{R}_{i-1}^2$, which can be obtained by differentiating Eq. (14.2) with respect to the incoming ray $\bar{R}_{i-1}$, to give

$$\frac{\partial^2 \mathrm{OPL_i}}{\partial \bar{R}_{i-1}^2} = \xi_{i-1} \frac{\partial^2 \lambda_i}{\partial \bar{R}_{i-1}^2}. \tag{15.22}$$

$\partial^2 \lambda_i/\partial \bar{R}_{i-1}^2$ is given in Eqs. (15.2) and (15.7) for flat and spherical boundary surfaces, respectively. As discussed in the following, and further addressed in Sect. 15.5, the Hessian matrix $\partial^2 \mathrm{OPL_i}/\partial \bar{X}_0^2$ given in Eq. (15.20) enables many wavefront aberration problems to be solved.

The problem of computing the wavefront aberration of an optical system by adding a reference sphere as a virtual boundary surface has been discussed in Sect. 14.3 of Chap. 14 (see Fig. 14.7). It was explained that for the modified system shown in Fig. 14.7, the wavefront aberration $W(\bar{X}_0)$ can be computed by Eq. (14.27). This equation can be rewritten as

$$\begin{aligned} W(\bar{X}_0) &= \mathrm{OPL}(\bar{P}_0, \bar{P}_n) - \mathrm{OPL}(\bar{P}_0, \bar{P}_{n/chief}) = \sum_{i=1}^{i=n} \mathrm{OPL_i} - \sum_{i=1}^{i=n} \mathrm{OPL_{i/chief}} \\ &= \sum_{i=1}^{i=n} \left(\mathrm{OPL_i} - \mathrm{OPL_{i/chief}} \right). \end{aligned} \tag{15.23}$$

The first-order change of the wavefront aberration, $\Delta W(\bar{X}_0)$, was derived in Eq. (14.29). In this section, the change of the wavefront aberration, $\Delta W(\bar{X}_0)$, is estimated by means of a Taylor series expansion containing up to the quadratic term. Consider the system shown in Fig. 14.8, in which two rays originate from point source $\bar{P}_0$, namely a reference source ray $\bar{R}_0 = \begin{bmatrix} \bar{P}_0 & \bar{\ell}_0 \end{bmatrix}^T$ with variable vector $\bar{X}_0 = \begin{bmatrix} P_{0x} & P_{0y} & P_{0z} & \alpha_0 & \beta_0 \end{bmatrix}^T$ and a neighboring source ray $\bar{R}_0' = \begin{bmatrix} \bar{P}_0 & \bar{\ell}_0' \end{bmatrix}^T$ with variable vector $\bar{X}_0' = \begin{bmatrix} P_{0x} & P_{0y} & P_{0z} & \alpha_0' & \beta_0' \end{bmatrix}^T$. If the wavefront aberration $W(\bar{X}_0)$ of the reference ray $\bar{R}_0$ is given, it is possible to estimate the wavefront aberration $W(\bar{X}_0')$ of the neighboring source ray $\bar{R}_0'$ by $W(\bar{X}_0') = W(\bar{X}_0) + \Delta W(\bar{X}_0)$, where

$$\Delta W(\bar{X}_0) = \left(\sum_{i=1}^{i=n} \frac{\partial \mathrm{OPL_i}}{\partial \bar{X}_0} \right) \Delta \bar{X}_0 + \frac{1}{2} \Delta \bar{X}_0^T \left(\sum_{i=1}^{i=n} \frac{\partial^2 \mathrm{OPL_i}}{\partial \bar{X}_0^2} \right) \Delta \bar{X}_0, \tag{15.24}$$

in which $\Delta \bar{X}_0 = \bar{X}_0' - \bar{X}_0 = \begin{bmatrix} 0 & 0 & 0 & \alpha_0' - \alpha_0 & \beta_0' - \beta_0 \end{bmatrix}^T$. Evaluating the Jacobian and Hessian matrices of Eq. (15.24) with respect to reference ray $\bar{R}_0$, the wavefront aberration $W(\bar{X}_0')$ of the neighboring ray $\bar{R}_0'$ can then be obtained as the sum of $W(\bar{X}_0)$ and $\Delta W(\bar{X}_0)$.

Example 15.1 Given two rays originating from the same point source (i.e., reference ray $\bar{R}_0$ and neighboring ray $\bar{R}'_0$, see Fig. 14.8), the wavefront of the neighboring ray $W(\bar{X}'_0)$ can be estimated with up to the quadratic term of the Taylor series expansion provided that the wavefront aberration of the reference ray $W(\bar{X}_0)$ is known. Curves B and C in Fig. 14.9 show the wavefront aberrations $W(\bar{X}'_0) = W(\bar{X}_0) + \Delta W(\bar{X}_0)$ obtained when estimating $\Delta W(\bar{X}_0)$ by Eqs. (14.29) and (15.24), respectively, and setting the reference ray $\bar{R}_0$ as the chief ray $\bar{R}_{0/chief}$. It is noted from Curve B that since $\partial OPL_i/\partial \bar{X}_0 = \bar{0}_{1\times 5}$ (i = 1 to i = n) for the reference ray $\bar{R}_{0/chief}$, the value of $W(\bar{X}'_0) = W(\bar{X}_0) + \Delta W(\bar{X}_0)$ obtained using Eq. (14.29) is equal to zero for all of the neighboring rays. In addition, it is seen that Curve C, in which $\Delta W(\bar{X}_0)$ in the wavefront aberration $W(\bar{X}'_0) = W(\bar{X}_0) + \Delta W(\bar{X}_0)$ is estimated using Eq. (15.24) with up to the quadratic term of the Taylor series expansion, is in good agreement with the solution obtained via raytracing for the close neighboring rays. However, the accuracy of the estimated solution reduces as the distance of the neighboring ray $\bar{R}'_0$ from the reference ray $\bar{R}_0$ increases.

15.5 Change of Wavefront Aberration Due to Translation of Point Source $\bar{P}_0$

In evaluating the change in the wavefront aberration caused by a translation of the point source, it is first necessary to determine the change in the variable vector $\Delta\bar{X}_{0/chief}$ of the chief ray when the point source is translated. It will be recalled that the chief ray of a point source $\bar{P}_0$ is a ray $\bar{R}_{0/chief} = \begin{bmatrix}\bar{P}_0 & \bar{\ell}_{0/chief}\end{bmatrix}^T$ which originates from $\bar{P}_0$ with a unit directional vector

$$\bar{\ell}_{0/chief} = \begin{bmatrix} \ell_{0x/chief} \\ \ell_{0y/chief} \\ \ell_{0z/chief} \\ 0 \end{bmatrix} = \begin{bmatrix} C\beta_{0/chief}C(90^\circ + \alpha_{0/chief}) \\ C\beta_{0/chief}S(90^\circ + \alpha_{0/chief}) \\ S\beta_{0/chief} \\ 0 \end{bmatrix} \tag{15.25}$$

and strikes the center of the aperture (see Fig. 15.2 and Eq. (2.3)). As discussed earlier in relation to Eq. (7.29), when the aperture is labeled as the gth boundary

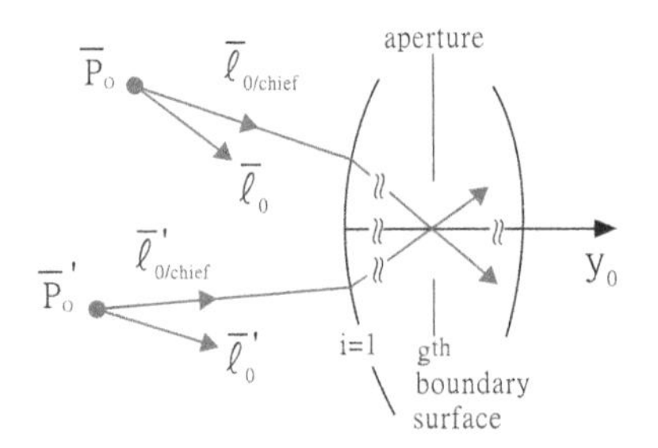

Fig. 15.2 Feasibility of estimating wavefront aberration via Taylor series expansion when point source $\bar{P}_0$ is translated to a neighboring point $\bar{P}'_0$

surface of the optical system, the change in the chief ray at the aperture, $\Delta\bar{R}_g$, caused by any change in the variable vector of the chief ray,

$$\Delta\bar{X}_{0/chief} = \begin{bmatrix} \Delta P_{0x} & \Delta P_{0y} & \Delta P_{0z} & \Delta\alpha_{0/chief} & \Delta\beta_{0/chief} \end{bmatrix}^T, \tag{15.26}$$

can be computed as

$$\Delta\bar{R}_g = \begin{bmatrix} \Delta P_{gx} \\ \Delta P_{gy} \\ \Delta P_{gz} \\ \Delta\ell_{gx} \\ \Delta\ell_{gy} \\ \Delta\ell_{gz} \end{bmatrix} = \bar{M}_g\bar{M}_{g-1}\cdots\bar{M}_2\bar{M}_1\bar{S}_0\,\Delta\bar{X}_{0/chief}$$
$$= \begin{bmatrix} c_{11} & c_{12} & c_{13} & c_{14} & c_{15} & c_{16} \\ c_{21} & c_{22} & c_{23} & c_{24} & c_{25} & c_{26} \\ c_{31} & c_{32} & c_{33} & c_{34} & c_{35} & c_{36} \\ c_{41} & c_{42} & c_{43} & c_{44} & c_{45} & c_{46} \\ c_{51} & c_{52} & c_{53} & c_{54} & c_{55} & c_{56} \\ c_{61} & c_{62} & c_{63} & c_{64} & c_{65} & c_{66} \end{bmatrix} \Delta\bar{X}_{0/chief}. \tag{15.27}$$

Taking the first and third components of Eq. (15.27), and setting $\Delta P_{gx} = \Delta P_{gz} = 0$, yields

$$\begin{bmatrix} \Delta P_{gx} \\ \Delta P_{gz} \end{bmatrix} = \begin{bmatrix} 0 \\ 0 \end{bmatrix} = \begin{bmatrix} c_{11} & c_{12} & c_{13} \\ c_{31} & c_{32} & c_{33} \end{bmatrix} \begin{bmatrix} \Delta P_{0x} \\ \Delta P_{0y} \\ \Delta P_{0z} \end{bmatrix} + \begin{bmatrix} c_{14} & c_{15} \\ c_{34} & c_{35} \end{bmatrix} \begin{bmatrix} \Delta\alpha_{0/chief} \\ \Delta\beta_{0/chief} \end{bmatrix}. \tag{15.28}$$

From Eq. (15.28), the change in the polar coordinates $\alpha_{0/chief}$ and $\beta_{0/chief}$ when the reference point source translates from $\bar{P}_0 = \begin{bmatrix} P_{0x} & P_{0y} & P_{0z} & 1 \end{bmatrix}^T$ to a neighboring point source $\bar{P}'_0 = \begin{bmatrix} P'_{0x} & P'_{0y} & P'_{0z} & 1 \end{bmatrix}^T$ can be estimated as

$$\begin{bmatrix} \Delta\alpha_{0/chief} \\ \Delta\beta_{0/chief} \end{bmatrix} = \left(\frac{-1}{(c_{14}c_{35} - c_{34}c_{15})} \begin{bmatrix} c_{35} & -c_{15} \\ -c_{34} & c_{14} \end{bmatrix} \begin{bmatrix} c_{11} & c_{12} & c_{13} \\ c_{31} & c_{32} & c_{33} \end{bmatrix} \right) \begin{bmatrix} \Delta P_{0x} \\ \Delta P_{0y} \\ \Delta P_{0z} \end{bmatrix}$$
$$= \begin{bmatrix} u_{41} & u_{42} & u_{43} \\ u_{51} & u_{52} & u_{53} \end{bmatrix} \begin{bmatrix} \Delta P_{0x} \\ \Delta P_{0y} \\ \Delta P_{0z} \end{bmatrix}, \tag{15.29}$$

where

$$\begin{bmatrix} \Delta P_{0x} \\ \Delta P_{0y} \\ \Delta P_{0z} \end{bmatrix} = \bar{P}'_0 - \bar{P}_0 = \begin{bmatrix} P'_{0x} - P_{0x} \\ P'_{0y} - P_{0y} \\ P'_{0z} - P_{0z} \end{bmatrix}.$$

The change in the variable vector $\Delta\bar{X}_{0/chief}$ when the point source is translated from $\bar{P}_0$ to $\bar{P}'_0$ can then be obtained as

$$\Delta\bar{X}_{0/chief} = \begin{bmatrix} \Delta P_{0x} \\ \Delta P_{0y} \\ \Delta P_{0z} \\ \Delta\alpha_{0/chief} \\ \Delta\beta_{0/chief} \end{bmatrix} = \begin{bmatrix} P'_{0x} - P_{0x} \\ P'_{0y} - P_{0y} \\ P'_{0z} - P_{0z} \\ \alpha'_{0/chief} - \alpha_{0/chief} \\ \beta'_{0/chief} - \beta_{0/chief} \end{bmatrix}$$
$$= \begin{bmatrix} 1 & 0 & 0 \\ 0 & 1 & 0 \\ 0 & 0 & 1 \\ u_{41} & u_{42} & u_{43} \\ u_{51} & u_{52} & u_{53} \end{bmatrix} \begin{bmatrix} P'_{0x} - P_{0x} \\ P'_{0y} - P_{0y} \\ P'_{0z} - P_{0z} \end{bmatrix} = \begin{bmatrix} 1 & 0 & 0 \\ 0 & 1 & 0 \\ 0 & 0 & 1 \\ u_{41} & u_{42} & u_{43} \\ u_{51} & u_{52} & u_{53} \end{bmatrix} \Delta\bar{P}_0. \quad (15.30)$$

From Eq. (15.30), $\Delta\bar{X}_{0/chief}$ can be computed as

$$\Delta\bar{X}_{0/chief} = \frac{\partial\bar{X}_{0/chief}}{\partial\bar{P}_0}\Delta\bar{P}_0 \quad (15.31)$$

with

$$\frac{\partial\bar{X}_{0/chief}}{\partial\bar{P}_0} = \begin{bmatrix} 1 & 0 & 0 \\ 0 & 1 & 0 \\ 0 & 0 & 1 \\ u_{41} & u_{42} & u_{43} \\ u_{51} & u_{52} & u_{53} \end{bmatrix}. \quad (15.32)$$

The existing literature on wavefront aberrations assumes the point source $\bar{P}_0$ to be fixed. As a result, the chief ray is unchanged (i.e., $\Delta\bar{X}_{0/chief} = \bar{0}_{5\times1}$). However, the following discussions propose an approach based on a Taylor series expansion for estimating the change in the wavefront aberration $W(\bar{X}'_0)$ of a neighboring ray when the point source $\bar{P}_0$ translates to a new point $\bar{P}'_0$. Consider the case shown in Fig. 15.2 involving four rays, namely two rays originating from reference point source $\bar{P}_0$ (i.e., a general ray $\bar{R}_0 = \begin{bmatrix} \bar{P}_0 & \bar{\ell}_0 \end{bmatrix}^T$ with variable vector $\bar{X}_0 = \begin{bmatrix} P_{0x} & P_{0y} & P_{0z} & \alpha_0 & \beta_0 \end{bmatrix}^T$ and a chief ray $\bar{R}_{0/chief} = \begin{bmatrix} \bar{P}_0 & \bar{\ell}_{0/chief} \end{bmatrix}^T$ with variable vector $\bar{X}_{0/chief} = \begin{bmatrix} P_{0x} & P_{0y} & P_{0z} & \alpha_{0/chief} & \beta_{0/chief} \end{bmatrix}^T$) and two rays originating from a neighboring point source $\bar{P}'_0$ (i.e., a general ray $\bar{R}'_0 = \begin{bmatrix} \bar{P}'_0 & \bar{\ell}'_0 \end{bmatrix}^T$ with variable vector $\bar{X}'_0 = \begin{bmatrix} P'_{0x} & P'_{0y} & P'_{0z} & \alpha'_0 & \beta'_0 \end{bmatrix}^T$

and a chief source ray $\bar{R}'_{0/\text{chief}} = \begin{bmatrix} \bar{P}'_0 & \bar{\ell}'_{0/\text{chief}} \end{bmatrix}^T$ with variable vector $\bar{X}'_{0/\text{chief}} = \begin{bmatrix} P'_{0x} & P'_{0y} & P'_{0z} & \alpha'_{0/\text{chief}} & \beta'_{0/\text{chief}} \end{bmatrix}^T$). When the point source translates from point $\bar{P}_0$ to point $\bar{P}'_0$, the resulting change in the wavefront aberration, $\Delta W(\bar{X}_0)$, can be estimated by a Taylor series expansion of either the first-order or up to the quadratic term as follows:

$$\begin{aligned} \Delta W(\bar{X}_0) &= \left(\frac{\partial OPL(\bar{P}_0, \bar{P}_n)}{\partial \bar{X}_0}\right)_{\bar{R}_0} \Delta\bar{X}_0 - \left(\frac{\partial OPL(\bar{P}_0, \bar{P}_n)}{\partial \bar{X}_0}\right)_{\bar{R}_{0/\text{chief}}} \Delta\bar{X}_{0/\text{chief}} \\ &= \left(\sum_{i=1}^{i=n} \frac{\partial OPL_i}{\partial \bar{X}_0}\right)_{\bar{R}_0} \Delta\bar{X}_0 - \left(\sum_{i=1}^{i=n} \frac{\partial OPL_i}{\partial \bar{X}_0}\right)_{\bar{R}_{0/\text{chief}}} \Delta\bar{X}_{0/\text{chief}}, \end{aligned} \tag{15.33}$$

$$\begin{aligned} \Delta W(\bar{X}_0) &= \left(\frac{\partial OPL(\bar{P}_0, \bar{P}_n)}{\partial \bar{X}_0}\right)_{\bar{R}_0} \Delta\bar{X}_0 - \left(\frac{\partial OPL(\bar{P}_0, \bar{P}_n)}{\partial \bar{X}_0}\right)_{\bar{R}_{0/\text{chief}}} \Delta\bar{X}_{0/\text{chief}} \\ &\quad + \frac{1}{2}(\Delta\bar{X}_0)^T \left(\frac{\partial^2 OPL(\bar{P}_0, \bar{P}_n)}{\partial \bar{X}_0^2}\right)_{\bar{R}_0} \Delta\bar{X}_0 - \frac{1}{2}(\Delta\bar{X}_{0/\text{chief}})^T \left(\frac{\partial^2 OPL(\bar{P}_0, \bar{P}_n)}{\partial \bar{X}_0^2}\right)_{\bar{R}_{0/\text{chief}}} \Delta\bar{X}_{0/\text{chief}} \\ &= \left(\sum_{i=1}^{i=n} \frac{\partial OPL_i}{\partial \bar{X}_0}\right)_{\bar{R}_0} \Delta\bar{X}_0 - \left(\sum_{i=1}^{i=n} \frac{\partial OPL_i}{\partial \bar{X}_0}\right)_{\bar{R}_{0/\text{chief}}} \Delta\bar{X}_{0/\text{chief}} + \frac{1}{2}(\Delta\bar{X}_0)^T \left(\sum_{i=1}^{i=n} \frac{\partial^2 OPL_i}{\partial \bar{X}_0^2}\right)_{\bar{R}_0} \Delta\bar{X}_0 \\ &\quad - \frac{1}{2}(\Delta\bar{X}_{0/\text{chief}})^T \left(\sum_{i=1}^{i=n} \frac{\partial^2 OPL_i}{\partial \bar{X}_0^2}\right)_{\bar{R}_{0/\text{chief}}} \Delta\bar{X}_{0/\text{chief}}, \end{aligned} \tag{15.34}$$

where $(\,)_{\bar{R}_0}$ and $(\,)_{\bar{R}_{0/\text{chief}}}$ indicate that the matrices within ( ) are evaluated in terms of the reference ray $\bar{R}_0$ and chief ray $\bar{R}_{0/\text{chief}}$, respectively, and

$$\Delta\bar{X}_0 = \begin{bmatrix} P'_{0x} - P_{0x} \\ P'_{0y} - P_{0y} \\ P'_{0z} - P_{0z} \\ \alpha'_0 - \alpha_0 \\ \beta'_0 - \beta_0 \end{bmatrix}. \tag{15.35}$$

Using Eq. (15.31), (15.33) and (15.34) can be rewritten respectively as

$$\Delta W(\bar{X}_0) = \left(\sum_{i=1}^{i=n} \frac{\partial OPL_i}{\partial \bar{X}_0}\right)_{\bar{R}_0} \Delta\bar{X}_0 - \left(\sum_{i=1}^{i=n} \frac{\partial OPL_i}{\partial \bar{X}_0}\right)_{\bar{R}_{0/\text{chief}}} \frac{\partial \bar{X}_{0/\text{chief}}}{\partial \bar{P}_0} \Delta\bar{P}_0, \tag{15.36}$$

Table 15.1 Comparison of estimated wavefront aberration for various translations of point source

	$\Delta P_{0y} = 6$		$\Delta P_{0y} = 12$		$\Delta P_{0y} = 18$	
	$W(\bar{X}'_0)$	Error (%)	$W(\bar{X}'_0)$	Error (%)	$W(\bar{X}'_0)$	Error (%)
Raytracing	0.010252		0.010865		0.011465	
Equation (14.29)	−5.990622	−∞	−11.99015	−∞	−17.98968	−∞
Equation (15.24)	−5.989751	−∞	−11.98920	−∞	−17.98866	−∞
Equation (15.36)	0.009378	−8.52	0.009852	−9.33	0.010325	−9.94
Equation (15.37)	0.010248	−0.04	0.010797	−0.63	0.011334	−1.15

$$\begin{aligned}\Delta W(\bar{X}_0) = {} & \left(\sum_{i=1}^{i=n} \frac{\partial OPL_i}{\partial \bar{X}_0}\right)_{\bar{R}_0} \Delta\bar{X}_0 - \left(\sum_{i=1}^{i=n} \frac{\partial OPL_i}{\partial \bar{X}_0}\right)_{\bar{R}_{0/chief}} \frac{\partial \bar{X}_{0/chief}}{\partial \bar{P}_0} \Delta\bar{P}_0 \\ & + \frac{1}{2}(\Delta\bar{X}_0)^T \left(\sum_{i=1}^{i=n} \frac{\partial^2 OPL_i}{\partial \bar{X}_0^2}\right)_{\bar{R}_0} \Delta\bar{X}_0 \\ & - \frac{1}{2}(\Delta\bar{P}_0)^T \left(\frac{\partial \bar{X}_{0/chief}}{\partial \bar{P}_0}\right)^T \left(\sum_{i=1}^{i=n} \frac{\partial^2 OPL_i}{\partial \bar{X}_0^2}\right)_{\bar{R}_{0/chief}} \frac{\partial \bar{X}_{0/chief}}{\partial \bar{P}_0} \Delta\bar{P}_0,\end{aligned} \quad (15.37)$$

where $\Delta\bar{P}_0 = \begin{bmatrix} P'_{0x} - P_{0x} & P'_{0y} - P_{0y} & P'_{0z} - P_{0z} \end{bmatrix}^T$ and $\Delta\bar{X}_0$ is given in Eq. (15.35). The wavefront aberration $W(\bar{X}'_0)$ of the neighboring ray $\bar{R}'_0$ can then be estimated as the sum of the wavefront aberration $W(\bar{X}_0)$ of the reference ray $\bar{R}_0 = \begin{bmatrix} \bar{P}_0 & \bar{\ell}_0 \end{bmatrix}^T$ and the change in the wavefront aberration $\Delta W(\bar{X}_0)$ given in Eq. (15.36) or (15.37).

Example 15.2 Table 15.1 compares the results obtained via raytracing and Eqs. (14.29), (15.24), (15.36) and (15.37), respectively, for $W(\bar{X}'_0)$ given a reference ray with a variable vector of $\bar{X}_0 = \begin{bmatrix} 0 & -507 & 0 & 0.7^\circ & -0.4^\circ \end{bmatrix}^T$ and a translated source ray with $\Delta\bar{X}_0 = \begin{bmatrix} 0 & \Delta P_{0y} & 0 & 0.2^\circ & 0.2^\circ \end{bmatrix}^T$, where $\Delta P_{0y} = 6$, 12 or 18, respectively. Note that the chief ray $\bar{R}'_{0/chief}$ is evaluated as $\bar{X}'_{0/chief} = \bar{X}_{0/chief} + \Delta\bar{X}_{0/chief}$ using Eq. (15.31). In general, the results show that Eqs. (15.36) and (15.37) yield accurate estimates of the waveform aberration for all values of the point source translation. However, Eqs. (14.29) and (15.24) fail to correctly estimate the wavefront aberration $W(\bar{X}'_0)$ given a translation of the point source $\bar{P}_0$ to a neighboring point source $\bar{P}'_0$ since they do not consider the effects of the translation motion of the point source.

15.6 Wavefront Shape Along Ray Path

As shown in Fig. 15.3, at any point on a surface $\bar{\Omega}$, a unit normal vector $\bar{n}_{\Omega}$ can be found which is at right angles to the surface. The intersection of a plane containing this normal vector $\bar{n}_{\Omega}$ and the surface forms a curve called a normal section with a curvature known as the normal curvature. For most points on most surfaces, different sections have different curvatures. The maximum and minimum values of these curvatures are called the principal curvatures of the surface and are designated as $\kappa_{\Omega 1}$ and $\kappa_{\Omega 2}$, respectively. The product of these two curvatures (i.e., $\kappa_{\Omega 1}\kappa_{\Omega 2}$) is referred to as the Gaussian curvature, where the sign of this curvature can be used to characterize the surface. In this section, a method is proposed for investigating the wavefront shape by determining the two principal curvatures, $\kappa_{\Omega 1}$ and $\kappa_{\Omega 2}$, and their directions.

Figure 15.4 illustrates a general ray (Eq. (2.1))

$$\bar{R}_i = \begin{bmatrix} \bar{P}_i & \bar{\ell}_i \end{bmatrix}^T = \begin{bmatrix} P_{ix} & P_{iy} & P_{iz} & \ell_{ix} & \ell_{iy} & \ell_{iz} \end{bmatrix}^T$$

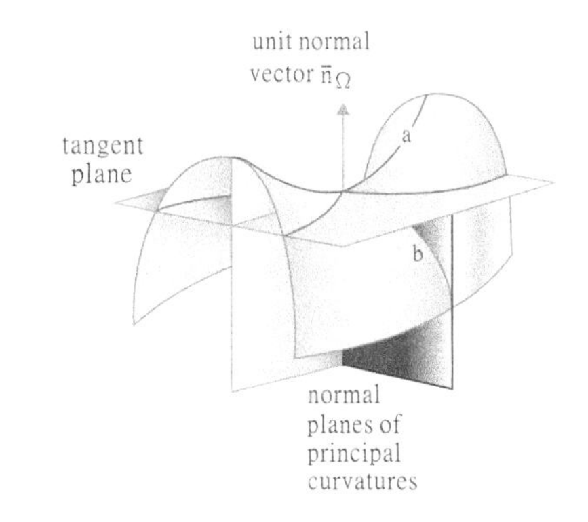

Fig. 15.3 Unit normal vector, tangent plane and normal planes at a point on a surface. Note that a and b are the intersection curves of the normal planes and the surface

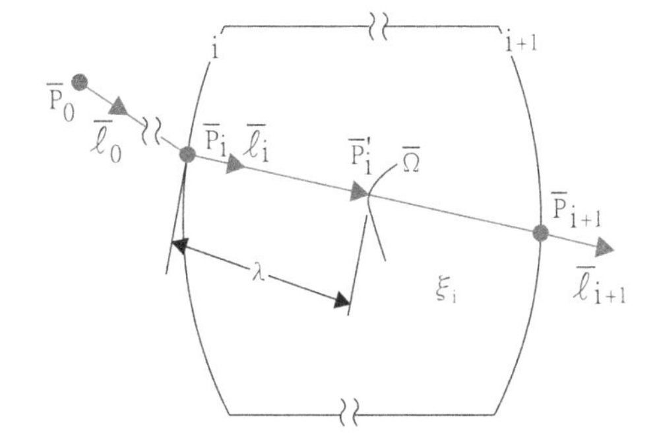

Fig. 15.4 Wavefront $\bar{\Omega}$, defined as loci of all points having the same OPL from point source $\bar{P}_0$

reflected/refracted at the ith boundary surface in an optical system. Any intermediate point $\bar{P}'_i$ along this ray as it travels from $\bar{P}_i$ to $\bar{P}_{i+1}$, where $\bar{P}_{i+1}$ is the incidence point on the (i + 1)th boundary surface, can be expressed as

$$\bar{P}'_i = \begin{bmatrix} P'_{ix} \\ P'_{iy} \\ P'_{iz} \\ 1 \end{bmatrix} = \begin{bmatrix} P_{ix} \\ P_{iy} \\ P_{iz} \\ 1 \end{bmatrix} + \lambda \begin{bmatrix} \ell_{ix} \\ \ell_{iy} \\ \ell_{iz} \\ 0 \end{bmatrix}, \tag{15.38}$$

where variable λ is the geometrical length from $\bar{P}_i$ to $\bar{P}'_i$. According to optical theory, the wavefront $\bar{\Omega}$ is the loci of all the points having the same OPL from the point source $\bar{P}_0$. Let $OPL(\bar{P}_0, \bar{P}'_i)$ be the OPL of the ray measured from point source $\bar{P}_0$ to intermediate point $\bar{P}'_i$ in Fig. 15.4. The condition for the wavefront $\bar{\Omega}$ having $OPL = OPL_{constant}$ is given as

$$\begin{aligned} OPL(\bar{P}_0, \bar{P}'_i) &= OPL_{constant} = \xi_0\lambda_1 + \xi_1\lambda_2 + \ldots + \xi_{i-1}\lambda_i + \xi_i\lambda \\ &= OPL_1 + OPL_2 + \ldots + OPL_i + \xi_i\lambda. \end{aligned} \tag{15.39}$$

Substituting λ in Eq. (15.39) into Eq. (15.38), the following expression is obtained for the wavefront $\bar{\Omega}$ with a constant optical path length $OPL(\bar{P}_0, \bar{P}'_i) = OPL_{constant}$:

$$\bar{\Omega} = \begin{bmatrix} \Omega_x \\ \Omega_y \\ \Omega_z \\ 1 \end{bmatrix} = \begin{bmatrix} P_{ix} \\ P_{iy} \\ P_{iz} \\ 1 \end{bmatrix} + \frac{[OPL_{constant} - (OPL_1 + OPL_2 + \ldots + OPL_i)]}{\xi_i} \begin{bmatrix} \ell_{ix} \\ \ell_{iy} \\ \ell_{iz} \\ 0 \end{bmatrix}. \tag{15.40}$$

In the following discussions, the first and second fundamental forms of the wavefront $\bar{\Omega}$ defined in Eq. (15.40) are used to investigate the wavefront principal curvatures and principal vectors, respectively. Since $\bar{R}_i = \begin{bmatrix} \bar{P}_i & \bar{\ell}_i \end{bmatrix}^T$ and $OPL_i = \xi_{i-1}\lambda_i$ in Eq. (15.40) are both functions of α_0 and β_0, the following quantities are required to investigate the shape of the wavefront $\bar{\Omega}$ using Eq. (15.40): (1) the Jacobian matrix $\partial\bar{R}_i/\partial\bar{X}_0$ given in Eq. (7.30); (2) the Hessian matrix $\partial^2\bar{R}_i/\partial\bar{X}_0^2$ derived in Eq. (15.13); (3) the Jacobian matrix $\partial OPL_i/\partial\bar{X}_0$ given in Eq. (14.5); and (4) the Hessian matrix $\partial^2 OPL_i/\partial\bar{X}_0^2$ derived in Eq. (15.20).

15.6.1 Tangent and Unit Normal Vectors of Wavefront Surface

The principal curvatures at any given point on a wavefront $\bar{\Omega}$ indicate the extent to which the wavefront bends in different directions at that point. In this section, the differential geometry-based method [3] is used to determine the principal curvatures and principal directions of the wavefront along the ray path through an optical system. In Eq. (15.40), the wavefront is defined in terms of the spherical coordinates α_0 and β_0. When the wavefront is regular (i.e., $(\partial\bar{\Omega}/\partial\alpha_0) \times (\partial\bar{\Omega}/\partial\beta_0) \neq \bar{0}$), the tangent vector $\bar{t}_\Omega$ and unit normal vector $\bar{n}_\Omega$ can be determined respectively as

$$\bar{t}_\Omega = \begin{bmatrix} t_{\Omega x} \\ t_{\Omega y} \\ t_{\Omega z} \\ 0 \end{bmatrix} = \frac{\partial\bar{\Omega}}{\partial\alpha_0} d\alpha_0 + \frac{\partial\bar{\Omega}}{\partial\beta_0} d\beta_0 = \begin{bmatrix} \partial\Omega_x/\partial\alpha_0 \\ \partial\Omega_y/\partial\alpha_0 \\ \partial\Omega_z/\partial\alpha_0 \\ 0 \end{bmatrix} d\alpha_0 + \begin{bmatrix} \partial\Omega_x/\partial\beta_0 \\ \partial\Omega_y/\partial\beta_0 \\ \partial\Omega_z/\partial\beta_0 \\ 0 \end{bmatrix} d\beta_0, \tag{15.41}$$

$$\bar{n}_\Omega = \begin{bmatrix} n_{\Omega x} \\ n_{\Omega y} \\ n_{\Omega z} \\ 0 \end{bmatrix} = \frac{\frac{\partial\bar{\Omega}}{\partial\alpha_0} \times \frac{\partial\bar{\Omega}}{\partial\beta_0}}{\left\| \frac{\partial\bar{\Omega}}{\partial\alpha_0} \times \frac{\partial\bar{\Omega}}{\partial\beta_0} \right\|} = \frac{1}{\left\| \frac{\partial\bar{\Omega}}{\partial\alpha_0} \times \frac{\partial\bar{\Omega}}{\partial\beta_0} \right\|} \begin{bmatrix} \frac{\partial\Omega_y}{\partial\alpha_0}\frac{\partial\Omega_z}{\partial\beta_0} - \frac{\partial\Omega_z}{\partial\alpha_0}\frac{\partial\Omega_y}{\partial\beta_0} \\ \frac{\partial\Omega_z}{\partial\alpha_0}\frac{\partial\Omega_x}{\partial\beta_0} - \frac{\partial\Omega_x}{\partial\alpha_0}\frac{\partial\Omega_z}{\partial\beta_0} \\ \frac{\partial\Omega_x}{\partial\alpha_0}\frac{\partial\Omega_y}{\partial\beta_0} - \frac{\partial\Omega_y}{\partial\alpha_0}\frac{\partial\Omega_x}{\partial\beta_0} \\ 0 \end{bmatrix}, \tag{15.42}$$

where the notation $\| \, \|$ denotes the magnitude of the corresponding vector, and

$$\frac{\partial\bar{\Omega}}{\partial\alpha_0} = \begin{bmatrix} \partial\Omega_x/\partial\alpha_0 \\ \partial\Omega_y/\partial\alpha_0 \\ \partial\Omega_z/\partial\alpha_0 \\ 0 \end{bmatrix} = \begin{bmatrix} \partial P_{ix}/\partial\alpha_0 \\ \partial P_{iy}/\partial\alpha_0 \\ \partial P_{iz}/\partial\alpha_0 \\ 0 \end{bmatrix} + \frac{[OPL_{constant} - (OPL_1 + OPL_2 + \ldots + OPL_i)]}{\xi_i} \begin{bmatrix} \partial\ell_{ix}/\partial\alpha_0 \\ \partial\ell_{iy}/\partial\alpha_0 \\ \partial\ell_{iz}/\partial\alpha_0 \\ 0 \end{bmatrix} - \frac{1}{\xi_i}\left(\frac{\partial OPL_1}{\partial\alpha_0} + \frac{\partial OPL_2}{\partial\alpha_0} + \ldots + \frac{\partial OPL_i}{\partial\alpha_0}\right) \begin{bmatrix} \ell_{ix} \\ \ell_{iy} \\ \ell_{iz} \\ 0 \end{bmatrix}, \tag{15.43}$$

$$\frac{\partial\bar{\Omega}}{\partial\beta_0}=\begin{bmatrix}\partial\Omega_x/\partial\beta_0\\ \partial\Omega_y/\partial\beta_0\\ \partial\Omega_z/\partial\beta_0\\ 0\end{bmatrix}=\begin{bmatrix}\partial P_{ix}/\partial\beta_0\\ \partial P_{iy}/\partial\beta_0\\ \partial P_{iz}/\partial\beta_0\\ 0\end{bmatrix}+\frac{[OPL_{constant}-(OPL_1+OPL_2+\ldots+OPL_i)]}{\xi_i}\begin{bmatrix}\partial\ell_{ix}/\partial\beta_0\\ \partial\ell_{iy}/\partial\beta_0\\ \partial\ell_{iz}/\partial\beta_0\\ 0\end{bmatrix}-\frac{1}{\xi_i}\left(\frac{\partial OPL_1}{\partial\beta_0}+\frac{\partial OPL_2}{\partial\beta_0}+\ldots+\frac{\partial OPL_i}{\partial\beta_0}\right)\begin{bmatrix}\ell_{ix}\\ \ell_{iy}\\ \ell_{iz}\\ 0\end{bmatrix}. \tag{15.44}$$

It is noted from Fig. 2.1 that when the numerator of $\bar{n}_\Omega$ shown in Eq. (15.42) is $(\partial\bar{\Omega}/\partial\alpha_0)\times(\partial\bar{\Omega}/\partial\beta_0)$, the component $n_{\Omega y}$ of $\bar{n}_\Omega$ is always positive. In other words, the angle between $\bar{n}_\Omega$ and the y_o axis is an acute angle (see Figs. 2.1 and 3.12) since α_0 is always confined in the range of $0°<\alpha_0+90°<180°$. This finding is important since the principal curvature is taken to be positive if the wavefront turns in the same direction as the chosen unit normal vector $\bar{n}_\Omega$, and negative otherwise. The magnitude of vector $(\partial\bar{\Omega}/\partial\alpha_0)\times(\partial\bar{\Omega}/\partial\beta_0)$ in Eq. (15.42) is calculated as

$$\left\|\frac{\partial\bar{\Omega}}{\partial\alpha_0}\times\frac{\partial\bar{\Omega}}{\partial\beta_0}\right\|=\sqrt{\left(\frac{\partial\Omega_y}{\partial\alpha_0}\frac{\partial\Omega_z}{\partial\beta_0}-\frac{\partial\Omega_z}{\partial\alpha_0}\frac{\partial\Omega_y}{\partial\beta_0}\right)^2+\left(\frac{\partial\Omega_z}{\partial\alpha_0}\frac{\partial\Omega_x}{\partial\beta_0}-\frac{\partial\Omega_x}{\partial\alpha_0}\frac{\partial\Omega_z}{\partial\beta_0}\right)^2+\left(\frac{\partial\Omega_x}{\partial\alpha_0}\frac{\partial\Omega_y}{\partial\beta_0}-\frac{\partial\Omega_y}{\partial\alpha_0}\frac{\partial\Omega_x}{\partial\beta_0}\right)^2}. \tag{15.45}$$

15.6.2 *First and Second Fundamental Forms of Wavefront Surface*

In differential geometry, the first fundamental form is the inner product on the tangent space of a surface. As such, it permits the curvature and metric properties of the wavefront (e.g., the length and area) to be calculated in a manner consistent with the ambient space. The first fundamental form of the wavefront given in Eq. (15.40) is expressed as

$$I_\Omega=\bar{t}_\Omega\cdot\bar{t}_\Omega=Ed\alpha_0^2+2Fd\alpha_0d\beta_0+Gd\beta_0^2, \tag{15.46}$$

in which the coefficients are given by

$$E = \frac{\partial \Omega_x}{\partial \alpha_0}\frac{\partial \Omega_x}{\partial \alpha_0} + \frac{\partial \Omega_y}{\partial \alpha_0}\frac{\partial \Omega_y}{\partial \alpha_0} + \frac{\partial \Omega_z}{\partial \alpha_0}\frac{\partial \Omega_z}{\partial \alpha_0}, \tag{15.47}$$

$$F = \frac{\partial \Omega_x}{\partial \alpha_0}\frac{\partial \Omega_x}{\partial \beta_0} + \frac{\partial \Omega_y}{\partial \alpha_0}\frac{\partial \Omega_y}{\partial \beta_0} + \frac{\partial \Omega_z}{\partial \alpha_0}\frac{\partial \Omega_z}{\partial \beta_0}, \tag{15.48}$$

$$G = \frac{\partial \Omega_x}{\partial \beta_0}\frac{\partial \Omega_x}{\partial \beta_0} + \frac{\partial \Omega_y}{\partial \beta_0}\frac{\partial \Omega_y}{\partial \beta_0} + \frac{\partial \Omega_z}{\partial \beta_0}\frac{\partial \Omega_z}{\partial \beta_0}. \tag{15.49}$$

The second fundamental form of a surface provides a measure of the change of the unit normal direction from one point on the wavefront to another. Together with the first fundamental form, it serves to determine the principal curvatures of the wavefront. The second fundamental form of the wavefront is defined as the projections of the second-order derivatives of $\bar{\Omega}$ onto its unit normal vector $\bar{n}_\Omega$, i.e.,

$$II_\Omega = L d\alpha_0^2 + 2M d\alpha_0 d\beta_0 + N d\beta_0^2, \tag{15.50}$$

where the coefficients are given by

$$\begin{aligned} L = \frac{\partial^2 \bar{\Omega}}{\partial \alpha_0^2} \cdot \bar{n}_\Omega = \frac{1}{\left\| \partial \bar{\Omega}/\partial \alpha_0 \times \partial \bar{\Omega}/\partial \beta_0 \right\|} & \left[\frac{\partial^2 \Omega_x}{\partial \alpha_0^2} \left(\frac{\partial \Omega_y}{\partial \alpha_0}\frac{\partial \Omega_z}{\partial \beta_0} - \frac{\partial \Omega_z}{\partial \alpha_0}\frac{\partial \Omega_y}{\partial \beta_0} \right) \right. \\ & \left. + \frac{\partial^2 \Omega_y}{\partial \alpha_0^2} \left(\frac{\partial \Omega_z}{\partial \alpha_0}\frac{\partial \Omega_x}{\partial \beta_0} - \frac{\partial \Omega_x}{\partial \alpha_0}\frac{\partial \Omega_z}{\partial \beta_0} \right) + \frac{\partial^2 \Omega_z}{\partial \alpha_0^2} \left(\frac{\partial \Omega_x}{\partial \alpha_0}\frac{\partial \Omega_y}{\partial \beta_0} - \frac{\partial \Omega_y}{\partial \alpha_0}\frac{\partial \Omega_x}{\partial \beta_0} \right) \right], \end{aligned} \tag{15.51}$$

$$\begin{aligned} M = \frac{\partial^2 \bar{\Omega}}{\partial \alpha_0 \partial \beta_0} \cdot \bar{n}_\Omega = \frac{1}{\left\| \partial \bar{\Omega}/\partial \alpha_0 \times \partial \bar{\Omega}/\partial \beta_0 \right\|} & \left[\frac{\partial^2 \Omega_x}{\partial \alpha_0 \partial \beta_0} \left(\frac{\partial \Omega_y}{\partial \alpha_0}\frac{\partial \Omega_z}{\partial \beta_0} - \frac{\partial \Omega_z}{\partial \alpha_0}\frac{\partial \Omega_y}{\partial \beta_0} \right) \right. \\ & \left. + \frac{\partial^2 \Omega_y}{\partial \alpha_0 \partial \beta_0} \left(\frac{\partial \Omega_z}{\partial \alpha_0}\frac{\partial \Omega_x}{\partial \beta_0} - \frac{\partial \Omega_x}{\partial \alpha_0}\frac{\partial \Omega_z}{\partial \beta_0} \right) + \frac{\partial^2 \Omega_z}{\partial \alpha_0 \partial \beta_0} \left(\frac{\partial \Omega_x}{\partial \alpha_0}\frac{\partial \Omega_y}{\partial \beta_0} - \frac{\partial \Omega_y}{\partial \alpha_0}\frac{\partial \Omega_x}{\partial \beta_0} \right) \right], \end{aligned} \tag{15.52}$$

$$\begin{aligned} N = \frac{\partial^2 \bar{\Omega}}{\partial \beta_0^2} \cdot \bar{n}_\Omega = \frac{1}{\left\| \partial \bar{\Omega}/\partial \alpha_0 \times \partial \bar{\Omega}/\partial \beta_0 \right\|} & \left[\frac{\partial^2 \Omega_x}{\partial \beta_0^2} \left(\frac{\partial \Omega_y}{\partial \alpha_0}\frac{\partial \Omega_z}{\partial \beta_0} - \frac{\partial \Omega_z}{\partial \alpha_0}\frac{\partial \Omega_y}{\partial \beta_0} \right) \right. \\ & \left. + \frac{\partial^2 \Omega_y}{\partial \beta_0^2} \left(\frac{\partial \Omega_z}{\partial \alpha_0}\frac{\partial \Omega_x}{\partial \beta_0} - \frac{\partial \Omega_x}{\partial \alpha_0}\frac{\partial \Omega_z}{\partial \beta_0} \right) + \frac{\partial^2 \Omega_z}{\partial \beta_0^2} \left(\frac{\partial \Omega_x}{\partial \alpha_0}\frac{\partial \Omega_y}{\partial \beta_0} - \frac{\partial \Omega_y}{\partial \alpha_0}\frac{\partial \Omega_x}{\partial \beta_0} \right) \right], \end{aligned} \tag{15.53}$$

in which,

$$\frac{\partial^2\bar{\Omega}}{\partial\alpha_0^2}=\begin{bmatrix}\partial^2\Omega_x/\partial\alpha_0^2\\ \partial^2\Omega_y/\partial\alpha_0^2\\ \partial^2\Omega_z/\partial\alpha_0^2\\ 0\end{bmatrix}=\begin{bmatrix}\partial^2 P_{ix}/\partial\alpha_0^2\\ \partial^2 P_{iy}/\partial\alpha_0^2\\ \partial^2 P_{iz}/\partial\alpha_0^2\\ 0\end{bmatrix}$$

$$+\frac{[OPL_{constant}-(OPL_1+OPL_2+\ldots+OPL_i)]}{\xi_i}\begin{bmatrix}\partial^2\ell_{ix}/\partial\alpha_0^2\\ \partial^2\ell_{iy}/\partial\alpha_0^2\\ \partial^2\ell_{iz}/\partial\alpha_0^2\\ 0\end{bmatrix} \tag{15.54}$$

$$-\frac{1}{\xi_i}\left(\frac{\partial^2 OPL_1}{\partial\alpha_0^2}+\frac{\partial^2 OPL_2}{\partial\alpha_0^2}+\ldots+\frac{\partial^2 OPL_i}{\partial\alpha_0^2}\right)\begin{bmatrix}\ell_{ix}\\ \ell_{iy}\\ \ell_{iz}\\ 0\end{bmatrix}.$$

The components of

$$\partial^2\bar{\Omega}/\partial\alpha_0\partial\beta_0=\begin{bmatrix}\partial^2\Omega_x/\partial\alpha_0\partial\beta_0 & \partial^2\Omega_y/\partial\alpha_0\partial\beta_0 & \partial^2\Omega_z/\partial\alpha_0\partial\beta_0 & 0\end{bmatrix}^T$$

and

$$\partial^2\bar{\Omega}/\partial\beta_0^2=\begin{bmatrix}\partial^2\Omega_x/\partial\beta_0^2 & \partial^2\Omega_y/\partial\beta_0^2 & \partial^2\Omega_z/\partial\beta_0^2 & 0\end{bmatrix}^T$$

in Eqs. (15.51), (15.52) and (15.53) can be obtained simply by replacing α_0^2 in Eq. (15.54) with $\alpha_0\beta_0$ and β_0^2, respectively.

15.6.3 Principal Curvatures of Wavefront

The principal curvatures of the wavefront at point $\bar{P}_i'$ in Fig. 15.4 can be determined in several but equivalent ways. One particularly elegant way involves solving the roots of the equation

$$\det\left(\begin{bmatrix}L-\kappa_\Omega E & M-\kappa_\Omega F\\ M-\kappa_\Omega F & N-\kappa_\Omega G\end{bmatrix}\right)=0. \tag{15.55}$$

If κ_Ω is one of the principal curvatures, Eq. (15.55) is not invertible. Thus, there exists a non-zero unit column vector $\bar{T}_\Omega=\begin{bmatrix}h_{\Omega 1} & h_{\Omega 2}\end{bmatrix}^T$ (denoted as the principal direction in differential geometry) with real number components such that

$$\begin{bmatrix} L - \kappa_\Omega E & M - \kappa_\Omega F \\ M - \kappa_\Omega F & N - \kappa_\Omega G \end{bmatrix} \begin{bmatrix} h_{\Omega 1} \\ h_{\Omega 2} \end{bmatrix} = 0. \tag{15.56}$$

Let $\kappa_{\Omega 1}$ and $\kappa_{\Omega 2}$ be the principal curvatures at any designated point on the wavefront $\bar{\Omega}$. The following properties can be derived: (1) $\kappa_{\Omega 1}$ and $\kappa_{\Omega 2}$ are real numbers; (2) if $\kappa_{\Omega 1} = \kappa_{\Omega 2} = \kappa_\Omega$, then every tangent vector to the wavefront is a principal vector; (3) if $\kappa_{\Omega 1} \neq \kappa_{\Omega 2}$, then any two non-zero principal tangent vectors $\bar{T}_{\Omega 1}$ and $\bar{T}_{\Omega 2}$ corresponding to $\kappa_{\Omega 1}$ and $\kappa_{\Omega 2}$, respectively, are perpendicular (note that orthogonality of the principal tangent vectors is not general, but is true if the second fundamental form is symmetric); (4) the Gaussian curvature $\kappa_{\Omega 1}\kappa_{\Omega 2}$ is given by $\kappa_{\Omega 1}\kappa_{\Omega 2} = (LN - M^2)/(EG - F^2)$; and (5) the principal directions are unchanged when the ray travels through a medium with a constant refractive index.

As described above, the curvature is taken to be positive if the wavefront turns in the same direction as the unit normal vector $\bar{n}_\Omega$. Thus, when looking in the direction of $\bar{n}_\Omega$ given in Eq. (15.42), all concave wavefronts (i.e., diverging fronts) have negative curvatures, while all convex wavefronts (i.e., converging fronts) have positive curvatures. Furthermore, the sign of the Gaussian curvature $\kappa_{\Omega 1}\kappa_{\Omega 2}$ can be used to characterize the surface. Specifically, if both principal curvatures have the same sign: $\kappa_{\Omega 1}\kappa_{\Omega 2} > 0$, then the Gaussian curvature is positive and the surface is said to have an elliptic point. At such points, the surface is dome like, locally lying on one side of its tangent plane. Furthermore, all of the sectional curvatures have the same sign. Conversely, if the principal curvatures have different signs: $\kappa_{\Omega 1}\kappa_{\Omega 2} < 0$, then the Gaussian curvature is negative and the surface is said to have a hyperbolic point. At such points, the surface is saddle shaped. Moreover, the sectional curvatures are equal to zero, giving the asymptotic directions. Finally, if one of the principal curvatures is zero: $\kappa_{\Omega 1}\kappa_{\Omega 2} = 0$, the Gaussian curvature is also zero and the surface is said to have a parabolic point.

Example 15.3 Consider the wavefront along the path of the chief ray originating from a point source $\bar{P}_0 = [0 \quad -507 \quad 170 \quad 1]^T$ in the axis-symmetrical system shown in Fig. 3.2. The principal radii of curvature ($1/\kappa_{\Omega 1}$ and $1/\kappa_{\Omega 2}$) of the wavefront at incidence points $\bar{P}_i$ (i = 1 to i = 10) are listed in Table 15.2. Moreover, the changes in $1/\kappa_{\Omega 1}$ and $1/\kappa_{\Omega 2}$ with the travel distance of the chief ray are illustrated in Fig. 15.5. As shown, the wavefront from point source $\bar{P}_0$ to point

Table 15.2 Principal radii of curvature of wavefront at each boundary surface encountered by the chief ray originating from $\bar{P}_0 = [0 \quad -507 \quad 170 \quad 1]^T$ in the axis-symmetrical system shown in Fig. 3.2

Incidence point $\bar{P}_i$	$\bar{P}_1$	$\bar{P}_2$	$\bar{P}_3$	$\bar{P}_4$	$\bar{P}_5$
Principal radii	107.59722	103.94575	103.94575	46.33882	43.00113
	108.45233	104.57090	104.57090	53.02069	49.68300
Incidence point $\bar{P}_i$	$\bar{P}_6$	$\bar{P}_7$	$\bar{P}_8$	$\bar{P}_9$	$\bar{P}_{10}$
Principal radii	34.59148	74.06272	−108.87190	164.39263	46.36329
	41.27335	96.17959	−22.70970	273.53115	57.16198

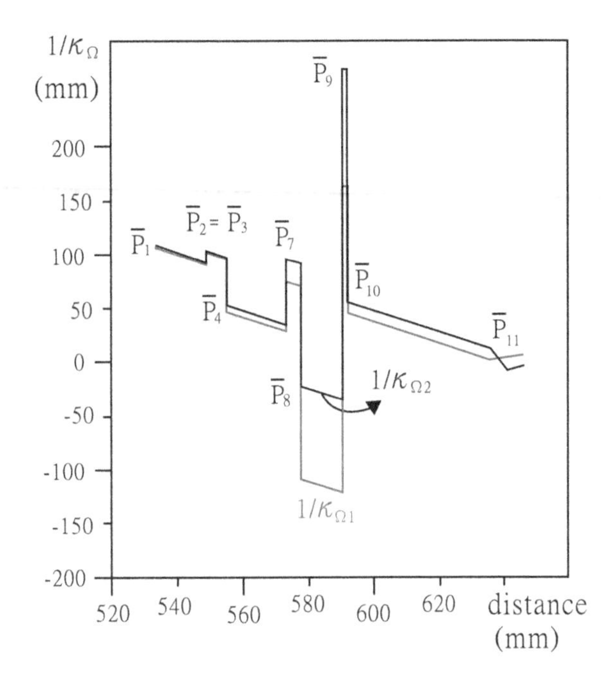

Fig. 15.5 Variation of principal radii $1/\kappa_{\Omega 1}$ and $1/\kappa_{\Omega 2}$ with travel distance of chief ray originating from point source $\bar{P}_0 = [0 \quad -507 \quad 170 \quad 1]^T$ in the axis-symmetrical system shown in Fig. 3.2

$\bar{P}_1$ is spherical. As a result, both principal radii of curvature are equal to the radius of the sphere and there are thus no distinguishable principal directions. As discussed above, diverging wavefronts have a negative curvature while converging wavefronts have a positive one. It is noted from Fig. 15.5 that the wavefront along the chief ray diverges from $\bar{P}_8$ to $\bar{P}_9$. It is also noted that the principal radii of curvature approach zero as the ray approaches the 11th boundary surface (i.e., the image plane); indicating a focusing of the rays originating from $\bar{P}_0$. Note that this finding is confirmed by the fact that the unit normal vector $\bar{n}_\Omega$ changes sign as the ray approaches $\bar{P}_{11}$ (see Fig. 15.6).

Example 15.4 Figure 15.7 shows a typical wavefront $\bar{\Omega}$ and corresponding principal radii for a ray traveling through a medium with a constant refractive index. Let $d\pi_i$ denote an element of the area of this wavefront at incidence point $\bar{P}_i$. All of the rays passing through $d\pi_i$ intersect some subsequent wavefront $\bar{\Omega}'$ within an elemental area $d\pi_i'$ at point $\bar{P}_i'$ (as shown in Fig. 15.4). Let $d\eta_1$ and $d\eta_2$ be the angles subtended by area $d\pi_i$ at the centers of the two principal curvatures, respectively. If B_i is the irradiance at incidence point $\bar{P}_i$ (as computed from Eq. (13.10) by replacing n with i), the irradiance B_i' at any intermediate point $\bar{P}_i'$ can be computed via the following energy conservation law:

$$\frac{B_i'}{B_i} = \frac{d\pi_i}{d\pi_i'} = \frac{d\eta_1 d\eta_2/(\kappa_1 \kappa_2)}{d\eta_1 d\eta_2/(\kappa_1' \kappa_2')} = \frac{\kappa_1' \kappa_2'}{\kappa_1 \kappa_2}. \tag{15.57}$$

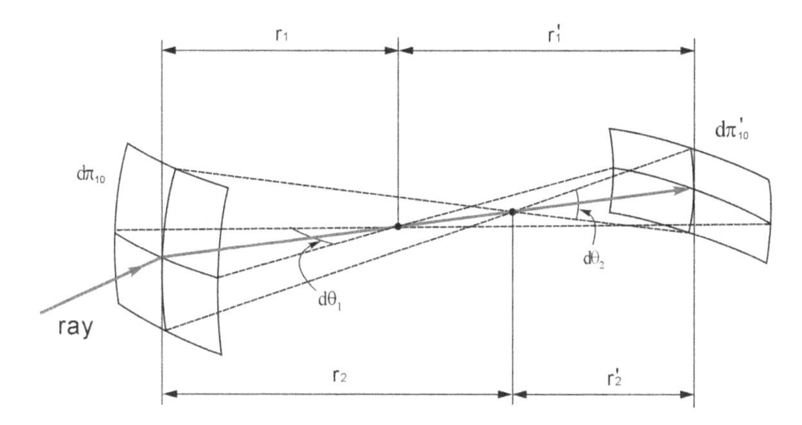

Fig. 15.6 Element of wavefront after emerging from 10th boundary surface (modified from Fig. 2 of [12]). Note that $d\pi_{10}$ and $d\pi'_{10}$ denote the infinitesimal areas of the wavefronts at points $\bar{P}_{10}$ and $\bar{P}'_{10}$, respectively

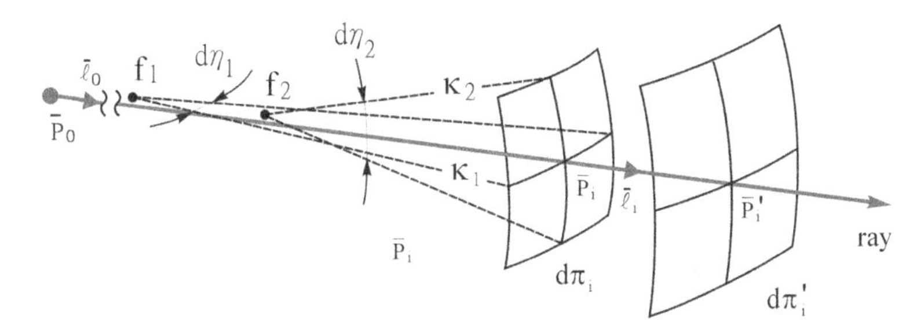

Fig. 15.7 Wavefront and principal radii (modified from Fig. 3 of [10])

Equation (15.57) shows that the intensity along the ray is proportional to the Gaussian curvature $\kappa_1\kappa_2$ of the wavefront. For example, the irradiance of any intermediate point along the chief ray between points $\bar{P}_8$ and $\bar{P}_9$ in Example 15.3 is equal to $B'_8 = B_8\kappa'_1\kappa'_2/(\kappa_1\kappa_2) = 15.457746\kappa'_1\kappa'_2$, where $B_8 = 0.006252$ when normalized such that the total flux from point source $\bar{P}_0$ is equal to one (as shown by Eq. (13.7)). Notably, this intensity law suggests the feasibility of deriving an algorithm to determine the irradiance by tracking the Gaussian curvature as the ray travels through the optical system.

Example 15.5 As shown in Table 15.2, the principal curvatures at the point of incidence of the chief ray on the 10th boundary surface in the system shown in Fig. 3.2 are $\kappa_1 = 1/46.36329$ and $\kappa_2 = 1/57.16198$, respectively. The corresponding principal unit tangent vectors are $\bar{T}_1 = \begin{bmatrix} 0 & 0.268532 & 0.963271 & 0 \end{bmatrix}^T$ and $\bar{T}_2 = \begin{bmatrix} -1 & 0 & 0 & 0 \end{bmatrix}^T$, respectively. It was shown by Kneisly [9] that the principal directions remain unchanged for a ray traveling through a medium with a constant refractive index. Furthermore, it was shown by Mitchell and Hanrahan [10] that when a ray travels from $\bar{P}_i$ (at which the principal curvatures of the wavefront are κ_1 and κ_2) to an intermediate point $\bar{P}'_i$ (see Figs. 15.4 and 15.6), the principal curvatures κ'_1 and κ'_2 at $\bar{P}'_i$ have the forms

$$\kappa'_1 = \frac{\kappa_1}{1 - \rho\kappa_1}, \tag{15.58}$$

$$\kappa'_2 = \frac{\kappa_2}{1 - \rho\kappa_2}. \tag{15.59}$$

Inspecting the above equations, it is noted that after point $\bar{P}_{10}$, the denominator of Eq. (15.58) or Eq. (15.59) becomes zero when the converging wavefront moves a distance $\rho_1 = 1/\kappa_1 = 46.36329$ or $\rho_2 = 1/\kappa_2 = 57.16198$, respectively. Hence, the irradiance tends to infinity (see Fig. 15.6). These positions of extremely high irradiance represent the caustic surfaces of the wavefront. The caustic positions can also be determined by solving Eqs. (15.43) and (15.44) subject to $\partial\bar{\Omega}/\partial\alpha_0 = 0$ and $\partial\bar{\Omega}/\partial\beta_0 = 0$, respectively; resulting in values of $\rho_1 = 46.36333$ and $\rho_2 = 57.16197$, respectively.

Example 15.6 The simplest approach for determining the irradiance in an optical system is the ray-counting method [11], in which multiple rays are traced and a count is made of the number of rays hitting each grid of a mesh on the image plane. However, the ray-counting method has a poor accuracy near the focal point of the optical system or in the vicinity of the caustic surfaces, where the irradiance is very high. This problem arises since when the ray converges toward a focus point, the tube area becomes very small and hence the irradiance is significantly increased (see Fig. 15.6). However, deriving the wavefront shape using the method proposed above involves the computation of local infinitesimal areas only; and hence the problem of poor accuracy is overcome. Figure 15.8a, b illustrate the variation of the irradiance along three different ray paths originating from point source $\bar{P}_0 = \begin{bmatrix} 0 & -507 & 170 & 1 \end{bmatrix}^T$ in the axis-symmetrical system shown in Fig. 3.2. It is noted in Fig. 15.8b that the irradiance increases rapidly between the 10th and 11th boundary surfaces. This is to be expected since all of the rays originating from $\bar{P}_0 = \begin{bmatrix} 0 & -507 & 170 & 1 \end{bmatrix}^T$ are focused on the 11th boundary surface (i.e., the image plane).

Example 15.7 Consider the wavefront along the path of a ray originating from point source $\bar{P}_0 = \begin{bmatrix} 0 & -40 & 0 & 1 \end{bmatrix}^T$ with $\alpha_0 = 4.5°$ and $\beta_0 = 1.5°$ in the non-axially symmetrical system shown in Fig. 3.14. The principal radii of curvature of the wavefront at incidence points $\bar{P}_i$ (i = 1 to i = 12) are listed in Table 15.3. In

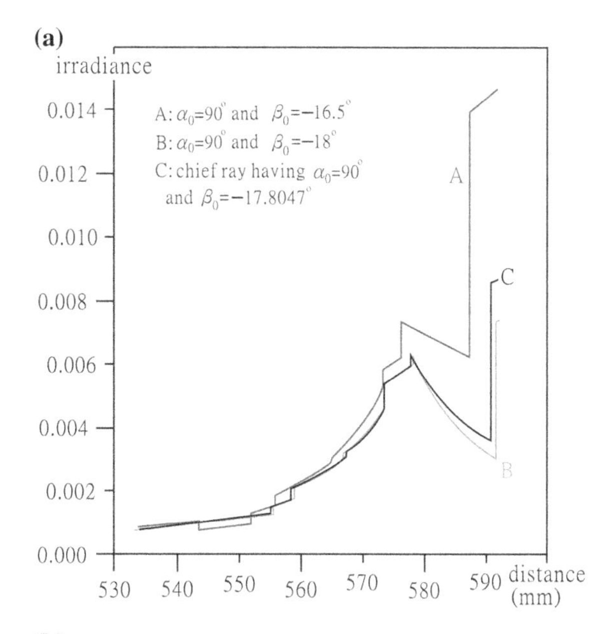

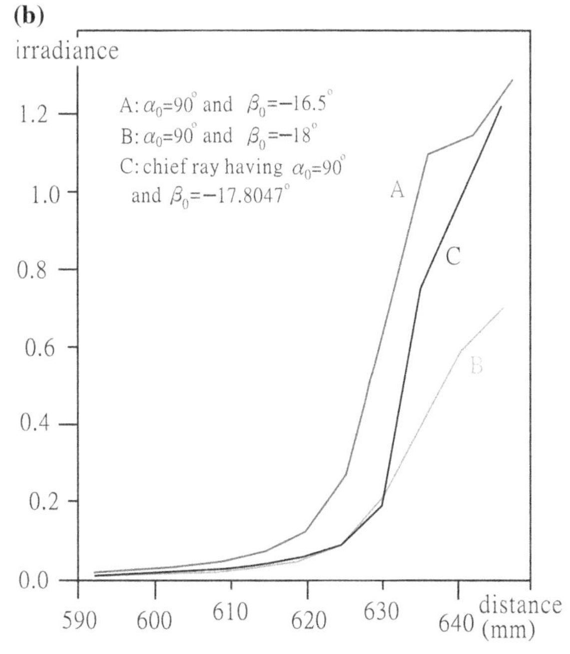

Fig. 15.8 Variation of irradiance with travel distance for three rays originating from point source $\bar{P}_0 = [0 \quad -507 \quad 170 \quad 1]^T$ in axis-symmetrical system shown in Fig. 3.2. **a** Irradiance of rays when traveling from 1st to 10th boundary surface. **b** Irradiance of rays when traveling from 10th boundary surface to image plane

addition, the variations of the principal radii of curvature with the travel distance of the ray are shown in Fig. 15.9. In general, the results presented in Table 15.3 and Fig. 15.9 indicate that the method proposed herein is applicable to both axis-symmetrical and non-axially symmetrical optical systems.

Table 15.3 Principal radii of curvature of wavefront at each boundary surface encountered by a ray originating from $\bar{P}_0 = [0 \quad -40 \quad 0 \quad 1]^T$ with $\alpha_0 = 4.5°$ and $\beta_0 = 1.5°$ in the non-axially symmetrical system shown in Fig. 3.14

Incidence point $\bar{P}_i$	$\bar{P}_1$	$\bar{P}_2$	$\bar{P}_3$	$\bar{P}_4$	$\bar{P}_5$	$\bar{P}_6$
Principal radii	−251.18806	320.94801	378.03749	−394.22296	223.84250	169.00521
	−227.35411	404.74967	486.96250	−229.40263	386.00136	292.72790
Incidence point $\bar{P}_i$	$\bar{P}_7$	$\bar{P}_8$	$\bar{P}_9$	$\bar{P}_{10}$	$\bar{P}_{11}$	$\bar{P}_{12}$
Principal radii	218.44317	−374.54701	165.92683	−166.82421	108.66737	42.56389
	378.88820	−173.51977	361.77838	−109.30423	204.95631	76.80608

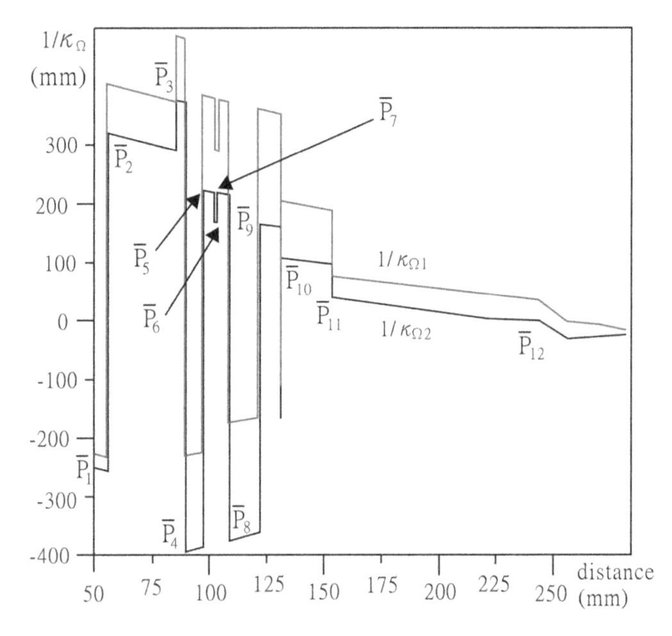

Fig. 15.9 Variation of principal radii, $1/\kappa_1$ and $1/\kappa_2$, with travel distance for a ray originating from point source $\bar{P}_0 = [0 \quad -40 \quad 0 \quad 1]^T$ with $\alpha_0 = 4.5°$ and $\beta_0 = 1.5°$ in non-axially symmetrical system shown in Fig. 3.16

Appendix 1

The explicit expression of $\partial^2\lambda_i/\partial\bar{R}_{i-1}^2$ when a ray hits a flat boundary surface $\bar{r}_i$ has the form (see Eq. (7.5))

$$
\begin{aligned}
\frac{\partial^2\lambda_i}{\partial\bar{R}_{i-1}^2} &= \frac{1}{E_i^2}\frac{\partial E_i}{\partial\bar{R}_{\underline{i-1}}}\frac{\partial D_i}{\partial\bar{R}_{i-1}} + \frac{1}{E_i^2}\frac{\partial D_i}{\partial\bar{R}_{\underline{i-1}}}\frac{\partial E_i}{\partial\bar{R}_{i-1}} - \frac{2D_i}{E_i^3}\frac{\partial E_i}{\partial\bar{R}_{\underline{i-1}}}\frac{\partial E_i}{\partial\bar{R}_{i-1}} \\
&= \frac{1}{E_i^2}\begin{bmatrix} 0 & 0 & 0 & J_{ix}J_{ix} & J_{ix}J_{iy} & J_{ix}J_{iz} \\ & 0 & 0 & J_{iy}J_{ix} & J_{iy}J_{iy} & J_{iy}J_{iz} \\ & & 0 & J_{iz}J_{ix} & J_{iz}J_{iy} & J_{iz}J_{iz} \\ & & & 0 & 0 & 0 \\ & \text{symm.} & & & 0 & 0 \\ & & & & & 0 \end{bmatrix} \\
&\quad - \frac{2D_i}{E_i^3}\begin{bmatrix} 0 & 0 & 0 & 0 & 0 & 0 \\ & 0 & 0 & 0 & 0 & 0 \\ & & 0 & 0 & 0 & 0 \\ & & & J_{ix}J_{ix} & J_{ix}J_{iy} & J_{ix}J_{iz} \\ & \text{symm.} & & & J_{iy}J_{iy} & J_{iy}J_{iz} \\ & & & & & J_{iz}J_{iz} \end{bmatrix}.
\end{aligned}
\tag{15.60}
$$

When $\bar{\ell}_i$ is the unit directional vector of the reflected ray at $\bar{r}_i$, $\partial^2\bar{\ell}_i/\partial\bar{R}_{i-1}^2$ is given by

$$
\frac{\partial^2\bar{\ell}_i}{\partial\bar{R}_{i-1}^2} = \bar{0}_{4\times6\times6}. \tag{15.61}
$$

When $\bar{\ell}_i$ is the unit directional vector of the refracted ray at $\bar{r}_i$, $\partial^2\bar{\ell}_i/\partial\bar{R}_{i-1}^2$ can be obtained by differentiating Eq. (7.10), to give

$$
\frac{\partial^2\bar{\ell}_i}{\partial\bar{R}_{i-1}^2} = \begin{bmatrix} \partial^2\ell_{ix}/\partial\bar{R}_{i-1}^2 \\ \partial^2\ell_{iy}/\partial\bar{R}_{i-1}^2 \\ \partial^2\ell_{iz}/\partial\bar{R}_{i-1}^2 \\ \bar{0} \end{bmatrix}, \tag{15.62}
$$

where

$$\frac{\partial^2 \ell_{ix}}{\partial \bar{R}_{i-1}^2} = \frac{s_i N_i^2 (1 - N_i^2) J_{ix}}{\sqrt{(1 - N_i^2 + N_i^2 E_i^2)^3}} \begin{bmatrix} 0 & 0 & 0 & 0 & 0 & 0 \\ & 0 & 0 & 0 & 0 & 0 \\ & & 0 & 0 & 0 & 0 \\ & & & J_{ix}J_{ix} & J_{ix}J_{iy} & J_{ix}J_{iz} \\ & \text{symm.} & & & J_{iy}J_{iy} & J_{iy}J_{iz} \\ & & & & & J_{iz}J_{iz} \end{bmatrix}, \tag{15.63}$$

$$\frac{\partial^2 \ell_{iy}}{\partial \bar{R}_{i-1}^2} = \frac{s_i N_i^2 (1 - N_i^2) J_{iy}}{\sqrt{(1 - N_i^2 + N_i^2 E_i^2)^3}} \begin{bmatrix} 0 & 0 & 0 & 0 & 0 & 0 \\ & 0 & 0 & 0 & 0 & 0 \\ & & 0 & 0 & 0 & 0 \\ & & & J_{ix}J_{ix} & J_{ix}J_{iy} & J_{ix}J_{iz} \\ & \text{symm.} & & & J_{iy}J_{iy} & J_{iy}J_{iz} \\ & & & & & J_{iz}J_{iz} \end{bmatrix}, \tag{15.64}$$

$$\frac{\partial^2 \ell_{iz}}{\partial \bar{R}_{i-1}^2} = \frac{s_i N_i^2 (1 - N_i^2) J_{iz}}{\sqrt{(1 - N_i^2 + N_i^2 E_i^2)^3}} \begin{bmatrix} 0 & 0 & 0 & 0 & 0 & 0 \\ & 0 & 0 & 0 & 0 & 0 \\ & & 0 & 0 & 0 & 0 \\ & & & J_{ix}J_{ix} & J_{ix}J_{iy} & J_{ix}J_{iz} \\ & \text{symm.} & & & J_{iy}J_{iy} & J_{iy}J_{iz} \\ & & & & & J_{iz}J_{iz} \end{bmatrix}. \tag{15.65}$$

Appendix 2

The Hessian matrix $\partial^2 \bar{R}_i / \partial \bar{R}_{i-1}^2$ of the reflected/refracted ray $\bar{R}_i = \begin{bmatrix} \bar{P}_i & \bar{\ell}_i \end{bmatrix}^T$ at a spherical boundary surface $\bar{r}_i$ with respect to the incoming ray $\bar{R}_{i-1}$ is composed of Eqs. (15.6), (15.10) and (15.11). However, the following terms are required before $\partial^2 \bar{R}_i / \partial \bar{R}_{i-1}^2$ can be determined.

(1) Determination of $\partial^2 \sigma_i / \partial \bar{R}_{i-1}^2$, $\partial^2 \rho_i / \partial \bar{R}_{i-1}^2$ and $\partial^2 \tau_i / \partial \bar{R}_{i-1}^2$:
Differentiating Eq. (7.57) of Appendix 1 in Chap. 7 with respect to incoming ray $\bar{R}_{i-1}$ yields

$$\frac{\partial^2 \bar{P}_i}{\partial \bar{R}_{i-1}^2} = {}^0\bar{A}_i \begin{bmatrix} \partial^2 \sigma_i / \partial \bar{R}_{i-1}^2 \\ \partial^2 \rho_i / \partial \bar{R}_{i-1}^2 \\ \partial^2 \tau_i / \partial \bar{R}_{i-1}^2 \\ \bar{0} \end{bmatrix}, \tag{15.66}$$

where $\partial^2\bar{P}_i/\partial\bar{R}_{i-1}^2$ is given in Eq. (15.6). The terms $\partial^2\sigma_i/\partial\bar{R}_{i-1}^2$, $\partial^2\rho_i/\partial\bar{R}_{i-1}^2$ and $\partial^2\tau_i/\partial\bar{R}_{i-1}^2$ in Eq. (15.66) can then be obtained as

$$\begin{bmatrix} \partial^2\sigma_i/\partial\bar{R}_{i-1}^2 \\ \partial^2\rho_i/\partial\bar{R}_{i-1}^2 \\ \partial^2\tau_i/\partial\bar{R}_{i-1}^2 \\ \bar{0} \end{bmatrix} = \left({}^0\bar{A}_i\right)^{-1}\frac{\partial^2\bar{P}_i}{\partial\bar{R}_{i-1}^2}. \tag{15.67}$$

(2) Determination of $\partial^2\alpha_i/\partial\bar{R}_{i-1}^2$ and $\partial^2\beta_i/\partial\bar{R}_{i-1}^2$:
$\partial^2\alpha_i/\partial\bar{R}_{i-1}^2$ can be obtained by differentiating Eq. (7.59) of Appendix 1 in Chap. 7 with respect to $\bar{R}_{i-1}$, to give

$$\begin{aligned} \frac{\partial^2\alpha_i}{\partial\bar{R}_{i-1}^2} &= \frac{1}{\sigma_i^2+\rho_i^2}\left(\frac{\partial\sigma_i}{\partial\bar{R}_{i-1}}\frac{\partial\rho_i}{\partial\bar{R}_{i-1}} + \sigma_i\frac{\partial^2\rho_i}{\partial\bar{R}_{i-1}^2} - \frac{\partial\rho_i}{\partial\bar{R}_{i-1}}\frac{\partial\sigma_i}{\partial\bar{R}_{i-1}} - \rho_i\frac{\partial^2\sigma_i}{\partial\bar{R}_{i-1}^2}\right) \\ &\quad - \frac{2}{\left(\sigma_i^2+\rho_i^2\right)^2}\left(\sigma_i\frac{\partial\sigma_i}{\partial\bar{R}_{i-1}} + \rho_i\frac{\partial\rho_i}{\partial\bar{R}_{i-1}}\right)\left(\sigma_i\frac{\partial\rho_i}{\partial\bar{R}_{i-1}} - \rho_i\frac{\partial\sigma_i}{\partial\bar{R}_{i-1}}\right). \end{aligned} \tag{15.68}$$

Similarly, $\partial^2\beta_i/\partial\bar{R}_{i-1}^2$ can be determined by differentiating Eq. (7.60) of Appendix 1 in Chap. 7 with respect to $\bar{R}_{i-1}$, to give

$$\begin{aligned} \frac{\partial^2\beta_i}{\partial\bar{R}_{i-1}^2} &= \frac{\sqrt{(\sigma_i^2+\rho_i^2)}}{(\sigma_i^2+\rho_i^2+\tau_i^2)}\frac{\partial^2\tau_i}{\partial\bar{R}_{i-1}^2} + \frac{1}{(\sigma_i^2+\rho_i^2+\tau_i^2)\sqrt{(\sigma_i^2+\rho_i^2)}}\left(\sigma_i\frac{\partial\sigma_i}{\partial\bar{R}_{i-1}} + \rho_i\frac{\partial\rho_i}{\partial\bar{R}_{i-1}}\right)\frac{\partial\tau_i}{\partial\bar{R}_{i-1}} \\ &\quad - \frac{2\sqrt{(\sigma_i^2+\rho_i^2)}}{(\sigma_i^2+\rho_i^2+\tau_i^2)^2}\left(\sigma_i\frac{\partial\sigma_i}{\partial\bar{R}_{i-1}} + \rho_i\frac{\partial\rho_i}{\partial\bar{R}_{i-1}} + \tau_i\frac{\partial\tau_i}{\partial\bar{R}_{i-1}}\right)\frac{\partial\tau_i}{\partial\bar{R}_{i-1}} \\ &\quad - \frac{1}{(\sigma_i^2+\rho_i^2+\tau_i^2)\sqrt{(\sigma_i^2+\rho_i^2)}}\frac{\partial\tau_i}{\partial\bar{R}_{i-1}}\left(\sigma_i\frac{\partial\sigma_i}{\partial\bar{R}_{i-1}} + \rho_i\frac{\partial\rho_i}{\partial\bar{R}_{i-1}}\right) \\ &\quad - \frac{\tau_i}{(\sigma_i^2+\rho_i^2+\tau_i^2)\sqrt{(\sigma_i^2+\rho_i^2)}}\left(\frac{\partial\sigma_i}{\partial\bar{R}_{i-1}}\frac{\partial\sigma_i}{\partial\bar{R}_{i-1}} + \sigma_i\frac{\partial^2\sigma_i}{\partial\bar{R}_{i-1}^2} + \frac{\partial\rho_i}{\partial\bar{R}_{i-1}}\frac{\partial\rho_i}{\partial\bar{R}_{i-1}} + \rho_i\frac{\partial^2\rho_i}{\partial\bar{R}_{i-1}^2}\right) \\ &\quad + \frac{2\tau_i}{(\sigma_i^2+\rho_i^2+\tau_i^2)^2\sqrt{(\sigma_i^2+\rho_i^2)}}\left(\sigma_i\frac{\partial\sigma_i}{\partial\bar{R}_{i-1}} + \rho_i\frac{\partial\rho_i}{\partial\bar{R}_{i-1}} + \tau_i\frac{\partial\tau_i}{\partial\bar{R}_{i-1}}\right)\left(\sigma_i\frac{\partial\sigma_i}{\partial\bar{R}_{i-1}} + \rho_i\frac{\partial\rho_i}{\partial\bar{R}_{i-1}}\right) \\ &\quad + \frac{\tau_i}{(\sigma_i^2+\rho_i^2+\tau_i^2)(\sigma_i^2+\rho_i^2)^{3/2}}\left(\sigma_i\frac{\partial\sigma_i}{\partial\bar{R}_{i-1}} + \rho_i\frac{\partial\rho_i}{\partial\bar{R}_{i-1}}\right)\left(\sigma_i\frac{\partial\sigma_i}{\partial\bar{R}_{i-1}} + \rho_i\frac{\partial\rho_i}{\partial\bar{R}_{i-1}}\right). \end{aligned} \tag{15.69}$$

(3) Determination of $\partial^2\bar{n}_i/\partial\bar{R}_{i-1}^2$:

One approach to compute $\partial^2\bar{n}_i/\partial\bar{R}_{i-1}^2$ is to differentiate Eq. (7.61) of Appendix 1 in Chap. 7 with respect to $\bar{R}_{i-1}$, to give

$$\frac{\partial^2\bar{n}_i}{\partial\bar{R}_{i-1}^2} = \begin{bmatrix} \partial^2 n_{ix}/\partial\bar{R}_{i-1}^2 \\ \partial^2 n_{iy}/\partial\bar{R}_{i-1}^2 \\ \partial^2 n_{iz}/\partial\bar{R}_{i-1}^2 \\ \bar{0} \end{bmatrix} = {}^0\bar{A}_i \frac{\partial^2({}^i\bar{n}_i)}{\partial\bar{R}_{i-1}^2}, \tag{15.70}$$

where $\partial^2({}^i\bar{n}_i)/\partial\bar{R}_{i-1}^2$ is obtained by differentiating Eq. (7.62) of Appendix 1 in Chap. 7 with respect to $\bar{R}_{i-1}$, to give

$$\begin{aligned} \frac{\partial^2({}^i\bar{n}_i)}{\partial\bar{R}_{i-1}^2} = {} & s_i \begin{bmatrix} -S\beta_i C\alpha_i \\ -S\beta_i S\alpha_i \\ C\beta_i \\ 0 \end{bmatrix} \frac{\partial^2\beta_i}{\partial\bar{R}_{i-1}^2} + s_i \begin{bmatrix} -C\beta_i S\alpha_i \\ C\beta_i C\alpha_i \\ 0 \\ 0 \end{bmatrix} \frac{\partial^2\alpha_i}{\partial\bar{R}_{i-1}^2} + s_i \begin{bmatrix} -C\beta_i C\alpha_i \\ -C\beta_i S\alpha_i \\ -S\beta_i \\ 0 \end{bmatrix} \frac{\partial\beta_i}{\partial\bar{R}_{\underline{i-1}}} \frac{\partial\beta_i}{\partial\bar{R}_{i-1}} \\ & + s_i \begin{bmatrix} S\beta_i S\alpha_i \\ -S\beta_i C\alpha_i \\ 0 \\ 0 \end{bmatrix} \frac{\partial\beta_i}{\partial\bar{R}_{\underline{i-1}}} \frac{\partial\alpha_i}{\partial\bar{R}_{i-1}} + s_i \begin{bmatrix} S\beta_i S\alpha_i \\ -S\beta_i C\alpha_i \\ 0 \\ 0 \end{bmatrix} \frac{\partial\alpha_i}{\partial\bar{R}_{\underline{i-1}}} \frac{\partial\beta_i}{\partial\bar{R}_{i-1}} \\ & + s_i \begin{bmatrix} -C\beta_i C\alpha_i \\ -C\beta_i S\alpha_i \\ 0 \\ 0 \end{bmatrix} \frac{\partial\alpha_i}{\partial\bar{R}_{\underline{i-1}}} \frac{\partial\alpha_i}{\partial\bar{R}_{i-1}}. \end{aligned} \tag{15.71}$$

(4) Determination of $\partial^2(C\theta_i)/\partial\bar{R}_{i-1}^2$:

The term $\partial^2(C\theta_i)/\partial\bar{R}_{i-1}^2$ is computed from Eq. (7.63) of Appendix 1 in Chap. 7 by noting that $\partial^2\ell_{i-1x}/\partial\bar{R}_{i-1}^2 = \partial^2\ell_{i-1y}/\partial\bar{R}_{i-1}^2 = \partial^2\ell_{i-1z}/\partial\bar{R}_{i-1}^2 = \bar{0}$, to give

$$\begin{aligned} \frac{\partial^2(C\theta_i)}{\partial\bar{R}_{i-1}^2} = {} & -\left(\ell_{i-1x}\frac{\partial^2 n_{ix}}{\partial\bar{R}_{i-1}^2} + \ell_{i-1y}\frac{\partial^2 n_{iy}}{\partial\bar{R}_{i-1}^2} + \ell_{i-1z}\frac{\partial^2 n_{iz}}{\partial\bar{R}_{i-1}^2}\right) \\ & -\left(\frac{\partial\ell_{i-1x}}{\partial\bar{R}_{\underline{i-1}}}\frac{\partial n_{ix}}{\partial\bar{R}_{i-1}} + \frac{\partial\ell_{i-1y}}{\partial\bar{R}_{\underline{i-1}}}\frac{\partial n_{iy}}{\partial\bar{R}_{i-1}} + \frac{\partial\ell_{i-1z}}{\partial\bar{R}_{\underline{i-1}}}\frac{\partial n_{iz}}{\partial\bar{R}_{i-1}}\right) \\ & -\left(\frac{\partial\ell_{i-1x}}{\partial\bar{R}_{i-1}}\frac{\partial n_{ix}}{\partial\bar{R}_{\underline{i-1}}} + \frac{\partial\ell_{i-1y}}{\partial\bar{R}_{i-1}}\frac{\partial n_{iy}}{\partial\bar{R}_{\underline{i-1}}} + \frac{\partial\ell_{i-1z}}{\partial\bar{R}_{i-1}}\frac{\partial n_{iz}}{\partial\bar{R}_{\underline{i-1}}}\right). \end{aligned} \tag{15.72}$$

References

1. Stavroudis ON (1972) The optics of rays, wavefronts, and caustics. Academic Press
2. Stavroudis ON (2006) The mathematics of geometrical and physical optics. Wiley-VCH Verlag
3. Pressley A (2001) Elementary differential geometry. The Springer Undergraduate Mathematics Series, p 123
4. Hoffnagle JA, Shealy DL (2011) Refracting the k-function: Stavroudis's solution to the eikonal equation for multielement optical systems. J Opt Soc Am A: 28:1312–1321
5. Shealy DL, Burkhard DG (1973) Caustic surfaces and irradiance for reflection and refraction from an ellipsoid, elliptic parabolic and elliptic cone. Appl Opt 12:2955–2959
6. Shealy DL, Burkhard DG (1976) Caustic surface merit functions in optical design. J Opt Soc Am 66:1122
7. Shealy DL (1976) Analytical illuminance and caustic surface calculations in geometrical optics. Appl Opt 15:2588–2596
8. Shealy DL, Hoffnagle JA (2008) Wavefront and caustics of a plane wave refracted by an arbitrary surace. J Opt Soc Am A: 25:2370–2382
9. Kneisly JA II (1964) Local curvature of wavefronts in an optical system. J Opt Soc Am 54:229–235
10. Mitchell DP, Hanrahan P (1992) Illumination from curved reflectors. In: Computer Graphics, vol 26, in Proceedings of SIGGRAPH, pp 283–291
11. Smith WJ (2001) Modern optical engineering, 3rd edn. Edmund Industrial Optics, Barrington, N.J
12. Burkhard DG, Shealy DL (1981) Simplified formula for the illuminance in an optical system. Appl Opt 20:897–909

Chapter 16
Hessian Matrices of Ray $\bar{R}_i$ with Respect to Incoming Ray $\bar{R}_{i-1}$ and Boundary Variable Vector $\bar{X}_i$

As discussed in the previous chapter, the study of optical systems basically involves either systems analysis or systems design. Systems design is the reverse problem of systems analysis. That is, it involves the determination of the system variable vector $\bar{X}_{sys}$ required to create an optical system which meets certain specifications. In systems design, optimization methods play a key role in tuning the system variable vector $\bar{X}_{sys}$ in such a way as to maximize the system performance, as defined by a merit function $\bar{\Phi}$. Optimization methods can be broadly classified as either direct search methods or derivative-based methods. Many studies have shown that cycling may occur when tuning the variables using a direct approach. Thus, derivative-based approaches are generally preferred. Of the various derivative-based approaches available, the Jacobian matrix $d\bar{R}_i/d\bar{X}_{sys}$ described in Chap. 7 and Hessian matrix $d^2\bar{R}_i/d\bar{X}_{sys}^2$ derived in this chapter provide a particularly efficient means of determining the search direction for many existing gradient-based optimization schemes, if $\bar{\Phi}$ is expressed in terms of a ray.

16.1 Hessian Matrix of a Ray with Respect to System Variable Vector

As shown in Eq. (7.1), a ray $\bar{R}_i$ is a recursive function of the given source ray $\bar{R}_0$. Its Hessian matrix $d^2\bar{R}_i/d\bar{X}_{sys}^2$ can be determined as (see Fig. 16.1 and Eq. (9) in [1])

P.D. Lin, *Advanced Geometrical Optics*, Progress in Optical Science and Photonics 4, DOI 10.1007/978-981-10-2299-9_16

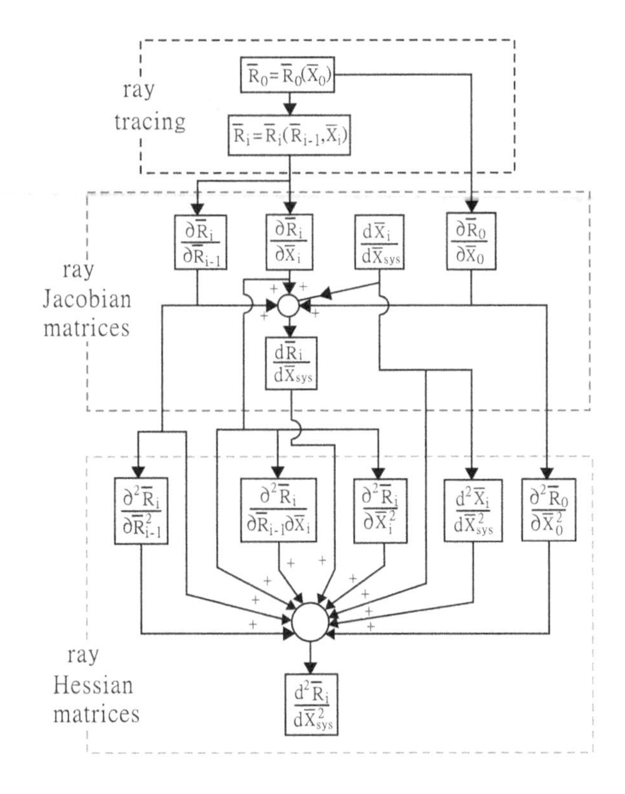

Fig. 16.1 Three-level hierarchical chart showing the determination of $d^2\bar{R}_i/d\bar{X}_{sys}^2$

$$\frac{d^2\bar{R}_i}{d\bar{X}_{sys}^2}=\left[\left(\frac{d\bar{R}_{i-1}}{d\bar{X}_{sys}}\right)^T \quad \left(\frac{d\bar{X}_i}{d\bar{X}_{sys}}\right)^T\right]\begin{bmatrix}\frac{\partial^2\bar{R}_i}{\partial\bar{R}_{i-1}^2} & \frac{\partial^2\bar{R}_i}{\partial\bar{R}_{i-1}\partial\bar{X}_i}\\ \frac{\partial^2\bar{R}_i}{\partial\bar{X}_i\partial\bar{R}_{i-1}} & \frac{\partial^2\bar{R}_i}{\partial\bar{X}_i^2}\end{bmatrix}\begin{bmatrix}\frac{d\bar{R}_{i-1}}{d\bar{X}_{sys}}\\ \frac{d\bar{X}_i}{d\bar{X}_{sys}}\end{bmatrix}+\left[\frac{\partial\bar{R}_i}{\partial\bar{R}_{i-1}} \quad \frac{\partial\bar{R}_i}{\partial\bar{X}_i}\right]\begin{bmatrix}\frac{d^2\bar{R}_{i-1}}{d\bar{X}_{sys}^2}\\ \frac{d^2\bar{X}_i}{d\bar{X}_{sys}^2}\end{bmatrix}. \tag{16.1}$$

It is noted from Eq. (16.1) that $d^2\bar{R}_i/d\bar{X}_{sys}^2$ must be determined sequentially from $i = 1$ to $i = n$ since $d^2\bar{R}_{i-1}/d^2\bar{X}_{sys}$ must be known in advance each time. Moreover, Eq. (16.1) shows that to determine $d^2\bar{R}_i/d\bar{X}_{sys}^2$, it is first necessary to obtain the following eight matrices:

(1) $d\bar{R}_{i-1}/d\bar{X}_{sys}$, which is given in Eq. (7.55) with $g = i - 1$;
(2) $d\bar{X}_i/d\bar{X}_{sys}$, which is derived in Sects. 8.2, 8.3 and 8.4 for a source ray, flat boundary surface and spherical boundary surface, respectively;
(3) $\partial\bar{R}_i/\partial\bar{R}_{i-1}$, which is determined in Sects. 7.2 and 7.3 for flat and spherical boundary surfaces, respectively;
(4) $\partial\bar{R}_i/\partial\bar{X}_i$, which is derived in Sects. 7.4 and 7.5 for flat and spherical boundary surfaces, respectively;

(5) $\partial^2\bar{R}_i/\partial\bar{R}_{i-1}^2$, which is determined in Sects. 15.1 and 15.2 for flat and spherical boundary surfaces, respectively;
(6) $\partial^2\bar{R}_i/\partial\bar{X}_i^2$, the Hessian matrix of ray $\bar{R}_i$ with respect to the boundary variable vector $\bar{X}_i$ (derived in Sects. 16.2 and 16.4 for flat and spherical boundary surfaces, respectively);
(7) $\partial^2\bar{R}_i/\partial\bar{R}_{i-1}\partial\bar{X}_i$, the Hessian matrix of ray $\bar{R}_i$ with respect to the incoming ray $\bar{R}_{i-1}$ and boundary variable vector $\bar{X}_i$ (determined in Sects. 16.3 and 16.5 for flat and spherical boundary surfaces, respectively);
(8) $d^2\bar{X}_i/d\bar{X}_{sys}^2$, the Hessian matrix of boundary variable vector $\bar{X}_i$ with respect to the system variable vector $\bar{X}_{sys}$ (addressed in Chap. 17).

In order to determine Eq. (16.1), the following sections of this chapter present a formal methodology for determining the two Hessian matrices $\partial^2\bar{R}_i/\partial\bar{R}_{i-1}\partial\bar{X}_i$ and $\partial^2\bar{R}_i/\partial\bar{X}_i^2$ for flat and spherical boundary surfaces.

16.2 Hessian Matrix $\partial^2\bar{R}_i/\partial\bar{X}_i^2$ for Flat Boundary Surface

The Hessian matrix $\partial^2\bar{R}_i/\partial\bar{X}_i^2$ of a ray $\bar{R}_i$ reflected/refracted at a flat boundary surface with respect to its boundary variable vector (Eq. 2.47)

$$\bar{X}_i = \begin{bmatrix} J_{ix} & J_{iy} & J_{iz} & e_i & \xi_{i-1} & \xi_i \end{bmatrix}^T$$

is defined as

$$\frac{\partial^2\bar{R}_i}{\partial\bar{X}_i^2} = \begin{bmatrix} \partial^2\bar{P}_i/\partial\bar{X}_i^2 \\ \partial^2\bar{\ell}_i/\partial\bar{X}_i^2 \end{bmatrix}_{6\times6\times6}. \tag{16.2}$$

It is noted from Eq. (16.2) that $\partial^2\bar{R}_i/\partial\bar{X}_i^2$ comprises two components, namely $\partial^2\bar{P}_i/\partial\bar{X}_i^2$ and $\partial^2\bar{\ell}_i/\partial\bar{X}_i^2$. The two components are separately discussed in the following sections.

16.2.1 Hessian Matrix of Incidence Point $\bar{P}_i$

The Hessian matrix $\partial^2\bar{P}_i/\partial\bar{X}_i^2$ of incidence point $\bar{P}_i$ can be obtained by differentiating Eq. (7.31) with respect to the boundary variable vector $\bar{X}_i$, to give

$$\frac{\partial^2 \bar{P}_i}{\partial \bar{X}_i^2} = \frac{\partial^2 \lambda_i}{\partial \bar{X}_i^2} \bar{\ell}_{i-1}. \tag{16.3}$$

Since $\partial^2 D_i/\partial \bar{X}_i^2 = \bar{0}_{6\times6}$ and $\partial^2 E_i/\partial \bar{X}_i^2 = \bar{0}_{6\times6}$, it follows from Eq. (7.32) that $\partial^2 \lambda_i/\partial \bar{X}_i^2$ has the form

$$\frac{\partial^2 \lambda_i}{\partial \bar{X}_i^2} = \frac{1}{E_i^2}\frac{\partial E_i}{\partial \bar{X}_{\underline{i}}}\frac{\partial D_i}{\partial \bar{X}_i} + \frac{1}{E_i^2}\frac{\partial D_i}{\partial \bar{X}_{\underline{i}}}\frac{\partial E_i}{\partial \bar{X}_i} - \frac{2D_i}{E_i^3}\frac{\partial E_i}{\partial \bar{X}_{\underline{i}}}\frac{\partial E_i}{\partial \bar{X}_i}. \tag{16.4}$$

The terms D_i, E_i, $\partial \bar{D}_i/\partial \bar{X}_i$ and $\partial \bar{E}_i/\partial \bar{X}_i$ in Eq. (16.4) are given by Eqs. (2.40), (2.41), (7.33) and (7.34), respectively.

16.2.2 Hessian Matrix of Unit Directional Vector $\bar{\ell}_i$ of Reflected Ray

For a reflected ray, the Hessian matrix $\partial^2 \bar{\ell}_i/\partial \bar{X}_i^2$ of the unit directional vector $\bar{\ell}_i$ can be obtained by differentiating Eq. (7.35) with respect to boundary variable vector $\bar{X}_i$. Since $\partial^2 \bar{n}_i/\partial \bar{X}_i^2 = \bar{0}$ and $\partial^2 (C\theta_i)/\partial \bar{X}_i^2 = \bar{0}$ (see Figs. (7.36) and (7.37)), it follows that

$$\frac{\partial^2 \bar{\ell}_i}{\partial \bar{X}_i^2} = 2\frac{\partial (C\theta_i)}{\partial \bar{X}_{\underline{i}}}\frac{\partial \bar{n}_i}{\partial \bar{X}_i} + 2\frac{\partial (C\theta_i)}{\partial \bar{X}_i}\frac{\partial \bar{n}_i}{\partial \bar{X}_{\underline{i}}}, \tag{16.5}$$

where $\partial \bar{n}_i/\partial \bar{X}_i$ and $\partial (C\theta_i)/\partial \bar{X}_i$ are given by Eqs. (7.36) and (7.37), respectively.

16.2.3 Hessian Matrix of Unit Directional Vector $\bar{\ell}_i$ of Refracted Ray

The Hessian matrix $\partial^2 \bar{\ell}_i/\partial \bar{X}_i^2$ of the unit directional vector $\bar{\ell}_i$ of a ray refracted at a flat boundary surface can be obtained directly by differentiating Eq. (7.38) with respect to $\bar{X}_i$. Since $\partial^2 (C\theta_i)/\partial \bar{X}_i^2 = \bar{0}$ and $\partial \ell_i/\partial \bar{X}_i = \bar{0}$, the following explicit expression for $\partial^2 \bar{\ell}_i/\partial \bar{X}_i^2$ can be obtained:

$$
\begin{aligned}
\frac{\partial^2 \bar{\ell}_i}{\partial \bar{X}_i^2} = {} & \left(N_i C\theta_i - \sqrt{1 - N_i^2 + (N_i C\theta_i)^2}\right) \frac{\partial^2 \bar{n}_i}{\partial \bar{X}_i^2} + \left(N_i - \frac{N_i^2 C\theta_i}{\sqrt{1 - N_i^2 + (N_i C\theta_i)^2}}\right) \frac{\partial (C\theta_i)}{\partial \bar{X}_{\underline{i}}} \frac{\partial \bar{n}_i}{\partial \bar{X}_i} \\
& + \left(C\theta_i - \frac{N_i\left((C\theta_i)^2 - 1\right)}{\sqrt{1 - N_i^2 + (N_i C\theta_i)^2}}\right) \frac{\partial N_i}{\partial \bar{X}_{\underline{i}}} \frac{\partial \bar{n}_i}{\partial \bar{X}_i} + \left(N_i - \frac{N_i^2 C\theta_i}{\sqrt{1 - N_i^2 + (N_i C\theta_i)^2}}\right) \frac{\partial (C\theta_i)}{\partial \bar{X}_i} \frac{\partial \bar{n}_i}{\partial \bar{X}_{\underline{i}}} \\
& + \left(1 - \frac{2 N_i C\theta_i}{\sqrt{1 - N_i^2 + (N_i C\theta_i)^2}} + \frac{N_i^3 C\theta_i\left((C\theta_i)^2 - 1\right)}{\sqrt{1 - N_i^2 + (N_i C\theta_i)^2}}\right) \frac{\partial N_i}{\partial \bar{X}_{\underline{i}}} \frac{\partial (C\theta_i)}{\partial \bar{X}_i} \bar{n}_i \\
& + \left(- \frac{N_i^2}{\sqrt{1 - N_i^2 + (N_i C\theta_i)^2}} + \frac{(N_i^2 C\theta_i)^2}{\left(1 - N_i^2 + (N_i C\theta_i)^2\right)^{3/2}}\right) \frac{\partial (C\theta_i)}{\partial \bar{X}_{\underline{i}}} \frac{\partial (C\theta_i)}{\partial \bar{X}_i} \bar{n}_i \\
& + \left(C\theta_i - \frac{N_i\left((C\theta_i)^2 - 1\right)}{\sqrt{1 - N_i^2 + (N_i C\theta_i)^2}}\right) \frac{\partial^2 N_i}{\partial \bar{X}_i^2} \bar{n}_i + \left(C\theta_i - \frac{N_i\left((C\theta_i)^2 - 1\right)}{\sqrt{1 - N_i^2 + (N_i C\theta_i)^2}}\right) \frac{\partial N_i}{\partial \bar{X}_i} \frac{\partial \bar{n}_i}{\partial \bar{X}_{\underline{i}}} \\
& + \left(\frac{1 - (C\theta_i)^2}{\sqrt{1 - N_i^2 + (N_i C\theta_i)^2}} - \frac{N_i^2\left(1 - (C\theta_i)^2\right)^2}{\left(1 - N_i^2 + (N_i C\theta_i)^2\right)^{3/2}}\right) \frac{\partial N_i}{\partial \bar{X}_{\underline{i}}} \frac{\partial N_i}{\partial \bar{X}_i} \bar{n}_i \\
& + \left(1 + \frac{N_i^3 C\theta_i\left((C\theta_i)^2 - 1\right)}{\left(1 - N_i^2 + (N_i C\theta_i)^2\right)^{3/2}}\right) \frac{\partial (C\theta_i)}{\partial \bar{X}_{\underline{i}}} \frac{\partial N_i}{\partial \bar{X}_i} \bar{n}_i + \frac{\partial^2 N_i}{\partial \bar{X}_i^2} \bar{\ell}_{i-1},
\end{aligned}
\tag{16.6}
$$

where $\bar{n}_i$, $C\theta_i$, $\partial N_i/\partial \bar{X}_i$, $\partial \bar{n}_i/\partial \bar{X}_i$ and $\partial(C\theta_i)/\partial \bar{X}_i$ are given in Eqs. (2.36), (2.44), (7.39), (7.36) and (7.37), respectively. The remaining term in Eq. (16.6), i.e., $\partial^2 N_i/\partial \bar{X}_i^2$, can be obtained by further differentiating Eq. (7.39) to give

$$
\frac{\partial^2 N_i}{\partial \bar{X}_i^2} = \begin{bmatrix} 0 & 0 & 0 & 0 & 0 & 0 \\ 0 & 0 & 0 & 0 & 0 & 0 \\ 0 & 0 & 0 & 0 & 0 & 0 \\ 0 & 0 & 0 & 0 & 0 & 0 \\ 0 & 0 & 0 & 0 & 0 & -1/\xi_i^2 \\ 0 & 0 & 0 & 0 & -1/\xi_i^2 & 2N_i/\xi_i^2 \end{bmatrix}. \tag{16.7}
$$

16.3 Hessian Matrix $\partial^2 \bar{R}_i/\partial \bar{X}_i \partial \bar{R}_{i-1}$ for Flat Boundary Surface

The Hessian matrix $\partial^2 \bar{R}_i/\partial \bar{X}_i \partial \bar{R}_{i-1}$ of a ray $\bar{R}_i$ reflected/refracted at a flat boundary surface with respect to the incoming ray $\bar{R}_{i-1}$ and boundary variable vector $\bar{X}_i$ has the form

$$\frac{\partial^2 \bar{R}_i}{\partial \bar{X}_i \partial \bar{R}_{i-1}} = \begin{bmatrix} \partial^2 \bar{P}_i / \partial \bar{X}_i \partial \bar{R}_{i-1} \\ \partial^2 \bar{\ell}_i / \partial \bar{X}_i \partial \bar{R}_{i-1} \end{bmatrix}_{6\times 6\times 6}. \tag{16.8}$$

It is noted that Eq. (16.8) is not a symmetrical matrix, and hence the sub-matrices (i.e., $\partial^2 \bar{P}_i / \partial \bar{X}_i \partial \bar{R}_{i-1}$ and $\partial^2 \bar{\ell}_i / \partial \bar{X}_i \partial \bar{R}_{i-1}$) are also non-symmetrical. It is further noted that Eq. (16.8) comprises two sub-matrices, i.e., $\partial^2 \bar{P}_i / \partial \bar{R}_{i-1} \partial \bar{X}_i$ and $\partial^2 \bar{\ell}_i / \partial \bar{R}_{i-1} \partial \bar{X}_i$. The former matrix is the Hessian matrix of the incidence point, while the latter matrix is the Hessian matrix of the reflected/refracted unit directional vector. The two matrices are formulated in the following sections.

16.3.1 Hessian Matrix of Incidence Point $\bar{P}_i$

The Hessian matrix $\partial^2 \bar{P}_i / \partial \bar{X}_i \partial \bar{R}_{i-1}$ of incidence point $\bar{P}_i$ on a flat boundary surface can be obtained by differentiating Eq. (7.4) with respect to the boundary variable vector $\bar{X}_i$. Since $\partial^2 \bar{P}_{i-1} / \partial \bar{X}_i \partial \bar{R}_{i-1} = \bar{0}$, $\partial^2 \bar{\ell}_{i-1} / \partial \bar{X}_i \partial \bar{R}_{i-1} = \bar{0}$ and $\partial \bar{\ell}_{i-1} / \partial \bar{X}_i = \bar{0}$, it follows that

$$\frac{\partial^2 \bar{P}_i}{\partial \bar{X}_i \partial \bar{R}_{i-1}} = \frac{\partial \lambda_i}{\partial \bar{X}_i} \frac{\partial \bar{\ell}_{i-1}}{\partial \bar{R}_{i-1}} + \frac{\partial^2 \lambda_i}{\partial \bar{X}_i \partial \bar{R}_{i-1}} \bar{\ell}_{i-1}, \tag{16.9}$$

where $\partial^2 \lambda_i / \partial \bar{X}_i \partial \bar{R}_{i-1}$ is obtained by differentiating Eq. (7.5) with respect to the boundary variable vector $\bar{X}_i$, yielding

$$\begin{aligned} \frac{\partial^2 \lambda_i}{\partial \bar{X}_i \partial \bar{R}_{i-1}} = & -\frac{1}{E_i} \frac{\partial^2 D_i}{\partial \bar{X}_i \partial \bar{R}_{i-1}} + \frac{1}{E_i^2} \frac{\partial E_i}{\partial \bar{X}_i} \frac{\partial D_i}{\partial \bar{R}_{i-1}} + \frac{D_i}{E_i^2} \frac{\partial^2 E_i}{\partial \bar{X}_i \partial \bar{R}_{i-1}} + \frac{1}{E_i^2} \frac{\partial D_i}{\partial \bar{X}_i} \frac{\partial E_i}{\partial \bar{R}_{i-1}} \\ & - \frac{2D_i}{E_i^3} \frac{\partial E_i}{\partial \bar{X}_i} \frac{\partial E_i}{\partial \bar{R}_{i-1}}. \end{aligned} \tag{16.10}$$

Terms D_i, E_i, $\partial D_i / \partial \bar{R}_{i-1}$, $\partial E_i / \partial \bar{R}_{i-1}$, $\partial D_i / \partial \bar{X}_i$ and $\partial E_i / \partial \bar{X}_i$ are given by Eqs. (2.40), (2.41), (7.8), (7.9), (7.33) and (7.34), respectively. The remaining terms in Eq. (16.10), i.e., $\partial^2 D_i / \partial \bar{X}_i \partial \bar{R}_{i-1}$ and $\partial^2 E_i / \partial \bar{X}_i \partial \bar{R}_{i-1}$, are obtained by differentiating Eqs. (7.8) and (7.9), respectively, with respect to $\bar{X}_i$, to obtain

$$\frac{\partial^2 D_i}{\partial \bar{X}_i \partial \bar{R}_{i-1}} = \begin{bmatrix} \bar{I}_{3\times 3} & \bar{0}_{3\times 3} \\ \bar{0}_{3\times 3} & \bar{0}_{3\times 3} \end{bmatrix}, \tag{16.11}$$

$$\frac{\partial^2 E_i}{\partial \bar{X}_i \partial \bar{R}_{i-1}} = \begin{bmatrix} \bar{0}_{3\times 3} & \bar{I}_{3\times 3} \\ \bar{0}_{3\times 3} & \bar{0}_{3\times 3} \end{bmatrix}. \tag{16.12}$$

16.3.2 Hessian Matrix of Unit Directional Vector $\bar{\ell}_i$ of Reflected Ray

The Hessian matrix $\partial^2\bar{\ell}_i/\partial\bar{X}_i\partial\bar{R}_{i-1}$ of the unit directional vector $\bar{\ell}_i$ of a reflected ray at a flat boundary surface can be determined by differentiating Eq. (7.10) with respect to the boundary variable vector $\bar{X}_i$. Since $\partial^2\bar{\ell}_{i-1}/\partial\bar{X}_i\partial\bar{R}_{i-1} = \bar{0}$, it follows that

$$\frac{\partial^2\bar{\ell}_i}{\partial\bar{X}_i\partial\bar{R}_{i-1}} = 2\frac{\partial^2(C\theta_i)}{\partial\bar{X}_i\partial\bar{R}_{i-1}}\bar{n}_i + 2\frac{\partial(C\theta_i)}{\partial\bar{R}_{i-1}}\frac{\partial\bar{n}_i}{\partial\bar{X}_i}, \tag{16.13}$$

where $\bar{n}_i$, $\partial\bar{n}_i/\partial\bar{X}_i$ and $\partial(C\theta_i)/\partial\bar{R}_{i-1}$ are given in Eqs. (2.36), (7.36) and (7.11), respectively. The remaining term in Eq. (16.13), i.e., $\partial^2(C\theta_i)/\partial\bar{X}_i\partial\bar{R}_{i-1}$, is obtained by differentiating Eq. (7.11) with respect to $\bar{X}_i$, to give

$$\frac{\partial^2(C\theta_i)}{\partial\bar{X}_i\partial\bar{R}_{i-1}} = s_i\frac{\partial^2 E_i}{\partial\bar{X}_i\partial\bar{R}_{i-1}} = s_i\begin{bmatrix}\bar{0}_{3\times3} & \bar{I}_{3\times3}\\ \bar{0}_{3\times3} & \bar{0}_{3\times3}\end{bmatrix}. \tag{16.14}$$

16.3.3 Hessian Matrix of Unit Directional Vector $\bar{\ell}_i$ of Refracted Ray

The Hessian matrix $\partial^2\bar{\ell}_i/\partial\bar{X}_i\partial\bar{R}_{i-1}$ of the unit directional vector $\bar{\ell}_i$ of a ray refracted at a flat boundary surface can be derived by differentiating Eq. (7.12) with respect to the boundary variable vector $\bar{X}_i$. Since $\partial^2\bar{\ell}_{i-1}/\partial\bar{X}_i\partial\bar{R}_{i-1} = \bar{0}$, it follows that

$$\begin{aligned}\frac{\partial^2\bar{\ell}_i}{\partial\bar{X}_i\partial\bar{R}_{i-1}} = {} & \left(N_i - \frac{N_i^2 C\theta_i}{\sqrt{1-N_i^2+(N_iC\theta_i)^2}}\right)\frac{\partial\bar{n}_i}{\partial\bar{X}_i}\frac{\partial(C\theta_i)}{\partial\bar{R}_{i-1}} + \left(N_i - \frac{N_i^2 C\theta_i}{\sqrt{1-N_i^2+(N_iC\theta_i)^2}}\right)\bar{n}_i\frac{\partial^2(C\theta_i)}{\partial\bar{X}_i\partial\bar{R}_{i-1}}\\ & + \left(1 - \frac{2N_iC\theta_i}{\sqrt{1-N_i^2+(N_iC\theta_i)^2}} - \frac{N_i^3C\theta_i\left(1-(C\theta_i)^2\right)}{\left(1-N_i^2+(N_iC\theta_i)^2\right)^{3/2}}\right)\frac{\partial N_i}{\partial\bar{X}_i}\bar{n}_i\frac{\partial(C\theta_i)}{\partial\bar{R}_{i-1}}\\ & + \left(\frac{-N_i^2}{\sqrt{1-N_i^2+(N_iC\theta_i)^2}} + \frac{N_i^4(C\theta_i)^2}{\left(1-N_i^2+(N_iC\theta_i)^2\right)^{3/2}}\right)\frac{\partial(C\theta_i)}{\partial\bar{X}_i}\bar{n}_i\frac{\partial(C\theta_i)}{\partial\bar{R}_{i-1}} + \frac{\partial N_i}{\partial\bar{X}_i}\frac{\partial\bar{\ell}_{i-1}}{\partial\bar{R}_{i-1}}.\end{aligned} \tag{16.15}$$

Terms $\bar{n}_i$, $C\theta_i$, $\partial N_i/\partial\bar{X}_i$, $\partial\bar{n}_i/\partial\bar{X}_i$, $\partial(C\theta_i)/\partial\bar{X}_i$, $\partial(C\theta_i)/\partial\bar{R}_{i-1}$, $\partial\bar{\ell}_{i-1}/\partial\bar{R}_{i-1}$ and $\partial^2(C\theta_i)/\partial\bar{X}_i\partial\bar{R}_{i-1}$ in Eq. (16.15) are obtained from Eqs. (2.36), (2.44), (7.39), (7.36), (7.37), (7.11), (7.7) and (16.14), respectively.

16.4 Hessian Matrix $\partial^2\bar{R}_i/\partial\bar{X}_i^2$ for Spherical Boundary Surface

This section determines the Hessian matrix $\partial^2\bar{R}_i/\partial\bar{X}_i^2$ of a reflected/refracted ray $\bar{R}_i$ with respect to the variable vector (Eq. 2.30)

$$\bar{X}_i = \begin{bmatrix} t_{ix} & t_{iy} & t_{iz} & \omega_{ix} & \omega_{iy} & \omega_{iz} & \xi_{i-1} & \xi_i & R_i \end{bmatrix}^T$$

of a spherical boundary surface. The two components of this Hessian matrix, namely $\partial^2\bar{P}_i/\partial\bar{X}_i^2$ and $\partial^2\bar{\ell}_i/\partial\bar{X}_i^2$, are discussed in the following sections.

16.4.1 Hessian Matrix of Incidence Point $\bar{P}_i$

The Hessian matrix $\partial^2\bar{P}_i/\partial\bar{X}_i^2$ of incidence point $\bar{P}_i$ on a spherical boundary surface can be obtained by differentiating Eq. (7.43) with respect to the spherical boundary variable vector $\bar{X}_i$. Since $\partial\bar{\ell}_{i-1}/\partial\bar{X}_i = \bar{0}$, it follows that

$$\frac{\partial^2\bar{P}_i}{\partial\bar{X}_i^2} = \bar{\ell}_{i-1}\frac{\partial^2\lambda_i}{\partial\bar{X}_i^2}. \tag{16.16}$$

Furthermore, since $\partial^2 D_i/\partial\bar{X}_i^2 = \bar{0}$, it follows from Eq. (7.44) that $\partial^2\lambda_i/\partial\bar{X}_i^2$ has the form

$$\begin{aligned}\frac{\partial^2\lambda_i}{\partial\bar{X}_i^2} = &\pm\frac{1}{\sqrt{D_i^2 - E_i}}\frac{\partial D_i}{\partial\bar{X}_{\underline{i}}}\frac{\partial D_i}{\partial\bar{X}_i} \pm \frac{D_i}{2\left(D_i^2 - E_i\right)^{3/2}}\left(-2D_i\frac{\partial D_i}{\partial\bar{X}_{\underline{i}}} + \frac{\partial E_i}{\partial\bar{X}_{\underline{i}}}\right)\frac{\partial D_i}{\partial\bar{X}_i}\\ &\pm\frac{-1}{2\sqrt{\left(D_i^2 - E_i\right)}}\frac{\partial^2 E_i}{\partial\bar{X}_i^2} \pm \frac{1}{4\left(D_i^2 - E_i\right)^{3/2}}\left(2D_i\frac{\partial D_i}{\partial\bar{X}_{\underline{i}}} - \frac{\partial E_i}{\partial\bar{X}_{\underline{i}}}\right)\frac{\partial E_i}{\partial\bar{X}_i},\end{aligned} \tag{16.17}$$

where D_i, E_i, $\partial D_i/\partial\bar{X}_i$ and $\partial E_i/\partial\bar{X}_i$ are given by Eqs. (2.17), (2.18), (7.45) and (7.46), respectively. Meanwhile, $\partial^2 E_i/\partial\bar{X}_i^2$ is obtained by differentiating (7.46) with respect to boundary variable vector $\bar{X}_i$, yielding

$$\frac{\partial^2 E_i}{\partial\bar{X}_i^2} = 2\begin{bmatrix} \bar{I}_{3\times3} & \bar{0}_{3\times5} & \bar{0}_{3\times1} \\ \bar{0}_{5\times3} & \bar{0}_{5\times5} & \bar{0}_{5\times1} \\ \bar{0}_{1\times3} & \bar{0}_{1\times5} & -1 \end{bmatrix}. \tag{16.18}$$

It is noted that Eq. (16.17) is a symmetrical matrix.

16.4.2 Hessian Matrix of Unit Directional Vector $\bar{\ell}_i$ of Reflected Ray

The Hessian matrix $\partial^2\bar{\ell}_i/\partial\bar{X}_i^2$ of the unit directional vector $\bar{\ell}_i$ of a ray reflected at a spherical boundary surface can be determined by differentiating Eq. (7.47) with respect to the boundary variable vector $\bar{X}_i$, to give

$$\frac{\partial^2\bar{\ell}_i}{\partial\bar{X}_i^2} = 2\frac{\partial^2(C\theta_i)}{\partial\bar{X}_i^2}\bar{n}_i + 2\frac{\partial(C\theta_i)}{\partial\bar{X}_i}\frac{\partial\bar{n}_i}{\partial\bar{X}_{\underline{i}}} + 2\frac{\partial(C\theta_i)}{\partial\bar{X}_{\underline{i}}}\frac{\partial\bar{n}_i}{\partial\bar{X}_i} + 2(C\theta_i)\frac{\partial^2\bar{n}_i}{\partial\bar{X}_i^2}. \tag{16.19}$$

Terms $\bar{n}_i$ and $C\theta_i$ are given in Eqs. (2.10) and (2.21), respectively, while $\partial\bar{n}_i/\partial\bar{X}_i$ and $\partial(C\theta_i)/\partial\bar{X}_i$ are given in Eqs. (7.78) and (7.80) of Appendix 2 in Chap. 7. Furthermore, $\partial^2\bar{n}_i/\partial\bar{X}_i^2$ and $\partial^2(C\theta_i)/\partial\bar{X}_i^2$ are obtained by further differentiating Eqs. (7.78) and (7.80) of Appendix 2 with respect to $\bar{X}_i$. (Note that the related derivations are given in Eqs. (16.42) and (16.44) in Appendix 1 in this chapter.)

16.4.3 Hessian Matrix of Unit Directional Vector $\bar{\ell}_i$ of Refracted Ray

The Hessian matrix $\partial^2\bar{\ell}_i/\partial\bar{X}_i^2$ of the unit directional vector of a ray refracted at a spherical boundary surface can be derived directly by differentiating Eq. (7.48) with respect to boundary variable vector $\bar{X}_i$, to give

$$\begin{aligned}
\frac{\partial^2\bar{\ell}_i}{\partial\bar{X}_i^2} &= \begin{bmatrix} \partial^2\ell_{ix}/\partial\bar{X}_i^2 \\ \partial^2\ell_{iy}/\partial\bar{X}_i^2 \\ \partial^2\ell_{iz}/\partial\bar{X}_i^2 \\ \bar{0} \end{bmatrix}_{4\times9\times9} \\
&= \frac{\partial^2 N_i}{\partial\bar{X}_i^2}(C\theta_i)\bar{n}_i + \frac{\partial N_i}{\partial\bar{X}_i}\frac{\partial(C\theta_i)}{\partial\bar{X}_{\underline{i}}}\bar{n}_i + \frac{\partial N_i}{\partial\bar{X}_i}(C\theta_i)\frac{\partial\bar{n}_i}{\partial\bar{X}_{\underline{i}}} + \frac{\partial N_i}{\partial\bar{X}_{\underline{i}}}\frac{\partial(C\theta_i)}{\partial\bar{X}_i}\bar{n}_i \\
&\quad + N_i\frac{\partial^2(C\theta_i)}{\partial\bar{X}_i^2}\bar{n}_i + N_i\frac{\partial(C\theta_i)}{\partial\bar{X}_i}\frac{\partial\bar{n}_i}{\partial\bar{X}_{\underline{i}}} + \frac{\partial N_i}{\partial\bar{X}_I}(C\theta_i)\frac{\partial\bar{n}_i}{\partial\bar{X}_i} + N_i\frac{\partial(C\theta_i)}{\partial\bar{X}_{\underline{i}}}\frac{\partial\bar{n}_i}{\partial\bar{X}_i} + N_i(C\theta_i)\frac{\partial^2\bar{n}_i}{\partial\bar{X}_i^2} \\
&\quad + \left(-\sqrt{1 - N_i^2 + (N_i)^2}\right)\frac{\partial^2\bar{n}_i}{\partial\bar{X}_i^2} + \left(\frac{N_i\left(1 - (C\theta_i)^2\right)}{\sqrt{1 - N_i^2 + (N_i C\theta_i)^2}}\right)\frac{\partial N_i}{\partial\bar{X}_{\underline{i}}}\frac{\partial\bar{n}_i}{\partial\bar{X}_i}
\end{aligned}$$

$$
\begin{aligned}
&+\left(\frac{-N_i^2C\theta_i}{\sqrt{1-N_i^2+(N_iC\theta_i)^2}}\right)\frac{\partial C\theta_i}{\partial \bar{X}_i}\frac{\partial \bar{n}_i}{\partial \bar{X}_i}+\left(\frac{-N_i^2C\theta_i}{\sqrt{1-N_i^2+(N_iC\theta_i)^2}}\right)\frac{\partial^2(C\theta_i)}{\partial \bar{X}_i^2}\bar{n}_i\\
&+\left(\frac{-N_i^2C\theta_i}{\sqrt{1-N_i^2+(N_iC\theta_i)^2}}\right)\frac{\partial (C\theta_i)}{\partial \bar{X}_i}\frac{\partial \bar{n}_i}{\partial \bar{X}_i}\\
&+\left(\frac{-N_i^2}{\sqrt{1-N_i^2+(N_iC\theta_i)^2}}+\frac{N_i^4(C\theta_i)^2}{\left(1-N_i^2+(N_iC\theta_i)^2\right)^{3/2}}\right)\frac{\partial (C\theta_i)}{\partial \bar{X}_i}\frac{\partial (C\theta_i)}{\partial \bar{X}_i}\bar{n}_i\\
&+\left(\frac{-2N_iC\theta_i}{\sqrt{1-N_i^2+(C\theta_i)^2}}-\frac{N_i^3C\theta_i\left(1-(C\theta_i)^2\right)}{\left(1-N_i^2+(N_iC\theta_i)^2\right)^{3/2}}\right)\frac{\partial N_i}{\partial \bar{X}_i}\frac{\partial (C\theta_i)}{\partial \bar{X}_i}\bar{n}_i+\frac{\partial^2 N_i}{\partial \bar{X}_i^2}\bar{\ell}_{i-1}.
\end{aligned}
\tag{16.20}
$$

Terms $\bar{n}_i$, $C\theta_i$ and $\partial N_i/\partial \bar{X}_i$ in Eq. (16.20) are given in Eqs. (2.10), (2.21) and (7.49), respectively, while $\partial \bar{n}_i/\partial \bar{X}_i$ and $\partial (C\theta_i)/\partial \bar{X}_i$ are given in Eqs. (7.78) and (7. 80) of Appendix 2 in Chap. 7. Meanwhile, terms $\partial^2 \bar{n}_i/\partial \bar{X}_i^2$ and $\partial^2 (C\theta_i)/\partial \bar{X}_i^2$ are obtained by further differentiating Eqs. (7.78) and (7.80) in Appendix 2 of Chap. 7 with respect to $\bar{X}_i$. (Note that the related derivations are presented in Eqs. (16.42) and (16.44) in Appendix 1 of this chapter.) The final term in Eq. (16.20), namely $\partial^2 N_i/\partial \bar{X}_i^2$, is determined by further differentiating Eq. (7.49) to give

$$
\frac{\partial^2 N_i}{\partial \bar{X}_i^2}=\begin{bmatrix}\bar{0}_{6\times 6} & \bar{0}_{6\times 1} & \bar{0}_{6\times 1} & \bar{0}_{6\times 1}\\ \bar{0}_{1\times 6} & 0 & -1/\xi_i^2 & 0\\ \bar{0}_{1\times 6} & -1/\xi_i^2 & 2N_i/\xi_i^2 & 0\\ \bar{0}_{1\times 6} & 0 & 0 & 0\end{bmatrix}. \tag{16.21}
$$

16.5 Hessian Matrix $\partial^2 \bar{R}_i/\partial \bar{X}_i \partial \bar{R}_{i-1}$ for Spherical Boundary Surface

As for the case of a flat boundary surface, the Hessian matrix $\partial^2 \bar{R}_i/\partial \bar{X}_i \partial \bar{R}_{i-1}$ and its sub-matrices for a spherical boundary surface are not symmetrical matrices. $\partial^2 \bar{R}_i/\partial \bar{X}_i \partial \bar{R}_{i-1}$ can be formulated as

$$
\frac{\partial^2 \bar{R}_i}{\partial \bar{X}_i \partial \bar{R}_{i-1}}=\begin{bmatrix}\partial^2 \bar{P}_i/\partial \bar{X}_i \partial \bar{R}_{i-1}\\ \partial^2 \bar{\ell}_i/\partial \bar{X}_i \partial \bar{R}_{i-1}\end{bmatrix}_{6\times 9\times 6}, \tag{16.22}
$$

where $\partial^2 \bar{P}_i/\partial \bar{X}_i \partial \bar{R}_{i-1}$ is the Hessian matrix of the incidence point and $\partial^2 \bar{\ell}_i/\partial \bar{X}_i \partial \bar{R}_{i-1}$ is the Hessian matrix of the reflected/refracted unit directional vector. The two matrices are separately derived in the following sections.

16.5.1 Hessian Matrix of Incidence Point $\bar{P}_i$

The Hessian matrix $\partial^2\bar{P}_i/\partial\bar{X}_i\partial\bar{R}_{i-1}$ of incidence point $\bar{P}_i$ on a spherical boundary surface can be obtained by differentiating Eq. (7.20) with respect to the boundary variable vector $\bar{X}_i$. Since $\partial^2\bar{P}_{i-1}/\partial\bar{X}_i\partial\bar{R}_{i-1} = \partial^2\bar{\ell}_{i-1}/\partial\bar{X}_i\partial\bar{R}_{i-1} = \bar{0}$ and $\partial\bar{\ell}_{i-1}/\partial\bar{X}_i = \bar{0}$, it follows that

$$\frac{\partial^2\bar{P}_i}{\partial\bar{X}_i\partial\bar{R}_{i-1}} = \frac{\partial\lambda_i}{\partial\bar{X}_i}\frac{\partial\bar{\ell}_{i-1}}{\partial\bar{R}_{i-1}} + \frac{\partial^2\lambda_i}{\partial\bar{X}_i\partial\bar{R}_{i-1}}\bar{\ell}_{i-1}, \tag{16.23}$$

where $\partial^2\lambda_i/\partial\bar{X}_i\partial\bar{R}_{i-1}$ is obtained by differentiating Eq. (7.21) with respect to $\bar{X}_i$, yielding

$$\begin{aligned}\frac{\partial^2\lambda_i}{\partial\bar{X}_i\partial\bar{R}_{i-1}} = & -\frac{\partial^2 D_i}{\partial\bar{X}_i\partial\bar{R}_{i-1}} \pm \frac{1}{2\sqrt{D_i^2 - E_i}}\left(2\frac{\partial D_i}{\partial\bar{X}_i}\frac{\partial D_i}{\partial\bar{R}_{i-1}} + 2D_i\frac{\partial^2 D_i}{\partial\bar{X}_i\partial\bar{R}_{i-1}} - \frac{\partial^2 E_i}{\partial\bar{X}_i\partial\bar{R}_{i-1}}\right)\\ & \pm\frac{1}{4\left(D_i^2 - E_i\right)^{3/2}}\left(2D_i\frac{\partial D_i}{\partial\bar{X}_i} - \frac{\partial E_i}{\partial\bar{X}_i}\right)\left(-2D_i\frac{\partial D_i}{\partial\bar{R}_{i-1}} + \frac{\partial E_i}{\partial\bar{R}_{i-1}}\right).\end{aligned} \tag{16.24}$$

Terms D_i, E_i, $\partial D_i/\partial\bar{R}_{i-1}$ $\partial E_i/\partial\bar{R}_{i-1}$, $\partial D_i/\partial\bar{X}_i$ and $\partial E_i/\partial\bar{X}_i$ in Eq. (16.24) are given by Eqs. (2.17), (2.18), (7.24), (7.25), (7.45) and (7.46), respectively. In addition, terms $\partial^2 D_i/\partial\bar{X}_i\partial\bar{R}_{i-1}$ and $\partial^2 E_i/\partial\bar{X}_i\partial\bar{R}_{i-1}$ are obtained by further differentiating Eq. (7.24) and Eq. (7.25) with respect to the spherical boundary variable vector $\bar{X}_i$, to give

$$\frac{\partial^2 D_i}{\partial\bar{X}_i\partial\bar{R}_{i-1}} = -\begin{bmatrix}\bar{0}_{3\times3} & \bar{I}_{3\times3}\\ \bar{0}_{6\times3} & \bar{0}_{6\times3}\end{bmatrix}, \tag{16.25}$$

$$\frac{\partial^2 E_i}{\partial\bar{X}_i\partial\bar{R}_{i-1}} = -2\begin{bmatrix}\bar{I}_{3\times3} & \bar{0}_{3\times3}\\ \bar{0}_{6\times3} & \bar{0}_{6\times3}\end{bmatrix}. \tag{16.26}$$

16.5.2 Hessian Matrix of Unit Directional Vector $\bar{\ell}_i$ of Reflected Ray

The Hessian matrix $\partial^2\bar{\ell}_i/\partial\bar{X}_i\partial\bar{R}_{i-1}$ of the unit directional vector $\bar{\ell}_i$ of a ray reflected at a spherical boundary surface can be determined by differentiating Eq. (7.26) with respect to boundary variable vector $\bar{X}_i$. Since $\partial^2\bar{\ell}_{i-1}/\partial\bar{X}_i\partial\bar{R}_{i-1} = \bar{0}$, then

$$\frac{\partial^2\bar{\ell}_i}{\partial\bar{X}_i\partial\bar{R}_{i-1}} = 2\frac{\partial(C\theta_i)}{\partial\bar{X}_i}\frac{\partial\bar{n}_i}{\partial\bar{R}_{i-1}} + 2C\theta_i\frac{\partial^2\bar{n}_i}{\partial\bar{X}_i\partial\bar{R}_{i-1}} + 2\frac{\partial^2(C\theta_i)}{\partial\bar{X}_i\partial\bar{R}_{i-1}}\bar{n}_i + 2\frac{\partial(C\theta_i)}{\partial\bar{R}_{i-1}}\frac{\partial\bar{n}_i}{\partial\bar{X}_i}. \tag{16.27}$$

Terms $\bar{n}_i$, $C\theta_i$, $\partial\bar{n}_i/\partial\bar{R}_{i-1}$, $\partial(C\theta_i)/\partial\bar{R}_{i-1}$, $\partial\bar{n}_i/\partial\bar{X}_i$ and $\partial(C\theta_i)/\partial\bar{X}_i$ are given in Eqs. (2.10), (2.21), (7.61), (7.63), (7.78) and (7.80) of the appendices in Chap. 7, respectively. In addition, $\partial^2\bar{n}_i/\partial\bar{X}_i\partial\bar{R}_{i-1}$ and $\partial^2(C\theta_i)/\partial\bar{X}_i\partial\bar{R}_{i-1}$ are given in Eqs. (16.50) and (16.52) of Appendix 2 in this chapter.

16.5.3 Hessian Matrix of Unit Directional Vector $\bar{\ell}_i$ of Refracted Ray

The Hessian matrix $\partial^2\bar{\ell}_i/\partial\bar{X}_i\partial\bar{R}_{i-1}$ of the unit directional vector $\bar{\ell}_i$ of a ray refracted at a spherical boundary surface can be derived by further differentiating Eq. (7.27) with respect to the boundary variable vector $\bar{X}_i$, to give

$$\begin{aligned}
\frac{\partial^2\bar{\ell}_i}{\partial\bar{X}_i\partial\bar{R}_{i-1}} &= \begin{bmatrix} \partial^2\ell_{ix}/\partial\bar{X}_i\partial\bar{R}_{i-1} \\ \partial^2\ell_{iy}/\partial\bar{X}_i\partial\bar{R}_{i-1} \\ \partial^2\ell_{iz}/\partial\bar{X}_i\partial\bar{R}_{i-1} \\ \bar{0} \end{bmatrix}_{4\times 9\times 6} \\
&= N_i\left(\frac{\partial^2 C\theta_i}{\partial\bar{X}_i\partial\bar{R}_{i-1}}\bar{n}_i + \frac{\partial C\theta_i}{\partial\bar{R}_{i-1}}\frac{\partial\bar{n}_i}{\partial\bar{X}_i} + \frac{\partial(C\theta_i)}{\partial\bar{X}_i}\frac{\partial\bar{n}_i}{\partial\bar{R}_{i-1}} + C\theta_i\frac{\partial^2\bar{n}_i}{\partial\bar{X}_i\partial\bar{R}_{i-1}}\right) + \frac{\partial N_i}{\partial\bar{X}_i}\left(\bar{n}_i\frac{\partial C\theta_i}{\partial\bar{R}_{i-1}} + C\theta_i\frac{\partial\bar{n}_i}{\partial\bar{R}_{i-1}}\right) \\
&\quad + \left(-\sqrt{1-N_i^2+(N_iC\theta_i)^2}\right)\frac{\partial^2\bar{n}_i}{\partial\bar{X}_i\partial\bar{R}_{i-1}} + \left(\frac{N_i\left(1-(C\theta_i)^2\right)}{\sqrt{1-N_i^2+(N_iC\theta_i)^2}}\right)\frac{\partial N_i}{\partial\bar{X}_i}\frac{\partial\bar{n}_i}{\partial\bar{R}_{i-1}} \\
&\quad + \left(\frac{-N_i^2C\theta_i}{\sqrt{1-N_i^2+(N_iC\theta_i)^2}}\right)\frac{\partial(C\theta_i)}{\partial\bar{X}_i}\frac{\partial\bar{n}_i}{\partial\bar{R}_{i-1}} \\
&\quad + \left(\frac{-N_i^2C\theta_i}{\sqrt{1-N_i^2+(N_iC\theta_i)^2}}\right)\frac{\partial\bar{n}_i}{\partial\bar{X}_i}\frac{\partial(C\theta_i)}{\partial\bar{R}_{i-1}} + \left(\frac{-N_i^2C\theta_i}{\sqrt{1-N_i^2+(N_iC\theta_i)^2}}\right)\bar{n}_i\frac{\partial^2(C\theta_i)}{\partial\bar{X}_i\partial\bar{R}_{i-1}} \\
&\quad - \left(\frac{N_i^2}{\sqrt{1-N_i^2+(N_iC\theta_i)^2}} + \frac{N_i^4(C\theta_i)^2}{\left(1-N_i^2+(N_iC\theta_i)^2\right)^{3/2}}\right)\frac{\partial(C\theta_i)}{\partial\bar{X}_i}\bar{n}_i\frac{\partial(C\theta_i)}{\partial\bar{R}_{i-1}} \\
&\quad - \left(\frac{2N_iC\theta_i}{\sqrt{1-N_i^2+(N_iC\theta_i)^2}} + \frac{N_i^3C\theta\left(1-(C\theta_i)^2\right)_i}{\left(1-N_i^2+(N_iC\theta_i)^2\right)^{3/2}}\right)\frac{\partial N_i}{\partial\bar{X}_i}\bar{n}_i\frac{\partial(C\theta_i)}{\partial\bar{R}_{i-1}} + \frac{\partial N_i}{\partial\bar{X}_i}\frac{\partial\bar{\ell}_{i-1}}{\partial\bar{R}_{i-1}}
\end{aligned} \tag{16.28}$$

Terms $\bar{n}_i$, $C\theta_i$, $\partial N_i/\partial\bar{X}_i$, $\partial\bar{n}_i/\partial\bar{R}_{i-1}$, $\partial(C\theta_i)/\partial\bar{R}_{i-1}$, $\partial\bar{n}_i/\partial\bar{X}_i$ and $\partial(C\theta_i)/\partial\bar{X}_i$ in Eq. (16.28) are given in Eqs. (2.10), (2.21), and (7.49), and (7.61), (7.63), (7.78) and (7.80) of the appendices in Chap. 7, respectively.

Appendix 1

To compute the Hessian matrix $\partial^2\bar{R}_i/\partial\bar{X}_i^2$ of a ray with respect to boundary variable vector $\bar{X}_i$ at a spherical boundary surface $\bar{r}_i$, it is first necessary to compute the following terms:

(1) Determination of $\partial^2({}^0\bar{A}_i)/\partial\bar{X}_i^2$:

$\partial^2({}^0\bar{A}_i)/\partial\bar{X}_i^2$ is a $4 \times 4 \times 9 \times 9$ matrix and can be obtained by further differentiating $\partial({}^0\bar{A}_i)/\partial\bar{X}_i$ shown in Eqs. (7.64) to (7.72) of Appendix 2 in Chap. 7 with respect to spherical boundary variable vector $\bar{X}_i$. A careful examination of Eqs. (7.64) to Eq. (7.72) of Appendix 2 in Chap. 7 shows that the non-zero components of $\partial^2({}^0\bar{A}_i)/\partial\bar{X}_i^2$ have the forms

$$\frac{\partial^2({}^0\bar{A}_i)}{\partial\omega_{ix}\partial\omega_{ix}} = \begin{bmatrix} 0 & -C\omega_{iz}S\omega_{iy}S\omega_{ix} + S\omega_{iz}C\omega_{ix} & -C\omega_{iz}S\omega_{iy}C\omega_{ix} - S\omega_{iz}S\omega_{ix} & 0 \\ 0 & -S\omega_{iz}S\omega_{iy}S\omega_{ix} - C\omega_{iz}C\omega_{ix} & -S\omega_{iz}S\omega_{iy}C\omega_{ix} + C\omega_{iz}S\omega_{ix} & 0 \\ 0 & -C\omega_{iy}S\omega_{ix} & -C\omega_{iy}C\omega_{ix} & 0 \\ 0 & 0 & 0 & 0 \end{bmatrix}, \tag{16.29}$$

$$\frac{\partial^2({}^0\bar{A}_i)}{\partial\omega_{iy}\partial\omega_{ix}} = \begin{bmatrix} 0 & C\omega_{iz}C\omega_{iy}C\omega_{ix} & -C\omega_{iz}C\omega_{iy}S\omega_{ix} & 0 \\ 0 & S\omega_{iz}C\omega_{iy}C\omega_{ix} & -S\omega_{iz}C\omega_{iy}S\omega_{ix} & 0 \\ 0 & -S\omega_{iy}C\omega_{ix} & S\omega_{iy}S\omega_{ix} & 0 \\ 0 & 0 & 0 & 0 \end{bmatrix}, \tag{16.30}$$

$$\frac{\partial^2({}^0\bar{A}_i)}{\partial\omega_{iz}\partial\omega_{ix}} = \begin{bmatrix} 0 & -S\omega_{iz}S\omega_{iy}C\omega_{ix} + C\omega_{iz}S\omega_{ix} & S\omega_{iz}S\omega_{iy}S\omega_{ix} + C\omega_{iz}C\omega_{ix} & 0 \\ 0 & C\omega_{iz}S\omega_{iy}C\omega_{ix} + S\omega_{iz}S\omega_{ix} & -C\omega_{iz}S\omega_{iy}S\omega_{ix} + S\omega_{iz}C\omega_{ix} & 0 \\ 0 & 0 & 0 & 0 \\ 0 & 0 & 0 & 0 \end{bmatrix}, \tag{16.31}$$

$$\frac{\partial^2({}^0\bar{A}_i)}{\partial\omega_{ix}\partial\omega_{iy}} = \begin{bmatrix} 0 & C\omega_{iz}C\omega_{iy}C\omega_{ix} & -C\omega_{iz}C\omega_{iy}S\omega_{ix} & 0 \\ 0 & S\omega_{iz}C\omega_{iy}C\omega_{ix} & -S\omega_{iz}C\omega_{iy}S\omega_{ix} & 0 \\ 0 & -S\omega_{iy}C\omega_{ix} & S\omega_{iy}S\omega_{ix} & 0 \\ 0 & 0 & 0 & 0 \end{bmatrix}, \tag{16.32}$$

$$\frac{\partial^2({}^0\bar{A}_i)}{\partial\omega_{iy}\partial\omega_{iy}} = \begin{bmatrix} -C\omega_{iz}C\omega_{iy} & -C\omega_{iz}S\omega_{iy}S\omega_{ix} & -C\omega_{iz}S\omega_{iy}C\omega_{ix} & 0 \\ -S\omega_{iz}C\omega_{iy} & -S\omega_{iz}S\omega_{iy}S\omega_{ix} & -S\omega_{iz}S\omega_{iy}C\omega_{ix} & 0 \\ S\omega_{iy} & -C\omega_{iy}S\omega_{ix} & -C\omega_{iy}C\omega_{ix} & 0 \\ 0 & 0 & 0 & 0 \end{bmatrix}, \tag{16.33}$$

$$\frac{\partial^2({}^0\bar{A}_i)}{\partial\omega_{iz}\partial\omega_{iy}} = \begin{bmatrix} S\omega_{iz}S\omega_{iy} & -S\omega_{iz}C\omega_{iy}S\omega_{ix} & -S\omega_{iz}C\omega_{iy}C\omega_{ix} & 0 \\ -C\omega_{iz}S\omega_{iy} & C\omega_{iz}C\omega_{iy}S\omega_{ix} & C\omega_{iz}C\omega_{iy}C\omega_{ix} & 0 \\ 0 & 0 & 0 & 0 \\ 0 & 0 & 0 & 0 \end{bmatrix}, \quad (16.34)$$

$$\frac{\partial^2({}^0\bar{A}_i)}{\partial\omega_{ix}\partial\omega_{iz}} = \begin{bmatrix} 0 & -S\omega_{iz}S\omega_{iy}C\omega_{ix} + C\omega_{iz}S\omega_{ix} & S\omega_{iz}S\omega_{iy}S\omega_{ix} + C\omega_{iz}C\omega_{ix} & 0 \\ 0 & C\omega_{iz}S\omega_{iy}C\omega_{ix} + S\omega_{iz}S\omega_{ix} & -C\omega_{iz}S\omega_{iy}S\omega_{ix} + S\omega_{iz}C\omega_{ix} & 0 \\ 0 & 0 & 0 & 0 \\ 0 & 0 & 0 & 0 \end{bmatrix}, \quad (16.35)$$

$$\frac{\partial^2({}^0\bar{A}_i)}{\partial\omega_{iy}\partial\omega_{iz}} = \begin{bmatrix} S\omega_{iz}S\omega_{iy} & -S\omega_{iz}C\omega_{iy}S\omega_{ix} & -S\omega_{iz}C\omega_{iy}C\omega_{ix} & 0 \\ -C\omega_{iz}S\omega_{iy} & C\omega_{iz}C\omega_{iy}S\omega_{ix} & C\omega_{iz}C\omega_{iy}C\omega_{ix} & 0 \\ 0 & 0 & 0 & 0 \\ 0 & 0 & 0 & 0 \end{bmatrix}, \quad (16.36)$$

$$\frac{\partial^2({}^0\bar{A}_i)}{\partial\omega_{iz}\partial\omega_{iz}} = \begin{bmatrix} -C\omega_{iz}C\omega_{iy} & -C\omega_{iz}S\omega_{iy}S\omega_{ix} + S\omega_{iz}C\omega_{ix} & -C\omega_{iz}S\omega_{iy}C\omega_{ix} - S\omega_{iz}S\omega_{ix} & 0 \\ -S\omega_{iz}C\omega_{iy} & -S\omega_{iz}S\omega_{iy}S\omega_{ix} - C\omega_{iz}C\omega_{ix} & -S\omega_{iz}S\omega_{iy}C\omega_{ix} + C\omega_{iz}S\omega_{ix} & 0 \\ 0 & 0 & 0 & 0 \\ 0 & 0 & 0 & 0 \end{bmatrix}. \quad (16.37)$$

(2) Determination of $\partial^2\sigma_i/\partial\bar{X}_i^2$, $\partial^2\rho_i/\partial\bar{X}_i^2$ and $\partial^2\tau_i/\partial\bar{X}_i^2$:

Differentiating Eq. (7.74) of Appendix 2 in Chap. 7 with respect to spherical boundary variable vector $\bar{X}_i$ yields

$$\frac{\partial^2\bar{P}_i}{\partial\bar{X}_i^2} = \frac{\partial^2({}^0\bar{A}_i)}{\partial\bar{X}_i^2}\begin{bmatrix}\sigma_i\\ \rho_i\\ \tau_i\\ 1\end{bmatrix} + \frac{\partial({}^0\bar{A}_i)}{\partial\bar{X}_i}\begin{bmatrix}\partial\sigma_i/\partial\bar{X}_{\underline{i}}\\ \partial\rho_i/\partial\bar{X}_{\underline{i}}\\ \partial\tau_i/\partial\bar{X}_{\underline{i}}\\ 0\end{bmatrix} + \frac{\partial({}^0\bar{A}_i)}{\partial\bar{X}_{\underline{i}}}\begin{bmatrix}\partial\sigma_i/\partial\bar{X}_i\\ \partial\rho_i/\partial\bar{X}_i\\ \partial\tau_i/\partial\bar{X}_i\\ 0\end{bmatrix} + {}^0\bar{A}_i\begin{bmatrix}\partial^2\sigma_i/\partial\bar{X}_i^2\\ \partial^2\rho_i/\partial\bar{X}_i^2\\ \partial^2\tau_i/\partial\bar{X}_i^2\\ 0\end{bmatrix}, \quad (16.38)$$

where $\partial^2\bar{P}_i/\partial\bar{X}_i^2$ is given in Eq. (16.16). $\partial^2\sigma_i/\partial\bar{X}_i^2$, $\partial^2\rho_i/\partial\bar{X}_i^2$ and $\partial^2\tau_i/\partial\bar{X}_i^2$ can then be determined from

$$\begin{bmatrix}\partial^2\sigma_i/\partial\bar{X}_i^2\\ \partial^2\rho_i/\partial\bar{X}_i^2\\ \partial^2\tau_i/\partial\bar{X}_i^2\\ 0\end{bmatrix} = ({}^0\bar{A}_i)^{-1}\left(\frac{\partial^2\bar{P}_i}{\partial\bar{X}_i^2} - \frac{\partial^2({}^0\bar{A}_i)}{\partial\bar{X}_i^2}\begin{bmatrix}\sigma_i\\ \rho_i\\ \tau_i\\ 1\end{bmatrix} - \frac{\partial({}^0\bar{A}_i)}{\partial\bar{X}_i}\begin{bmatrix}\partial\sigma_i/\partial\bar{X}_{\underline{i}}\\ \partial\rho_i/\partial\bar{X}_{\underline{i}}\\ \partial\tau_i/\partial\bar{X}_{\underline{i}}\\ 0\end{bmatrix} - \frac{\partial({}^0\bar{A}_i)}{\partial\bar{X}_{\underline{i}}}\begin{bmatrix}\partial\sigma_i/\partial\bar{X}_i\\ \partial\rho_i/\partial\bar{X}_i\\ \partial\tau_i/\partial\bar{X}_i\\ 0\end{bmatrix}\right). \quad (16.39)$$

(3) Determination of $\partial^2\alpha_i/\partial\bar{X}_i^2$ and $\partial^2\beta_i/\partial\bar{X}_i^2$:

$\partial^2\alpha_i/\partial\bar{X}_i^2$ and $\partial^2\beta_i/\partial\bar{X}_i^2$ can be obtained by differentiating Eqs. (7.76) and (7.77) of Appendix 2 in Chap. 7, respectively, with respect to spherical boundary variable vector $\bar{X}_i$, yielding

$$\begin{aligned}\frac{\partial^2\alpha_i}{\partial\bar{X}_i^2} &= \frac{1}{\sigma_i^2+\rho_i^2}\left[\sigma_i\frac{\partial^2\rho_i}{\partial\bar{X}_i^2} + \frac{\partial\sigma_i}{\partial\bar{X}_{\underline{i}}}\frac{\partial\rho_i}{\partial\bar{X}_i} - \rho_i\frac{\partial^2\sigma_i}{\partial\bar{X}_i^2} - \frac{\partial\rho_i}{\partial\bar{X}_{\underline{i}}}\frac{\partial\sigma_i}{\partial\bar{X}_i}\right] \\ &\quad - \frac{2}{\left(\sigma_i^2+\rho_i^2\right)^2}\left[\sigma_i\frac{\partial\sigma_i}{\partial\bar{X}_{\underline{i}}} + \rho_i\frac{\partial\rho_i}{\partial\bar{X}_{\underline{i}}}\right]\left(\sigma_i\frac{\partial\rho_i}{\partial\bar{X}_i} - \rho_i\frac{\partial\sigma_i}{\partial\bar{X}_i}\right),\end{aligned} \tag{16.40}$$

$$\begin{aligned}\frac{\partial^2\beta_i}{\partial\bar{X}_i^2} &= \frac{\sqrt{(\sigma_i^2+\rho_i^2)}}{(\sigma_i^2+\rho_i^2+\tau_i^2)}\frac{\partial^2\tau_i}{\partial\bar{X}_i^2} - \frac{2\sqrt{(\sigma_i^2+\rho_i^2)}}{(\sigma_i^2+\rho_i^2+\tau_i^2)^2}\left(\sigma_i\frac{\partial\sigma_i}{\partial\bar{X}_{\underline{i}}} + \rho_i\frac{\partial\rho_i}{\partial\bar{X}_{\underline{i}}} + \tau_i\frac{\partial\tau_i}{\partial\bar{X}_{\underline{i}}}\right)\frac{\partial\tau_i}{\partial\bar{X}_i} \\ &\quad + \frac{1}{(\sigma_i^2+\rho_i^2+\tau_i^2)\sqrt{(\sigma_i^2+\rho_i^2)}}\left(\sigma_i\frac{\partial\sigma_i}{\partial\bar{X}_{\underline{i}}} + \rho_i\frac{\partial\rho_i}{\partial\bar{X}_{\underline{i}}}\right)\frac{\partial\tau_i}{\partial\bar{X}_i} \\ &\quad - \frac{\tau_i}{(\sigma_i^2+\rho_i^2+\tau_i^2)\sqrt{(\sigma_i^2+\rho_i^2)}}\left(\frac{\partial\sigma_i}{\partial\bar{X}_{\underline{i}}}\frac{\partial\sigma_i}{\partial\bar{X}_i} + \sigma_i\frac{\partial^2\sigma_i}{\partial\bar{X}_i^2} + \frac{\partial\rho_i}{\partial\bar{X}_{\underline{i}}}\frac{\partial\rho_i}{\partial\bar{X}_i} + \rho_i\frac{\partial^2\rho_i}{\partial\bar{X}_i^2}\right) \\ &\quad - \frac{1}{(\sigma_i^2+\rho_i^2+\tau_i^2)\sqrt{(\sigma_i^2+\rho_i^2)}}\frac{\partial\tau_i}{\partial\bar{X}_i}\left(\sigma_i\frac{\partial\sigma_i}{\partial\bar{X}_i} + \rho_i\frac{\partial\rho_i}{\partial\bar{X}_i}\right) \\ &\quad + \frac{2\tau_i}{(\sigma_i^2+\rho_i^2+\tau_i^2)^2\sqrt{(\sigma_i^2+\rho_i^2)}}\left(\sigma_i\frac{\partial\sigma_i}{\partial\bar{X}_{\underline{i}}} + \rho_i\frac{\partial\rho_i}{\partial\bar{X}_{\underline{i}}} + \tau_i\frac{\partial\tau_i}{\partial\bar{X}_{\underline{i}}}\right)\left(\sigma_i\frac{\partial\sigma_i}{\partial\bar{X}_i} + \rho_i\frac{\partial\rho_i}{\partial\bar{X}_i}\right) \\ &\quad + \frac{\tau_i}{(\sigma_i^2+\rho_i^2+\tau_i^2)(\sigma_i^2+\rho_i^2)^{3/2}}\left(\sigma_i\frac{\partial\sigma_i}{\partial\bar{X}_{\underline{i}}} + \rho_i\frac{\partial\rho_i}{\partial\bar{X}_{\underline{i}}}\right)\left(\sigma_i\frac{\partial\sigma_i}{\partial\bar{X}_i} + \rho_i\frac{\partial\rho_i}{\partial\bar{X}_i}\right).\end{aligned} \tag{16.41}$$

(4) Determination of $\partial^2\bar{n}_i/\partial\bar{X}_i^2$:

One approach to compute $\partial^2\bar{n}_i/\partial\bar{X}_i^2$ is to differentiate Eq. (7.78) of Appendix 2 in Chap. 7 with respect to spherical boundary variable vector $\bar{X}_i$, to give

$$\frac{\partial^2\bar{n}_i}{\partial\bar{X}_i^2} = \frac{\partial^2({}^0\bar{A}_i)}{\partial\bar{X}_i^2}{}^i\bar{n}_i + \frac{\partial({}^0\bar{A}_i)}{\partial\bar{X}_i}\frac{\partial({}^i\bar{n}_i)}{\partial\bar{X}_{\underline{i}}} + \frac{\partial({}^0\bar{A}_i)}{\partial\bar{X}_{\underline{i}}}\frac{\partial({}^i\bar{n}_i)}{\partial\bar{X}_i} + {}^0\bar{A}_i\frac{\partial^2({}^i\bar{n}_i)}{\partial\bar{X}_i^2}, \tag{16.42}$$

where

$$\frac{\partial^2({}^{i}\bar{n}_i)}{\partial\bar{X}_i^2} = s_i \begin{bmatrix} -S\beta_i C\alpha_i \\ -S\beta_i S\alpha_i \\ C\beta_i \\ 0 \end{bmatrix} \frac{\partial^2\beta_i}{\partial\bar{X}_i^2} + s_i \begin{bmatrix} -C\beta_i S\alpha_i \\ C\beta_i C\alpha_i \\ 0 \\ 0 \end{bmatrix} \frac{\partial^2\alpha_i}{\partial\bar{X}_i^2}$$
$$+ s_i \left(\begin{bmatrix} -C\beta_i C\alpha_i \\ -C\beta_i S\alpha_i \\ -S\beta_i \\ 0 \end{bmatrix} \frac{\partial\beta_i}{\partial\bar{X}_i} + \begin{bmatrix} S\beta_i S\alpha_i \\ -S\beta_i C\alpha_i \\ 0 \\ 0 \end{bmatrix} \frac{\partial\alpha_i}{\partial\bar{X}_i} \right) \frac{\partial\beta_i}{\partial\bar{X}_i} \tag{16.43}$$
$$+ s_i \left(\begin{bmatrix} S\beta_i S\alpha_i \\ -S\beta_i C\alpha_i \\ 0 \\ 0 \end{bmatrix} \frac{\partial\beta_i}{\partial\bar{X}_i} + \begin{bmatrix} -C\beta_i C\alpha_i \\ -C\beta_i S\alpha_i \\ 0 \\ 0 \end{bmatrix} \frac{\partial\alpha_i}{\partial\bar{X}_i} \right) \frac{\partial\alpha_i}{\partial\bar{X}_i}.$$

(5) Determination of $\partial^2(C\theta_i)/\partial\bar{X}_i^2$:

The term $\partial^2(C\theta_i)/\partial\bar{X}_i^2$ can be computed from Eq. (7.80) of Appendix 2 in Chap. 7 as

$$\frac{\partial^2(C\theta_i)}{\partial\bar{X}_i^2} = -\left(\ell_{i-1x}\frac{\partial^2 n_{ix}}{\partial\bar{X}_i^2} + \ell_{i-1y}\frac{\partial^2 n_{iy}}{\partial\bar{X}_i^2} + \ell_{i-1z}\frac{\partial^2 n_{iz}}{\partial\bar{X}_i^2}\right), \tag{16.44}$$

where the three components of $\partial^2\bar{n}_i/\partial\bar{X}_i^2$ are given in Eq. (16.42) of this appendix.

Appendix 2

To compute the ray Hessian matrix $\partial^2\bar{R}_i/\partial\bar{X}_i\partial\bar{R}_{i-1}$ for a spherical boundary surface $\bar{r}_i$, it is first necessary to determine the following terms:

(1) Determination of $\partial^2({}^{0}\bar{A}_i)/\partial\bar{X}_i\bar{R}_{i-1}$, a matrix with dimensions $4 \times 4 \times 9 \times 6$:

${}^{0}\bar{A}_i$ is not a function of the incoming ray $\bar{R}_{i-1}$. Therefore, the following equation is obtained:

$$\frac{\partial^2({}^{0}\bar{A}_i)}{\partial\bar{X}_i\partial\bar{R}_{i-1}} = \begin{bmatrix} \bar{0} & \bar{0} & \bar{0} & \bar{0} \\ \bar{0} & \bar{0} & \bar{0} & \bar{0} \\ \bar{0} & \bar{0} & \bar{0} & \bar{0} \\ \bar{0} & \bar{0} & \bar{0} & \bar{0} \end{bmatrix}_{4\times4\times9\times6}. \tag{16.45}$$

(2) Determination of $\partial^2\sigma_i/\partial\bar{X}_i\partial\bar{R}_{i-1}$, $\partial^2\rho_i/\partial\bar{X}_i\partial\bar{R}_{i-1}$ and $\partial^2\tau_i/\partial\bar{X}_i\partial\bar{R}_{i-1}$:

Differentiating Eq. (7.57) of Appendix 1 in Chap. 7 with respect to spherical boundary variable vector $\bar{X}_i$ yields

$$\frac{\partial^2 \bar{P}_i}{\partial \bar{X}_i \partial \bar{R}_{i-1}} = \frac{\partial ({}^0\bar{A}_i)}{\partial \bar{X}_i} \begin{bmatrix} \partial \sigma_i / \partial \bar{R}_{i-1} \\ \partial \rho_i / \partial \bar{R}_{i-1} \\ \partial \tau_i / \partial \bar{R}_{i-1} \\ \bar{0} \end{bmatrix} + {}^0\bar{A}_i \begin{bmatrix} \partial^2 \sigma_i / \partial \bar{X}_i \partial \bar{R}_{i-1} \\ \partial^2 \rho_i / \partial \bar{X}_i \partial \bar{R}_{i-1} \\ \partial^2 \tau_i / \partial \bar{X}_i \partial \bar{R}_{i-1} \\ \bar{0} \end{bmatrix}, \tag{16.46}$$

where $\partial^2 \bar{P}_i / \partial \bar{X}_i \partial \bar{R}_{i-1}$ is given in Eq. (16.23). $\partial^2 \sigma_i / \partial \bar{X}_i \partial \bar{R}_{i-1}$, $\partial^2 \rho_i / \partial \bar{X}_i \partial \bar{R}_{i-1}$ and $\partial^2 \tau_i / \partial \bar{X}_i \partial \bar{R}_{i-1}$ can then be determined from

$$\begin{bmatrix} \partial^2 \sigma_i / \partial \bar{X}_i \partial \bar{R}_{i-1} \\ \partial^2 \rho_i / \partial \bar{X}_i \partial \bar{R}_{i-1} \\ \partial^2 \tau_i / \partial \bar{X}_i \partial \bar{R}_{i-1} \\ \bar{0} \end{bmatrix} = \left({}^0\bar{A}_i\right)^{-1} \left(\frac{\partial^2 \bar{P}_i}{\partial \bar{X}_i \partial \bar{R}_{i-1}} - \frac{\partial ({}^0\bar{A}_i)}{\partial \bar{X}_i} \begin{bmatrix} \partial \sigma_i / \partial \bar{R}_{i-1} \\ \partial \rho_i / \partial \bar{R}_{i-1} \\ \partial \tau_i / \partial \bar{R}_{i-1} \\ \bar{0} \end{bmatrix} \right). \tag{16.47}$$

(3) Determination of $\partial^2 \alpha_i / \partial \bar{X}_i \partial \bar{R}_{i-1}$ and $\partial^2 \beta_i / \partial \bar{X}_i \partial \bar{R}_{i-1}$:

$\partial^2 \alpha_i / \partial \bar{X}_i \partial \bar{R}_{i-1}$ and $\partial^2 \beta_i / \partial \bar{X}_i \partial \bar{R}_{i-1}$ can be obtained by differentiating Eqs. (7.59) and (7.60) of Appendix 1 in Chap. 7, respectively, with respect to boundary variable vector $\bar{X}_i$, to give

$$\begin{aligned} \frac{\partial^2 \alpha_i}{\partial \bar{X}_i \partial \bar{R}_{i-1}} &= \frac{1}{\sigma_i^2 + \rho_i^2} \left[\frac{\partial \sigma_i}{\partial \bar{X}_i} \frac{\partial \rho_i}{\partial \bar{R}_{i-1}} + \sigma_i \frac{\partial^2 \rho_i}{\partial \bar{X}_i \partial \bar{R}_{i-1}} - \frac{\partial \rho_i}{\partial \bar{X}_i} \frac{\partial \sigma_i}{\partial \bar{R}_{i-1}} - \rho_i \frac{\partial^2 \sigma_i}{\partial \bar{X}_i \partial \bar{R}_{i-1}} \right] \\ &\quad - \frac{2}{\left(\sigma_i^2 + \rho_i^2\right)^2} \left(\sigma_i \frac{\partial \sigma_i}{\partial \bar{X}_i} + \rho_i \frac{\partial \rho_i}{\partial \bar{X}_i} \right) \left(\sigma_i \frac{\partial \rho_i}{\partial \bar{R}_{i-1}} - \rho_i \frac{\partial \sigma_i}{\partial \bar{R}_{i-1}} \right), \end{aligned} \tag{16.48}$$

$$\begin{aligned} \frac{\partial^2 \beta_i}{\partial \bar{X}_i \partial \bar{R}_{i-1}} &= \frac{\sqrt{(\sigma_i^2 + \rho_i^2)}}{(\sigma_i^2 + \rho_i^2 + \tau_i^2)} \frac{\partial^2 \tau_i}{\partial \bar{X}_i \partial \bar{R}_{i-1}} + \frac{1}{(\sigma_i^2 + \rho_i^2 + \tau_i^2)\sqrt{(\sigma_i^2 + \rho_i^2)}} \left(\sigma_i \frac{\partial \sigma_i}{\partial \bar{X}_i} + \rho_i \frac{\partial \rho_i}{\partial \bar{X}_i} \right) \frac{\partial \tau_i}{\partial \bar{R}_{i-1}} \\ &\quad - \frac{2\sqrt{(\sigma_i^2 + \rho_i^2)}}{\left(\sigma_i^2 + \rho_i^2 + \tau_i^2\right)^2} \left(\sigma_i \frac{\partial \sigma_i}{\partial \bar{X}_i} + \rho_i \frac{\partial \rho_i}{\partial \bar{X}_i} + \tau_i \frac{\partial \tau_i}{\partial \bar{X}_i} \right) \frac{\partial \tau_i}{\partial \bar{R}_{i-1}} \\ &\quad - \frac{\tau_i}{(\sigma_i^2 + \rho_i^2 + \tau_i^2)\sqrt{(\sigma_i^2 + \rho_i^2)}} \left(\frac{\partial \sigma_i}{\partial \bar{X}_i} \frac{\partial \sigma_i}{\partial \bar{R}_{i-1}} + \sigma_i \frac{\partial^2 \sigma_i}{\partial \bar{X}_i \partial \bar{R}_{i-1}} + \frac{\partial \rho_i}{\partial \bar{X}_i} \frac{\partial \rho_i}{\partial \bar{R}_{i-1}} + \rho_i \frac{\partial^2 \rho_i}{\partial \bar{X}_i \partial \bar{R}_{i-1}} \right) \\ &\quad - \frac{1}{(\sigma_i^2 + \rho_i^2 + \tau_i^2)\sqrt{(\sigma_i^2 + \rho_i^2)}} \left(\frac{\partial \tau_i}{\partial \bar{X}_i} \right)^T \left(\sigma_i \frac{\partial \sigma_i}{\partial \bar{R}_{i-1}} + \rho_i \frac{\partial \rho_i}{\partial \bar{R}_{i-1}} \right) \\ &\quad + \frac{2\tau_i}{\left(\sigma_i^2 + \rho_i^2 + \tau_i^2\right)^2 \sqrt{(\sigma_i^2 + \rho_i^2)}} \left(\sigma_i \frac{\partial \sigma_i}{\partial \bar{X}_i} + \rho_i \frac{\partial \rho_i}{\partial \bar{X}_i} + \tau_i \frac{\partial \tau_i}{\partial \bar{X}_i} \right) \left(\sigma_i \frac{\partial \sigma_i}{\partial \bar{R}_{i-1}} + \rho_i \frac{\partial \rho_i}{\partial \bar{R}_{i-1}} \right) \\ &\quad + \frac{\tau_i}{(\sigma_i^2 + \rho_i^2 + \tau_i^2)\left(\sigma_i^2 + \rho_i^2\right)^{3/2}} \left(\sigma_i \frac{\partial \sigma_i}{\partial \bar{X}_i} + \rho_i \frac{\partial \rho_i}{\partial \bar{X}_i} \right) \left(\sigma_i \frac{\partial \sigma_i}{\partial \bar{R}_{i-1}} + \rho_i \frac{\partial \rho_i}{\partial \bar{R}_{i-1}} \right). \end{aligned} \tag{16.49}$$

(4) Determination of $\partial^2\bar{n}_i/\partial\bar{X}_i\partial\bar{R}_{i-1}$:

$\partial^2\bar{n}_i/\partial\bar{X}_i\partial\bar{R}_{i-1}$ can be obtained by differentiating Eq. (7.61) of Appendix 1 in Chap. 7 with respect to $\bar{X}_i$, to give

$$\frac{\partial^2\bar{n}_i}{\partial\bar{X}_i\partial\bar{R}_{i-1}} = \begin{bmatrix} \partial^2 n_{ix}/\partial\bar{X}_i\partial\bar{R}_{i-1} \\ \partial^2 n_{iy}/\partial\bar{X}_i\partial\bar{R}_{i-1} \\ \partial^2 n_{iz}/\partial\bar{X}_i\partial\bar{R}_{i-1} \\ \bar{0} \end{bmatrix} = \frac{\partial({}^0\bar{A}_i)}{\partial\bar{X}_i}\frac{\partial({}^i\bar{n}_i)}{\partial\bar{R}_{i-1}} + {}^0\bar{A}_i\frac{\partial^2({}^i\bar{n}_i)}{\partial\bar{X}_i\partial\bar{R}_{i-1}}, \tag{16.50}$$

where $\partial^2({}^i\bar{n}_i)/\partial\bar{X}_i\partial\bar{R}_{i-1}$ is obtained by differentiating Eq. (7.62) of Appendix 1 in Chap. 7, to give

$$\begin{aligned}\frac{\partial^2({}^i\bar{n}_i)}{\partial\bar{X}_i\partial\bar{R}_{i-1}} &= s_i\begin{bmatrix} -S\beta_i C\alpha_i \\ -S\beta_i S\alpha_i \\ C\beta_i \\ 0 \end{bmatrix}\frac{\partial^2\beta_i}{\partial\bar{X}_i\partial\bar{R}_{i-1}} + s_i\begin{bmatrix} -C\beta_i S\alpha_i \\ C\beta_i C\alpha_i \\ 0 \\ 0 \end{bmatrix}\frac{\partial^2\alpha_i}{\partial\bar{X}_i\partial\bar{R}_{i-1}} \\ &+ s_i\left(\begin{bmatrix} -C\beta_i C\alpha_i \\ -C\beta_i S\alpha_i \\ -S\beta_i \\ 0 \end{bmatrix}\frac{\partial\beta_i}{\partial\bar{X}_i} + \begin{bmatrix} S\beta_i S\alpha_i \\ -S\beta_i C\alpha_i \\ 0 \\ 0 \end{bmatrix}\frac{\partial\alpha_i}{\partial\bar{X}_i}\right)\frac{\partial\beta_i}{\partial\bar{R}_{i-1}} \\ &+ s_i\left(\begin{bmatrix} S\beta_i S\alpha_i \\ -S\beta_i C\alpha_i \\ 0 \\ 0 \end{bmatrix}\frac{\partial\beta_i}{\partial\bar{X}_i} + \begin{bmatrix} -C\beta_i C\alpha_i \\ -C\beta_i S\alpha_i \\ 0 \\ 0 \end{bmatrix}\frac{\partial\alpha_i}{\partial\bar{X}_i}\right)\frac{\partial\alpha_i}{\partial\bar{R}_{i-1}}.\end{aligned} \tag{16.51}$$

(5) Determination of $\partial^2(C\theta_i)/\partial\bar{X}_i\partial\bar{R}_{i-1}$:

$\partial^2(C\theta_i)/\partial\bar{X}_i\partial\bar{R}_{i-1}$ can be computed from Eq. (7.63) of Appendix 1 in Chap. 7 as

$$\begin{aligned}\frac{\partial^2(C\theta_i)}{\partial\bar{X}_i\partial\bar{R}_{i-1}} &= -\left(\ell_{i-1x}\frac{\partial^2 n_{ix}}{\partial\bar{X}_i\partial\bar{R}_{i-1}} + \ell_{i-1y}\frac{\partial^2 n_{iy}}{\partial\bar{X}_i\partial\bar{R}_{i-1}} + \ell_{i-1z}\frac{\partial^2 n_{iz}}{\partial\bar{X}_i\partial\bar{R}_{i-1}}\right) \\ &- \left(\frac{\partial\ell_{i-1x}}{\partial\bar{R}_{i-1}}\frac{\partial n_{ix}}{\partial\bar{X}_i} + \frac{\partial\ell_{i-1y}}{\partial\bar{R}_{i-1}}\frac{\partial n_{iy}}{\partial\bar{X}_i} + \frac{\partial\ell_{i-1z}}{\partial\bar{R}_{i-1}}\frac{\partial n_{iz}}{\partial\bar{X}_i}\right),\end{aligned} \tag{16.52}$$

where the three components of $\partial\bar{n}_i/\partial\bar{X}_i$ in Eq. (16.52) are given in Eq. (7.78) of Appendix 2 in Chap. 7, while the three components of $\partial^2\bar{n}_i/\partial\bar{X}_i\partial\bar{R}_{i-1}$ are given in Eq. (16.50) of this appendix.

Reference

1. Lin PD (2014) The derivative matrices of a skew-ray for spherical boundary surfaces and their applications in system analysis and design. Appl Opt 53:3085–3100

Chapter 17
Hessian Matrix of Boundary Variable Vector $\bar{X}_i$ with Respect to System Variable Vector $\bar{X}_{sys}$

As discussed in Chap. 8, the system variable vector $\bar{X}_{sys}$ for a general optical problem may not simply be the collection of all the components of the constituent boundary variable vectors $\bar{X}_i$ (i = 0 to i = n). As a result, computing the Hessian matrix $d^2\bar{X}_i/d\bar{X}_{sys}^2$ of the boundary variable vector $\bar{X}_i$ with respect to the system variable vector $\bar{X}_{sys}$ is highly challenging. In the following discussions, the notation $d^2\bar{X}_i/d\bar{X}_{sys}^2 = H_i(u, v, w)$ is used to indicate the Hessian matrix of the uth boundary variable in $\bar{X}_i$ with respect to the vth and wth variables in $\bar{X}_{sys}$, respectively. In analyzing and designing practical optical systems, it is necessary to determine the Hessian matrix of three different types of boundary variable vector, namely the source variable vector $\bar{X}_0$ given in Eq. (2.4), the boundary variable vector $\bar{X}_i$ of a spherical boundary surface given in Eq. (2.30), and the boundary variable vector $\bar{X}_i$ of a flat boundary surface given in Eq. (2.47).

17.1 Hessian Matrix $\partial^2\bar{X}_0/\partial\bar{X}_{sys}^2$ of Source Ray

The Jacobian matrix of source variable vector (Eq. 2.4)

$$\bar{X}_0 = \begin{bmatrix} P_{0x} & P_{0y} & P_{0z} & \alpha_0 & \beta_0 \end{bmatrix}^T$$

P.D. Lin, *Advanced Geometrical Optics*, Progress in Optical Science and Photonics 4, DOI 10.1007/978-981-10-2299-9_17

with respect to system variable vector $\bar{X}_{sys}$ is given in Eq. (8.1), with the non-zero components given in Eq. (8.2). Equations (8.2), (8.3), (8.4) and (8.5) yield the following expressions:

$$\frac{\partial^2 \bar{X}_0}{\partial \bar{X}_{sys} \partial \bar{X}_0} = \bar{0}_{5 \times q_{sys} \times 5}, \tag{17.1}$$

$$\frac{\partial^2 \bar{X}_0}{\partial \bar{X}_{sys} \partial \bar{X}_\xi} = \bar{0}_{5 \times q_{sys} \times q_\xi}, \tag{17.2}$$

$$\frac{\partial^2 \bar{X}_0}{\partial \bar{X}_{sys} \partial \bar{X}_R} = \bar{0}_{5 \times q_{sys} \times q_R}, \tag{17.3}$$

$$\frac{\partial^2 \bar{X}_0}{\partial \bar{X}_{sys} \partial \bar{X}_{rest}} = \bar{0}_{5 \times q_{sys} \times q_{rest}}. \tag{17.4}$$

Therefore, the Hessian matrix $d^2\bar{X}_0/d\bar{X}_{sys}^2$ of source ray vector $\bar{X}_0$ with respect to $\bar{X}_{sys}$ is a null matrix, i.e.,

$$\frac{d^2 \bar{X}_0}{d\bar{X}_{sys}^2} = H_0(u, v, w) = \bar{0}_{5 \times q_{sys} \times q_{sys}}. \tag{17.5}$$

17.2 Hessian Matrix $\partial^2\bar{X}_i/\partial\bar{X}_{sys}^2$ for Flat Boundary Surface

The Jacobian matrix of flat boundary variable vector (Eq. 2.47)

$$\bar{X}_i = [J_{ix} \quad J_{iy} \quad J_{iz} \quad e_i \quad \xi_{i-1} \quad \xi_i]^T$$

with respect to system variable vector $\bar{X}_{sys}$ is given in Eq. (8.6). The Hessian matrix $d^2\bar{X}_i/d\bar{X}_{sys}^2$ for a flat boundary surface comprises twelve sub-matrices, namely

$$\frac{d^2\bar{X}_i}{d\bar{X}_{sys}^2} = H_i(u,v,w) = \begin{bmatrix} \partial^2 J_{ix}/\partial \bar{X}_{sys}^2 \\ \partial^2 J_{iy}/\partial \bar{X}_{sys}^2 \\ \partial^2 J_{iz}/\partial \bar{X}_{sys}^2 \\ \partial^2 e_i/\partial \bar{X}_{sys}^2 \\ \partial^2 \xi_{i-1}/\partial \bar{X}_{sys}^2 \\ \partial^2 \xi_i/\partial \bar{X}_{sys}^2 \end{bmatrix}$$
$$= \begin{bmatrix} \frac{\partial^2(J_{ix},J_{iy},J_{iz})}{\partial \bar{X}_{sys}\partial \bar{X}_0} & \frac{\partial^2(J_{ix},J_{iy},J_{iz})}{\partial \bar{X}_{sys}\partial \bar{X}_\xi} & \frac{\partial^2(J_{ix},J_{iy},J_{iz})}{\partial \bar{X}_{sys}\partial \bar{X}_R} & \frac{\partial^2(J_{ix},J_{iy},J_{iz})}{\partial \bar{X}_{sys}\partial \bar{X}_{rest}} \\ \frac{\partial^2 e_i}{\partial \bar{X}_{sys}\partial \bar{X}_0} & \frac{\partial^2 e_i}{\partial \bar{X}_{sys}\partial \bar{X}_\xi} & \frac{\partial^2 e_i}{\partial \bar{X}_{sys}\partial \bar{X}_R} & \frac{\partial^2 e_i}{\partial \bar{X}_{sys}\partial \bar{X}_{rest}} \\ \frac{\partial^2(\xi_{i-1},\xi_i)}{\partial \bar{X}_{sys}\partial \bar{X}_0} & \frac{\partial^2(\xi_{i-1},\xi_i)}{\partial \bar{X}_{sys}\partial \bar{X}_\xi} & \frac{\partial^2(\xi_{i-1},\xi_i)}{\partial \bar{X}_{sys}\partial \bar{X}_R} & \frac{\partial^2(\xi_{i-1},\xi_i)}{\partial \bar{X}_{sys}\partial \bar{X}_{rest}} \end{bmatrix}. \tag{17.6}$$

Equation (17.6) can be evaluated via the following three-step procedure [1]:

(1) Initialize all components of $H_i(u,v,w)$ to zero.
(2) Determine $H_i(u,w,v)$ by differentiating $J_i(u,v)$ with respect to the wth component of system variable vector $\bar{X}_{sys}$. Since the $J_i(u,v)$ terms of Eqs. (8.7)–(8.17) are all constants, the following results are obtained:

$$\frac{\partial^2(J_{ix}, J_{iy}, J_{iz})}{\partial \bar{X}_{sys}\partial \bar{X}_0} = \bar{0}_{3\times q_{sys}\times 5}, \tag{17.7}$$

$$\frac{\partial^2 e_i}{\partial \bar{X}_{sys}\partial \bar{X}_0} = \bar{0}_{1\times q_{sys}\times 5}, \tag{17.8}$$

$$\frac{\partial^2(\xi_{i-1}, \xi_i)}{\partial \bar{X}_{sys}\partial \bar{X}_0} = \bar{0}_{2\times q_{sys}\times 5}, \tag{17.9}$$

$$\frac{\partial^2(J_{ix}, J_{iy}, J_{iz})}{\partial \bar{X}_{sys}\partial \bar{X}_\xi} = \bar{0}_{3\times q_{sys}\times q_\xi}, \tag{17.10}$$

$$\frac{\partial^2 e_i}{\partial \bar{X}_{sys}\partial \bar{X}_\xi} = \bar{0}_{1\times q_{sys}\times q_\xi}, \tag{17.11}$$

$$\frac{\partial^2(\xi_{i-1}, \xi_i)}{\partial \bar{X}_{sys}\partial \bar{X}_\xi} = \bar{0}_{2\times q_{sys}\times q_\xi}, \tag{17.12}$$

$$\frac{\partial^2(J_{ix}, J_{iy}, J_{iz})}{\partial \bar{X}_{sys}\partial \bar{X}_R} = \bar{0}_{3\times q_{sys}\times q_R}, \tag{17.13}$$

$$\frac{\partial^2 e_i}{\partial \bar{X}_{sys} \partial \bar{X}_R} = \bar{0}_{1\times q_{sys}\times q_R}, \tag{17.14}$$

$$\frac{\partial^2 (\xi_{i-1}, \xi_i)}{\partial \bar{X}_{sys} \partial \bar{X}_R} = \bar{0}_{2\times q_{sys}\times q_R}, \tag{17.15}$$

$$\frac{\partial^2 (\xi_{i-1}, \xi_i)}{\partial \bar{X}_{sys} \partial \bar{X}_{rest}} = \bar{0}_{2\times q_{sys}\times q_{rest}}. \tag{17.16}$$

(3) To determine $d^2(J_{ix}, J_{iy}, J_{iz})/d\bar{X}_{sys}^2$ and $d^2 e_i/d\bar{X}_{sys}^2$, it is first necessary to obtain $d^2(\bar{r}_i)/d\bar{X}_{sys}^2$, i.e., the second-order derivative matrix of $\bar{r}_i$ with respect to variables x_v and x_w in $\bar{X}_{sys}$. The required matrix can be obtained by further differentiating Eq. (8.24) with respect to $\bar{X}_{sys}$, to give

$$\frac{d^2(\bar{r}_i)}{d\bar{X}_{sys}^2} = \frac{d^2({}^0\bar{A}'_{ej})}{d\bar{X}_{sys}^2}({}^{ej}\bar{r}_i) + \frac{d({}^0\bar{A}'_{ej})}{d\bar{X}_{sys}}\frac{d({}^{ej}\bar{r}_i)}{d\underline{\bar{X}_{sys}}} + \frac{d({}^0\bar{A}'_{ej})}{d\underline{\bar{X}_{sys}}}\frac{d({}^{ej}\bar{r}_i)}{d\bar{X}_{sys}} + ({}^0\bar{A}'_{ej})\frac{d^2({}^{ej}\bar{r}_i)}{d\bar{X}_{sys}^2}. \tag{17.17}$$

The detailed derivations of $d({}^0\bar{A}'_{ej})/d\bar{X}_{sys}$ and $d({}^{ej}\bar{r}_i)/d\bar{X}_{sys}$ are presented in Eqs. (8.77) and (8.83) of Appendix 3 in Chap. 8. Meanwhile, $d^2({}^0\bar{A}'_{ej})/d\bar{X}_{sys}^2$ and $d^2({}^{ej}\bar{r}_i)/d\bar{X}_{sys}^2$ are determined using Eqs. (17.62) and (17.66) in Appendix 3 of this chapter.

Example 17.1 Consider the generic prism shown in Fig. 8.2 by setting j = 1. All of the components in matrix $d^2\bar{X}_i/d\bar{X}_{sys}^2 = H_i(u,v,w)$ are equal to zero if ${}^0\bar{A}_{e1} = \bar{I}_{4\times 4}$ and $d^2({}^{e1}\bar{r}_i)/d\bar{X}_{sys}^2 = \bar{0}$ (i = 1 to i = 3). Consequently, the required derivatives of ${}^0\bar{A}_{e1}$ in Eq. (17.17) with j = 1 are all zero matrices.

Example 17.2 Consider the corner-cube mirror system shown in Fig. 3.19. The Hessian matrix $d^2\bar{R}_3/d\bar{X}_{sys}^2$ of the system illustrates the cross-coupling effects of $\bar{X}_0$, $\bar{X}_{e1}$ and $\bar{X}_i$ (i = 1 to i = 3) (given in Example 3.17) on the exit ray. As discussed Sect. 1.10, the Hessian matrix is essentially a 3-D table of values, and is therefore not easily presented in a conventional 2-D table format. However, the graphical representation suggested by Olson [2] provides a good intuitive understanding, as shown in Fig. 17.1. The results presented in Fig. 17.1 show that none of the variables in the system vector $\bar{X}_{e1}$ have a cross-coupling effect on the exit ray $\bar{\ell}_3$. Thus, provided that the flat boundary surfaces are perfectly fabricated, the corner-cube mirror preserves a retro-reflective property irrespective of its alignment with respect to the world coordinate frame $(xyz)_0$.

Fig. 17.1 Use of Hessian matrix $d^2\bar{R}_3/d\bar{X}_{sys}^2$ in determining cross-coupling effects of system variable vector on unit directional vector ℓ_{3z} of exit ray in corner-cube mirror system shown in Fig. 3.19

(a)

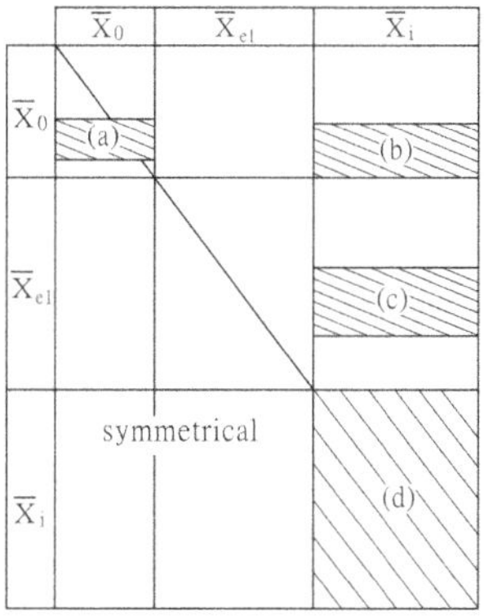

	$\bar{X}_0$	$\bar{X}_{e1}$	$\bar{X}_i$
$\bar{X}_0$	(a)		(b)
$\bar{X}_{e1}$			(c)
$\bar{X}_i$	symmetrical		(d)

	P_{0x}	P_{0y}	P_{0z}	α_0	β_0
β_0					-0.71

(b)

	J_{1x}	J_{1y}	e_1	J_{2x}	J_{2z}	e_2	J_{3y}	J_{3z}	e_3
α_0	-0.71	1.22			1.22			-0.71	
β_0	1.22	0.71			0.71			1.22	

(c)

	J_{1x}	J_{1y}	e_1	J_{2x}	J_{2z}	e_2	J_{3y}	J_{3z}	e_3
ω_{e1x}				1.22			1.22		
ω_{e1y}				0.71			0.71		
ω_{e1z}	0.71	-1.22			-1.22			0.71	

(d)

	J_{1x}	J_{1y}	e_1	J_{2x}	J_{2z}	e_2	J_{3y}	J_{3z}	e_3
J_{1x}	-2.83							-2.83	
J_{1y}		-2.83			-2.83				
e_1									
J_{2x}									
J_{2z}				1.22	-2.83			-1.41	
e_2			symmetrical						
J_{3y}								-0.71	
J_{3z}								-2.83	
e_3									

17.3 Design of Optical Systems Possessing Only Flat Boundary Surfaces

In this section, the validity of the proposed Hessian matrix as a tool for the design of optical systems consisting of only flat boundary surfaces is demonstrated using the two surface tracking systems shown in Figs. 17.2 and 17.3, respectively [3].

Example 17.3 As shown in Fig. 17.2, the tracking system comprises one laser diode and two first-surface-mirrors. The laser diode emits a source ray $\bar{R}_0 = \begin{bmatrix} \bar{P}_0 & \bar{\ell}_0 \end{bmatrix}^T$ in order to track the target point $\bar{P}_{3/\text{target}} = \begin{bmatrix} P_{3x/\text{target}} & P_{3y/\text{target}} & P_{3z/\text{target}} & 1 \end{bmatrix}^T$ on a 3-D surface by means of two intermediate reflection processes at mirrors $\bar{r}_1$ and $\bar{r}_2$, respectively. Note that the boundary surface of the two mirrors are given as

$$\bar{r}_1 = \begin{bmatrix} J_{1x} & J_{1y} & J_{1z} & e_1 \end{bmatrix}^T = \begin{bmatrix} -0.17299 & -0.01513 & 0.98481 & 0 \end{bmatrix}^T,$$

$$\bar{r}_2 = \begin{bmatrix} J_{2x} & J_{2y} & J_{2z} & e_2 \end{bmatrix}^T = \begin{bmatrix} 0.20679 & 0.96297 & -0.17299 & 0 \end{bmatrix}^T.$$

Note also that $P_{3z/\text{target}}$ is a function of $P_{3x/\text{target}}$ and $P_{3y/\text{target}}$ for a 3-D surface. Consequently, two design variables (denoted as $\bar{X}_{sys} = \begin{bmatrix} x_1 & x_2 \end{bmatrix}^T$) are sufficient to track $\bar{P}_{3/\text{target}}$.

In designing the tracking system shown in Fig. 17.2, the first step is to formulate the problem as an optimization problem with the following merit function:

$$\bar{\Phi} = \frac{1}{2}\left[(P_{3x} - P_{3x/\text{target}})^2 + (P_{3y} - P_{3y/\text{target}})^2\right].$$

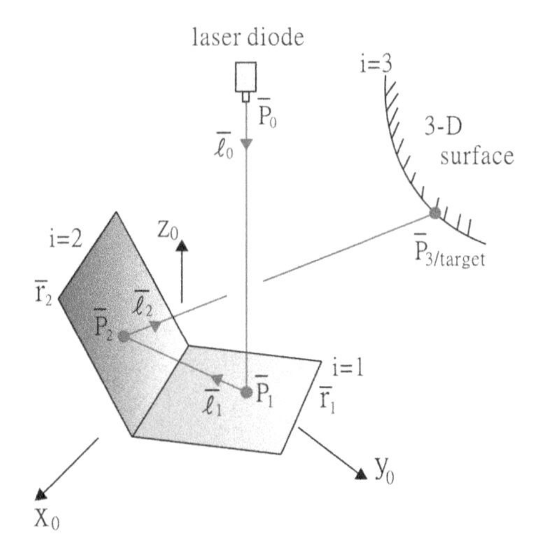

Fig. 17.2 General laser surface tracking system

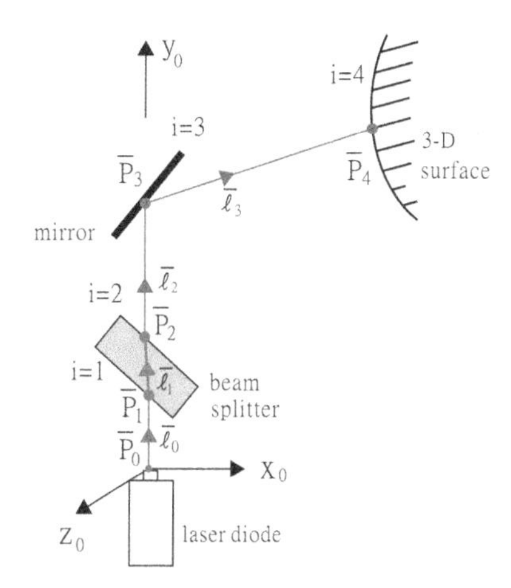

Fig. 17.3 Leica LTD 500 laser tracking system

It is seen that the merit function is an implicit function of $\bar{X}_{sys}$. In addition, it is noted that the desired solutions of $P_{3x} = P_{3x/target}$ and $P_{3y} = P_{3y/target}$ are obtained when the $\bar{X}_{sys}$ value is found such that $\bar{\Phi} = 0$.

In the following discussions, the validity of the Hessian matrix as a design tool is demonstrated using two common optimization approaches, namely the steepest-descent method and the classical Newton method [4]. Note that the step size is set to unity in both methods. The steepest-descent method is one of the simplest and most commonly used optimization methods since the search direction $\Delta\bar{X}_{sys}$ is evaluated simply by taking the Jacobian of the merit function (Eq. 1.75). By contrast, in the classical Newton method, the search direction $\Delta\bar{X}_{sys}$ is evaluated in terms of both the Jacobian matrix and the Hessian matrix (see Eq. 1.76). For the merit function defined above, the Jacobian and Hessian matrices are given respectively as

$$\frac{d\bar{\Phi}}{d\bar{X}_{sys}} = (P_{3x} - P_{3x/target})\frac{dP_{3x}}{d\bar{X}_{sys}} + (P_{3y} - P_{3y/target})\frac{dP_{3y}}{d\bar{X}_{sys}},$$

$$\frac{d^2\Phi}{d\bar{X}^2_{sys}} = (P_{3x} - P_{3x/target})\frac{d^2P_{3x}}{d\bar{X}^2_{sys}} + (P_{3y} - P_{3y/target})\frac{d^2P_{3y}}{d\bar{X}^2_{sys}} + \frac{dP_{3x}}{d\bar{X}_{\underline{sys}}}\frac{dP_{3x}}{d\bar{X}_{sys}} + \frac{dP_{3y}}{d\bar{X}_{\underline{sys}}}\frac{dP_{3y}}{d\bar{X}_{sys}}.$$

In the following discussions, two alternative designs (designated as Design#1 and Design#2, respectively) are considered for tracking the target point $\bar{P}_{3/target} = \begin{bmatrix} -300 & 200 & P_{3z/target} & 1 \end{bmatrix}^T$ in Fig. 17.2.

Design #1 The laser diode is mounted on two mutually perpendicular slides such that the emitted laser ray has the form

Table 17.1 Merit function and system variable values as function of number of iterations for Design #1 laser tracking system given use of steepest-descent optimization method

Iteration counter	$\bar{\Phi}$	X_1 (mm)	X_2 (mm)
1	$>10^{10}$	55250	55250
10	456400	995.118	1508.098
20	86.651	102.706	184.764
30	0.016	90.392	166.541
36	0.000	90.234	166.306

$$\bar{R}_0 = \begin{bmatrix} x_1 & x_2 & 500 & C(-45^\circ)C(210^\circ) & C(-45^\circ)S(210^\circ) & S(-45^\circ) \end{bmatrix}^T.$$

In other words, the laser diode is mounted on a horizontal plane at a height of 500 mm above the x_0y_0 plane with its unit directional vector $\bar{\ell}_0$ fixed at $\alpha_0 = 120^\circ$ and $\beta_0 = -45^\circ$ (see Eq. 2.3). Table 17.1 lists the values of the merit function $\bar{\Phi}$ (and corresponding values of x_1 and x_2) as a function of the iteration counter k given an initial guess of the system vector as $\bar{X}_{sys} = \begin{bmatrix} 55250 & 55250 \end{bmatrix}^T$ and the use of the steepest-descent optimization method. It is observed that the merit function has a value of $\bar{\Phi} \cong 0$ after k = 36 iterations. By contrast, the classical Newton optimization method converges to the desired solution in just one iteration; irrespective of the initial guess made for $\bar{X}_{sys}$. In other words, the classical Newton optimization method yields a highly effective solution for the considered design problem when implemented using the search direction given in Eq. (1.76).

Design #2. The laser diode is fixed at $\bar{P}_0 = \begin{bmatrix} 100 & 200 & 500 & 1 \end{bmatrix}^T$ with a unit directional vector of

$$\bar{\ell}_0 = \begin{bmatrix} Cx_2C(90^\circ + x_1) & Cx_2S(90^\circ + x_1) & Sx_2 & 0 \end{bmatrix}^T,$$

where angles x_1 and x_2 are set by two servomotors. In this particular example, the steepest-descent method requires significant computational effort to converge. Table 17.2 gives the value of the merit function $\bar{\Phi}$ as a function of the number of iterations when using the classical Newton optimization method with an initial guess of $\bar{X}_{sys} = \begin{bmatrix} 130^\circ & -55^\circ \end{bmatrix}^T$. It is seen that the merit function converges to $\bar{\Phi} = 0$ after just eight iterations. In other words, the effectiveness of the classical Newton method based on the Jacobian and Hessian matrices of $\bar{\Phi}$ is once again confirmed.

Example 17.4 The Leica LTD 500 laser tracking system [5] is designed for the precise position measurement of large or distant objects and is used for such applications as aligning aircraft wings or bridge girders during assembly. The tracking system comprises two servomotors driving a mirror; with one motor rotating the mirror horizontally and the other orienting the mirror vertically. Together, the two motors are capable of directing the laser beam at a target retro-reflector (see Fig. 17.3). Angle encoders on the two motors provide the horizontal and vertical angles of the laser beam, while an interferometer (not shown in

Table 17.2 Merit function and system variable values as function of number of iterations for Design #2 laser tracking system given use of classical Newton optimization method

Iteration counter	$\bar{\Phi}$	X_1	X_2
1	14304.380	130.000°	−55.000°
3	25587.429	127.769°	−28.928°
5	161.329	122.400°	−42.425°
8	0.000	122.474°	−43.654°

Fig. 17.3) indicates the distance to the retro-reflector. The vectors of the source ray $\bar{R}_0$, two flat boundary surfaces of the beam-splitter (i.e., $\bar{r}_1$ and $\bar{r}_2$), and driven mirror $\bar{r}_3$ are given respectively as

$$\bar{R}_0 = \begin{bmatrix} 0 & 0 & 0 & 0 & 1 & 0 \end{bmatrix}^T,$$
$$\bar{r}_1 = \begin{bmatrix} J_{1x} & J_{1y} & J_{1z} & e_1 \end{bmatrix}^T = \begin{bmatrix} 1/\sqrt{2} & 1/\sqrt{2} & 0 & -35.35534 \end{bmatrix}^T,$$
$$\bar{r}_2 = \begin{bmatrix} J_{2x} & J_{2y} & J_{2z} & e_2 \end{bmatrix}^T = \begin{bmatrix} 1/\sqrt{2} & 1/\sqrt{2} & 0 & -37.35534 \end{bmatrix}^T,$$
$$\bar{r}_3 = \begin{bmatrix} J_{3x} & J_{3y} & J_{3z} & e_3 \end{bmatrix}^T = \begin{bmatrix} C(x_2)C(x_1) & C(x_2)S(x_1) & S(x_2) & -100C(x_2)S(x_1) \end{bmatrix}^T.$$

Assume that the laser tracking system is to measure target point $\bar{P}_{4/target} = \begin{bmatrix} 20 & 130 & P_{4z/target} & 1 \end{bmatrix}^T$ on the 3-D surface shown in Fig. 17.3. Thus, the merit function used to determine the optimal system variable vector $\bar{X}_{sys} = \begin{bmatrix} x_1 & x_2 \end{bmatrix}^T$ has the form

$$\bar{\Phi} = \frac{1}{2}\left[(P_{4x} - 20)^2 + (P_{4y} - 130)^2\right].$$

As in the previous example, the optimization problem can be solved using both the steepest-descent method and the classical Newton method. In both methods, let the step size be set to unity and the initial guess be taken as $\bar{X}_{sys} = \begin{bmatrix} 87° & 3° \end{bmatrix}^T$. It can be shown that the steepest-descent method fails to converge within 500 iterations. By contrast, the classical Newton method converges to a solution of $\bar{X}_{sys} = \begin{bmatrix} 46.846° & -46.340° \end{bmatrix}^T$ after 30 iterations.

17.4 Hessian Matrix $\partial^2\bar{X}_i/\partial\bar{X}_{sys}^2$ for Spherical Boundary Surface

Let $d^2\bar{X}_i/d\bar{X}_{sys}^2 = [H_i(u, v, w)]$ denote the Hessian matrix of the uth variable in the boundary variable vector (Eq. 2.30)

$$\bar{X}_i = \begin{bmatrix} t_{ix} & t_{iy} & t_{iz} & \omega_{ix} & \omega_{iy} & \omega_{iz} & \xi_{i-1} & \xi_i & R_i \end{bmatrix}^T$$

of a spherical boundary surface with respect to the vth and wth variables in the system variable vector $\bar{X}_{sys}$, respectively. Note that $d^2\bar{X}_i/d\bar{X}_{sys}^2$ is a $9 \times q_{sys} \times q_{sys}$ matrix and can be partitioned into the following sub-matrices:

$$\frac{d^2\bar{X}_i}{d\bar{X}_{sys}^2} = H_i(u,v,w) = \begin{bmatrix} \partial^2 t_{ix}/\partial\bar{X}_{sys}^2 \\ \partial^2 t_{iy}/\partial\bar{X}_{sys}^2 \\ \partial^2 t_{iz}/\partial\bar{X}_{sys}^2 \\ \partial^2 \omega_{ix}/\partial\bar{X}_{sys}^2 \\ \partial^2 \omega_{iy}/\partial\bar{X}_{sys}^2 \\ \partial^2 \omega_{iz}/\partial\bar{X}_{sys}^2 \\ \partial^2 \xi_{i-1}/\partial\bar{X}_{sys}^2 \\ \partial^2 \xi_i/\partial\bar{X}_{sys}^2 \\ \partial^2 R_i/\partial\bar{X}_{sys}^2 \end{bmatrix}$$
$$= \begin{bmatrix} \frac{\partial^2(t_{ix},t_{iy},t_{iz})}{\partial\bar{X}_{sys}\partial\bar{X}_0} & \frac{\partial^2(t_{ix},t_{iy},t_{iz})}{\partial\bar{X}_{sys}\partial\bar{X}_\xi} & \frac{\partial^2(t_{ix},t_{iy},t_{iz})}{\partial\bar{X}_{sys}\partial\bar{X}_R} & \frac{\partial^2(t_{ix},t_{iy},t_{iz})}{\partial\bar{X}_{sys}\partial\bar{X}_{rest}} \\ \frac{\partial^2(\omega_{ix},\omega_{iy},\omega_{iz})}{\partial\bar{X}_{sys}\partial\bar{X}_0} & \frac{\partial^2(\omega_{ix},\omega_{iy},\omega_{iz})}{\partial\bar{X}_{sys}\partial\bar{X}_\xi} & \frac{\partial^2(\omega_{ix},\omega_{iy},\omega_{iz})}{\partial\bar{X}_{sys}\partial\bar{X}_R} & \frac{\partial^2(\omega_{ix},\omega_{iy},\omega_{iz})}{\partial\bar{X}_{sys}\partial\bar{X}_{rest}} \\ \frac{\partial^2(\xi_{i-1},\xi_i)}{\partial\bar{X}_{sys}\partial\bar{X}_0} & \frac{\partial^2(\xi_{i-1},\xi_i)}{\partial\bar{X}_{sys}\partial\bar{X}_\xi} & \frac{\partial^2(\xi_{i-1},\xi_i)}{\partial\bar{X}_{sys}\partial\bar{X}_R} & \frac{\partial^2(\xi_{i-1},\xi_i)}{\partial\bar{X}_{sys}\partial\bar{X}_{rest}} \\ \frac{\partial^2 R_i}{\partial\bar{X}_{sys}\partial\bar{X}_0} & \frac{\partial^2 R_i}{\partial\bar{X}_{sys}\partial\bar{X}_\xi} & \frac{\partial^2 R_i}{\partial\bar{X}_{sys}\partial\bar{X}_R} & \frac{\partial^2 R_i}{\partial\bar{X}_{sys}\partial\bar{X}_{rest}} \end{bmatrix} \tag{17.18}$$

The Hessian matrix $H_i(u,v,w)$ can be determined via the following three-step procedure [6]:

(1) Initialize the components of $H_i(u,v,w)$ to zero.
(2) Determine $H_i(u,v,w)$ by differentiating $J_i(u,v)$ with respect to the wth component of vector $\bar{X}_{sys}$. Since the $J_i(u,v)$ terms of Eqs. (8.26)–(8.39) are all constants, the following results of Eq. (17.18) are obtained:

$$\frac{\partial^2(t_{ix}, t_{iy}, t_{iz})}{\partial\bar{X}_{sys}\partial\bar{X}_0} = \bar{0}_{3\times q_{sys}\times 5}, \tag{17.19}$$

$$\frac{\partial^2(\omega_{ix}, \omega_{iy}, \omega_{iz})}{\partial\bar{X}_{sys}\partial\bar{X}_0} = \bar{0}_{3\times q_{sys}\times 5}, \tag{17.20}$$

$$\frac{\partial^2(\xi_{i-1}, \xi_i)}{\partial\bar{X}_{sys}\partial\bar{X}_0} = \bar{0}_{2\times q_{sys}\times 5}, \tag{17.21}$$

$$\frac{\partial^2 R_i}{\partial \bar{X}_{sys} \partial \bar{X}_0} = \bar{0}_{1 \times q_{sys} \times 5}, \tag{17.22}$$

$$\frac{\partial^2 (t_{ix}, t_{iy}, t_{iz})}{\partial \bar{X}_{sys} \partial \bar{X}_\xi} = \bar{0}_{3 \times q_{sys} \times q_\xi}, \tag{17.23}$$

$$\frac{\partial^2 (\omega_{ix}, \omega_{iy}, \omega_{iz})}{\partial \bar{X}_{sys} \partial \bar{X}_\xi} = \bar{0}_{3 \times q_{sys} \times q_\xi}, \tag{17.24}$$

$$\frac{\partial^2 (\xi_{i-1}, \xi_i)}{\partial \bar{X}_{sys} \partial \bar{X}_\xi} = \bar{0}_{2 \times q_{sys} \times q_\xi}, \tag{17.25}$$

$$\frac{\partial^2 R_i}{\partial \bar{X}_{sys} \partial \bar{X}_\xi} = \bar{0}_{1 \times q_{sys} \times q_\xi}, \tag{17.26}$$

$$\frac{\partial^2 (\omega_{ix}, \omega_{iy}, \omega_{iz})}{\partial \bar{X}_{sys} \partial \bar{X}_R} = \bar{0}_{3 \times q_{sys} \times q_R}, \tag{17.27}$$

$$\frac{\partial^2 (\xi_{i-1}, \xi_i)}{\partial \bar{X}_{sys} \partial \bar{X}_R} = \bar{0}_{2 \times q_{sys} \times q_R}, \tag{17.28}$$

$$\frac{\partial^2 R_i}{\partial \bar{X}_{sys} \partial \bar{X}_R} = \bar{0}_{1 \times q_{sys} \times q_R}, \tag{17.29}$$

$$\frac{\partial^2 (\xi_{i-1}, \xi_i)}{\partial \bar{X}_{sys} \partial \bar{X}_{rest}} = \bar{0}_{2 \times q_{sys} \times q_{rest}}, \tag{17.30}$$

$$\frac{\partial^2 R_i}{\partial \bar{X}_{sys} \partial \bar{X}_{rest}} = \bar{0}_{1 \times q_{sys} \times q_{rest}}, \tag{17.31}$$

(3) Determine $\partial^2(t_{ix}, t_{iy}, t_{iz})/\partial \bar{X}_{sys} \partial \bar{X}_R$, $\partial^2(t_{ix}, t_{iy}, t_{iz})/\partial \bar{X}_{sys} \partial \bar{X}_{rest}$ and $\partial^2(\omega_{ix}, \omega_{iy}, \omega_{iz})/\partial \bar{X}_{sys} \partial \bar{X}_{rest}$.

To establish these terms, it is first necessary to obtain $d^2({}^0A_i)/d\bar{X}_{sys}^2 = [\partial^2({}^0A_i)/\partial x_w \partial x_v]$ by differentiating Eq. (8.44), to give

$$\frac{d^2({}^0\bar{A}_i)}{d\bar{X}_{sys}^2} = \frac{d^2({}^0\bar{A}_{ej})}{d\bar{X}_{sys}^2} {}^{ej}\bar{A}_i + {}^0\bar{A}_{ej} \frac{d^2({}^{ej}\bar{A}_i)}{d\bar{X}_{sys}^2} + \frac{d({}^0\bar{A}_{ej})}{d\bar{X}_{sys}} \frac{d({}^{ej}\bar{A}_i)}{d\bar{X}_{\underline{sys}}} + \frac{d({}^0\bar{A}_{ej})}{d\bar{X}_{\underline{sys}}} \frac{d({}^{ej}\bar{A}_i)}{d\bar{X}_{sys}}, \tag{17.32}$$

where $d^2({}^0\bar{A}_{ej})/d\bar{X}_{sys}^2$ and $d^2({}^{ej}\bar{A}_i)/d\bar{X}_{sys}^2$ are the second-order derivative matrices of ${}^0\bar{A}_{ej}$ and ${}^{ej}\bar{A}_i$, respectively, with respect to $\bar{X}_{sys}$. Note that terms $d^2({}^0\bar{A}_{ej})/d\bar{X}_{sys}^2$

Table 17.3 $H_i(9, v, w)$ for system shown in Fig. 3.12. Note that radius R_i is not a component of $\bar{X}_i$ for i = 2, 3, 5 and 6 since $\bar{r}_2$, $\bar{r}_3$, $\bar{r}_5$ and $\bar{r}_6$ in Fig. 3.12 are flat boundary surfaces

H_i (u, v, w) \ $\bar{X}_i$	i = 1	i = 2	i = 3	i = 4	i = 5	i = 6	= 7
$\bar{X}_{sys}$	R_i u = 9			R_i u = 9			R_i u = 9
$\bar{X}_0$	0			0			0
$\bar{X}_\xi$	0			0			0
$\bar{X}_R$	0			0			0
$\bar{X}_{rest}$	0			0			0

and $d^2({}^{ej}\bar{A}_i)/d\bar{X}_{sys}^2$ in Eq. (17.32) are determined in Appendixes 1 and 2 in this chapter, respectively. The Hessian matrix of the six pose variables (i.e., t_{ix}, t_{iy}, t_{iz}, ω_{ix}, ω_{iy}, and ω_{iz}) with respect to the system variable vector $\bar{X}_{sys}$ (i.e., $d^2t_{ix}/d\bar{X}_{sys}^2$, $d^2t_{iy}/d\bar{X}_{sys}^2$, $d^2t_{iz}/d\bar{X}_{sys}^2$, $d^2\omega_{ix}/d\bar{X}_{sys}^2$, $d^2\omega_{iy}/d\bar{X}_{sys}^2$ and $d^2\omega_{iz}/d\bar{X}_{sys}^2$) can then be determined using the method shown in Appendix 4 in this chapter.

Example 17.5 The Hessian matrices $H_i(9, v, w)$ (i = 1 to i = 7 and v = 1 to v = 17) for the cat's eye system shown in Fig. 3.12 can be obtained by further differentiating the equations given in Table 8.4 with respect to the system variable vector $\bar{X}_{sys}$. The corresponding results are listed in Table 17.3. The non-zero values are as follows: $H_1(9, 10, 10) = 0$, $H_4(9, 11, 11) = 0$ and $H_7(9, 10, 10) = 0$.

Example 17.6 The Hessian matrices $H_i(9, v, w)$ (i = 1 to i = 5 and v = 1 to v = 12) for the retro-reflector system shown in Fig. 3.13 are presented in Table 17.4 (see also Table 8.5).

Example 17.7 The non-zero values of $\partial^2(t_{ix}, t_{iy}, t_{iz})/\partial\bar{X}_R^2$, $\partial^2(t_{ix}, t_{iy}, t_{iz})/\partial\bar{X}_{rest}^2$ and $\partial^2(\omega_{ix}, \omega_{iy}, \omega_{iz})/\partial\bar{X}_{rest}^2$ for the system shown in Fig. 3.12 are given as follows:

Table 17.4 $H_i(9, v, w)$ for system shown in Fig. 3.13

H_i (u, v, w) \ $\bar{X}_i$	i = 1	i = 2	i = 3	i = 4	i = 5
$\bar{X}_{sys}$	R_i u = 9	R_i u = 9	R_i u = 9	R_i u = 9	R_i u = 9
$\bar{X}_0$	0	0	0	0	0
$\bar{X}_\xi$	0	0	0	0	0
$\bar{X}_R$	0	0	0	0	0
$\bar{X}_{rest}$	0	0	0	0	0

$$
\begin{aligned}
H_3(1,15,16) = {} & H_3(1,16,15) = -H_3(2,15,15) = -H_3(2,17,17) = \\
& H_3(4,12,17) = -H_3(4,14,15) = -H_3(4,15,14) = \\
& H_3(4,17,12) = H_5(1,15,16) = H_5(1,16,15) = \\
& -H_5(2,15,15) = -H_5(2,17,17) = \\
& H_5(4,12,17) = -H_5(4,14,15) = -H_5(4,15,14) = \\
& H_5(4,17,12) = 1,\ H_3(4,17,17) = \\
& H_3(4,15,15) = H_5(4,15,15) = H_5(4,17,17) = 0.1.
\end{aligned}
$$

Example 17.8 The non-zero values of $\partial^2(t_{ix}, t_{iy}, t_{iz})/\partial\bar{X}_R^2$, $\partial^2(t_{ix}, t_{iy}, t_{iz})/\partial\bar{X}_{rest}^2$ and $\partial^2(\omega_{ix}, \omega_{iy}, \omega_{iz})/\partial\bar{X}_{rest}^2$ for the system shown in Fig. 3.13 are given as

$$
H_1(2,8,8) = H_2(2,9,9) = H_4(2,9,9) = H_5(2,8,8) = H_3(2,10,10) = 0.
$$

17.5 Design of Retro-reflectors

In this section, the validity of the proposed methodology is further demonstrated using the two retro-reflector systems shown in Figs. 3.12 and 3.13, respectively. In laser interferometer systems, retro-reflectors are required to provide a high degree of parallelism of the reflected ray; particularly when performing large-range measurements. To achieve this goal, the proposed methodology is used to optimize the system variable vector of the two retro-reflectors in such a way as to minimize the divergence angle of the reflected rays. It is shown numerically in the following examples that the tuned retro-reflectors greatly improve the parallelism of the reflected rays.

Ideally a retro-reflector should return the source rays $\bar{R}_0$ which are incident parallel to the optical axis along exactly the same direction along which they originally came (i.e., $\bar{\ell}_{n/ideal} = \begin{bmatrix} 0 & -1 & 0 & 0 \end{bmatrix}^T$). In designing the retro-reflectors shown in Figs. 3.12 and 3.13, the first step is to formulate the design task as an optimization problem with the following merit function:

$$
\bar{\Phi} = 10^6 \left\{ \sum_1^3 0.5 \left[\ell_{nx}^2 + (\ell_{ny} + 1)^2 + \ell_{nz}^2 \right] \right\}. \tag{17.33}
$$

Equation (17.33) is used here to compute the deviation between the actual unit directional vector $\bar{\ell}_n = \begin{bmatrix} \ell_{nx} & \ell_{ny} & \ell_{nz} & 0 \end{bmatrix}^T$ and the ideal unit directional vector $\bar{\ell}_{n/ideal} = \begin{bmatrix} 0 & -1 & 0 & 0 \end{bmatrix}^T$ of three rays originating from three different point sources (i.e., $\bar{P}_0 = \begin{bmatrix} 0 & -10 & P_{0z/1} & 1 \end{bmatrix}^T$, $\bar{P}_0 = \begin{bmatrix} 0 & -10 & P_{0z/2} & 1 \end{bmatrix}^T$ and $\bar{P}_0 = \begin{bmatrix} 0 & -10 & P_{0z/3} & 1 \end{bmatrix}^T$) each with a unit directional vector

$\bar{\ell}_0 = [0 \quad 1 \quad 0 \quad 0]^T$. It is noted that $\bar{\Phi}$ is deliberately scaled up by a factor of 10^6 to decrease the number of leading zeros in the calculated result. Clearly, the numerical value of $\bar{\Phi}$ depends on the system variable vector $\bar{X}_{sys}$. In addition, the ideal solution of $\bar{\ell}_{n/ideal} = [0 \quad -1 \quad 0 \quad 0]^T$ is obtained when the value of the system variable vector $\bar{X}_{sys}$ is found such that $\bar{\Phi} = 0$.

For the merit function defined in Eq. (17.33), the Jacobian and Hessian matrices are given respectively as

$$\frac{d\bar{\Phi}}{d\bar{X}_{sys}} = 10^6 \sum_1^3 \left\{ \ell_{nx} \frac{d\ell_{nx}}{d\bar{X}_{sys}} + (\ell_{ny} + 1) \frac{d\ell_{ny}}{d\bar{X}_{sys}} + \ell_{nz} \frac{d\ell_{nz}}{d\bar{X}_{sys}} \right\}, \tag{17.34}$$

and

$$\begin{aligned} \frac{d^2\bar{\Phi}}{d\bar{X}^2_{sys}} = 10^6 \sum_1^3 \Bigg\{ & \ell_{nx} \frac{d^2\ell_{nx}}{d\bar{X}^2_{sys}} + (\ell_{ny} + 1) \frac{d^2\ell_{ny}}{d\bar{X}^2_{sys}} + \ell_{nz} \frac{d^2\ell_{nz}}{d\bar{X}^2_{sys}} \\ & + \frac{d\ell_{nx}}{d\underline{\bar{X}_{sys}}} \frac{d\ell_{nx}}{d\bar{X}_{sys}} + \frac{d\ell_{ny}}{d\underline{\bar{X}_{sys}}} \frac{d\ell_{ny}}{d\bar{X}_{sys}} + \frac{d\ell_{nz}}{d\underline{\bar{X}_{sys}}} \frac{d\ell_{nz}}{d\bar{X}_{sys}} \Bigg\}. \end{aligned} \tag{17.35}$$

Example 17.9 Cat's eye retro-reflector composed of two hemispheres.

In this example, the use of the Hessian matrix as a design tool is demonstrated using three different optimization methods (denoted as Methods 1–3, respectively). Method 3 is the classical Newton's method (p. 460 of [4]), and needs both the Jacobian matrix and the Hessian matrix to find the minima of the merit function $\bar{\Phi}$. By contrast, Methods 1 and 2 are both quasi-Newton methods (p. 466 of [4]), and use only the Jacobian matrix of the merit function to generate an approximation of the Hessian matrix. The difference between the two methods lies in the technique used to obtain the Jacobian matrix. Specifically, Method 1 uses the finite difference (FD) approach to estimate the matrix, whereas Method 2 uses the user-supplied Jacobian matrix given by Eq. (17.34). Notably, the Hessian matrix is not computed in either method since it is simply estimated by analyzing successive Jacobian matrices.

Assuming that the retro-reflector shown in Fig. 3.12 is axis-symmetrical, misalignment errors need not be considered (i.e., $t_{e2x} = t_{e2z} = \omega_{e2x} = \omega_{e2y} = \omega_{e2z} = 0$). Furthermore, the refractive indices ξ_{air} and ξ_{glue} are given as $\xi_{air} = 1$ and $\xi_{glue} = 1.2$, respectively. If the radii (i.e., R_1 and R_4) of the two hemispheres are tuned simultaneously in the optimization process, the volume of the retro-reflector may be scaled up when minimizing the merit function Φ. This is clearly impractical for a real-world design problem, in which the volume of the retro-reflector system is fixed. Thus, one of the length dimensions of $\bar{X}_{sys}$ must be prescribed in advance (e.g., $R_4 = -60$ in the present case). As a result, the system variable vector reduces to $\bar{X}_{sys} = [\xi_{e1} \quad \xi_{e2} \quad R_1 \quad v_2]^T$.

Table 17.5 Variation of merit function $\bar{\Phi}$ and divergence angle Γ with iteration number when using three different optimization methods to design retro-reflector system shown in Fig. 3.12

Iteration counter	Method 1		Method 2		Method 3	
	Φ	Γ	Φ	Γ	Φ	Γ
1	3840.574	4.44°	3840.574	4.44°	3840.574	4.44°
2	1363.109	2.23°	1363.109	2.23°	100.226	0.80°
3	494.771	1.25°	494.771	1.25°	28.738	0.24°
7	15.439	0.18°	15.438	0.18°	13.644	0.17°
12	15.439	0.18°	15.400	0.18°		
30			13.644	0.17°		
Time consumed (sec.)	0.02		1.75		9.65	
No. of iteration	12		30		7	
No. of function evaluation	72		65		7	

Table 17.5 lists the values of the merit function $\bar{\Phi}$ for three rays with heights $P_{0z/1} = 5$, $P_{0z/2} = 10$ and $P_{0z/3} = 15$, respectively, given an initial guess, upper bound and lower bound of the system variable vector as follows:

$$\bar{X}_{sys/initial\ guess} = \begin{bmatrix} 1.6 & 1.7 & 1/30 & 0 \end{bmatrix}^T,$$
$$\bar{X}_{sys/upper} = \begin{bmatrix} 1.8 & 1.8 & 1/20 & 0.1 \end{bmatrix}^T,$$
$$\bar{X}_{sys/lower} = \begin{bmatrix} 1.2 & 1.2 & 1/40 & 0 \end{bmatrix}^T.$$

Method 3, which uses the proposed method to determine the Jacobian and Hessian matrices, converges to a final solution of $\bar{X}_{sys} = \begin{bmatrix} 1.6879 & 1.8 & 1/40 & 0 \end{bmatrix}^T$ after just 7 iterations (see Table 17.5). Moreover, the merit function $\bar{\Phi}$ and divergence angle Γ (i.e., the angle between $\bar{\ell}_7$ and $\bar{\ell}_0$ with height $P_{0z/3}$) are reduced from $\bar{\Phi} = 3840.574$ to $\bar{\Phi} = 13.644$ and $\Gamma = 4.44°$ to $\Gamma = 0.17°$, respectively. By contrast, Method 1, which uses the FD method to determine the Jacobian matrix, fails to converge to the optimal solution even after 72 function calls. Similarly, Method 2 fails to converge even after 65 function calls. In other words, the proposed methodology provides a more effective means of determining the search direction than the two quasi-Newton methods. However, it should be noted that Method 3 is significantly slower than Methods 1 or 2 due to the time required to compute the Jacobian and Hessian matrices defined in Eqs. (17.34) and (17.35), respectively.

Example 17.10 Cat's eye retro-reflector with lens-mirror configuration.

Table 17.6 Variation of merit function $\bar{\Phi}$ and divergence angle Γ with iteration number when using two different optimization methods to design the retro-reflector system shown in Fig. 3.12

Iteration counter	Steepest descent method	Classical Newton method	
		$\bar{\Phi}$	Γ
1	712.8158	712.8158	1.906°
2		343.6206	1.312°
5		49.9745	0.516°
10		5.6896	0.159°
12		14.2015	0.296°
17		0.7179	0.006°
No. of iteration		17	

For the retro-reflector system shown in Fig. 3.13, the primary lens is separated from the secondary mirror by a distance v_2. In optimizing the system, variables ξ_{air} and q_{e1} are set as $\xi_{air} = 1$ and $q_{e1} = 6.5$, respectively, in order to avoid scaling the system dimensions up or down when tuning the variables. The system variable vector is thus reduced to $\bar{X}_{sys} = [\,\xi_{e1} \quad R_1 \quad R_2 \quad R_3 \quad v_2\,]^T$. Let the retro-reflector system be optimized using two optimization methods, namely the steepest-descent method (see Eq. (1.75) and the classical Newton method (see Eq. (1.76)), with a unit step size in both cases. Table 17.6 shows the values of the merit function $\bar{\Phi}$ evaluated for three rays with heights of $P_{0z/1} = 5$, $P_{0z/2} = 15$ and $P_{0z/3} = 25$, respectively, given the use of the two different methods and an initial guess of $\bar{X}_{sys/initial\ guess} = [\,1.5 \quad 182 \quad -182 \quad -200 \quad 200\,]^T$ in both cases. It is seen that for the steepest-descent method, the optimization procedure terminates at the second iteration since the outcome of Eq. (1.75) leads to impossible results. By contrast, the classical Newton method converges to a final solution of

$$\bar{X}_{sys} = [\,1.578 \quad 103.9846 \quad -246.7281 \quad -3.2694 \quad 122.6542\,]^T.$$

The literature contains many search direction determination methods [4]. Gradient-based methods (which use the Jacobian and/ or Hessian matrices to determine the search direction) include the steepest-descent method, the classical Newton method, the conjugate gradient method, and the modified Newton method [4]. In practice, it is extremely difficult to predict which search direction determination method is best suited to any particular optical design optimization task. However, the examples presented above show that irrespective of the method chosen (i.e., the steepest-descent method or the classic Newton method in the present case), it is important that the Jacobian and/ or Hessian matrices of the merit function are known.

Appendix 1

It is noted from Eqs. (17.17) and (17.32) that to compute $d^2(\bar{r}_i)/d\bar{X}_{sys}^2$ and $d^2(^0\bar{A}_i)/d\bar{X}_{sys}^2$, it is first necessary to determine $d^2(^0\bar{A}_{ej})/d\bar{X}_{sys}^2$ by differentiating $d(^0\bar{A}_{ej})/d\bar{X}_{sys}$ in Eq. (8.48) of Appendix 1 in Chap. 8 with respect to $\bar{X}_{sys}$, i.e.,

$$\frac{d^2(^0\bar{A}_{ej})}{d\bar{X}_{sys}^2} = \left[\frac{\partial^2(^0\bar{A}_{ej})}{\partial x_w \partial x_v}\right] = \begin{bmatrix} \partial^2 I_{ejx}/\partial x_w\partial x_v & \partial^2 J_{ejx}/\partial x_w\partial x_v & \partial^2 K_{ejx}/\partial x_w\partial x_v & \partial^2 t_{ejx}/\partial x_w\partial x_v \\ \partial^2 I_{ejy}/\partial x_w\partial x_v & \partial^2 J_{ejy}/\partial x_w\partial x_v & \partial^2 K_{ejy}/\partial x_w\partial x_v & \partial^2 t_{ejy}/\partial x_w\partial x_v \\ \partial^2 I_{ejz}/\partial x_w\partial x_v & \partial^2 J_{ejz}/\partial x_w\partial x_v & \partial^2 K_{ejz}/\partial x_w\partial x_v & \partial^2 t_{ejz}/\partial x_w\partial x_v \\ 0 & 0 & 0 & 0 \end{bmatrix}. \tag{17.36}$$

Let ω_{ejx}, ω_{ejy} and ω_{ejz} be expressed as linear combinations of the components of $\bar{X}_{sys}$, as shown in Eqs. (8.45), (8.46) and (8.47) of Appendix 1 in Chap. 8. Therefore, the components of Eq. (17.36) of this appendix are given by

$$\begin{aligned}\frac{\partial^2 I_{ejx}}{\partial x_w \partial x_v} = &-b_v\left(b_w C\omega_{ejy} C\omega_{ejz} - c_w S\omega_{ejy} S\omega_{ejz}\right)\\ &- c_v\left(-b_w S\omega_{ejy} S\omega_{ejz} + c_w C\omega_{ejy} C\omega_{ejz}\right),\end{aligned} \tag{17.37}$$

$$\begin{aligned}\frac{\partial^2 I_{ejy}}{\partial x_w \partial x_v} = &-b_v\left(b_w C\omega_{ejy} S\omega_{ejz} + c_w S\omega_{ejy} C\omega_{ejz}\right)\\ &- c_v\left(b_w S\omega_{ejy} C\omega_{ejz} + c_w C\omega_{ejy} S\omega_{ejz}\right),\end{aligned} \tag{17.38}$$

$$\frac{\partial^2 I_{ejz}}{\partial x_w \partial x_v} = b_v b_w S\omega_{ejy}, \tag{17.39}$$

$$\begin{aligned}\frac{\partial^2 J_{ejx}}{\partial x_w \partial x_v} = &\ a_v\left(-a_w S\omega_{ejx} S\omega_{ejy} C\omega_{ejz} + b_w C\omega_{ejx} C\omega_{ejy} C\omega_{ejz} - c_w C\omega_{ejx} S\omega_{ejy} S\omega_{ejz}\right)\\ &+ b_v\left(a_w C\omega_{ejx} C\omega_{ejy} C\omega_{ejz} - b_w S\omega_{ejx} S\omega_{ejy} C\omega_{ejz} - c_w S\omega_{ejx} C\omega_{ejy} S\omega_{ejz}\right)\\ &- c_v\left(a_w C\omega_{ejx} S\omega_{ejy} S\omega_{ejz} + b_w S\omega_{ejx} C\omega_{ejy} S\omega_{ejz} + c_w S\omega_{ejx} S\omega_{ejy} C\omega_{ejz}\right)\\ &+ a_v\left(a_w C\omega_{ejx} S\omega_{ejz} + c_w S\omega_{ejx} C\omega_{ejz}\right) - c_v\left(-a_w S\omega_{ejx} C\omega_{ejz} - c_w C\omega_{ejx} S\omega_{ejz}\right),\end{aligned} \tag{17.40}$$

$$\frac{\partial^2 J_{ejy}}{\partial x_w \partial x_v} = a_v\left(-a_w S\omega_{ejx} S\omega_{ejy} S\omega_{ejz} + b_w C\omega_{ejx} C\omega_{ejy} S\omega_{ejz} + c_w C\omega_{ejx} S\omega_{ejy} C\omega_{ejz}\right)$$
$$+ b_v\left(a_w C\omega_{ejx} C\omega_{ejy} S\omega_{ejz} - b_w S\omega_{ejx} S\omega_{ejy} S\omega_{ejz} + c_w S\omega_{ejx} C\omega_{ejy} C\omega_{ejz}\right)$$
$$+ c_v\left(a_w C\omega_{ejx} S\omega_{ejy} C\omega_{ejz} + b_w S\omega_{ejx} C\omega_{ejy} C\omega_{ejz} - c_w S\omega_{ejx} S\omega_{ejy} S\omega_{ejz}\right)$$
$$- a_v\left(a_w C\omega_{ejx} C\omega_{ejz} - c_w S\omega_{ejx} S\omega_{ejz}\right) - c_v\left(-a_w S\omega_{ejx} S\omega_{ejz} + c_w C\omega_{ejx} C\omega_{ejz}\right), \tag{17.41}$$

$$\frac{\partial^2 J_{ejz}}{\partial x_w \partial x_v} = a_v\left(-a_w S\omega_{ejx} C\omega_{ejy} - b_w C\omega_{ejx} S\omega_{ejy}\right) - b_v\left(a_w C\omega_{ejx} S\omega_{ejy} + b_w S\omega_{ejx} C\omega_{ejy}\right), \tag{17.42}$$

$$\frac{\partial^2 K_{ejx}}{\partial x_w \partial x_v} = -a_v\left(a_w C\omega_{ejx} S\omega_{ejy} C\omega_{ejz} + b_w S\omega_{ejx} C\omega_{ejy} C\omega_{ejz} - c_w S\omega_{ejx} S\omega_{ejy} S\omega_{ejz}\right)$$
$$+ b_v\left(-a_w S\omega_{ejx} C\omega_{ejy} C\omega_{ejz} - b_w C\omega_{ejx} S\omega_{ejy} C\omega_{ejz} - c_w C\omega_{ejx} C\omega_{ejy} S\omega_{ejz}\right)$$
$$- c_v\left(-a_w S\omega_{ejx} S\omega_{ejy} S\omega_{ejz} + b_w C\omega_{ejx} C\omega_{ejy} S\omega_{ejz} + c_w C\omega_{ejx} S\omega_{ejy} C\omega_{ejz}\right)$$
$$- a_v\left(-a_w S\omega_{ejx} S\omega_{ejz} + c_w C\omega_{ejx} C\omega_{ejz}\right) - c_v\left(a_w C\omega_{ejx} C\omega_{ejz} - c_w S\omega_{ejx} S\omega_{ejz}\right), \tag{17.43}$$

$$\frac{\partial^2 K_{ejy}}{\partial x_w \partial x_v} = -a_v\left(a_w C\omega_{ejx} S\omega_{ejy} S\omega_{ejz} + b_w S\omega_{ejx} C\omega_{ejy} S\omega_{ejz} + c_w S\omega_{ejx} S\omega_{ejy} C\omega_{ejz}\right)$$
$$+ b_v\left(-a_w S\omega_{ejx} C\omega_{ejy} S\omega_{ejz} - b_w C\omega_{ejx} S\omega_{ejy} S\omega_{ejz} + c_w C\omega_{ejx} C\omega_{ejy} C\omega_{ejz}\right)$$
$$+ c_v\left(-a_w S\omega_{ejx} S\omega_{ejy} C\omega_{ejz} + b_w C\omega_{ejx} C\omega_{ejy} C\omega_{ejz} - c_w C\omega_{ejx} S\omega_{ejy} S\omega_{ejz}\right)$$
$$- a_v\left(-a_w S\omega_{ejx} C\omega_{ejz} - c_w C\omega_{ejx} S\omega_{ejz}\right) + c_v\left(a_w C\omega_{ejx} S\omega_{ejz} + c_w S\omega_{ejx} C\omega_{ejz}\right), \tag{17.44}$$

$$\frac{\partial^2 K_{ejz}}{\partial x_w \partial x_v} = -a_v\left(a_w C\omega_{ejx} C\omega_{ejy} - b_w S\omega_{ejx} S\omega_{ejy}\right) - b_v\left(-a_w S\omega_{ejx} S\omega_{ejy} + b_w C\omega_{ejx} C\omega_{ejy}\right), \tag{17.45}$$

$$\frac{\partial^2 t_{ejx}}{\partial x_w \partial x_v} = \frac{\partial^2 t_{ejx}}{\partial x_w \partial x_v}, \tag{17.46}$$

$$\frac{\partial^2 t_{ejy}}{\partial x_w \partial x_v} = \frac{\partial^2 t_{ejy}}{\partial x_w \partial x_v}, \tag{17.47}$$

$$\frac{\partial^2 t_{ejz}}{\partial x_w \partial x_v} = \frac{\partial^2 t_{ejz}}{\partial x_w \partial x_v}. \tag{17.48}$$

Note that Eqs. (17.46), (17.47) and (17.48) indicate that $d^2t_{ejx}/d\bar{X}_{sys}^2$, $d^2t_{ejy}/d\bar{X}_{sys}^2$ and $d^2t_{ejz}/d\bar{X}_{sys}^2$ can be obtained simply by differentiating their corresponding expressions (e.g., $\partial^2 t_{e1x}/\partial x_6^2 = \partial^2 t_{e1y}/\partial x_7^2 = \partial^2 t_{e1z}/\partial x_8^2 = 0$ in the example shown in Fig. 3.13).

Appendix 2

The components of the Jacobian matrix $d({}^{ej}\bar{A}_i)/d\bar{X}_{sys}$ are listed in Eqs. (8.65)–(8.73) of Appendix 2 in Chap. 8. The Hessian matrix $d^2({}^{ej}\bar{A}_i)/d\bar{X}_{sys}^2$ can be determined as

$$\frac{d^2({}^{ej}\bar{A}_i)}{d\bar{X}_{sys}^2} = \left[\frac{\partial^2({}^{ej}\bar{A}_i)}{\partial x_w \partial x_v}\right]_{4\times 4\times q_{sys}\times q_{sys}} = \begin{bmatrix} \partial^2({}^{ej}I_{ix})/\partial x_w\partial x_v & \partial^2({}^{ej}J_{ix})/\partial x_w\partial x_v & \partial^2({}^{ej}K_{ix})/\partial x_w\partial x_v & \partial^2({}^{ej}t_{ix})/\partial x_w\partial x_v \\ \partial^2({}^{ej}I_{iy})/\partial x_w\partial x_v & \partial^2({}^{ej}J_{iy})/\partial x_w\partial x_v & \partial^2({}^{ej}K_{iy})/\partial x_w\partial x_v & \partial^2({}^{ej}t_{iy})/\partial x_w\partial x_v \\ \partial^2({}^{ej}I_{iz})/\partial x_w\partial x_v & \partial^2({}^{ej}J_{iz})/\partial x_w\partial x_v & \partial^2({}^{ej}K_{iz})/\partial x_w\partial x_v & \partial^2({}^{ej}t_{iz})/\partial x_w\partial x_v \\ \bar{0} & \bar{0} & \bar{0} & \bar{0} \end{bmatrix}. \tag{17.49}$$

Let ${}^{ej}\omega_{ix}$, ${}^{ej}\omega_{iy}$ and ${}^{ej}\omega_{iz}$ be expressed as linear combinations of the components of $\bar{X}_{sys}$, as shown in Eqs. (8.61), (8.62) and (8.63) of Appendix 2 in Chap. 8. Therefore, the components of Eq. (17.49) of this appendix can be computed from Eq. (8.65)–(8.73) of Appendix 2 in Chap. 8 as

$$\begin{aligned}\frac{\partial^2({}^{ej}I_{ix})}{\partial x_w \partial x_v} = &-g_v\left(g_w C({}^{ej}\omega_{iy})C({}^{ej}\omega_{iz}) - h_w S({}^{ej}\omega_{iy})S({}^{ej}\omega_{iz})\right) \\ &- h_v\left(-g_w S({}^{ej}\omega_{iy})S({}^{ej}\omega_{iz}) + h_w C({}^{ej}\omega_{iy})C({}^{ej}\omega_{iz})\right),\end{aligned} \tag{17.50}$$

$$\begin{aligned}\frac{\partial^2({}^{ej}I_{iy})}{\partial x_w \partial x_v} = &-g_v\left(g_w C({}^{ej}\omega_{iy})S({}^{ej}\omega_{iz}) + h_w S({}^{ej}\omega_{iy})C({}^{ej}\omega_{iz})\right) \\ &+ h_v\left(-g_w S({}^{ej}\omega_{iy})C({}^{ej}\omega_{iz}) - h_w C({}^{ej}\omega_{iy})S({}^{ej}\omega_{iz})\right),\end{aligned} \tag{17.51}$$

$$\frac{\partial^2({}^{ej}I_{iz})}{\partial x_w \partial x_v} = g_v g_w S({}^{ej}\omega_{iy}), \tag{17.52}$$

$$\begin{aligned}\frac{\partial^2({}^{ej}J_{ix})}{\partial x_w \partial x_v} = &\rho_v\left(-\rho_w S({}^{ej}\omega_{ix})S({}^{ej}\omega_{iy})C({}^{ej}\omega_{iz}) + g_w C({}^{ej}\omega_{ix})C({}^{ej}\omega_{iy})C({}^{ej}\omega_{iz}) - h_w C({}^{ej}\omega_{ix})S({}^{ej}\omega_{iy})S({}^{ej}\omega_{iz})\right) \\ &+ g_v\left(\rho_w C({}^{ej}\omega_{ix})C({}^{ej}\omega_{iy})C({}^{ej}\omega_{iz}) - g_w S({}^{ej}\omega_{ix})S({}^{ej}\omega_{iy})C({}^{ej}\omega_{iz}) - h_w S({}^{ej}\omega_{ix})C({}^{ej}\omega_{iy})S({}^{ej}\omega_{iz})\right) \\ &- h_v\left(\rho_w C({}^{ej}\omega_{ix})S({}^{ej}\omega_{iy})S({}^{ej}\omega_{iz}) + g_w S({}^{ej}\omega_{ix})C({}^{ej}\omega_{iy})S({}^{ej}\omega_{iz}) + h_w S({}^{ej}\omega_{ix})S({}^{ej}\omega_{iy})C({}^{ej}\omega_{iz})\right) \\ &+ \rho_v\left(\rho_w C({}^{ej}\omega_{ix})S\omega_{ejz} + h_w S({}^{ej}\omega_{ix})C({}^{ej}\omega_{iz})\right) - h_v\left(-\rho_w S({}^{ej}\omega_{ix})C\omega_{ejz} - h_w C({}^{ej}\omega_{ix})S({}^{ej}\omega_{iz})\right),\end{aligned} \tag{17.53}$$

$$\begin{aligned}\frac{\partial^2(^{ej}J_{iy})}{\partial x_w \partial x_v} = {} & \rho_v\left(-\rho_w S(^{ej}\omega_{ix})S(^{ej}\omega_{iy})S(^{ej}\omega_{iz}) + g_w C(^{ej}\omega_{ix})C(^{ej}\omega_{iy})S(^{ej}\omega_{iz}) + h_w C(^{ej}\omega_{ix})S(^{ej}\omega_{iy})C(^{ej}\omega_{iz})\right)\\ & + g_v\left(\rho_w C(^{ej}\omega_{ix})C(^{ej}\omega_{iy})S(^{ej}\omega_{iz}) - g_w S(^{ej}\omega_{ix})S(^{ej}\omega_{iy})S(^{ej}\omega_{iz}) + h_w S(^{ej}\omega_{ix})C(^{ej}\omega_{iy})C(^{ej}\omega_{iz})\right)\\ & + h_v\left(\rho_w C(^{ej}\omega_{ix})S(^{ej}\omega_{iy})C(^{ej}\omega_{iz}) + g_w S(^{ej}\omega_{ix})C(^{ej}\omega_{iy})C(^{ej}\omega_{iz}) - h_w S(^{ej}\omega_{ix})S(^{ej}\omega_{iy})S(^{ej}\omega_{iz})\right)\\ & - \rho_v\left(\rho_w C(^{ej}\omega_{ix})C(^{ej}\omega_{iz}) - h_w S(^{ej}\omega_{ix})S(^{ej}\omega_{iz})\right) - h_v\left(-\rho_w S(^{ej}\omega_{ix})S(^{ej}\omega_{iz}) + h_w C(^{ej}\omega_{ix})C(^{ej}\omega_{iz})\right),\end{aligned} \tag{17.54}$$

$$\begin{aligned}\frac{\partial^2(^{ej}J_{iz})}{\partial x_w \partial x_v} = {} & \rho_v\left(-\rho_w S(^{ej}\omega_{ix})C(^{ej}\omega_{iy}) - g_w C(^{ej}\omega_{ix})S(^{ej}\omega_{iy})\right)\\ & - g_v\left(\rho_w C(^{ej}\omega_{ix})S(^{ej}\omega_{iy}) + g_w S(^{ej}\omega_{ix})C(^{ej}\omega_{iy})\right),\end{aligned} \tag{17.55}$$

$$\begin{aligned}\frac{\partial^2(^{ej}K_{ix})}{\partial x_w \partial x_v} = {} & -\rho_v\left(\rho_w C(^{ej}\omega_{ix})S(^{ej}\omega_{iy})C(^{ej}\omega_{iz}) + g_w S(^{ej}\omega_{ix})C(^{ej}\omega_{iy})C(^{ej}\omega_{iz}) - h_w S(^{ej}\omega_{ix})S(^{ej}\omega_{iy})S(^{ej}\omega_{iz})\right)\\ & + g_v\left(-\rho_w S(^{ej}\omega_{ix})C(^{ej}\omega_{iy})C(^{ej}\omega_{iz}) - g_w C(^{ej}\omega_{ix})S(^{ej}\omega_{iy})C(^{ej}\omega_{iz}) - h_w C(^{ej}\omega_{ix})C(^{ej}\omega_{iy})S(^{ej}\omega_{iz})\right)\\ & - h_v\left(-\rho_w S(^{ej}\omega_{ix})S(^{ej}\omega_{iy})S(^{ej}\omega_{iz}) + g_w C(^{ej}\omega_{ix})C(^{ej}\omega_{iy})S(^{ej}\omega_{iz}) + h_w C(^{ej}\omega_{ix})S(^{ej}\omega_{iy})C(^{ej}\omega_{iz})\right)\\ & - \rho_v\left(-\rho_w S(^{ej}\omega_{ix})S\omega_{ejz} + h_w C(^{ej}\omega_{ix})C(^{ej}\omega_{iz})\right) - h_v\left(\rho_w C(^{ej}\omega_{ix})C\omega_{ejz} - h_w S(^{ej}\omega_{ix})S(^{ej}\omega_{iz})\right),\end{aligned} \tag{17.56}$$

$$\begin{aligned}\frac{\partial^2(^{ej}K_{iy})}{\partial x_w \partial x_v} = {} & -\rho_v\left(\rho_w C(^{ej}\omega_{ix})S(^{ej}\omega_{iy})S(^{ej}\omega_{iz}) + g_w S(^{ej}\omega_{ix})C(^{ej}\omega_{iy})S(^{ej}\omega_{iz}) + h_w S(^{ej}\omega_{ix})S(^{ej}\omega_{iy})C(^{ej}\omega_{iz})\right)\\ & + g_v\left(-\rho_w S(^{ej}\omega_{ix})C(^{ej}\omega_{iy})S(^{ej}\omega_{iz}) - g_w C(^{ej}\omega_{ix})S(^{ej}\omega_{iy})S(^{ej}\omega_{iz}) + h_w C(^{ej}\omega_{ix})C(^{ej}\omega_{iy})C(^{ej}\omega_{iz})\right)\\ & + h_v\left(-\rho_w S(^{ej}\omega_{ix})S(^{ej}\omega_{iy})C(^{ej}\omega_{iz}) + g_w C(^{ej}\omega_{ix})C(^{ej}\omega_{iy})C(^{ej}\omega_{iz}) - h_w C(^{ej}\omega_{ix})S(^{ej}\omega_{iy})S(^{ej}\omega_{iz})\right)\\ & - \rho_v\left(-\rho_w S(^{ej}\omega_{ix})C(^{ej}\omega_{iz}) - h_w C(^{ej}\omega_{ix})S(^{ej}\omega_{iz})\right) + h_v\left(\rho_w C(^{ej}\omega_{ix})S(^{ej}\omega_{iz}) + h_w S(^{ej}\omega_{ix})C(^{ej}\omega_{iz})\right),\end{aligned} \tag{17.57}$$

$$\begin{aligned}\frac{\partial^2(^{ej}K_{iz})}{\partial x_w \partial x_v} = {} & -\rho_v\left(\rho_w C(^{ej}\omega_{ix})C(^{ej}\omega_{iy}) - g_w S(^{ej}\omega_{ix})S(^{ej}\omega_{iy})\right)\\ & - g_v\left(-\rho_w S(^{ej}\omega_{ix})S(^{ej}\omega_{iy}) + g_w C(^{ej}\omega_{ix})C(^{ej}\omega_{iy})\right),\end{aligned} \tag{17.58}$$

$$\frac{\partial^2(^{ej}t_{ix})}{\partial x_w \partial x_v} = \frac{\partial^2(^{ej}t_{ix})}{\partial x_w \partial x_v}, \tag{17.59}$$

$$\frac{\partial^2(^{ej}t_{iy})}{\partial x_w \partial x_v} = \frac{\partial^2(^{ej}t_{iy})}{\partial x_w \partial x_v}, \tag{17.60}$$

$$\frac{\partial^2(^{ej}t_{iz})}{\partial x_w \partial x_v} = \frac{\partial^2(^{ej}t_{iz})}{\partial x_w \partial x_v}. \tag{17.61}$$

Note that Eqs. (17.59), (17.60) and (17.61) indicate that $d^2(^{ej}t_{ix})/d\bar{X}_{sys}^2$, $d^2(^{ej}t_{iy})/d\bar{X}_{sys}^2$ and $d^2(^{ej}t_{iz})/d\bar{X}_{sys}^2$ can be obtained simply by differentiating their corresponding expressions. For example, $\partial^2(^{e1}t_{2y})/\partial x_{11}^2 = 0$, $\partial^2(^{e1}t_{2y})/\partial x_{25}^2 = 0$ and $\partial^2(^{e1}t_{2y})/\partial x_{12}^2 = 0$ for $^{e1}\bar{A}_2$ in Example 3.3.

Appendix 3

It is noted from Eq. (17.17) that to compute $d^2(\bar{r}_i)/d\bar{X}^2_{sys}$, it is first necessary to have the numerical values of ${}^0\bar{A}'_{ej}$ (the known pose matrix of an element in a given system), ${}^{ej}\bar{r}_i$ (the known expression of a flat boundary surface in a given system), $d({}^0\bar{A}'_{ej})/d\bar{X}_{sys}$ (Eq. (8.77) of Appendix 3 in Chap. 8), $d({}^{ej}\bar{r}_i)/d\bar{X}_{sys}$ (Eq. (8.83) of Appendix 3 in Chap. 8), $d^2({}^0\bar{A}'_{ej})/d\bar{X}^2_{sys}$ and $d^2({}^{ej}\bar{r}_i)/d\bar{X}^2_{sys}$. The two required terms, $d^2({}^0\bar{A}'_{ej})/d\bar{X}^2_{sys}$ and $d^2({}^{ej}\bar{r}_i)/d\bar{X}^2_{sys}$, are presented in the following.

(1) Determination of $d^2({}^0\bar{A}'_{ej})/d\bar{X}^2_{sys}$:

$d^2({}^0\bar{A}'_{ej})/d\bar{X}^2_{sys}$ can be obtained directly by differentiating Eq. (8.77) of Appendix 3 in Chap. 8, to give

$$\frac{d^2({}^0\bar{A}'_{ej})}{d\bar{X}^2_{sys}} = \begin{bmatrix} \partial^2 I_{ejx}/\partial x_w \partial x_v & \partial^2 J_{ejx}/\partial x_w \partial x_v & \partial^2 K_{ejx}/\partial x_w \partial x_v & \bar{0} \\ \partial^2 I_{ejy}/\partial x_w \partial x_v & \partial^2 J_{ejy}/\partial x_w \partial x_v & \partial^2 K_{ejy}/\partial x_w \partial x_v & \bar{0} \\ \partial^2 I_{ejz}/\partial x_w \partial x_v & \partial^2 J_{ejz}/\partial x_w \partial x_v & \partial^2 K_{ejz}/\partial x_w \partial x_v & \bar{0} \\ \partial^2 f_{ejx}/\partial x_w \partial x_v & \partial^2 f_{ejy}/\partial x_w \partial x_v & \partial^2 f_{ejz}/\partial x_w \partial x_v & \bar{0} \end{bmatrix}, \quad (17.62)$$

with

$$\begin{aligned} \frac{\partial^2 f_{ejx}}{\partial x_w \partial x_v} = &-\left(\frac{\partial^2 I_{ejx}}{\partial x_w \partial x_v} t_{ejx} + \frac{\partial^2 I_{ejy}}{\partial x_w \partial x_v} t_{ejy} + \frac{\partial^2 I_{ejz}}{\partial x_w \partial x_v} t_{ejz} + I_{ejx} \frac{\partial^2 t_{ejx}}{\partial x_w \partial x_v} + I_{ejy} \frac{\partial^2 t_{ejy}}{\partial x_w \partial x_v} + I_{ejz} \frac{\partial^2 t_{ejz}}{\partial x_w \partial x_v}\right), \\ &-\left(\frac{\partial I_{ejx}}{\partial x_v}\frac{\partial t_{ejx}}{\partial x_w} + \frac{\partial I_{ejy}}{\partial x_v}\frac{\partial t_{ejy}}{\partial x_w} + \frac{\partial I_{ejz}}{\partial x_v}\frac{\partial t_{ejz}}{\partial x_w} + \frac{\partial I_{ejx}}{\partial x_w}\frac{\partial t_{ejx}}{\partial x_v} + \frac{\partial I_{ejy}}{\partial x_w}\frac{\partial t_{ejy}}{\partial x_v} + \frac{\partial I_{ejz}}{\partial x_w}\frac{\partial t_{ejz}}{\partial x_v}\right) \end{aligned} \quad (17.63)$$

$$\begin{aligned} \frac{\partial^2 f_{ejy}}{\partial x_w \partial x_v} = &-\left(\frac{\partial^2 J_{ejx}}{\partial x_w \partial x_v} t_{ejx} + \frac{\partial^2 J_{ejy}}{\partial x_w \partial x_v} t_{ejy} + \frac{\partial^2 J_{ejz}}{\partial x_w \partial x_v} t_{ejz} + J_{ejx} \frac{\partial^2 t_{ejx}}{\partial x_w \partial x_v} + J_{ejy} \frac{\partial^2 t_{ejy}}{\partial x_w \partial x_v} + J_{ejz} \frac{\partial^2 t_{ejz}}{\partial x_w \partial x_v}\right), \\ &-\left(\frac{\partial J_{ejx}}{\partial x_v}\frac{\partial t_{ejx}}{\partial x_w} + \frac{\partial J_{ejy}}{\partial x_v}\frac{\partial t_{ejy}}{\partial x_w} + \frac{\partial J_{ejz}}{\partial x_v}\frac{\partial t_{ejz}}{\partial x_w} + \frac{\partial J_{ejx}}{\partial x_w}\frac{\partial t_{ejx}}{\partial x_v} + \frac{\partial J_{ejy}}{\partial x_w}\frac{\partial t_{ejy}}{\partial x_v} + \frac{\partial J_{ejz}}{\partial x_w}\frac{\partial t_{ejz}}{\partial x_v}\right) \end{aligned} \quad (17.64)$$

$$\begin{aligned} \frac{\partial^2 f_{ejz}}{\partial x_w \partial x_v} = &-\left(\frac{\partial^2 K_{ejx}}{\partial x_w \partial x_v} t_{ejx} + \frac{\partial^2 K_{ejy}}{\partial x_w \partial x_v} t_{ejy} + \frac{\partial^2 K_{ejz}}{\partial x_w \partial x_v} t_{ejz} + K_{ejx} \frac{\partial^2 t_{ejx}}{\partial x_w \partial x_v} + K_{ejy} \frac{\partial^2 t_{ejy}}{\partial x_w \partial x_v} + K_{ejz} \frac{\partial^2 t_{ejz}}{\partial x_w \partial x_v}\right), \\ &-\left(\frac{\partial K_{ejx}}{\partial x_v}\frac{\partial t_{ejx}}{\partial x_w} + \frac{\partial K_{ejy}}{\partial x_v}\frac{\partial t_{ejy}}{\partial x_w} + \frac{\partial K_{ejz}}{\partial x_v}\frac{\partial t_{ejz}}{\partial x_w} + \frac{\partial K_{ejx}}{\partial x_w}\frac{\partial t_{ejx}}{\partial x_v} + \frac{\partial K_{ejy}}{\partial x_w}\frac{\partial t_{ejy}}{\partial x_v} + \frac{\partial K_{ejz}}{\partial x_w}\frac{\partial t_{ejz}}{\partial x_v}\right) \end{aligned} \quad (17.65)$$

(2) Determination of $d^2({}^{ej}\bar{r}_i)/d\bar{X}^2_{sys}$:

$d^2({}^{ej}\bar{r}_i)/d\bar{X}^2_{sys}$ can be obtained by differentiating Eq. (8.83) of Appendix 3 in Chap. 8, to give

$$\frac{d^2({}^{ej}\bar{r}_i)}{d\bar{X}_{sys}^2} = \begin{bmatrix} d^2({}^{ej}J_{ix})/d\bar{X}_{sys}^2 \\ d^2({}^{ej}J_{iy})/d\bar{X}_{sys}^2 \\ d^2({}^{ej}J_{iz})/d\bar{X}_{sys}^2 \\ d^2({}^{ej}e_i)/d\bar{X}_{sys}^2 \end{bmatrix}, \tag{17.66}$$

where $d^2({}^{ej}J_{ix})/d\bar{X}_{sys}^2$, $d^2({}^{ej}J_{iy})/d\bar{X}_{sys}^2$, $d^2({}^{ej}J_{iz})/d\bar{X}_{sys}^2$ are given, respectively, by Eqs. (17.53), (17.54), (17.55) of Appendix 2 in this chapter. $d^2({}^{ej}e_i)/d\bar{X}_{sys}^2$ can be obtained directly by differentiating Eq. (8.84) of Appendix 3 in Chap. 8, to give

$$\begin{aligned} \frac{d^2({}^{ej}e_i)}{d\bar{X}_{sys}^2} = -\Bigg(& \frac{d^2({}^{ej}J_{ix})}{d\bar{X}_{sys}^2}({}^{ej}t_{ix}) + \frac{d({}^{ej}J_{ix})}{d\bar{X}_{sys}}\frac{d({}^{ej}t_{ix})}{d\bar{X}_{sys}} + \frac{d({}^{ej}J_{ix})}{d\bar{X}_{sys}}\frac{d({}^{ej}t_{ix})}{d\bar{X}_{sys}} + ({}^{ej}J_{ix})\frac{d^2({}^{ej}t_{ix})}{d\bar{X}_{sys}^2} \\ & + \frac{d({}^{ej}J_{iy})}{d\bar{X}_{sys}^2}({}^{ej}t_{iy}) + \frac{d({}^{ej}J_{iy})}{d\bar{X}_{sys}}\frac{d({}^{ej}t_{iy})}{d\bar{X}_{sys}} + \frac{d({}^{ej}J_{iy})}{d\bar{X}_{sys}}\frac{d({}^{ej}t_{iy})}{d\bar{X}_{sys}} + ({}^{ej}J_{iy})\frac{d^2({}^{ej}t_{iy})}{d\bar{X}_{sys}^2} \\ & + \frac{d({}^{ej}J_{iz})}{d\bar{X}_{sys}^2}({}^{ej}t_{iz}) + \frac{d({}^{ej}J_{iz})}{d\bar{X}_{sys}}\frac{d({}^{ej}t_{iz})}{d\bar{X}_{sys}} + \frac{d({}^{ej}J_{iz})}{d\bar{X}_{sys}}\frac{d({}^{ej}t_{iz})}{d\bar{X}_{sys}} + ({}^{ej}J_{iz})\frac{d^2({}^{ej}t_{iz})}{d\bar{X}_{sys}^2}\Bigg). \end{aligned} \tag{17.67}$$

Having obtained ${}^{0}\bar{A}_{ej}$, ${}^{ej}\bar{r}_i$, $d({}^{0}\bar{A}_{ej})/d\bar{X}_{sys}$, $d({}^{ej}\bar{r}_i)/d\bar{X}_{sys}$, $d^2({}^{0}\bar{A}'_{ej})/d\bar{X}_{sys}^2$, and $d^2({}^{ej}\bar{r}_i)/d\bar{X}_{sys}^2$, the Hessian matrix $d^2(\bar{r}_i)/d\bar{X}_{sys}^2$ can be computed numerically from Eq. (17.17).

Appendix 4

$d({}^{0}\bar{A}_i)/d\bar{X}_{sys}$ and $d^2({}^{0}\bar{A}_i)/d\bar{X}_{sys}^2$ can be computed from Eqs. (8.44) and (17.32), respectively, as

$$\frac{d({}^{0}\bar{A}_i)}{d\bar{X}_{sys}} = \begin{bmatrix} \partial I_{ix}/\partial x_v & \partial J_{ix}/\partial x_v & \partial K_{ix}/\partial x_v & \partial t_{ix}/\partial x_v \\ \partial I_{iy}/\partial x_v & \partial J_{iy}/\partial x_v & \partial K_{iy}/\partial x_v & \partial t_{iy}/\partial x_v \\ \partial I_{iz}/\partial x_v & \partial J_{iz}/\partial x_v & \partial K_{iz}/\partial x_v & \partial t_{iz}/\partial x_v \\ 0 & 0 & 0 & 0 \end{bmatrix}, \tag{17.68}$$

$$\frac{d^2({}^0\bar{A}_i)}{d\bar{X}_{sys}^2} = \begin{bmatrix} \partial I_{ix}^2/\partial x_w \partial x_v & \partial J_{ix}^2/\partial x_w \partial x_v & \partial K_{ix}^2/\partial x_w \partial x_v & \partial t_{ix}^2/\partial x_w \partial x_v \\ \partial I_{iy}^2/\partial x_w \partial x_v & \partial J_{iy}^2/\partial x_w \partial x_v & \partial K_{iy}^2/\partial x_w \partial x_v & \partial t_{iy}^2/\partial x_w \partial x_v \\ \partial I_{iz}^2/\partial x_w \partial x_v & \partial J_{iz}^2/\partial x_w \partial x_v & \partial K_{iz}^2/\partial x_w \partial x_v & \partial t_{iz}^2/\partial x_w \partial x_v \\ \bar{0} & \bar{0} & \bar{0} & \bar{0} \end{bmatrix}. \tag{17.69}$$

As described in the following, the Hessian matrices of the six pose variables (i.e., t_{ix}, t_{iy}, t_{iz}, ω_{ix}, ω_{iy} and ω_{iz}) for a spherical boundary surface $\bar{r}_i$ can be obtained by further differentiating the equations given in Appendix 4 in Chap. 8. For example, the Hessian matrix $d^2\omega_{iz}/d\bar{X}_{sys}^2 = [\partial^2\omega_{iz}/\partial x_w \partial x_v]$ can be obtained by differentiating Eq. (8.85) of Appendix 4 in Chap. 8, to give

$$\frac{\partial^2\omega_{iz}}{\partial x_w \partial x_v} = \frac{F\frac{\partial E}{\partial x_w} - E\frac{\partial F}{\partial x_w}}{F^2}, \tag{17.70}$$

where

$$\frac{\partial F}{\partial x_w} = 2\left(I_{ix}\frac{\partial I_{ix}}{\partial x_w} + I_{iy}\frac{\partial I_{iy}}{\partial x_w}\right), \tag{17.71}$$

$$\frac{\partial E}{\partial x_w} = \frac{\partial I_{ix}}{\partial x_w}\frac{\partial I_{iy}}{\partial x_v} + I_{ix}\frac{\partial^2 I_{iy}}{\partial x_w \partial x_v} - \frac{\partial I_{iy}}{\partial x_w}\frac{\partial I_{ix}}{\partial x_v} - I_{iy}\frac{\partial^2 I_{ix}}{\partial x_w \partial x_v}. \tag{17.72}$$

Furthermore, differentiating Eq. (8.88) of Appendix 4 in Chap. 8 yields

$$\frac{\partial^2\omega_{iy}}{\partial x_w \partial x_v} = \frac{L\left(\frac{\partial G}{\partial x_w} + \frac{\partial H}{\partial x_w}\right) - (G+H)\frac{\partial L}{\partial x_w}}{L^2}, \tag{17.73}$$

where

$$\frac{\partial L}{\partial x_w} = 2I_{iz}\frac{\partial I_{iz}}{\partial x_w} + 2(I_{ix}C\omega_{iz} + I_{iy}S\omega_{iz})\left[\left(\frac{\partial I_{ix}}{\partial x_w}C\omega_{iz} + \frac{\partial I_{iy}}{\partial x_w}S\omega_{iz}\right) + (-I_{ix}S\omega_{iz} + I_{iy}C\omega_{iz})\frac{\partial \omega_{iz}}{\partial x_w}\right]. \tag{17.74}$$

$$\begin{aligned} \frac{\partial G}{\partial x_w} = {} & \frac{\partial I_{iz}}{\partial x_w}\left[(-I_{ix}S\omega_{iz} + I_{iy}C\omega_{iz})\frac{\partial \omega_{iz}}{\partial x_v} + \left(\frac{\partial I_{ix}}{\partial x_v}C\omega_{iz} + \frac{\partial I_{iy}}{\partial x_v}S\omega_{iz}\right)\right] \\ & + I_{iz}\left[\left(-\frac{\partial I_{ix}}{\partial x_w}S\omega_{iz} + \frac{\partial I_{iy}}{\partial x_w}C\omega_{iz}\right)\frac{\partial \omega_{iz}}{\partial x_v} + (-I_{ix}C\omega_{iz} - I_{iy}S\omega_{iz})\frac{\partial \omega_{iz}}{\partial x_w}\frac{\partial \omega_{iz}}{\partial x_v}\right. \\ & + (-I_{ix}S\omega_{iz} + I_{iy}C\omega_{iz})\frac{\partial^2\omega_{iz}}{\partial x_w \partial x_v} + \left(\frac{\partial^2 I_{ix}}{\partial x_w \partial x_v}C\omega_{iz} + \frac{\partial^2 I_{iy}}{\partial x_w \partial x_v}S\omega_{iz}\right) \\ & \left. + \left(-\frac{\partial I_{ix}}{\partial x_v}S\omega_{iz} + \frac{\partial I_{iy}}{\partial x_v}C\omega_{iz}\right)\frac{\partial \omega_{iz}}{\partial x_w}\right], \end{aligned} \tag{17.75}$$

$$\frac{\partial H}{\partial x_w} = -\left(I_{ix}C\omega_{iz} + I_{iy}S\omega_{iz}\right)\frac{\partial^2 I_{iz}}{\partial x_w \partial x_v} - \left(\frac{\partial I_{ix}}{\partial x_w}C\omega_{iz} + \frac{\partial I_{iy}}{\partial x_w}S\omega_{iz}\right)\frac{\partial I_{iz}}{\partial x_v} - \left(-I_{ix}S\omega_{iz} + I_{iy}C\omega_{iz}\right)\frac{\partial \omega_{iz}}{\partial x_w}\frac{\partial I_{iz}}{\partial x_v}. \tag{17.76}$$

Finally, differentiating Eq. (8.92) of Appendix 4 in Chap. 8 gives

$$\frac{\partial^2 \omega_{ix}}{\partial x_w \partial x_v} = \frac{M\left(\frac{\partial P}{\partial x_w}T + P\frac{\partial T}{\partial x_w} - Q\frac{\partial U}{\partial x_w} - U\frac{\partial Q}{\partial x_w}\right) - (PT - UQ)\frac{\partial M}{\partial x_w}}{M^2}, \tag{17.77}$$

where

$$\begin{aligned}\frac{\partial M}{\partial x_w} &= 2\left(K_{ix}S\omega_{iz} - K_{iy}C\omega_{iz}\right)\left[\left(\frac{\partial K_{ix}}{\partial x_w}S\omega_{iz} - \frac{\partial K_{iy}}{\partial x_w}C\omega_{iz}\right) + \left(K_{ix}C\omega_{iz} + K_{iy}S\omega_{iz}\right)\frac{\partial \omega_{iz}}{\partial x_w}\right] \\ &\quad + 2\left(-J_{ix}S\omega_{iz} + J_{iy}C\omega_{iz}\right)\left[\left(-\frac{\partial J_{ix}}{\partial x_w}S\omega_{iz} + \frac{\partial J_{iy}}{\partial x_w}C\omega_{iz}\right) + \left(-J_{ix}C\omega_{iz} - J_{iy}S\omega_{iz}\right)\frac{\partial \omega_{iz}}{\partial x_w}\right],\end{aligned} \tag{17.78}$$

$$\frac{\partial P}{\partial x_w} = -\frac{\partial J_{ix}}{\partial x_w}S\omega_{iz} + \frac{\partial J_{iy}}{\partial x_w}C\omega_{iz} + \left(-J_{ix}C\omega_{iz} - J_{iy}S\omega_{iz}\right)\frac{\partial \omega_{iz}}{\partial x_w}, \tag{17.79}$$

$$\begin{aligned}\frac{\partial T}{\partial x_w} &= \left(\frac{\partial K_{ix}}{\partial x_w}C\omega_{iz} + \frac{\partial K_{iy}}{\partial x_w}S\omega_{iz}\right)\frac{\partial \omega_{iz}}{\partial x_v} + \left(-K_{ix}S\omega_{iz} + K_{iy}C\omega_{iz}\right)\frac{\partial \omega_{iz}}{\partial x_w}\frac{\partial \omega_{iz}}{\partial x_v} \\ &\quad + \left(K_{ix}C\omega_{iz} + K_{iy}S\omega_{iz}\right)\frac{\partial^2 \omega_{iz}}{\partial x_w \partial x_v} + \left(C\omega_{iz}\frac{\partial K_{ix}}{\partial x_v} + S\omega_{iz}\frac{\partial K_{iy}}{\partial x_v}\right)\frac{\partial \omega_{iz}}{\partial x_w} \\ &\quad + \left(S\omega_{iz}\frac{\partial^2 K_{ix}}{\partial x_w \partial x_v} - C\omega_{iz}\frac{\partial^2 K_{iy}}{\partial x_w \partial x_v}\right),\end{aligned} \tag{17.80}$$

$$\frac{\partial U}{\partial x_w} = \left(K_{ix}C\omega_{iz} + K_{iy}S\omega_{iz}\right)\frac{\partial \omega_{iz}}{\partial x_w} + \left(\frac{\partial K_{ix}}{\partial x_w}S\omega_{iz} - \frac{\partial K_{iy}}{\partial x_w}C\omega_{iz}\right), \tag{17.81}$$

$$\begin{aligned}\frac{\partial Q}{\partial x_w} &= \left(-\frac{\partial J_{ix}}{\partial x_w}C\omega_{iz} - \frac{\partial J_{iy}}{\partial x_w}S\omega_{iz}\right)\frac{\partial \omega_{iz}}{\partial x_v} + \left(J_{ix}S\omega_{iz} - J_{iy}C\omega_{iz}\right)\frac{\partial \omega_{iz}}{\partial x_w}\frac{\partial \omega_{iz}}{\partial x_v} \\ &\quad + \left(-J_{ix}C\omega_{iz} - J_{iy}S\omega_{iz}\right)\frac{\partial^2 \omega_{iz}}{\partial x_w \partial x_v} + \left(-S\omega_{iz}\frac{\partial^2 J_{ix}}{\partial x_w \partial x_v} + C\omega_{iz}\frac{\partial^2 J_{iy}}{\partial x_w \partial x_v}\right) \\ &\quad - \left(C\omega_{iz}\frac{\partial J_{ix}}{\partial x_v} + S\omega_{iz}\frac{\partial J_{iy}}{\partial x_v}\right)\frac{\partial \omega_{iz}}{\partial x_w}.\end{aligned} \tag{17.82}$$

The Hessian matrices of the three translation pose variables (i.e., $d^2t_{ix}/d\bar{X}_{sys}^2 = [\partial^2 t_{ix}/\partial x_w \partial x_v]$, $d^2t_{iy}/d\bar{X}_{sys}^2 = [\partial^2 t_{iy}/\partial x_w \partial x_v]$ and $d^2t_{iz}/d\bar{X}_{sys}^2 = [\partial^2 t_{iz}/\partial x_w \partial x_v]$) can be determined by differentiating Eqs. (8.98), (8.99) and (8.100) of Appendix 4 in Chap. 8, to give

$$\begin{aligned}\frac{\partial^2 t_{ix}}{\partial x_w \partial x_v} = {} & \frac{\partial^2 I_{ejx}}{\partial x_w \partial x_v}\left({}^{ej}t_{ix}\right) + \frac{\partial^2 J_{ejx}}{\partial x_w \partial x_v}\left({}^{ej}t_{iy}\right) + \frac{\partial^2 K_{ejx}}{\partial x_w \partial x_v}\left({}^{ej}t_{iz}\right) + \frac{\partial^2 \left(t_{ejx}\right)}{\partial x_w \partial x_v} + I_{ejx}\frac{\partial^2 \left({}^{ej}t_{ix}\right)}{\partial x_w \partial x_v} \\ & + J_{ejx}\frac{\partial^2 \left({}^{ej}t_{iy}\right)}{\partial x_w \partial x_v} + K_{ejx}\frac{\partial^2 \left({}^{ej}t_{iz}\right)}{\partial x_w \partial x_v} + \frac{\partial I_{ejx}}{\partial x_v}\frac{\partial \left({}^{ej}t_{ix}\right)}{\partial x_w} + \frac{\partial J_{ejx}}{\partial x_v}\frac{\partial \left({}^{ej}t_{iy}\right)}{\partial x_w} + \frac{\partial K_{ejx}}{\partial x_v}\frac{\partial \left({}^{ej}t_{iz}\right)}{\partial x_w} \\ & + \frac{\partial I_{ejx}}{\partial x_w}\frac{\partial \left({}^{ej}t_{ix}\right)}{\partial x_v} + \frac{\partial J_{ejx}}{\partial x_w}\frac{\partial \left({}^{ej}t_{iy}\right)}{\partial x_v} + \frac{\partial K_{ejx}}{\partial x_w}\frac{\partial \left({}^{ej}t_{iz}\right)}{\partial x_v},\end{aligned} \tag{17.83}$$

$$\begin{aligned}\frac{\partial^2 t_{iy}}{\partial x_w \partial x_v} = {} & \frac{\partial^2 I_{ejy}}{\partial x_w \partial x_v}\left({}^{ej}t_{ix}\right) + \frac{\partial^2 J_{ejy}}{\partial x_w \partial x_v}\left({}^{ej}t_{iy}\right) + \frac{\partial^2 K_{ejy}}{\partial x_w \partial x_v}\left({}^{ej}t_{iz}\right) + \frac{\partial^2 \left(t_{ejy}\right)}{\partial x_w \partial x_v} + I_{ejy}\frac{\partial^2 \left({}^{ej}t_{ix}\right)}{\partial x_w \partial x_v} \\ & + J_{ejy}\frac{\partial^2 \left({}^{ej}t_{iy}\right)}{\partial x_w \partial x_v} + K_{ejy}\frac{\partial^2 \left({}^{ej}t_{iz}\right)}{\partial x_w \partial x_v} + \frac{\partial I_{ejy}}{\partial x_v}\frac{\partial \left({}^{ej}t_{ix}\right)}{\partial x_w} + \frac{\partial J_{ejy}}{\partial x_v}\frac{\partial \left({}^{ej}t_{iy}\right)}{\partial x_w} + \frac{\partial K_{ejy}}{\partial x_v}\frac{\partial \left({}^{ej}t_{iz}\right)}{\partial x_w} \\ & + \frac{\partial I_{ejy}}{\partial x_w}\frac{\partial \left({}^{ej}t_{ix}\right)}{\partial x_v} + \frac{\partial J_{ejy}}{\partial x_w}\frac{\partial \left({}^{ej}t_{iy}\right)}{\partial x_v} + \frac{\partial K_{ejy}}{\partial x_w}\frac{\partial \left({}^{ej}t_{iz}\right)}{\partial x_v},\end{aligned} \tag{17.84}$$

$$\begin{aligned}\frac{\partial^2 t_{iz}}{\partial x_w \partial x_v} = {} & \frac{\partial^2 I_{ejz}}{\partial x_w \partial x_v}\left({}^{ej}t_{ix}\right) + \frac{\partial^2 J_{ejz}}{\partial x_w \partial x_v}\left({}^{ej}t_{iy}\right) + \frac{\partial^2 K_{ejz}}{\partial x_w \partial x_v}\left({}^{ej}t_{iz}\right) + \frac{\partial^2 \left(t_{ejz}\right)}{\partial x_w \partial x_v} + I_{ejz}\frac{\partial^2 \left({}^{ej}t_{ix}\right)}{\partial x_w \partial x_v} \\ & + J_{ejz}\frac{\partial^2 \left({}^{ej}t_{iy}\right)}{\partial x_w \partial x_v} + K_{ejz}\frac{\partial^2 \left({}^{ej}t_{iz}\right)}{\partial x_w \partial x_v} + \frac{\partial I_{ejz}}{\partial x_v}\frac{\partial \left({}^{ej}t_{ix}\right)}{\partial x_w} + \frac{\partial J_{ejz}}{\partial x_v}\frac{\partial \left({}^{ej}t_{iy}\right)}{\partial x_w} + \frac{\partial K_{ejz}}{\partial x_v}\frac{\partial \left({}^{ej}t_{iz}\right)}{\partial x_w} \\ & + \frac{\partial I_{ejz}}{\partial x_w}\frac{\partial \left({}^{ej}t_{ix}\right)}{\partial x_v} + \frac{\partial J_{ejz}}{\partial x_w}\frac{\partial \left({}^{ej}t_{iy}\right)}{\partial x_v} + \frac{\partial K_{ejz}}{\partial x_w}\frac{\partial \left({}^{ej}t_{iz}\right)}{\partial x_v}.\end{aligned} \tag{17.85}$$

References

1. Lin PD (2013) Analysis and design of prisms using the derivatives of a ray, Part II: the derivatives of boundary variable vector with respect to system variable vector. Appl Opt 52:4151–4162
2. Olson C, Youngworth RN (2008) Alignment analysis of optical systems using derivative information. Proc of SPIE 7068:1–10
3. Lin PD (2013) Design of optical systems using derivatives of rays: derivatives of variable vector of spherical boundary surfaces with respect to system variable vector. Appl Opt 52:7271–7287
4. Arora JS (2012) Introduction to optimum design, 3rd edn. Elsevier Inc., p 482
5. Leica Inc., Geodesy and Industrial Systems Center, Norcross, GA
6. Lin PD (2013) Analysis and design of prisms using the derivatives of a ray, Part i: the derivatives of a ray with respect to boundary variable vector. Appl Opt 52:4137–4150

Chapter 18
Hessian Matrix of Optical Path Length

Chapter 14 presented a method for determining the Jacobian matrix of the optical path length (OPL) of a skew ray. In this chapter, the method is extended to the Hessian matrix of OPL in a non-axially symmetrical optical system. The proposed method facilitates the cross-sensitivity analysis of the OPL with respect to arbitrary system variables and provides an ideal basis for automatic optical system design applications in which the merit function is defined in terms of wavefront aberrations.

18.1 Determination of Hessian Matrix of OPL

As stated in Sect. 14.2, the notation $\mathrm{OPL}(\bar{P}_g, \bar{P}_h)$, where g and h are two positive integers and $0 \le g < h \le n$, represents the OPL between two incidence points $\bar{P}_g$ and $\bar{P}_h$ of a ray as it propagates through an optical system containing n boundary surfaces. It will be recalled that $\mathrm{OPL}(\bar{P}_g, \bar{P}_h)$ can be obtained simply by summing OPL_i (see Eq. (14.13)). Consequently, the Hessian matrix of $\mathrm{OPL}(\bar{P}_g, \bar{P}_h)$ with respect to the system variable vector $\bar{X}_{sys}$ is given as

$$\frac{\partial^2 \mathrm{OPL}(\bar{P}_0, \bar{P}_n)}{\partial \bar{X}_{sys}^2} = \sum_{i=1}^{i=n} \frac{\partial^2 \mathrm{OPL}_i}{\partial \bar{X}_{sys}^2}. \tag{18.1}$$

The Hessian matrix of the OPL derived in this chapter provides an efficient means of determining the search direction in gradient-based optimization schemes in which the merit function $\bar{\Phi}$ is defined in terms of wavefront aberrations. To

P.D. Lin, *Advanced Geometrical Optics*, Progress in Optical Science and Photonics 4, DOI 10.1007/978-981-10-2299-9_18

obtain the Hessian matrix, the following equation is first required to compute Eq. (18.1):

$$\frac{d^2OPL_i}{d\bar{X}_{sys}^2} = \left[\left(\frac{d\bar{R}_{i-1}}{d\bar{X}_{sys}}\right)^T \quad \left(\frac{d\bar{X}_i}{d\bar{X}_{sys}}\right)^T\right] \begin{bmatrix} \frac{\partial^2 OPL_i}{\partial \bar{R}_{i-1}\partial \bar{R}_{i-1}} & \frac{\partial^2 OPL_i}{\partial \bar{R}_{i-1}\partial \bar{X}_i} \\ \frac{\partial^2 OPL_i}{\partial \bar{X}_i \partial \bar{R}_{i-1}} & \frac{\partial^2 OPL_i}{\partial \bar{X}_i \partial \bar{X}_i} \end{bmatrix} \begin{bmatrix} \frac{d\bar{R}_{i-1}}{d\bar{X}_{sys}} \\ \frac{\partial \bar{X}_i}{d\bar{X}_{sys}} \end{bmatrix} + \begin{bmatrix} \frac{\partial OPL_i}{\partial \bar{R}_{i-1}} & \frac{\partial OPL_i}{\partial \bar{X}_i} \end{bmatrix} \begin{bmatrix} \frac{d^2\bar{R}_{i-1}}{d\bar{X}_{sys}^2} \\ \frac{d^2\bar{X}_i}{d\bar{X}_{sys}^2} \end{bmatrix}. \tag{18.2}$$

Figure 18.1 presents a hierarchical chart with three levels (i.e., OPL_i from ray tracing, OPL_i Jacobian matrix, and OPL_i Hessian matrix) showing the calculations involved in computing $d^2OPL_i/d\bar{X}_{sys}^2$. It is noted that the equations become increasingly sophisticated and advanced toward the lower levels. It is also noted from Eq. (18.2) that a total of nine terms are required to determine $d^2OPL_i/d\bar{X}_{sys}^2$. Six of these terms have already been presented in previous chapters, namely $d\bar{R}_{i-1}/d\bar{X}_{sys}$ (Eqs. (7.55) with $g = i - 1$), $d\bar{X}_i/d\bar{X}_{sys}$ (Eqs. 8.6 and 8.25), $\partial OPL_i/\partial \bar{R}_{i-1}$ (Eq. 14.2), $\partial OPL_i/\partial \bar{X}_i$ (Eq. 14.6), $d^2\bar{X}_i/d\bar{X}_{sys}^2$ (Eqs. 17.6 and 17.18), and $d^2\bar{R}_i/d\bar{X}_{sys}^2$

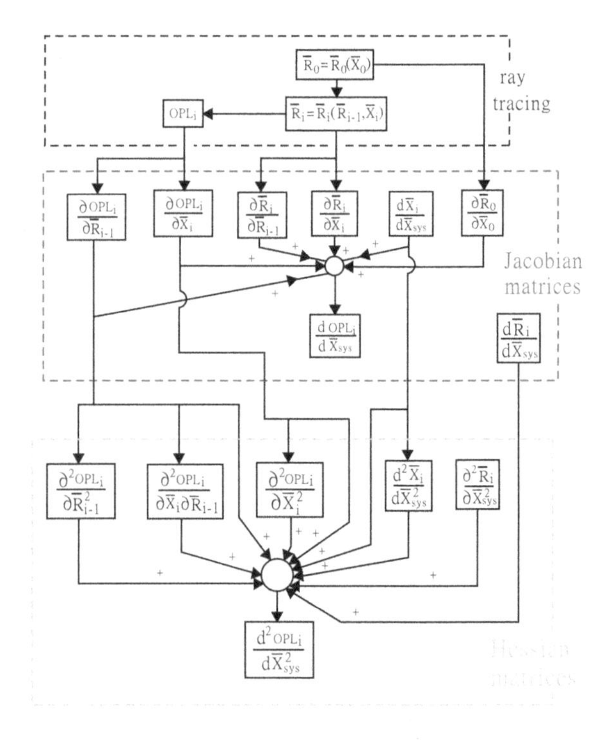

Fig. 18.1 Hierarchical chart showing three levels involved in determining OPL_i and its Jacobian and Hessian matrices

(Eq. (16.1)). Consequently, the following sections present derivations only for $\partial^2 OPL_i/\partial\bar{R}_{i-1}^2$, $\partial^2 OPL_i/\partial\bar{X}_i\partial\bar{R}_{i-1}$ and $\partial^2 OPL_i/\partial\bar{X}_i^2$.

18.1.1 Hessian Matrix of OPL_i with Respect to Incoming Ray $\bar{R}_{i-1}$

The Hessian matrix of OPL_i with respect to the incoming ray $\bar{R}_{i-1}$ (i.e., $\partial^2 OPL_i/\partial\bar{R}_{i-1}^2$) can be obtained directly by differentiating Eq. (14.2) with respect to $\bar{R}_{i-1}$. Since $\partial\xi_{i-1}/\partial\bar{R}_{i-1} = \bar{0}$, the Hessian matrix is obtained as

$$\frac{\partial^2 OPL_i}{\partial\bar{R}_{i-1}^2} = \xi_{i-1}\frac{\partial^2\lambda_i}{\partial\bar{R}_{i-1}^2}, \tag{18.3}$$

where $\partial^2\lambda_i/\partial\bar{R}_{i-1}^2$ is given in Eqs. (15.2) and (15.7) for flat and spherical boundary surfaces, respectively.

18.1.2 Hessian Matrix of OPL_i with Respect to $\bar{X}_i$ and $\bar{R}_{i-1}$

The Hessian matrix of OPL_i with respect to the boundary variable vector $\bar{X}_i$ and incoming ray $\bar{R}_{i-1}$ can be obtained for both spherical and flat boundary surfaces by differentiating Eq. (14.2) with respect to $\bar{X}_i$, to give

$$\frac{\partial^2 OPL_i}{\partial\bar{X}_i\partial\bar{R}_{i-1}} = \frac{\partial\xi_{i-1}}{\partial\bar{X}_i}\frac{\partial\lambda_i}{\partial\bar{R}_{i-1}} + \xi_{i-1}\frac{\partial^2\lambda_i}{\partial\bar{X}_i\partial\bar{R}_{i-1}}. \tag{18.4}$$

Terms $\partial\xi_{i-1}/\partial\bar{X}_i$ and $\partial\lambda_i/\partial\bar{R}_{i-1}$ are given in Chap. 14 (Eqs. (14.7) and (14.8)) and Chap. 7 (Eqs. (7.5) and (7.21)), respectively. Meanwhile, term $\partial^2\lambda_i/\partial\bar{X}_i\partial\bar{R}_{i-1}$ is given in Chap. 16 (Eqs. (16.10) and (16.24)).

18.1.3 Hessian Matrix of OPL_i with Respect to Boundary Variable Vector $\bar{X}_i$

The Hessian matrix of OPL_i with respect to the boundary variable vector $\bar{X}_i$ can be obtained directly by differentiating Eq. (14.6) with respect to $\bar{X}_i$. Since

$\partial^2\xi_{i-1}/\partial\bar{X}_i^2 = \bar{0}$, the following equation is obtained for both spherical and flat boundary surfaces:

$$\frac{\partial^2 OPL_i}{\partial\bar{X}_i^2} = \frac{\partial\lambda_i}{\partial\bar{X}_i}\frac{\partial\xi_{i-1}}{\partial\bar{X}_i} + \frac{\partial\xi_{i-1}}{\partial\bar{X}_i}\frac{\partial\lambda_i}{\partial\bar{X}_i} + \xi_{i-1}\frac{\partial^2\lambda_i}{\partial\bar{X}_i^2}. \tag{18.5}$$

Terms $\partial\lambda_i/\partial\bar{X}_i$ (Eqs. 7.32 and 7.44), $\partial\xi_{i-1}/\partial\bar{X}_i$ (Eqs. (14.7) and (14.8)) and $\partial^2\lambda_i/\partial\bar{X}_i^2$ (Eqs. (16.4) and (16.17)) have all been presented in previous chapters.

18.2 System Analysis Based on Jacobian and Hessian Matrices of Wavefront Aberrations

The theory developed in this chapter is applicable to both axis-symmetrical systems and non-axially symmetrical systems. However for convenience, the following discussions consider only the axis-symmetrical system shown in Fig. 3.2 for illustration purposes. Figure 14.11 shows the modified system obtained for Fig. 3.2 when introducing an additional reference sphere into the system. (Note that the reference sphere is labeled as i = 11 and the image plane is thus marked as i = 12.) The figure shows the path of a general ray $\bar{R}_0 = \begin{bmatrix}\bar{P}_0 & \bar{\ell}_0\end{bmatrix}^T$ originating from point source $\bar{P}_0$ with unit directional vector $\bar{\ell}_0$ and then propagating through the system. Figure 14.8 provides a simplified view of the system in which most of the boundary surfaces are excluded so as to show more clearly the reference sphere and image plane by setting n = 11. It will be recalled that in wavefront aberration computations, the reference sphere $\bar{r}_{ref}$ may not pass through the center of the exit pupil [1]. However, it is possible to define the wavefront aberration of a system by using an arbitrary reference sphere $\bar{r}_{ref}$ of radius $\bar{R}_{ref}$ centered at $\bar{P}_{12/chief}$ (i.e., the incidence point of chief ray at on the image plane). As shown in Fig. 14.8, the chief ray $\bar{R}_{0/chief} = \begin{bmatrix}\bar{P}_0 & \bar{\ell}_{0/chief}\end{bmatrix}^T$ originating from point source $\bar{P}_0$ is incident on the reference sphere at point $\bar{P}_{11/chief}$. As shown in Eq. (14.24) by setting n = 11, the wavefront aberration $W(\bar{X}_0)$ of the general ray $\bar{R}_0$ is equal to the difference between $OPL(\bar{P}_0, \bar{P}_{11})$ and $OPL(\bar{P}_0, \bar{P}_{11/chief})$, where $\bar{P}_{11}$ and $\bar{P}_{11/chief}$ are the incidence points of the general ray $\bar{R}_0$ and chief ray $\bar{R}_{0/chief}$, respectively, on the reference sphere. The Hessian matrix $d^2W(\bar{X}_0)/d\bar{X}_{sys}^2$ of the wavefront aberration $W(\bar{X}_0)$ with respect to $\bar{X}_{sys}$ can be obtained from Eq. (14.28) as

$$\frac{d^2W(\bar{X}_0)}{d\bar{X}_{sys}^2} = \frac{d^2OPL(\bar{P}_0, \bar{P}_n)}{d\bar{X}_{sys}^2} - \frac{d^2OPL(\bar{P}_0, \bar{P}_{n/chief})}{d\bar{X}_{sys}^2}. \tag{18.6}$$

The Hessian matrix of a function is usually used in large-scale optimization problems with Newton-type methods since it provides the coefficient of the quadratic term in the local Taylor series expansion of the function. The change in the wavefront aberration $\Delta W(\bar{X}_0)$ resulting from changes in one or more of the system variables can be estimated by a Taylor series expansion of either a first-order or up to the quadratic term, i.e.,

$$\Delta W(\bar{X}_0) = \frac{dW(\bar{X}_0)}{d\bar{X}_{sys}} \Delta\bar{X}_{sys}, \tag{18.7}$$

$$\Delta W(\bar{X}_0) = \frac{dW(\bar{X}_0)}{d\bar{X}_{sys}} \Delta\bar{X}_{sys} + \frac{1}{2}\left(\Delta\bar{X}_{sys}^T \frac{d^2W(\bar{X}_0)}{d\bar{X}_{sys}^2} \Delta\bar{X}_{sys}\right). \tag{18.8}$$

Example 18.1 To demonstrate the use of the Hessian matrix in estimating the wavefront aberration, consider the optical system shown in Fig. 14.11. Assume that a source ray $\bar{R}_0$ originating from $\bar{P}_0 = [0 \quad -507 \quad 170 \quad 1]^T$ is incident on the aperture at $(x_5, z_5) = (0, 5)$. Furthermore, assume that one of the system variables (say, v_4) deviates from its nominal value while the remaining variables remain unchanged. The second column in Table 18.1 shows the numerical results obtained for $\Delta W(\bar{X}_0)$ using a raytracing approach. Let these results be taken as the mathematically-exact numeric values of $\Delta W(\bar{X}_0)$ for different values of Δv_4. The third and fourth columns in Table 18.1 show the estimated values of $\Delta W(\bar{X}_0)$ obtained from Eq. (18.7) and the corresponding errors relative to the raytracing values, respectively. It is seen that the estimation errors range from −1.34 % to −25.40 %. In other words, the accuracy of Eq. (18.7) is highly sensitive to the value of Δv_4. By contrast, the estimation errors associated with Eq. (18.8) are both far smaller than those associated with Eq. (18.7) and more robust toward changes in the value of Δv_4 (see the fifth column in Table 18.1).

Table 18.1 Comparison of exact and estimated changes in wavefront aberration due to changes in variable v_4

Δv_4	ΔW from raytracing	Equation (18.7)		Equation (18.8)	
		ΔW	Error%	ΔW	Error%
0.1	0.00988	0.00974	−1.34	0.00987	−0.03
0.2	0.02003	0.01948	−2.72	0.02000	−0.13
0.4	0.04127	0.03897	−5.58	0.04104	−0.56
0.6	0.06394	0.05845	−8.57	0.06312	−1.27
0.8	0.07588	0.06820	−10.12	0.07455	−1.74
1.2	0.14313	0.11691	−18.32	0.13558	−5.27
1.6	0.20896	0.15587	−25.40	0.18907	−9.52

18.3 System Design Based on Jacobian and Hessian Matrices of Wavefront Aberrations

One of the most important applications of Eq. (18.1) is that of facilitating automatic optical design systems based on a merit function defined in terms of the wavefront aberration. To demonstrate the use of the proposed Hessian matrix of the OPL in providing an ideal search direction for such optimization methods, let the system shown in Fig. 14.11 again be taken for illustration purposes. For simplicity, assume that only the separations of the three elements in the system (i.e., j = 4, j = 5 and j = 6) are adjustable. In other words, the design problem imitates the practical case where it is necessary to determine the positions of multiple commercial elements so as to achieve a specified position of the system aperture (i.e., v_3). The system variable vector thus reduces to $\bar{X}_{sys} = [v_4 \quad v_5 \quad v_6]^T$. In optimizing the element positions, it is first necessary to define a merit function with which to characterize the defects of the system. Assume that $\bar{P}_0$ is a point source which emits a large number of rays into space. To calculate the wavefront aberration, it is necessary to trace a subset of these rays (say, f) from point $\bar{P}_0$ through the system. In general, the merit function is evaluated for several point sources (say, q) in the field of view. Let the merit function $\bar{\Phi}$ and its Jacobian matrix $\partial\bar{\Phi}/\partial\bar{X}_{sys}$ have the forms given in Eqs. (14.30) and (14.31), respectively. The Hessian matrix $\partial^2\bar{\Phi}/\partial\bar{X}_{sys}^2$ can then be obtained from Eq. (14.31) as

$$\begin{aligned}\frac{\partial^2\bar{\Phi}}{\partial\bar{X}_{sys}^2} &= 2\sum_1^q\sum_1^m\left(\frac{\partial W}{\partial\bar{X}_{sys}}\frac{\partial W}{\partial\bar{X}_{sys}} + W\frac{\partial^2 W}{\partial\bar{X}_{sys}^2}\right)\\ &= 2\sum_1^q\sum_1^m\left\{\left[OPL(\bar{P}_0,\bar{P}_n) - OPL(\bar{P}_0,\bar{P}_{n/chief})\right]\left[\frac{\partial^2 OPL(\bar{P}_0,\bar{P}_n)}{\partial\bar{X}_{sys}^2} - \frac{\partial^2 OPL(\bar{P}_0,\bar{P}_{n/chief})}{\partial\bar{X}_{sys}^2}\right]\right.\\ &\left.+\left[\frac{\partial OPL(\bar{P}_0,\bar{P}_n)}{\partial\bar{X}_{sys}} - \frac{\partial OPL(\bar{P}_0,\bar{P}_{n/chief})}{\partial\bar{X}_{sys}}\right]\left[\frac{\partial OPL(\bar{P}_0,\bar{P}_n)}{\partial\bar{X}_{sys}} - \frac{\partial OPL(\bar{P}_0,\bar{P}_{n/chief})}{\partial\bar{X}_{sys}}\right]\right\}.\end{aligned} \tag{18.9}$$

Example 18.2 In this example, the benefit of the Hessian matrix as a design and analysis tool is demonstrated using the steepest-descent method and the modified Newton method. It will be recalled that the steepest-descent method is one of the simplest and most commonly used optimization methods since its search direction $\Delta\bar{X}_{sys}$ is simply evaluated by taking the Jacobian matrix of the merit function (Eq. 1.75). By contrast, in the modified Newton method, the search direction $\Delta\bar{X}_{sys}$ is evaluated in terms of both the Jacobian matrix and the Hessian matrix (Eq. 1.76). Let the merit function defined in Eq. (14.30) be evaluated for three (q = 3) point sources ($\bar{P}_0 = [0 \quad -507 \quad u \quad 1]^T$, u = 5,15,25). Furthermore, let each of these point sources generate m = 17 rays (i.e., one chief ray and 16 general rays evenly distributed over the aperture j = 3 with a radius of 13.716 mm). Table 18.2 lists the

Table 18.2 Convergence of system variable values and merit function values for optical system shown in Fig. 14.11 given use of steepest descent method and modified Newton method

f	Steepest decent method				Modified Newton method			
	v_4	v_5	v_6	$\bar{\Phi}$	v_4	v_5	v_6	$\bar{\Phi}$
1	8.000	3.000	40.000	1.012646	8.000	3.000	40.000	1.012646
2	8.184	2.995	40.073	0.973706	15.544	7.338	54.082	0.113918
3	8.365	2.991	40.145	0.935976	14.629	8.191	49.618	0.002934
4	8.543	2.987	40.216	0.899436	14.542	8.331	48.870	0.000824
5	8.719	2.984	40.287	0.864063	14.527	8.343	48.803	0.000796
6	8.891	2.981	40.356	0.829837	14.520	8.343	48.805	0.000794
7	9.061	2.978	40.424	0.796733	14.514	8.341	48.818	0.000793
30	12.209	2.981	41.704	0.291153				
70	15.044	3.008	42.885	0.041500				
90	15.705	3.0106	43.164	0.015505				
120	16.229	3.0113	43.389	0.004259				
213	16.616	3.013	43.562	0.001439				

values of the merit function $\bar{\Phi}$ (and corresponding values of $\bar{X}_{sys}$) at various iteration counts f given an initial guess of $\bar{X}_{sys} = [8 \quad 3 \quad 40]^T$. It is seen that the merit function has a value of $\Phi = 0.001439$ after 213 iterations when the steepest-decent optimization method is used. However, the modified Newton method converges to $\Phi = 0.000793$ in just seven iterations. In other words, the modified Newton method, based on the Jacobian and Hessian matrices given in Eqs. (14.31) and (18.9), respectively, provides a highly effective approach for solving the current design problem. However, it should be noted that the method incurs a significantly longer CPU time since it requires additional computations to obtain the Hessian matrices.

Reference

1. Milkŝ A (2002) Dependence of the wave-front aberration on the radius of the reference sphere. J Opt Soc Am 19:1187–1190

VITA

Psang Dain Lin
Distinguished Professor
National Cheng Kung University
Department of Mechanical Engineering
Tainan, Taiwan 70101
Phone: (886) –6-2080563
Website: http://140.116.31.121/
E-mail: pdlin@mail.ncku.edu.tw

Education:

July 1979, BS, National Cheng Kung University, Department of Mechanical Engineering, Taiwan.

July 1984, MS, National Cheng Kung University, Department of Mechanical Engineering, Taiwan.

July 1989, Ph.D., Northwestern University, Department of Mechanical Engineering, USA.

P.D. Lin, *Advanced Geometrical Optics*, Progress in Optical Science and Photonics 4, DOI 10.1007/978-981-10-2299-9

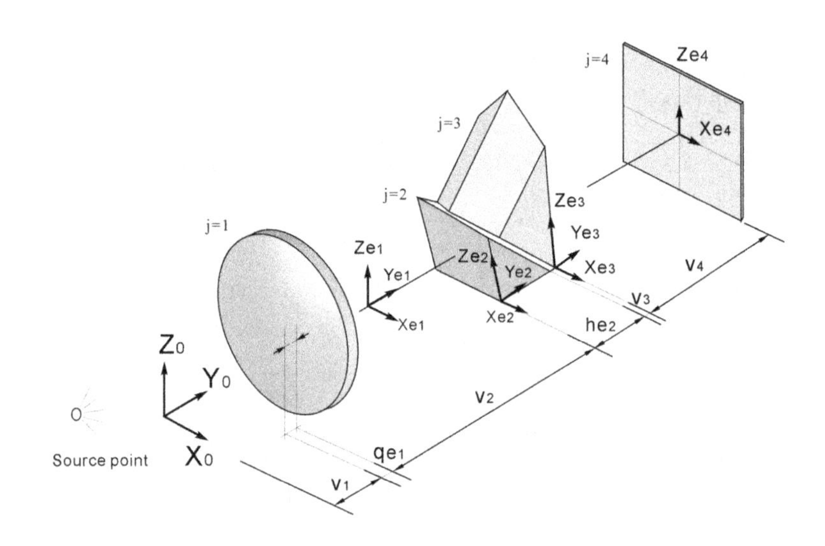

j=4
Ze4
Xe4
j=3
j=2
Ze3
Ye3
j=1
Ze1
Ze2
Ye2
Xe3
Ye1
V4
Xe1
Xe2
V3
he2
Z0
Y0
V2
O
Source point
X0
qe1
V1

CPSIA information can be obtained
at www.ICGtesting.com
Printed in the USA
LVHW08s0019200918
590745LV00003B/21/P